# Priming and Pretreatment of Seeds and Seedlings

Mirza Hasanuzzaman • Vasileios Fotopoulos
Editors

# Priming and Pretreatment of Seeds and Seedlings

## Implication in Plant Stress Tolerance and Enhancing Productivity in Crop Plants

Springer

*Editors*
Mirza Hasanuzzaman
Department of Agronomy
Sher-e-Bangla Agricultural University
Dhaka, Bangladesh

Vasileios Fotopoulos
Department of Agricultural Sciences,
Biotechnology and Food Science
Cyprus University of Technology
Limassol, Cyprus

ISBN 978-981-13-8624-4 ISBN 978-981-13-8625-1 (eBook)
https://doi.org/10.1007/978-981-13-8625-1



This Springer imprint is published by the registered company Springer Nature Singapore Pte Ltd.
The registered company address is: 152 Beach Road, #21-01/04 Gateway East, Singapore 189721, Singapore

# Preface

Climate change-related environmental conditions, such as drought, salinity, and heat, are among the most devastating abiotic stresses for crop yield. Their impacts are further exacerbated due to anthropogenic contribution, thus generating a major threat to global food security. Simultaneously, agriculture is one of the top water-consuming sectors globally (69% of the water withdrawal), and agricultural intensification is expected to increase the demand for water worldwide. Drought, salt, and heat tolerance improvement in major crops is a valuable tool to reduce crop yield losses, to increase agricultural water use efficiency, and to cultivate areas that were unable to be cultivated due to low precipitation and/or salt accumulation.

Different methodologies have been employed for enhancing climate change-related stress tolerance in plants and therefore increase crop water use efficiency; some are particularly time-consuming (e.g., conventional breeding), and others are currently unacceptable in many countries around the world (e.g., plant genetic modification). Priming is a rapidly emerging field in plant stress physiology and crop stress management. Plants treated with certain natural or synthetic compounds (i.e., chemical agents) and/or biological agents (such as PGPB and AMF) prior stress events show enhanced tolerance when exposed to sub-optimal abiotic conditions. There are different kinds of priming, like hydropriming, halopriming, magnetopriming, and chemical priming, which have become very popular in the last couple of decades. Recently, seedling pretreatments with a low dose of antioxidants, trace elements, amino acids, phytohormones, and other signaling molecules were found to provide enhance stress tolerance under stressful conditions. Stress impacts on plant growth and yield in primed plants are remarkably reduced in comparison with non-primed plants. However, further research is needed to better understand how plants better adapt to multiple environmental constraints after priming and establish stress management practices based on biological and chemical priming, while very few studies were dedicated on seed priming against combined abiotic stresses to date.

The editors and contributing authors hope that this book will provide an up-to-date view on the exciting field of seed/plant priming and lead to new discussions and efforts to further develop this multifaceted technology. Therefore, the present volume would be a valuable source of scientific information to advanced students, early-stage researchers, faculty, and scientists involved in agriculture, plant sciences, biotechnology, plant breeding, and related areas.

We, the editors, would like to give special thanks to the authors for their outstanding and timely work in producing such fine chapters. We are highly thankful to Ms. Mei Hann Lee, editor, *Life Sciences*, Springer, Japan, for her prompt responses during the acquisition. We are also thankful to Selvakumar Rajendran, project coordinator of this book, and all other editorial staff for their precious help in formatting and incorporating editorial changes in the manuscripts. Special thanks to Dr. Md. Mahabub Alam, Mr. Abdul Awal Chowdhury Masud, and Khursheda Parvin of Sher-e-Bangla Agricultural University, Bangladesh, for their generous help in formatting the manuscripts. The editors and contributing authors hope that this book will include a practical update on our knowledge for the role of plant nutrients in abiotic stress tolerance.

Dhaka, Bangladesh — Mirza Hasanuzzaman
Lemesos, Cyprus — Vasileios Fotopoulos

# Contents

# Contributors


**Mazhar M. F. Abdalla** Agronomy Department, Faculty of Agriculture, Cairo University, Giza, Egypt

**Magdi T. Abdelhamid** Botany Department, National Research Centre, Giza, Egypt

**Maqsood Ahmad** Mirpur University of Science and Technology (MUST), Mirpur, Pakistan

**Muhammad Ahmad** Center for Health & Bioresources, Austrian Institute of Technology, Vienna, Austria

**Niaz Ahmad** Department of Soil Science, Bahauddin Zakariya University, Multan, Punjab, Pakistan

**Riaz Ahmad** Department of Horticulture, Bahauddin Zakariya University, Multan, Punjab, Pakistan

**Shakeel Ahmad** Department of Agronomy, Bahauddin Zakariya University, Multan, Punjab, Pakistan

**Muhammad Akmal** Institute of Soil Science, Pir Mehr Ali Shah Arid Agriculture University, Rawalpindi, Punjab, Pakistan

**Iftikhar Ali** Institute of Soil and Environmental Sciences, Gomal University, Dera Ismail Khan, Pakistan

**Muhammad Arif Ali** Department of Soil Science, Bahauddin Zakariya University, Multan, Punjab, Pakistan

**Qasim Ali** College of Agriculture, Bahauddin Zakariya University, Layyah, Pakistan

**Sajid Ali** Department of Horticulture, Bahauddin Zakariya University, Multan, Punjab, Pakistan

**Shafaqat Ali** Department of Environmental Sciences and Engineering, Government College University Faisalabad, Faisalabad, Pakistan

**Anjali Anand** Division of Plant Physiology, ICAR-Indian Agricultural Research Institute, New Delhi, India

**Muhammad Akbar Anjum** Department of Horticulture, Bahauddin Zakariya University, Multan, Punjab, Pakistan

**Shakeel Ahmad Anjum** Department of Agronomy, University of Agriculture Faisalabad, Faisalabad, Punjab, Pakistan

**Muhammad Saleem Arif** Department of Environmental Sciences and Engineering, Government College University Faisalabad, Faisalabad, Pakistan

**Muhammad Arshad** Institute of Environmental Sciences and Engineering, School of Civil and Environmental Engineering, National University of Sciences and Technology, Islamabad, Pakistan

**Muhammad Arslan Ashraf** Department of Botany, Government College University Faisalabad, Faisalabad, Pakistan

**Sana Ashraf** Institute of Soil and Environmental Sciences, University of Agriculture, Faisalabad, Pakistan

**Rashida Parveen** Department of Plant Pathology, Bahauddin Zakariya University, Multan, Punjab, Pakistan

**Jay Prakash Awasthi** Plant Molecular Biotechnology Laboratory, Department of Life Science and Bioinformatics, Assam University, Silchar, Assam, India

**Muhammad Azam** Institute of Horticultural Sciences, University of Agriculture Faisalabad, Faisalabad, Pakistan

**Aditya Banerjee** Post Graduate Department of Biotechnology, St. Xavier's College (Autonomous), Kolkata, West Bengal, India

**Muhammad Bilal** Department of Environmental Sciences, University of COMSATS, Abbottabad, Pakistan

**Pankaj Borgohain** Plant Molecular Biotechnology Laboratory, Department of Life Science and Bioinformatics, Assam University, Silchar, Assam, India

**K. Chakraborty** Department of Crop Physiology & Biochemistry, ICAR-National Rice Research Institute, Cuttack, India

**Bhaben Chowardhara** Plant Molecular Biotechnology Laboratory, Department of Life Science and Bioinformatics, Assam University, Silchar, Assam, India

**Darwish S. Darwish** Agronomy Department, Faculty of Agriculture, Cairo University, Giza, Egypt

**Farah Deeba** Department of Biotechnology and Genetic Engineering, Kohat University of Science and Technology, Kohat, Pakistan

**Sanjenbam Sanjibia Devi** Plant Molecular Biotechnology Laboratory, Department of Life Science and Bioinformatics, Assam University, Silchar, Assam, India

**Shaghef Ejaz** Department of Horticulture, Bahauddin Zakariya University, Multan, Punjab, Pakistan

**Mahmoud H. El Kholy** Soil and Water Research Institute, Agricultural Research Center, Giza, Egypt

**Raafat R. El-Masry** Botany Department, National Research Centre, Giza, Egypt

**Ayman EL Sabagh** Department of Agronomy, Faculty of Agriculture, Kafrelsheikh University, Kafrelsheikh, Egypt

**Vasileios Fotopoulos** Department of Agricultural Sciences, Biotechnology and Food Science, Cyprus University of Technology, Limassol, Cyprus

**Dibakar Ghosh** ICAR Directorate of Weed Research, Jabalpur, Madhya Pradesh, India

**Idrees Haider** Department of Soil Science, Bahauddin Zakariya University, Multan, Punjab, Pakistan

**Mirza Hasanuzzaman** Department of Agronomy, Sher-e-Bangla Agricultural University, Dhaka, Bangladesh

**Hafiz Athar Hussain** Institute of Environment and Sustainable Development in Agriculture, Chinese Academy of Agricultural Sciences, Beijing, China

**Haziq Hussain** Department of Biotechnology and Genetic Engineering, Kohat University of Science and Technology, Kohat, Pakistan

**Iqbal Hussain** Department of Botany, Government College University Faisalabad, Faisalabad, Pakistan

**Sadam Hussain** Department of Agronomy, University of Agriculture Faisalabad, Faisalabad, Punjab, Pakistan

**Saddam Hussain** Department of Agronomy, University of Agriculture Faisalabad, Faisalabad, Punjab, Pakistan

**Sajid Hussain** College of Agriculture, Bahauddin Zakariya University, Layyah, Pakistan

**Sajjad Hussain** Department of Horticulture, Bahauddin Zakariya University, Multan, Punjab, Pakistan

**Ehab Awad-Allah Ibrahim** Cross Pollinated Vegetable Crops Research Department, Horticulture Research Institute, Agricultural Research Center, Giza, Egypt

**Muhammad Ijaz** College of Agriculture, Bahauddin Zakariya University, Layyah, Pakistan

**Fariha Ilyas** Department of Soil Science, Bahauddin Zakariya University, Multan, Punjab, Pakistan

**Mazhar Iqbal** Department of Environmental Sciences, Quaid-i-Azam University, Islamabad, Pakistan

**Muhammad Iqbal** Department of Botany, Government College University Faisalabad, Faisalabad, Pakistan

**Mehwish Jan** Department of Horticulture, Bahauddin Zakariya University, Multan, Punjab, Pakistan

**Muhammad Kamran** Department of Agronomy, University College of Agriculture, University of Sargodha, Sargodha, Pakistan

**Saradia Kar** Plant Molecular Biotechnology Laboratory, Department of Life Science and Bioinformatics, Assam University, Silchar, Assam, India

**Ryota Kataoka** Faculty of Life & Environmental Sciences, Department of Environmental Sciences, University of Yamanashi, Yamanashi, Japan

**Muhammad Fasih Khalid** Department of Horticulture, Bahauddin Zakariya University, Multan, Punjab, Pakistan

**Abdul Khaliq** Department of Agronomy, University of Agriculture Faisalabad, Faisalabad, Punjab, Pakistan

**Imran Khan** Department of Agronomy, University of Agriculture Faisalabad, Faisalabad, Punjab, Pakistan

**Muhammad Daud Khan** Department of Biotechnology and Genetic Engineering, Kohat University of Science and Technology, Kohat, Pakistan

**Samiya Mahmood Khan** Department of Plant Pathology, Bahauddin Zakariya University, Multan, Pakistan

**Nicholas Emmanuel Korres** Department of Crop, Soil, and Environmental Sciences, University of Arkansas, Fayetteville, AR, USA

**Dong Jin Lee** Department of Crop Science and Biotechnology, Dankook University, Chungnam, Republic of Korea

**Ahmad Mahmood** Faculty of Life & Environmental Sciences, Department of Environmental Sciences, University of Yamanashi, Yamanashi, Japan

**Abdul Majeed** Department of Botany, Government Degree College Pabbi, Nowshera, Khyber Pakhtunkhwa, Pakistan

**Sajid Masood** State Key Laboratory of Soil and Sustainable Agriculture, Institute of Soil Science, Chinese Academy of Sciences, Nanjing, China

**Komal Mazhar** Department of Forestry and Range Management, University of Agriculture Faisalabad, Faisalabad, Pakistan

**Debojyoti Moulick** Department of Environmental Science, University of Kalyani, Nadia, West Bengal, India

Plant Molecular Biotechnology Laboratory, Department of Life Science, Assam University, Silchar, Assam, India

**Zahir Muhammad** Department of Botany, University of Peshawar, Peshawar, Khyber Pakhtunkhwa, Pakistan

**A. K. Mukherjee** Department of Crop Protection, ICAR-National Rice Research Institute, Cuttack, India

**Ghulam Mustafa** Department of Soil Science, Bahauddin Zakariya University, Multan, Punjab, Pakistan

**Faisal Nadeem** Department of Agronomy, Faculty of Agriculture, University of Agriculture, Faisalabad, Pakistan

**Muhammad Nafees** Department of Horticultural Sciences, University college of Agriculture and Environmental Sciences, The Islamia University Bahawalpur, Bahawalpur, Punjab, Pakistan

**Aamir Nawaz** Department of Horticulture, Bahauddin Zakariya University, Multan, Punjab, Pakistan

**Ahmad Nawaz** College of Agriculture, Bahauddin Zakariya University, Layyah, Pakistan

**Abdul-Sattar Nizami** Center of Excellence in Environmental Studies (CEES), King Abdulaziz University, Jeddah, Saudi Arabia

**Shinya Oba** Plant Production Control Laboratory, Faculty of Applied Biological Sciences, Gifu University, Gifu, Japan

**Elsayed Omer** Medicinal and Aromatic Plants Research Department, National Research Centre, Giza, Egypt

**Sanjib Kumar Panda** Plant Molecular Biotechnology Laboratory, Department of Life Science and Bioinformatics, Assam University, Silchar, Assam, India

**Ragab Ragab** Centre for Ecology and Hydrology (CEH), Wallingford, UK

**Muhammad Junaid Rao** Key Laboratory of Horticultural Plant Biology (Ministry of Education), Huazhong Agricultural University, Wuhan, People's Republic of China

**Rizwan Rasheed** Department of Botany, Government College University Faisalabad, Faisalabad, Pakistan

**Abdul Rehman** College of Agriculture, Bahauddin Zakariya University, Layyah, Pakistan

Department of Crop Science and Biotechnology, Dankook University, Chungnam, Republic of Korea

**Rehmanullah** Department of Botany, University of Peshawar, Peshawar, Khyber Pakhtunkhwa, Pakistan

**Muhammad Riaz** Department of Environmental Sciences and Engineering, Government College University Faisalabad, Faisalabad, Pakistan

**Aryadeep Roychoudhury** Post Graduate Department of Biotechnology, St. Xavier's College (Autonomous), Kolkata, West Bengal, India

**Abdul Saboor** Department of Soil Science, Bahauddin Zakariya University, Multan, Punjab, Pakistan

**Bedabrata Saha** Plant Molecular Biotechnology Laboratory, Department of Life Science and Bioinformatics, Assam University, Silchar, Assam, India

**Smita Sahoo** Plant Molecular Biotechnology Laboratory, Department of Life Science and Bioinformatics, Assam University, Silchar, Assam, India

**Subhas Chandra Santra** Department of Environmental Science, University of Kalyani, Nadia, West Bengal, India

**Muhammad Saqib** Department of Horticulture, Bahauddin Zakariya University, Multan, Punjab, Pakistan

**Muhammad Sarfraz** College of Agriculture, Bahauddin Zakariya University, Layyah, Pakistan

**R. K. Sarkar** Department of Crop Physiology & Biochemistry, ICAR-National Rice Research Institute, Cuttack, India

**Taskeen Sarwar** College of Agriculture, Bahauddin Zakariya University, Layyah, Pakistan

**Abdul Sattar** College of Agriculture, Bahauddin Zakariya University, Layyah, Pakistan

**Muhammad Shahid** Department of Environmental Sciences, COMSATS University Islamabad, Vehari Campus, Pakistan

**Babar Shahzad** Tasmanian Institute of Agriculture, University of Tasmania, Hobart, TAS, Australia

**Muhammad Taimoor Shakeel** Department of Plant Pathology, Bahauddin Zakariya University, Multan, Punjab, Pakistan

**Anket Sharma** Plant Stress Physiology Laboratory, Department of Botanical & Environmental Sciences, Guru Nanak Dev University, Amritsar, India

**Poornima Sharma** Division of Plant Physiology, ICAR-Indian Agricultural Research Institute, New Delhi, India

**Ahmad Sher** College of Agriculture, Bahauddin Zakariya University, Layyah, Pakistan

**Muhammad Sohail** Department of Horticulture, Bahauddin Zakariya University, Multan, Punjab, Pakistan

**Bhaben Tanti** Department of Botany, Gauhati University, Guwahati, Assam, India

**Mohsin Tanveer** Tasmanian Institute of Agriculture, University of Tasmania, Hobart, TAS, Australia

**Meenakshi Thakur** College of Horticulture and Forestry, Neri, Hamirpur, Dr. Y.S. Parmar University of Horticulture and Forestry, Nauni, Solan, Himachal Pradesh, India

**Aman Ullah** Department of Agronomy, Faculty of Agriculture, University of Agriculture, Faisalabad, Pakistan

**Muhammad Waqas** Department of Botanical and Environmental Sciences, Kohat University of Science and Technology, Kohat, Pakistan

**Allah Wasaya** College of Agriculture, Bahauddin Zakariya University, Layyah, Pakistan

**Tauqeer Ahmad Yasir** College of Agriculture, Bahauddin Zakariya University, Layyah, Pakistan

**Sumaira Zafar** Department of Horticulture, Bahauddin Zakariya University, Multan, Punjab, Pakistan

**Iqra Zakir** Department of Agronomy, Bahauddin Zakariya University, Multan, Punjab, Pakistan

# About the Editors

**Mirza Hasanuzzaman** is professor of Agronomy at Sher-e-Bangla Agricultural University in Dhaka. He received his PhD on "Plant Stress Physiology and Antioxidant Metabolism" from Ehime University, Japan, with a scholarship from the Japanese Government (MEXT). Later, he completed his postdoctoral research at the Center of Molecular Biosciences, University of the Ryukyus, Japan, as a recipient of the Japan Society for the Promotion of Science (JSPS) Postdoctoral Fellowship. He was also the recipient of the Australian Government's Endeavour Research Fellowship for postdoctoral research as an Adjunct Senior Researcher at the University of Tasmania, Australia. His current work is focused on the physiological and molecular mechanisms of environmental stress tolerance. He has published over 80 articles in peer-reviewed journals, has edited 8 books, and has written 30 book chapters on important aspects of plant physiology, plant stress tolerance, and crop production. According to Scopus®, His publications have received about 3,800 citations with an *h*-index of 31. He is an editor and reviewer for more than 50 peer-reviewed international journals and was a recipient of the "Publons Peer Review Award 2017 and 2018." He has been honored by different authorities for his outstanding performance in different fields, like research and education, and has received the World Academy of Science Young Scientist Award (2014).

**Vasileios Fotopoulos** is associate professor in Structural and Functional Plant Biology and head of the CUT Plant Stress Physiology Group established in 2008. His main scientific research focuses on the study of nitro-oxidative signaling cascades involved in the plant's response to stress factors, while emphasis is being given in the development of chemical treatment technologies towards the amelioration of abiotic stress factors and promotion of plant growth. To date, he is the author of 62 scientific papers published in peer-reviewed journals (h-index=23, Source: Scopus), as well as 7 book chapters. He currently serves as associate editor in *BMC Plant Biology*, *Gene*, and five other journals. He has also been assigned to evaluate competitive research proposals from different countries (France, Belgium, Poland, Chile, Latvia, Greece, Italy, Portugal, Israel, Qatar, Austria, Cyprus, Denmark), EU proposals (EUROSTARS), while he is an active review panel member for COST Actions (EU) in the field of "Natural Sciences." Finally, he has acted as examiner of MSc theses/PhD dissertations from Italy, South Africa, and the Netherlands.

# Methods of Seed Priming

**Ahmad Sher, Taskeen Sarwar, Ahmad Nawaz, Muhammad Ijaz, Abdul Sattar, and Shakeel Ahmad**

## Contents



**Abstract** The food security of world population depends on our limited agricultural land and the reproductive capacity of field crops. Hence, plant scientist are putting their efforts to increase the crop yield by using the existing resources. Seed priming is a very promising, efficient, and low-cost approach to increase the germination, the growth, as well as the productive capability of crops. Water, inorganic salts, sugars, solid medium with water and nutrients, beneficial microbes, micronutrients, hormones, rhizobacteria, and organic sources are used as priming agents for seeds. Seed priming is not only used for improving the plant growth and yield but it also increases the abiotic stress tolerance. This chapter will focus on classification of seed priming by using different priming agents.

A. Sher · T. Sarwar · A. Nawaz (✉) · M. Ijaz · A. Sattar
College of Agriculture, Bahauddin Zakariya University,
Layyah, Pakistan

S. Ahmad (✉)
Department of Agronomy, Bahauddin Zakariya University, Multan, Punjab, Pakistan
e-mail: shakeelahmad@bzu.edu.pk



M. Hasanuzzaman, V. Fotopoulos (eds.), *Priming and Pretreatment of Seeds and Seedlings*, https://doi.org/10.1007/978-981-13-8625-1_1

**Keywords** Seed treatment · Hydropriming · Halopriming · Pre-treatment · Seedling vigor

# 1 Introduction

Seed priming is the control hydration of seeds in water or a solution of low osmotic potential to initiate the germination metabolism without radical protrusion. Many studies have reported that seed priming improves the stand establishment and productivity of field crops (Farooq et al. 2005, 2006a, b, c, d, 2007, 2008a, b, 2010). There are several types of seed priming including (i) hydropriming, (ii) halopriming, (iii) osmopriming, (iv) solid matrix priming, (v) biopriming, (vi) nutripriming, and (vii) seed priming with hormones, plant growth regulators, and other organic sources. Each of this seed priming technique has been discussed in detail below.

# 2 Hydropriming

In hydropriming, the seeds are simply soaked in water prior to sowing for a defined period of time depending upon the radical protrusion time of each plant species. Aeration may or may not be provided to the seeds. It is a very simple, low-cost technology as simple water is used in this priming technique. Hydropriming is followed by surface drying or redrying of seeds to their original weight. Many studies have reported that hydropriming may improve the stand establishment, seedling vigor, and productivity of field crops under optimal and suboptimal conditions. For example, Roy and Srivastava (1999) found that hydropriming of wheat seeds improved the seed germination in a saline soil. In another study, seed emergence and plant growth at vegetative and reproductive stages were improved in maize due to hydropriming (Nagar et al. 1998). Many studies have reported that hydropriming of seeds improved the seed germination and seedling emergence of rice (Basra et al. 2005; Farooq et al. 2006d), maize (Mohammadi et al. 2008; Dezfuli et al. 2008), wheat (Harris et al. 2001; Afzal et al. 2007; Nawaz et al. 2017), safflower (Bastia et al. 1999), chickpea (Harris et al. 1999; Kaur et al. 2002; Kamithi et al. 2016), lentil (Ghassemi et al. 2008), safflower (Ashrafi and Razmjoo 2010), mountain rye (Ansari and Zadeh 2012), and pearl millet (Kumar et al. 2002). In conclusion, hydropriming improves seed germination, seedling emergence, and productivity of field crops.

## 3 Halopriming

Halopriming involves the soaking of seeds in the aerated solutions of inorganic salts (potassium nitrate, sodium chloride, calcium sulfate, and calcium chloride) of variable concentration. Various studies have reported that halopriming improved the stand establishment, seedling growth, and productivity of diverse crop species under optimal and suboptimal conditions. For example, Khan et al. (2009) found a significant improvement in the stand establishment and seedling growth of hot pepper due to seed priming with sodium chloride under saline soil conditions. In another study, the priming of wheat seeds with sodium chloride and potassium chloride was useful for improvement in its performance in saline soil (Iqbal et al. 2006). Under salt stress, priming of sorghum seeds with potassium nitrate or calcium chloride improved the activity of amylase and proteases (Kadiri and Hussaini 1999) and increased the soluble sugars, free amino acids, and proteins in pigeon pea (Jyotsna and Srivastava 1998). In another study, Sedghi et al. (2010) found that seed priming with sodium chloride or potassium nitrate improved the stand establishment and seedling growth of sunflower. Improvement in α-amylase and dehydrogenase activities due to seed priming with potassium nitrate solution in muskmelon has also been reported (Singh et al. 1999). The priming of wheat seeds with calcium chloride improved the performance of wheat under saline stress (Roy and Srivastava 1999). Likewise, seed priming of millet and sorghum with calcium chloride and potassium nitrate improved the activity of α-amylase and proteases under salt stress (Kadiriand Hussaini 1999). Chang-Zheng et al. (2002) also reported that seed priming with mixed salt solution improved the seed germination and activities of α- and p-amylase and root dehydrogenase in rice under salt stress. In conclusion, halopriming improves the stand establishment, germination metabolism, and productivity of many crops under optimal and suboptimal conditions.

## 4 Osmopriming

In osmopriming, the seeds are soaked in aerated solution of sugars (sorbitol, mannitol, etc.) or polyethylene glycol (PEG), followed by surface drying or redrying to their original weight. It is also known as osmotic priming or osmo-conditioning. Many studies have reported that osmopriming improves the stand establishment and seedling/crop growth under optimal and suboptimal conditions. For example, osmopriming (20% PEG-8000) of ryegrass and sorghum seeds improved the germination rate of both crops under suboptimal conditions (Hur 1991). Salehzade et al. (2009) also found that osmopriming (with PEG-8000) of wheat seeds improved the stand establishment. In another study, the osmopriming (with PEG-As) enhanced the ATPase activity in peanut with substantial improvement in RNA syntheses and activity of acid phosphatase in the cotyledon and embryonic axis (Nawaz et al. 2013). In tomato, osmopriming (with PEG 6000) improved the seedling emergence, seedling vigor, stay green, and leaf growth as compared to unprimed seeds (Pradhan et al. 2015). In rice under salt stress, seed priming with mannitol (1, 2, and 3%) improved the seedling dry weight, the membrane stability, and stay green with a simultaneous reduction in the $Na^+/K^+$ (Theerakulpisut et al. 2017). Improvement in stand establishment of sorghum and peanut owing to mannitol seed priming has also been reported (Safiatou 2012). Many other studies have reported that seed priming with PEG improved germination in bell pepper (Thakur et al. 1997), soybean (Gongping et al. 2000; Sadeghi et al. 2011), tomato (Nagarajan and Pandita 2001), muskmelon (Nascimento 2003), bitter gourd (Pandita and Nagarajan 2004; Thirusenduraselvi and Jerlin 2009), cauliflower (Pallavi et al. 2006), safflower (Nabizadeh et al. 2012), barley (Shukla et al. 2016), and wheat (Azadi et al. 2013). In conclusion, osmopriming improves the germination metabolism in field and horticultural crops under a diverse array of environmental conditions.

## 5 Solid Matrix Priming

Matrix priming, also known as the solid matric conditioning, is accomplished with the controlled and limited hydration, as in hydropriming and osmopriming. However, the matrix priming utilizes the solid medium (matrix which delivers water and nutrients to seed prior to emergence of radical including vermiculite and diatomaceous and water-absorbent polymer) for seed priming purpose (Taylor and Harman 1990). These solid matrix materials have low bulk density and low osmotic potential and high water-holding capacity. Khan (1992) recommended the use of matrix priming due to the good water-holding capacity of the solid matrix and its ease of removal from the seed. The matrix priming has the ability to provide a good amount of oxygen to the seed during the process of hydration. In a study, the germination metabolism was improved by the matrix priming in *Helichrysum bracteatum* L. (Grzesik and Nowak 1998). Ilyas et al. (2002) found that matrix conditioning of hot pepper seeds improved the protein level and fruit quality. In another study on pepper and tomato by Kubik et al. (1988), it was found that solid matrix priming was much useful for improvement in germination. In conclusion, solid matrix priming might be a good option to boost up the seed germination in different crop species, especially the horticultural crops.

## 6 Biopriming

It is a newly emerged technique of seed treatment which integrates the physiological (seed hydration) and biological (seed inoculation with the beneficial organisms) mechanisms (Reddy 2012; Rakshit et al. 2015). Biopriming has been recently used as an alternate method of managing the soil- and seed-borne pathogens. Callan et al. (1990) first introduced the procedure of biopriming. Like other seed priming techniques, radical protrusion in not allowed in the biopriming. The exudates released from the seed may serve as source of energy and nutrients to the biocontrol agents during biopriming (Wright et al. 2003), thus facilitating the proliferation and the colonization of these biocontrol agents over the surface of seeds which facilitate the nutrient/water uptake. According to Rakshit et al. (2015), the biopriming with beneficial microbes offers an innovative crop protection toll by improving the seed quality, seedling vigor, and plant ability to withstand the suboptimal growth conditions, thus ensuring the sustainable crop production. Mahmood et al. (2016) reviewed that biopriming speeds up germination, ensures the uniformity of seedling emergence, and thus improves the crop yields and quality. Recently, the nutrient-solubilizing bacteria have been used as seed priming agent along with the nutrient media to enhance the nutrient uptake in crops. For example, Rehman et al. (2018a, b) found that *Pseudomonas*-aided zinc application or Zn seed priming with endophytic bacteria improved the productivity and biofortification of bread wheat. In conclusion, biopriming with beneficial microbes may be useful to combat the disease problem and improve the bioavailability of the micronutrients to the crops.

## 7 Nutripriming

Recently, the priming of seeds in different micronutrient soltutions is getting momentum in order to improve the micronutrient availability into plants and their final assimilation in the seed (biofortification) to reduce the malnutrition. Many studies have reported that seed priming with zinc, boron (B), and magnesium at pre-optimized rates improved the performance of different field crops owing to an improvement in seed germination, growth, and yield parameters. For example, Rehman (2012) conducted a field study to evaluate the effect of seed priming with boron in increasing the chickpea productivity. Seeds of two chickpea genotypes Punjab-2008 and T-3009 were soaked in aerated solutions of different B concentrations viz. 0.1, 0.01, 0.001% B. Seed priming in 0.001% B solution improved the stand establishment and uniformity. Seed priming in 0.001% B solution and hydropriming improved the leaf emergence, leaf elongation, plant height, leaf area index, crop and leaf growth rate, reduced time to flowering, caused early maturity, improved grain filling rate and duration, and grain enrichment with boron. In another study, boron nutripriming improved the germination and early seedling growth of rice in Punjab, Pakistan (Farooq et al. 2011). Various other studies have reported an improvement in the stand establishment and productivity of field crops due to nutripriming (Ullah et al. 2017; Iqbal et al. 2017; Rehman et al. 2015, 2018a, b; Farooq et al. 2018). In conclusion, nutripriming improves stand establishment, growth, productivity, and grain biofortification of field crops.

## 8 Seed Priming with Plant Growth Regulators, Hormones, and Other Organic Sources

Seed priming with hormones, plant growth-promoting rhizobacteria, and other organics sources has been reported to improve the stand establishment, growth, and productivity of field and horticultural crops under optimal and suboptimal conditions. For example, improvement in seed germination in the *Podophyllum hexandrum* (Nautiyal et al. 1987; Nadeem et al. 2000; Thakur et al. 2010), onion (Nalini et al. 2001), okra (Singh et al. 2004), bell pepper (Yogananda et al. 2004), brinjal (Kumar 2005), sunflower (Singh et al. 2006), and sesame (Kyuak et al. 1995) has been reported due to seed priming with gibberellic acid. In a study, Zhang et al. (2007) found that seed priming of lucerne seeds with brassinolide at an optimized concentration increased the seed germination and seedling growth under salt stress.

Seed priming with hormones to improve the crop performance under stressful environments has also attained greater attention in recent past (Sneideris et al. 2015). The hormonal priming generally consists of priming of seeds with different chemicals including hydrochloric acid, hypochlorite and other natural substances. In a study, the seed priming with ascorbic acid and salicylic acid (500 ppm each) improved the seedling growth and seedling dry weight under saline and non-saline conditions (Afzal et al. 2006). In pepper, Khan et al. (2009) found that seed priming with salicylic acid and acetyl salicylic acid improved the uniformity of seedling emergence under saline conditions. In addition to these chemicals, seed priming with ethylene has been reported to improve the seed germination of lettuce under high temperature stress (Nascimento et al. 2004). Afzal et al. (2005, 2006) also reported that hormonal priming in spring wheat with ascorbic acid, salicylic acid, and abscisic acid was useful for alleviation of salinity stress. Basra et al. (2006) found that seed priming in aerated solutions of either, ascorbate or salicylic acid (at concentrations of 10 and 20 ppm for 48 h) improved the stand establishment and seedling dry weight in rice. In conclusion, seed priming with plant growth regulators, hormones, and other organic sources is a viable option to improve the seed germination, growth, and productivity of crops.

## References

Afzal I, Basra SMA, Ahmad N, Farooq M (2005) Optimization of hormonal priming techniques for alleviation of salinity stress in wheat (*Triticum aestivum* L.). Caderno de Pesquisa Sér. Bio, Santa Cruz do Sul 17(1):95–109

Afzal I, Basra SMA, Farooq M, Nawaz A (2006) Alleviation of salinity stress in spring wheat by hormonal priming with ABA, salicylic acid and ascorbic acid. Int J Agric Biol 8(1):23–28

Afzal I, Basra SMA, Ahmad N, Lodhi TE (2007) Counteraction of salinity stress on wheat plants by pre-sowing seed treatments. Pak J Agric Sci 44:104–109

Ansari O, Zadeh FS (2012) Osmo and hydro priming improvement germination characteristics and enzyme activity of Mountain Rye (*Secale montanum*) seeds under drought stress. J Stress Physiol Biochem 8(4):253–261

Ashrafi E, Razmjoo K (2010) Effect of priming on seed germination and field emergence of safflower (*Carthamus tinctorius* L.). Seed Sci Technol 38(3):675–681
Azadi MS, Younesi E, Tabatabaei SA (2013) Seed germination, seedling growth and enzyme activity of wheat seed primed under drought and different temperature conditions. J Stress Physiol Biochem 9(4):1–4
Basra SMA, Farooq M, Tabassum R (2005) Physiological and biochemical aspects of seed vigor enhancement treatments in fine rice (*Oryza sativa* L.). Seed Sci Technol 33:623–628
Basra SM, Farooq M, Wahid A, Khan MB (2006) Rice seed invigoration by hormonal and vitamin priming. Seed Sci Technol 34(3):753–758
Bastia DK, Rout AK, Mohanty SK, Prusty AM (1999) Effect of sowing date sowing methods and seed soaking on yield and oil content of rainfed safflower grown in Kalahandi, Orissa. Indian J Agron 44:621–623
Callan NW, Mathre D, Miller JB (1990) Bio-priming seed treatment for biological control of *pythium ultimum* preemergence damping-off in sh2 sweet corn. Plant Dis 74(5):368–372
Chang-Zheng H, Jin H, Zhi-Yu Z, Song-Lin R, Wen-Jian S (2002) Effect of seed priming with mixed-salt solution on germination and physiological characteristics of seedling in rice (*Oryza sativa* L.) under stress conditions. J Zhejiang Univ (Agric Life Sci) 28:175–178
Dezfuli PM, Zadeh FS, Janmohammadi M (2008) Influence of priming techniques on seed germination behavior of maize inbred lines (*Zea mays* L.). J Agric Biol Sci 3(3):22–25
Farooq M, Basra SMA, Hafeez K, Ahmad N (2005) Thermal hardening: a new seed vigor enhancement tool in rice. J Integr Plant Biol 47:187–193
Farooq M, Basra SMA, Wahid A (2006a) Priming of field-sown rice seed enhances germination, seedling establishment, allometry and yield. Plant Growth Regul 49:285–294
Farooq M, Basra SMA, Tabassum R, Afzal I (2006b) Enhancing the performance of direct seeded fine rice by seed priming. Plant Prod Sci 9:446–456
Farooq M, Basra SMA, Khalid M, Tabassum R, Mehmood T (2006c) Nutrient homeostasis, reserves metabolism and seedling vigor as affected by seed priming in coarse rice. Can J Bot 84:1196–1202
Farooq M, Basra SMA, Afzal I, Khaliq A (2006d) Optimization of hydropriming techniques for rice seed invigoration. Seed Sci Technol 34:507–512
Farooq M, Basra SMA, Ahmad N (2007) Improving the performance of transplanted rice by seed priming. Plant Growth Regul 51:129–137
Farooq M, Basra SMA, Asad SA (2008a) Comparison of conventional puddling and dry tillage in rice–wheat system. Paddy Water Environ 6:397–404
Farooq M, Basra SMA, Rehman H, Saleem BA (2008b) Seed priming enhances the performance of late sown wheat (*Triticumaestivum* L.) by improving chilling tolerance. J Agron Crop Sci 194:55–60
Farooq M, Basra SMA, Wahid A, Ahmad N (2010) Changes in nutrient-homeostasis and reserves metabolism during rice seed priming: consequences for germination and seedling growth. Agric Sci China 9:101–108
Farooq M, Atique-ur-Rehman AT, Habib M (2011) Boron nutripriming improves the germination and early seedling growth of rice (*Oryza sativa* L.). J Plant Nutr 34(10):1507–1515
Farooq M, Ullah A, Rehman A, Nawaz A, Nadeem A, Wakeel A, Nadeem F, Siddique KH (2018) Application of zinc improves the productivity and biofortification of fine grain aromatic rice grown in dry seeded and puddled transplanted production systems. Field Crops Res 216:53–62
Ghassemi GK, Aliloo AA, Valizadeh M, Moghaddam M (2008) Effect of hydro and osmo-priming on seed germination and Field Emergence of Lentil (*Lens culinaris* Medik.). Not Bot Horti Agrobot Cluj 36(1):29–33
Gongping GU, Guo Rong WU, Chang Mei L, Chang Fang Z (2000) Effects of PEG priming on vigour index and activated oxygen metabolism in soybean seedlings. Chin J Oil Crop Sci 22:26–30

Grzesik M, Nowak J (1998) Effects of matriconditioning and hydro-priming on *Helichrysumbracteatum* L. seed germination, seedling emergence and stress tolerance. Seed Sci Technol 26:363–376

Harris D, Joshi A, Khan PA, Gothakar P, Sodhi PS (1999) On-farm seed priming in semi-arid agriculture: Development and evaluation in corn, rice and chickpea in India using participatory methods. Exp Agric 35:15–29

Harris D, Raghuwanshi BS, Gangwar JS, Singh SC, Joshi KD, Rashid A, Hollington PA (2001) Participatory evaluation by farmers of on-farm seed priming in wheat in India, Nepal and Pakistan. Exp Agric 37:403–415

Hur SN (1991) Effect of osmoconditioning on the productivity of Italian ryegrass and sorghum under suboptimal conditions. Korean J Anim Sci 33:101–105

Ilyas S, Sutariati GAK, Suwarno FC, Sudarsono (2002) Matriconditioning improves the quality and protein level of medium vigor hot pepper seed. Seed Technol 65–75

Iqbal M, Ashraf M, Jamil A, Rehman S (2006) Does seed priming induce changes in the levels of some endogenous plant hormones in hexaploidwheat plants under salt stress. J Int Plant Biol 48:181–189

Iqbal S, Farooq M, Cheema SA, Afzal I (2017) Boron seed priming improves the seedling emergence, growth, grain yield and grain biofortification of bread wheat. Int J Agri Biol 19(1):177–182

Jyotsna V, Srivastava AK (1998) Physiological basis of salt stress resistance in pigeonpea (Cajanuscajan L.)-II. Pre-sowing seed soaking treatment in regulating early seedling metabolism during seed germination. Plant Physiol Biochem 25:89–94

Kadiri M, Hussaini MA (1999) Effect of hardening pretreatments on vegetative growth, enzyme activities and yield of *Pennisetumamericanum* and *Sorghum bicolor* L. Global J Pure Appl Sci 5:179–183

Kamithi KD, Wachira F, Kibe AM (2016) Effects of different priming methods and priming durations on enzyme activities in germinating chickpea (Cicerarietinum L.). Am J Nat Appl Sci 1:1):1–1):9

Kaur S, Gupta AK, Kaur N (2002) Effect of osmo and hydropriming of chickpea seeds on seedling growth and carbohydrate metabolism under water deficit stress. Plant Growth Regul 37:17–22

Khan AA (1992) Preplant physiological seed conditioning. Hortic Rev 14:131–181

Khan HA, Ayub CM, Pervez MA, Bilal RM, Shahid MA, Ziaf K (2009) Effect of seed priming with NaCl on salinity tolerance of hot pepper (*Capsicum annuum* L.) at seedling stage. Soil Environ 28:81–87

Kubik KK, Eastin JA, Eastin JD, Eskridge KM (1988) Solid matrix priming of tomato and pepper. In: Proceedings of International Conference on Stand Establishment in Horticultural crops, Lancaster, pp 86–96

Kumar S (2005) Influence of pre-sowing seed treatment and seed pelleting on storability in brinjal (Solanummelongena L.). MSc (Agri) Thesis, University of Agriculture Science, Dharwad, India

Kumar A, Gangwar JS, Prasad SC, Harris D (2002) On-farm seed priming increases yield of direct-sown finger millet in India. Int Sorghum Millets News 43:90–92

Kyuak H, Hopper NW, Bridham RD (1995) Effects of temperature and presoaking on germination, root length and shoot length of sesame (*Sesamumindicum* L.). Environ Exp Bot 35(3):345–351

Mahmood A, Turgay OC, Farooq M, Hayat R (2016) Seed biopriming with plant growth promotingrhizobacteria: a review. FEMS Microbiology Ecol 92(8). PMID 2722222, https://doi.org/10.1093/fersec/fiw112

Mohammadi MJ, Dezfuli PM, Sharifzadeh F (2008) Seed invigoration techniques to improve germination and early growth of inbred line of maize under salinity and drought stress. J Appl Plant Physiol 34:215–226

Nabizadeh E, Yousefi F, Gerami F (2012) Osmopriming effects on safflower seed germination and seedling growth. Eur J Exp Biol 2(3):623630

Nadeem M, Palni LMS, Purohit AN, Pandey H, Nandi SK (2000) Propagation and conservation of Podophyllum hexandrum Royle: an important medicinal herb. Nandi Biol Conserv 92:121–129
Nagar R, Dadlani PM, Sharma SP (1998) Effect of hydro-priming on field emergence and crop growth of maizegenotypes. Seed Sci Res 26:1–5
Nagarajan S, Pandita VK (2001) Improvement in germination characteristics in artificially aged seeds of tomato by osmoconditioning. Seed Res 29(2):136–140
Nalini T, Poonam S, Lal C, Katiyar PK, Vaish CP (2001) Effect of presowing seed treatments on germination growth and yield of onion. Seed Res 29(2):238–239
Nascimento WM (2003) Muskmelon seed germination and seedling development in response to seed priming. Sci Agric 60(1):71–75
Nascimento WM, Cantliffe DJ, Huber DJ (2004) Ethylene evolution and endo-beta-mannanase activity during lettuce seed germination at high temperature. Sci Agric 61(2):156–163
Nautiyal MC, Rawat AS, Bhadula SK, Purohit AN (1987) Seed germination in Podophyllum hexandrum. Seed Res 15:206–209
Nawaz J, Hussain M, Jabbar A, Nadeem GA, Sajid M, Subtain MU, Shabbir I (2013) Seed priming a technique. Int J Agric Crop Sci 6(20):1373–1381
Nawaz A, Farooq M, Ahmad R, Basra SMA, Lal R (2016) Seed priming improves stand establishment and productivity of no till wheat grown after direct seeded aerobic and transplanted flooded rice. Eur J Agron 76:130–137
Pallavi HM, Kalappa VP, Ramegowda (2006) Application of vigour tests and seed invigoration in cauliflower. XII Nation Seed Seminiferous Prosperity Qual Seed 24(26):101
Pandita VK, Nagarajan S (2004) Effect of pre-sowing treatments in improving emergence of bitter gourd seedlings. Indian J Hortic 61(3):280–281
Pradhan N, Prakash P, Manimurugan C, Tiwari SK, Sharma RP, Singh PM (2015) Screening of tomato genotypes using osmopriming with PEG 6000 under salinity stress. Res Environ Life Sci 8(2):245–250
Rakshit A, Singh HB, Sen A (2015) Nutrient use efficiency: from basics to advances. Springer, New Delhi
Reddy PP (2012) Bio-priming of seeds. In: Recent advances in crop protection. Springer, New Delhi, pp 83–90
Rehman A (2012) Evaluating the potential of seed priming with boron for improving chickpea productivity. M.Sc. (Hons.) Dissertation. Department of Agronomy, University of Agriculture, Faisalabad
Rehman A, Farooq M, Ahmad R, Basra SMA (2015) Seed priming with zinc improves the germination and early seedling growth of wheat. Seed Sci Technol 43(2):262–268
Rehman A, Farooq M, Naveed M, Nawaz A, Shahzad B (2018a) Seed priming of Zn with endophytic bacteria improves the productivity and grain biofortification of bread wheat. Eur J Agron 94:98–107
Rehman A, Farooq M, Naveed M, Ozturk L, Nawaz A (2018b) Pseudomonas-aided zinc application improves the productivity and biofortification of bread wheat. Crop Pasture Sci 69(7):659–672
Roy NK, Srivastava AK (1999) Effect of presoaking seed treatment on germination and amylase activity of wheat (*Triticumaestivum* L.) under salt stress conditions. Rachis 18:46–51
Sadeghi H, Khazaei F, Yari L, Sheidaei S (2011) Effect of seed osmopriming on seed germination behavior and vigour of soyabean (*Glycine max* L.). Agric Biol Sci 6:39–43
Safiatou SS (2012) Effect of different seed priming methods on germination, seedling establishment and vigour in sorghum (*Sorghum bicolor* (L.) Moench.) andbambara groundnut (*Vignasubterrenea* (L.) Verdc.). Doctoral dissertation
Salehzade H, Izadkhah Shishvan M, Chiyasi M (2009) Effect of seed priming on germination and seedling growth of wheat (*Triticum aestivum* L.). J Biol Sci 4:629–631
Sedghi M, Ali N, Esmaielpour B (2010) Effect of seed priming on germination and seedling growth of two medicinal plants under salinity. J Emir Food Agric 22:130–113

Shukla N, Verma Y, Shukla PK, Misra P (2016) Effect of polyethylene glycol (PEG) 6000 on seed priming in drought tolerant and sensitive barley (*Hordeumvulgare* L.) seeds. Int J Plant Sci 11(1):75–78

Singh G, Gill SS, Sandhu KK (1999) Improved performance of muskmelon (*Cucumis melo*) seeds with osmoconditioning. Acta Agrobot 52(1–2):121–137

Singh J, Kanwar JS, Geeta B (2004) Seed vigour as influenced by different seed priming treatments in okra (*Abelmoschusesculentus* L. Moench). Seed Res 32(2):122–125

Singh P, Kumar A, Bajpai VP, Singh CB, Jatar AL (2006) Effect of pretreatment on partially aged seeds of sunflower (Heliatnhusannuus L.) for germination, vigour and seedling growth. National Seed Seminar, Abstract XII. Angrau, Hyderabad, p 130

Sneideris LC, Gavassi MA, Campos ML, D'Amico-Damiao V, Carvalho RF (2015) Effects of hormonal priming on seed germination of pigeon pea under cadmium stress. An Acad Bras Cienc 87(3):1847–1852

Taylor AG, Harman GE (1990) Concepts and technologies of selected seed treatments. Annu Rev Phytopathol 28:321–329

Thakur A, Thakur PS, Bhardwaj J (1997) The influence of osmoconditioning on germination potential and seedling performance of bell pepper. Seed Res 25(1):25–30

Thakur A, Thakur PS, Dutt V, Thakur CL (2010) Conservation of Podophyllum hehandrum – an endangered medicinal herb through seeds. Indian J Plant Physiol 15:110–116

Theerakulpisut P, Kanawapee N, Panwong B (2017) Seed priming alleviated salt stress effects on rice seedlings by improving $Na^{+}/K^{+}$ and maintaining membrane integrity. Int J Plant Biol 7:6402

Thirusenduraselvi D, Jerlin R (2009) Osmopriming of seeds to improve the performance of bitter gourd cv. CO-1. Int J Plant Sci 4(1):182–187

Ullah A, Farooq M, Nadeem A, Rehman A, Asad SA, Nawaz A (2017) Manganese nutrition improves the productivity and grain biofortification of fine grain aromatic rice in conventional and conservation production systems. Paddy Water Environ 15(3):563–572

Wright B, Rowse HR, Whipps JM (2003) Application of beneficial microorganisms to seeds during drum priming. Biomed Sci Technol 13(6):599–614

Yogananda DK, Vyakarnahal BS, Shekhargouda M (2004) Effect of seed invigoration with growth regulations and micronutrients on germination and seedling vigour of bell pepper cv. California Wonder. Karnataka J Agric Sci 17(4):811–813

Zhang S, Hu J, Zhang Y, Xie XJ, Knapp A (2007) Seed priming with brassinolide improves lucerne (*Medicago sativa* L.) seed germination and seedling growth in relation to physiological changes under salinity stress. Aust J Agric Res 5(8):811–815

# Advances in the Concept and Methods of Seed Priming

**Muhammad Waqas, Nicholas Emmanuel Korres, Muhammad Daud Khan, Abdul-Sattar Nizami, Farah Deeba, Iftikhar Ali, and Haziq Hussain**

## Contents



---

M. Waqas (✉)
Department of Botanical and Environmental Sciences, Kohat University of Science and Technology, Kohat, Pakistan

N. E. Korres
Department of Crop, Soil, and Environmental Sciences, University of Arkansas, Fayetteville, AR, USA

M. D. Khan · F. Deeba · H. Hussain
Department of Biotechnology and Genetic Engineering, Kohat University of Science and Technology, Kohat, Pakistan

A.-S. Nizami
Center of Excellence in Environmental Studies (CEES), King Abdulaziz University, Jeddah, Saudi Arabia

I. Ali
Institute of Soil and Environmental Sciences, Gomal University, Dera Ismael Khan, Pakistan



M. Hasanuzzaman, V. Fotopoulos (eds.), *Priming and Pretreatment of Seeds and Seedlings*, https://doi.org/10.1007/978-981-13-8625-1_2

**Abstract** The critical stages during the growth of crops are the uniform seed germination, early seedling growth, and uniform plant stand. Low crop yield is attributed to uneven seed germination and seedling growth. Therefore, the quality of seed can be improved through priming in addition to the field management techniques for better seed germination. Priming is a physiological technique of seed hydration and drying to enhance the pregerminative metabolic process for rapid germination, seedling growth, and final yield under normal as well as stressed conditions. The primed seeds show faster and uniform seed germination due to different enzyme activation, metabolic activities, biochemical process of cell repair, protein synthesis, and improvement of the antioxidant defense system as compared to unprimed seeds. There are many techniques of seed priming which are broadly divided into conventional methods (hydro-priming, osmo-priming, nutrient priming, chemical priming, bio-priming, and priming with plant growth regulators) and advanced methods (nano-priming and priming with physical agents). However, priming is strongly affected by various factors such as temperature, aeration, light, priming duration, and seed characteristics. This chapter highlights the priming mechanism and the available technologies as a tool for superficial seed germination and crop stand. An experiment with reference to the importance of priming toward vigor seed germination and seedling growth was conducted, and its results have been added in this chapter.

**Keywords** Seed priming · Germination · Antioxidant defense system · Metabolic activities · Crop growth

## 1 Introduction

Seed treatment before sowing is the foundation for activation of seed resources that in combination with external ingredients could contribute to the efficient plant growth and high yield. Various physiological and non-physiological techniques are available for enhancing seed performance as well as to combat environmental constraints. The physiological treatments for improving seed germination and stand establishment are composed of seed hydration techniques such as humidification, wetting, and presoaking. The other techniques for promoting germination are comprised of chemical treatments, seed inoculation with beneficial microbes, and seed coating. Seed priming is a physiological method of controlled hydration and drying to enhance sufficient pre-germinative metabolic process for rapid germination (Dawood 2018). This is one of the economic and feasible technologies for uniform seed development in most of the field crops. Other benefits include efficient nutrients uptake, water use efficiency, release photo- and thermo-dormancy, maturity, and crop yield (Hill et al. 2008; Bagheri 2014; Lara et al. 2014; Dutta 2018). However, many factors affect the performance of seed priming such as plant species, priming duration, temperature, priming media, and their concentration and storage conditions. The treated seed with a proper reagent can germinate better, for instance, seed treatment with inorganic salts (KCl and $KNO_3$) enhanced the germination and growth rates. The KCl improved the starch and protein contents, whereas $KNO_3$ increased the fruit size and quality (Singh et al. 2015a, b).

The other priming reagents involved in breaking seed dormancy are gibberellic acid (GA) and cytokinins (Assefa et al. 2010). Priming induces a set of biochemical changes such as enzyme activation, metabolism of germination inhibitors, repair of cell damages, and imbibition to promote germination (Farooq et al. 2010). Priming also modifies biochemical and physiological nature of embryo and affects the release of substances during germination phase II that activates the production of hydrolytic enzymes. These substances make high-energy compounds and essential chemicals for the germinating seedlings available (Renugadevi and Vijayageetha 2006). Therefore, the positive effects of seed priming are highly attributed to various biochemical phenomena such as improvement of the antioxidant defense system and restoration of metabolic activities through the synthesis of proteins and nucleic acids (RNA and DNA) (Di Girolamo and Barbanti 2012).

## 2 History of Seed Priming

Various seed treatment techniques are introduced and examined for uniform germination under different environmental conditions. Evenari (1984) reported that the efforts for improving seed germination and growth are dated back to ancient Greeks. Theophrastus (371–287 B.C.), during an investigation, observed that cucumber seeds when soaked in water result in faster and uniform germination as compared to unprimed seeds (Theophrastus, Enquiry into Plants, Book VII, I.6). Likewise, the Roman naturalist Gaius Plinius Secundus (23–79 A.D.) in his Encyclopedia reported the positive effects of presoaking of cucumber seeds in honey and water for seed germination (Gaius 1949–1954). Afterward, in 1539–1619, the French botanist Oliver de Serres reported about the seed soaked in manure water for 2 days and then dried before sowing as an effective way of seed treatment for better crop growth.

Charles Darwin tested osmo-priming on lettuce and cress seeds in seawater and observed high germination in the treated seeds as compared to nonprimed seed (Darwin 1855). The modern concept of seed priming is presented by Ells (1963), who highlighted the critical parameters related to seed treatment. He observed high germination rates when seeds were treated with a specific nutrient solution. Koehler (1967) reported that treatment with salt solution promotes RNA accumulation that, in turn, enhances other physiological process and results in high seed emergence. May et al. (1962) stated that seed drying for certain time at specific level after priming exerts beneficial effect and leads to fast germination under normal as well as stressful conditions (Berrie and Drennan 1971). Heydecker et al. (1973) used organic chemical polyethylene glycol (PEG) H–(O–$CH_2$–$CH_2$)n–OH, a high molecular weight compound, for seed pretreatment to boost germination and avoid several problems associated with salts treatment like hardening. The priming technology, so far in research and development, has been adopted as a novel technique for getting a uniform crop standby several seed and agricultural companies.

## 3 Phenomenon of Seed Priming

After sowing, seeds remain in the soil for a certain period to absorb water and some essential nutrients for their growth. Seed priming is a technique to reduce this time and makes the germination quickly and uniformly. In addition to hydration, priming also reduces the sensitivity of seed to external environmental factors (Afzal et al. 2016). Priming promotes seed germination under three stages such as imbibition, germination, and growth (Fig. 1). During the imbibition stage, the water uptake promotes protein synthesis and respiratory activities through messenger ribonucleic acid (mRNA). The second stage is related to the initiation of different physiological activities related to germination such as protein synthesis, mitochondria synthesis, and alteration in soluble sugars (Varier et al. 2010). The critical factor during seed priming is the controlled water uptake during the second stage, before the emergence and growth of radical from the seed coat during the last stage. The second

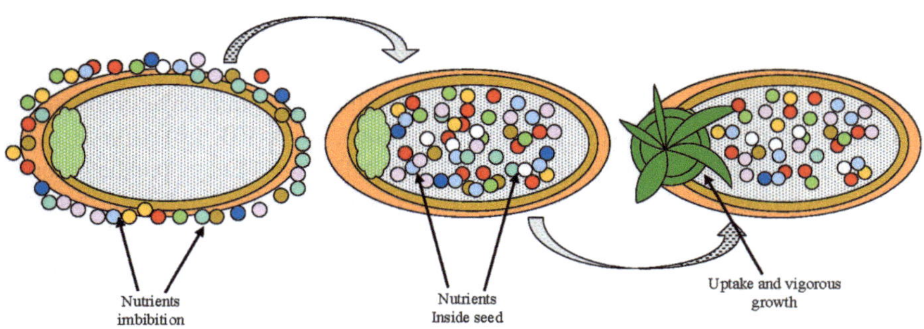


**Fig. 1** General phenomenon of seed priming

stage (germination) is much sensitive to environmental factors than the third stage (Côme and Thévenot 1982). Therefore, during priming, the seeds that have passed through the second stage could germinate under variant environmental conditions as compared to unprimed seeds (Corbineau and Côme 2006).

# 4 Methods of Seed Priming

There are several techniques of seed priming that are broadly divided into conventional and advanced methods. The traditional techniques are comprised of hydro-priming, osmo-priming, nutrient priming, chemical priming, bio-priming, seed priming with plant growth regulators, and priming with plant extracts, whereas the advanced techniques of seed priming include seed priming through nanoparticles and priming through physical agents (Table 1). These techniques are described below.

## *4.1 Conventional Seed Priming Methods*

### 4.1.1 Hydro-priming

Hydro-priming is a simple and economical technique in which seeds are soaked in water for a specific period and dried to a certain moisture level before sowing (Singh et al. 2015a, b). This technology is useful in areas with adverse environmental conditions including high heat and drought stress. Hydro-priming improves the water uptake efficiency and seed hydration under such conditions (McDonald 2000). However, it is critical to maintaining optimum humidity and temperature to avoid radical projection, as hydro-priming may result in noncontrolled water uptake (Taylor et al. 1998). The key technique of hydro-priming is "drum priming" invented by Rowse (1991). Drum priming is comprised of a drum containing seed lot connected to a boiler producing vapors. The vapors upon entering the drum condense into water droplets. During treatment, the increase in relative seed mass is measured along with strict control of the volume of the water and time required for seed hydration (Warren and Bennett 1997). Various research studies have explained that during drought stress conditions, hydro-priming increases the germination and seedling growth by 3–4 times as compared to nonpriming (Kaur et al. 2002). Likewise, Sung and Chiu (1995) reported that watermelon seeds when subjected to hydro-priming result in fast germination and seedling growth. In addition to several food crops such as *Allium porrum*, coriander, pyrethrum, and wheat, hydro-priming resulted in a uniform early germination in many desert plants like desert cacti (Dubrovsky 1996).

**Table 1** The effect of conventional and advance seed priming techniques on crops growth and development

| Crop | Priming solutions | Priming duration (hours) | Observations | Studies |
|---|---|---|---|---|
| Pepper | Marigold flowers petal extract | 24 | 10% high seed germination<br>15% high seedling emergence rate<br>Reduced the MGT up to 40%<br>High germination index<br>High seedling weight | Mavi (2016) |
| Wheat | Water, KCl, $CaCl_2$ | 12, 14, 24 | Improved crop performance under chilling temperature<br>Reduced time to start emergence by ~16%<br>Reduced time to 50% germination and MET by ~17 and 33%, respectively<br>Increase the plant height, fertile tillers, and straw yield<br>Enhanced the grain yield by ~12 | Farooq et al. (2008) |
| Rice | Water | 24 | Improved seedling growth by enhancing germination index, seedling vigor index, and germination energy<br>Reduced mean germination time<br>Increased panicle number($m^{-2}$)<br>Improved crop growth and final yield | Mahajan et al. (2011) |
| Marigold | Solution Zn and Mn | 12 | Enhanced the germination rate up to 93%<br>Increased SVI by 18.5%<br>Flower yield was increased >63%<br>~50% increase in essential oil production | Mirshekari et al. (2012) |
| Barley | Solution of Zn and P | 12 | Increased the germination rate from 65 to 95%<br>50% germination was achieved after 2 days<br>High nutrient accumulation without affecting germination<br>Increased the plant height, root, and shoot biomass<br>Stimulate root growth by 27%<br>Increased the water use efficiency by 44% under drought conditions | Ajouri et al. (2004) |

(continued)

**Table 1** (continued)

| Crop | Priming solutions | Priming duration (hours) | Observations | Studies |
|---|---|---|---|---|
| Linseed | Salicylic acid, $CaCl_2$ and moringa leaf extract | 12 | Reduced time to 50% germination and mean germination time<br>Enhanced the GI and final germination<br>Increased the fresh and dry weight of root and shoot<br>Increase in seed weight by 9.30%, biological yield by 34.16%, seed yield by 39.49%, harvest index by 4.12%, and oil contents by 13.39% | Rehman et al. (2014) |
| Cabbage | Urea | 8 | Primed seed showed high germination rate<br>Enhanced antioxidative enzyme activity<br>Improved soluble sugar and proline content | Yan (2015) |
| Tomato | Plants extracts | 48 | Maximum germination up to 90%<br>Perk up seed vigor index, seedling length, and fresh and dry biomass<br>Suppress plant disease and increased the survival rate up to 68% | Prabha et al. (2016) |
| Rice | Polyamine | 48 | Increased seed germination and seedling length<br>Affect fresh and dry weight<br>Enhanced net photosynthesis, proline production, water use efficiency<br>Improved anthocyanins, soluble phenolics, and membrane properties | Farooq et al. (2009) |
| Mung bean | Water and polyethylene glycol | 6, 12 | Earlier emergence and flowering as compared to unprimed<br>Rapid physiological maturity<br>Increased the grain $pod^{-1}$ by 14%, grain weight by 3.5%, and grain yield by 12% | Khan et al. (2008) |
| Rice | Selenium | 24 | Triggered germination rate and early seedling growth<br>Greater α-amylase activity; enhanced activities of enzymatic antioxidants, peroxidase, catalase, and glutathione peroxidase; and higher soluble sugars were observed as compared to unprimed<br>High chlorophyll content<br>Greater membrane stability, high starch metabolism, and increased activities of antioxidant | Khaliq et al. (2015) |

(continued)

**Table 1** (continued)

| Crop | Priming solutions | Priming duration (hours) | Observations | Studies |
|---|---|---|---|---|
| Fennel | Salicylic acid | | Enhanced germination rate<br>Increased seed stamina index, seedling fresh and dry weight | Farahbakhsh (2012) |
| Rice | Water and $KH_2PO_4$ | 24 | Priming with both water and P significantly enhanced seed emergence and seedling growth<br>Better performance for plant parameters<br>Increased shoot biomass and root length<br>High P concentration in shoot due to priming with P solution | Pame et al. (2015) |
| Peanut | ZnO nanoparticles | 3 | Increased seed germination rate<br>Improved stem and root growth<br>Increased pod yield per plant by 34%<br>Enhanced growth and crop yield | Prasad et al. (2012) |
| Wheat | Silver nanoparticles | – | Increased percent germination<br>Increased plumule and radicle length | Salehi and Tamaskani (2008) |
| Rice | Calcium-phosphate nanoparticles | 48 | Improved seedling growth<br>Stimulate metabolites and enzymes related with antioxidative responses | Upadhyaya et al. (2017) |
| Rice | SiO2 | 6 | Nanoparticles showed no toxic effect on seedling growth<br>Improved root length, seedling dry weight, and root volume | Adhikari et al. (2013) |

### 4.1.2 Osmo-priming

The method of osmo-priming is patented by Heydecker et al. (1973). It is a widely used commercial technique in which seeds are hydrated to a controlled level to allow pregermination metabolic activities (Halmer 2004). During the process, seeds are exposed to a controlled level of imbibition because of excess water entry to seed resulting in reactive oxygen species (ROS) accumulation as well as oxidative damage of cellular components such as proteins, lipid membranes, and nucleic acids. Osmo-priming through a delayed water entry to seed reduces the ROS accumulation and thus protects the cell from oxidative injury. Osmotica such as PEG [H-(O-$CH_2$-$CH_2$)n-OH], sugar, mannitol ($C_6H_{14}O_6$), and sorbitol ($C_6H_{14}O_6$) are added to the solution for lowering water uptake. In addition, different salts such as $NaNO_3$, $MgCl_2$, NaCl, and $KNO_3$ are used for osmo-priming. Singh et al. (2014) experimented on osmo-priming using cowpea. They used $KNO_3$ as priming solution with three levels of time durations (6, 8, and 10 h). Their results showed that in comparison to unprimed treatments, osmo-priming was proved superior in terms of all germination and growth parameters. Furthermore, osmo-priming with $KNO_3$ showed greater results than hydro-priming for all tested parameters. Osmo-priming is technically and financially more feasible as compared to hydro-priming, because osmo-primed seed results in quicker germination with low cost and better water conservation, thus providing a promising alternative to the farmers (Moradi and Younesi 2009). Jett et al. (1996) stated that the controlled seed hydration in osmo-priming preserves plasma membrane and causes quicker germination. However, during solution selection for osmo-priming, the morphology of seed should be considered as the semipermeable outer layers in some seeds are highly sensitive that affect the efficacy of priming (Pill 1995). The semipermeability of this outer layer is due to the presence of amorphous tissue between seed coat and pericarp that inhibits solute exchange, thus controlling priming agent and water to enter to the seed (Zhou et al. 2013). The internal osmotic equilibrium and nutritional balance of the seed will be disturbed if the solution is not properly selected according to the permeability of the seed due to the penetration of ions released from priming solution (salts) (Bradford 1995).

### 4.1.3 Nutrient Priming

The saturation of seeds with a certain concentration of nutrients for a specific period before sowing is known as nutrient priming (Shivay et al. 2016). Priming of crop seeds with either micro- or macronutrients enhances the nutrient substances and augments the germination, sprout (seedling) development, and water uptake efficiency. Micronutrient seed priming is a well-known technique to increase the osmosis for water regulation in seeds during the germination period (Singh 2007). For instance, seed primed with sodium molybdate dihydrate (0.02% and 0.04%) for 5 h improved the yield of mung bean (Umair et al. 2011). Likewise, macronutrient seed priming is the most effective technique (Rakshit et al. 2013). For example, potassium (K) is a mineral nutrient, and soaking of crop seeds with K increases the tolerance possibility of plant life against the different environmental stress conditions (Cakmak 2005). Grain yield of wheat and chickpea can be improved by priming with Zn solutions (Arif et al. 2007). Likewise, in moderately Zn-deficient soils such as alkaline soil, Zn priming is helpful to mimic Zn deficiency in plants (Harris et al. 2008). It also improves crop appearance, growth, yield, and nutrient absorption (Shivay et al. 2016). Chickpea seeds primed in 0.05% solution of $ZnSO_4.7H_2O$ (zinc sulfate heptahydrate) give a 19% high seeds production and have 29% more seed's Zn as compared to that of nonprimed chickpea seeds (Harris et al. 2008). Besides, mineral nutrient priming increases plants' tolerance to various environmental stress conditions (Marschner 1995). Ascorbic acid is also an essential vitamin nutrient and has been used for seed priming because it is an antioxidant in nature. Seeds primed with elevated concentrations of internal plant ascorbate retain the antioxidant capability of plants and protect them from oxidative stress damage (Zhou et al. 2009). Tavili et al. (2009) reported that seed treatment with ascorbic acid increases the germination rate of *Agropyron elongatum* grown in salt stress conditions.

### 4.1.4 Chemical Priming

Numerous chemicals are in use to soak a variety of crops seeds before germination. Natural and synthetic chemicals like choline, chitosan, putrescine, ethanol, paclobutrazol, $ZnSO_4$, $KH_2PO_4$, $CuSO_4$, and Se are used in seed priming to enhance growth and tolerance in crop plants (Jisha et al. 2013). Pretreatment of seeds with these chemicals increases crop plants growth, and plants attain resistance to various abiotic stresses. Priming of *Salvia* L. and *Capsicum annuum* L. seeds and other crop species with butenolide compound promotes seedling vitality and emergence (Demir et al. 2012). The improved seedling appearance, because of butenolide pretreatment, lessens the chance of plant pathogens' attack. Similarly, presoaking rice seeds with (1% or 5%) ethanol solution results in more rapidly and consistent germination rate and high leaf numbers (Farooq et al. 2006). Putrescine is another chemical compound that can be used for seed treatment. Soaking seeds of tobacco in this compound solution develop the cold stress tolerance at the stage of germination and seedling growth through the regulation of antioxidant system (Xu et al. 2011). Also presoaking with a paclobutrazol compound developed salt stress tolerance for *Catharanthus roseus* because of antioxidant system regulation (Jaleel et al. 2007). Shahrokhi et al. (2011) described that seed priming of turf grass with this compound in drought stress affects the plant physiology, though it is associated with the concentration of the paclobutrazol solution and the temperament of the cultivar. In addition, chitosan is a large-sized cationic polysaccharide molecule generally obtained during waste materials of seafood processing. Priming with chitosan increased disease resistance, the rate, and percentage of germination and the lipolytic activity of lipase, GA3, and indole-3-acetic acid (IAA) quantity and as well enhanced the quality of seeds for crop plants (Shao et al. 2005). Wheat seeds pretreatment in chitosan solution stimulated resistance to several diseases of crops and improved seed quality (Reddy et al. 1999). Seeds layered with chitosan increased seed germination rate and tolerance in seedlings of hybrid rice during stress condition (Ruan and Xue 2002). Maize seeds when soaked in different acidic solutions of chitosan increased the vitality of seedlings (Shao et al. 2005). Furthermore, in cold stress conditions, it enhanced the maize seeds germination velocity and hence benefited the seedlings growth (Guan et al. 2009).

### 4.1.5 Bio-priming

It is a seed-presoaking technique along with the inoculation of beneficial microorganisms. It combines both the biological agent (microorganisms) and physiological soaking (seed hydration) phase. Callan and Coworkers first depicted the bio-priming in 1990 for the biological management of *Pythium* pre-emergence of sh2 sweet corn. Incorporated imbibitions with a biocontrol mediator at certain temperature enhance fortification. Additionally, seed priming along with the beneficial microorganisms possibly will promote the maturity of the crop plants, mainly if the inoculated microorganisms colonize the rhizosphere of the plant and maintain plant physiology and plant growth for a longer period (Bennett and Whipps 2008). Since it is a biological approach to use both bacteria and fungi, competitors will also counter to both soil and seeds endured pathogens (Afzal et al. 2016). Callan et al. (1990) reported that bio-priming involves the varnishing of seeds by bacterial biocontrol negotiator like *Pseudomonas aureofaciens* Kluyver AB254 strain and hydrating at 23 °C for 20 h in damped vermiculite or on damped germination blotters in self-sealing plastic bags. The leakage of seeds release during the period of bio-priming could provide nutrients and strength for inoculated biocontrol agents (Wright et al. 2003). These flattering environments contribute to the migration and propagation of inoculated biocontrol mediators above the surface of the seed and assist water and nutrients uptake throughout the bio-priming period. Priming by a diverse group of beneficial microbes could not only augment the seed's nature but also boost seedling strength and capability to combat both biotic and abiotic stresses (Rakshit et al. 2015). The microbes mostly designed for bio-priming of seeds belong to *Pseudomonas* spp., *Enterobacter* spp., *Trichoderma* spp., and *Bacillus* spp. (Raj et al. 2004). For vegetable seeds, adequate bio-priming remedies were accomplished with *Trichoderma harzianum* strain, followed by *Trichoderma pseudokoningii*, *Bacillus* spp., *Gliocladium* spp., and *Pseudomonas fluorescens* (Ilyas 2006). Recently, bio-priming used as a substitute to control several soil- and seed-borne pathogens. For instance, mutual response of both *Trichoderma harzianum* and *Pseudomonas fluorescens*, when applied on pepper seeds as bio-priming agent, results in a significant growth of seedlings (Kumar et al. 2010; Reddy 2012). Various rhizobacterial inoculants are used as priming agents to control pathogenic fungal strains and to enhance crop yield. Most of the rhizospheric bacterial strains boost plant growth and physiology and thus are called plant growth-promoting rhizobacteria (PGPR) (Tonelli et al. 2011). Within the roots of tomato and rice plants, mycorrhizal fungi activate the aggregation of several transcripts and proteins that also predicts function in the plant defense mechanism (Pozo and Azcon-Aguilar 2007).

### 4.1.6 Priming with Plant Growth Regulators (PGR)

Seed treatment with plant growth regulators (PGR) is known to mitigate the harmful effects of several environmental stresses (Bahrani and Pourreza 2012; Jisha et al. 2013). Mendoza et al. (2002) reported that priming pepper seeds in salicylic acid protected the seedlings from adverse effects of a chill. Bell pepper primed with $GA_3$ (200 ppm) showed a higher rate of several physiological parameters including germination, shoot root length, and seedling vigor indices as compared to the control (Yogananda et al. 2004). In this context, indole-3-acetic acid (IAA), one of the prime auxins in plants, regulated cell division, enhanced photosynthetic activities, and activated the translocation of carbohydrates that enhance root initiation, flowering, and fruit setting and ripening (MacDonald 1997; Awan et al. 1999; Naeem et al. 2004). Similarly, gibberellins having antagonistic effects with abscisic acid (ABA) regulate seed germination and plant growth. Abu-Muriefah (2017) stated that the improved seed germination due to $GA_3$ priming might be due to its effect on stored food within seeds. He further described that $GA_3$ is known to stimulate the synthesis and production of the hydrolases resulting in the germination of seeds. Certain hydrolase enzymes are involved in seed germination process that makes the endosperm accessible to the embryo. In addition, cytokinins, ethylene, abscisic acid, and salicylic acid are some of the other plant growth promoter hormones that promote growth and regulate plant responses under various stress conditions. Similarly, certain new compounds such as $KNO_3$ (3%), $KH_2PO_4$ (2%), and PEG solutions (10%) are now being used as priming agents showing enhanced germination as compared to nonprimed seeds (Korkmaz and Korkmaz 2009; Ozbay and Susluoglu 2016).

### 4.1.7 Priming with Plant Extract

Allelochemicals such as phenolic compounds, terpenoids, flavonoids, saponins, alkaloids, and steroids may inhibit or stimulate plant growth (Narwal 1994). Saponins can enhance nutrient absorption as they are readily soluble in water. Alkaloids, saponins, and phenolic compounds present in the leaves of various plants are involved in the production of antioxidant activities and protect the plants against pathogens (Satish et al. 2007). Embryo and other associated structures are generally assumed to be activated by certain physiologically active substances that result in more water absorption and eventually in higher vigor index due to the development of an efficient roots system (Rangaswamy et al. 1993). Some plants are rich in saponin and alkaloids such as *Chlorophytum* leaves, while others are rich in terpenoids, steroids, flavonoids, and antiquinone such as neem leaves (Raphael 2012; Chakraborthy et al. 2014). Dawood et al. (2012) reported that amalgamation of fenugreek seeds (10%) or guava leaves or lantana leaves (20%) into the soil significantly enhanced carbohydrates and photosynthetic pigments of leaf tissues in sunflower. Similarly, the reduction in mortality and high seedling vigor in tomato was reported by priming tomato seeds with *Azadirachta*, *Chlorophytum*, and *Vinca* (Prabha et al. 2016).

## *4.2 Advanced Methods of Seed Priming*

### 4.2.1 Seed Priming Through Nanoparticles

Nanotechnology utilizes nanoparticles less than 100 nm in size, and it has a promising role in transforming food production and agriculture (Fraceto et al. 2016). The excessive use of chemical fertilizers can be reduced by utilizing nanomaterials in agriculture (Upadhyaya et al. 2017). In this context, priming seeds with nanoparticles has been reported to enhance seed germination and vigor in many crops. Ghafari and Razmjoo (2013) reported that seed priming with calcium-phosphate, $SiO_2$, ZnO, and Ag nanoparticles enhanced germination and seedling development. The mechanism behind high seed germination in nano-priming is the greater penetration via seed coat that improves nutrient and water uptake efficiency of the seed (Dutta 2018).

### 4.2.2 Seed Priming Through Physical Agents

The magnetic field, UV radiation, gamma radiation, X-rays, and microwaves are some of the physical agents that are used for seed priming (Bilalis et al. 2012). Priming with magnetic field has been reported to improve germination rate, vigor, and seedling biomass as well as tolerance to various environmental stresses. The tolerances to different stresses and improved germination rate have been attributed to a reduction in reactive oxygen species (ROS) with increasing activities of antioxidant enzymes (Bhardwaj et al. 2012; Araujo et al. 2016). However, the effects of ionizing radiation such as gamma (γ) rays are dose and intensity dependent. These rays interact with cellular components directly and are reported to improve the germination at lower doses (less than 10 Gy). Certain changes in hormonal network of plant cells take place that in turn trigger the antioxidative capacity and lead to early dormancy breaking and improved germination (Qi et al. 2015). The application of mechanical waves (ultrasound) is another physical method of priming having a frequency in the range of 20–100 kHz. In ultrasound priming, mechanical pressure is imposed on seed coat that increases the seed's porosity known as acoustic cavitation and activation of enzymatic and other biological reactions due to greater water uptake in the seed. Thus, in ultrasonic priming, mass transfer of the absorbed water is enhanced that allows it to react freely with the cell embryo.

## 5 Factors Affecting Seed Priming

Seed priming is highly affected by various biotic and abiotic factors such as aeration, temperature, time, and seed quality. Among these, aeration is the most important and effective factor affecting seed respiration, seed viability, and seed emergence/germination (Bujalski and Nienow 1991; Heydecker et al. 1973). Heydecker and Coolbear (1977) and Bujalski et al. (1989) reported the impact of aeration in seed priming by observing enhanced germination percentage in aerated PEG solution treatment as compared to nonaerated. Similarly, the temperature is another important factor influencing the germination of seeds. Basra et al. (2005) reported that optimum temperature ranges from 15 to 30 °C for most of the seed germination. On the other hand, McDonald (2000) reported slow germination at a lower priming temperature. Wahid et al. (2008) documented a range of 15–20 °C for seed priming, and the duration of priming may extend from almost 8 h to 14 days based on plant species, osmotic solution, osmotic potential, and temperature (Finch-Savage et al. 1991). Seed quality is another important factor in seed germination, and a viable and vigorous seed is the first most necessary for seed priming (Cantliffe 1987). Other seed characteristics also play a role in seed priming and germination process. For instance, Patanè et al. (2008) reported osmo-priming with PEG solution unsuitable for sorghum seeds priming. Sorghum is rich in tannin that could be removed with the solution during treatment and hence leads to lower seed germination. In this regard, Passam et al. (1989) stated that the salt solution is effective as compared to mannitol and PEG solutions. Similarly, O'Sullivan and Bouw (1984) reported $KNO_3$ and $K_3PO_4$ as an effective priming solution in pepper seeds as compared to PEG.

# 6 Seed Priming: Physiological Basis and Plant Response

## *6.1 Occurrence of Seed Germination and Seedling Growth*

In vegetable crops, high yield and growth are primarily associated with seedling health and early emergence which induces a potential to cope with various biotic and abiotic stresses. These all result in high yield and quality crops (Cantliffe 2003). Seed priming is one of the primitive techniques used to enhance early seed emergence and initiates several processes involved in seed germination (Asgedom and Becker 2001). Therefore, seed priming boosts the imbibition and metabolic processes resulting in enhanced seed germination, germination uniformity, and seedling growth and development in both normal and stress conditions (Ansari et al. 2012; Dey et al. 2014; Nayban et al. 2017). Several studies have reported that seed priming can enhance early seed emergence and growth in stress conditions as compared to untreated seeds (Bradford 1986; Chen et al. 2012). Yogananda et al. (2004) observed the increased physiological response of bell pepper treated with $GA_3$ and $KNO_3$ solution. Similarly, Yadav et al. (2011) also reported cold and salt stress tolerance in primed seeds of pepper with 100% survival. In another study, Ahmadvand et al. (2012) evaluated the effect of $KNO_3$-primed seeds of soybean in various cultivars and reported a significant increase in germination and seed emergence followed by increased physiological parameters like fresh weight, dry weight, root length, and shoot length.

## *6.2 Crop Nutrition and Yield*

Crop nutrient deficiency known as hidden hunger is a global dilemma resulting in low yields and poor-quality products (Dey et al. 2014). Seed priming is an approach to provide nutrients to the seed emergence and activate various biochemical processes necessary for seed germination. This not only provides the nutrients to establish seedling growth and emergence but also helps in enrichment of grain nutrient status (Singh et al. 2015a, b) and has been proved an economical method of nutrients application as compared to soil application and has been found economical as compared to soil application (Slaton et al. 2001). Ajouri et al. (2004) reported Zn as an effective priming agent in barley (*Hordeum vulgare* L.). Another study was done to assess the Zn content of early grown radicles and coleoptiles and was found much higher up to 200 mg $kg^{-1}$. The study concluded that Zn might be involved in various processes during early seed development (Cakmak 2005). Additionally, higher contents of Zn may enhance seed resistance to soil-borne diseases and ensure crop growth and development (Marschner 1995). Several studies by Khan et al. (2008), Bhowmick et al. (2010), Umair et al. (2013), and Bhowmick (2013) have reported seed priming as a suitable strategy for increased crop yields. Harris (2006) reported that increased crop yield might be attributed to crop density and individual crop performance. Srivastava and Bose (2012) studied the effect of ($Mg(NO_3)_2$ and $KNO_3$) salts as a priming agent for rice crop and reported increased physiological response to primed seeds as compared to untreated. Similarly, Arif et al. (2008) also studied the effect of priming seeds in soybean and reported priming increased seed emergence and establishment and attributed all this to the early metabolic activities as it activates radical protrusion. However, they reported that extended priming duration might decrease crop yield and 6 h was found as a suitable duration for soybean seed priming.

## *6.3 Seed Priming for Stress Management*

Field crops are subject to various types of biotic and abiotic stress like herbivores, pathogens attack as well as cold, heat, heavy metal, nutrient deficiency, salinity, and drought (Fedoroff et al. 2010). Among these stresses, salinity, drought, and temperature cause low growth and development (Jaleel et al. 2007; Thakur et al. 2010). These types of stress induce ROS production in a plant cell that results in cell injury and ultimately plants failure. In order to combat these ROS, plants have a self-defense mechanism that eliminates ROS species and protects plant cells from damage (Baxter et al. 2013). The intensity of stress injury depends upon various stress-related factors like stress tolerance, timing, and intensity (Niinemets 2009). Seed priming is considered one of the most promising techniques to enhance seed resistance to all these biotic and abiotic stresses (Van Hulten et al. 2006). Uchida et al. (2002) reported the importance of compounds used in seed priming technique as they alleviate the deteriorating effects of both biotic and abiotic stresses concerning plant growth. Kibinza et al. (2011) explained that germination percentages were improved by priming that was due to the considerable drop in $H_2O_2$ accumulation. Catalase activity was restored by priming that protects the stressed seeds from reactive oxygen species. Furthermore, Kester et al. (1997) reported that protein content in various plant tissues might be increased by seed priming via improved performance of protein synthesis system and increased production of L-isoaspartyl methyltransferase enzyme that plays an important role in repairing plant tissue proteins. Priming also enhances seed germination by enhancing the activities of protease and amylase that hydrolyze protein and starch into simple forms to make them available for the embryo (Miransari and Smith 2014). Sajedi et al. (2011) stated that ROS produced under $PbCl_2$ stress may be reduced by hormonal priming. Similarly, in another study, Pereira et al. (2009) showed improved seedling emergence by osmo-priming on carrot seed germination under extreme temperature conditions and water stress.

## 7 Assessment of Priming Effects on Plant Growth and Development

### *7.1 Seed Priming Using Compost Extract for Improving Germination Parameters*

Compost tea, a liquid extract, is obtained when compost is steeped in water for a period with the aim of transferring soluble organic matters (OM), beneficial microbes, and nutrients into the solution (Mohd Din et al. 2017). The application of this microbial- and nutrient-rich tea is known for stimulating the plant's growth and soil fertility (Ahmad et al. 2017). Compost tea is prepared aerobically (aerated tea) or anaerobically (nonaerated tea); however, little evidence exists as regards which method either aerated or nonaerated is more beneficial for agricultural purposes (Amos 2017). There have been numerous studies for evaluating the potential of compost tea in suppressing the plant diseases (Mengesha et al. 2017). However, there are limited studies on the effect of compost tea on crop growth and development (Kim et al. 2015). Therefore, an assessment that was carried out aims to examine the stimulatory and inhibitory effects of compost teas on seed germination and seedling growth through a novel method of seed priming in compost tea, since no previous study has been conducted.

### *7.2 Preparation and Characteristics of Compost Tea*

The compost used for compost teas production was prepared from food waste using an in-vessel compost bioreactor (Waqas et al. 2017). Two different types of compost teas, i.e., aerated and nonaerated, were prepared by mixing compost and water at 1:10 ratio (i.e., 1 kg dry compost in 10 L distilled water). The mixture was steeped at 25 °C for 72 h. A standard brewing method was used for steeping and extraction period during which the compost was in contact with water (Scheuerell and Mahaffee 2006). The aeration, in aerated compost tea, was maintained on a continuous basis by stirring the solution through a mechanical agitator with 80 revolutions per minute (RPM) throughout the extraction period of 72 h. The schematic diagram of the bioreactor used for making the aerated compost tea is shown in Fig. 2. Nonaerated compost tea was prepared by using a standard method of bucket fermentation (Diver 2002). This approach is referred to as the European-style or European method for compost tea preparation and dates back to hundreds of years

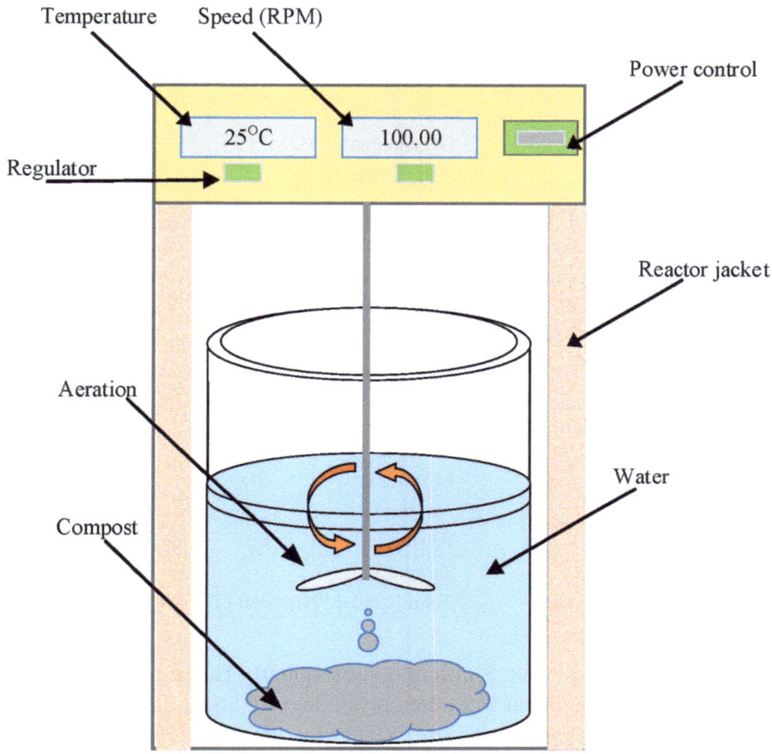


**Fig. 2** Schematic diagram of reactor used for producing aerated tea

(Brinton et al. 2004). During this process, the mixture was initially stirred and then left undisturbed at 25 °C for 3 days (Weltzien 1991). After the designated steeping time, the compost extracts were filtered through a muslin cloth. Different physio-chemical characteristics of the compost tea were determined by following the standard methods of analysis, and the results are presented in Table 2. For experimentation, the compost tea was diluted using distilled water. The dilution concentrations were 0% (100% extract), 25% (75% extract, 25% water), 50% (50% extract, 50% water), 75% (25% extract, 75% water), and 100% (100% water). The 100% compost tea diluted solution was used as a control treatment for comparing the concentration effects of both compost teas.

## *7.3 Seed Priming and Experimental Setup*

Seed priming in the compost tea was introduced by soaking the mung bean seeds in compost tea for a definite period of time. Seeds' surfaces were disinfected to avoid any bacterial or fungal contamination by imbibing them in a mixture of ethanol and

**Table 2** Characteristics of compost and compost tea through aerated and nonaerated fermentation

| Parameters | Compost |
|---|---|
| pH | 8.02 |
| Electrical conductivity (EC) (mS $cm^{-1}$) | 3.36 |
| Moisture content (MC) (%) | 39.31 |
| Organic matter (OM) (%) | 56.81 |
| Ash content (%) | 43.13 |
| Carbon (%) | 31.61 |
| $NH_4$-N (mg $kg^{-1}$) | 168.32 |
| $NO_3$-N (mg $kg^{-1}$) | 117.91 |
| Nitrification index (NI) | 1.42 |
| | Compost tea |
| pH | 8.94 |
| EC (mS $cm^{-1}$) | 4.48 |
| Ca (mg $L^{-1}$) | 17.97 |
| K (mg $L^{-1}$) | 52.33 |
| Mg (mg $L^{-1}$) | 19.41 |
| Na (mg $L^{-1}$) | 173.13 |
| Fe (mg $L^{-1}$) | 5.31 |
| Zn (mg $L^{-1}$) | 0.14 |
| Total Kjeldahl Nitrogen (TKN) (%) | 0.47 |

distilled water (70:30 v/v) for 5 min and subsequently rinsing them with distilled water. Ten seeds of mung bean were then primed/soaked over dampened filter paper to assess the compost tea phytotoxicity. 20 ml of a diluted solution of each compost tea was used as a priming solution in 15 cm petri dish (Fisher brand, Fisher Scientific, Waltham MA). The petri dishes were incubated for 24 h at 25 °C in the dark and were then covered with a plastic wrapping to avoid any water loss during the priming (Mavi 2014). After priming, the seeds were washed with distilled (for consistency reasons) water and dried back to about 14% grain moisture at 36 °C in drying oven (Pame et al. 2015). A factorial experiment in a totally randomized design was used to investigate the effects of compost tea aeration methods (aerated vs. nonaerated) and priming treatments (primed vs. unprimed) on a lot of ten mung bean seeds at five different dilution concentrations of 0%, 25%, 50%, 75%, and 100%. The tea dilution at 100% solution was referred to control treatment as it contained 0% compost tea and 100% distilled water. Three replicates were used for each treatment combination. Ten seeds of mung bean were homogenously distributed on two layers of sterile Whatman™ filter paper in each petri dish. For unprimed seeds, 5 ml of each diluted solution of compost tea was applied to moisten the filter paper, whereas the petri dishes containing primed seed in compost tea were applied with only distilled water when required. The petri dishes were monitored daily, and an equal amount of tea solutions/distilled water was supplied when necessary to keep the moisture of seeds or seedlings at adequate levels. Petri dishes were incubated in a growth chamber at 27 °C for 14 days. During experimentation, different germination and growth parameters were determined through standard formulas, and the obtained data were subjected to analysis of variance (ANOVA) and least significant difference (LSD) test at 5% probability level (Steel et al. 1997).

## *7.4 Effects on Germination Parameters*

### 7.4.1 Germination Rate and Germination Index (GI)

The ANOVA results indicated that compost teas' dilution and seed priming with compost tea significantly affected (state the level of significance, e.g., $\alpha = 0.05$) the seed germination and germination index. No significant differences on seed germination rate and germination index were observed for the aeration system, i.e., aerated and nonaerated teas exert the same effect on germination. The germination rate of the seeds exposed to aerated compost tea was equal to 86.3%, like nonaerated tea (86%). Among priming effects, the highest stimulation in germination rate was recorded for primed seeds, resulting in 91.6% germination, whereas that of unprimed seeds was recorded at 80.6% (Table 3). Similarly, among the tea dilution concentrations, the highest germination rate (94.2%) was observed in dilution of tea to 50%. However, a drastic reduction in germination rate was observed with increasing the tea concentration, and it was noticed that at 0% dilution (100% tea concentration), the lowest germination rate of 72.5% was observed (Table 3). Similarly, the two-way interaction between the aeration system with priming and dilution and the three-way combined interaction of aeration, dilution, and priming showed a nonsignificant effect on germination rate.

Similarly, the results showed that the maximum GI of 4.5 was estimated for primed seed, whereas the lowest GI of 2.5 was recorded for unprimed seed (Table 3). It was observed that priming had an advantageous effect in terms of seed germina-

**Table 3** Effect of aeration, dilution, and priming on the germination and growth parameters of mung bean

| Compost tea | | G% | GI | MGT | SVI | Root length | Shoot length |
|---|---|---|---|---|---|---|---|
| Aeration | Aerated | 86.3 a | 3.6 a | 6.1 a | 221.7 a | 2.4 a | 5.6 a |
| | Nonaerated | 86 a | 3.4 a | 5.9 a | 135.3 b | 1.5 b | 5.2 a |
| Priming | Primed | 91.6 a | 4.5 a | 4.8 b | 234.1 a | 2.5 a | 7.1 a |
| | Unprimed | 80.6 b | 2.5 b | 7.1 a | 122.9 b | 1.5 b | 3.8 b |
| Dilution | 0% | 72.5 c | 2.8 c | 6.3 ab | 73.2 c | 0.9 c | 3.3 c |
| | 25% | 84.2 b | 3.3 b | 6.2 ab | 158.9 b | 1.8 b | 4.9 b |
| | 50% | 94.2 a | 3.8 a | 5.6 bc | 241 a | 2.5 a | 4.7 b |
| | 75% | 88.3 ab | 3.8 a | 5.3 c | 243.4 a | 2.7 a | 7.2 a |
| | Control | 91.6 ab | 3.6 ab | 6.5 a | 176.1 b | 1.9 b | 6.7 a |

Means of the same category followed by different letters are significantly different at $P \leq 0.05$ level using LSD test

*G%* germination percentage, *GI* germination index, *MGT* mean germination time, *SVI* seed vigor index

**Table 4** Interaction of aeration and dilution toward GI, SVI, and seedling growth of mung bean

| Aeration | Dilution | GI | SVI | Root length | Shoot length |
|---|---|---|---|---|---|
| Aerated | 0% | 3.1 e | 80 cd | 1.05 de | 3.5 d |
| | 25% | 3.5 bcde | 192 b | 2.2 b | 5.6 bc |
| | 50% | 4 a | 336 a | 3.4 a | 5.6 bc |
| | 75% | 3.9 ab | 320.7 a | 3.6 a | 7.7 a |
| | Control | 3.4 cde | 179.8 b | 2.1 bc | 6.8 ab |
| Nonaerated | 0% | 2.5 f | 66.3 d | 0.9 e | 3.1 d |
| | 25% | 3.2 de | 125.8 bcd | 1.4 cde | 4.2 cd |
| | 50% | 3.6 abcd | 146 bc | 1.6 bcd | 3.9 d |
| | 75% | 3.8 abc | 166.2 b | 1.9 bc | 7.1 ab |
| | Control | 3.9 ab | 172.3 b | 1.8 bc | 6.6 ab |

Means of the same category followed by different letters are significantly different at $P \leq 0.05$ level using LSD test
*GI* germination index, *SVI* seed vigor index

tion parameters. Similarly, the mean data for tea dilution showed that the maximum GI of 3.8 in relation to control treatment was recorded at a dilution of tea to 50%, which was similar to 75% diluted solution with GI of 3.8 (Table 3). Moreover, like germination rate, reduction in values of GI was also observed in the low diluted solutions of compost tea. The results showed that the minimum GI values of 2.8 and 3.3 were recorded at 0 and 25% diluted solutions, respectively (Table 3). In addition, the ANOVA results showed significant effects of the interaction of aeration with dilution ratios. The interaction data revealed that the maximum GI value of 4 was recorded in aerated tea at 50% dilution followed by dilution of tea to 75% with GI of 3.9 in the same tea (Table 4). Similarly, for nonaerated compost tea, the values for same tea diluted concentration (50 and 75%) were observed to be lower than aerated tea. For nonaerated compost tea, the results showed that the maximum GI value of 3.9 had been recorded for control treatment followed by 75% and 50% dilution with the GI of 3.8 and 3.6, respectively. However, for both aerated and nonaerated teas, the lowest value for GI was observed for dilution of teas to 25% and 0% (Table 4).

### 7.4.2 Mean Germination Time (MGT) and Seed Vigor Index (SVI)

MGT showed the rapidity of germination; hence, the lower the value of MGT, the earlier is the germination. The ANOVA results showed highly significant variation for MGT in dilution and priming of compost tea. The mean data in Table 3 revealed that the rapid germination was recorded for primed seeds with least MGT value of 4.8, whereas the higher MGT value of 7.1 for unprimed depicts the delayed seed germination. For dilution of tea, the lowest MGT value of 5.3 was noticed at 75% tea dilution. However, this value for MGT was statistically near to dilution of tea to 50% (5.6). Similarly, the maximum MGT value of 6.5 was computed for control treatment that showed low rapidity of seed germination. For aeration, the value for MGT in aerated tea was 6.1, which was statistically similar to the MGT value of 5.9

for nonaerated tea (Table 3). This nonsignificant difference showed that the aeration system did not affect MGT of the tested seeds. The overall results revealed that priming reduced the MGT by 31.2% in comparison with unpriming. Similarly, compared to control, the solution containing 25% and 50% compost tea reduced the MGT by 18.4 and 12.9%, respectively.

SVI expressed the level of seed performance and activity during the germination and seedling emergence. It was found that SVI was significantly affected by aeration, tea dilution, and priming. Moreover, the results indicated that the two-way interaction between aeration with dilution and priming also had significant effects toward SVI. The results showed that among the aeration systems, the utmost SVI of 221.7 was recorded for aerated tea, leaving behind nonaerated tea with the value of 135.3 (Table 3). Similarly, among the priming effects, the highest values were recorded for primed seed that was 234.1 whereas the least values (122.9) were estimated for unprimed seeds. In addition, compost tea dilution showed high variation toward SVI. The highest SVI (243.4) was observed for dilution of tea to 75% that was statistically like 50% diluted tea (241), whereas the least SVI (73.2) was recorded for dilution of tea to 0% (Table 3). Furthermore, the mean data for the two-way interaction of aeration with dilution showed that the maximum SVI (336) was observed for aerated tea diluted at 50% (Table 4). This value was found statistically similar to 75% tea dilution with the SVI value of 320.7. Similarly, the lowest SVI of 66.3 was recorded at 0% tea dilution in nonaerated (Table 4). Furthermore, the mean data for the interaction of priming with dilution showed that the utmost values for SVI (353.5) were observed for primed seeds with a dilution of tea to 50% followed by 25 and 75% diluted tea solutions with SVI values of 272 and 270.3, respectively (Table 5), whereas the minimum SVI (28.2 and 45.8) was noticed for dilution of tea to 0 and 25% in unprimed seeds, respectively (Table 5). The combined interaction of aeration, dilution, and priming showed that highest SVI (526.7) was noticed in primed seeds with the dilution of aerated compost tea to 50%. Similarly, the second highest peak value of 370.7 for SVI also observed for primed seed with aerated tea at tea dilution

**Table 5** Interaction of priming and dilution toward SVI and seedling growth of mung bean

| Priming | Dilution | SVI | Root length | Shoot length |
|---|---|---|---|---|
| Primed | 0% | 118.2 d | 1.5 e | 6 c |
| | 25% | 272 b | 2.9 ab | 8.6 a |
| | 50% | 353.5 a | 3.6 a | 7.7 ab |
| | 75% | 270.3 b | 2.8 bc | 7.7 ab |
| | Control | 156.7 cd | 1.6 de | 5.4 c |
| Unprimed | 0% | 28.2 e | 0.4 f | 0.6 d |
| | 25% | 45.8 e | 0.6 f | 1.3 d |
| | 50% | 128.5 d | 1.4 e | 1.8 d |
| | 75% | 216.5 bc | 2.7 bc | 6.2 bc |
| | Control | 195.5 c | 2.3 cd | 9.1 a |

Means of the same category followed by different letters are significantly different at $P \leq 0.05$ level using LSD test
*SVI* seed vigor index

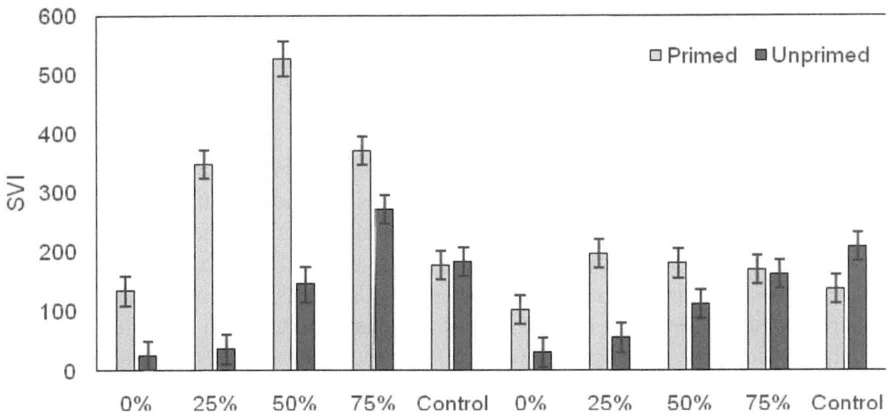


**Fig. 3** Interaction of aeration, dilution, and priming on seed vigor index of mung bean. Vertical bars represent ± standard error of the mean

to 75%. Conversely, the lowest SVI (25.7 and 30.7) was noticed for unprimed seeds at 0% tea dilution in both the aerated and nonaerated teas, respectively (Fig. 3).

### 7.4.3 Effects on Root and Shoot Length

The tea dilution and priming and their combined interaction affected significantly (a = 0.05) mung bean root and shoot length. However, the effect of aeration on the shoot length was observed nonsignificant. The results showed that the maximum mean root length (2.4 cm) was recorded for aerated tea. Similarly, the mean data for priming depict that the utmost root (2.5 cm) and shoot length (7.1 cm) were observed for primed seed, whereas for unprimed seed, the observed root and shoot length were 1.5 and 3.8 cm, respectively (Table 3). In addition, the mean data for compost tea dilution showed that the maximum root length (2.7 and 2.5 cm) was observed for dilution of teas to 75 and 50%, whereas the highest shoot length (7.2 cm) was noticed for dilution of tea to 75%. Correspondingly, the minimum root (0.9 cm) and shoot length (3.3 cm) was estimated for dilution of tea to 0% (Table 3). The interaction of aeration with dilution showed that among the aerated and nonaerated compost teas, the highest root length (3.6 cm) was recorded for tea dilution to 75% that was statistically similar to dilution of tea to 50% (3.4 cm) (Table 4). Similarly, the highest shoot length (7.7 cm) was recorded in aerated compost tea at 75% dilution, whereas the least values for root and shoot length (0.9 and 3.1 cm) were observed in nonaerated compost tea at 0% dilution followed by 25% dilution of the same tea (Table 4). As the ANOVA results also showed significant differences among the interaction between priming and tea dilution, the utmost root length (3.6 cm) was observed for primed seed in compost tea at 50% dilution, whereas for shoot length, the maximum value (8.6 cm) was recorded at 25% dilution. The bare minimum root and shoot length (0.4 and 0.6 cm) were calculated for unprimed seeds treated with compost tea at 0% dilution followed with 25% dilution in the same tea (Table 5).

The combined interaction of aeration, tea dilution, and priming revealed that the maximum seedling root length (5.3 cm) was recorded for primed seed at 50% diluted solution of the aerated tea (Fig. 4). Similarly, the highest shoot length of 10 and 9.2 cm was recorded for seeds primed with dilution of aerated tea to 25 and 50%, respectively (Fig. 5), whereas the maximum inhibition in root and shoot length was resulted by 100% tea concentration (0% dilution). The recorded root and shoot length at the respective 0% dilution were 0.4 and 0.73 cm in aerated and 0.5 and 0.6 cm in nonaerated compost tea (Figs. 4 and 5).

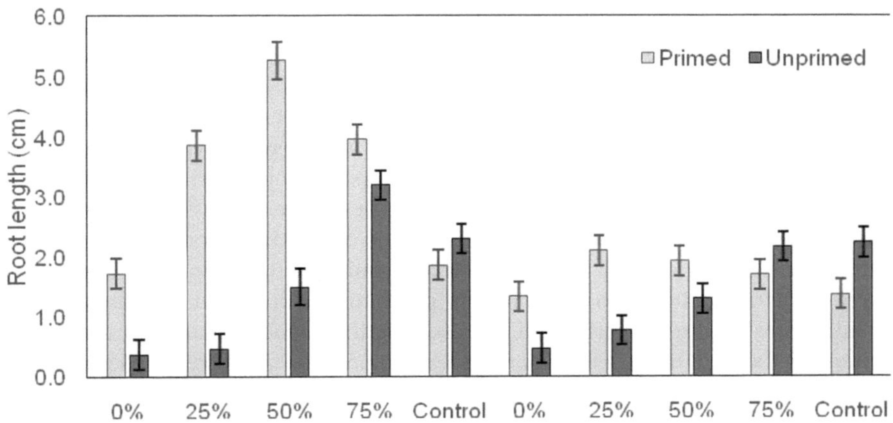


**Fig. 4** Interaction of aeration, dilution, and priming on root length of mung bean. Vertical bars represent ± standard error of the mean

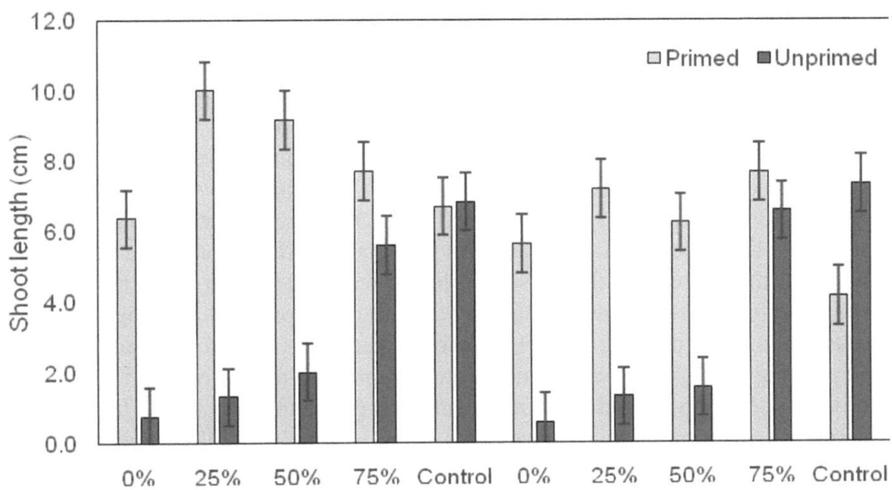


**Fig. 5** Interaction of aeration, dilution, and priming on shoot length of mung bean. Vertical bars represent ± standard error of the mean

## 8 Limitations and Perspective in Seed Priming Technology

Seed priming has been developed as a promising technology for superficial crop stand in a variety of environmental conditions. However, many protocols such as seed desiccation (redrying) after priming may affect different physiochemical process which reduces seed longevity and viability (Heydecker and Gibbins 1977; Halmer 2004). Other conditions for posttreatment such as storage temperature, air composition, and moisture also negatively affect seed viability (Schwember and Bradford 2005). Similarly, the prolonged seed treatment during priming may also cause loss of seed tolerance to desiccation (Sliwinska and Jendrzejczak 2002). Priming itself in certain circumstances may also cause different problems. For instance, all priming protocols may not lead to significant germination and growth where inappropriate priming conditions may cause degradation of the protective proteins (Capron et al. 2000). Hence, it is critical to select specific priming protocol for different plants about germination and growth in different environmental conditions. Thus, for filling the gap and successful application of priming technology, detailed studies focusing on treatment technologies, gene expressions, and molecular mechanisms need to be fully explored (Araujo et al. 2016). Correspondingly, the advanced methods of seed priming such as priming with nanoparticles may also have deleterious effects on environment, plant, and human health. In this regard, solid studies need to be performed for resolving the impact of nanomaterials when enter the food chain by using them in agriculture. Extensive researches are still required for each priming technology in terms of optimal dose, exposure time, and dose rate that could affect plant growth and development.

## 9 Conclusions

Seed priming is the physiological process of controlled seed hydration to enhance sufficient pregerminative metabolic process, efficient nutrient uptake and water use efficiency, breaking dormancy, timely maturity, and crop yield. During imbibition, the water uptake promotes protein synthesis and respiratory activities by using extant messenger ribonucleic acid (mRNA) with the initiation of different physiological activities related to germination. This technology has been found to be the most feasible and economical for uniform seed emergence in most of the field crops. There are many well-developed seed-priming techniques such as hydro-priming, osmo-priming, nutrient priming, chemical priming, bio-priming, priming with plant growth regulators, priming with plant extracts, seed priming through nanoparticles, and priming through physical agents. However, priming technology still has several limitations. The prolonged seed treatment during priming may cause loss of seed tolerance to desiccation that reduces seed viability. Similarly, all priming protocols may not lead to significant germination and growth where inappropriate priming conditions may cause degradation in the protective proteins. Hence, extensive researches are required in selecting specific priming protocol for different plants regarding germination and growth under various environmental conditions.

## References

Abu-Muriefah SS (2017) Phytohormonal priming improves germination and antioxidant enzymes of soybean (*Glycine max*) seeds under lead (Pb) stress. Biosci Res 14(1):42–56

Adhikari T, Kundu S, Rao AS (2013) Impact of SiO2 and Mo nano particles on seed germination of rice (Oryza sativa L.). Int J Agric Food Sci Technol 4(8):809–816

Afzal I, Rehman HU, Naveed M, Basra SMA (2016) Recent advances in seed enhancements. In New challenges in seed biology-basic and translational research driving seed technology. InTech, pp 47–74

Ahmad I, Basra SMA, Akram M, Wasaya A, Ansar M, Hussain S, Iqbal A, Hussain SA (2017) Improvement of antioxidant activities and yield of spring maize through seed priming and foliar application of plant growth regulators under heat stress conditions. Semina: Ciências Agrárias 38:47–56

Ahmadvand G, Soleimani F, Saadatian B, Pouya M (2012) Effect of seed priming with potassium nitrate on germination and emergence traits of two soybean cultivars under salinity stress conditions. Am Eurasian J Agric Environ Sci 12:769–774

Ajouri A, Asgedom H, Becker M (2004) Seed priming enhances germination and seedling growth of barley under conditions of P and Zn deficiency. J Plant Nutr Soil Sci 167(5):630–636

Amos D (2017) Aerated compost tea (ACT) to improve soil biology and to act as a biofertiliser/biofungicide, pp 1–2. http://orgprints.org/31042/. Accessed 2 June 2017

Ansari O, Chogazardi H, Sharifzadeh F, Nazarli H (2012) Seed reserve utilization and seedling growth of treated seeds of mountain rye (*Secale montanum*) as affected by drought stress. Cercetari Agronomice in Moldova 45(2):43–48

Araujo SDS, Paparella S, Dondi D, Bentivoglio A, Carbonera D, Balestrazzi A (2016) Physical methods for seed invigoration: advantages and challenges in seed technology. Front Plant Sci 7:646

Arif M, Waqas M, Nawab K, Shahid M (2007) Effect of seed priming in Zn solutions on chickpea and wheat. Afr. Crop Sci Conf Proc 8:237–240

Arif M, Jan MT, Marwat KB, Khan MA (2008) Seed priming improves emergence and yield of soybean. Pak J BotPak J Bot 40(3):1169–1177

Asgedom H, Becker M (2001) Effects of seed priming with nutrient solutions on germination, seedling growth and weed competitiveness of cereals in Eritrea. In: Proceedings of Deutscher Tropentag. University of Bonn & ATSAF, Margraf Pub. Press, Weickersheim, p 282

Assefa MK, Hunje R, Koti RV (2010) Enhancement of seed quality in soybean following priming treatment. Karnataka J Agric Sci 23:787–789

Awan IU, Baloch MS, Sadozai NS, Sulemani MZ (1999) Stimulatory effect of GA3 and IAA on ripening process, kernel development and quality of rice. Pakistan J Biol Sci(Pakistan) 2:410–412

Bagheri MZ (2014) The effect of maize priming on germination characteristics, catalase and peroxidase enzyme activity and total protein content under salt stress. Int J Biosci 4(2):104–112

Bahrani A, Pourreza J (2012) Gibberellic acid and salicylic acid effects on seed germination and seedlings growth of wheat (*Triticum aestivum* L.) under salt stress condition. World Appl Sci J 18(5):633–641

Basra SMA, Farooq M, Tabassam R, Ahmad N (2005) Physiological and biochemical aspects of pre-sowing seed treatments in fine rice (*Oryza sativa* L.). Seed Sci Technol 33(3):623–628

Baxter A, Mittler R, Suzuki N (2013) ROS as key players in plant stress signalling. J Exp Bot 65(5):1229–1240

Benavides-Mendoza A, Ramirez-Rodriguez H, Robledo-Torres V, Hernandez-Davila J, Ramirez-Mezquitic JG, Bacopulos-Te-llez E, Bustamante-Garcia MA (2002) Seed treatment with salicylates modifies stomatal distribution, stomatal density and the tolerance to cold stress in pepper seedlings. In: Proceedings of the 16th international pepper conference (Tampico, Tamaulipas, Mexico (Vol. 2), November

Bennett AJ, Whipps JM (2008) Dual application of beneficial microorganisms to seed during drum priming. Appl Soil Ecol 38(1):83–89
Berrie AMM, Drennan DSH (1971) The effect of hydration-dehydration on seed germination. New Phytol 70(1):135–142
Bhardwaj J, Anand A, Nagarajan S (2012) Biochemical and biophysical changes associated with magnetopriming in germinating cucumber seeds. Plant Physiol Biochem 57:67–73
Bhowmick MK (2013) Seed priming and foliar nutrition towards enhancing Lathyrus productivity under rice-utera system. In: Proceedings of the Centenary Session, 100th Indian Science Congress, Section of Agriculture and Forestry Sciences, 3–7 Jan 2013, Kolkata, West Bengal, India, p 219
Bhowmick M K, Biswas PK, Sen P, Bhattacharyya P (2010) Studies on seed priming, row spacing and foliar nutrition in chickpea under rainfed conditions in West Bengal, India. In: Proceedings of the international seminar on climate change and environmental challenges of 21st century, December, pp 7–9
Bilalis DJ, Katsenios N, Efthimiadou A, Karkanis A, Efthimiadis P (2012) Investigation of pulsed electromagnetic field as a novel organic pre-sowing method on germination and initial growth stages of cotton. Electromagn Biol Med 31(2):143–150
Bradford KJ (1986) Manipulation of seed water relations via osmotic priming to improve germination under stress conditions. HortScience (USA)
Bradford KJ (1995) Water relations in seed germination. Seed Dev Germination 1(13):351–396
Brinton W, Storms P, Evans E, Hill J (2004) Compost teas: microbial hygiene and quality in relation to method of preparation. Biodynamics. 249:36–45
Bujalski W, Nienow AW (1991) Large-scale osmotic priming of onion seeds: a comparison of different strategies for oxygenation. Sci Hortic 46(1–2):13–24
Bujalski W, Nienow AW, Gray D (1989) Establishing the large scale osmotic priming of onion seeds by using enriched air. Ann Appl Biol 115(1):171–176
Cakmak I (2005) The role of potassium in alleviating detrimental effects of abiotic stresses in plants. J Plant Nutr Soil Sci 168(4):521–530
Callan NW, Mathre D, Miller JB (1990) Bio-priming seed treatment for biological control of *Pythium ultimum* pre-emergence damping-off in sh-2 sweet corn. Plant Dis 74:368–372
Cantliffe DJ (1987) Priming of lettuce for early and uniform emergence under conditions of environmental stress. Acta Hortic 122:29–38
Cantliffe DJ (2003) Seed enhancements. Acta Hortic 607:53–59
Capron I, Corbineau F, Dacher F, Job C, Côme D, Job D (2000) Sugar beet seed priming: effects of priming conditions on germination, solubilization of 11-S globulin and accumulation of LEA proteins. Seed Sci Res 10(3):243–254
Chakraborthy GS, Aeri V, Verma P, Singh S (2014) Phytochemical and antimicrobial studies of *Chlorophytum borivilianum*. Pharmacophore 5:258–261
Chen K, Fessehaie A, Arora R (2012) Dehydrin metabolism is altered during seed osmopriming and subsequent germination under chilling and desiccation in Spinaciaoleracea L. cv. Bloomsdale: possible role in stress tolerance. Plant Sci 183:27–36
Côme D, Thévenot C (1982) Environmental control of embryo dormancy and germination. In: The physiology and biochemistry of seed development, dormancy and germination, pp 271–298
Corbineau F, Côme D (2006) Priming: a technique for improving seed quality. Seed testing international. ISTA News Bulletin No 132:38–40
Darwin C (1855) Effect of salt-water on the germination of seeds. Gardeners Chron Agric Gaz 47:773
Dawood MG (2018) Stimulating plant tolerance against abiotic stress through seed priming. In: Advances in seed priming. Springer, Singapore, pp 147–183
Dawood MG, El-Awadi ME, El-Rokiek KG (2012) Physiological impact of fenugreek, guava and lantana on the growth and some chemical parameters of sunflower plants and associated weeds. J Ame Sci 8(6):166–174

Demir I, Ozuaydın I, Yasar F, Van Staden J (2012) Effect of smoke-derived butenolide priming treatment on pepper and salvia seeds in relation to transplant quality and catalase activity. S Afr J Bot 78:83–87

Dey S, Pramanik K, Mukherjee S, Poddar S, Barik K (2014) Effect of seed priming and straw mulching on growth and productivity of chickpea (*Cicer arietinum* L.) under rainfed condition. SATSA Mukhaptra Annu Tech Issue 18:126–131

Di Girolamo G, Barbanti L (2012) Treatment conditions and biochemical processes influencing seed priming effectiveness. Ital J Agron 7(2):25

Diver S (2002) Notes on compost teas: a supplement to the ATTRA publication: compost teas for plant disease control. Appropriate Technology Transfer for Rural Areas (ATTRA), National Sustainable Agriculture Information Service, AR. (National Center for Appropriate Technology, Montana). Retrieved March, 5, 2012

Dubrovsky JG (1996) Seed hydration memory in Sonoran Desert cacti and its ecological implication. Am J Bot 83(5):624–632

Dutta P (2018) Seed priming: new vistas and contemporary perspectives. In: Advances in seed priming. Springer, Singapore, pp 3–22

Ells JE (1963) The influence of treating tomato seed with nutrient solutions on emergence rate and seedling growth. In Proc Amer Soc Hort Sci 83:684–687

Evenari M (1984) Seed physiology: its history from antiquity to the beginning of the 20th century. Bot Rev 50(2):119–142

Farahbakhsh H (2012) Germination and seedling growth in un-primed and primed seeds of Fenel as affected by reduced water potential induced by NaCl. Int Res J Appl Basic Sci 3(4):737–744

Farooq M, Basra SM, Rehman H, Mehmood T (2006) Germination and early seedling growth as affected by pre-sowing ethanol seed treatments in fine rice. Int J Agric BiolInt J Agric Biol 8:19–22

Farooq M, Basra SMA, Rehman H, Saleem BA (2008) Seed priming enhances the performance of late sown wheat (*Triticum aestivum* L.) by improving chilling tolerance. J Agron Crop Sci 194(1):55–60

Farooq M, Wahid A, Lee DJ (2009) Exogenously applied polyamines increase drought tolerance of rice by improving leaf water status, photosynthesis and membrane properties. Acta Physiol Plant 31(5):937–945

Farooq M, Basra SM, Wahid A, Ahmad N (2010) Changes in nutrient-homeostasis and reserves metabolism during rice seed priming: consequences for seedling emergence and growth. Agric Sci China 9(2):191–198

Fedoroff NV, Battisti DS, Beachy RN, Cooper PJ, Fischhoff DA, Hodges CN et al (2010) Radically rethinking agriculture for the 21st century. Science 327(5967):833–834

Finch-Savage WE, Gray D, Dickson GM (1991) The combined effects of osmotic priming with plant growth regulator and fungicide soaks on the seed quality of five bedding plant species. Seed Sci Technol 19(2):495–503

Fraceto LF, Grillo R, de Medeiros GA, Scognamiglio V, Rea G, Bartolucci C (2016) Nanotechnology in agriculture: which innovation potential does it have? Front Environ Sci 4:20

Gaius PS (1949) Naturalishistoria, vol. IV–VII, Books 12–27 (trans: Rackham H, Jones WHS, Eichholz DE). Harvard University Press, Massachussets and William Heinemann, London

Ghafari H, Razmjoo J (2013) Effect of foliar application of nano-iron oxidase, iron chelate and iron sulphate rates on yield and quality of wheat. Int JAgron Plant Prod 4(11):2997–3003

Guan YJ, Hu J, Wang XJ, Shao CX (2009) Seed priming with chitosan improves maize germination and seedling growth in relation to physiological changes under low temperature stress. J Zhejiang Univ Sci B 10(6):427–433

Halmer P (2004) Methods to improve seed performance in the field. In: Handbook of seed physiology, pp 125–65

Harris D (2006) Development and testing of "on-farm" seed priming. Adv Agron 90:129–178

Harris D, Rashid A, Miraj G, Arif M, Yunas M (2008) 'On-farm'seed priming with zinc in chickpea and wheat in Pakistan. Plant Soil 306(1–2):3–10

Heydecker W, Coolbear P (1977) Seed treatments for improved performance-survey and attempted prognosis. Seed Sci Technol 5:353–425

Heydecker W, Gibbins BM (1977) The'priming'of seeds. In: Symposium on seed problems in horticulture, vol 83, pp 213–224

Heydecker W, Higgins J, Gulliver RL (1973) Accelerated germination by osmotic seed treatment. Nature 246(5427):42

Hill H, Bradford KJ, Cunningham J, Taylor AG (2008) Primed lettuce seeds exhibit increased sensitivity to moisture during aging. Acta Hortic 782:135–141

Ilyas S (2006) Seed treatments using matriconditioning to improve vegetable seed quality. J Agronomi Indonesia 34(2)

Jaleel CA, Gopi R, Manivannan P, Panneerselvam R (2007) Responses of antioxidant defense system of *Catharanthus roseus* (L.) G. Don. Topaclobutrazol treatment under salinity. Acta Physiol Plant 29(3):205–209

Jett LW, Welbaum GE, Morse RD (1996) Effects of matric and osmotic priming treatments on broccoli seed germination. J Am Soc Hortic Sci 121(3):423–429

Jisha KC, Vijayakumari K, Puthur JT (2013) Seed priming for abiotic stress tolerance: an overview. Acta Physiol Plant 35(5):1381–1396

Kaur S, Gupta AK, Kaur N (2002) Effect of osmo-and hydropriming of chickpea seeds on seedling growth and carbohydrate metabolism under water deficit stress. Plant Growth Regul 37(1):17–22

Kester ST, Geneve RL, Houtz RL (1997) Priming and accelerated ageing affect L-isoaspartyl methyltransferase activity in tomato (*Lycopersicon esculentum* Mill.) seed. J Exp Bot 48(4):943–949

Khaliq A, Aslam F, Matloob A, Hussain S, Geng M, Wahid A, ur Rehman H (2015) Seed priming with selenium: consequences for emergence, seedling growth, and biochemical attributes of rice. Biol Trace Elem Res 166(2):236–244

Khan A, Khalil SK, Khan AZ, Marwat KB, Afzal A (2008) The role of seed priming in semi-arid area for mung bean phenology and yield. Pak J Bot 40(6):2471–2480

Kibinza S, Bazin J, Bailly C, Farrant JM, Corbineau F, El-Maarouf-Bouteau H (2011) Catalase is a key enzyme in seed recovery from ageing during priming. Plant Sci 181(3):309–315

Kim MJ, Shim CK, Kim YK, Hong SJ, Park JH, Han EJ, Kim JH, Kim SC (2015) Effect of aerated compost tea on the growth promotion of lettuce, soybean, and sweet corn in organic cultivation. Plant Pathol J 31(3):259–268

Koehler DE (1967) Thesis. Purdue University, USA

Korkmaz A, Korkmaz Y (2009) Promotion by 5-aminolevulenic acid of pepper seed germination and seedling emergence under low-temperature stress. Sci Hortic 119(2):98–102

Kumar S, Arya MC, Sinfh R (2010) Management of sweet pepper diseases and growth promotion by *Pseudomonas fluorescens* and *Trichoderma harzianum* in mid hills of Central Himalayas, India. Indian Phytopathol 63(2):181–186

Lara TS, Lira JMS, Rodrigues AC, Rakocevic M, Alvarenga AA (2014) Potassium nitrate priming affects the activity of nitrate reductase and antioxidant enzymes in tomato germination. J Agric Sci 6(2):72

MacDonald H (1997) Auxin perception and signal transduction. Physiol Plant 100:423–430

Mahajan G, Sarlach RS, Japinder S, Gill MS (2011) Seed priming effects on germination, growth and yield of dry direct-seeded rice. J Crop Improv 25(4):409–417

Marschner H (1995) Mineral nutrition of higher plants, 2nd edn. Academic, San Diego. Mineral nutrition of high plant. Academic Press, pp: 330–355

Mavi K (2014) Use of extract from dry marigold (*Tagates*spp.) flowers to prime eggplant (*Solanum melongena*L.) seeds. Acta Sci Pol-Hortorumcultus 13:3–12

Mavi K (2016) The effect of organic priming with Marigold herbal tea on seeds quality in Aji pepper (*Capsicum baccatum* var. pendulum Willd.). Mustafa Kemal Üniversitesi Ziraat Fakültesi Dergisi 21(1):31–39

May LH, Milthorpe EJ, Milthorpe FL (1962) Pre-sowing hardening of plants to drought. In: Field crop abstracts, vol 15, pp 93–98

McDonald MB (2000) Seed priming. In: Black M, Bewley JD (eds) Seed technology and its biological basis. Sheffield Academic Press, Sheffield, pp 287–325

Mengesha WK, Powell SM, Evans KJ, Barry KM (2017) Diverse microbial communities in non-aerated compost teas suppress bacterial wilt. World J Microbiol Biotechnol 33(3):49–55

Miransari M, Smith DL (2014) Plant hormones and seed germination. Environ Exp Bot 99:110–121

Mirshekari B, Baser S, Allahyari S, Hamedanlu N (2012) On-farm seed priming with Zn+ Mn is an effective way to improve germination and yield of marigold. Afr J Microbiol Res 6(28):5796–5800

Mohd Din ARJ, Cheng KK, Sarmidi MR (2017) Assessment of compost extract on yield and phytochemical contents of Pak Choi (*Brassica Rapa* cv. Chinensis) grown under different fertilizer strategies. Commun Soil Sci Plant Anal 48(3):274–284

Moradi A, Younesi O (2009) Effects of osmo-and hydro-priming on seed parameters of grain sorghum (*Sorghum bicolor* L.). Aust J Basic Appl Sci 3(3):1696–1700

Naeem M, Bhatti IRAM, Ahmad RH, Ashraf MY (2004) Effect of some growth hormones (GA3, IAA and kinetin) on the morphology and early or delayed initiation of bud of lentil (*Lens culinaris* Medik). Pak J Bot 36(4):801–809

Narwal SS (1994) Allelopathy in crop production. Scientific Publishers, Jodhpur, p 288

Nayban G, Mandal AK, De BK (2017) Seed priming: a low-cost climate-resilient tool for improving germination, growth and productivity of mungbean. SATSA Mukhaptra Annu Tech Issue 21:162–172

Niinemets Ü (2009) Mild versus severe stress and BVOCs: thresholds, priming and consequences. Trends Plant Sci 15(3):145–153

O'Sullivan JOHN, Bouw WJ (1984) Pepper seed treatment for low-temperature germination. Can J Plant Sci 64(2):387–393

Ozbay N, Susluoglu Z (2016) Assessment of growth regulator prohexadione calcium as priming agent for germination enhancement of pepper at low temperature. JAPS: J Anim Plant Sci 26(6):1652–1658

Pame AR, Kreye C, Johnson D, Heuer S, Becker M (2015) Effects of genotype, seed P concentration and seed priming on seedling vigor of rice. Exp Agric 51(03):370–381

Passam HC, Karavites PI, Papandreou AA, Thanos CA, Georghiou K (1989) Osmo-conditioning of seeds in relation to growth and fruit yield of aubergine, pepper, cucumber and melon in unheated greenhouse cultivation. Sci Hortic 38(3–4):207–216

Patanè C, Cavallaro V, D'Agosta G, Cosentino SL (2008) Plant emergence of PEG-osmoprimed seeds under suboptimal temperatures in two cultivars of sweet sorghum differing in seed tannin content. J Agron Crop Sci 194(4):304–309

Pereira MD, Dias DCFDS, Dias LADS, Araújo EF (2009) Primed carrot seeds performance under water and temperature stress. Sci Agric 66(2):174–179

Pill WG (1995) Low water potential and presowing germination treatments to improve seed quality. In: Seed quality, pp 319–359

Pozo MJ, Azcón-Aguilar C (2007) Unraveling mycorrhiza-induced resistance. Curr Opin Plant Biol 10(4):393–398

Prabha D, Negi S, Kumari P, Negi YK, Chauhan JS (2016) Effect of seed priming with some plant leaf extract on seedling growth characteristics and root rot disease in tomato. Int J Agric Syst 4(1):46–51

Prasad TNVKV, Sudhakar P, Sreenivasulu Y, Latha P, Munaswamy V, Reddy KR et al (2012) Effect of nanoscale zinc oxide particles on the germination, growth and yield of peanut. J Plant Nutr 35(6):905–927

Qi W, Zhang L, Wang L, Xu H, Jin Q, Jiao Z (2015) Pretreatment with low-dose gamma irradiation enhances tolerance to the stress of cadmium and lead in Arabidopsis thaliana seedlings. Ecotoxicol Environ Saf 115:243–249

Raj SN, Shetty NP, Shetty HS (2004) Seed bio-priming with Pseudomonas fluorescens isolates enhances growth of pearl millet plants and induces resistance against downy mildew. Int J Pest Manag 50(1):41–48

Rakshit A, Pal S, Rai S, Rai A, Bhowmick MK, Singh HB (2013) Micronutrient seed priming: a potential tool in integrated nutrient management. SATSA Mukhaptra Annu Tech Issue 17:77–89
Rakshit A, Sunita K, Pal S, Singh A, Singh HB (2015) Bio-priming mediated nutrient use efficiency of crop species. In: Nutrient use efficiency: from basics to advances. Springer, New Delhi, pp 181–191
Rangaswamy A, Purushothaman S, Devasenapathy P (1993) Seed hardening in relation to seedling quality characters of crops. Madras Agric J 80(9):535–537
Raphael E (2012) Phytochemical constituents of some leaves extract of *Aloe vera* and *Azadirachta indica* plant species. Glob Adv Res J Environ Sci Toxicol 1(2):014–017
Reddy PP (2012) Bio-priming of seeds. In: Recent advances in crop protection. Springer, New Delhi, pp 83–90
Reddy MVB, Arul J, Angers P, Couture L (1999) Chitosan treatment of wheat seeds induces resistance to *Fusarium graminearum* and improves seeds quality. J Agric Food Chem 47:67–72
Rehman H, Nawaz Q, Basra SMA, Afzal I, Yasmeen A (2014) Seed priming influence on early crop growth, phenological development and yield performance of linola (*Linum usitatissimum* L.). J Integr Agric 13(5):990–996
Renugadevi J, Vijayageetha V (2006) Organic seed fortification in cluster bean (*Cyamopsistetra gonoloba* L.) TAUB. In: International conference on indigenous vegetables and legumes. Prospectus for fighting poverty, hunger and malnutrition, vol 752, December, pp 335–337)
Rowse HR (1991) Methods of priming seeds. UK Patent, 2(192), 781
Ruan SL, Xue QZ (2002) Effects of chitosan coating on seed germination and salt-tolerance of seedlings in hybrid rice (*Oryza sativa* L.). Acta Agron Sin 28:803–808
Sajedi NA, Ardakani MR, Madani H, Naderi A, Miransari M (2011) The effects of selenium and other micronutrients on the antioxidant activities and yield of corn (Zea mays L.) under drought stress. Physiol Mol Biol Plants 17(3):215–222
Salehi M, Tamaskani F (2008) Pretreatment effect of nanosilver on germination and seedling growth of wheat under salt stress. In: Proceeding of 11th Iranian congress in seed sciences and echnology. Gorgan, Iran
Satish S, Mohana DC, Ranhavendra MP, Raveesha KA (2007) Antifungal activity of some plant extracts against important seed borne pathogens of Aspergillus sp. Int J Agric Technol 3(1):109–119
Scheuerell SJ, Mahaffee WF (2006) Variability associated with suppression of graymold (*Botrytis cinerea*) on geranium by foliar applications of nonaerated and aerated compost teas. Plant Dis 90:1201–1208
Schwember AR, Bradford KJ (2005) Drying rates following priming affect temperature sensitivity of germination and longevity of lettuce seeds. Hort Sci 40(3):778–781
Shahrokhi M, Tehranifar A, Hadizadeh H, Selahvarzi Y (2011) Effect of drought stress and paclobutrazol-treated seeds on physiological response of *Festuca arundinacea* L. Master and *Lolium perenne* L. Barrage. J Biol Environ Sci 5(14):77–85
Shao CX, Hu J, Song WJ, Hu WM (2005) Effects of seed priming with chitosan solutions of different acidity on seed germination and physiological characteristics of maize seedling. J Zhejiang Univ (Agric Life Sci) 31(6):705–708
Shivay YS, Singh U, Prasad R, Kaur R (2016) Agronomic interventions for micronutrient biofortification of pulses. Indian J Agron 61(4th IAC Special Issue):161–172
Singh MV (2007) Efficiency of seed treatment for ameliorating zinc deficiency in crops. Zinc Crops 24–26
Singh A, Dahiru R, Musa M, Sani Haliru B (2014) Effect of Osmopriming duration on germination, emergence, and early growth of Cowpea (*Vigna unguiculata* (L.) Walp.) in the Sudan Savanna of Nigeria. Int J Agron 2014:1–4
Singh H, Jassal RK, Kang JS, Sandhu SS, Kang H, Grewal K (2015a) Seed priming techniques in field crops-a review. Agric Rev 36(4):251–264
Singh U, Praharaj CS, Shivay YS, Kumar L, Singh SS (2015b) Ferti-fortification: an agronomic approach for micronutrient enrichment of pulses. In Pulses: challenges and opportunities under changing climatic scenario, In: Proceedings of the national conference on "Pulses: challenges and opportunities under changing climatic scenario, vol. 29, pp 208–222

Slaton NA, Wilson CE, Ntamatungiro S, Norman RJ, Boothe DL (2001) Evaluation of zinc seed treatments for rice. Agron J 93(1):152–157

Sliwinska E, Jendrzejczak E (2002) Sugar-beet seed quality and DNA synthesis in the embryo in relation to hydration-dehydration cycles. Seed Sci Technol 30(3):597–608

Srivastava AK, Bose B (2012) Effect of nitrate seed priming on phenology, growth rate and yield attributes in rice (*Oryza sativa* L.). Vegetos Int J Plant Res 25(2):174–181

Steel RGD, Torrie JH, Dickey D (1997) Principles and procedures of statistics: a biometrical approach, 3rd edn. McGraw Hill Book Company, New York, pp 172–177

Sung JM, Chiu KY (1995) Hydration effect on seedling emergence strength of watermelon seeds differing in ploidy. Plant Sci 110(1):21–26

Tavili A, Zare S, Enayati A (2009) Hydropriming, ascorbic and salicylic acid influence on germination of *Agropyro nelongatum* host. Seeds under salt stress. Res J Seed Sci 2(1):16–22

Taylor AG, Allen PS, Bennett MA, Bradford KJ, Burris JS, Misra MK (1998) Seed enhancements. Seed Sci Res 8(2):245–256

Thakur P, Kumar S, Malik JA, Berger JD, Nayyar H (2010) Cold stress effects on reproductive development in grain crops: an overview. Environ Exp Bot 67(3):429–443

Tonelli ML, Furlan A, Taurian T, Castro S, Fabra A (2011) Peanut priming induced by biocontrol agents. Physiol Mol Plant Pathol 75(3):100–105

Uchida A, Jagendorf AT, Hibino T, Takabe T, Takabe T (2002) Effects of hydrogen peroxide and nitric oxide on both salt and heat stress tolerance in rice. Plant Sci 163(3):515–523

Umair A, Ali S, Hayat R, Ansar M, Tareen MJ (2011) Evaluation of seed priming in mung bean (*Vigna radiata*) for yield, nodulation and biological nitrogen fixation under rainfed conditions. Afr J Biotechnol 10(79):18122–18129

Umair A, Ali S, Sarwar M, Bashir K, Tareen MJ, Malik MA (2013) Assessment of some priming techniques in mungbean (*Vigna radiata*): a green house study. Pak J Agric Res 26:4

Upadhyaya H, Begum L, Dey B, Nath PK, Panda SK (2017) Impact of calcium phosphate nanoparticles on rice plant. J Plant Sci Phytopathol 1:1–10

Van Hulten M, Pelser M, Van Loon LC, Pieterse CM, Ton J (2006) Costs and benefits of priming for defense in Arabidopsis. Proc Natl Acad Sci 103(14):5602–5607

Varier A, Vari AK, Dadlani M (2010) The subcellular basis of seed priming. Curr Sci 99:450–456

Wahid A, Noreen A, Basra SM, Gelani S, Farooq M (2008) Priming-induced metabolic changes in sunflower (Helianthus annuus) achenes improve germination and seedling growth. Bot Stud 49(4):343–350

Waqas M, Nizami AS, Aburiazaiza AS, Barakat MA, Ismail IMI, Rashid MI (2017) Optimization of food waste compost with the use of biochar. J Environ Manag. https://doi.org/10.1016/j.jenvman.2017.06.015

Warren JE, Bennett MA (1997) Seed hydration using the drum priming system. Hort Sci 32(7):1220–1221

Weltzien HC (1991) Biocontrol of foliar fungal disease with compost extracts. In: Andrews JH, Hirano SS (eds) Microbial ecology of leaves. Springer-Verlag, New York, pp 430–450

Wright B, Rowse H, Whipps JM (2003) Microbial population dynamics on seeds during drum and steeping priming. Plant Soil 255(2):631–640

Xu S, Hu J, Li Y, Ma W, Zheng Y, Zhu S (2011) Chilling tolerance in *Nicotiana tabacum* induced by seed priming with putrescine. Plant Growth Regul 63(3):279–290

Yadav PV, Kumari M, Ahmed Z (2011) Seed priming mediated germination improvement and tolerance to subsequent exposure to cold and salt stress in capsicum. Res J Seed Sci 4(3):125–136

Yan M (2015) Seed priming stimulate germination and early seedling growth of Chinese cabbage under drought stress. S Afr J Bot 99:88–92

Yogananda DK, Vyakarnahal BS, Shekhargouda M (2004) Effect of seed invigoration with growth regulations and micronutrients on germination and seedling vigour of bell pepper cv. California Wonder. Karnataka J Agri Sci 17(4):811–813

Zhou ZS, Guo K, Elbaz AA, Yang ZM (2009) Salicylic acid alleviates mercury toxicity by preventing oxidative stress in roots of *Medicago sativa*. Environ Exp Bot 65(1):27–34

Zhou J, Wang Y, Jahufer Z (2013) Location and chemical composition of semi-permeable layer of forage seeds. Bangladesh J Bot 42(1):23–30

# Physiological, Biochemical, and Molecular Aspects of Seed Priming

**Sadam Hussain, Saddam Hussain, Abdul Khaliq, Shafaqat Ali, and Imran Khan**

## Contents



**Abstract** Seed priming is a low-cost and effective strategy to enhance seed germination, vigor index, and yield in many field crops mainly under adverse environmental conditions. Seed priming is known to trigger the normal metabolic developments during early stage of germination, before the radicle protrusion. Higher, faster, and synchronized germination of primed seeds largely occurs due to enzyme activation, reduced imbibition time, metabolic reparation during imbibition, buildup of germination-promoting metabolites, and osmotic adjustment. Moreover, plants emerging from primed seeds exhibit faster activation of cellular defense systems, which trigger tolerance against consequent exposure to environmental stresses in the field. Several seed priming approaches including hydropriming,

S. Hussain · S. Hussain (✉) · A. Khaliq · I. Khan
Department of Agronomy, University of Agriculture Faisalabad, Faisalabad, Punjab, Pakistan

S. Ali
Department of Environmental Sciences and Engineering, Government College University Faisalabad, Faisalabad, Pakistan



M. Hasanuzzaman, V. Fotopoulos (eds.), *Priming and Pretreatment of Seeds and Seedlings*, https://doi.org/10.1007/978-981-13-8625-1_3

nutrient priming, hormonal priming, chemical priming, osmopriming, and redox priming can be effectively used under different environmental conditions. The present chapter provides an overview of the physiological, biochemical, and molecular changes modulated by seed priming, which enhance seed germination and plant growth. Moreover, it discusses the possible mechanisms associated with seed priming-induced abiotic stress tolerance in plants.

**Keywords** Abiotic stresses · Antioxidants · Seed priming · Osmotic adjustment · Physiological and molecular mechanisms

## Abbreviations

| | |
|---|---|
| AQP | Aquaporin |
| AsA | Ascorbic acid |
| ATP | Adenosine triphosphate |
| BER | Base excision repair |
| CAT | Catalase |
| DEPs | Differentially expressed proteins |
| DHN | Dehydrins |
| GA | Gibberellin |
| GR | Glutathione reductase |
| $H_2O_2$ | Hydrogen peroxide |
| HR | Homologous recombination |
| LEA | Late embryogenesis abundant |
| MT | Metallothionein |
| NaCl | Sodium chloride |
| NER | Nucleotide excision repair |
| OA | Osmotic adjustment |
| OP | Osmotic potential |
| PAs | Polyamines |
| PEG | Polyethylene glycol |
| POD | Peroxidases |
| Pro | Proline |
| ROS | Reactive oxygen species |
| RWC | Relative water contents |
| SOD | Superoxide dismutase |

## 1 Introduction

Changes in climatic patterns have triggered the occurrence of biotic and abiotic stresses. These stresses are generally interrelated and cause undesirable morphological, biochemical, physiological, and molecular changes that affect the plant growth and development and ultimately the seed yield (Hussain et al. 2018a). Seed priming is an effective strategy to impart the abiotic stress tolerance besides strengthening the defense line of crop plants. In seed priming, presowing treatments are applied that control the hydration level within seed and allow pregerminative metabolic processes (physiochemical) to proceed while preventing radical emergence (Hussain et al. 2015; Lutts et al. 2016; Zheng et al. 2016; Hussain et al. 2018a). This phenomenon is described by an enhanced activation of plant defense mechanisms (Anderson et al. 2017). Stimuli from abiotic cues can prompt the priming establishment by acting as warning signals (Mauch-Mani et al. 2017). Upon stimulus perception, changes generally occur in the plant at the physiological, biochemical as well as at molecular levels.

Seed priming promotes the germination-related functions and enhances germination rate, vigor index, root length, photosynthetic efficacy, biomass production, and many other growth attributes in plants (Hussain et al. 2016a). Moreover, seed priming improves the plant biochemical status by induction and de novo synthesis of hydrolyses activity (e.g., α-amylase) and soluble sugar levels during germination process while nitrate reductase (NR) activity and N content in the growing seedlings under normal condition with respect to untreated seeds (Sharma and Bose 2006; Anaytullah and Bose 2007). Further, the definite role of priming in enhancing germination and growth of different crop plants under different abiotic stresses has been well established (Kaya et al. 2006; Zhang et al. 2006; Zhuo et al. 2009; Jisha et al. 2013; Paparella et al. 2015; Wang et al. 2016; Zheng et al. 2016; Hussain et al. 2018a, b). Seedlings emerged from primed seeds showed better antioxidative defense system owing to enhanced activities/levels of glutathione reductase (GR), catalase (CAT), peroxidases (POD), superoxide dismutase (SOD), and ascorbic acid (AsA), and regulation of stress protein like aquaporins (AQP), dehydrins (DHN), and late embryogenesis abundant (LEA) proteins (Wechsberg et al. 1994; Fashui 2002; Bolkhina et al. 2003; Anaytullah et al. 2012; Chen et al. 2013). In seed priming, seeds are hydrated in different types of solutions for the initiation of certain metabolic processes (Hussain et al. 2015) such as protein synthesis, and synthesis or repair of mitochondria, which permits the initial germination (Bray et al. 1989; Jisha et al. 2013; Paparella et al. 2015).

This chapter sums up the recent work accomplished on seed priming regarding its role in modulation of physio-biochemical and molecular mechanisms during germination and postgermination phases. It also discusses the basis of seed priming-induced enhancement of abiotic stress responses in crop plants during their different development stages.

# 2 Physiological, Biochemical, and Molecular Aspects of Seed Priming

## *2.1 Pregerminative Metabolism*

At physiological level, priming treatments inflict various metabolic alterations in seed with the start of imbibition process (Bray et al. 1989; Bray 1995; Paparella et al. 2015). Because of rehydration in seed priming, major cellular processes in seed such as the de novo synthesis of proteins and nucleic acids, adenosine triphosphate (ATP) production, activation of antioxidants accumulation of phospholipids and sterols, and DNA repair mechanisms are regulated (Paparella et al. 2015). Seeds are commonly exposed to various abiotic stresses during the initial stage of their germination, and thus, the oxidative damage of nucleic acid, lipids, and proteins is not an unusual phenomenon (Kranner et al. 2010). Within this framework, the seed repair response is important for conserving seed vigor and efficient germination (Oge` et al. 2008). Proper repair of DNA damage allows the embryo cells to recommence cell cycle development and DNA replication occurs; nevertheless, oxidative injury because of defective DNA repair mechanisms leads to cell death (Kranner et al. 2010; Balestrazzi et al. 2011b; Waterworth et al. 2011; Ventura et al. 2012).

Priming treatments lead to an increased solubilization of storage proteins in seed, enhanced antioxidative activity (Bourgne et al. 2000; Randhir and Shetty 2005), and reduced lipid peroxidation rate in plants (Yeh et al. 2005). A proteome analysis of *Arabidopsis* seed germination after priming identified various storage proteins (12S-cruciferin β-subunits) which specifically appear during seed priming (hydro- and osmopriming) (Gallardo et al. 2004). Sung and Chang (1993) reported that other reserve mobilization enzymes such as isocitrate lyase (for mobilization of lipids) and carbohydrates (-amylases) are also triggered during seed-priming treatments. In primed seeds, a significantly higher number of mitochondria were also observed in osmoprimed leek cells (Corbineau et al. 2000). During priming treatments, upregulation of α- and β-tubulin subunits proteins, which are important in cell division, was also noted by Varier et al. (2010). Repair mechanism of DNA is an important constituent of the "pregerminative metabolism," which is activated with the start of seed imbibition, accompanied by uncontrolled buildup of reactive oxygen species (ROS) (Paparella et al. 2015). Numerous studies have specified that the major DNA repair processes such as nucleotide- (NER) and base excision repair (BER) are stimulated during the early seed imbibition phase in order to sustain the genome integrity (Balestrazzi et al. 2012; Cordoba-Canero et al. 2014; Paparella et al. 2015).

The increased activities of antioxidant enzymes permit the control accumulation of ROS during water uptake by seed (imbibition process) (Bailly et al. 2000; Hsu et al. 2003). Scavenging ROS in seed is described as antioxidant potential of the seed, and essential requirement to enhance germination under environmental extremities (Liu et al. 2007). The antioxidative response of seed can be assessed by monitoring the expression profiles of genes encoding enzymatic antioxidant such as

SOD, which is important in scavenging of superoxide radicals (Yao et al. 2012; Paparella et al. 2015). Likewise, some other antioxidants such as CAT, APX, and GR were enhanced at transcript as well as enzyme activity level, representing the activation of antioxidant defense system (Chen et al. 2014; Macovei et al. 2014; Paparella et al. 2015). In short, seed priming activates various protective functions owing to physiological and metabolic alterations during pregerminative phase of seed, which provide the way for faster germination and vigorous seedling establishment.

## 2.2 *Better Imbibition and Vigorous Seedling Growth*

Seed priming may persuade structural and ultrastructural changes in seed that facilitates and speed up the process of imbibition, thus leading to uniform seedling emergence and better seedling stand establishment (Galhaut et al. 2014). The first phase of germination (imbibitions phase), which starts with the uptake of water by seed, is similar in both primed and nonprimed seed by comparing water uptake and time. During phase II (lag phase) of germination in primed seeds, hydration treatment permits controlled uptake of water by seed, whether alive or dead, and stimulates the pregerminative metabolic process, but emergence of radicle is prevented, characterized by the extended lag phase. Phase III (radical protrusion phase) is also similar in primed and nonprimed seeds and covers the germination and postgermination phases (Rajjou et al. 2012). Primed seed also showed the higher activities of many enzymes for metabolism of proteins (proteases), carbohydrates (α and β amylases), and lipids (isocitrate lyase) involved in mobilization of stored reserves in seed (Varier et al. 2010; Sisodia et al. 2018). These enzymes play pivot role in breakdown of the macromolecules for embryo growth and development that exerts positive influence on early and better seedling emergence (Varier et al. 2010). Seed priming can break seed dormancy, curtail seedling emergence time, improve seedling vigor, and leads to better germination and growth of plants (Mondal et al. 2011; Srivastava and Bose 2012). Seed priming with bioflavonoid enhanced both root and shoot elongation and increased the levels of photosynthetic pigments, flavonoids, and phenolics (Singh et al. 2016). Bose et al. (2018) stated that seed priming with polyethylene glycol significantly improved the levels of photosynthetic pigments under abiotic stress conditions.

## 2.3 *Osmotic Adjustment*

Seed priming has been reported to shorten the time of imbibition (phase I) and lag phases (phase II) (Khan et al. 2009; Ibrahim 2016). The embryo swelling inside the primed seed speeds up the germination process through facilitating the absorption of water (Elouaer and Hannachi 2012; Ibrahim 2016). Accumulation of osmolytes (e.g., sugars) is a crucial mechanism of plant stress tolerance (Ibrahim 2016), which reduces the osmotic potential (OP) of cell and allows osmotic adjustment (OA) under adverse environmental conditions (Matias et al. 2015; Kubala et al. 2015b; Ibrahim 2016). In priming, enhancement of physiological as well as biochemical changes takes place in seed during suspension of germination by negligible matric and low OP of the imbibing medium. Nonpenetrating organic solutes or salts in solid matrices (matriconditioning) and liquid medium (osmoconditioning) are used to generate the water potential equilibrium between seed and osmotic medium needed for conditioning. In *Vigna radiata* (mungbean), priming of seeds with sub-lethal dose of sodium chloride (halopriming) ameliorated the toxic effects of salinity stress by accumulation of osmolytes and enhancement of antioxidant defense system for OA (Saha et al. 2010). γ-Aminobutyric acid (GABA) treatment was helpful in maintaining an ideal OP in plants, which make them able to survive under osmotic stress conditions, without disturbing the normal functions of the cell. Lower OP with less change in leaf water content showed better OA while encountering osmotic stress (Hinckley et al. 1980). Higher proline (Pro) accumulation is also a common response of plants to osmotic stress (Ashraf and Foolad 2007). Proline regulates the cellular redox potential, stabilizes the macromolecules and subcellular structures, and triggers the stress-responsive genes/proteins (Szabados and Savouré 2010). During osmotic stress, accumulation of Pro occurs mainly due to reduced degradation and increased synthesis (Verbruggen and Hermans 2008), and partially occurs due to abscisic acid (ABA) and hydrogen peroxide ($H_2O_2$) signaling (Rejeb et al. 2014). While studying on *Zea mays* seedlings, Yang et al. (2009) demonstrated that the $H_2O_2$ led to Pro accumulation and upregulated the expression of $\Delta^1$-pyrroline-5-carboxylate synthetase gene in radicle and coleoptile. Although accumulation of Pro has been reported in priming treated seeds (Farhoudi et al. 2011), little is known regarding the involvement of $H_2O_2$ and putative gene expression in this process.

## 2.4 Membrane Properties

Auto-oxidation of storage metabolites reserves during seed storage may lead to lipid peroxidation, which may further disrupt cellular compartmentalization and membranes (Rakshit and Singh 2018). Long-term storage of seed affects the activities and functions of various enzymes and cellular organelles, ultimately reducing the viability and vigor of seedling (Rakshit and Singh 2018). During seed priming, initial phase (phase I or imbibition phase) is characterized by repairing of mitochondria and DNA, enhanced respiration and energy metabolism, cell cycle initiation, hormone signaling, gene transcription and gene translation, and regulation of stress-responsive genes (DHN, LEA, AQP, etc.), activation of priming memory, ROS signaling, and regulation of antioxidant enzymes (Rakshit and Singh 2018). During lag phase, activation of priming memory and synthesis of protein through mRNA are recruited upon rehydration. In postgermination phase, mobilization of stored reserve and elongation of radicle cells occur, and after rupture of seed outer covering (seed coat), radicle emerges out (Chen and Arora 2013). Seed priming treatments trigger the active absorption of ions with higher ATP availability, and reduce the metabolites leakage via repair of deteriorated seed parts leading to improved growth of embryo (Dahal et al. 1990). Seed priming improves the cell membrane integrity, negates the oxidative damage, and counteracts lipid peroxidation, and these attributes are directly linked with repair of biochemical lesions (Villiers and Edgcumbe 1975), maintenance of seed viability (Basu et al. 1973), activation of enzymes (Sananda and Bose 2012), and improved rate of germination (Lee et al. 1998). It has also been reported that menadione sodium bisulfite, a novel priming agent (Prasad et al. 1994), was capable of inducing resistance against oxidative stress in *Arabidopsis* (Borges et al. 2009). Priming with polyamines has also been reported to stabilize the bilayer surface, retard the membrane deterioration (Basra et al. 1994), scavenge the free radicals, and protect the macromolecules and membranes from oxidative damages (Roberts et al. 1986; Besford et al. 1993) under adverse environmental conditions.

## *2.5 Antioxidant Defense System*

In primed seedlings, different antioxidants such as POD, APX, SOD, and CAT have been known to play an important role in enhancing stress tolerance (Bolkhina et al. 2003). These antioxidants may guard the cellular membranes against the harmful effects of ROS such as $H_2O_2$, hydroxyl radicals, superoxide radicals, and singlet oxygen (Posmyk et al. 2001). Enhanced activities of CAT, SOD, and POD in seedlings emerged from primed seeds have been reported under normal and stress conditions (Goswami et al. 2013; Zheng et al. 2016; Hussain et al. 2016a, b).

Plant antioxidant systems scavenge the excessive ROS production persuaded by various stresses and play an important role during seed storage, germination, and development (Bailly 2004; De Tullio and Arrigoni 2003). Antioxidant system constitutes both enzymatic (such as APX, CAT, and SOD) and nonenzymatic compounds (e.g., GSH and AsA). Each antioxidant usually has a specific function, for instance, CAT degrades the $H_2O_2$ into water and oxygen, while APX-induced catalysis of $H_2O_2$ is reliant on AsA-GSH cycle. Here, AsA acts as election donor to stimulate $H_2O_2$ degradation by the APX, while GSH and its enzymes (monodehydroascorbate reductase; MDHAR, GR, dehydroascorbate reductase; DHAR) are accountable for AsA regeneration. Moreover, various antioxidants can have different functions at a similar developmental phase. For instance, CAT was regulated in dormant mature as well as germinating seeds, whereas APX was not noticed in physiologically quiescent seeds (De Tullio and Arrigoni 2003). Seed priming modifies the ROS accumulation and alters the expression of genes and enzymes of antioxidative defense system. Kibinza et al. (2011) noted that sunflower priming showed enhanced expression of gene encoding CAT and demonstrated that CAT is important enzyme that play role in recovery of vigor in the aged seeds.

## 2.6 Changes in Metabolic Events

Several metabolic events, such as DNA replication (Lanteri et al. 1994); synthesis of RNA, DNA, and proteins (Bray 1995; Bailly et al. 2000); and accumulation of beta-tubulin (De Castro et al. 1995), are triggered with seed priming (Paparella et al. 2015). In sunflower (*Helianthus annuus*), Chojnowski et al. (1997) noted that priming of seed increased the respiration rate and conversion of ACC (1-aminocyclopropane 1-carboxylic acid) to the ethylene. Understanding the responses of biochemical or molecular markers is of a significant interest for evaluating the efficacy of seed priming. In tomato (*Solanum lycopersicum*) seeds, a significant positive correlation was noted between DNA replication and osmotic treatment (osmoconditioning) (Bailly et al. 2000). Job et al. (1997) found that an 11-S globulin beta-chain was a good indicator among seed priming-induced effects in sugar beet (*Beta vulgaris*). Seed-priming treatments have been reported to improve the germination rate of aged seeds (Bailly et al. 1996, 2000), which was mainly attributed to lower lipid peroxidation rate, and restoration of antioxidative defense mechanisms, e.g., increased activities of GR and CAT (Bailly et al. 1996, 2000). The CAT enzyme controls the lipid peroxidation rate by scavenging of $H_2O_2$, while GR is known to produce glutathione.

Several other metabolic and cellular events such as including induction of stress-responsive proteins (LEAs and HSPs), cell division and elongation, $H^+$-ATPase activity, plasma membrane fluidity, and changes in proteome and transcriptome have been demonstrated to understand the basis of priming-induced abiotic stress tolerance in plants (Gallardo et al. 2001; Zhuo et al. 2009).

## 2.7 Hormonal Balance and Regulation

The antagonistic roles of different plant hormones like GA (gibberellin) and ABA in regulating seed germination are well understood (Bewley 1997). However, molecular and physiological mechanisms underlying the influence of seed priming treatments on germination in relation to plant hormones are poorly illustrated. The GA is crucial for seed germination, as evident by the inability of GA-deficient *Arabidopsis* (ga1-3) and tomato (gib-1) mutants to germinate without the exogenous application of GA (Groot and Karrssen 1987).

There are two distinct development programs involved in seed maturation and germination: shift from quiescent state to germination and switch from embryo development to maturation (Kermode 1990). Both steps are affected by the coordination between growth promoter (GA) and growth retardants (ABA) (Nambara et al. 2010). The interaction between ABA and GA is controlled by regulation of genes involved in signaling pathways and biosynthesis of these hormones and metabolism of germination-related proteins, stress proteins, and storage compounds (van der Geest 2002; Weiss and Ori 2007). Weitbrecht et al. (2011) have reviewed the role of ABA/GA in regulation of seed germination. During the embryogenesis, accumulation of ABA inhibits vivipary and modulates seed maturation. While GA antagonizes ABA during germination and development, therefore, GA biosynthesis signaling is generally reduced in mature seeds but increased during germination. El-Araby et al. (2006) reported that seed-priming treatments enhance GA/ABA ratio which, in turn, regulates seed germination. In primed seeds, the possible basis of regulation of hormones in relation to germination includes different GA-induced germination-related processes, mainly including weakening of endosperm, reserve mobilization, and elongation of embryo cell (Chen et al. 2001; Chen and Bradford 2000; Nonogaki et al. 2000; Chen et al. 2002; Sung et al. 2008). Only few studies exist on interaction of ABA and GA during seed priming, and its contribution in seed priming-induced regulation of germination. The available studies have outlined a differential response of seed priming (e.g., increased ABA signaling during seed priming) to ABA-GA interaction (Lopez-Molina et al. 2002; Catusse et al. 2011).

Ethylene also affects the germination process through enhancing speed and rate of germination (Siriwitayawan et al. 2003). The relationship between priming and ethylene has been reported in various plant species, such as peanut, cucumber, tomato, and lettuce (Cantliffe et al. 2000; Habdas et al. 1998; Siriwitayawan et al. 2003). Two different mechanisms are suggested regarding the role of ethylene in seed priming: (1) higher ethylene production enhances the activity of endo-β-mannanase, which can help germination at high temperature via weakening of endosperm, and (2) ethylene can modulate OA by regulating the osmotic adjustment in primed seeds and increasing germination even under stressful conditions (Cantliffe et al. 2000).

## 2.8 *Aquaporins and Tonoplast Intrinsic Proteins*

Aquaporins (AQPs) have significant role in kinetic exchange of water. Recently, different studies have demonstrated the better water uptake in primed seeds because of faster imbibition with respect to nonprimed seeds (Galhaut et al. 2014; Kubala et al. 2015a), and the ectopic expression of AQPs genes (SoPIP2; 1) was enhanced in primed seeds (Chen et al. 2013). The transmembrane transport of water through the regulation of AQPs enhanced the capacity of seed to absorb and transport water to different plant tissues. It was observed that during imbibition process, the transport of water and adequate water supply for the embryo through AQPs might be one of the major components mediated in primed seeds, that impacts the rate of germination as well as stress resistance (Chen et al. 2013; Wojtyla et al. 2016).

Increased germination potential and vigor of seeds under different priming techniques have been well reported (Hu et al. 2005; Pant and Bose 2016). In *Arabidopsis* model plant, proteomic analysis in primed and unprimed seeds identified 1300 proteins, and an abundant change was recorded in 74 proteins prior to emergence, and some new proteins were identified during dehydration (Gallardo et al. 2001). In wheat seeds, proteomics analysis during artificial aging and seed priming (hydropriming) showed 162 differentially expressed proteins (DEPs) responsible for energy supply, metabolism, and stress responses (Lv et al. 2016), while 531 DEPs were noted in primed seeds compared to nonprimed seeds, and various upregulated DEPs are involved in process of energy supply (such as tricarboxylic acid cycle, glycolysis, fatty acid oxidation), anabolic processes (such as synthesis of various fats and amino acids), and cell division and growth (Lv et al. 2016).

## *2.9 Dehydrins (Late Embryogenesis Abundant Proteins)*

DHNs (group 2 LEA protein) (Candat et al. 2014) are associated with enhanced stress tolerance in primed seeds (Wechsberg et al. 1994). Degradation of DHN was noted during priming (osmopriming) in *Pisum sativum* L. (peas) and *Beta vulgaris* L. (beet) (Capron et al. 2000; Baker et al. 1995; Gumilevskaya and Azarkovich 2010), indicating that primed seeds may reduce DHNs than unprimed seeds. A reduction in the abundance of DHNs in primed seeds was noted in *Pisum sativum* (peas), *Spinacia oleracea* (spinach), and *Beta vulgaris* (beet) compared to unprimed seeds (Baker et al. 1995; Chen et al. 2012a, b). Chen et al. (2012b) found four DHN-like proteins in spinach seeds during early phase of osmopriming and they demonstrated that these proteins were degraded progressively to a lower level in primed seeds, compared with untreated seeds.

## *2.10 Reactive Oxygen Species: Key Signaling Molecules in Priming*

Seed priming alters the accumulation of ROS in plants (Pal et al. 2013). Several evidence indicate that ROS, although toxic at higher concentration, could also assist as signal molecules to regulate hormone signaling, programmed cell death, plant growth and development, and tolerance against various abiotic stress factors (Mittler et al. 2004). ROS also contribute in sensing of the external environment and thus control the plant development (Smirnoff 2005). A complex system is present for ROS metabolism that interacts closely with hormonal signaling systems and assists plants in regulation of developmental proceedings as well as abiotic stress responses (El-Maarouf-Bouteau and Bailly 2008). These authors noted that light or moderate accumulation of ROS is beneficial for germination, as it may activate signaling cascades for seed germination, promote the weakening of endosperm, and mediate programmed cell death in the aleurone layer (Bailly et al. 2008; El-Maarouf-Bouteau and Bailly 2008). Under priming treatments, such beneficial accumulation of ROS may occur from partial hydration and dehydration of seed. Bailly et al.

(2008) reported the decreased ROS accumulation during seed development of in sunflower; however, their levels were increased rapidly with seed hydration because of regulation of various ROS-producing metabolisms mainly including lipid catabolism and respiration. During seed-priming process, the enhanced ROS production can be a natural response that occurred due to the partial hydration of seed, rather than an indicator of stress. Nevertheless, $H_2O_2$ accumulation was enhanced by PEG exposure (>24 h) in *Medicago* (Balestrazzi et al. 2011a). Likewise, a combined application of GA and PEG persuaded high accumulation of ROS in *Larix deciduas* (Larch) and *Pinus sylvestris* (pine) seeds (Naglreiter et al. 2005). Apart from the positive role of ROS, an excessive generation of ROS can be injurious to germination of seed because these may cause seed aging and cellular perturbations (McDonald 1999). It was suggested that the accumulation of ROS should be controlled tightly in order to play the role of positive regulator of germination (Bailly et al. 2008), particularly during osmopriming where less availability of the water can cause more production of ROS.

## 2.11 Activation of DNA Repair Pathways

DNA repair and defense mechanisms are stimulated in seeds upon rehydration to decrease the growth inhibition during development of seedling. During maturation and storage conditions, seeds are exposed to various adverse conditions, which cause DNA oxidation, loss of D NA integrity, and hinder the cell cycle (Bray and West 2005). To increase the chances of successful germination and preserve the seed vigor, DNA repair processes must be kept at proper level in embryo. DNA repair mechanisms mainly including BER and NER together with antioxidants are regarded as key pregerminative metabolic processes (Huang et al. 2007; Wojtyla et al. 2016). In barrel clover (*Medicago truncatula*) seeds, the TYROSYL-DNA PHOSPHODIESTERASE (MtTdp1α; GB# FJ858738) and MtTdp1β (GB# BT006446.1) genes encoding α- and β-tyrosyl-DNA phosphodiesterase1 and DNA repair genes MtTop1α (GB# CA919655) and MtTop1β (GB# CX526330) were upregulated rapidly during the imbibition process (Macovei et al. 2010; Wojtyla et al. 2016). The increased expressions of MtTop1 α and MtTop1 β and as well as for MtTdp1 α and MtTdp1 β were also noted during PEG priming of *Medicago truncatula* seeds (Balestrazzi et al. 2011a). In *Arabidopsis* seed, enhanced expression of AtOGG1 gene (DNA glycosylase) was also recorded during desiccation and water uptake (Wojtyla et al. 2016). The PEG priming of *M. truncatula* seeds also showed the involvement of AtTFIIS and MtTFIIS in DNA repair mechanisms during seed imbibition (Macovei et al. 2011b; Wojtyla et al. 2016), but Balestrazzi et al. (2011a) demonstrated that higher regulation of TFII-S was not observed in PEG-primed seeds. DNA repair mechanisms (NER, BER, HR, etc.) are now regarded as pregerminative metabolic process, important for recrudescence of cell cycle activity (Huang et al. 2007; Wojtyla et al. 2016). The repair processes of DNA should be regarded as the most valuable and essential process activated in the process of seed priming (Macovei et al. 2011a; Ventura et al. 2012; Rajjou et al. 2012). Prior to replication in primed seeds, repair of DNA damage occurs primarily through DNA synthesis (Varier et al. 2010). In chick pea seeds, the role of DNA repair during seed priming was suggested based on the expression analysis of genes involved in DNA repair directly or indirectly (Sharma and Maheshwari 2015; Wojtyla et al. 2016). Seed priming also regulates the expression of some genes/proteins which play essential role for cell division. In *Arabidopsis thaliana* seeds, both osmoprimed (−0.75 MPa PEG 6000) and hydroprimed resulted in accumulation of tubulin subunits (α tubulin and β tubulin) (Gallardo et al. 2001; Kubala et al. 2015a). In osmoprimed seeds of *Lycopersicon esculentum*, DNA replication and synchronization of cell cycle were also recorded (Özbingöl et al. 1999; Wojtyla et al. 2016). Moreover, osmopriming (1.2 MPa PEG) in *Brassica napus* seeds enhanced the regulation of genes/proteins involved in cell cycle, such as microtubule motor activity proteins (At3g45850, At4g39050), proteins associated with microtubule (At5g55230, At1g24764), cell division control protein 48 homolog C (At3g01610; CDC48C), and tubulin subunits (beta- and gamma-tubulin) (Kubala et al. 2015a; Wojtyla et al. 2016). The above discussion partially confirmed that different priming techniques promote higher and faster activation of pregerminative metabolism, such as DNA processing during replication and repair mechanisms, and transcriptional activity.

## 3 Seed Priming and Abiotic Stress Tolerance in Plants

Crop plants often pass through a period of abiotic stresses during their life cycle under natural environments which adversely affect their growth and productivity (Hussain et al. 2018a). Different stresses such as drought, heat, cold, and salinity are threatening the crop yields. To counter the effects of these abiotic stresses, plants modulate various physio-biochemical processes. It has been well understood that most of the abiotic stresses decrease the uptake during first phase of germination. Less water availability under abiotic stresses affects the process of the cell elongation, leading to reduced embryo growth followed by poor emergence of seedling. Seed priming, a pragmatic technique, has been successfully employed to achieve the proper stand establishment under normal and stressful conditions (Jafar et al. 2012; Hussain et al. 2016a, b). Recently, several reports have described the beneficial impacts of priming under different abiotic stresses in various field crops (Jisha et al. 2013; Hussain et al. 2016a, b; Hussain et al. 2018a). Plants emerged from the primed seeds showed vigorous head start and higher stress tolerance mainly because of more effective energy metabolism, OA, quick cellular defense systems, enlarged embryo, and enhanced enzymatic activation (Jisha et al. 2013). In PEG-primed seeds, Pant and Bose (2016) noted that the germination, vigor, and relative water content were enhanced with respect to nonprimed seeds. Under cold stress, seed priming with salicylic acid (SA) + $H_2O_2$ combination significantly enhanced germination and seedling growth, which was closely related with ROS detoxification, hormonal metabolism, metabolites accumulation, and efficient energy supply (Li et al. 2017). In tobacco, seed priming enhanced the cold tolerance during germination and growth of seedling by regulation of antioxidative defense in plant tissues (Xu et al. 2011). Guan et al. (2009) demonstrated that seed priming improved the germination speed and seedling growth of maize under low temperature. Similarly, Elkoca et al. (2007) suggested the seed priming (hydropriming and osmopriming) for better seedling emergence and vigorous growth of chickpea under chilling stress. Srivastava et al. (2010) reported the beneficial influence of various seed-priming techniques to increase salinity and drought tolerance in Indian mustard. Kumar et al. (2016) stated that hydro- and halopriming ($Mg(NO_3)_2$ and $Ca(NO_3)_2$) can alleviate the adversities of heavy metal ($HgCl_2$) stress in wheat by enhancing the germination, seedling length, soluble sugar contents, and α amylase activity.

## 4 Conclusion

Seed priming offers a smart, innovative, realistic, and effective option for achieving faster and uniform emergence, vigorous stand establishment, and higher productivity in crop plants under normal and stressful conditions. Vigorous start and greater stress tolerance in plants raised from primed seeds primarily occur because of more effective energy metabolism, OA, embryo enlargement, enhanced activation of different enzymes, increased DNA and RNA synthesis, and quick cellular defense responses. Seed priming can stimulate various signaling cascades during early growth phase and leads to faster and efficient defense responses in plants. The exact molecular mechanisms behind seed priming are not fully explored; it is postulated that sensitization of seed was related with higher buildup of signaling proteins. Under abiotic stress conditions, activation of signaling proteins amplifies signal transduction, thus resulting in rapid activation of plant defense responses. Priming helps repair some of the damages caused by seed erosion, which leads to increased vigor of primed seeds. Seed priming also affects the metabolic phase of germination and thus triggers the early replication of DNA. It is desirable to develop the appropriate seed-priming approaches for different plant species to overcome the challenges of the environmental extremities. Although the beneficial effects of seed priming in enhancing abiotic stress tolerance in plants are widely reported, the exact mechanism behind such effects is poorly known. Therefore, the imminent goals of agricultural researchers should be focused on identifying the novel proteins/genes and transcription factors, which are regulated due to seed priming under different abiotic stresses.

## References


Anaytullah, Bose B (2007) Nitrate-hardened seeds increase germination, amylase activity and proline content in wheat seedlings at low temperature. Physiol Mol Biol Plants 13:199–207

Anaytullah, Srivastava AK, Bose B (2012) Impact of seed hardening treatment with nitrate salts on nitrogen and antioxidant defense metabolisms in Triticum aestivum L under different sowing conditions. Vegetos 25:292–299

Anderson AJ, McLean JE, Jacobson AR, Britt DW (2017) CuO and ZnO nanoparticles modify interkingdom cell signaling processes relevant to crop production. J Agric Food Chem 66(26):6513–6524

Ashraf MFMR, Foolad M (2007) Roles of glycine betaine and proline in improving plant abiotic stress resistance. Environ Exp Bot 59(2):206–216

Bailly C (2004) Active oxygen species and antioxidant in seed biology. Seed Sci Res 14:93–107

Bailly C, Benamar A, Corbineau F, Côme D (1996) Changes in malondialdehyde content and in superoxide dismutase, catalase and glutathione reductase activities in sunflower seeds as related to deterioration during accelerated aging. Physiol Plant 97:104–110

Bailly C, Benamar A, Corbineau F, Come D (2000) Antioxidant systems in sunflower (*Helianthus annuus* L.) seeds as affected by priming. Seed Sci Res 10:35–42

Bailly C, El-Maarouf-Bouteau H, Corbineau F (2008) From intracellular signaling networks to cell death: the dual role of reactive oxygen species in seed physiology. C R Biol 331:806–814

Baker EH, Bradford KJ, Bryant JA, Rost TL (1995) A comparison of desiccation related proteins (dehydrin and QP47) in peas (*Pisum sativum*). Seed Sci Res 5:185–193

Balestrazzi A, Confalonieri M, Macovei A, Carbonera D (2011a) Seed imbibition in Medicago truncatula Gaertn. Expression profiles of DNA repair genes in relation to PEG-mediated stress. J Plant Physiol 168:706–713

Balestrazzi A, Confalonieri M, Macovei A, Dona` M, Carbonera D (2011b) Genotoxic stress and DNA repair in plants: emerging functions and tools for improving crop productivity. Plant Cell Rep 30:287–295

Balestrazzi A, Confalonieri M, Dona` M, Carbonera D (2012) Genotoxic stress, DNA repair, and crop productivity. In: Tuteja N, Gill SS (eds) Crop improvement under adverse conditions. Springer, Berlin, pp 153–169

Basra AS, Singh B, Malik CP (1994) Priming-induced changes in polyamine levels in relation to vigor of aged onion seeds. Bot Bull Acad Sinica 35:19–23

Basu RN, Chattopadhyay K, Pal P (1973) Maintenance of seed viability in rice (*Oryza sativa* L.) and Jute (*Corchorus capsularis* L. and *C. olitorius* L.). Indian Agric 18:76–79

Besford RT, Richardson CM, Campos JL, Tiburcio AF (1993) Effects of polyamines on stabilization of molecular complexes in thylakoid membranes of osmotically stressed oat leaves. Planta 189:201–206

Bewley JD (1997) Seed germination and dormancy. Plant Cell 9:1055–1066

Bolkhina O, Virolainen E, Fagerstedt K (2003) Antioxidants, oxidative damage and oxygen deprivation stress: a review. Ann Bot 91:179–194

Borges AA, Dobon A, Expósito-Rodríguez M, Jiménez-Árias D, Borges-Pérez A, CasañasSánchez V, Pérez JA, Luis JC, Tornero P (2009) Molecular analysis of menadione-induced resistance against biotic stress in Arabidopsis. Plant Biotechnol J 7:744–762

Bose B, Kumar M, Singhal RK, Mondal S (2018) Impact of seed priming on the modulation of Physico-chemical and molecular processes during germination, growth, and development of crops. In: Advances in seed priming. Springer, Singapore, pp 23–40

Bourgne S, Job C, Job D (2000) Sugar beet seed priming: solubilization of the basic subunit of 11-S globulin in individual seeds. Seed Sci Res 10:153–156

Bray CM (1995) Biochemical processes during the osmopriming of seeds. In: Kigel J, Galili G (eds) Seed development and germination. Marcel Dekker, New York, pp 767–789

Bray CM, West CE (2005) DNA repair mechanisms in plants: crucial sensors and effectors for the maintenance of genome integrity. New Phytol 168:511–528

Bray CM, Davison PA, Ashraf M, Taylor MR (1989) Biochemical events during osmopriming of leek seed. Ann Appl Biol 102:185–193

Candat A, Paszkiewicz G, Neveu M, Gautier R, Logan DC, Avelange-Macherel MH, Macherel D (2014) The ubiquitous distribution of late embryogenesis abundant proteins across cell compartments in Arabidopsis offers tailored protection against abiotic stress. Plant Cell 26:114

Cantliffe DJ, Sung Y, Nascimento WM (2000) Lettuce seed germination. Hortic Rev 24:229–275

Capron I, Corbineau F, Dacher F, Job C, Côme D, Job D (2000) Sugar beet seed priming: effects of priming conditions on germination, solubilization of 11-S globulin and accumulation of LEA proteins. Seed Sci Res 10:243–254

Catusse J, Meinhard J, Job C, Strub J, Fischer U, Pestsova E, West-Hoff P, Van Dorsselaer A, Job D (2011) Proteomics reveals potential biomarkers of seed vigor in sugarbeet. Proteomics 11:1569–1580

Chen K, Arora R (2013) Priming memory invokes seed stress-tolerance. Environ Exp Bot 94:33–45

Chen F, Bradford KJ (2000) Expression of an expansin is associated with endosperm weakening during tomato seed germination. Plant Physiol 124(3):1265–1274

Chen F, Dahal P, Bradford KJ (2001) Two tomato expansin genes show divergent expression and localization in embryos during seed development and germination. Plant Physiol 127(3):928–936

Chen F, Nonogaki H, Bradford KJ (2002) A gibberellin-regulated xyloglucanendotransglycosylase gene is expressed in the endosperm cap during tomato seed germination. J Exp Bot 53:215–223

Chen H, Chu P, Zhou Y, Li Y, Liu J, Ding Y, Tsang EW, Jiang L, Wu K, Huang S (2012a) Overexpression of AtOGG1, a DNA glycosylase/AP lyase, enhances seed longevity and abiotic stress tolerance in Arabidopsis. J Exp Bot 63(11):4107–4121

Chen K, Fessehaie A, Arora R (2012b) Dehydrin metabolism is altered during seed osmopriming and subsequent germination under chilling and desiccation in Spinacia oleracea L. cv. Bloomsdale: possible role in stress tolerance. Plant Sci 183:27–36

Chen K, Fessehaie A, Arora R (2013) Aquaporin expression during seed osmopriming and post-priming germination in spinach. Biol Plantarum 57(1):193–198

Chen C, Letnik I, Hacham Y, Dobrev P, Ben-Daniel BH, Vankova R, Amir R, Miller G (2014) ASCORBATE PEROXIDASE6 protects Arabidopsis desiccating and germinating seeds from stress and mediates cross talk between reactive oxygen species, abscisic acid, and auxin. Plant Physiol 166:370–383

Chojnowski M, Corbineau F, Côme D (1997) Physiological and biochemical changes induced in sunflower seeds by osmopriming and subsequent drying, storage and aging. Seed Sci Res 7:323–331

Corbineau F, Ozbingol N, Vineland D, Come D (2000) Improvement of tomato seed germination by osmopriming as related to energy metabolism. In: Black M, Bradford KJ, Vasquez-Ramos J (eds) Seed biology advances and applications: proceedings of the sixth international workshop on seeds. CABI Cambridge, Merida, pp 467–474

Cordoba-Canero D, Rolda n´-Arjona T, Ariza RR (2014) Arabidopsis ZDP DNA 30-phosphatase and ARP endonuclease function in 8-oxoG repair initiated by FPG and OGG1 DNA glycosylases. Plant J 79:824–834

Dahal P, Bradford KJ, Jones RA (1990) Effects of priming and endosperm integrity on seed germination rates of tomato genotypes II. J Exp Bot 41:1441–1453

De Castro R, Zheng X, Bergervoet JHW, De Vos CHR, Bino RJ (1995) β -tubulin accumulation and DNA replication in imbibing tomato seeds. Plant Physiol 109:499–504

De Tullio MC, Arrigoni O (2003) The ascorbic acid system in seeds: to protect and to serve. Seed Sci Res 13:249–260

El-Araby MM, Moustafa SMA, Ismail AI, Hegazi AZA (2006) Hormone and phenol levels during germination and osmopriming of tomato seeds, and associated variations in protein patterns and anatomical seed features. Acta Agronomica Hungarica 54(4):441–457

Elkoca E, Haliloglu K, Esitken A, Ercisli S (2007) Hydro-and osmopriming improve chickpea germination. Acta Agric Scand Sect B Soil Plant Sci 57(3):193–200

El-Maarouf-Bouteau H, Bailly C (2008) Oxidative signaling in seed germination and dormancy. Plant Signal Behav 3:175–182
Elouaer MA, Hannachi C (2012) Seed priming to improve germination and seedling growth of safflower (*Carthamustinctorius*) under salt stress. Eur J Biosci 6:76–84
Farhoudi R, Saeedipour S, Mohammadreza D (2011) The effect of NaCl seed priming on salt tolerance, antioxidant enzyme activity, proline and carbohydrate accumulation of Muskmelon (*Cucumismelo* L.) under saline conditions. Afr J Agric Res 6:136–370
Fashui H (2002) Study on the mechanism of cerium nitrate effects on germination of aged rice seed. Biol Trace Elem Res 87:191–200
Galhaut L, Lespinay A, Walker DJ, Bernal MP, Correal E, Lutts S (2014) Seed priming of *Trifoliumrepens* L. improved germination and early seedling growth on heavy metal contaminated soil. Water Air Soil Pollut 225:1–15
Gallardo K, Job C, Groot SPC, Puype M, Demol H, Vandekerekhove J et al (2001) Proteomic analysis of Arabidopsis seed germination and priming. Plant Physiol 126:835–848
Gallardo K, Job C, Groot SPC, Puype M, Demol H, Vandekerekhove J, Job D (2004) Proteomics of Arabidopsis seed germination and priming. In: Nicholas G (ed) The biology of seeds: recent advances. CABI, Cambridge, pp 199–209
Goswami A, Banerjee R, Raha S (2013) Drought resistance in rice seedlings conferred by seed priming: role of the anti-oxidant defense mechanisms. Protoplasma 250:1115–1129
Groot SPC, Karrssen CM (1987) Gibberellins regulate seed germination in tomato by weakening: a study with gibberellin-deficient mutants. Planta 171:525–531
Guan YJ, Hu J, Wang XJ, Shao CX (2009) Seed priming with chitosan improves maize germination and seedling growth in relation to physiological changes under low temperature stress. J Zhejiang Univ Sci B 10(6):427–433
Gumilevskaya NA, Azarkovich MI (2010) Identification and characterization of dehydrins in horse chestnut recalcitrant seeds. Russ J Plant Physiol 57:859–864
Habdas H, Szafirowska A, Sokolowska A, Grzesik M (1998) Cytological and physiological effects of matriconditioning on low viable cucumber seed germination. In: XXV IntHort congress, part 7: quality of horticultural products, pp 113–120
Hinckley TM, Durhme F, Hinckley AR, Richter H (1980) Water relations of drought hardy shrubs: osmotic potential and stomatal reactivity. Plant Cell Environ 3:131–140
Hsu CC, Chen CL, Chen JJ, Sung JM (2003) Accelerated agingenhanced lipid peroxidation in bitter gourd seeds and effects of priming and hot water soaking treatments. Sci Hortic 98:201–212
Hu J, Zhu ZY, Song WJ, Wang JC, Hu WM (2005) Effects of sand priming on germination and field performance in direct-sown rice (*Oryza sativa* L.). Seed Sci andTech 33(1):243–248
Huang Z, Boubriak I, Osborne DJ, Dong M, Gutterman Y (2007) Possible role of pectin-containing mucilage and dew in repairing embryo DNA of seeds adapted to desert conditions. Ann Bot 101:277–283
Hussain S, Zheng M, Khan F, Khaliq A, Fahad S, Peng S et al (2015) Benefits of rice seed priming are offset permanently by prolonged storage and the storage conditions. Sci Rep 5:8101. https://doi.org/10.1038/srep08101
Hussain S, Khan F, Hussain HA, Nie L (2016a) Physiological and biochemical mechanisms of seed priming-induced chilling tolerance in Rice cultivars. Front Plant Sci 7:116. https://doi.org/10.3389/fpls.2016.00116
Hussain S, Khan F, Cao W, Wu L, Geng M (2016b) Seed priming alters the production and detoxification of reactive oxygen intermediates in rice seedlings grown under sub-optimal temperature and nutrient supply. Front Plant Sci 7:439. https://doi.org/10.3389/fpls.2016.00439
Hussain HA, Hussain S, Khaliq A, Ashraf U, Anjum SA, Men S, Wang L (2018a) Chilling and drought stresses in crop plants: implications, cross talk, and potential management opportunities. Front Plant Sci 9:393
Hussain S, Khaliq A, Tanveer M, Matloob A, Hussain HA (2018b) Aspirin priming circumvents the salinity-induced effects on wheat emergence and seedling growth by regulating starch metabolism and antioxidant enzyme activities. Acta Physiol Plant 40(4):68

Ibrahim EA (2016) Seed priming to alleviate salinity stress in germinating seeds. J Plant Physiol 192:38–46

Jafar MZ, Farooq M, Cheema MA, Afzal I, Basra SMA, Wahid MA, Aziz T, Shahid M (2012) Improving the performance of wheat by seed priming under saline conditions. J Agron Crop Sci 198(1):38–45

Jisha KC, Vijayakumari K, Puthur JT (2013) Seed priming for abiotic stress tolerance: an overview. Acta Physiol Plant 35(5):1381–1396

Job C, Kersulec A, Ravasio L, Chareyre S, Pépin R, Job D (1997) The solubilization of the basic subunit of sugar beet seed 11-S globulin during priming and early germination. Seed Sci Res 7:225–243

Kaya MD, Okcu G, Atak M, Cikili Y, Kolsarici O (2006) Seed treatments to overcome salt and drought stress during germination in sunflower (*Helianthus annuus* L.). Eur J Agron 24:291–295

Kermode AR (1990) Regulatory mechanisms involved in transition from seed development to germination Crit. Rev Plant Sci 9:155–195

Khaliq A, Aslam F, Matloob A, Hussain S, Geng M, Wahid A, ur Rehman H (2015) Seed priming with selenium: consequences for emergence, seedling growth, and biochemical attributes of rice. Biol Trace Elem Res 166(2):236–244

Khan HA, Ayub CM, Pervez MA, Bilal RM, Shahid MA, Ziaf K (2009) Effect of seed priming with NaCl on salinity tolerance of hot pepper (*Capsicum annuum* L.) at seedling stage. Soil Environ 28:81–87

Kibinza S, Bazin J, Bailly C, Farrant JM, Corbineau F, El-Maarouf-Bouteau H (2011) Catalase is a key enzyme in seed recovery from ageing during priming. Plant Sci 181:309–315

Kranner I, Beckett RP, Minibayeva FV, Seal CE (2010) What is stress? Concepts, definitions and applications in seed science. New Phytol 188:655–673

Kubala S, Garnczarska M, Wojtyla Ł, Clippe A, Kosmala A, Zmienk´o A et al (2015a) Deciphering priming-induced improvement of rapeseed (*Brassica napus* L.) germination through an integrated transcriptomic and proteomic approach. Plant Sci 231:94–113

Kubala S, Wojtyla Ł, Quinet M, Lechowska K, Lutts S, Garnczarska M (2015b) Enhanced expression of the proline synthesis gene P5CSA in relation to seed osmopriming improvement of *Brassica napus* germination under salinity stress. Plant Sci 183:1–12

Kumar M, Pant B, Mondal S, Bose B (2016) Hydro and halo priming: influenced germination responses in wheat Var-HUW-468 under heavy metal stress. Acta Physiol Plant 38:217

Lanteri S, Saracco F, Kraak HL, Bino RJ (1994) The effects of priming on nuclear replication activity and germination of pepper (*Capsicum annuum*) and tomato (*Lycopersicon esculentum*) seeds. Seed Sci Res 4:81–87

Lee SS, Kim JH, Hong SB, Yun SH, Park EH (1998) Priming effect of rice seeds on seedling establishment under adverse soil conditions. Korean J Crop Sci 43:194–198

Li Z, Xu J, Gao Y, Wang C, Guo G, Luo Y, Huang Y, Hu W, Sheteiwy MS, Guan Y, Hu J (2017) The synergistic priming effect of exogenous salicylic acid and $H_2O_2$ on chilling tolerance enhancement during maize (*Zea mays* L.) seed germination. Front Plant Sci 8:1153

Liu X, Xing D, Li L, Zhang L (2007) Rapid deterioration of seed vigour based on the level of superoxide generation during early imbibition. Photochem Photobiol Sci 6:767–774

Lopez-Molina L, Mongrand S, McLachlin DT, Chait BT, Chua NH (2002) ABI5 acts downstream of ABI3 to execute an ABA-dependent growth arrest during germination. Plant J 32(3):317–328

Lutts S, Benincasa P, Wojtyla L, Kubala S, Pace R, Lechowska K, Quinet M, Garnczarska M (2016) Seed priming: new comprehensive approaches for an old empirical technique. In: New challenges in seed biology-basic and translational research driving seed technology. InTech

Lv Y, Zhang S, Wang J, Hu Y (2016) Quantitative proteomic analysis of wheat seeds during artificial ageing and priming using the isobaric tandem mass tag labeling. PLoS One 11(9):0162851

Macovei A, Balestrazzi A, Confalonieri M, Carbonera D (2010) The Tdp1; 1; (Tyrosyl-DNA phosphodiesterase) gene family in barrel medic (*Medicago truncatula* Gaertn.): bioinformatic investigation and expression profiles in response to copper- and PEG-mediated stress. Planta 232:393–407

Macovei A, Balestrazzi A, Confalonieri M, Buttafava A, Carbonera D (2011a) The TFIIS and TFIIS-like genes from Medicago truncatula are involved in oxidative stress response. Gene 470:20–30
Macovei A, Balestrazzi A, Confalonieri M, Fae` M, Carbonera D (2011b) New insights on the barrel medic MtOGG1 and MtFPG functions in relation to oxidative stress response in planta and during seed imbibition. Plant Physiol Biochem 49:1040–1050
Macovei A, Garg B, Raikwar S, Balestrazzi A, Carbonera D, Buttafava A, Bremont JFJ, Gill SS, Tuteja N (2014) Synergistic exposure of rice seeds to different doses of gamma-ray and salinity stress resulted in increased antioxidant enzyme activities and gene-specific modulation of TC-NER pathway. Biomed Res Int 2014:676–934
Matias JR, Ribeiro RC, Aragão CA, AraújoGgl DBF (2015) Physiological changes in osmo and hydroprimed cucumber seeds germinated in biosaline water. J Seed Sci 37:7–15
Mauch-Mani B, Baccelli I, Luna E, Flors V (2017) Defense priming: an adaptive part of induced resistance. Annu Rev Plant Biol 68:485–512
McDonald MB (1999) Seed deterioration: physiology, repair and assessment. Seed Sci Technol 27:177–237
Mittler R, Vanderauwera S, Gollery M, Van Breusegem F (2004) Reactive oxygen gene network of plants. Trends Plant Sci 9(10):490–498
Mondal S, Vijai P, Bose B (2011) Role of seed hardening in rice variety Swarna (MTU 7029). Res J Seed Sci 4:157–165
Naglreiter C, Reichenauer TG, Goodman BA, Bolhàr-Nordenkampf HR (2005) Free radical generation in Pinussylvestris and Larix decidua seeds primed with polyethylene glycol or potassium salt solutions. Plant Physiol Biochem 43(2):117–123
Nambara E, Okamoto M, Tatematsu K, Yano R, Seo M, Kamiya Y (2010) Abscisic acid and the control of seed dormancy and germination. Seed Sci Res 20(2):55–67
Nonogaki H, Gee OH, Bradford KJ (2000) A germination-specific endo-β-mannanase gene is expressed in the micropylar endosperm cap of tomato seeds. Plant Physiol 123(4):1235–1246
Oge` L, Bourdais G, Bove J, Collet B, Godin B, Granier F, Boutin JP, Job D, Jullien M, Grappin P (2008) Protein repair L-isoaspartyl methyltransferase 1 is involved in both seed longevity and germination vigor in Arabidopsis. Plant Cell 20:3022–3037
Özbingöl N, Corbineau F, Groot SPC, Bino RJ, Côme D (1999) Activation of the cell cycle in tomato (Lycopersicon esculentum mill) seeds during osmoconditioning as related to temperature and oxygen. Ann Bot 84:245–251
Pal M, Gondor OK, Janda T (2013) Role of salicylic acid in acclimation to low temperature. Acta Agron Hung 61:161–172
Pant B, Bose B (2016) Mitigation of the influence of PEG-6000 imposed water stress on germination of halo primed rice seeds. Int J Agric Environ Biotechnol 9:275
Paparella S, Araújo SS, Rossi G, Wijayasinghe M, Carbonera D, Balestrazzi A (2015) Seed priming: state of the art and new perspectives. Plant Cell Rep 34:1281–1293
Posmyk MM, Corbineau F, Vinel D, Bailly C, Côme D (2001) Osmoconditioning reduces physiological and biochemical damage induced by chilling in soybean seeds. Physiol Plant 111(4):473–482
Prasad TK, Anderson MD, Martin BA, Stewart CR (1994) Evidence for chilling induced oxidative stress in maize seedlings and a regulatory role for hydrogen peroxide. Plant Cell 6:65–74
Rajjou L, Duval M, Gallardo K, Catusse J, Bally J, Job C, Job D (2012) Seed germination and vigor. Annu Rev Plant Biol 63:507–533
Rakshit A, Singh HB (2018) Advances in seed priming. Nature Springer
Randhir R, Shetty K (2005) Developmental stimulation of total phenolics and related antioxidant activity in light- and dark-germinated corn by natural elicitors. Process Biochem 40:1721–1732
Rejeb KB, Abdelly C, Savouré A (2014) How reactive oxygen species and proline face stress together. Plant Physiol Biochem 80:278–284

Roberts DR, Dumbroff EB, Thompson JE (1986) Exogenous polyamines alter membrane fluidity in bean leaves—a basis for their potential misinterpretation of their true physiological role. Planta 167:395–401

Saha P, Chatterjee P, Biswas AK (2010) NaCl pretreatment alleviates salt stress by enhancement of antioxidant defense system and osmolyte accumulation in mungbean (*Vignaradiata* L. Wilczek). Indian J Exp Biol 48:593–560

Sananda M, Bose B (2012) Kinetics studies on α-amylase extracted from germinating wheat endosperm of primed and non-primed seeds. Indian J Agric Biochem 25:137–141

Sharma MK, Bose B (2006) Effect of seed hardening with nitrate salts on seedling emergence, plant growth and nitrate assimilation of wheat (*Triticum aestivum* L.). Physiol Mol Biol Plants 12:173–176

Sharma SN, Maheshwari A (2015) Expression patterns of DNA repair genes associated with priming small and large chickpea (*Cicer arietinum*) seeds. Seed Sci Technol 43:250–261

Singh A, Gupta R, Pandey R (2016) Rice seed priming with picomolarrutin enhances rhiospheric Bacillus subtilis CIM colonization and plant growth. PLoS One 11(1):e0146013

Siriwitayawan G, Geneve RL, Downie AB (2003) Seed germination of ethylene perception mutants of tomato and Arabidopsis. Seed Sci Res 13:303–314

Sisodia A, Padhi M, Pal AK, Barman K, Singh AK (2018) Seed priming on germination, growth and flowering in flowers and ornamental trees. Springer Nature

Smirnoff N (2005) Antioxidants and reactive oxygen species in plants. Preface. Blackwell Publishing, Ltd., Oxford. https://doi.org/10.1002/9780470988565

Srivastava AK, Bose B (2012) Effect of nitrate seed priming on phenology, crop growth rate and yield attributes in rice (*Oryza sativa* L.). Vegetos 25:174–181

Srivastava AK, Lokhande VH, Patade VY, Suprasanna P, Sjahril R, D'Souza SF (2010) Comparative evaluation of hydro-, chemo-, and hormonal-priming methods for imparting salt and PEG stress tolerance in Indian mustard (*Brassica juncea* L.). Acta Physiol Plant 32:1135–1144

Sung FJM, Chang YH (1993) Biochemical activities associated with priming of sweet corn seeds to improve vigor. Seed Sci Technol 21:97–105

Sung Y, Cantliffe DJ, Nagata RT, Nascimento WM (2008) Structural changes in lettuce seed during germination at high temperature altered by genotype, seed maturation temperature, and seed priming. J Am Soc Hort Sci 133(2):300–311

Szabados L, Savouré A (2010) Proline: a multifunctional amino acid. Trends Plant Sci 5:89–97

Van der Geest AHM (2002) Seed genomics: germinating opportunities. Seed Sci Res 12:145–153

Varier A, Vari AK, Dadlani M (2010) The sub cellular basis of seed priming. Curr Sci 99:450–456

Ventura L, Dona` M, Macovei A, Carbonera D, Buttafava A, Mondoni A, Rossi G, Balestrazzi A (2012) Understanding the molecular pathways associated with seed vigor. Plant Physiol Biochem 60:196–206

Verbruggen N, Hermans C (2008) Proline accumulation in plants: a review. Amino Acids 35:753–759

Villiers TA, Edgcumbe DJ (1975) On the cause of seed determination in dry storage. Seed Sci Technol 3:761–774

Wang W, Chen Q, Hussain S, Mei J, Dong H, Peng S, Huang L, Cui K, Nie L (2016) Pre-sowing seed treatments in direct-seeded early rice: consequences for emergence, seedling growth and associated metabolic events under chilling stress. Sci Rep. https://doi.org/10.1038/srep19637

Waterworth WM, Drury GE, Bray CM, West CE (2011) Repairing breaks in the plant genome: the importance of keeping it together. New Phytol 192:805–822

Wechsberg GE, Probert RJ, Bray CM (1994) The relationship between 'dehydrin like' protein sand seed longevity in *Ranunculus sceleratus* L. J Exp Bot 45:1027–1030

Weiss D, Ori N (2007) Mechanisms of cross talk between gibberellin and other hormones. Plant Physiol 144(3):1240–1246

Weitbrecht K, Müller K, Leubner-Metzger G (2011) First off the mark: early seed germination. J Exp Bot 62(10):3289–3309

Wojtyla Ł, Lechowska K, Kubala S, Garnczarska M (2016) Molecular processes induced in primed seeds—increasing the potential to stabilize crop yields under drought conditions. J Plant Physiol 203:116–126
Xu S, Hu J, Li Y, Ma W, Zheng Y, Zhu S (2011) Chilling tolerance in *Nicotiana tabacum* induced by seed priming with putrescine. Plant Growth Regul 63:279–290
Yang S-H, Lan S-S, Gong M (2009) Hydrogen peroxide-induced proline and metabolic pathway of its accumulation in maize seedlings. J Plant Physiol 166:1694–1699
Yao Z, Liu L, Gao F, Rampitsch C, Reinecke DM, Ozga JA, Ayele BT (2012) Developmental and seed aging mediated regulation of antioxidative genes and differential expression of proteins during pre- and post-germinative phases in pea. J Plant Physiol 15:1477–1488
Yeh YM, Chiu KY, Chen CL, Sung JM (2005) Partial vacuum extends the longevity of primed bitter gourd seeds by enhancing their anti-oxidative activities during storage. Sci Hortic 104:101–112
Zhang S, Hu J, Liu N, Zhu Z (2006) Pre-sowing seed hydration treatment enhances the cold tolerance of direct-sown rice. Seed Sci Technol 34(3):593–601
Zheng M, Tao Y, Hussain S, Jiang Q, Peng S, Huang J et al (2016) Seed priming in dry direct-seeded rice: consequences for emergence, seedling growth and associated metabolic events under drought stress. Plant Growth Regul 78:167–178
Zhuo J, Wang W, Lu Y, Sen W, Wang X (2009) Osmopriming-regulated changes of plasma membrane composition and function were inhibited by phenylarsin oxide in soybean seeds. J Integr Plant Biol 9:858–867

# Fundamental Processes Involved in Seed Priming

**Ehab Awad-Allah Ibrahim**

## Contents



---

E. A.-A. Ibrahim (✉)
Cross Pollinated Vegetable Crops Research Department, Horticulture Research Institute, Agricultural Research Center, Giza, Egypt



M. Hasanuzzaman, V. Fotopoulos (eds.), *Priming and Pretreatment of Seeds and Seedlings*, https://doi.org/10.1007/978-981-13-8625-1_4

**Abstract** Seed priming is an effective tool to enhance germination of seeds and subsequent growth of seedlings under both ideal and adverse conditions. The positive effects of seed priming are due to varied physiological, biochemical, and molecular changes. These improvements include activation of enzymes that are involved in cellular metabolism, metabolism of inhibitors, breaking seed dormancy, and water imbibition. Environmental stresses can adversely affect growth of plant at the physiological, biochemical, and molecular levels. The plants grown from the primed seeds tend to have better tolerance to abiotic stresses. Seed priming enhances the metabolic processes within the germinating seed, leading to buildup of abiotic stress tolerance. However, physiological, biochemical, and cellular mechanisms of this phenomenon are not well understood. This chapter deals with recent progress in understanding the roles of seed priming in various physiological, biochemical, and molecular mechanisms during seed germination and post-germination stages as well as how it assists in regulating plant tolerance to abiotic stresses. Further, factors affecting seed priming and mechanisms of abiotic stress tolerance with priming agents were explained in the current state research work.

**Keywords** Seed priming · Abiotic stress · Physiological basis · Biochemical basis · Seed germination · Priming agents

## Abbreviations

| | |
|---|---|
| ABA | Abscisic acid |
| AgNPs | Biocompatible silver nanoparticles |
| APX | Ascorbate peroxidase |
| AsA | Ascorbic acid |
| ATP | Adenosine triphosphate |
| BABA | Beta-aminobutyric acid |
| CAT | Catalase |
| FC | Field capacity |
| GA3 | Gibberellin |
| GB | Glycine betaine |

| | |
|---|---|
| GPOX | Guaiacol peroxidase |
| GR | Glutathione reductase |
| GSH | Reduced glutathione |
| HSPs | Heat shock proteins |
| IAA | Indole-3-acetic acid |
| JA | Jasmonic acid |
| LEA | Late embryogenic abundance |
| MDA | Malondialdehyde |
| NR | Nitrate reductase |
| PAL | Phenylalanine ammonia-lyase |
| PAs | Polyamines |
| PEG | Polyethylene glycol |
| POX | Peroxidase |
| PPO | Polyphenol oxidase |
| ROIs | Reactive oxygen intermediates |
| ROS | Reactive oxygen species |
| SA | Salicylic acid |
| SOD | Superoxide dismutase |
| XOD | Xanthine oxide |

## 1 Introduction

Abiotic stresses such as salinity, drought, and temperature severity are the most significant factors that adversely affect the crop production (Jakab et al. 2005; Jaleel et al. 2009; Thakur et al. 2010). These abiotic stresses are responsible for either inhibition or delayed germination of seed and establishment of seedling (Almansouri et al. 2001). Germination is a crucial stage of plant's life that is highly responsive to change of environmental conditions. It is controlled by physiological, biochemical, and molecular processes related to embryo development (Bewley 1997; Rajjou et al. 2012; Bewley et al. 2013; Lutts et al. 2016). Abiotic stresses cause a series of changes in several physiological, biochemical, and molecular processes in cells (Xiong and Zhu 2002). They disturb the metabolic balance of plant cells, which may result in oxidative stress (Shafi et al. 2009), membrane damage (Korkmaz et al. 2010), reduction of cellular respiration (Sugie et al. 2006), and reactive oxygen species production (Suzuki and Mittler 2006).

Seed priming is a promising strategy used to improve performance of seed by enhancing the rate of germination and emergence uniformity that results in better and faster development of seedling (Cramer 2002; Basra et al. 2004; Castañares and Bouzo 2018; Sohail et al. 2018). Several seed priming methods have been developed to improve seed germination and seedling growth. Seed priming with several agents has developed as a promising approach to enhance plant stress tolerance. Seeds can be primed with different chemicals, ions, organic compounds, hormones, and antioxidants (Ghassemi-Golezai and Esmaeilpour 2008; Eskandari 2013; Jisha et al.

2013; Nawaz et al. 2013; Ibrahim 2016; Masondo et al. 2018). The seed priming success is strongly correlated with environmental variables during the seed priming process, plant species, seed lot and vigor, and priming method (Parera and Cantliffe 1994; Corbineau and Come 2006; Farooq et al. 2012; Maiti and Pramanik 2013).

Seed priming stimulates many of the physiological, biochemical, and molecular mechanisms that are involved in the early stages of germination, and it can induce abiotic stress tolerance mechanisms that improve growth during and after seedling establishment. Seed priming activates a series of physiological, biochemical, cellular, systemic, and molecular modulations that enhances plant growth under abiotic stress conditions (Varier et al. 2010; Eisvand et al. 2010; Chen and Arora 2013; Siri et al. 2013; Ibrahim 2016; Lutts et al. 2016; Pal et al. 2017).

Seed priming improves the germination rate and uniformity by metabolic repair during imbibition, osmotic adjustment, and stimulation of pre-germinative metabolic processes in the primed seeds. These include effects on early reserve mobilization, repair mechanisms, and DNA processing during replication (Bradford 1986; Bray et al. 1989; Ibrahim 2016; Lutts et al. 2016). Seed priming also promotes many of germination-related activities such as energy metabolism, reserve mobilization, endosperm weakening, respiration, and supporting the expansion of embryo cells (Pandita et al. 2007; Chen and Arora 2011; Varier et al. 2010). Seed priming improves germination by the regulation of DNA repair pathways, degrading enzymes, catalase, and other antioxidant-scavenging enzymes, the de novo synthesis of proteins and nucleic acids, and the accumulation of phospholipids and sterols (Afzal et al. 2002; Chen 2011; Rajjou et al. 2012; Kubala et al. 2015a; Paparella et al. 2015). Moreover, proper seedling development from primed seed may be due to induced plasma membrane fluidity, cell division, elongation, and abiotic stress-responsive proteins (Ibrahim 2016).

Seed priming also promotes the specific defense mechanism-related abiotic stress responses. It increases the accumulation of protective proteins such as heat shock proteins (HSPs) and late embryogenesis abundant, enhances the antioxidant scavenging enzymes such as SOD, GPX and CAT, activates the membrane efflux pumps, and upregulates the genes-encoding peroxiredoxin (Li et al. 2005; Varier et al. 2010; Catusse et al. 2011; Chen 2011). Seed priming also regulates stress tolerance by increasing of protein synthesis potential and posttranslational modifications and by keeping the optimum quotient for the translational turnover (Kubala et al. 2015b).

## 2 The Physiology of Seed Germination

The germination of seeds is the series of steps that progress to the radical protrusion. Dry mature seeds have very low water content (5–15%) with metabolic activities at a standstill. The start of metabolism in dry mature seeds and growth of embryo depend on suitable environment conditions such as suitable temperature, water, and oxygen. Sufficient water uptake (imbibition) is the most important factor to start the germination and the metabolic activities in the embryo. Water uptake in germinating

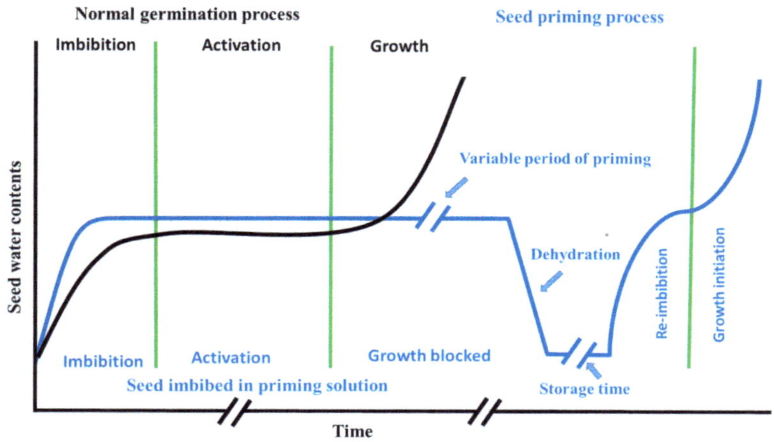


**Fig. 1** Schematic representation of the water uptake within the seed during standard germination and seed priming process. (Rajjou et al. 2012)

seed is normally a triphasic process, with a rapid initial water uptake (phase I, i.e., imbibition), followed by an activation phase with slight change in water content (phase II), and a further increase in water uptake (phase III) that occurs as the embryo axis elongates and with resumption of growth (Bewley 1997; Rajjou et al. 2012; Bewley et al. 2013; Lutts et al. 2016). Before the end of activation phase, germination stays a reversible process, and the seeds can be dehydrated again and stay viable during storing and able to subsequently reinitiate germination under suitable conditions. When water goes into dry seeds, the germination process of seed that has no problem with dormancy and coat permeability occurs in three phases (Fig. 1) (Bewely and Black 1994; Bewley et al. 2013; Eskandari 2013). The germination process can be divided into the following three phases:

## 2.1 *Initial Fast Imbibition*

When a dry seed is soaked in water, the solutes inside the seed cells decrease the turgor potential and lead to rapid uptake of water (phase I), and as the seed water potential increases during imbibition, the seed water content increases. The cell wall resistance to expansion results in a turgor potential increase in the cell. During seed imbibition, proteins are synthesized in the embryo using existing and new messenger ribonucleic acid (mRNA), and DNA and mitochondria are repaired and synthesized (McDonald 2000; Varier et al. 2010; Rajjou et al. 2012; Rosental and Nonogaki 2014). When the water potential of the cells in a seed increases, the water uptake slows, and the seed enters lag phase (phase II).

## 2.2 *Starting of Metabolic Processes in the Seed (Lag Phase)*

Phase II is known as the lag phase, in which the water potential of seed is in balance with that of the seed environment. During this phase, the seed absorbs a small or negligible amount of water over a relatively long period, but significant metabolic activity is associated with activation of enzymes and increased respiration. Therefore, the seed can complete all its physiological pre-germination processes and get ready for radicle emergence. The major metabolic changes related to germination such as new mitochondria and protein synthesis are initiated during this phase. Thus, it is also called an activation phase (Di Girolamo and Barbanti 2012a, b; Rajjou et al. 2012; Rosental and Nonogaki 2014). Phases I and II are the most delicate stages for the germination process and are essential to successful seed priming (Bewley 1997).

## 2.3 *Subsequent Radicle Emergence and Resumption of Growth*

When the pre-germination embryonic processes are completed in phase II, the seed transfers to phase III, which results in radicle protrusion through the seed coat, and water and oxygen absorption rapidly increases (Bradford 1986; Welbaum et al. 1998). Also, this phase increases root growth, and its emergence coincides with its characteristic division and enlargement of cells (Bradford 1995; Rajjou et al. 2012).

# 3 The Benefits of Seed Priming

Seed germination and stand establishment are critical factors in the life cycle of plant under both normal and stress conditions. Many methods to modify the processes involved in germination have been performed to enhance the behavior of seeds. Seed priming is the most commonly used method to enhance seed germination potential and increased abiotic stress tolerance (Bewley et al. 2013; Paparella et al. 2015; Ibrahim 2016). The valuable effects of seed priming with different methods and agents have been found in many crops (Maiti and Pramanik 2013; Ibrahim 2016; Lutts et al. 2016). The beneficial effects are not exclusive of seed germination and can be extended to later stages (Chen and Arora 2013). Seed priming is adopted as a method to improve performance of seed germination especially when applied under diverse environmental conditions or to poor-quality seeds. The seed priming benefits have also been found for subsequent seedling growth, resulting in a stronger crop stand (Nerson 2007; Dursun and Ekinci 2010; Sharma et al. 2014) and better ability to compete with weeds and to tolerate infection of pathogens (Ellis and Butcher 1988; Hill et al. 2008) and improve establishment of plants under abiotic stress conditions (Soeda et al. 2005; Nerson 2007; Khan et al. 2009a, b; Patade et al. 2009; Bewley et al. 2013; Maiti and Pramanik 2013; Paparella et al. 2015). Seed priming can break or alleviate photo- and thermo-dormancy in plants such as lettuce and endive (Bewley et al. 2013). The primed seeds give earlier and more uniform germination, more vigorous seedlings, and great tolerance to adverse conditions (Cramer 2002; Ozbay and Susluoglu 2016). Seed priming promotes the proliferation of root hairs that may lead to more efficient nutrient and water uptake. Moreover, it reduces several seed-borne pathogens, including *Xanthomonas campestris* in *Brassica* seed and *Septoria* in celery (Durner 2013). Seed priming is a low-cost, simple, and low-risk technique that can be a useful technology for farmers.

## 4 Mechanism of Seed Priming

Seed priming is a pre-sowing method that hydrates partially the seed in a specific environment for a certain period, followed by seed drying. It permits pre-germinative metabolic processes to the point of germination process to start while preventing radicle emergence (Bradford 1986; Dell Aquila and Tritto 1991; Pill 1995; Basra et al. 2003; Giri and Schillinger 2003; Chen and Arora 2013; Paparella et al. 2015; Ibrahim 2016). Seed priming allows some pre-germinative processes to occur without actual germination. Soaking of seeds into water stimulates a variety of metabolic changes necessary for germination. These changes include water imbibition, breaking seed dormancy, enzyme activation, etc. (Ajouri et al. 2004). In primed seeds, moistures are controlled at a level just less than what is required for actual germination, but just suitable to allow germinative metabolic processes to start. Primed seeds complete phase I (hydration) and II (lag phase) without entering the initial growth phase (phase-III) (Bradford 1986; Basra et al. 2003). Although the germination is not completed, the seed storage reserves are converted into a form that nourishes the embryo during germination. These changes allow the treated seeds to move to the third stage of germination, which involves radicle protrusion as soon as favorable environmental conditions are found (Ibrahim 2016).

Dehydration of seed after priming (drying-back) is essential to allow storage of seeds. After soaking period, seeds are rinsed with water and re-dried to levels compatible with storage. This keeps the positive effects of seed priming, without the quality loss that is caused by rapid deterioration of seed (Parera and Cantliffe 1992; Halmer 2004; Varier et al. 2010; Ratikanta 2011; Di Girolamo and Barbanti 2012a, b). Seed drying after the start of radicle emergence causes injury to vigor and viability of seed. When primed seed is in phase III, it tolerates a return to the initial moisture without damage (Fig. 1) (Taylor et al. 1998; Di Girolamo and Barbanti 2012a, b; Rajjou et al. 2012; Ibrahim 2016).

Primed seeds can be stored until time of sowing. The seed desiccation tolerance and longevity are two essential traits for primed seed (Ellis and Hong 1994; Hay and Probert 1995; Gurusinghe and Bradford 2001). The longevity of primed seed depends on the dehydration and storage conditions. A rapid dehydration may alter the soluble carbohydrate content. This decreases the tolerance to dehydration and seed longevity. Conversely, a slow dehydration may increase subsequent longevity (Bruggink et al. 1999; Gurusinghe and Bradford 2001). The primed seed can be stored for a period without harm. Primed seed often has a shorter life span than non-primed seed and should be stored under optimal conditions prior to planting.

## 5 Factors Affecting Seed Priming

The effectiveness of the seed priming is mainly influenced by various environmental variables during the seed priming process such as osmotic potential of priming solution, priming duration and ambient temperature, priming method applied, seed lot and vigor, plant species, and the storage condition of primed seeds. Therefore, the best results from seed priming are obtained if the priming technique is optimized (Parera and Cantliffe 1994; Corbineau and Come 2006; Farooq et al. 2012; Maiti and Pramanik 2013). These factors are often tested in specific experiments. Factors to consider are as follows:

### *5.1 Kind of Priming*

Each of the priming methods has its advantages and disadvantages and may have different effects (Tzortzakis 2009). Hydro-priming and solid matrix with calcium aluminum silicate (1:1:0.4, seed/water/solid matrix) increased significantly the germination and seedling vigor of okra (Sharma et al. 2014) in comparison with halo-priming and solid matrix methods. Halo-priming of hot pepper seeds showed a better performance than hydro- and unprimed seeds under low (3 dS $m^{-1}$) and moderate (5 dS $m^{-1}$) salt stress conditions (Amjad et al. 2007). At the same osmotic potential of NaCl and PEG, salt (NaCl) stress had a lower influence on cowpea seed germination and seedling growth than did the water stress simulated by the PEG (Murillo-Amador et al. 2002). Moreover, germination percentage of leek and carrot seeds primed with salt solution such as monopotassium phosphate ($KH_2PO_4$) was lower than those primed with polyethylene glycol (PEG 6000 molecular weight) (Brocklehurst et al. 1984). The positive effects of $KNO_3$ priming were more than NaCl priming, which are similar to those found for watermelon (*Citrullus lanatus* (Thunb.) Mansf) (Demir and Van de Venter 1999; Demir and Oztokat 2003), cucumber (*Cucumis sativus* L.) (Ghassemi-Golezani and Esmaeilpour 2008), and muskmelon (*Cucumis melo* L.) (Bradford et al. 1988). According to Alevarado and Bradford (1988) and Bellti et al. (1993), the superiority of seed priming with $KNO_3$ to NaCl is related to more nitrogen and potassium accumulation in seeds primed with $KNO_3$. The salt solution of potassium nitrate (KNO3) treatment showed better performance of tomato seeds than the PEG treatment (Frett et al. 1991; Lara et al. 2014). Priming tomato seeds with $KNO_3$ treatment gave the best results in comparison with water and vitamin C, gibrescol (GA3), as well as PEG 8000 treatment (Bocian and Holubowiczs 2008).

Moreover, seed priming of cucumber seeds with ethrel (100 ppm) or $KNO_3$ (1%) showed the better performance than GA3 (100 ppm) (Sowmya et al. 2013). In bottle gourd, GA3 was the best for seed osmopriming followed by $KH_2PO_4$ and $Na_2HPO_4$ (Das et al. 2014). Priming the hot pepper seeds in acetylsalicylic acid was found to be superior over salicylic acid treatment (Khan et al. 2009b). Cowpea seeds primed with GA3 showed better performance compared to indole-3-acetic acid (IAA) treatment (Saeedipour 2013). Solid matrix priming had a greater effect on seed germination than osmotic priming, particularly when PEG is used as the osmotic (Khan 1992). Muskmelon seed priming with salt solutions enhanced faster germination and reduced the germination percentage than priming with PEG or mannitol (Nascimento et al. 2004). Imbibition of muskmelon seed in salt solutions allows ions to penetrate into the seed and reduces seed germination; however, the seed is not damaged because some ions like $K^+$ are excluded from the embryo by the perisperm envelope (Welbaum and Bradford 1989). Seed priming with potassium nitrate ($KNO_3$), PEG, or NaCl improved germination and establishment of seedling; seed priming with $KNO_3$ had the most favorable effects in tomato (Govinden-Soulange and Levantard 2008; Lara et al. 2014). Priming of rice seeds with 5-aminolevulinic acid had higher germination and growth of seedling than those obtained by hydro-priming (Kanto et al. 2015).

## *5.2 Temperature*

Both seed priming and germination require similar temperature and other environmental factors. Each crop has a different optimum temperature for germination. In this respect, the optimum temperature for priming and germination of tomato seeds was 27–28 °C (Ozbingol et al. 1998). However, priming of onion seeds at 20 °C gave better results (Haigh et al. 1986; Barbara 2015). This may be due to the fact that each plant has different optimal temperatures for seed priming. Above and sub-optimal temperature delays germination due to a longer lag phase (Bradford 1995; Copeland and McDonald 2001). Low temperatures during seed imbibition can delay the physiological and biochemical processes of germination and take a longer time to achieve the same results, even though the seeds absorb a huge amount of water (Bradford 1986; Lee et al. 1998; McDonald 2000). A long seed priming allows radical protrusion and causes permanent damage during dehydration due to a lower moisture (Parera and Cantliffe 1994; Di Girolamo and Barbanti 2012a, b).

## *5.3 Oxygen Availability*

Most seeds can germinate in an atmosphere that is composed of about 20% of oxygen, but in seed priming, they can even germinate at lesser levels of oxygen (Copeland and McDonald 2001). Seed priming decreases the seed's sensitivity to lower oxygen levels. Seeds require a sufficient amount of oxygen (more than 10%)

during seed priming (Bujalski et al. 1989; Ozbingol et al. 1998; Farooq et al. 2012). At lower levels of oxygen, anaerobic respiration is taking place in some seeds, leading to an increase in ethanol to toxicity level (Furutani et al. 1986).

Aeration of soaking solution improves water uptake and decreases the time needed for seed priming (Yeoung et al. 1995; Ozbingol et al. 1998; Nascimento 2003). This treatment promotes faster germination of seed and accelerates growth of seedling in many crops (Nelson and Govers 1986; Bradford et al. 1988; Ozbingol et al. 1998; Nascimento 2003; Demir and Okcu 2004). On the other hand, seed soaking in aerated solution for a long time had deleterious effects on seed germination because of increased metabolism and salt penetration into the seeds (Welbaum and Bradford 1989; Nascimento 2003). Because of the low solubility of oxygen (50%) and mobility (10%), the relative oxygen availability to the seeds is decreased by 5%. Therefore, aeration system is needed to support seed respiration especially in a PEG solution (Bujalski et al. 1989; Bujalski and Nienow 1991). Primed seeds in an aerated PEG solution show more improved percentage of germination than in non-aerated solution (Heydecker and Coolbear 1977; Bujalski et al. 1989; Pereira et al. 2009). However, Cantliffe et al. (1981) found that there is no difference in lettuce seed germination between aerated and non-aerated solution of $K_3PO_4$. Moreover, soaking seeds in aerated PEG or salt solution gives more benefits than moist solid media (Akers and Holley 1986; Bujalski et al. 1989).

## 5.4 Osmotic Potential and Solution Concentration

The osmotic potential is zero in distilled water (0 MPa), and negative in priming solutions (> −2.0 MPa), but can reach to as very low as −350 to −50 MPa in air-dry seed because of the high solute content in cytoplasm. When seed is soaked in priming solution, water will move from the solution where it has a higher water potential into the seed where it has a lower water potential until equilibrium is reached. The imbibition of seed during soaking is directly related to the seed size and coat permeability and hydratable substrate content (Bradford 1995; Bewley et al. 2013). Concentration and water potential of priming solution are particularly important factors that determine the effectiveness of seed priming. High concentrations of ions in priming solution have detrimental effects on seed germination. They can be easily absorbed and accumulated in seed, resulting in inhibition of germination because nutritional imbalance and cytotoxic effects (Bradford 1995; Paparella et al. 2015). If the concentration is below the minimum threshold, priming treatment fails to initiate the expected physiological and biochemical processes of germination. On the other hand, if the concentration of the priming solution is too high or the duration of process is too long, seed germination will either be retarded or hampered completely as a result of toxicity (Sadeghi et al. 2009). The priming solution concentration also affects the seed performance under drought or salt stress. Priming with salt solutions at high concentration requires long-term treatment period and reduces the percentage of germination. Even though using low concentration of salt solutions requires short-term treatment period, it enhances the performance of the primed seeds, thus increasing the germination rate, improving the seedling vigor, and improving salt stress tolerance (Kaymak et al. 2008; Khan et al. 2009a; Nakaune et al. 2012; Saranya et al. 2017).

Various studies have shown that increasing of agent (e.g., NaCl, $CaCl_2$, $KNO_3$, and micronutrients) concentration in priming solutions might reduce germination indices and seedling growth in many crops. These reductions may be due to the greater osmotic pressures and ion toxicity (Bonilla et al. 2004; Nascimento et al. 2004; Esmaielpour et al. 2006; Diniz et al. 2009; Neamatollahi et al. 2009; Kaveh et al. 2011; Memon et al. 2013; Ebrahimi et al. 2014; Singh et al. 2014).

## 5.5 *Duration of Treatment*

The duration of treatments is important for some plants (Bradford 1986; Bocian and Holubowicz 2008). Controlled imbibition of water that forms the basis of seed priming technology is inherently affected by soaking duration. Priming duration along with the optimum concentration of any priming media used is a very important factor that determines the germination success and seedling establishment. This is because water imbibition of the seeds during priming process is directed up to the saturation level of the seeds (Ghassemi-Golezani and Esmaeilpour 2008). Seed priming before radicle emergence can be beneficial. Nevertheless, further extended priming duration may have negative effects on seed germination due to death of the seedling on drying (Ajouri et al. 2004). The ideal duration of seed priming differs according to the priming agent type, the solution osmotic potential, species, and priming temperature. Jeong et al. (2000) studied the effect of duration and temperature of seed priming on seed germination of many vegetables. Priming chemicals were −0.50 MPa PEG 8000 for carrot, 50 mM $K_3PO_4$ for lettuce, 200 mM $KH_2PO_4$ for onion, and 100 mM $Ca(NO_3)_2$ for Welsh onion seeds. Priming temperature varied from 10 °C to 25 °C, and priming duration differed from 12 h to 5 days, depending on the crop. The seed treatment, priming duration, and priming temperature had significant effects on percentage germination of carrot, lettuce, and onion seeds. Nevertheless, the priming duration and temperature had no effects on percentage germination of Welsh onion seeds. Singh et al. (2014) found that priming onion seeds for 24 h significantly improved seed germination (81.6%) as compared to 12 h, 36 h, or 48 h duration, and increasing priming duration from 2 days to 6 days led to a significant decrease in seed germination. Moreover, priming for 36 h gave the fastest germination time for maize seeds among 12 h to 48 h priming duration (Hacisalihoglu et al. 2018).

Generally, the success of seed priming is influenced by the duration of priming and vegetable species. Hydro-priming for 7 and 14 h can be successfully applied to enhanced seed and seedling vigor of bean (Ghassemi-Golezani et al. 2010). In addition, hydro-priming for 96 h was more active in increasing seed germination of onion than 48 h (Caseiro et al. 2004). Saeedipour (2013) found that the phytohormone seed priming for 6 h was more effective than 12 h in cowpea seed germination. Govinden-Soulange and Levantard (2008) found that tomato can be emerged earlier by priming seed in PEG for 2 days than 7 days. Moreover, priming of pepper seeds in PEG for 6 days resulted in increasing number of abnormal seedlings (Cantliffe et al. 1981).

## 5.6 Seed Quality

Seed quality is the key factor affecting the seed priming processing. Use of vigorous and pathogen-free seed is recommended to get a good priming result, e.g., the percentage and rate of emergence (Cantliffe et al. 1981; Cantliffe and Elballa 1994; Wheeler and Ellis 1994). The use of good seeds is essential for seed priming. There are differences in seed vigor between the seed lots; therefore, the seed priming response may vary among seed lots. To achieve maximum viability and vigor of seeds, seed harvest at a time as close as the physiological maturity stage is needed (Brocklehurst and Dearman 1983; Bocian and Holubowicz 2008). Using seeds harvested at suitable time and appropriate agent with precise concentration for priming can increase seed germination performance, especially under drought and salt conditions (Kazemi and Eskandari 2012). Melon seeds were more responsive to osmopriming with $KNO_3$ when harvested at 40 days after anthesis than 60 days (Welbaum and Bradford 1990). Watermelon seeds of 20 days after anthesis had the maximum benefit of priming such as emergence percentages and rate and weight of seedling compared to 30 and 40 days after anthesis (Demir and Mavi 2004).

## 5.7 Light

Light may be necessary in some priming treatments if the species require light for germination. Photoblastic seeds such as lettuce and celery that need light to germinate may be exposed to light during seed priming to decrease dormancy (Khan et al. 1978).

## 5.8 Dehydration after Priming

Dehydration after priming by removing a large quantity of moisture from the seed is a key to successful long-term seed storage, so the seeds should be dehydrated to their original moisture level. A rapid dehydration after seed priming leads to loss of the advancement obtained by priming. It may alter the soluble carbohydrate content, which in turn reduces desiccation tolerance and longevity of storage (Gurusinghe and Bradford 2001). On the contrary, a slow drying-back may increase the accumulation of late embryogenesis abundant proteins that offer a beneficial mechanism to improve seed longevity. However, fast re-drying at high temperatures may prompt the synthesis of heat shock proteins that are valuable to seed longevity (Gurusinghe et al. 2002). Drying seeds slowly after priming leads to regular loss of water that increases tolerance to dehydration damage by giving time for protective mechanisms to work. Moreover, a slow drying-back increased the gene expression involved in defense against abiotic stresses and DNA protection in seeds of *Brassica oleracea* (Soeda et al. 2005). Schwember and Bradford (2005) found that the fast or slow drying-back after priming lettuce seeds had lower seed longevity than non-primed seeds.

The conditions immediately after seed priming may affect seed longevity. Different kinds of abiotic stresses have been used before dehydration to reserve the benefits of priming and to maintain seed quality. This procedure may be seen as a stimulating cross-tolerance to keep priming benefits and promote the defense mechanisms that prevent the loss of seed viability during storage. In this respect, submission of mild water stress or heat shock before re-drying increased the mean rate of germination and reduced the mean time of germination (Genoud and Metraux 1999; Nascimento and West 2000; Kranner et al. 2010; Lira et al. 2015).

### *5.9 Storage Condition*

The longevity of primed seed is affected by many factors such as oxygen, relative air humidity, seed moisture content, and temperature (Rajjou and Debeaujon 2008; Wang et al. 2018). The lowest seed germination and growth seedling attributes were found when primed rice seeds were stored at 25 °C, while no significant reduction was observed during storage of seed at −4 °C (Hussain et al. 2015). The relative air humidity during storage might be more essential than the storage temperature in determining the longevity of seed. In general, low temperature, low relative humidity, and vacuum probably establish optimum conditions for storage of the primed seeds (Wang et al. 2018).

## 6 Physiological, Biochemical, and Molecular Responses to Seed Priming

The period between seed planting and seedling emergence plays a crucial role in stand establishment of crops (Hubbard et al. 2012). Seed priming has been used to improve germination rate and uniformity (Gupta et al. 2008; Girolamo and Barbanti 2012; Jisha et al. 2013). Seed priming following re-desiccation of the seed activates some processes stimulating germination (Asgedom and Becker 2001). Primed seeds usually have more and uniform germination due to decreased lag time of imbibition, activation of enzyme, enhanced germination metabolism, improved repair processes, and osmotic adjustment (Lee and Kim 2000; Farooq et al. 2006; Hussain et al. 2015). Seed priming enhances performance of seed and offers earlier and uniform germination by physiological, biochemical, molecular, and cellular changes (Fig. 2) (Cramer 2002; Beckers and Conrath 2007; Di Girolamo and Barbanti 2012a, b; Siri et al. 2013; Manonmani et al. 2014).

Phase I is characterized by the priming memory activation. During this phase, DNA and mitochondria are repaired, proteins are synthesized, stress response genes are stimulated, and signaling pathways are regulated. Phase II is the most critical phase, whereas the major metabolic and cellular activities are increased, resulting in germination initiation. In this phase, a series of proteins such as those related to ROS scavenging and signaling, storage, and stress response are synthesized in the embryo using existing and new messenger ribonucleic acid (mRNA), and the mitochondria and DNA are also synthesized and repaired. In Phase III, water and oxygen absorption increases, major reserve material mobilization occurs, embryonic axis elongates, and finally radicle protrudes from the coat of seed to complete germination and to start the post-germination stage (McDonald 2000; Varier et al. 2010; Rajjou et al. 2012; Chen and Arora 2013; Rosental and Nonogaki 2014).

The improvements conferred by seed priming are largely related to modulate the germination metabolic activities within the seed needed for breaking seed dormancy, enzyme activation, antioxidant activities, nucleic acid (DNA and RNA) and protein synthesis, repair processes, transcription and translation, DNA replication, germination metabolism, increasing the amount of ATP and the number of mitochondria, repairing deteriorated seed parts, and decreasing metabolite leakage. In general, seed priming activates a series of biochemical, physiological, and metabolic changes that are related to improve germination performance and vigorous growth as well as increase abiotic stress tolerance (Bray et al. 1989; Smith and Comb 1991; Bewley and Black 1994; McDonald 2000; Afzal et al. 2002; Netondo et al. 2004; Di Girolamo and Barbanti 2012a, b; Manonmani et al. 2014; Paparella et al. 2015; Ibrahim 2016; Wojtyla et al. 2016).

The benefits of seed priming include the upregulation of proteins and genes involved in cell division, cell wall modification, transcription, reserve mobilization, translation, oxidative stress response, and water transport, wherein DNA and membrane repair is better than normal imbibition (Dell Aquila and Tritto 1991; Giri and Schillinger 2003; Rajjou et al. 2012; Bewley et al. 2013; Kubala et al. 2015b).

The improved germination conferred by seed priming is associated with higher protein abundance involved in the oxidative stress tolerance, targeted proteolysis and posttranslational processing capacity, removal or destruction of dormancy blocks, repair of membranes, immature embryo development, and alteration of the tissues covering the embryo and general pre-germination metabolism enhancement (Dell Aquila and Tritto 1991; Giri and Schillinger 2003; Rajjou et al. 2012; Bewley et al. 2013; Kubala et al. 2015b).

Seed priming upregulates genes encoding proteins that are involved in weakening of endosperm, such as xyloglucan endotransglucosylase endo-mannanase and expansin (Nakaune et al. 2012). The improved activity and expression of these pro-

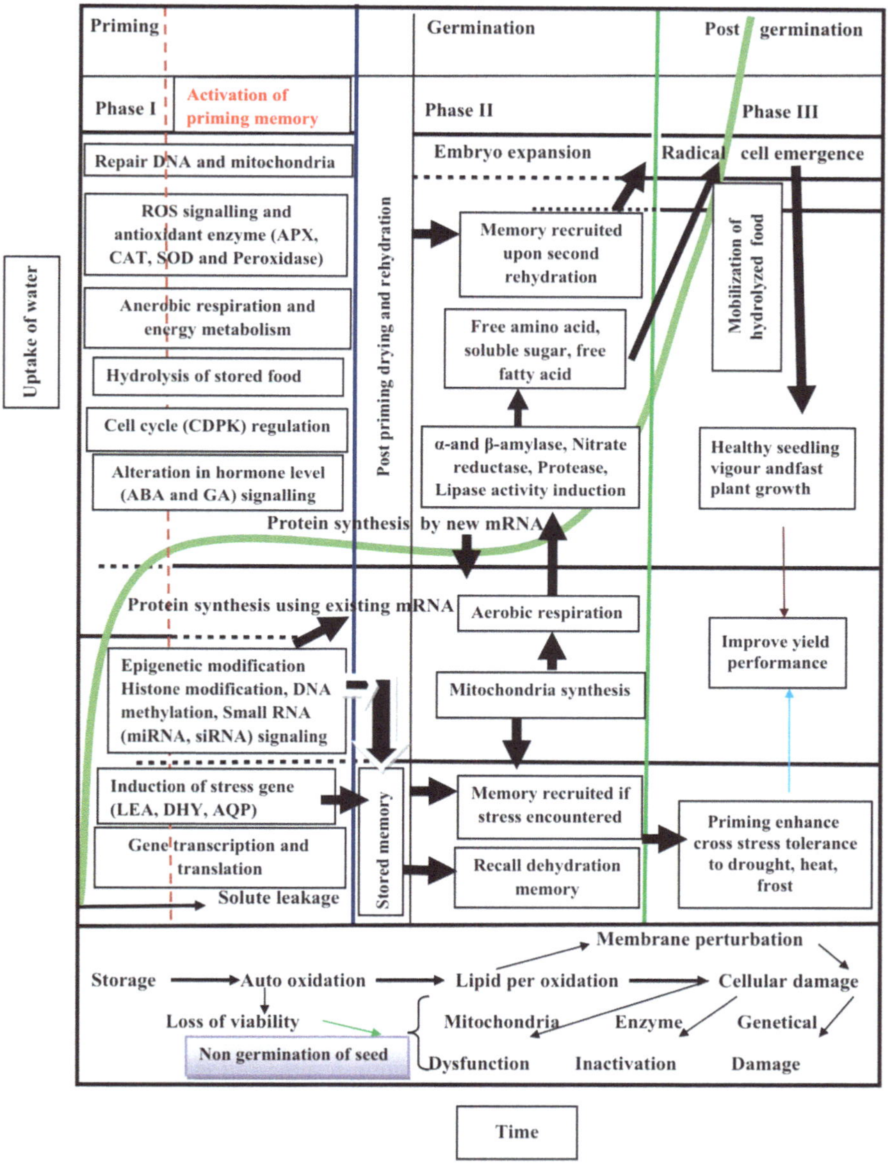


**Fig. 2** Model for stress memory and its relation to signaling of abiotic stress in primed seed. (Chen and Arora 2013; Bose et al. 2018)

teins might decrease the physical tolerance to protrusion of the radicle via shrinkage of the endosperm shrinkage, cell wall dissolution, and cytoplasm condensation (Sung et al. 2008). Certain storage proteins such as cruciferin and globulins are identified only during seed priming and not during seed germination. Similarly, stress-related proteins such as HSPs are specifically synthesized during osmopriming and not during imbibition in water. These proteins protect the cell from moisture stress during osmopriming, and they work as molecular chaperones and protect naturally damaged proteins (Afzal et al. 2002; Varier et al. 2010).

ROS at low concentration act as signaling molecules. Abiotic stress conditions lead to an excessive ROS production and accumulation, resulting in oxidative damage. They have a detrimental effect on molecules at high levels. They lead to DNA oxidation, which results in strand breaks in DNA, leading to germination failure. Seed priming stimulates the ROS generation at low levels, which reflects the strong activation of cell division (Bray and West 2005; Jisha and Puthur 2014; Wojtyla et al. 2016).

The primed seeds enhance the activity of most important enzymes that are implicated in cellular metabolism. These enzymes include acid phosphatase, esterase, glyoxysome enzymes, and enzymes that are involved in the mobilization of stored reserves such as α and β amylases, and isocitrate lyase (Lee and Kim 2000; Lin and Sung 2001; Varier et al. 2010; Di Girolamo and Barbanti 2012a, b). Moreover, seed priming increases the antioxidant enzyme activities (Afzal et al. 2002; Sedghi et al. 2014; Sharma et al. 2014) as well as the accumulation of non-enzymatic antioxidants (e.g., AsA, GSH, proline, and soluble sugars) (Gong et al. 2005; Sharma and Dubey 2007; Gong et al. 2008; Farhoudi et al. 2011; Zhu et al. 2011; Nawaz et al. 2012; Ibrahim 2016). These increases can lead to optimizing defense mechanisms through decreasing the production of ROS (e.g., $O_2$ and $H_2O_2$) and increasing the scavenging of ROS during seed germination (Noctor and Foyer 1998; Vranova et al. 2002; McDonald 2000; Hsu et al. 2003; Kibinza et al. 2011; Ansari and Sharif-Zadeh 2012; Sedghi et al. 2014; Paparella et al. 2015).

The ability to repair DNA during drying-back of seeds is associated with the germination progress. The pre-germinative metabolic processes that are essential to recover from DNA damage are ROS scavenging and mechanisms of DNA repair such as homologous recombination nucleotide, nonhomologous end-joining, and base excision repair (Huang et al. 2008).

The molecular responses to seed priming involve a DNA repair mechanism, and gene expression for syntheses of new RNAs and proteins (Varier et al. 2010). DNA repair mechanisms must be kept at suitable levels in the embryo to increase the chances of successful germination (Huang et al. 2008; Wojtyla et al. 2016). The activated DNA repair mechanisms during imbibition resume cell cycle progression of embryonic cells and undergo replication of the DNA and then enhance the germination efficiency, but defective repair of DNA leads to increased cell death (Waterworth et al. 2010; Kranner et al. 2010; Balestrazzi et al. 2011). DNA repair is the most valuable and vital process stimulated by seed priming, and it determines the potential benefits of seed priming (Rajjou et al. 2012; Ventura et al. 2012; Wojtyla et al. 2016). The increase of germination with seed priming may be associated with upregulated specific genes such as germination-associated, antioxidant, and plant growth-promoting genes (Sadeghi et al. 2011; Sharma et al. 2015; Wojtyla et al. 2016).

Seed priming induces the expression of 952 genes and 75 proteins required to promote pre-germinative metabolism including transcriptional activity, DNA replication, and repair mechanisms. Moreover, it induces expression of gene-coded proteins involved in regulation and maintenance of the cell cycle (Kubala et al. 2015a), genes involved directly and indirectly in DNA repair (Sharma and Maheshwari 2015), and APX, CAT, and SOD genes (Kubala et al. 2013).

## 7 Stresses-Induced Metabolic Changes in Germinating Seeds

Stress conditions affected negatively on seed germination in many plants (Patade et al. 2011; Ansari and Sharif-Zadeh 2012). Germination of seed and growth of seedling are the most sensitive stages to abiotic stress in the biological life cycle of the plants (Patade et al. 2011; Hasan et al. 2013; Fazlali et al. 2013). Even when high-quality seeds could be at hand, various abiotic stresses can be delayed or prevent seed germination (Haghighi et al. 2012). In the early phase of seed germination, abiotic stresses can cause wide oxidative damage to nucleic acids, proteins, and lipids (Kranner et al. 2010; Ventura et al. 2012; Paparella et al. 2015; Ibrahim 2016).

Drought, salinity, and high temperature are the major abiotic stresses that adversely affect the development, growth, and yields of crops. They cause a variety of biochemical, physiological, and metabolic changes in plants that include ROS generation that can induce membrane lipid peroxidation, leading to membrane damage, severe damage in DNA, and leakage of cell and photosynthetic component destruction (Hasanuzzaman et al. 2013; Jisha and Puthur 2014; Sheteiwy et al. 2018). Salinity affects seed germination by making osmotic potential that prevents water uptake during imbibition or by toxic effects of sodium and chloride ions on embryo viability (Munns 2002; Khajeh-Hosseini et al. 2003; Jahromi et al. 2008; Daszkowska-Golec 2011; Sheteiwy et al. 2018). The toxic effects of ions disrupt the enzymes and other macromolecule structures, damage organelles of cell and plasma membrane, inhibit protein synthesis, and uncouple respiration and photosynthesis (Feng et al. 2002; Parida and Das 2005; Panda and Khan 2009). Moreover, salinity stress alters hormone balance in plant tissue. Excessive salt is related to reductions in auxin, gibberellins, salicylic acid, and cytokines and an increase in abscisic acid (ABA) and jasmonic acid (JA) in the plant tissues (Javid et al. 2011; Miransari and Smith 2014). Seed germination is inhibited by salinity-induced disturbance of the metabolic processes such as decrease in water uptake and changes in the stored reserve mobilization, affecting the structural organization of proteins (Foolad and Lin 1997; Machado Neto et al. 2004).

Plants response to salinity and drought stresses shows similar physiological mechanisms. The plant cell perceives both stresses as water deprivation (Tavili et al. 2011). Meanwhile, cold stress can disturb the metabolic balance of the cell, resulting in a reduction of cellular respiration (Sugie et al. 2006) and membrane damage (Korkmaz et al. 2010). Exposure of germinating seeds to various abiotic stresses resulted in oxidative stress as indicated by overproduction of ROS hydrogen peroxide, superoxide, and hydroxyl radicals, which are very reactive and may cause damage to cells through oxidation of nucleic acids, proteins, and lipids (Pastori and Foyer 2002; Apel and Hirt 2004; Bandeoglu et al. 2004; Ashraf and Foolad 2005; Suzuki and Mittler 2006; Munns and Tester 2008).

Plants have an antioxidant system to scavenge the toxic ROS. The efficient running of the antioxidant defense system is vital for plants to tolerate adverse abiotic stresses (Gong et al. 2005). ROS play two contrary roles in plants: at low concentrations, they act as signaling molecules that regulate the defense response. Meanwhile, excess cellular levels of ROS cause damage to proteins, membrane lipids, nucleic acids, carbohydrates, pigments, chloroplasts, and enzymes that ultimately results in oxidative stress (Pastori and Foyer 2002; Apel and Hirt 2004; Bandeoglu et al. 2004; Smirnoff 2005; del Río and Puppo 2009; Liu et al. 2010; Goud and Kachole 2011; Suzuki et al. 2012; Yi et al. 2014). ROS damage depends on the equilibrium between production and scavenging of ROS at a suitable time and place, and this equilibrium may be disturbed by abiotic stresses (Gratao et al. 2005).

When plant grows under normal conditions, the ROS production and scavenging are in a dynamic balance. But when plants are exposed to various abiotic stresses, the ROS generation and accumulation can disrupt this delicate balance (Neill et al. 2002; Ashraf and Harris 2004; Munns and Tester 2008; Das et al. 2015; Sheteiwy et al. 2018).

ROS provokes damage to cell macromolecules (proteins, nucleic acids, and lipids) and ultimately to cellular structure or even damage to the antioxidative system (McDonald 1999; Bandeoglu et al. 2004; Ashraf and Foolad 2005; Gratao et al. 2005; Munns and Tester 2008; Hela et al. 2011; Kaewnaree et al. 2011; Manaa et al. 2013).

Under abiotic stress, peroxidation of lipid may be one of the most vital molecular mechanisms involved in inhibiting germination of seed (Yang et al. 2010). Abiotic stress-induced generation of ROS increases the contents of malondialdehyde (MDA) that is the most mutagenic product of lipid peroxidation and could also cause damage to the cell by reacting with cellular macromolecules (Xu et al. 2015; Ali et al. 2017; Sheteiwy et al. 2018). Several studies have reported an increase in MDA contents under various abiotic stresses (Khan and Panda 2008; Ali et al. 2017; Sheteiwy et al. 2018). It is also possible to provide an indication of the damage of membrane at the cellular level under abiotic stress (Weisany et al. 2012). Therefore, its content reflects oxidative damage in the plant tissue (Gong et al. 2008).

The damage degree depends on the balance between ROS formation and its removal by the antioxidative scavenging systems that provide crucial protection against oxidative damage, and it appears to be a relatively more important stress-tolerant trait. Stress-induced ROS accumulation is countered by enzymatic antioxidant systems. The antioxidative system comprises numerous enzymes including the antioxidant enzymes (such as SOD and CAT) and nonenzymatic compounds with low molecular mass antioxidants such as AsA, glutathione, thioredoxin, and carotenoids (Sharma and Dubey 2007; Gong et al. 2008; Zhu et al. 2011). Thus, the antioxidant capacity of plants that assist in ROS scavenging and cell protection is related to their salt tolerance (Gong et al. 2005; Gupta et al. 2010; Das et al. 2015;

Ibrahim 2016). Superoxide is converted into hydrogen peroxide ($H_2O_2$) by the action of SOD, and finally to molecular oxygen and water by CAT, thus preventing the formation of extremely dangerous hydroxyl radical from $H_2O_2$. This plays many essential roles within the seed during cell membrane alteration (Goel et al. 2003; Wang et al. 2009; Gill and Tuteja 2010; Kibinza et al. 2011).

# 8 The Mechanisms of Seed Priming-Induced Abiotic Stress Tolerance

Adaptation to abiotic stresses includes complex processes involving various changes such as the induction-specific genes, elevated antioxidative activities, transient increases in plant regulators levels, protective proteins, suppression energy-consuming pathways, and osmolytes accumulation (Bartels and Sunkar 2005).

Exogenous application of some agents through seed priming can enhance seed germination performance and seedling establishment of many plants under abiotic stress conditions. Seed priming is one of the useful physiological approaches that can be used to tolerate adverse abiotic stresses such as drought, salinity, and chilling in various plant species (Fig. 3) (Sivritepe et al. 2005; Ashraf and Foolad 2005; Farooq et al. 2008a; Patade et al. 2009; Bayat and Sepehri 2012; Kazemi and Eskandari 2012; Khan et al. 2012; Hussain et al. 2016b). The beneficial impacts of seed priming may be very clear mainly under abiotic stress conditions (Ashraf and Foolad 2005; Beckers and Conrath 2007; Chen 2011; Chen and Arora 2013; Ibrahim 2016; Masondo et al. 2018).

Abiotic stress tolerance assimilated by seed priming is likely reached via two strategies. In the first strategy, seed priming stimulates the pre-germination metabolic processes such as enhancement in the energy metabolism, early mobilization of seed food reserves, elongation of embryo cell, and endosperm weakening (Corbineau et al. 2000; Benamar et al. 2003; Pandita et al. 2007; Sung et al. 2008; Li et al. 2010; Sun et al. 2010; Chen and Arora 2011), which provide the conversion of quiescent seeds into germinating state and lead to increased germination. In the second strategy, seed priming imposes biotic stresses on seeds, which represses the protrusion of the radicle but supports stress responses, inducing cross-tolerance to abiotic stresses, activation of enzymes, and osmotic adjustment. These stress tolerance strategies form a "priming memory" in germinating primed seeds that can be recruited following stress exposure and mediate more stress tolerance (Bruce et al. 2007; Chen and Arora 2013; Hussain et al. 2016b; Ibrahim 2016).

Seed priming is a pre-germination exposure to stress that leaves stress memory in the seeds to tolerate future stresses in their environments (Fig. 2) (Beckers and Conrath 2007; Tanou et al. 2012). The seed priming process involves prior exposure to some factors that empower the plants to overcome the future stress exposure. It permits accelerated defense mechanism against adverse environments by inducing the increase of enzymatic and nonenzymatic antioxidant potential, the accumulation of inactive forms of signaling proteins and essential transcription factors, and the

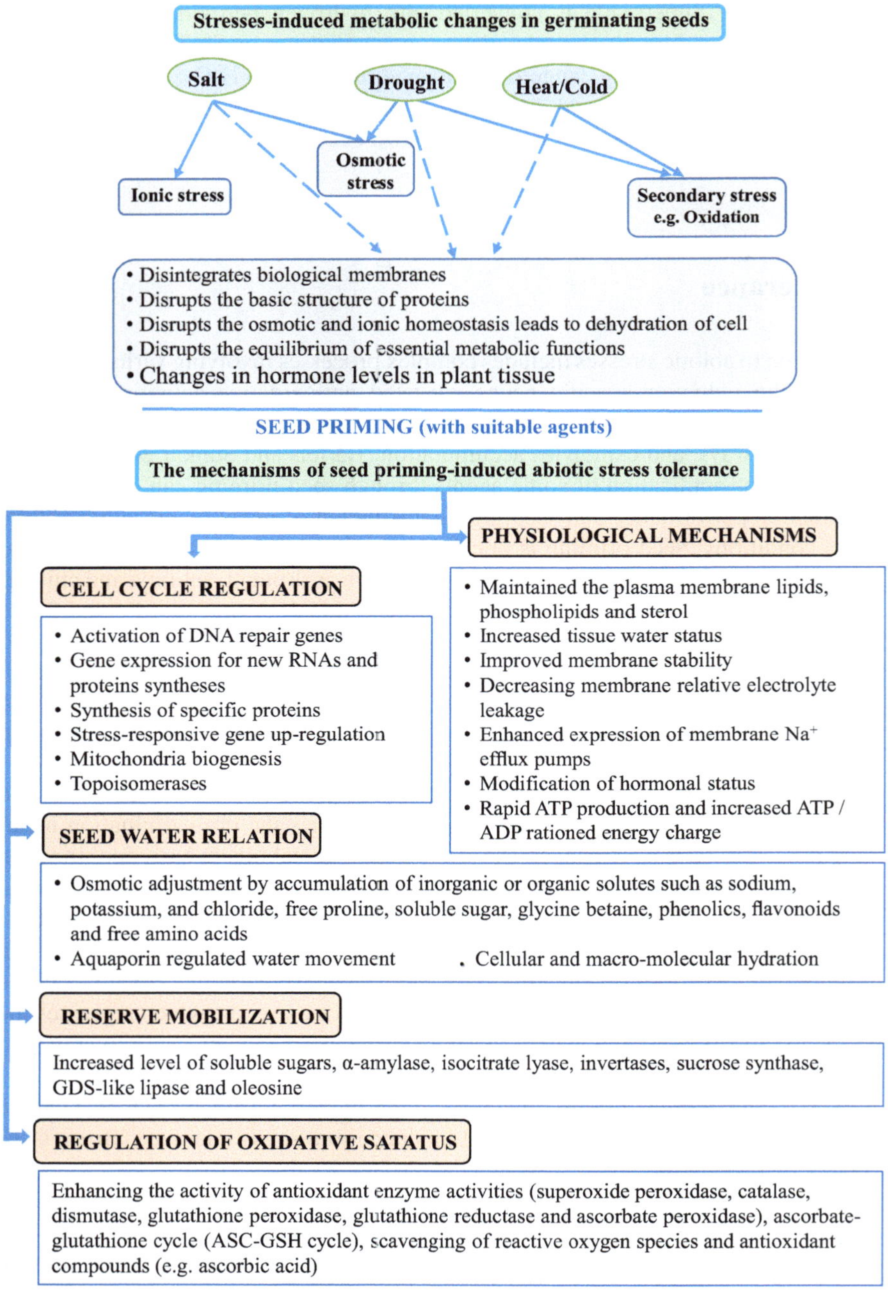


**Fig. 3** The adverse effects of abiotic stresses and mechanisms of seed priming to induce abiotic stresses tolerance in germinating seeds

occurrence of epigenetic changes as well as inducing of jasmonate-linked defense responses to the epigenetic changes (Rowse 1996; Farooq et al. 2006, 2009; Beckers and Conrath 2007; Bruce et al. 2007; Tanou et al. 2012; Kasote et al. 2019).

Seed priming improves the protein synthesis performance in the plant tissue; prompts the specific protein synthesis such as heat shock proteins (HSPs) that increase heat and low temperature stress tolerance; and increases L-isoaspartyl protein methyltransferase enzyme, which repairs tissue protein (Kester et al. 1997; Ji et al. 2017).

Seed priming increases abiotic stress tolerance by enhancing the accumulation of osmotic substances such as proline, soluble sugars, and soluble proteins that regulate water potential and increasing the protective activities of enzymes such as catalase (CAT), superoxide dismutase (SOD), and peroxidase (POX) that defend cells against oxidative stress under abiotic stresses (Mittal and Dubey 1995; Bohnert and Shen 1999; Kazemi and Eskandari 2012; Joshi et al. 2013; Hussain et al. 2016a, b; Ibrahim 2016; Pal et al. 2017).

Plants overcome stress-induced osmotic effects such as salinity and drought through accumulation of high concentrations of organic or inorganic solutes in the cytoplasm (de Azevedo Neto et al. 2005; Munns 2005; Gonçalves et al. 2007; Farhoudi et al. 2011; Farouk 2011; Zhu et al. 2011; Joshi et al. 2013; Matias et al. 2015; Ibrahim 2016).

Under abiotic stress conditions, the antioxidant system is considered as one of the key defense systems in germinating primed seeds to control ROS accumulation under abiotic stresses (McDonald 2000; Hsu et al. 2003; Gong et al. 2005; Gupta et al. 2010; Das et al. 2015; Paparella et al. 2015; Ibrahim 2016; Banerjee and Roychoudhury 2018). Seed priming increases the activities of antioxidant enzymes such as CAT, POX, SOD, and GR and increases the level of nonenzymatic antioxidants such as ascorbic acid. These actions are involved in the ROS scavenging and protection of seed through the reduction of superoxide and hydrogen peroxide production (Noctor and Foyer 1998; McDonald 2000; Afzal et al. 2002; Vranova et al. 2002; Hsu et al. 2003; Gong et al. 2008; Zhu et al. 2011; Ansari and Sharif-Zadeh 2012; Nawaz et al. 2012; Sharma et al. 2014; Sedghi et al. 2014; Matias et al. 2015; Paparella et al. 2015; Ibrahim 2016; Pal et al. 2017; Kasote et al. 2019). Moreover, seed priming also prompts the accumulation of photoprotective pigments such as anthocyanin that enhances scavenging of ROS and plant protection (Banerjee and Roychoudhury 2016).

Seed priming reduces the accumulation of MDA that is increased under abiotic stresses and used as marker of oxidative stress. This reduction is associated with better membrane repair and inductive responses of antioxidant enzymes during seed priming process (Nawaz et al. 2012; Das and Roychoudhury 2014; Dong et al. 2014; Paparella et al. 2015; Banerjee and Roychoudhury 2018).

Seed priming activates multiple stress-responsive genes (Wojtyla et al. 2016; Paul and Roychoudhury 2017). It has a higher BnPIP1 and Bn-TIP2 gene expression in seeds during germination under osmotic and salt stresses. BnPIP1 provides an adequate water supply for storage and nutrient metabolism at the early stage of germination, leading to increased germination and tolerance to salt and osmotic stresses. The induction of Bn-TIP2 may be correlated with protrusion of radicle and enlargement or proliferation of the cell in later phases of germination (Gao et al. 1999). Chilling and drought tolerance in spinach primed seeds are enhanced by the expression of four aquaporin genes (SoPIP1;1, SoPIP1;2, SoPIP2;1, and SoıTIP) that contribute to water transport across vacuolar membranes and plasma to enable water supply to expanding tissues. Moreover, some AQP genes exhibited greater chilling and drought tolerance in primed seeds (Chen et al. 2013).

The beneficial effects of seed priming under salinity stress have been manifested in various plants like tomato (Cayuela et al. 1996; Cuartero et al. 2006), melon (*Cucumis melo* L.) (Sivritepe et al. 2003), hot pepper (Khan et al. 2009a, b), lettuce (Nasri et al. 2011), and okra (Dkhil et al. 2014). The positive effects of seed priming under extreme temperature stress were found in carrot (Pill and Finch-Savage 1988), tomato (Pill et al. 1991), watermelon (Demir and Oztokat 2003), muskmelon (Nascimento 2003), and asparagus (Bittencourt et al. 2004). Seed priming of carrot in −1.0 and −1.2 MPa PEG 6000 improved seed performance and seedling emergence under higher and suboptimal temperatures (Pereira et al. 2009). The beneficial effect of seed priming on the performance of onion seed at suboptimal temperatures is probably associated with the increase of endo-mannanase activity in seeds before radicle protrusion; this enzyme is involved in endosperm cell wall hydrolysis (Kępczyńska et al. 2003). Seed priming alleviated thermoinhibition in lettuce by increasing maximum germination temperature (Schwember and Bradford 2010). Moreover, seed priming may be beneficial for maize genotypes with known sensitivity to low temperature (Hacisalihoglu et al. 2018).

# 9 Mechanisms of Abiotic Stress Tolerance with Priming Agents

## 9.1 *Priming Methods and Agents*

Several seed priming methods have been extensively used and developed to increase the synchrony and speed of seed germination (Ghassemi-Golezani and Esmaeilpour 2008; Eskandari 2013; Jisha et al. 2013; Nawaz et al. 2013; Lutts et al. 2016). Osmopriming, halo priming, and osmo-conditioning are the most commonly used priming methods. Seed priming is the process of soaking of seed in aerated low-water potential solutions to permit seed imbibition but prevents the protrusion of radicle. Different kinds of agents are used to make low-water potential solutions (Bray 1995; Chen and Arora 2011). The critical success factor in seed priming is to let sufficient water to stimulate metabolic processes, but not enough to allow a further expansion and growth of the cell. The vital goal of osmopriming is to minimize the oxidative stress mediated by ROS with delaying entry of water. Thus, the water potential of the osmotic agent is an important consideration when designing a method (Taylor et al. 1998; Paparella et al. 2015). The osmopriming process is similar to an extended early seed imbibition, which sets in motion a gradual progress essential for activities of pre-germinative metabolism. Thus, using osmopriming as a model is particularly helpful to study the seed conversion from a dry and metabolically quiescent state to a hydrated and metabolically active state (Chen and Arora 2011).

Seed priming with many natural or synthetic compounds has been researched in various crops (Table 1). The commonly used agents in seed priming are polyethylene glycol, nutrients, inorganic salts, biostimulants, and pure water (Horii et al. 2007; Butler et al. 2009; Hussain et al. 2015; Masondo et al. 2018). Moreover, it has also been stated that seed priming with plant growth regulators, polyamines, and certain other organic sources enhances seeds' performance in many crops. Different types of priming agents have different properties and effectiveness. Therefore, each plant species needs to be optimized for the priming solutions (Horii et al. 2007).

## 9.2 *Polyethylene Glycol*

Polyethylene glycol (PEG) is the most common substance used to control water potential in primed seed due to its nontoxic nature and large molecular size (6000–8000 mw) that impedes its penetration into the seed (Thomas et al. 2000). Osmopriming with PEG results in increasing the seed germination potential and strengthening the antioxidant system, which ultimately leads to increase in germinating seed tolerance to abiotic stresses (Chen and Arora 2011; Mouradi et al. 2016). Germination has been reported to be enhanced by priming with polyethylene glycol at −1.5 MPa for 7 days in onion seeds (Dorna et al. 2013), at −1.25 MPa for 2 days in tomato (Govinden-Soulange and Levantard 2008), and at 0.5 MPa in sweet corn (Ghiyasi et al. 2008). Seed priming with PEG at −0.6 MPa for 24 h at 25 °C enhanced the activities of PO and CAT as well as reduced the MDA content and the electrolyte leakage in germinating seed of *Moroccan alfalfa* under water deficit (Mouradi et al. 2016). In addition, PEG priming increased proline content in seedlings of sunflower under salt stress (Moghanibashi et al. 2013). Siri et al. (2013) found that the seed priming with PEG improved the sweet pepper quality and the membrane repair, which results in lipid peroxidation reduction and enhancement of antioxidant system synthesis responsible for removing ROS. Moreover, gene expression of antioxidant enzymes was downregulated with PEG priming under nano-ZnO stress (Salah et al. 2015).

**Table 1** Various seed priming methods and agents applied in various crops to ameliorate abiotic stress tolerance and the main results observed

| Crop | Methods and agents of seed priming | Stress type | Main results of seed priming | References |
|---|---|---|---|---|
| *Brassica juncea* | Halopriming $CaCl_2$ | Salt | Increased total soluble sugars, proline, and glycine betaine accumulation as well as decreased MDA and phenolic levels | Srivastava et al. (2010) |
| *Brassica napus* | Osmopriming PEG | Salt | Induced increase in proline content, enhanced the *P5CSA* gene upregulation and enzyme activities, reduced $H_2O_2$ accumulation and downregulation of the PDH gene | Kubala et al. (2015b) |
| *Brassica napus* | Osmopriming Ascorbic acid | Drought | Enhanced the activity of CAT and POX | Razaji et al. (2014) |
| *Brassica oleracea* | Hormone priming Methyl jasmonate (25 μmol $L^{-1}$) | Salt | Increased the glucobrassicin, indolic glucosinolate, anthocyanin, neo-glucobrassicin, and chlorogenic acid derivative contents | Hassini et al. (2017) |
| *Cajanus cajan* | Hormone priming Jasmonic acid | Copper stress | Increased the SOD and POX activities | Poonam et al. (2013) |
| *Capsicum annuum* | Halopriming KCl (10 mM) | Salt | Increased proline accumulation | Aloui et al. (2014) |
| *Capsicum annuum* | Chemical priming (thiourea) Hydropriming | Cold | Increased expression of *PROX1*, *CaWRKY30*, *osmotin*, SOD, CAH, and Cu/Zn genes | Patade et al. (2012) |
| *Capsicum annuum* | Osmopriming Glycine betaine (10 mM) | Salt | Decreased MDA levels and enhanced content of proline and activity of SOD enzyme | Korkmaz and Şirikçi (2011) |
| *Cicer arietinum* | Osmopriming Chitosan (0.1% and 0.2%) | Salt | Decreased Na and malondialdehyde level and increased proline and total carbohydrate content and K/Na ratio | Mahdavi et al. (2015) |
| *Cucumis melo* | Halopriming NaCl (100 mM) | Salt | Increased antioxidants, soluble carbohydrates, and proline content | Farhoudi et al. (2011) |
| *Cucumis sativus* | Halopriming $CaCl_2$ (2 mM) | Salt | Increased the proline accumulation | Joshi et al. (2013) |

| | | | | |
|---|---|---|---|---|
| *Cucumis sativus* | Hormone priming<br>Melatonin (1 μM) | Salt | Promoted catabolism of ABA and biosynthesis of gibberellic acid and recharged the antioxidant system | Zhang et al. (2014) |
| *Cucurbita pepo* | Osmopriming<br>Ascorbic acid (0.5 mM) | Salt | Enhanced the catalase and peroxidase activities | Fazlali et al. (2013) |
| *Helianthus annuus* | Hormone priming<br>Jasmonic acid | Chilling | Improved content of carbohydrate | Gornik and Lahuta (2017) |
| *Jatropha curcas* | Osmopriming<br>Nitric oxide (Na-nitroprusside) | Salt | Increased the ascorbate and glutathione accumulation in the endosperm-embryo axis and enhanced the activities of CAT and GR | Gadelha et al. (2017) |
| *Medicago sativa* | Osmopriming (PEG) | Salt | Changed patterns of 84 proteins | Yacoubi et al. (2013) |
| *Medicago sativa* | Osmopriming<br>Mannitol (4%) | Salt | Increased the antioxidant enzyme activities and reduced electrolyte leakage and MDA contents | Amooaghaie (2011) |
| *Oryza sativa* | Hormonal priming<br>Salicylic acid (2 mM) | | Reduced $Na^+$ accumulation, increased content of $K^+$ and $Ca^+$, enhanced enzyme activities that involved in assimilation of secondary metabolism and decreased ROS and MDA contents | Sheteiwy et al. (2018) |
| *Oryza sativa* | Osmopriming<br>Spermidine (5 mM) | Salt | Upregulated several transcription factors and stress-responsive genes and increased the expression of membrane $Na^+$ efflux pumps | Paul et al. (2017) |
| *Oryza sativa* | Osmopriming<br>BABA | Salt | Reduced MDA content and lipid peroxidation and increased the activities of antioxidant enzyme | Jisha and Puthur (2016) |
| *Oryza sativa* | Hormonal priming<br>Salicylic acid (100 mg $L^{-1}$)<br>Chemical priming of Se (60 μM) | Chilling | Reduced levels of ROIs and increased the activities of MAO and XOD | Hussain et al. (2016a) |
| *Oryza sativa* | Osmopriming<br>PEG (10%)<br>$KNO_3$ (0.5%)<br>Spermidine (0.5 mmol $L^{-1}$) | Drought | Increased the α-amylase activity, antioxidant enzyme activity, and lipid peroxidation | Zheng et al. (2016) |

(continued)

**Table 1** (continued)

| Crop | Methods and agents of seed priming | Stress type | Main results of seed priming | References |
|---|---|---|---|---|
| *Oryza sativa* | Osmopriming Polyamines | Drought | Improved net water use efficiency, leaf water status, photosynthesis, free proline production, soluble phenolics, anthocyanins, membrane properties, and activity of CAT and reduced $H_2O_2$ level | Farooq et al. (2009) |
| *Phaseolus vulgaris* | Hormonal priming ABA (30 ppm) | Salt | Increased phosphatidylcholine/phosphatidylethanolamine ratio and confers root cell membrane protection | Salama et al. (2015) |
| *Silybum marianum* | Halopriming $KNO_3$ (0.25 mM) | Salt | Improved peroxidase activity | Zavariyan et al. (2015) |
| *Smooth vetch* | Hormone priming Salicylic acid (0.5 mM) | Water deficiency | Increased the proline and glycine betaine accumulation | Namdari and Baghbani (2017) |
| *Solanum lycopersicum* | Halopriming NaCl (300 mM) | Salt | Promoted germination and endosperm weakening and increased gibberellin content | Nakaune et al. (2012) |
| *Sorghum bicolor* | Osmopriming (PEG) | Excessive soil moisture drought | Enhanced APX, CAT, POX, and SOD activities, increased soluble sugars, compatible solutes, proline, and free amino acids level, and decreased electrolyte leakage and content of MDA | Zhang et al. (2015) |
| *Spinacia oleracea* | Osmopriming (PEG) | Drought chilling | Increased the SoPIP1;1, SoPIP2;1, and SoıTIP expression | Chen et al. (2013) |
| *Spinacia oleracea* | Osmopriming (PEG) | Desiccation chilling | Increased 30, 26, 19 kD dehydrin-like proteins and dehydrin (CAP85) accumulation as well as stimulated gene expression | Chen et al. (2012b) |
| *Spinacia oleracea* | Osmopriming PEG | Desiccation chilling | Decreased the accumulation of MDA | Chen and Arora (2011) |
| *Stevia rebaudiana Bertoni* | Nutripriming Fe, B, Se | Salt | Increased antioxidant activity of the enzyme | Shahverdi et al. (2017) |

| | | | | |
|---|---|---|---|---|
| *Triticum aestivum* | Halopriming<br>Silicon priming<br>($Na_2SiO_3$) (30 mM) | Salt | Increased accumulation of $Ca^{2+}$ and $K^+$ ions | Azeem et al. (2015) |
| *Triticum aestivum* | Osmopriming<br>Choline (5 mM) | Salt | Increased accumulation of glycine betaine and maintained osmotic potential, minimized toxic elements ($Na^+$, $Cl^-$), maintained beneficial elements ($K^+$, $Ca^{2+}$), and reduced oxidative stress | Salama et al. (2011) |
| *Triticum durum* | Hormone priming<br>Salicylic acid<br>(0.1 mM) | Salt | Enhanced phenylalanine ammonia lyase, peroxidase activities, and accumulation of flavonoids and phenolics | Karadag and Yucel (2017) |
| *Triticum durum* | Hormone priming<br>Salicylic acid<br>(0.5 mM)<br>Calcium $Ca^{2+}$ (50 mM) | Salt | Increased total carotenoid, flavonoid, and phenol accumulation and enhanced ascorbic acid oxidase and phenylalanine ammonia lyase activities and increased the levels of lipid peroxidation | Yucel and Heybet (2016) |
| *Triticum durum* | Osmopriming<br>Ascorbic acid | Salt | Increased proteins associated with storage, metabolism, defense, and protein destination | Fercha et al. (2014) |
| *Triticum durum* | Osmopriming<br>Ascorbic acid | Salt | Changed 72 protein accumulations that are involved in metabolism, protein synthesis, defense response, and proteolysis | Fercha et al. (2013) |
| *Vicia faba* | Hormonal priming<br>Salicylic acid<br>(0.2 mM) | Salt | Enhanced CAT, POX, GR, and APX activities | Azooz (2009) |
| *Vicia faba* | Hormone priming<br>Melatonin (100 mM and 500 mM) | Salt | Increased total carbohydrates, photosynthetic efficiency, indole acetic acid, total phenols, and $K^+$ and $Ca^{2+}$ levels | Dawood and El-Awadi (2015) |
| *Vigna radiata* | Osmopriming<br>BABA (1 mM) | Drought | Increased total protein proline and total carbohydrate accumulation, increased the superoxide dismutase, nitrate reductase, and guaiacol peroxidase activities, and reduced MDA level | Jisha and Puthur (2016) |
| *Zea mays* | Hormone priming<br>Melatonin (0.8 mM) | Salt | Improved the activities of CAT, SOD, and PAL, content of relative water, proline, and total phenolic and decreased membrane relative electrolyte leakage, $Na^+$ content, and lipid peroxidation product | Jiang et al. (2016) |

## 9.3 Inorganic Salts

The other substances used to control water potential in seed germination are inorganic salts such as NaCl, $CaCl_2$, $NaNO_3$, $KNO_3$, $K_3PO_4$, $KH_2PO_4$, $MgSO_4$, $MnSO_4$, $CuSO_4$, $ZnSO_4$, and $MgCl_2$ as well as organic molecules such as glycerol, sorbitol, proline, mannitol, butenolide, putrescine, choline, chitosan, and paclobutrazol (Shao et al. 2005; Su et al. 2006; Foti et al. 2008; Hasanuzzaman et al. 2010; Demir et al. 2012; Hameed et al. 2013) that are inexpensive and easy to remove and aerate compared to PEG (Farooq et al. 2005; Paparella et al. 2015).

Seed priming with inorganic salts improves the antioxidant enzyme activities involved in seed germination and alters the mobilization of organic substances to different parts of the embryo. Seed priming with $ZnSO_4$ had higher catalase, superoxide dismutase, and peroxidase activities compared to hydro-primed plants (Aboutalebian and Nazari 2017). This treatment also alleviated drought stress by increasing the antioxidants in the seedlings of *N. sativa* (Fallah et al. 2018). Seed priming with NaCl increased the content of proline and soluble carbohydrate as well as enhanced antioxidant enzyme activities and reduced damage of seed membrane (Farhoudi et al. 2011). Seed priming with NaCl improved the germination performance of tomato by increasing content of gibberellin via activation of gibberellin biosynthetic gene and genes involved in endosperm cap weakening (Nakaune et al. 2012). Seed priming of amaranth with NaCl decreased MDA production and increased cell membrane stability as well as improved tolerance to salinity by promoting $K^+$ and $Ca^{2+}$ accumulation (Omami 2005). Seed priming with sodium nitroprusside increased the antioxidant enzyme activity and the contents of ascorbic acid, Pro, and total phenolics, resulting in alleviation of the adverse effects of salinity stress (Ali et al. 2017). Under suboptimal temperature conditions, seed priming with calcium nitrate improved germination rate and emergence of seedling (Batista et al. 2016). Seed priming with $KNO_3$ improved the salt tolerance of lettuce seeds by increasing the activities of acid phytase and phosphatase in the roots, cotyledons, and shoots (Nasri et al. 2011). The beneficial impact of calcium chloride priming on inducing salinity tolerance may be due to the antioxidant system activation and enhanced proline accumulation in cucumber seedlings (Joshi et al. 2013). Priming of sunflower seeds with $KNO_3$ improved the salinity tolerance by enhancing POX enzyme activity and decreasing MDA concentration in seedling (Farhoudi 2012). Seed priming with $KNO_3$ solution decreased the activity of peroxidase in *Silybum marianum* (Zavariyan et al. 2015). Pigeon pea seeds treated with $CaC1_2$ or $KNO_3$ showed improvements in soluble sugars, free amino acids, and proteins during germination under salinity stress (Jyotsna and Srivastava 1998). Pepper seedlings grown from primed seed with KCl, NaCl, and CaC12 had improved proline accumulation, water content, soluble sugar, polyphenols, soluble proteins, carotenoid content, and biomass (Aloui et al. 2014; Hassen et al. 2017).

Plant species reveal differences in response to PEG or salts, because they differ in the presence of a selective semipermeable layer, which surrounds the embryo and allows water absorption, preventing salt diffusion. Plants that do not have this layer such as cabbage and broccoli absorb ions and cause embryonic damages under osmopriming. Plants that have this layer such as tomato, pepper, lettuce, and melon may be safely subjected to soaking in solutions of inorganic salts (Welbaum and Bradford 1990; Taylor et al. 1997; Welbaum et al. 1998).

## 9.4 Fertilizers

Different fertilizers are used to improve seed priming efficiency. In this respect, soaking okra seeds in a single superphosphate or diammonium phosphate solution induced a higher germination percentage and accelerates seedling growth (Shah et al. 2011). In seed priming with nutrients, seeds are primed in a mixture of water and dissolved nutrients instead of pure water (Arif et al. 2005). The mineral-nutrient status of plants plays a vital role in enhancing plant tolerance to abiotic stresses (Marschner 1995).

## 9.5 Hormonal Priming

The seeds could be primed with plant growth chemicals or phytohormones, and the priming process is called hormonal priming. Seed priming with plant growth regulators improves ulates germination and emergence of various crops (Gao et al. 2002; Carvalho et al. 2011). Plant hormones play a key role in plant physiology and development by generating and transmitting different kinds of signals between and within the cells; the endogenous levels of phytohormones undergo significant changes under abiotic stresses (Iqbal and Ashraf 2013). Plant hormones play a vital role in modulating physiological and molecular responses to abiotic stresses, which eventually lead to overcome the adverse environmental conditions (Pandey 2017; Ku et al., 2018). Seed priming with phytohormones can increase seed germination by activating some enzymes such as amylase and protease, which hydrolyzed molecules of starch and protein into simple available form for embryo uptake (Miransari and Smith 2014). Abiotic stress conditions can reduce the levels of most growth upregulators; however, the exogenous application of hormones may overcome their deficiency (Yurekli et al. 2004; Akram et al. 2012; Babu et al. 2012). Salinity stress is associated with an increase in ABA and JA in the plant tissues and decreases in auxin, cytokinin, gibberellins, and salicylic acid. These changes are an initial process controlling the decrease of plant growth due to stress and can be alleviated by the exogenous application of plant growth regulators. Seed priming with several plant growth regulators such as salicylic acid and JA have been used to accelerate the seed germination, enhance the seedling establishment, and enhance the abiotic stress tolerance (Korkmaz et al. 2004; Farooq et al. 2008a; Krantev et al. 2008; Javid et al. 2011). Moreover, these treatments maintain the levels of IAA and cytokinin in the plant tissues that increase cell division (Sakhabutdinova et al. 2003). Priming of maize seeds with hormones induced abiotic stress tolerance by improving the oxidative stress tolerance (Afzal et al. 2008; Farooq et al. 2008b).

### 9.5.1 Gibberellins

Seed priming with gibberellins (GAs) improves the germination rate by increasing the amino acid content in the embryo, which releases enzymes to digest the endospermic starch when there is renewal of embryo growth at germination (Sedghi et al. 2014). Gibberellins play an important role in many essential processes of plant development and growth such as germination, stem elongation, and leaf expansion (Razem et al. 2006). Gibberellic acid (GA3) activates a-amylase involved in starch breakdown, which is stored mainly in the endosperm, and is degraded for embryo growth during germination (Taiz and Zeiger 2006). Moreover, gibberellins increase seed germination through inhibiting the activity of ABA and the related biosynthesis pathways that reduce ABA amounts as well as activating the ABA-catabolizing enzymes (Atia et al. 2009; Miransari and Smith 2014; Younesi and Moradi 2015). Moreover, seed priming using gibberellic acid solution enhanced the seed performance under both saline and osmotic stresses (Daur 2018).

Seed priming with GA3 improved germination performance and establishment of seedling, prepared a suitable metabolic reaction before germination, and improved salt tolerance in many plants by increasing the antioxidant enzyme activities and reducing the damage of membrane as estimated using electrolyte leakage and MDA biomarkers (Sedghi et al. 2010; Hela et al. 2011; Khan et al. 2012; Iqbal and Ashraf 2013; Younesi and Moradi 2015). On the other hand, seed priming with GA3 increased the seed germination rate and alleviated the adverse effects of drought stress in the early stages of seedling growth by enhancing antioxidants and proline content in the seedlings (Fallah et al. 2018). Primed *Agropyron elongatum* seeds with gibberellin and ABA increased activities of CAT and SOD under drought stress in comparison with unprimed seeds (Eisvand et al. 2010).

### 9.5.2 Salicylic Acid

Salicylic acid (SA) and acetyl salicylic acid are known to offer protection against abiotic stresses by triggering the potential to tolerate stress (Senaratna et al. 2000). Salicylicate is an endogenous growth regulator of phenolic nature that contributes in the regulation of physiological processes (Raskin 1992). Salicylicate cooperates with other signaling pathways including those regulated by ethylene and JA (Szalai et al. 2000; Ding and Wang 2003). It also increases the tolerance to osmotic stress and high or low temperature through the activation of guaiacol peroxidase and glutathione reductase (Borsani et al. 2001; He et al. 2002; Kang and Saltveit 2002). The beneficial impact of seeds primed with SA on seedling growth under salinity stress may be due to enhanced oxygen uptake and amylase activity and increased soluble sugar, protein, and free amino acid contents, as well as improved mobilization of nutrients from the cotyledons to the embryonic axis (Kathiresan et al. 1984; Zhang et al. 1999).

Salicylic acid priming at lower concentrations has been effectively used in enhancing hydrolase activities and that led to increase in a breakdown of food reserve and consequential early germination of the seeds (Sakhabutdinova et al. 2003). Priming with SA has also been found to increase germination parameters of melon by improving biochemical activities of the seeds during priming process (Basra et al. 2007).

Seed priming with SA recorded the highest mean activity of CAT enzyme and total phenol content in embryonic axis and cotyledon (Pal et al. 2017) and decreased ROS and MDA contents under salinity stress (Sheteiwy et al. 2018). Seeds of faba bean (*Vicia faba* L.) primed with SA enhanced salt tolerance by increasing CAT, POX, GR, and APX activities (Azooz 2009). Maize seed primed with SA increased seed germination and growth of seedling under both normal and low temperature conditions by enhancement of antioxidant system synthesis (Farooq et al. 2008a, b). Meanwhile, priming with SA increased growth of rice seedling under low temperature condition via enhancing the content of soluble sugar, proline, and protein (Pouramir-Dashtmian et al. 2014). Lada et al. (2004) indicated that hormonal (salicylic acid and gibberellin) priming increased germination of seed carrot in low temperature (5 °C). Moreover, seed priming with SA enhanced temperature stress tolerance in spring maize by an earlier emergence, increasing tissue water status, as well as improving membrane stability (Rehman et al. 2012). Sunflower seed priming with SA acid increased growth, content of carbohydrate, and chilling tolerance (Gornik and Lahuta 2017). Fennel seeds primed with salicylic acid also had better seed germination under low water potential (Farahbakhsh 2012).

Seed priming with SA and sodium hydrosulfide reduced methionine, proline, and arginine accumulation and increased nitric oxide and glycine betaine contents and regulated the expression of *ZmSAMD* and *ZmACS6* genes (genes participating in methionine metabolism) in maize plants under lead (Pb) stress (Zanganeh et al. 2018). However, seed priming with SA caused higher proline and glycine betaine (GB) accumulation under all available water levels except 100% FC and also higher total soluble sugars and trehalose under severe water deficit (25% FC). Salicylic acid primed seeds had higher content of relative water especially under higher water deficiency levels (Namdari and Baghbani 2017).

### 9.5.3 Jasmonic Acid

Jasmonic acid (JA) applications alleviate abiotic stresses in plants through activating the antioxidant enzymes (Yoon et al. 2009; Yan et al. 2013). JA pretreatment improved metal stress tolerance by increasing phytochelatin accumulation (Maksymiec et al. 2007). Seed priming with JA enhanced seedling growth and activity of SOD and POX, increased the accumulation of chlorophyll and carotenoid, and regulated the toxic effects of copper stress (Poonam et al. 2013). Seed priming with JA enhanced growth, content of carbohydrate, and chilling tolerance in sunflower (Gornik and Lahuta 2017).

## 9.6 Polyamines

Polyamines (PAs) are known to have positive effects on development and growth of plant (Watson and Malmberg 1998). They include spermidine, spermine, and putrescine. Polyamines are small ubiquitous compounds that have been involved in the regulation of many metabolic processes including protein synthesis, transcription, RNA modification, and enzyme activity modulation (Takahashi and Kakehi 2010) as well as a variety of stress responses in plants (Bouchereau et al. 1999). Polyamines bind to cations and can be placed in the context of anionic membrane components such as phospholipids, thereby acting as cellular protectors by stabilizing membranes under high stress conditions and protecting plants against various abiotic stresses (Basra et al. 1994; Bouchereau et al. 1999; Kusano et al. 2008). The effect of exogenous PAs on seed germination depends on the type and concentration of polyamine as well as the embryo dormancy state (Farooq et al. 2011). Seed priming with PAs had the beneficial effects on germination and seedling vigor of different crops (Afzal et al. 2009; Khan et al. 2012).

Seed priming with spermine improved salt stress tolerance in rice seedlings by minimizing oxidative damages; modulating antioxidants, osmolytes, and photosynthesis systems; increasing the activation of antioxidant enzymes; and reducing $H_2O_2$ and MDA levels (Paul and Roychoudhury 2016). Seed priming with spermidine (Spd) induced water stress tolerance in white clover by enhancing the activity of antioxidant enzymes, ascorbate-glutathione cycle, and the scavenging of ROS and decreasing lipid peroxidation levels, as well as improving cell membrane stability and starch metabolism (Li et al. 2014). Seed priming with Spd enhanced chilling tolerance in rice through improving the antioxidant enzyme activities; increasing glycine betaine, total phenolics, and flavonoid content; and upregulation of genes encoding enzymes that are involved in biosynthesis of PAs (Sheteiwy et al. 2017).

## 9.7 Ascorbic Acid

Seed priming with ascorbic acid alleviated salinity stress in pumpkin seedlings through increasing the catalase and peroxidase activity and reducing the respective enzymes' activity, which is possibly caused by the free radical neutralization resulting from salt stress (Fazlali et al. 2013). On the other hand, priming rapeseed seeds with ascorbic acid increased germination percentage, length of root and shoot, seedling vigor, and activity of CAT and POX under water stress condition (Razaji et al. 2014). Wheat seed primed with ascorbic acid had more emergence, growth, and water statue of seedlings under drought condition (Farooq et al. 2013).

## 9.8 Brassinosteroids

The plant hormone brassinosteroids regulate a wide range of physiological processes. *Medicago sativa* seeds primed with brassinosteroids increased salinity stress tolerance by enhancing antioxidant enzyme activity (POX, SOD, and CAT) (Zhang et al. 2007).

## 9.9 Beta-Aminobutyric Acid

Beta-aminobutyric acid (BABA) is identified as an effective inducer of abiotic stress tolerance in plants. This is done by interacting with several hormones like abscisic acid, salicylic acid, and ethylene or by producing signaling processes cascade mediated through $H_2O_2$ (Jakab et al. 2005; Zimmerli et al. 2008; Zhong et al. 2014). Priming rice grain with BABA decreased content of MDA by decreasing the biomembrane's lipid peroxidation and enhancing the activities of antioxidant enzyme, thus increasing the salt stress tolerance of rice seedlings (Jisha and Puthur 2016). Moreover, seed priming with beta-aminobutyric acid increased the total carbohydrate, proline, and protein accumulation and enhanced nitrate reductase activity, as well as enhanced the activities of antioxidant enzymes and reduced MDA content in green gram (Jisha and Puthur 2015).

## 9.10 Choline

Seed priming with choline improved salt tolerance in wheat plants by minimizing toxic elements ($Na^+$, $Cl^-$), maintaining beneficial elements ($K^+$, $Ca^{2+}$), increasing accumulation of glycine betaine, and decreasing oxidative stress, which reflect in decreasing lipid peroxidation. Moreover, choline priming alleviated salt stress injury by maintaining the plasma membrane lipids, phospholipids, and sterol (Salama et al. 2011; Salama and Mansour 2015). Moreover, choline is a precursor for phosphatidylcholine, which is a major membrane lipid in eukaryotic cells (Rathinasabapathi et al. 2000). Exogenous application of choline improved salt tolerance by increasing glycine betaine accumulation that synthesizes from choline via two-step oxidation (Sakamoto and Murata 2000; Cha-um et al. 2006; Su et al. 2006; Chen and Murata 2008; Salama et al. 2011).

## 9.11 5-Aminolevulinic Acid

Seed priming with 5-aminolevulinic acid increased germination and growth of seedling by enhancing the antioxidant enzyme activities (Kanto et al. 2015). Seed priming with 5-aminolevulinic acid increased chilling tolerance in two rice cultivars through improving the antioxidant enzyme activities (POD, SOD, APX, and GPX); increasing glycine betaine, total phenolics, and flavonoid content; and upregulation of genes encoding enzymes that are involved in biosynthesis of PAs (Sheteiwy et al. 2017).

## 9.12 Glycine Betaine

Priming of pepper seeds with glycine betaine (GB) increased the salt tolerance by enhancing the SOD enzyme activity and increasing proline content as well as reducing MDA content (Korkmaz and Sirikci 2011). The beneficial impact of glycine betaine priming on inducing salt tolerance in safflower may be due to the activation of SOD and CAT as well as the prevention of cell membrane damage by reducing oxidation of membrane lipid and improving the ion homeostasis (Alasvandyari et al. 2017). Seed treatment with GB also regulates chilling stress tolerance in cotton seedling by increasing antioxidant enzyme activities and decreasing $H_2O_2$ content and cell membrane damage (Cheng et al. 2018). Seed soaking can overcome the injurious effects of water stress on canola plants by increasing proline, IAA, and total soluble sugar content and activating the antioxidant enzymes as well as decreasing MDA and $H_2O_2$ (Dawood and Sadak 2014).

## 9.13 Melatonin

Seed priming with melatonin alleviated salt stress effects on maize by increasing the activities of antioxidant enzymes and increasing relative water, total phenolic, and proline content, as well as reducing $Na^+$ content, lipid peroxidation level, and membrane relative electrolyte leakage (Jiang et al. 2016). Seed priming with melatonin alleviated salt stress in faba bean seedlings by increasing the content of $K^+$, $Ca^{+2}$, indole acetic acid, total carbohydrate, pigments and total phenolic, and $K^+/Na^+$ and $Ca^{+2}/Na^+$ ratios (Dawood and EL-Awadi 2015).

## 9.14 Chitosan

Solid matrix priming with chitosan enhanced the speed of seed germination and seedling growth under salinity stress (Sen and Mandal 2016). Chitosan priming improved wheat seed germination and vigor of seedling under osmotic stress made by PEG (Hameed et al. 2014). Seed priming with chitosan had positive effects on seed germination and vigor of seedlings due to the resynthesis and activation of some enzymes, the reserve mobilization of food material, and the synthesis of DNA and RNA as well as activation of germination-responsive genes (Sadeghi et al. 2011). Seed priming with chitosan reduced the mean germination time; enhanced the germination index, seedling vigor, and activity of POX and CAT; and reduced the content of MDA and relative permeability of plasma membrane as well as increased the soluble sugars and proline concentrations in the maize seedlings under low temperature stress (Guan et al. 2009).

## *9.15 Butenolide*

Butenolide improved seedling emergence, vigor, and growth in pepper (Demir et al. 2012). Seed priming with butenolide stimulates many of the physiological, biochemical, and molecular mechanisms that are involved in germination and seedling growth such as increasing replicated DNA (Jain and van Staden 2007) and stimulation of the cell division cycle and mobilization of storage (Soos et al. 2009) as well as activation of enzymes that hydrolyze storage reserves (Bailly et al. 2001). Also, it can alleviate the adverse effects of high temperature and low osmotic potential on germinating seeds (Ghebrehiwot et al. 2008).

## *9.16 Biocompatible Silver Nanoparticles*

Biocompatible silver nanoparticles (AgNPs) can be applied as nanopriming agent. Seed priming with AgNPs can improve seed germination performance in at least three probable ways. Firstly, it can internalize seed coat and support water uptake, resulting in stimulation of seed metabolic activity and starch hydrolysis during initial imbibition (phase I). Secondly, it could increase α-amylase activity to fasten starch hydrolysis, leading to increased available soluble sugar content that enhances the growth of embryo and seedlings. Thirdly, they could mediate ROS generation, especially $H_2O_2$ that could act as signal molecules for promoting the upregulation of aquaporin genes that stimulate germination process and participating in loosening of cell wall and weakening of endosperm (Mahakham et al. 2017). The role of seed priming with AgNPs in alleviating the adverse effects of salinity stress has been reported (Abou-Zeid and Ismail 2018).

# 10 Conclusion

Abiotic stresses such as salinity, drought, and extreme temperature are the main challenges in agriculture. These stresses often result in different kinds of damage and activate different cellular responses and signaling pathways. Seed priming is a useful method to overcome the adverse effects of abiotic stresses in crops. Seed priming enhances cell signaling pathways and activities of plant defense-related responses. It has been detected that certain physiological, biochemical, and molecular changes are prompted by seed priming. The beneficial effects of seed priming are associated with the mobilization of reserve food, the repair of mitochondria and membranes, the repair and building up of nucleic acid, and the activation of genes involved in enzyme synthesis. Seed priming also regulates DNA replication, translation, and transcription. It will be necessary to develop suitable priming methods and agents for different plants to enhance abiotic stress tolerance in a sustainable manner. Moreover, future research efforts should focus on the molecular, biochemical, and physiological changes induced by seed priming under various abiotic stresses.

# References

Aboutalebian MA, Nazari S (2017) Seedling emergence and activity of some antioxidant enzymes of canola (*Brassica napus*) can be increased by seed priming. J Agric Sci 155(10):1541–1552

Abou-Zeid HM, Ismail GSM (2018) The role of priming with biosynthesized silver nanoparticles in the response of *Triticum aestivum* L. to salt stress. Egypt J Bot 58(1):73–85

Afzal I, Basra SMA, Ahmad N, Cheema MA, Warraich EA, Khaliq A (2002) Effect of priming and growth regulator treatment on emergence and seedling growth of hybrid maize *(Zea mays)*. Int J Agric Biol 4:303–306

Afzal I, Basra SMA, Shahid M, Farooq M, Saleem M (2008) Priming enhances germination of spring maize (*Zea mays* L.) under cool conditions. Seed Sci Technol 36:497–503

Afzal I, Munir F, Ayub CM, Basra SMA, Hameed A, Nawaz A (2009) Changes in antioxidant system, germination capacity and vigour of tomato seeds in response of priming with polyamines. Seed Sci Technol 37:765–770

Ajouri A, Haben A, Becker M (2004) Seed priming enhances germination and seedling growth of barley under conditions of P and Zn deficiency. J Plant Nutr Soil Sci 167:630–636

Akers SW, Holley KE (1986) SPS: a system for priming seed using aerated polyethylene glycol or salt solutions. HortScience 21:529–531

Akram NA, Ashraf M, Al-Qurainy F (2012) Aminolevulinic acid-induced regulation in some key physiological attributes and activities of antioxidant enzymes in sunflower (*Helianthus annuus* L.) under saline regimes. Sci Hortic 142:143–148

Alasvandyari F, Mahdavi B, Hosseini Shahab M (2017) Glycine betaine affects the antioxidant system and ion accumulation and reduces salinity-induced damage in safflower seedlings. Arch Biol Sci 69(1):139–147

Alevarado AD, Bradford KJ (1988) Priming and storage of tomato (*Lycopersicon esculentum*) seeds, effect of storage temperature on germination rate and viability. Seed Sci Technol 16:601–612

Ali Q, Daud MK, Haider MZ, Ali S, Rizwan M, Aslam N, Noman A, Iqbal N, Shahzad F, Deeba F, Ali I, Zhu SJ (2017) Seed priming by sodium nitroprusside improves salt tolerance in wheat

(*Triticum aestivum* L.) by enhancing physiological and biochemical parameters. Plant Physiol Biochem 119:50–58
Almansouri M, Kinet JM, Lutts S (2001) Effect of salt and osmotic stresses on germination in durum wheat (*Triticum durum*). Plant Soil 231:243–254
Aloui H, Souguir M, Latique S, Hannachi C (2014) Germination and growth in control and primed seeds of pepper as affected by salt stress. Cercetari Agron Moldova 47(3):83–95
Amjad M, Ziaf K, Iqbal Q, Ahmad I, Riaz MA (2007) Effect of seed priming on seed vigor and salt tolerance in hot pepper. Pak J Agric Sci 44(3):408–419
Amooaghaie R (2011) The effect of hydro and osmopriming on alfalfa seed germination and antioxidant defenses under salt stress. Afr J Biotechnol 10(33):6269–6275
Ansari O, Sharif-Zadeh F (2012) Osmo and hydro priming improvement germination characteristics and enzyme activity of Mountain Rye (*Secale montanum*) seeds under drought stress. J Stress Physiol Biochem 8(4):253–261
Apel K, Hirt H (2004) Reactive oxygen species: metabolism, oxidative stress, and signal transduction. Annu Rev Plant Biol 55:373–399
Arif M, Ali S, Shah A, Javeed N, Rashid A (2005) Seed priming maize for improving emergence and seedling growth. Sarhad J Agric 21:539–543
Asgedom H, Becker M (2001) Effects of seed priming with nutrient solutions on germination, seedling growth and weed competitiveness of cereals in Eritrea. In: Proceedings Deutscher Tropentag, University of Bonn and ATSAF, Magrraf Publishers Press, Weickersheim, p 282
Ashraf M, Foolad MR (2005) Pre-sowing seed treatment-a shotgun approach to improve germination, plant growth, and crop yield under saline and non-saline conditions. Adv Agron 88:223–271
Ashraf M, Harris PJC (2004) Potential biochemical indicators of salinity tolerance in plants. Plant Sci 166(1):3–16
Atia A, Debez A, Barhoumi Z, Smaoui A, Abdelly C (2009) ABA, GA3 and nitrate may control seed germination of *Crithmum maritimum* (*Apiaceae*) under saline conditions. C R Biol 332:704–710
Azeem M, Iqbal N, Kausar S, Javed MT, Akram MS, Sajid MA (2015) Efficacy of silicon priming and fertigation to modulate seedling's vigor and ion homeostasis of wheat (*Triticum aestivum* L.) under saline environment. Environ Sci Pollut Res Int 22:14367–14371
Azooz MM (2009) Salt stress mitigation by seed priming with salicylic acid in two faba bean genotypes differing in salt tolerance. Int J Agric Biol 11:343–350
Babu MA, Singh D, Gothandam KM (2012) The effect of salinity on growth, hormone and mineral elements in leaf and fruit of tomato cultivar PKM1. J Anim Plant Sci 22:159–164
Bailly C, Audigier C, Ladonne F, Wagner MH, Coste F (2001) Changes in oligosaccharides content and antioxidant enzyme activities in developing bean seeds as related to acquisition of drying tolerance and seed quality. J Exp Bot 52:701–708
Balestrazzi A, Confalonieri M, Macovei A, Carbonera D (2011) Seed imbibition in Medicago truncatula Gaertn: expression profiles of DNA repair genes in relation to PEG-mediated stress. J Plant Physiol 168:706–713
Bandeoglu E, Eyidogan F, Yücel M, Öktem HA (2004) Antioxidant responses of shoots and roots of lentil to NaCl-salinity stress. Plant Growth Regul 42:69–77
Banerjee A, Roychoudhury A (2016) Plant responses to light stress: oxidative damages, photoprotection and role of phytohormones. In: Ahammed GJ, Yu J-Q (eds) Plant hormones under challenging environmental factors. Springer Nature, Dordrecht, pp 181–213
Banerjee A, Roychoudhury A (2018) Seed priming Technology in the Amelioration of salinity stress in plants. In: Rakshit A, Singh H (eds) Advances in seed priming. Springer, Singapore
Barbara J (2015) Improving onion seed germination using priming treatments. Infrastruct Ecol Rural Areas 4:1437–1447
Bartels D, Sunkar R (2005) Drought and salt tolerance in plants. Crit Rev Plant Sci 24:23–58
Basra AS, Singh B, Malik CP (1994) Priming-induced changes in polyamine levels in relation to vigor of aged onion seeds. Bot Bull Acad Sin 35(19):23

Basra SMA, Farooq M, Khaliq A (2003) Comparative study of pre-sowing seed enhancement treatments in indica rice (*Oryza sativa* L.). Pak J Life Soc Sci 1:5–9

Basra SMA, Farooq M, Hafeez K, Ahmad N (2004) Osmohardening a new technique for rice seed invigoration. Int Rice Res Notes 29:80–81

Basra SMA, Farooq M, Rehman H, Saleem BA (2007) Improving the germination and early seedling growth in melon (*Cucumis melo* L.) by pre-sowing Salicylicate treatments. Int J Agric Biol 9(4):550–554

Batista TB, Cardoso ED, Binotti FFS, Costa E, Sá ME (2016) Priming and stress under high humidity and temperature on the physiological quality of Brachiaria brizantha cv MG-5 seeds. Acta Sci 38(1):123–127

Bayat S, Sepehri A (2012) Paclobutrazol and salicylic acid application ameliorates the negative effect of water stress on growth and yield of maize plants. Int J Res Agric Sci 8:127–139

Beckers GJM, Conrath U (2007) Priming for stress resistance: from the lab to the field. Curr Opin Plant Biol 10:1–7

Bellti P, Lanteris S, Lotito S (1993) Priming of Papaver nudicaule seeds for germination at low temperature. Hortic Sci 4:163–165

Benamar A, Tallon C, Macherel D (2003) Membrane integrity and oxidative properties of mitochondria isolated from imbibing pea seeds after priming or accelerated ageing. Seed Sci Res 13:35–45

Bewley JD (1997) Seed germination and dormancy. Plant Cell 9:1055–1066

Bewley JD, Black M (1994) Seeds: physiology of development and germination, 2nd edn. Plenum Press, New York, p 455

Bewley JD, Bradford KJ, Hilhorst HWM, Nonogaki H (2013) Seeds: physiology of development, germination and dormancy, 3rd edn. Springer, New York

Bittencourt MLC, Dias DCFS, Dias LAS, Araújo EF (2004) Effects of priming on asparagus seed germination and vigour under water and temperature stress. Seed Sci Technol 32:607–616

Bocian S, Holubowicz R (2008) Effect of different ways of priming tomato (*Lycopersicon esculentum* MILL) seeds on their quality. Pol J Nat Sci 23(4):729–739

Bohnert HJ, Shen B (1999) Transformation and compatible solutes. Sci Hortic 78:237–260

Bonilla I, El-Hamdaoui A, Bolaños L (2004) Boron and calcium increase *Pisum sativum* seed germination and seedling development under salt stress. Plant Soil 267:97–107

Borsani O, Valpuesta V, Botella MA (2001) Evidence for a role of salicylic acid in the oxidative damage generated by NaCl and osmotic stress in *Arabidopsis* seedlings. Plant Physiol 126:1024–1030

Bose B, Kumar M, Singhal RK, Mondal S (2018) Impact of seed priming on the modulation of physico-chemical and molecular processes during germination, growth, and development of crops. In: Rakshit A, Singh H (eds) Advances in seed priming. Springer, Singapore

Bouchereau A, Aziz A, Larther F, Martin-Tanguy J (1999) Polyamines and environmental challenges: recent development. Plant Sci 140:103–125

Bradford KJ (1986) Manipulation of seed water relations via osmotic priming to improve germination under stress conditions. HortScience 21:1105–1112

Bradford KJ (1995) Water relations in seed germination. In: Kigel J, Galili G (eds) Seed development and germination. Marcel Dekker, New York, pp 351–396

Bradford KJ, May DM, Hoyle BJ, Sibinski S, Scott SJ, Tyler KB (1988) Seed and soil treatments to improve emergence of muskmelon from cold or crusted soils. Crop Sci 28:1001–1005

Bray CM (1995) Biochemical processes during the osmopriming of seeds. In: Kigel J, Galili G (eds) Seed development and germination. Marcel Dekker, New York, pp 767–789

Bray CM, West CE (2005) DNA repair mechanisms in plants: crucial sensors and effectors for the maintenance of genome integrity. New Phytol 168:511–528

Bray CM, Davison PA, Ashraf M, Taylor RM (1989) Bio-chemical events during osmopriming of leek seed. Ann Appl Biol 102:185–193

Brocklehurst PA, Dearman J (1983) Interactions between seed priming treatments and nine lots of carrot, celery and onion. I. Laboratory germination. Ann Appl Biol 120:577–584

Brocklehurst PA, Dearman J, Drew RLK (1984) Effects of osmotic priming on seed germination and seedling growth in leek. Sci Hortic 24(3):201–210

Bruce TJA, Matthes MC, Napier JA, Pickett JA (2007) Stressful memories of plants: evidence and possible mechanisms. Plant Sci 173:603–608

Bruggink GT, Ooms JJJ, van der Toorn P (1999) Induction of longevity in primed seeds. Seed Sci Res 9:49–53

Bujalski W, Nienow AW (1991) Large-scale osmotic priming of onion seeds: a comparison of different strategies for oxygenation. Sci Hortic 46:13–24

Bujalski W, Nienow AW, Gray D (1989) Establishing the large scale osmotic priming of onion seeds using enriched air. Ann Appl Biol 115:171–176

Butler L, Hay F, Ellis R, Smith R, Murray T (2009) Priming and re-drying improve the survival of mature seeds of *Digitalis purpurea* during storage. Ann Bot 103:1261–1270

Cantliffe DJ, Elballa M (1994) Improved germination of carrot at stressful high temperature by seed priming. Proc Fla State Hortic Soc 107:121–128

Cantliffe DJ, Elbala M, Guedes AC, Odell GB, Perkinsveazie P, Schultheis JR, Seale DN, Shuler KD, Tanne I, Watkins JT (1981) Improving stand establishment of direct-seeded vegetables in Florida. In: Proceedings of Florida State Horticutural Society, vol 100, pp 213–216

Carvalho RF, Piotto FA, Schmidt D, Peters LP, Monteiro CC, Azevedo RA (2011) Seed priming with hormones does not alleviate induced oxidative stress in maize seedlings subjected to salt stress. Sci Agric 68:598–602

Caseiro R, Bennett MA, Marcos-Filho J (2004) Comparison of three priming techniques for onion seed lots differing in initial seed quality. Seed Sci Technol 32:365–375

Castañares JL, Bouzo CA (2018) Effect of different priming treatments and priming durations on melon germination behavior under suboptimal conditions. Open Agric 3(1):386–392

Catusse J, Meinhard J, Job C, Strub JM, Fischer U, Pestova E, Westohoff P, Van Dorselaer A, Job D (2011) Proteomics reveals potential biomarkers of seed vigor in sugar beet. Proteomics 11:1569–1580

Cayuela E, Perez-Alfocea F, Caro M, Bolarin MC (1996) Priming of seeds with NaCl induces physiological changes in tomato plants grown under salt stress. Physiol Plant 96:231–236

Cha-um S, Supaibulwatana K, Kirdmanee C (2006) Water relation, photosynthetic ability and growth of Thai jasmine rice (*Oryza sativa* L. ssp. *Indica* cv KDML105) to salt stress by application of exogenous glycinebetaine and choline. J Agron Crop Sci 192:25–36

Chen K (2011) Antioxidants and dehydrin metabolism associated with osmopriming-enhanced stress tolerance of germinating spinach (*Spinacia oleracea* L. cv. Bloomsdale) seeds PhD. Iowa State University, Ames, Iowa, USA

Chen K, Arora R (2011) Dynamics of the antioxidant system during seed osmopriming, post-priming germination, and seedling establishment in spinach (*Spinacia oleracea*). Plant Sci 180:212–220

Chen K, Arora R (2013) Priming memory invokes seed stress-tolerance. Environ Exp Bot 94:33–45

Chen THH, Murata N (2008) Glycinebetaine: an effective protectant against abiotic stress in plants. Trend Plant Sci 13:499–505

Chen K, Fessehaie A, Arora R (2013) Aquaporin expression during seed osmopriming and post-priming germination in spinach. Biol Plant 57:193–198

Cheng C, Pei LM, Yin TT, Zhang KW (2018) Seed treatment with glycine betaine enhances tolerance of cotton to chilling stress. J Agric Sci 156(3):323–333

Copeland LO, McDonald MB (2001) Principles of seed science and technology, 4th edn. Kluwer Academic Publishers, Dordrecht, p 488

Corbineau F, Come D (2006) Priming: a technique for improving seed quality. Seed Test Int 132:38–40

Corbineau F, Ozbincol N, Vinol D, Come D (2000) Improvement in tomato seed germination by osmopriming as related to energy metabolism. In: Black M, Bradford KJ, Vazquez-Ramos J (eds) Seed biology: advances and application. CABI, Oxon, pp 449–466

Cramer GR (2002) Sodium-calcium interactions under salinity stress in Läuchli a, Lüttge u, salinity Envirnoment-plants-molecules. Kluwer Academic Publishers, Dordrecht, pp 205–227

Cuartero J, Bolarin MC, Asins MJ, Moreno V (2006) Increasing salt tolerance in tomato. J Exp Bot 57:1045–1058

Das K, Roychoudhury A (2014) Reactive oxygen species (ROS) and response of antioxidants as ROS-scavengers during environmental stress in plants. Front Environ Sci 2:53
Das S, Dash FM, Nandi AK, Senapati N, Sarkar S, Pandey G (2014) Seed quality index an estimate used to predict response of bottle gourd seeds (*Lagenaria siceraria* (Mol) Standl) to hydro- and osmo-priming. Acta Adv Agric Sci 2(12):1–10
Das P, Nutan KK, Singla-Pareek SL, Pareek A (2015) Oxidative environment and redox homeostasis in plants: dissecting out significant contribution of major cellular organelles. Front Environ Sci 2:70
Daszkowska-Golec A (2011) Arabidopsis seed germination under abiotic stress as a concert of action of phytohormones. OMICS 15:763–774
Daur I (2018) Effects of hydro and hormonal priming on quinoa (*Chenopodium quinoa* Willd.) seed germination under salt and drought stress. Pak J Bot 50(5):1669–1673
Dawood MG, EL-Awadi ME (2015) Alleviation of salinity stress on *Vicia faba* L. plants via seed priming with melatonin. Acta Biol Colomb 20:223–235
Dawood MG, Sadak MS (2014) Physiological role of glycinebetaine in alleviating the deleterious effects of drought stress on canola plants (Brassica napus L.). Middle East J Agric Res 3:943–954
de Azevedo Neto AD, Prisco JT, Enéas-Filho J, Medeiros JV, Gomes-Filho E (2005) Hydrogen peroxide pre-treatment induces salt-stress acclimation in maize plants. J Plant Physiol 162(10):1114–1122
del Río LA, Puppo A (2009) Reactive oxygen species in plant signaling. Springer, Dordrecht/Heidelberg/London/New York
Dell Aquila A, Tritto V (1991) Germination and biochemical activities in wheat seeds following delayed harvesting, aging and osmotic. Seed Sci Technol 19:73–82
Demir I, Mavi K (2004) The effect of priming on seedling emergence of differentially matured watermelon (*Citrullus lanatus* (Thunb) Matsum and Nakai) seeds. Sci Hortic 102:467–473
Demir I, Okcu G (2004) Aerated hydration treatment for improved germination and seedling growth in aubergine (*Solanum melongena*) and pepper (*Capsicum annum*). Ann Appl Biol 144:121–123
Demir I, Oztokat C (2003) Effect of salt priming on germination and seedling growth at low temperature in watermelon seed during development. Seed Sci Technol 31:765–770
Demir I, Van de Venter HA (1999) The effect of priming treatments on the performance of watermelon (*Citrullus lanatus* (Thunb) Matsum and Nakai) seeds under temperature and osmotic stress. Seed Sci Technol 27:871–875
Demir I, Ozuaydın I, Yasar F, Staden JV (2012) Effect of smoke-derived butenolide priming treatment on pepper and salvia seeds in relation to transplant quality and catalase activity. S Afr J Bot 78:83–87
Di Girolamo G, Barbanti L (2012a) Treatment conditions and biochemical processes influencing seed priming effectiveness. Ital J Agron 7(e25):e178–e188
Di Girolamo G, Barbanti L (2012b) Treatment conditions and biochemical processes influencing seed priming effectiveness. Ital J Agron 7:8–18
Ding CK, Wang C (2003) The dual effects of methyl salicylate on ripening and expression of ethylene biosynthetic genes in tomato fruit. Plant Sci 164:589–596
Diniz KA, Silva PA, Oliveira JA, Evangelista JRE (2009) Sweet pepper seed responses to inoculation with microorganisms and coating with micronutrients, amino acids and plant growth regulators. Sci Agric 66:293–297
Dkhil BB, Issa A, Denden M (2014) Germination and seedling emergence of primed okra (*Abelmoschus esculentus* L.) seeds under salt stress and low temperature. Am J Plant Physiol 9:38–45
Dong L, Hao Z, Li Z, Zhu J, Wang Q (2014) Enhancement of welsh onion (*Allium fistulosum* L.) seed vigor by $KNO_3$ priming. J Agric Sci Technol 16:1345–1353
Dorna H, Jarosz M, Szopinska D, Szulc I, Rosinska A (2013) Germination vigour and health of primed *Allium cepa* L. seeds after storage. Acta Sci Pol Hortorum Cultus 12:43–58
Durner EF (2013) Principles of horticultural physiology. Cabi Publishing, Wallingford, p 416

Dursun A, Ekinci M (2010) Effects of different priming treatments and priming durations on germination percentage of parsley (*Petroselinum crispum* L.) seeds. Agric Sci 1:17–23

Ebrahimi R, Ahmadizadeh M, Rahbarian P (2014) Enhancing stand establishment of tomato cultivars under salt stress condition. Southwest J Hortic Biol Environ 5(1):19–42

Eisvand HR, Tavakkol-Afshari R, Sharifzadeh F, Maddah-Aref H, Hesamzadeh-Hejaz SM (2010) Effects of hormonal priming and drought stress on activity and isozyme profles of antioxidant enzymes in deteriorated seed of tall wheatgrass (*Agropyron elongatum* host). Seed Sci Technol 38:280–297

Ellis RH, Butcher PD (1988) The effects of priming and 'natural' differences in quality amongst onion seed lots on the response of the rate of germination to temperature and the identification of the characteristics under genotypic control. J Exp Bot 39:935–950

Ellis R, Hong TD (1994) Desiccation tolerance and potential longevity of developing seeds of rice (*Oryza sativa* L.). Ann Bot 73:501–506

Eskandari H (2013) Effects of priming technique on seed germination properties, emergence and field performance of crops: a review. Int J Agron Plant Prod 4:454–458

Esmaielpour B, Ghassemi-Golezani K, Rahimzadeh Khoei F, Gregoorian V, Toorchi M (2006) The effect of NaCl priming on cucumber seedling growth under salinity stress. J Food Agric Environ 4(2):347–349

Fallah S, Somayeh M, Mohammad P (2018) Seed priming improves seedling emergence and reduces oxidative stress in *Nigella sativa* under soil moisture stress. J Plant Nutr 41(1):29–40

Farahbakhsh H (2012) Germination and seedling growth in unprimed and primed seeds of fennel as affected by reduced water potential induced by NaCl. Int Res J Appl Basic Sci 3:737–744

Farhoudi R (2012) Evaluation effect of KNO3 seed priming on seedling growth and cell membrane damage of sunflower (*Helianthus annus*) under salt stress. Am Eurasian J Agric Environ Sci 12(3):384–388

Farhoudi R, Saeedipour S, Mohammadreza D (2011) The effect of NaCl seed priming on salt tolerance, antioxidant enzyme activity, proline and carbohydrate accumulation of muskmelon (*Cucumis melo* L.) under saline condition. Afr J Agric Res 6(6):1363–1370

Farooq M, Basra SMA, Saleem BA, Nafees M, Chishti SA (2005) Enhancement of tomato seed germination and seedling vigor by osmopriming. Pak J Agric 42(3–4):36–41

Farooq M, Basra SMA, Khalid A, Tabassum R, Mehmood T (2006) Nutrient homeostasis, reserves metabolism and seedling vigor as affected by seed priming in coarse rice. Can J Bot 84:1196–1202

Farooq M, Aziz T, Basra SMA, Cheema MA, Rehman H (2008a) Chilling tolerance in hybrid maize induced by seed priming with salicylic acid. J Agron Crop Sci 194:161–168

Farooq M, Basra SMA, Rehman H, Hussain M (2008b) Seed priming with polyamines improves the germination and early seedling growth in fine rice. J New Seeds 9:145–155

Farooq M, Basra SMA, Wahid A, Ahmad N, Saleem BA (2009) Improving the drought tolerance in rice (*Oryza sativa* L.) by exogenous application of salicylic acid. J Agron Crop Sci 195:237–246

Farooq M, Aziz T, Rehman H, Rehman A, Cheema SA, Aziz T (2011) Evaluating surface drying and re-drying for wheat seed priming with polyamines: effects on emergence, early seedling growth and starch metabolism. Acta Physiol Plant 33:1707–1713

Farooq M, Wahid A, Siddique KHM (2012) Micronutrient application through seed treatments – a review. J Soil Sci Plant Nutr 12(1):125–142

Farooq M, Irfan M, Aziz T, Ahmad I, Cheema SA (2013) Seed priming with ascorbic acid improves drought resistance of wheat. J Agron Crop Sci 199:12–22

Farouk S (2011) Ascorbic acid and α-tocopherol minimize salt-induced wheat leaf senescence. J Stress Physiol Biochem 7(3):58–79

Fazlali R, Asli DE, Moradi P (2013) The effect of seed priming by ascorbic acid on bioactive compounds of naked seed pumpkin (*Cucurbita pepo var styriaca*) under salinity stress. Int J Farm Allied Sci 2(17):587–590

Feng G, Zhang FS, Li XL, Tian CY, Tang C, Rengel Z (2002) Improved tolerance of maize plants to salt stress by *arbuscular mycorrhiza* is related to higher accumulation of soluble sugars in roots. Mycorrhiza 12:185–190

Fercha A, Capriotti AL, Caruso G, Cavaliere C, Gherroucha H, Samperi R, Stampachiacchiere S, Lagana A (2013) Gel-free proteomics reveal potential biomarkers of priming-induced salt tolerance in durum wheat. J Proteomics 91:486–499

Fercha A, Capriotti AL, Caruso G, Cavaliere C, Samperi R, Stampachiacchiere S, Laganà A (2014) Comparative analysis of metabolic proteome variation in ascorbate-primed and unprimed wheat seeds during germination under salt stress. J Proteomics 108:238–257

Foolad MR, Lin GY (1997) Genetic potential for salt tolerance during germination in *Lycopersicon* species. HortScience 32:296–300

Foti R, Aburenia K, Tigerea A, Gotosab J, Gerec J (2008) The efficacy of different seed priming osmotica on the establishment of maize (*Zea mays* L.) caryopses. J Arid Environ 72:1127–1130

Frett JJ, Pill WG, Morneau DC (1991) A comparison of priming agents for tomato and asparagus seeds. HortScience 26:1158–1159

Furutani SC, Zandstra BH, Price HC (1986) The effects of osmotic solute composition and duration and temperature of priming on onion seed germination. Seed Sci Technol 14:545–551

Gadelha CG, Miranda RS, Alencar NLM, Costa JH, Prisco JT, Gomes-Filho E (2017) Exogenous nitric oxide improves salt tolerance during establishment of Jatropha curcas seedlings by ameliorating oxidative damage and toxic ion accumulation. J Plant Physiol 212:69–79

Gao YP, Young L, Bonham-Smith P, Gusta LV (1999) Characterization and expression of plasma and tonoplast membrane aquaporins in primed seed of *Brassica napus* during germination under stress conditions. Plant Mol Biol 40:635–644

Gao YP, Bonham-Smith PC, Gusta LV (2002) The role of peroxiredoxin antioxidant and calmodulin in ABA-primed seeds of *Brassica napus* exposed to abiotic stresses during germination. J Plant Physiol 159:951–958

Genoud T, Métraux J (1999) Crosstalk in plant cell signaling: structure and function of the genetic network. Trends Plant Sci 4(12):503–507

Ghassemi-Golezani K, Esmaeilpour B (2008) The effect of salt priming on the performance of differentially matured cucumber (*Cucumis sativus*) seeds. Notulae Bot Hortic Agrobot Cluj-Napoca 36(2):67–70

Ghassemi-Golezani K, Jabbarpour S, Zehtab-Salmasi S, Mohammadi A (2010) Response of winter rapeseed (*Brassica napus* L.) cultivars to salt priming of seeds. Afr J Agric Res 5:1089–1094

Ghebrehiwot HM, Kulkarni MG, Kirkman KP, van Staden J (2008) Smoke-water and a smokeisolated butenolide improve germination and seedling vigor of *Eragrostis tef* (Zucc.) trotter under high temperature and low osmotic potential. J Agron Crop Sci 194:270–277

Gill SS, Tuteja N (2010) Reactive oxygen species and antioxidant machinery in abiotic stress tolerance in crop plants. Plant Physiol Biochem 48:909–930

Giri GS, Schillinger WF (2003) Seed priming winter wheat for germination, emergence and yield. Crop Sci 43:2135–2141

Goel A, Goel AK, Sheoran IS (2003) Changes in oxidative stress enzymes during artificial aging in cotton (*Gossypium hirsutum* L.) seeds. J Plant Physiol 160:1093–1100

Gonçalves JF, Becker AG, Cargnelutti D, Tabaldi LA, Pereira LB, Battisti V, Spanevello RM, Morsch VM, Nicoloso FT, Schetinger MRC (2007) Cadmium toxicity causes oxidative stress and induces response of the antioxidant system in cucumber seedlings. Braz J Plant Physiol 19(3):223–232

Gong HZ, Chen K, Wang S, Zhang C (2005) Silicon alleviates oxidative damage of wheat plants in pots under drought. Plant Sci 169:313–321

Gong HJ, Chen KM, Zhao ZG, Chen GC, Zhou WJ (2008) Effects of silicon on defense of wheat against oxidative stress under drought at different developmental stages. Biol Plant 52(3):592–596

Gornik K, Lahuta BL (2017) Application of phytohormones during seed hydropriming and heat shock treatment on sunflower (*Helianthus annuus* L.) chilling resistance and changes in soluble carbohydrates. Acta Physiol Plant 39:118

Goud PB, Kachole MS (2011) Effect of exogenous hydrogen peroxide on peroxidase and polyphenol oxidase activities in *Cajanus cajan* (L.) Millsp detached leaves. Int J Curr Res 3:61–65

Govinden-Soulange J, Levantard M (2008) Comparative studies of seed priming and pelleting on percentage and meantime to germination of seeds of tomato (*Lycopersicon esculentum* Mill). Afr J Agric Res 3(10):725–731
Gratao PL, Polle A, Lea PJ, Azevedo RA (2005) Making the life of heavy metal-stressed plants a little easier. Funct Plant Biol 32:481–494
Guan Y, Hu J, Wangi X, Shao C (2009) Seed priming with chitosan improves maize germination and seedling growth in relation to physiological changes under low temperature stress. J Zhejiang Univ Sci B 10(6):427–433
Gupta A, Dadlani M, Arun Kumar MB, Roy M, Naseem M, Choudhary VK, Maiti RK (2008) Seed priming: the aftermath. Int J Agric Environ Biotechnol 1:199–209
Gupta A, Zhuo J, Zha J, Reddy S, Olp J, Pai A (2010) Effect of different intravenous iron preparations on lymphocyte intracellular reactive oxygen species generation and subpopulation survival. BMC Nephrol 17:11–16
Gurusinghe SH, Bradford KJ (2001) Galactosyl-sucrose oligosaccharides and potential longevity of primed seeds. Seed Sci Res 11:121–133
Gurusinghe S, Powell ALT, Bradford KJ (2002) Enhanced expression of BiP is associated with treatments that extend storage longevity of primed tomato seeds. J Am Soc Hortic Sci 127:528–534
Hacisalihoglu G, Kantanka S, Miller N, Gustin JL, Settles AM (2018) Modulation of early maize seedling performance via priming under sub-optimal temperatures. PLoS One 13(11):e0206861
Haghighi M, Afifipour Z, Mozafarian M (2012) The alleviation effect of silicon on seed germination and seedling growth of tomato under salinity stress. Veget Crops Res Bull 76:119–126
Haıgh AM, Barlow EWR, Mılthrope F, Sınclaır PJ (1986) Field emergence of tomato (*Lycopersicon esculentum*), carrot (*Daucus carota*) and onion (*Allium cepa*) seeds primed in an aerated salt solution. J Am Soc Hortic Sci 111:660–665
Halmer P (2004) Methods to improve seed performance in the field. In: Benech-Arnold RL, Sánchez RA (eds) Seed physiology: applications to agriculture. Food Product Press, New York, pp 125–166
Hameed A, Sheikh MA, Jamil A, Maqsood S, Basra A (2013) Seed priming with sodium silicate enhances seed germination and seedling growth in wheat (*Triticum aestivum* L.) under water deficit stress induced by polyethylene glycol. Pak J Life Soc Sci 11(1):19–24
Hameed A, Sheikh MA, Hameed A, Farooq T, Basra SMA, Jamil A (2014) Chitosan seed priming improves seed germination and seedling growth in wheat (*Triticum aestivum* L.) under osmotic stress induced by polyethylene glycol. Philipp Agric Sci 97:294–299
Hasan J, Crawford RJ, Ivanova EP (2013) Antibacterial surfaces: the quest for a new generation of biomaterials. Trends Biotechnol 31:295–304
Hasanuzzaman M, Anwar Hossain M, Fujita M (2010) Selenium in higher plants: physiological role, antioxidant metabolism and abiotic stress tolerance. J Plant Sci 5:354–375
Hasanuzzaman M, Nahar K, Alam MM, Roychowdhury R, Fujita M (2013) Physiological, biochemical, and molecular mechanisms of heat stress tolerance in plants. Int J Mol Sci 14(5):9643–9684
Hassen A, Elouaer MA, Hannachi C (2017) Seed priming to improve seedling growth of pepper cultivars exposed to salt concentrations. Int J Veg Sci 23(6):489–507
Hassini I, Baenas N, Moreno DA, Carvajal M, Boughanmi N, Martinez Ballesta MDC (2017) Effects of seed priming, salinity and methyl jasmonate treatment on bioactive composition of *Brassica oleracea* var. *capitata* (white and red varieties) sprouts. J Sci Food Agric 97:2291–2299
Hay FR, Probert RJ (1995) The effect of different drying conditions and maturity on desiccation tolerance and seed longevity in *Digitalis purpurea* L. Ann Bot 76:739–647
He YL, Liu YL, Chen Q, Bian AH (2002) Thermotolerance related to antioxidation induced by salicylic acid and heat hardening in tall fescue seedlings. J Plant Physiol Mol Biol 28(2):89–95
Hela M, Nawel N, Imen T, Hanen Z, Imen BS, Raouia BM, Olfa B, Rym K, Mouhiba BNA, Abdelali H, Lachaâl M, Ouerghi Z (2011) Salt stress induced changes in germination, lipid per-

oxidation and antioxidant activities in lettuce (*Lactuca sativa* L.) seedlings. Afr J Biotechnol 10(65):14498–14506

Heydecker W, Coolbear T (1977) Seed treatments for improved performance survey and attempted prognosis. Seed Sci Technol 5:353–425

Hill H, Bradford KJ, Cunningham J, Taylor AG (2008) Primed lettuce seeds exhibit increased sensitivity to moisture during aging. Acta Hortic 782:135–141

Horii A, McCue P, Shetty K (2007) Seed vigour studies in corn, soybean and tomato in response to fish protein hydrolysates and consequences on phenolic-linked responses. Bioresour Technol 98:2170–2177

Hsu CC, Chen CL, Chen JJ, Sung JM (2003) Accelerated aging enhanced lipid peroxidation in bitter gourd seeds and effects of priming and hot water soaking treatments. Sci Hortic 98:201–212

Huang Z, Boubriak I, Osborne DJ, Dong M, Gutterman Y (2008) Possible role of pectin-containing mucilage and dew in repairing embryo DNA of seeds adapted to desert conditions. Ann Bot 101:277–283

Hubbard M, Germida J, Vujanovic V (2012) Fungal endophytes improve wheat seed germination under heat and drought stress. Botany 90:137–149

Hussain S, Zheng M, Khan F, Khaliq A, Fahad S, Peng S, Huang J, Cui K, Nie L (2015) Benefits of rice seed priming are offset permanently by prolonged storage and the storage conditions. Sci Rep 5:1–12

Hussain S, Khan F, Cao W, Wu L, Geng M (2016a) Seed priming alters the production and detoxification of reactive oxygen intermediates in rice seedlings grown under sub-optimal temperature and nutrient supply. Front Plant Sci 7:439

Hussain S, Khan F, Hussain HA, Nie L (2016b) Physiological and biochemical mechanisms of seed priming-induced chilling tolerance in rice cultivars. Front Plant Sci 7:116

Ibrahim EA (2016) Seed priming to alleviate salinity stress in germinating seeds. J Plant Physiol 192:38–46

Iqbal M, Ashraf M (2013) Gibberellic acid mediated induction of salt tolerance in wheat plants: growth, ionic partitioning, photosynthesis, yield and hormonal homeostasis. Environ Exp Bot 86:76–85

Jahromi F, Aroca R, Porcel R, Ruiz-Lozano JM (2008) Influence of salinity on the *in vitro* development of *Glomus intraradices* and on the *in vivo* physiological and molecular responses of mycorrhizal lettuce plants. Microb Ecol 55:45–53

Jain N, Van Staden J (2007) The potential of the smoke-derived compound, 3-methyl-2H-furo[2,3-c]pyran-2-one, as a priming agent for tomato seeds. Seed Sci Res 17:175–181

Jakab G, Ton J, Flors V, Zimmerli L, Metraux JP, Mauch-Mani B (2005) Enhancing *Arabidopsis* salt and drought stress tolerance by chemical priming for its abscisic acid responses. Plant Physiol 139(1):267–274

Jaleel CA, Riadh K, Gopi R, Manivannan P, Ines J, Al-Juburi HJ, Chang-Xing Z, Hong-Bo S, Panneerselvam R (2009) Antioxidant defense responses: physiological plasticity in higher plants under abiotic constraints. Acta Physiol Plant 31(3):427–436

Javid MG, Sorooshzadeh A, Moradi F, Sanavy SAMM, Allahdadi I (2011) The role of phytohormones in alleviating salt stress in crop plants. Aust J Crop Sci 6:726–734

Jeong Y, Kim JC, Lie J-C, Jeong YO, Cho JL (2000) Effect of priming duration and temperature on germinability of carrot, lettuce, onion and Welsh onion seeds. Korean J Hortic Sci Technol 18(3):327–333

Ji CY, Jin R, Xu Z, Kim HS, Lee C-J, Kang L, Kim SE, Lee HU, Lee JS, Kang CH, Chi YH, Lee SY, Xie Y, Li H, Ma D, Kwak SS (2017) Overexpression of *Arabidopsis* P3B increases heat and low temperature stress tolerance in transgenic sweet potato. BMC Plant Biol 17(1):139

Jiang X, Li H, Song X (2016) Seed priming with melatonin effects on seed germination and seedling growth in maize under salinity stress. Pak J Bot 48(4):1345–1352

Jisha KC, Puthur JT (2014) Halopriming of seeds imparts tolerance to NaCl and PEG induced stress in *Vigna radiata* (L.) Wilczek varieties. Physiol Mol Biol Plants 20(3):303–312

Jisha KC, Puthur JT (2015) Seed priming with BABA (β-amino butyric acid): a cost-effective method of abiotic stress tolerance in *Vigna radiata* (L.) Wilczek. Protoplasma 253(2):277–289

Jisha KC, Puthur JT (2016) Seed priming with beta-amino butyric acid improves abiotic stress tolerance in rice seedlings. Rice Sci 23(5):242–254
Jisha KC, Vijayakumari K, Puthur JT (2013) Seed priming for abiotic stress tolerance: an overview. Acta Physiol Plant 35:1381–1396
Joshi N, Jain A, Arya K (2013) Alleviation of salt stress in *Cucumis sativus* L. through seed priming with calcium chloride. Indian J Appl Res 3(11):22–25
Jyotsna V, Srivastava AK (1998) Physiological basis of salt stress resistance in pigeon pea (*Cajanus cajan* L.) II pre-sowing seed soaking treatment in regulating early seedling metabolism during seed germination. Plant Physiol Biochem 25:89–94
Kaewnaree P, Vichitphan S, Klanrit P, Siri B, Vichitphan K (2011) Effect of accelerated aging process on seed quality and biochemical changes in sweet pepper (*Capsicum annuum* Linn.) seeds. Biotechnology 2:175–182
Kang HM, Saltveit ME (2002) Chilling tolerance of maize, cucumber and rice seedlings leaves and roots are differently affected by salicylic acid. Physiol Plant 115:571–576
Kanto U, Jutamanee K, Osotsapar Y, Chai-arree W, Jattupornpong S (2015) Promotive effect of priming with 5-aminolevulinic acid on seed germination capacity, seedling growth and antioxidant enzyme activity in rice subjected to accelerated ageing treatment. Plant Prod Sci 18:443–454
Karadag B, Yucel NC (2017) Salicylic acid and fish flour pre-treatments affect wheat phenolic and flavonoid compounds, lipid peroxidation levels under salt stress. Cereal Res Commun 45:192–201
Kasote DM, Lee JHJ, Jayaprakasha GK, Patil BS (2019) Seed priming with iron oxide nanoparticles modulate antioxidant potential and defense-linked hormones in watermelon seedlings. ACS Sustain Chem Eng 7(5):5142–5151
Kathiresan K, Kalyani V, Gnanarethium JL (1984) Effect of seed treatments on field emergence, early growth and some physiological processes of sunflower (*Helianthus annuus* L.). Field Crops Res 9:255–259
Kaveh H, Nemati H, Farsi M, Jartoodeh SV (2011) How salinity affect germination and emergence of tomato lines. J Biol Environ Sci 15(3):159–163
Kaymak HC, Goveng I, Yarali F, Donmez MF (2008) The effects of biopriming with PGPR on germination of radish (*Raphanus sativus* L.) seeds under saline conditions. Turk J Agric 33:173–179
Kazemi K, Eskandari H (2012) Does priming improve seed performance under salt and drought stress? J Basic Appl Sci Res 2(4):3503–3507
Kępczyńska E, Piękna-Grochala J, Kępczyński J (2003) Effects of matriconditioning on onion seed germination, seedling emergence and associated physical and metabolic events. J Plant Growth Regul 41:269
Kester ST, Geneve RL, Houtz RL (1997) Priming and accelerated ageing effect L-isoaspartyl methyltransferase activity in tomato (*Lycopersicon esculentum* L.) seed. J Exp Bot 48:943–949
Khajeh-Hosseini M, Powell AA, Bingham IJ (2003) The interaction between salinity stress and seed vigor during germination of soyabean seeds. Seed Sci Technol 31:715–725
Khan AA (1992) Preplant physiological seed conditioning. In: Janick J (ed) Horticultural reviews, vol 13. Wiley, Oxford, pp 131–181
Khan MH, Panda SK (2008) Alterations in root lipid peroxidation and antioxidative responses in two rice cultivars under NaCl-salinity stress. Acta Physiol Plant 30:91–89
Khan AA, Tao AL, Knypl JS, Borkowska B, Powell LE (1978) Osmotic conditioning of seeds: physiological and biochemical changes. Acta Hortic 83:267–278
Khan HA, Ayub CM, Pervez MA, Bilal RM, Shahid MA, Ziaf K (2009a) Effect of seed priming with NaCl on salinity tolerance of hot pepper (*Capsicum annuum* L.) at seedling stage. Soil Environ 28:81–87
Khan HA, Pervez MA, Ayub CM, Ziaf K, Bilal RM, Shahid MA, Akhtar MA (2009b) Hormonal priming alleviates salt stress in hot pepper (*Capsicum annuum* L.). Soil Environ 28:130–135
Khan HA, Ziaf K, Amjad M, Iqbal Q (2012) Exogenous application of polyamines improves germination and early seedling growth of hot pepper. Chilean J Agric Res 72(3):429–433

Kibinza S, Bazin J, Bailly C, Farrant JM, Corbineau F, El-Maarouf-Bouteau H (2011) Catalase is a key enzyme in seed recovery from ageing during priming. Plant Sci 181:309–315
Korkmaz A, Sirikci R (2011) Improving salinity tolerance of germinating seeds by exogenous application of glycinebetaine in pepper. Seed Sci Technol 39:377–388
Korkmaz A, Tiryaki I, Nas MN, Ozbay N (2004) Inclusion of plant growth regulators into priming solution improves low-temperature germination and emergence of watermelon seeds. Can J Plant Sci 84:1161–1165
Korkmaz A, Korkmaz Y, Demirkiran AZ (2010) Enhancing chilling stress tolerance of pepper seedlings by exogenous application of 5-aminolevulinic acid. Environ Exp Bot 67:495–501
Kranner I, Beckett RP, Minibayeva FV, Seal CE (2010) What is stress? Concepts, definitions and applications in seed science. New Phytol 188:655–673
Krantev A, Yordanova R, Janda T, Szalai G, Popova L (2008) Treatment with salicylic acid decreases the effect of cadmium on photosynthesis in maize plants. J Plant Physiol 165:920–931
Ku YS, Sintaha M, Cheung MY, Lam HM (2018) Plant hormone signaling crosstalks between biotic and abiotic stress responses. Int J Mol Sci 19(10):3206
Kubala S, Wojtyla Ł, Garnczarska M (2013) Seed priming improves salt stress tolerance during germination by modulation of antioxidative capacity. Biotechnologia 94:223
Kubala S, Garnczarska M, Wojtyla Ł, Clippe A, Kosmala A, Żmieńko A, Lutts S, Quinet M (2015a) Deciphering priming-induced improvement of rapeseed (*Brassica napus* L.) germination through an integrated transcriptomic and proteomic approach. Plant Sci 231:94–113
Kubala S, Wojtyla Ł, Quinet M, Lechowska K, Lutts S, Garnczarska M (2015b) Enhanced expression of the proline synthesis gene *P5CSA* in relation to seed osmopriming improvement of Brassica napus germination under salinity stress. J Plant Physiol 183:1–12
Kusano T, Berberich T, Tateda C, Takahashi Y (2008) Polyamines: essential factors for growth and survival. Planta 228:367–381
Lada R, Stiles A, Surette MA, Caldwell C, Nowak J, Sturz AV, Blake TJ (2004) Stand establishment technologies for processing carrots. Acta Hortic (631):105–116
Lara TS, Lira JMS, Rodrigues AC, Rakocevic M, Alvarenga AA (2014) Potassium nitrate priming affects the activity of nitrate reductase and antioxidant enzymes in tomato germination. J Agric Sci 6(2):72–80
Lee SS, Kim JH (2000) Total sugars, α-amylase activity, and germination after priming of normal and aged rice seeds. Korean J Crop Sci 45:108–111
Lee SS, Kim JH, Hong SB, Yuu SH, Park EH (1998) Priming effect of rice seeds on seedling establishment under adverse soil conditions. Korean J Crop Sci 43:194–198
Li F, Wu X, Tsang E, Cutler AJ (2005) Transcriptional profling of imbibed *Brassica napus* seed. Genomics 86:718–730
Li Z, Lu GY, Zhang XK, Zou CS, Cheng Y, Zheng PY (2010) Improving drought tolerance of germinating seeds by exogenous application of gibberellic acid ($GA_3$) in rapeseed (*Brassica napus* L.). Seed Sci Technol 38:432–440
Li Z, Peng Y, Zhang XQ, Pan MH, Ma X, Huang LK et al (2014) Exogenous spermidine improves water stress tolerance of white clover (*Trifolium repens* L.) involved in antioxidant defence, gene expression and proline metabolism. Plant Omics 7:517–526
Lin JM, Sung JM (2001) Pre-sowing treatments for improving emergence of bitter gourd seedlings under optimal and sub-optimal temperatures. Seed Sci Technol 29:39–50
Lira JMS, Lara TS, Rodrigues AC, Dousseau S, Magalhães MM, Alvarenga AA (2015) Cross-tolerance mechanism induction in melon seeds by priming prior drying. Ciênc Agrotecnol 39(2):131–137
Liu Y, Ye N, Liu R, Chen M, Zhang J (2010) $H_2O_2$ mediates the regulation of ABA catabolism and GA biosynthesis in Arabidopsis seed dormancy and germination. J Exp Bot 61:2979–2990
Lutts S, Benincasa P, Wojtyla L, Kubala S, Pace R, Lechowska K, Quinet M, Garnczarska M (2016) Seed priming: new comprehensive approaches for an old empirical technique. In: Araujo S, Balestrazzi A (eds) New challenges in seed biology – basic and translational research driving seed technology, IntechOpen, pp 1–47

Machado Neto NB, Saturnino SM, Bomfim DC, Custodio CC (2004) Water stress induced by mannitol and sodium chloride in soybean cultivars. Braz Arch Biol Technol 47(4):521–529
Mahakham W, Sarmah AK, Maensiri S, Theerakulpisut P (2017) Nanopriming technology for enhancing germination and starch metabolism of aged rice seeds using phytosynthesized silver nanoparticles. Sci Rep 7:8263
Maiti R, Pramanik K (2013) Vegetable seed priming: a low cost, simple and powerful techniques for farmers' livelihood. Int J Bioresour Stress Manag 4(4):475–481
Maksymiec W, Wojcik M, Krupa Z (2007) Variation in oxidative stress and photochemical activity in *Arabidopsis thaliana* leaves subjected to cadmium and excess copper in the presence or absence of jasmonate and ascorbate. Chemosphere 66:421–427
Manaa A, Mimouni H, Wasti S, Gharbi E, Aschi-Smiti S, Faurobert M, Ahmed HB (2013) Comparative proteomic analysis of tomato (*Solanum lycopersicum*) leaves under salinity stress. Plant Omics J 6:268–277
Manonmani V, Begum MAJ, Jayanthi M (2014) Halo priming of seeds. Res J Seed Sci 7:1–13
Masondo NA, Kulkarni MG, Finnie JF, Van Staden J (2018) Influence of biostimulants-seed-priming on *Ceratotheca triloba* germination and seedling growth under low temperatures, low osmotic potential and salinity stress. Ecotoxicol Environ Saf 147:43–48
Matias JR, Ribeiro RC, Aragão CA, Araújo GGL, Dantas BF (2015) Physiological changes in osmo and hydroprimed cucumber seeds germinated in biosaline water. J Seed Sci 37(1):07–15
McDonald MB (1999) Seed deterioration: physiology, repair and assessment. Seed Sci Technol 27:177–237
McDonald MB (2000) Seed priming. In: Black M, Bewley JD (eds) Seed technology and its biological basis. Sheffield Academic Press, Sheffield, pp 287–325
Memon NN, Gandahi MB, Pahoja VM, Sharif N (2013) Response of seed priming with boron on germination and seedling sprouts of broccoli. Int J Agri Sci Res 3(2):183–194
Miransari M, Smith DL (2014) Plant hormones and seed germination. Environ Exp Bot 99:110–121
Mittal R, Dubey RS (1995) Influence of sodium chloride salinity on polyphenol oxidase, indole 3-acetic acid oxidase and catalase activities in rice seedlings differing in salt tolerance. Trop Sci 35:141–149
Moghanibashi M, Karimmojeni H, Nikneshan P (2013) Seed treatment to overcome drought and salt stress during germination of sunflwer (*Helianthus annuus* L.). J Agrobiol 30:89–96
Mouradi M, Bouizgaren A, Farissi M, Makoudi B, Kabbadj A, Very AA, Sentenac H, Qaddoury A, Ghoulam C (2016) Osmopriming improvesseeds germination, growth, antioxidant responses and membrane stabilityduring early stage of Moroccan alfalfa populations under water deficit. Chilean J Agric Res 76(3):265–272
Munns R (2002) Comparative physiology of salt and water stress. Plant Cell Environ 25:239–250
Munns R (2005) Genes and salt tolerance: bringing them together. New Phytol 167:645–663
Munns R, Tester M (2008) Mechanisms of salinity tolerance. Annu Rev Plant Biol 59:651–681
Murillo-Amador B, López-Aguilar R, Kaya C, Larrinaga-Mayoral J, Flores-Hernández A (2002) Comparative effects of NACl and polyethylene glycol on germination, emergence and seedling growth of cowpea. J Agron Crop Sci 188:235–247
Nakaune M, Hanada A, Yin YG, Matsukura C, Yamaguchi S (2012) Molecular and physiological dissection of enhanced seed germination using short-term low-concentration salt seed priming in tomato. Plant Physiol Biotechnol 52:28–37
Namdari A, Baghbani A (2017) Consequences of seed priming with salicylic acid and hydro priming on *Smooth Vetch* seedling growth under water deficiency. J Agric Sci 9(12):259
Nascimento WM (2003) Muskmelon seed germination and seedling development in response to seed priming. Sci Agric 60(1):71–75
Nascimento WM, West SH (2000) Drying during muskmelon (*Cucumis melo* L.) seed priming and its effects on seed germination and deterioration. Seed Sci Tecnhol 28(1):211–215
Nascimento WM, Souza DE, Aragão FA (2004) Muskmelon seed priming in relation to seed vigor. Sci Agric 61(1):114–117

Nasri N, Kaddour R, Mahmoudi H, Baatour O, Bouraoui N, Lachaâl M (2011) The effect of osmopriming on germination, seedling growth and phosphatase activities of lettuce under saline condition. Afr J Biotechnol 10(65):14366–14372

Nawaz A, Amjad M, Jahangir MM, Khan SM, Cui H, Hu J (2012) Induction of salt tolerance in tomato (*Lycopersicon esculentum* Mill.) seeds through sand priming. Aust J Crop Sci 6(7):1199–1203

Nawaz J, Hussain M, Jabbar A, Nadeem GA, Sajid M, Subtain MU, Shabbir I (2013) Seed priming a technique. Int J Agric Crop Sci 6(20):1373–1381

Neamatollahi E, Bannayan M, Ghanbari A, Haydari M, Ahmadian A (2009) Does hydro and osmo-priming improve fennel (*Foeniculum vulgare*) seeds germination and seedlings growth? Notulae Bot Hortic Agrobotanici ClujNapoca 37(2):190–194

Neill SJ, Desikan R, Clarke A, Hancock JT (2002) Nitric oxide is a novel component of abscisic acid signalling in stomatal guard cells. Plant Physiol 128:13–16

Nelson H, Govers A (1986) Salt priming of muskmelon seeds for low temperature germination. Sci Hortic 28:85–91

Nerson H (2007) Seed production and germinability of cucurbit crops. Seed Sci Biotechnol 1(1):1–10

Netondo GW, Onyango JC, Beck E (2004) Sorghum and salinity: II gas exchange and chlorophyll fluorescence of sorghum under salt stress. Crop Sci 44:806–811

Noctor G, Foyer CH (1998) Ascorbate and glutathione: keeping active oxygen under control. Annu Rev Plant Physiol Plant Mol Biol 49:249–279

Omami EN (2005) Salt tolerance of amaranth as affected by seed priming, MSc thesis. University of Pretoria, South Africa

Ozbay N, Susluoglu Z (2016) Assessment of growth regulator prohexadione calcium as priming agent for germination enhancement of pepper at low temperature. J Anim Plant Sci 26(6):1652–1658

Ozbingol N, Corbineau F, Côme D (1998) Responses of tomato seeds to osmoconditioning as related to temperature and oxygen. Seed Sci Res 8(3):377–384

Pal A, Ali MA, Pal AK (2017) Effect of seed priming on reserve mobilization, water uptake and antioxidative enzyme activities in germinating seeds of groundnut under salinity stress. Int J Agric Sci 9(36):4542–4545

Panda SK, Khan MH (2009) Growth, oxidative damage and antioxidant responses in greengram (*Vigna radiata* L.) under short-term salinity stress and its recovery. J Agron Crop Sci 195:442–454

Pandey GK (2017) Mechanism of hormone signaling under stress, vols 1 and 2. Wiley-Blackwell, Hoboken

Pandita VK, Anand A, Nagarajan S (2007) Enhancement of seed germination in hot pepper following presowing treatments. Seed Sci Technol 35:282–290

Paparella S, Araújo SS, Rossi G, Wijayasinghe M, Carbonera D, Balestrazzi A (2015) Seed priming: state of the art and new perspectives. Plant Cell Rep 34(8):1281–1293

Parera CA, Cantliffe DJ (1992) Enhanced emergence and seedling vigor in shrunken-2 sweet corn via seed disinfection and solid matrix priming. J Am Soc Hortic Sci 117:400–403

Parera CA, Cantliffe DJ (1994) Pre-sowing seed priming. Hortic Rev 16:109–141

Parida AK, Das AB (2005) Salt tolerance and salinity effect of plants: a review. Ecotoxicol Environ Saf 60:324–349

Pastori GM, Foyer CH (2002) Common components, networks and pathways of cross-tolerance to stress the central role of 'redox' and abscisic-acid-mediated controls. Plant Physiol 129:460–468

Patade VY, Sujata B, Suprasanna P (2009) Halopriming imparts tolerance to salt and PEG induced drought stress in sugarcane. Agric Ecosyst Environ 134:24–28

Patade VY, Maya K, Zakwan A (2011) Seed priming mediated germination improvement and tolerance to subsequent exposure to cold and salt stress in capsicum. Res J Seed Sci 4(3):125–136

Patade VY, Khatri D, Manoj K, Kumari M, Ahmed Z (2012) Cold tolerance in thiourea primed capsicum seedlings is associated with transcript regulation of stress responsive genes. Mol Biol Rep 39:10603–10613
Paul S, Roychoudhury A (2016) Seed priming with spermine ameliorates salinity stress in the germinated seedlings of two rice cultivars differing in their level of salt tolerance. Trop Plant Res 3(3):616–633
Paul S, Roychoudhury A (2017) Effect of seed priming with spermine/spermidine on transcriptional regulation of stress-responsive genes in salt-stressed seedlings of an aromatic rice cultivar. Plant Gene 11:133–142
Pereira MD, Dias DCFS, Dias LAS, Araújo EF (2009) Primed carrot seeds performance under water and temperature stress. Sci Agric 66(2):174–179
Pill WG (1995) Low water potential and presowing germination treatments to improve seed quality. In: Basra S (ed) Seed quality basic mechanisms and agricultural implications. Food Products Press, London, pp 319–360
Pill WG, Finch-Savage WE (1988) Effects of combining priming and plant growth regulator treatments on the synchronization of carrot seed germination. Ann Appl Biol 113:383–389
Pill WG, Frett JJ, Morneau DC (1991) Germination and seedling emergence of primed tomato and asparagus seeds under adverse conditions. Hortic Sci 26:1160–1162
Poonam S, Kaur H, Geetika S (2013) Effect of Jasmonic acid on photosynthetic pigments and stress markers in *Cajanus cajan* (L.) Millsp seedlings under copper stress. Am J Plant Sci 4:817–823
Pouramir-Dashtmian F, Khajeh-Hosseini M, Esfahani M (2014) Improving rice seedling physiological and biochemical processes under low temperature by seed priming with salicylic acid. Int J Plant Anim Environ Sci 4(2):565
Rajjou L, Debeaujon I (2008) Seed longevity: survival and maintenance of high germination ability of dry seeds. C R Biol 331:796–805
Rajjou L, Duval M, Gallardo K, Catusse J, Bally J, Job C, Job D (2012) Seed germination and vigor. Annu Rev Plant Biol 63:507–533
Raskin I (1992) Role of salicylic acid in plants. Annu Rev Plant Physiol Plant Mol Biol 43:439–463
Rathinasabapathi B, Sigua C, Ho J, Gage DA (2000) Osmoprotectant β-alanine betaine synthesis in the Plumbaginaceae: *S*-adenosyl-l-methionine dependent *N*- methylation of β-alanine to its betaine is via *N*-methyl and *N,N*-dimethyl β-alanines. Physiol Plant 109:225–231
Ratikanta KM (2011) Seed priming: an efficient farmers' technology to improve seedling vigour, seedling establishment and crop productivity. Int J Bioresour Stress Manag 2(3):297
Razaji A, Farzanian M, Sayfzadeh S (2014) The effects of seed priming by ascorbic acid on some morphological and biochemical aspects of rapeseed (*Brassica napus* L.) under drought stress condition. Int J Biosci 4(1):432–442
Rehman H, Afzal I, Farooq M, Aziz T, Ahmad SM (2012) Improving temperature stress resistance in spring maize by seed priming. In: Proceedings of 3rd international conference 'Frontiers in agriculture'. Dankook International Cooperation on Agriculture, Dankook University, Cheonansi, Republic of Korea
Rosental L, Nonogaki H, Fait A (2014) Activation and regulation of primary metabolism during seed germination. Seed Sci Res 24:1–15
Rowse HR (1996) Drum priming: a non-osmotic method of priming seeds. Seed Sci Technol 24(2):281–294
Sadeghi S, Rahnavard A, Ashrafi ZY (2009) Study of respond seeds wheat (*Triticum aestivum* L.) to osmotic priming, temperatures and local seed masses. Bot Res Int 2(2):69–73
Sadeghi H, Khazaei F, Yari L, Sheidaei S (2011) Effect of seed osmopriming on seed germination behaviour and vigor of soybean (*Glycine max* L.). J Agric Biol Sci 6:39–43
Saeedipour S (2013) Effect of phytohormone seed priming on germination and seedling growth of cowpea (*Vigna sinensis* L.) under different duration of treatment. Int J Biosci 3(12):187–192
Sakamoto A, Murata N (2000) Genetic engineering of glycinebetaine synthesis in plants: current status and implications for enhancement of stress tolerance. J Exp Bot 51:81–88

Sakhabutdinova R, Fatkhutdinova DR, Bezrukova MV, Shakirova FM (2003) Salicylic acid prevents the damaging action of stress factors on wheat plants. Bulg J Plant Physiol Special Issue:314–319

Salama KHA, Mansour MMF (2015) Choline priming-induced plasma membrane lipid alterations contributed to improved wheat salt tolerance. Acta Physiol Plant 37:1–7

Salama KHA, Mansour MMF, Hassan NS (2011) Choline priming improves salt tolerance in wheat (*Triticum aestivum* L.). Aust J Basic Appl Sci 5:126–132

Salama KHA, Ahmed HFS, El-Araby MMA (2015) Interaction of exogenous abscisic acid and salinity on the lipid root plasma membrane of Phaseolus vulgaris L. Egypt J Exp Biol (Bot) 11:189–196

Saranya N, Renugadevi J, Raja K, Rajashree V, Hemalatha G (2017) Seed priming studies for vigour enhancement in onion CO onion. J Pharmacogn Phytochem 6(3):77–82

Schwember AR, Bradford KJ (2005) Drying rates following priming affect temperature sensitivity of germination and longevity of lettuce seeds. Hortic Sci 40:778–781

Schwember AR, Bradford KJ (2010) A genetic locus and gene expression patterns associated with the priming effect on lettuce seed germination at elevated temperatures. Plant Mol Biol 73:105–118

Sedghi M, Nemati A, Esmaielpour B (2010) Effect of seed priming on germination and seedling growth of two medicinal plants under salinity. Emirates J Food Agric 22(2):130–139

Sedghi M, Amanpour-Balaneji B, Bakhshi J (2014) Physiological enhancement of medicinal pumpkin seeds (*Cucurbita pepo* var styriaca) with different priming methods. Iran J Plant Physiol 5(1):1209–1215

Sen SK, Mandal P (2016) Solid matrix priming with chitosan enhances seed germination and seedling invigoration in mung bean under salinity stress. J Cent Eur Agric 17(3):749–762

Senaratna T, Touchel D, Bumm E, Dixon K (2000) Acetyl salicylic acid induces multiple stress tolerance in bean and tomato plants. Plant Growth Regul 30:157–161

Shafi M, Bakht J, Hassan MJ, Raziuddin M, Zhang GP (2009) Effect of cadmium and salinity stresses on growth and antioxidant enzyme activities of wheat (*Triticum aestivum* L.). Bull Environ Contam Toxicol 82:772–776

Shah AR, Ara N, Shafi G (2011) Seed priming with phosphorus increased germination and yield of okra. Afr J Agric Res 6(16):3859–3876

Shahverdi MA, Omidi H, Tabatabaei SJ (2017) Effect of nutri-priming on germination indices and physiological characteristics of *stevia* seedling under salinity stress. J Seed Sci 39(4):353–362

Shao HB, Liang ZS, Shao MA (2005) Changes of anti-oxidative enzymes and MDA content under soil water deficits among 10 wheat (*Triticum aestivum* L.) genotypes at maturation stage. Colloids Surf B Biointerfaces 45(1):7–13

Sharma P, Dubey RS (2007) Involvement of oxidative stress and role of antioxidative defense system in growing rice seedlings exposed to toxic concentrations of aluminum. J Plant Cell Rep 26:2027–2038

Sharma SN, Maheshwari A (2015) Expression patterns of DNA repair genes associated with priming small and large chickpea (*Cicer arietinum*) seeds. Seed Sci Technol 43:250–261

Sharma AD, Rathore SVS, Srinivasan K, Tyagi RK (2014) Comparison of various seed priming methods for seed germination, seedling vigour and fruit yield in okra (*Abelmoschus esculentus* L. Moench). Sci Hortic 165(22):75–81

Sharma KK, Singh US, Sharma P, Kumar A, Sharma L (2015) Seed treatments for sustainable agriculture – a review. J Appl Nat Sci 7:521–539

Sheteiwy M, Shen H, Xu J, Guan Y, Song W, Hu J (2017) Seed polyamines metabolism induced by seed priming with spermidine and 5-aminolevulinic acid for chilling tolerance improvement in rice (*Oryza sativa* L.) seedlings. Environ Exp Bot 137:58–72

Sheteiwy MS, An J, Yin M, Jia X, Guan Y, He F, Hu J (2018) Cold plasma treatment and exogenous salicylic acid priming enhances salinity tolerance of *Oryza sativa* seedlings. Protoplasma 2018:1–21

Singh PK, Pandita VK, Tomar BS, Seth R (2014) Germination and field emergence in osmotic and solid matrix priming in onion (*Allium cepa*(. Indian J Agric Sci 84(12):1561–1564

Siri B, Vichitphan K, Kaewnaree P, Vichitphan S, Klanrit P (2013) Improvement of quality, membrane integrity and antioxidant systems in sweet pepper (*Capsicum annuum* Linn) seeds affected by osmopriming. Aust J Crop Sci 7(13):2068–2073
Sivritepe N, Sivritepe HO, Eris A (2003) The effects of NaCl priming on salt tolerance in melon seedlings grown under saline conditions. Sci Hortic 97:229–237
Sivritepe HO, Sivritepe N, Eris A, Turhan E (2005) The effects of NaCl pretreatments on salt tolerance of melons grown under long-term salinity. Sci Hortic 106:568–581
Smirnoff N (2005) Ascorbate, tocopherol and carotenoids: metabolism, pathway engineering and functions. In: Smirnoff N (ed) Antioxidants and reactive oxygen species in plants. Blackwell, Oxford, pp 53–86
Smith PT, Comb BG (1991) Physiological and enzymatic activity of pepper seeds (*Capsicum annuum*) during priming. Physiol Plant 82:433–439
Soeda Y, Konings MCJM, Vorst O, van Houwelingen AMML, Stoopen GM, Maliepaard CA, Kodde J, Bino RJ, Groot SPC, van der Geest AHM (2005) Gene expression programs during *Brassica oleracea* seed maturation, osmopriming, and germination are indicators of progression of the germination process and the stress tolerance level. Plant Physiol 137:354–368
Sohail SA, Chaurasia AK, Bara BM (2018) Effect of different seed priming methods on germination and vigour of Kabuli Chickpea (*Cicer kabulium* L.) seeds. Int J Curr Microbiol App Sci 7(8):1396–1404
Soos V, Juhasz A, Light ME, Van Staden J, Balazs E (2009) Smoke-water- induced changes of expression pattern in Grand Rapids lettuce achenes. Seed Sci Res 19:37–49
Sowmya KJ, Rame G, Bhanuprakash K, Yogeesha HS, Puttaraju TB, Channakeshava BC (2013) Enhancement of seed quality through Chemopriming in cucumber (*Cucumis sativus* L.). Mysore J Agric Sci 47(1):22–30
Srivastava AK, Lokhande VH, Patade VY, Suprasanna P, Sjahril R, D'Souza SF (2010) Comparative evaluation of hydro-, chemo-, and hormonal-priming methods for imparting salt and PEG stress tolerance in Indian mustard (*Brassica juncea* L.). Acta Physiol Plant 32:1135–1144
Su J, Hirji R, Zhang L, He C, Selvaraj G, Wu R (2006) Evaluation of the stress-inducible production of choline oxidase in transgenic rice as a strategy for producing the stress-protectant glycine betaine. J Exp Bot 57:1129–1135
Sugie A, Naydenov N, Mizuno N, Nakamura C, Takumi S (2006) Overexpression of wheat alternative oxidase gene *Waox1a* alters respiration capacity and response to reactive oxygen species under low temperature in transgenic *Arabidopsis*. Genes Genet Syst 81:349–354
Sun YY, Sun YJ, Wang MT, Li XY, Guo X, Hu R, Jun MA (2010) Effects of seed priming on germination and seedling growth under water stress in rice. Acta Agron Sin 36(11):1931–1940
Sung Y, Cantliffe DJR, Nagata T, Nascimento WM (2008) Structural changes in lettuce seed during germination at high temperature altered by genotype, seed maturation temperature, and seed priming. J Am Soc Hortic Sci 133:167–311
Suzuki N, Mittler R (2006) Reactive oxygen species and temperature stresses: a delicate balance between signaling and destruction. Physiol Plant 126:45–51
Suzuki N, Koussevitzky S, Mittler R, Miller G (2012) ROS and redox signaling in the response of plants to abiotic stress. Plant Cell Environ 35:259–270
Szalai G, TariJanda LT, Pestenacz A, Paldi E (2000) Effects of cold acclimation and salicylic acid on changes in ACC and MACC contents in maize during chilling. Biol Plant 43:637–640
Takahashi T, Kakehi JI (2010) Polyamines: ubiquitous polycations with unique roles in growth and stress responses. Ann Bot 105:1–6
Tanou G, Fotopoulos V, Molassiotis A (2012) Priming against environmental challenges and proteomics in plants: update and agricultural perspectives. Front Plant Sci 3:216
Tavili A, Zare S, Moosavi SA, Enayati A (2011) Effects of seed priming on germination characteristics of Bromus species under salt and drought conditions Amer-Eurasian. J Agric Environ Sci 10:163–168

Taylor AG, Beresniewicz MM, Goffnet MC (1997) Semipermeable layer in seeds. In: Ellis RH, Black M, Murdoch AJ, Hong TD (eds) Basic and applied aspects of seeds biology. Kluwer Academic Publishing, Dordrecht, pp 429–436

Taylor AG, Allen PS, Bennet MA, Bradford KJ, Burris JS, Misra MK (1998) Seed enhancements. Seed Sci Res 8:245–256

Thakur P, Kumar S, Malik JA, Berger JD, Nayyar H (2010) Cold stress effects on reproductive development in grain crops: an overview. Environ Exp Bot 67(3):429–443

Thomas UC, Varughese K, Thomas A, Sadanandan S (2000) Seed priming for increased vigour, viability and productivity of upland rice. Leisa India 4:14

Tzortzakis NG (2009) Effect of pre-sowing treatment on seed germination and seedling vigour in endive and chicory. Hortic Sci 36(3):117–125

Varier A, Vari AK, Dadlani M (2010) The sub cellular basis of seed priming. Curr Sci 99:450–456

Ventura L, Donà M, Macovei A, Carbonera D, Buttafava A, Mondoni A et al (2012) Understanding the molecular pathways associated with seed vigor. Plant Physiol Biochem 60:196–206

Vranova E, Inze D, Van Brensegem F (2002) Signal transduction during oxidative stress. J Exp Bot 53(372):1227–1236

Wang WB, Kim YH, Lee HS, Kim KY, Deng XP, Kwak SS (2009) Analysis of antioxidant enzyme activity during germination of alfalfa under salt and drought stresses. Plant Physiol Biochem 47:570–577

Wang W, He A, Peng S, Huang J, Cui K, Nie L (2018) The effect of storage condition and duration on the deterioration of primed rice seeds. Front Plant Sci 9:172

Waterworth WM, Masnavi G, Bhardwaj RM, Jiang Q, Bray CM, West CE (2010) A plant DNA ligase is an important determinant of seed longevity. Plant J 63:848–886

Watson MB, Malmberg RL (1998) Arginine decarboxylase (polyamine synthesis) mutants of Arabidopsis thaliana exhibit altered root growth. Plant J 13:231–239

Weisany W, Sohrabi Y, Heidari G, Siosemardeh A, Ghassemi-Golezani K (2012) Changes in antioxidant enzymes activity and plant performance by salinity stress and zinc application in soybean (*Glycine max* L.). Plant Omics J 5:60–67

Welbaum GE, Bradford KJ (1989) Water relations of seed development and germination in muskmelon (*Cucumis melo* L.) IV characteristics of the perisperm during seed development. Plant Physiol 92:1038–1045

Welbaum GE, Bradford KJ (1990) Water relations of seed development and germination in muskmelon (*Cucumis melo* L.) IV characteristics of the perisperm during seed development. Plant Physiol 92:1038–1045

Welbaum GE, Shen Z, Oluoch MO, Jett LM (1998) The evolution and effects of priming vegetable seeds. Seed Technol 20:209–235

Wheeler TR, Ellis RH (1994) Effects of seed quality and temperature on pre-emergence root growth of seedlings on onion (*Allium cepa* L.). Seed Sci Technol 22:141–148

Wojtyla Ł, Lechowska K, Kubala S, Garnczarska M (2016) Molecular processes induced in primed seeds increasing the potential to stabilize crop yields under drought conditions. J Plant Physiol 203:116–126

Xiong L, Zhu JK (2002) Molecular and genetic aspects of plant responses to osmotic stress. Plant Cell Environ 25:131–139

Xu H, Wei Y, Zhu Y, Lian L, Xie H, Cai Q et al (2015) Antisense suppression of Lox3 gene expression in rice endosperm enhances seed longevity. Plant Biotechnol J 13:526–539

Yacoubi R, Job C, Belghazi M, Chaibi W, Job D (2013) Proteomic analysis of the enhancement of seed vigour in osmoprimed alfalfa seeds germinated under salinity stress. Seed Sci Res 23:99–110

Yan Z, Chen J, Li X (2013) Methyl jasmonate as modulator of Cd toxicity in *Capsicum frutescens* var fasciculatum seedlings. Ecotoxicol Environ Saf 98:203–209

Yang Y, Yao Y, Zhang X (2010) Comparison of growth and physiological responses to severe drought between two altitudinal *Hippoph aerhamnoides* populations. Silva Fennica 44:603–614

Yeoung YR, Wilson DO, Murray GA (1995) Oxygen regulates imbibition of muskmelon seeds. Seed Sci Technol 23:843–850

Yi H, Liu X, Yi M, Chen G (2014) Dual role of hydrogen peroxide in *Arabidopsis* guard cells in response to sulfur dioxide. Adv Toxicol 407368

Yoon JY, Hamayun M, Lee SK, Lee IJ (2009) Methyl jasmonate alleviated salinity stress in soybean. J Crop Sci Biotechnol 12:63–68

Younesi O, Moradi A (2015) Effect of priming of seeds of *Medicago sativa* 'bami' with gibberellic acid on germination, seedlings growth and antioxidant enzymes activity under salinity stress. J Hortic Res 22:167–174

Yucel NC, Heybet EH (2016) Salicylic acid and calcium treatments improves wheat vigor, lipids and phenolics under high salinity. Acta Chim Slov 63:738–746

Yurekli F, Porgali ZB, Turkan I (2004) Variations in abscisic acid, indole-3-acetic acid, gibberellic acid and zeatin concentrations in two bean species subjected to salt stress. Acta Bio Cracov Ser Bot 46:201–212

Zanganeh R, Jamei R, Rahmani F (2018) Impacts of seed priming with salicylic acid and sodium hydrosulfide on possible metabolic pathway of two amino acids in maize plant under lead stress. Mol Biol Res Commun 7(2):83–88

Zavariyan A, Rad M, Asghari M (2015) Effect of seed priming by potassium nitrate on germination and biochemical indices in *Silybum marianum* L. under salinity stress. Int J Life Sci 9:23–29

Zhang S, Gao J, Song J, Zhang SG, Gao JY, Song JZ (1999) Effects of salicylic acid and aspirin on wheat seed germination under salt stress. Plant Physiol 35:29–32

Zhang S, Hu J, Zhang Y, Xie XJ, Knapp A (2007) Seed priming with brassinolide improves lucerne (Medicago sativa L.) seed germination and seedling growth in relation to physiological changes under salinity stress. Aust J Agric Res 58(8):811

Zhang HJ, Zhang N, Yang RC, Wang L, Sun QQ, Li DB, Cao YY, Weeda S, Zhao B, Ren S, Guo YD (2014) Melatonin promotes seed germination under high salinity by regulating antioxidant systems, ABA and $GA_4$ interaction in cucumber (*Cucumis sativus* L.). J Pineal Res 57:269–279

Zhang F, Yu J, Johnston CR, Wang Y, Zhu K, Lu F, Zhang Z, Zou J (2015) Seed priming with polyethylene glycol induces physiological changes in sorghum (*Sorghum bicolor* L. Moench) seedlings under suboptimal soil moisture environments. PLoS One 10:e0140620

Zheng M, Tao Y, Hussain S, Jiang Q, Peng S, Huang J, Cui K, Nie L (2016) Seed priming in dry direct-seeded rice: consequences for emergence, seedling growth and associated metabolic events under drought stress. Plant Growth Regul 78:167–178

Zhong YP, Wang B, Yan JH, Cheng LJ, Yao LM, Xiao L, Wu TL (2014) DL-β-amino butyric acid induced resistance in soybean against *Aphis glycines* Matsumura (*Hemiptera: Aphididae*). PLoS One 9(1):1–11

Zhu S, Zhang X, Luo T, Liu Q, Tang Z, Jing Z (2011) Effects of NaCl stress on seed germination, early seedling growth and physiological characteristics of cauliflower (*Brassica oleracea* L. var *botrytis* L.). Afr J Biotechnol 10(78):17940–17947

Zimmerli L, Hou BH, Tsai CH, Jakab G, Mauch-Mani B, Omerville S (2008) The xenobiotic beta-amino butyric acid enhances *Arabidopsis* thermo-tolerance. Plant J 53(1):144–156

# Seedling Pretreatment: Methods and Protocols

**Abdul Rehman, Komal Mazhar, Muhammad Ijaz, Qasim Ali, and Shakeel Ahmad**

## Contents



**Abstract** In the perspective of seed technology, physical methods that were generally used to increase the plant production provide advantages as compared to conventional methods that use chemical substances. The impacts of physical treatments for invigoration in seeds tended to be addressed at different levels that include changes in the gene expression, the structural and morphological aspects, and metabolite or protein accumulation. The physical methods used for the pre-sowing treatments are irradiation with microwaves (MWs), magneto-priming, and ionizing radiations (IRs). "Magneto-priming" depends upon the use of magnetic fields and portrayed as a cheap, noninvasive, and eco-friendly technique with demonstrated gainful impacts on vigor, crop yield, and seed germination. Radiation such as gamma rays and X-rays has been generally viewed as a powerful contrivance in

A. Rehman (✉) · M. Ijaz · Q. Ali
College of Agriculture, Bahauddin Zakariya University, Multan, Pakistan
e-mail: abdulpbg@bzu.edu.pk

K. Mazhar
Department of Forestry and Range Management, University of Agriculture Faisalabad, Faisalabad, Pakistan

S. Ahmad (✉)
Department of Agronomy, Bahauddin Zakariya University, Multan, Punjab, Pakistan
e-mail: shakeelahmad@bzu.edu.pk



M. Hasanuzzaman, V. Fotopoulos (eds.), *Priming and Pretreatment of Seeds and Seedlings*, https://doi.org/10.1007/978-981-13-8625-1_5

food technology and agricultural sciences. Positive impacts on seedling vigor, health, and germination were aroused by ultraviolet radiations (UV) that include UV-C and UV-A. Among all the identified physical treatments, a connected and basic research is yet expected to characterize the genotype, optimal dose required, and irradiation conditions that all depend on environment. Electron spin resonance has a huge prospective in regard to seed technology not correctly scrutinized to a point in time to end the treatment. Physical methods for the seed invigoration were described in the current chapter as well as emphasized to address the needs of seed technologists, trade makers, and producers by using the physical methods.

**Keywords** Seedling hardening · Chemical priming · Hydropriming · Phytohormones · Cross-tolerance

## 1 Introduction

The greatest difficulties that humanity is confronting are enhancing the supportability of agriculture while decreasing its ecological effect, to meet the food demand of the developing worldwide population (Edmondson et al. 2014). The impression of agricultural manageability depends on creative innovations and new methods with no antagonistic consequences on nonmarket goods.

Seeds with high vigor are intermediary of sustainable productivity and establishment of crop. The estimated seed market of the European Union in 2012 stretched up to 7 billion Euros as compared to 20% of world market, positioning third after China and the USA. Chemical compounds were generally used in modern agriculture, while the utilization of physical methods may speak in this monetarily focused on setting, inventive biotech/molecular tools, products, and treatments which are critical to accelerate the consolidation procedure of the seed industry.

Despite the extensive role of chemical methods in modern agriculture, the physical methods can also be able to play an alternative role to boost up the production in agriculture and can also be used for the storage and plant protection (Aladjadjiyan 2012). More benefits were obtained by adopting physical methods for seed invigoration as compared to the extensively used chemical treatments that involve the use of chemical substances. One of the advantages of physical methods is that they decrease the use of fertilizers, thereby lessening the contamination of the raw materials produced on the farm. Another benefit provided by the physical method is the disinfection of the seed during storage and before sowing (Aladjadjiyan 2012). Conceivable methodologies include ionizing radiation (IR), electromagnetic waves (EW), ultrasound waves (UW), and magnetic fields (MF). Extensively studied topic of the twentieth century is the treatment of the food using irradiation to ensure the greater shelflife and microbiological security; as a result, there is more concern towards the use of irradiation for the security of the food supply (Farkas and Mohácsi-Farkas 2011). The space research has shown great interest in studying the potential effect of irradiation on living organisms. Experiments revealed that Gamma rays and X-rays impede the growth of crop plant (Wolff et al. 2014; Arena et al. 2014). Because of such restriction, the cultivation of higher plants in space was impossible without new agriculture-related technologies that help to understand the possible effect of space factors on the environment but also on the development of greenhouse (De Micco et al. 2014). There is also a need to address the problems related to the development and the growth of plants in such extraterrestrial environment (Wolff et al. 2014).

Effect of physical treatments for invigoration in seeds tended to be addressed at different levels that include changes in the gene expression. Plant system serves as exceptional representations of the biological response to various factors related to environment that include the mutation in the cell due to certain chemical and physical substances (Macovei et al. 2014; Zaka et al. 2002; Confalonieri et al. 2014; Balestrazzi et al. 2011).

Effect of physical treatment on plants depends upon various factors such as MF or radiation (complete dose, its rate and type) and plant characteristics, i.e., cultivar, species, age, ploidy, and complexity of the target tissue or organ (De Micco et al. 2014). Recently, many interesting reports were published in toxological sciences about the usage of very minute concentation of toxins for the purpose of stimulation which is called hormesis (Belz and Piepho 2012; Calabrese and Baldwin 2000). It has been stated that germination, production of the photosynthetic pigment, and plant growth were stimulated by doing seed irradiation (Macovei et al. 2014; Kovács and Keresztes 2002; De Micco et al. 2014). However, the reports also stated the disastrous impacts of these irradiations were also seen in case of long exposure.

Priming is a well-recognized treatment that enhances the quality of the seed by temporarily stimulating the pre-germinative metabolism that in turn activates the processes related to DNA repair and antioxidant functions (Paparella et al. 2015). Priming of seed has arisen as an effective method to increase seed germination and seed vigor; seed priming also proved to be an effective method during field establishment and for seed growth during adversative environmental condition (Hussain et al. 2015; Ventura et al. 2012). By using treatments like osmopriming and hydropriming, the low vigor seed can be cured. In the perspective of seed technology, physical methods were proved to be more efficient when compared to the methods that were conventionally used (Bilalis et al. 2012). Extensively used physical method in the field of agriculture for pre-sowing treatment of seed was the usage of magnetic field (Aguilar et al. 2009). Magnetic field technique was generally noninvasive, cheap, and eco-friendly (Efthimiadou et al. 2014; Bilalis et al. 2012). This review will also address other physical treatments like UV radiation, microwave radiation, gamma rays, and X-rays.

The current manuscript addresses some of the extensively used physical methods of seed treatment and highlighted the possible advantages and disadvantages of these physical methods. Also, the future perception related to the use of the technique will also be discussed to meet the needs of the trade makers, consumers, and seed producers.

## 2 Magneto-Priming: An Appropriate Invigoration Protocol

The effects produced by the magnetic field in living organism are extensively studied. Magnetic field of the Earth is (50 μT) as stated by Belyavskaya (2004). The unit of magnetic flux density 1 tesla (T) is equal to 1 kg $s^{-2}$ $A^{-1}$; here, A stands for Ampere that is defined as electric current flowing with a charge of 1 coulomb per second. For the first time, the impacts of magnetic field on plants were described by Krylov and Tarakanova (1960); the phenomenon of magnetotropism was described as the auxin-like impact produced on the germinating seed by magnetic field. The techniques used in agriculture for the purpose of seed priming are electromagnetic field (EMF) and static magnetic field (SMF) that produced positive effects on crop yield, seed vigor, and seed germination (Baby et al. 2011). Static magnetic field (SMF) is considerably different from electromagnetic field (EMF). Static magnetic field can only be produced by a permanent magnet as produced by earth magnetic field or by the help of industrial processes; electromagnetic field is produced by electrically charged entities being extended indeterminately in space (Mitchell and Cambrosio 1997).

In the perspective of seed technology, the possible effects exerted by magnetic field were studied in order to derive the full potential of the magnetic field for the new seed treatments. The beneficial impacts produced by the magnetic field in order to improve the accumulation of the biomass and germination parameters were discussed for a wide range of plants by Teixeira da Silva and Dobránszki (2015). The current studies tested range of the magnetic field from 0 to 300 mT. Seeds that were primed with magnets show improved vigor, germination rates, root development, and seedling biomass. Tolerance to abiotic and biotic stress was increased by using magnetic field treatment (De Souza et al. 2006; Anand et al. 2012). Bhardwaj et al. (2012) described the enhanced antioxidant activities of catalase (CAT), glutathione reductase (GR), and superoxide dismutase (SOD) in magneto-primed cucumber (*Cucumis sativus* L.) seeds. Soybean (*Glycine max* (L.) Merr. var. Js-335) seeds that were magneto-primed produce lower amount of superoxide radicals ( $O_2^{\bullet-}$ ) (Baby et al. 2011). Therefore, magnetic field treatments have the added potentiality to be used for diminishing the disease-induced or drought effects on the productivity of crop.

On the basis of underlying assumptions, a tremendous interest was generated to study the biochemical, physiological, and molecular mechanisms involved to gradually improve the performance of the plants with magneto-primed seed. (Baby et al. 2011; Vashisth and Nagarajan 2010; Anand et al. 2012; Javed et al. 2011; Table 1) like sunflower (*Helianthus annuus* L.) seeds. Magnetic field acts as a priming agent in such a way to enhance the performance of the seeds. Seeds that were treated with magnetic field show enhanced protease, dehydrogenase, and alpha-amylase activities as compared to nontreated seeds. However, higher activity level of hydrolyzing enzymes was seen in seeds treated with static magnetic field (SMF) that related to the seedling vigor, germination, and rooting traits.

A very little information is available about the mechanism of plant's responce against magnetic field and their signal transduction pathway. Blue light photorecep-

**Table 1** Magnetic fields (MFs) effects on seed and seedling performance

| Species | Magnetic fields (MFs) intensity | Effects described | References |
|---|---|---|---|
| *Quercus suber* | 0.015 mT | Seedling biomass and sprouting improved | Celestino et al. (2000) |
| *Vicia faba* | 0.1 mT | Seedling growth improved | Rajendra et al. (2005) |
| *Solanum lycopersicum* | 100 and 170 mT | Increased biotic stress tolerance with improved growth and biomass | De Souza et al. (2006) |
| *Helianthus annuus* | 50 and 200 mT | Rate of germination increased Enhanced biomass accumulation and seedling length | Vashisth and Nagarajan (2010) |
| *Glycine max* | 200 and 150 mT | Improved efficiency for biomass accumulation and light harvesting | Baby et al. (2011) and Shine et al. (2011) |
| *Tagetes patula* | 100 mT | Enhanced germination, seedling emergence, and metabolism of starch | Afzal et al. (2012) |
| *Zea mays* | 100 and 200 mT | Enhanced photosynthesis, water status of leaf and seedling growth drought stress | Anand et al. (2012) |
| *Cucumis sativus* | 200 mT | Enhanced seedling emergence, germination, starch and anti-oxidative metabolism | Bhardwaj et al. (2012) |
| *Vigna radiate* | 5 mT | Enhanced seedling vigor, germination, anti-oxidative, and starch metabolism | Reddy et al. (2012) |
| *Triticum aestivum* | 30 mT | Neither seed germination improved for seedling growth. Under soil flooding, enhanced antioxidant potential | Balakhnina et al. (2015) |

tors–cryptochromes were responsible for the signaling of magnetic field in plants (Ahmad et al., 2007). However, there is still a need to fully understand the mechanisms involved in magnetobiology and the possible genotoxic impacts involved in the usage of magnetic fields (Ghodbane et al. 2013). In order to increase the seed vigor, seed germination, and photosynthetic capacity of plants primed with magnetic fields, there is still the need to fully understand the mechanisms lying behind the process.

In horticultural crop, environmental factors like humidity, light, and temperature are associated with good performance of seed, but the impact produced upon the application of environmental factors with magnetic field still remains unclear. By using the resulting knowledge, new seed treatments that were also locally adapted can be developed. Poinapen et al. (2013) exhibited that how the collective effect of magnetic field and environmental factors affect seed performance and viability in tomato (*Solanum lycopersicum* L.) in laboratory. It was shown that, during early stages of seed germination, relative humidity played a crucial role in modulating the performance of magneto-primed seeds.

# 3 Ionizing Radiation Treatments

## *3.1 Gamma Radiation, Most Promising Technique*

Recent studies are usually based on ionizing radiation in agriculture for food security. Gamma (γ) radiation are irradiation of high energy type, so it penetrates into plant and foods easily (Jayawardena and Peiris 1988). Application of γ-rays is done usually using cobalt-60 (Moussa 2006) in the form of Gray (Gy), i.e., 1 Gy is equal to the amount of absorbed irradiation per kilogram. But when biological material was used for treatment then amount of irradiation count in sievert units. Dose rate can also be used as starting treatments. For high sterilization conditions, γ-rays are used such as microbial experiments or insect-pest pathogenic treatments. They are also used as pest prevention agent. In food security, γ-rays help to cease sprouting of vegetables and delay ripening of fruits by stopping the important enzymes and, thus, increase shelf life (Moussa 2006; Mokobia and Anomohanran 2005). Furthermore, γ-rays also cause mutation and are helpful tools in plant breeding for the purpose of new traits and varieties' improvement (Irfaq and Nawab 2001).

Among the diverse radiobiology features, the topic that is gaining remarkable attention is the depiction of the gamma rays' effects on seeds. Commencing study mainly focused on the impact of low-dose rate on seeds and their germination percentage. Gamma rays directly affect the proteins, nucleic acid, and cell membranes at any stage (Kovács and Keresztes 2002). However, an unintended action is described by the production of reactive oxygen species (ROS) as of water radiolysis (Esnault et al. 2010; Borzouei et al. 2010). Reactive oxygen species have the ability to damage and diffuse the cellular organelles and macromolecules. Nonetheless, strength of γ-rays depends upon dose rate, duration, and intensity. Gamma rays when delivered enhance the establishment of seedling and germination percentage and thus act as a priming agent.

In seed technology field, gamma irradiations are used for seed invigoration and as dynamic force for inspecting the molecular mechanism which involves the activation of seed response to the available physical treatment. After that, the effect of gamma rays was studied on seeds of *Arabidopsis thaliana* for evaluating the biological responses (Qi et al. 2015). Germination index was stimulated when seed is irradiated with total doses that were lower than 100 Gy. Similarly, the growth rate of seedling, root length, and weight were found to be significantly increased as compared to control. Among the tested doses of irradiation, *A. thaliana* seeds with 50 Gy dose showed the high positive impacts in all growth parameters. With γ-rays, it was also proved that it enhances all the seedling parameters along with germination performance in crops as well as woody plants. Maity et al. (2005) examined γ-ray-induced impacts on *Phaseolus mungo* L. and *Oryza sativa* L. cv-2233 seeds. Range of the radiation applied is 50–350 Gy dose. Irradiation with less doses helps to enhance the morphological traits, i.e., shoot number, number of seed per panicle, plant height, and panicle length. However, high-dose exposure is reported to bring negative effects on the seedling growth parameters. For example, in rice, the greater stimulatory dose matched to 50 Gy, while *P. mungo* showed improvement with 200 Gy radiations. Their effects on maize were also considered (Marcu et al. 2013).

A test was conducted to compare radiation sensitivity of plant germination rate and photosynthetic pigment contents among different seeds conditions. The triggering effects of gamma rays were also observed at 2–30 Gy as low doses and 70 Gy as high doses presented to be detrimental to plant response. Through all research, it was clear that low dose of irradiation has a constructive output on yield traits in crops, as well on the seedling establishment and germination. Two pre-sowing treatments, i.e., gamma radiation (300, 400, and 500 Gy) and simple hydropriming, were applied on okra and activated the metabolic reaction. Both treatments were applied for the improvement of seedling conditions and germination rate. Application of γ-rays with 400 Gy showed superior results in germination, quality of seeds, and yield along with photosynthetic capacity (Hegazi and Hamideldin 2010). Hence, results showed that the efficacy of dose and radiation on germination and yield traits relents on seed species.

γ-rays treatments could not only be used for crops species but also for the conservation of native species. For this purpose, seeds of *Moluccella laevis* L. were tested on seedling stage (Minisi et al. 2013). Corresponding to previous explanations, γ-rays treated at 5 Gy as minute dose increase the percentage of germination than untreated samples or those treated with up to 20 Gy. Low-dose γ-rays also have the ability to stimulate seed vigor, germination, and growth of seedling in deadly nightshade (Abdel-Hady et al. 2008), wild oat (*Avena fatua* L.) (Maherchandani 1975; Majeed et al. 2010), garden cress (*Lepidium sativum* L.) (Dubey et al. 2007), and okra (*Abelmoschus esculentus* L. Monch.) (Moussa 2006). Collective evidences were similar in all these works that the less number of doses of γ-rays resulted in advantageous action in seeds that are physically treated, which hysterics the definition of hormesis (Luckey 1980, 2006). The threshold of low dose is typically placed just above the background of natural radiation levels (2.4 millisievert per year) and demarcated by the border between bionegative and biopositive effects. Hormesis signifies an adaptive retort of organisms and cells to intermittent and moderate stresses, thus provoking the initiation of repair mechanism and cellular defense against the stress-causing agent, or else missing in the lack of stress (Calabrese and Baldwin 2000; Mattson 2008).

Molecular and biological processes which can be benifited by radiation hormesis are still under argued. Various researchers pointed out reactive oxygen species production in seed in response to γ-rays, subsequently they acted as signaling molecules and antioxidant responses in stress. Resultantly, radiation-treated plant can tolerate the common stresses, i.e., water loss, temperature variation, and light intensity during growth (Gicquel et al. 2012). Abdel-Hady et al. (2008) and Sjodin (1962) appealed that low-dose gamma irradiation enhanced enzyme activity as well as proteins and nucleic acid synthesis in seeds.

Constructed on this, it is clearly described that significant increase is achieved in seed metabolism which further breaks dormancy and improves germination of seeds. Study was also conducted for the evaluation of chronic and severe gamma irradiation effects on genetic stability and expression in seeds (Kovalchuk et al. 2007). In retort to acute radiation, the plant metabolism is moved to activation of mechanisms for pro-survival, immediate repair of the damage, and perhaps the inhibition of cell differentiation/cell division. In distinction, chronic irradiation activated an alternative response, based on gene expression that involved in general stress and metabolism of nucleic is motivated. Furthermore, the output of the following study also expresses the variation in methylation pattern and chromatin modification, which are interlinked by chronic radiation.

Through all these researches, it was obvious that efficient results were attained by low-dose treatment of γ radiations in seed technology with low-dose treatment which gives efficient results. Nonetheless, additional studies are desirable to comprehend the molecular base evaluation, essential in the enhanced growth/response detected in exposed seeds. In spite of the advantageous features cited, γ-irradiation treatments in a seed equipment perception are subject to the proper formation of γ radiation conveniences able to graft at an industrial scale. Solicitation of this seed invigoration method still involves a widespread research imperative to describe the optimum conditions (dose rate, total dose) which should be contingent to genotype, crop species, and environmental background.

### *3.2 How Are Seedling Development and Seed Germination Affected by X-Rays*

Impact of X-rays on living organisms has not yet been completely assumed. The wavelength of X-rays from 0.01–10nm in range of electromagnetic spectrum and carry energies between 120 eV-120 KeV. In agriculture, Lenient X-rays from 0.12 to 12 keV are considered best for agriculture products (Kotwaliwale et al. 2014). In contrast, in humans, X-rays have broad-spectrum use in medical as well as in research on plants (Beard et al. 1958).

According to De Micco et al. (2014) and Einset and Collins (2015) after the 1960s, various researches have been published on the seed treated with X-ray radiations. Cumulative X-ray irradiation doses have been found to have a negative impact on root growth and germination of date palm seeds (Al-Enezi et al. 2012). On the contrary, researchers also found out stimulatory impact on growth of leaf of x-ray irradiation with 0.56 Gy radiation level. However, in tomato, with treatments of X-rays on seeds, significant increases in anatomical and ecophysiological characteristics are reported (De Micco et al. 2014). The complete results presented that development and germination of functional leaves were not dependent on doses.

In addition, during seed treatment with high dose of X-ray radiation, there are minor changes in leaves structure that structurally resulted in poor photosynthetic efficiency. The documentation of candidate species that were radiation-resistant can be a significant achievement in agricultural research (Arena et al. 2014).

Previously, very little information have been presented about the effects of X-rays on crops due to focusing only on plant physiology. New evidences related to physiological and molecular mechanisms essential for the resistance of plant tissues to X-rays could be revealed by using profiling techniques that were globally used (e.g., Omics).

## 3.3 Seeds Treatment with Ultraviolet Radiation

With the diminution of stratosphere ozone layer, assessment of solar ultraviolet (UV) radiation to the earth is more. Numerous studies showed the effect of UV radiation on plant and ecosystem (Kovács and Keresztes 2002; Heisler et al. 2003). A few studies reported the impact of UV radiation on seed germination and its biology (Table 2).

Since radiation is exceptionally destructive to organisms, it is not recommended for an organism under natural conditions (Hollósy 2002). Furthermore, this UV-C radiation is able to penetrate rapidly into plant tissues and is nonionizing, which maintains its prospective as a germicidal agent. Cabbage is treated with low dose of UV-C and resisted to black rot (Brown et al. 2001). Again, UV-C also has great impact on growth parameters and quality in green house. Pre-sowing treatment of lettuce with UV-C resulted in high salt stress. So, the consequences exhibited that seedlings treated with UV-C were competent to alleviate the effect of extreme salinity; it might be possible due to the enhanced free radical foraging activity in leaf tissues (Ouhibi et al. 2014).

In spite of UV-B radiation containing only 1.5% of the total spectrum, physiology of crops is also affected by radiation, including damaging of DNA, protein, and membranes resulting in restriction of growth of plant and thus reduction in photosynthesis (Choudhary and Agrawal 2014; Hideg et al. 2013). Their impacts are studied in mash bean for seedling growth, germination, and development

**Table 2** Ultraviolet radiations (UVs) effects on seed and seedling performance

| Species | UV type | Effects | References |
|---|---|---|---|
| *Carthamus tinctorius* | UV-B | Hampered seedling growth with improved rate of germination | Farokh et al. (2010) |
| *Arachis hypogaea* | UV-C | Improved seedling vigor and germination rate. Tolerance in seedlings against root infecting fungi | Siddiqui et al. (2011) |
| *Vigna radiata* | UV-C; UV-A | Improved seedling vigor and germination rate. Tolerance in seedlings against root-infecting fungi | Hamid and Jawaid (2011) and Siddiqui et al. (2011) |
| *Vigna mungo* | UV-B | Reduced germination in hydrated seeds, suppressed photosynthetic pigments, also decrease in root and shoot development | Shaukat et al. (2013) |
| *Phaseolus vulgaris* | UV-C | Improved rate of germination as well as bioactive molecules in seed | Guajardo-Flores et al. (2014) |
| *Lactuca sativa* | UV-C | Improved stress tolerance against NaCl | Ouhibi et al. (2014) |

(Shaukat et al. 2013). In contrast, authors described improved results in germination rate with UV-B light but not for germination percentage which remained same as earlier. Besides, many harmful results are clear using UV-B light, i.e., decreased length of root and shoot. Biochemically, it increases the amount of total soluble phenols and activity of tyrosine ammonia lyase.

UV-A radiation signifies about 6.3% of the received solar radiation and is considered as less dangerous (Hollósy 2002). Hamid and Jawaid (2011) examined the impacts of UV-A radiation on the seeds of mung bean and the results were auspicious and exhibited the improved results in germination rate and seedling performance with root–shoots length and leaf area. This is the only study which is accessible to the researcher for further research. However, a deeper insight is still requested to elucidate about the molecular mechanisms underlying the improvements occurring in the UV-treated seeds.

## 3.4 Microwaves Potentialities in Seed Technology

The main component of electromagnetic spectrum is microwave having frequency ranging from 300 MHz-300 GHz. So the electromagnetic radiation has both thermal and nonthermal effects in biological systems (Banik et al. 2003). Furthermore, different biological effects are caused by microwave such as wave forms, frequencies, duration of exposures, and modulation (Vian et al. 2006). Meanwhile, various studies show that a very limited literature is available on the effects of microwaves on humans and animals (Jayasanka and Asaeda 2014). When living tissue absorbed MW radiation, it brings dipole rotation in electron orbit and, hence, leads to the fast and selective heating. Interestingly, present studies focus on 2.45 Ghz radiations, which is occupied by molecule of water in living cells (Iuliana et al. 2013).

In seed technology before sowing and storing of seed, nonlethal MWs treatments have been widely used (Knox et al. 2013). Remarkably, growth of weeds can be prevented by MWs treatments (Sahin 2014). So, MW treatment is an alternative way of chemical control of weeds because it has no hazardous effect on environment; therefore, it is valid on greenhouse of horticulture and some ornamentals nurseries.

Nowadays, big problem in the field of agriculture is poor seed germination, but the seed invigoration treatment is the best method for enhancement rate of germination (Table 3). Effects of wireless routers, MWs, and mobile devices, were also studied in three aromatic species, i.e., Parsley, celery, and dill (Soran et al. 2014). So the seeds of these three species were bare and used to assess the effects on leaf anatomy, volatile emissions, and essential oil content. Results revealed effect of MWs on three species-induced chemical and structural modifications.

**Table 3** Microwaves (MWs) effects on seed and seedling performance

| Species | MWs intensity | Effects | References |
|---|---|---|---|
| *Triticum aestivum* | 2.45 GHz | Reduced seedborne infestation of *Fusarium graminearum* (wheat head blight fungus) and also reduction in seed vigor | Reddy et al. (1998) |
| *Raphanus sativus* | 10.5 and 12.5 GHz | Rate of germination is reduced and also decrease in rate of hypocotyl growth | Scialabba and Tamburello (2002) |
| *Lens culinaris* | 2.45 GHz | No effect on rate of germination. Length of seedling is stimulated 30 s exposure | Aladjadjiyan (2010) |
| *Solanum tuberosum* | 2.45–54 GHz | Highest improved biomass accumulation in seed potato germs | Jakubowski (2010) |
| *Hordeum vulgare* | 2.45 GHz | Improved vigor and seedling germination | Iuliana et al. (2013) |
| *Oryza sativa* | 2.45 GHz | Improved germination rate, similarly, an increase in the length of primary root and shoot length | Talei et al. (2013) |

# 4 Effects of Physical Seed Invigoration Studies by Physical Methods

Through electron paramagnetic resonance (EPR), which analysis of short lived species, i.e., reactive oxygen species for their quantitative and qualitative, can be measured? This potent tool, presently known as one of the most sensitive and specific, upswings from the novel studies in quantum mechanics supported by Zeeman (1897). That becomes necessary to the research based on biological system with transition metals and organic radicals (Sahu et al. 2013).

Electron paramagnetic resonance active system comprises a molecular orbit containing single unpaired electron. In the absenteeism of MF, same energy is possessed by the two quantum states. When a magnetic field is activated, increase in the energy state with decrease in the energy of $-1/2$ state depends upon the magnetic field strength. Single unpaired electrons were able to change their orbit state, and thus, actions related to the energy absorption are supervised and then converted into a spectrum (Weil and Bolton 2006). By using trace molecules, i.e., spin probes, EPR resolution can be further improved, which are artificially presented in the biological systems that are targeted. With the ability of stable diamagnetic tracer, molecules are capable to fix the momentary reactive oxygen species and instigate long-lived paramagnetic spin adducts, i.e., radical species. The second gather in cells during hours or minutes enables detection through EPR (Sahu et al. 2013). Electron paramagnetic resonance spectroscopy is presently used *in planta* to dichotomize and comprehend the varieties of functions played by reactive oxygen species in stress response and plant growth regulation mechanism (Steffen-Heins and Steffens 2015).

The prospective of this instrument for the purpose of high-resolution summarizing of radicals in seeds is still abundantly exploited. EPR is ominously added to comprehend the role of free radicals in seed weakening. In soybean seeds, an unknown organic radical was identified through electron paramagnetic resonance and then further confirmed and dated in dried seeds of maize (Leprince et al. 1995). Primarily at the seed level, effect of temperature and oxygen on reactive oxygen species (ROS)-induced wound was also exposed by EPR spectra. A fascinating feature of seed practicability discovered by electron paramagnetic resonance is connected to variations in thickness and movement of molecules stirring when the cytoplasm arrives in a shiny state. Meanwhile, when seeds are stored under low temperature, the distinctiveness takes place and it has been supervised by the help of spin probes. The molecular mobility of spin probes can be slow or fast reliant on their dissemination among different seed constituents. Nitroxide spin probes, undeniably, were effectively used to inaugurate an association between occurrence of intracellular glasses, molecular mobility, and stability of seed storage (Hoekstra and Leprince 1998; Buitink et al. 1998, 1999). Temperature-dependent and moisture changes stirring in pea and *Impatiens* seeds are connected through rate of aging and that is verified in cytosol because of the revolving motion of a spin probe (Buitink et al. 2000).

Mostly, scientists used a nitroxide spin probe in wheat embryos to check if solid EPR was efficaciously used for efficient examination of seeds for plasma membranes reliability (Smirnov et al. 1992). So, result revealed greater than 13% moisture in the semipermeable cell wall of seed. Furthermore, study on seeds of wheat also validated the impacts of aging of the embryo (Golovina et al. 1997). Also, it reveals that permeability of plasma membrane improved more quickly during seed aging. The strategic role of permeability of membrane precision during imbibition of seeds of neem (*Azadirachta indica* L.) was also examined by spin probe technique using nitroxyl (Sacandè et al. 2001).

Species profiles of paramagnetic are developing as favorable to screen against stress tolerance in grains. Large values of semiquinone with carbohydrate radicals were associated with seeds that are water stress sensitive and also linked with greater starch contents by biochemical tests. Researches also revealed that stress tolerance can be identified by EPR radical profiles. Likewise, Kurdziel et al. (2015) used EPR to examine wheat grains of four genotypes through free radical profiles in embryo, endosperm, and seed coat and found various levels of water stress tolerance. Drought-sensitive genotypes having more carbohydrates are identified as selectable marker under water stress tolerance. EPR was also used to examine the effect of ozone application on barley, oat, and wheat grains against oxidative stress tolerance (Labanowska et al. 2016). Electron paramagnetic resonance exposed that the number of paramagnetic species like Cu (II), Fe (III), and Mn (II), altered upon exposure of ozone, is also dependent on the tolerance behavior of particular genotype. Furthermore, the patterns of organic radicals that are stable (carbohydrate semiquinone and phenoxyl) suggestively changed ozone treatment as given, presenting improvement of paramagnetic species of the cultivars that are tolerant. The information assimilated over EPR constructed studies in seeds or food processing as revealed previously, sustenance the method to inspect the impacts of physical invigoration methods for seed treatments. Marcu et al. (2013) used EPR to study the biochemical changes induced in maize seeds when irradiated with cumulative doses of $\gamma$-radiation.

## 5 Conclusion

Seeds with extraordinary potency are substitution of crop formation and regular output. By using physical methods, seed vigor and germination are enhanced because physical method is eco-friendly and provides benefit on large scale. The presence of suitable services to achieve the physical treatment on the preemergence metabolism is obstructing its efficiency on priming treatment applications. It is also exhibited during evaluation that radiation treatment has a very little effect on biochemical changes or gene expression. Hence, the deficiency of biomarkers related to appropriate time or radiation dose restricts the application of seed procedures at large scale. Nonetheless, scientists from industry and academia are presently attentive on incapacitating these problems. Recent improvements on high-output methods (e.g., Omics) as compared with latest discharge of genomic properties on crops are predictable to lead. Additional studies are looked for to identify the molecular players generated during the riposte of seed invigoration treatments, specifically in radio-tolerant cultivars or species. In comparison, by intensifying the total cultivars analyzed with each changed method, it will be conceivable to recognize best appropriate goals for a physical treatment that is specific in stopping harmful conditions. The environmental situations in moderating the reaction of seed to the given radioactivity behaviors could not be deserted, establishing itself as an auspicious research opportunity. This information is vital to improve new plans to enterprise new treatments to modify and progress seed invigoration and germination. A combined method is required to speed up translational and basic research related to seed expertise, thus, at the end, fabricating strategies for the seed operatives.

## References

Abdel-Hady MS, Okasha EM, Soliman SSA, Talaat M (2008) Effect of gamma radiation and gibberellic acid on germination and alkaloid production in *Atropa belladonna* L. Aust J Basic Appl Sci 2:401–405

Afzal I, Mukhtar K, Qasim M, Basra SMA, Shahid M, Haq Z (2012) Magnetic stimulation of marigold seed. Int Agrophys 26:335–339. https://doi.org/10.2478/v10247-012-0047-1

Aguilar CH, Dominguez-Pacheco A, Carballo AC, Cruz-Orea A, Ivanov R, Bonilla JLL, Montañez JPV (2009) Alternating magnetic field irradiation effects. Acta Agrophysica 14:7–17

Ahmad M, Galland P, Ritz T, Wiltschko R, Wiltschko W (2007) Magnetic intensity affects cryptochrome-dependent responses in *Arabidopsis thaliana*. Planta 225:615–624. https://doi.org/10.1007/s00425-006-0383

Al-Enezi N, Al-Bahrany A, Al-Khayri J (2012) Effect of X-irradiation on date palm seed germination and seedling growth. Emir J Food Agric 24:415–424

Aladjadjiyan A (2010) Effect of microwave irradiation on seeds of lentils (Lens Culinaris, Med.). Rom J Biophys 20:213–221

Aladjadjiyan A (2012) Physical factors for plant growth stimulation improve food quality. In: Aladjadjiyan A (ed) Food production – approaches, challenges and tasks. InTech, Rijeka, pp 145–168

Anand A, Nagarajan S, Verma APS, Joshi DK, Pathak PC, Bhardwaj J (2012) Pre-treatment of seeds with static magnetic field ameliorates soil water stress in seedlings of maize (*Zea mays* L.). Indian J Biochem Biophys 49:63–70
Arena C, De Micco V, Macaeva E, Quintens R (2014) Space radiation effects on plant and mammalian cells. Acta Astronaut 104:419–431. https://doi.org/10.1016/j.actaastro.2014.05.005
Baby SM, Narayanaswamy GK, Anand A (2011) Superoxide radical production and performance index of photosystem II in leaves from magnetoprimed soybean seeds. Plant Signal Behav 6:1635–1637. https://doi.org/10.4161/psb.6.11.17720
Balakhnina T, Bulak P, Nosalewicz M, Pietruszewski S, Włodarczyk T (2015) The influence of wheat *Triticum aestivum* L. seed pre-sowing treatment with magnetic fields on germination, seedling growth, and antioxidant potential under optimal soil watering and flooding. Acta Physiol Plant 37:59. https://doi.org/10.1007/s11738-015-1802-2
Balestrazzi A, Confalonieri M, Macovei A, Carbonera D (2011) Seed imbibition in *Medicago truncatula* Gaertn: expression profiles of DNA repair genes in relation to PEG-mediated stress. J Plant Physiol 168:706–713. https://doi.org/10.1016/j.jplph.2010.10.008
Banik S, Bandyopadhyay S, Ganguly S (2003) Bioeffects of microwave – a brief review. Bioresour Technol 87:155–159. https://doi.org/10.1016/S0960-8524(02)00169-4
Beard BH, Haskins FA, Gardner CO (1958) Comparison of effects of X-rays and thermal neutrons on dormant seeds of barley, maize, mustard, and safflower. Genetics 43:728–736
Belyavskaya NA (2004) Biological effects due to weak magnetic field on plants. Adv Space Res 34:1566–1574. https://doi.org/10.1016/j.asr.2004.01.021
Belz RG, Piepho H-P (2012) Modeling effective dosages in hormetic dose-response studies. PLoS One 7:e33432. https://doi.org/10.1371/journal.pone.0033432
Bhardwaj J, Anand A, Nagarajan S (2012) Biochemical and biophysical changes associated with magnetopriming in germinating cucumber seeds. Plant Physiol Biochem 57:67–73. https://doi.org/10.1016/j.plaphy.2012.05.008
Bilalis DJ, Katsenios N, Efthimiadou A, Karkanis A, Efthimiadis P (2012) Investigation of pulsed electromagnetic field as a novel organic pre-sowing method on germination and initial growth stages of cotton. Electromagn Biol Med 31:143–150. https://doi.org/10.3109/15368378.2011.624660
Borzouei A, Naseriyan B, Majdabadi A, Kafi M, Khazaei H (2010) Effects of gamma radiation on germination and physiological aspects of wheat (*Triticum aestivum* L.) seedlings. Pak J Bot 42:2281–2290
Brown JE, Lu TY, Stevens C, Khan VA, Lu JY, Wilson CL, Collins DJ, Wilson MA, Lgwegbe ECK, Chalutz E, Droby S (2001) The effect of low dose ultraviolet light-C seed treatment on induced resistance in cabbage to black rot (*Xanthomonas campestris* pv. campestris). Crop Prot 20:873–883. https://doi.org/10.1016/S0261-2194(01)00037-0
Buitink J, Claessens MAE, Hemminga MA, Hoekstra FA (1998) Influence of water content and temperature on molecular mobility and intracellular glasses in seeds and pollen. Plant Phys 118:531–554. https://doi.org/10.1104/pp.118.2.531
Buitink J, Hemminga MA, Hoekstra FA (1999) Characterization of molecular mobility in seed tissues: an electron paramagnetic resonance spin probe study. Biophys J 76:3315–3322. https://doi.org/10.1016/S0006-3495(99)77484-9
Buitink J, Leprince O, Hemminga MA, Hoekstra FA (2000) Molecular mobility in the cytoplasm: an approach to describe and predict lifespan of dry germplasm. Proc Natl Acad Sci U S A 97:2385–2390. https://doi.org/10.1073/pnas.040554797
Calabrese EJ, Baldwin LA (2000) Radiation hormesis: its historical foundations as a biological hypothesis. Hum Exp Toxicol 19:41–75. https://doi.org/10.1191/096032700678815585
Celestino C, Picazo ML, Toribio M (2000) Influence of chronic exposure to an electromagnetic field on germination and early growth of *Quercus suber* seeds: preliminary study. Electromagn Biol Med 19:115–120
Confalonieri M, Faè M, Balestrazzi A, Donà M, Macovei A, Valassi A, Giraffa G, Carbonera D (2014) Enhanced osmotic stress tolerance in *Medicago truncatula* plants overexpressing the

DNA repair gene MtTdp2α (tyrosyl-DNA phosphodiesterase 2). Plant Cell Tissue Organ Cult 116:187–203. https://doi.org/10.1007/s11240-013-0395-y

Choudhary KK, Agrawal SB (2014) Ultraviolet-B induced changes in morphological, physiological and biochemical parameters of two cultivars of pea (*Pisum sativum* L.). Ecotoxicol Environ Saf 100:178–187. https://doi.org/10.1016/j.ecoenv.2013.10.032

De Micco V, Paradiso R, Aronne G, De Pascale S, Quarto M, Arena C (2014) Leaf anatomy and photochemical behaviour of *Solanum lycopersicum* L. plants from seeds irradiated with low-LET ionising radiation. Sci World J 2014:428141. https://doi.org/10.1155/2014/428141

De Souza A, Garcí D, Sueiro L, Gilart F, Porras E, Licea L (2006) Pre-sowing magnetic treatments of tomato seeds increase the growth and yield of plants. Bioelectromagnetics 27:247–257. https://doi.org/10.1002/bem.20206

Dubey AK, Yadav JR, Singh B (2007) Studies on induced mutations by gamma irradiation in okra (*Abelmoschus esculentus* (L.) Monch.). Progress Agric 7:46–48

Edmondson JL, Davies ZG, Gaston KJ, Leake JR (2014) Urban cultivation in allotments maintains soil qualities adversely affected by conventional agriculture. J Appl Ecol 51:880–889. https://doi.org/10.1111/1365-2664.12254

Efthimiadou A, Katsenios N, Karkanis A, Papastylianou P, Triantafyllidis V, Travlos I, Bilalis DJ (2014) Effects of presowing pulsed electromagnetic treatment of tomato seed on growth, yield, and lycopene content. Sci World J 2014:369745. https://doi.org/10.1155/2014/369745

Einset J, Collins AR (2015) DNA repair after X-irradiation: lessons from plants. Mutagenesis 30:45–50. https://doi.org/10.1093/mutage/geu054

Esnault M-A, Legue F, Chenal C (2010) Ionizing radiation: advances in plant response. Environ Exp Bot 68:231–237. https://doi.org/10.1016/j.envexpbot.2010.01.007

Farkas J, Mohácsi-Farkas C (2011) History and future of food irradiation. Trends Food Sci Technol 22:121–126. https://doi.org/10.1016/j.tifs.2010.04.002

Farokh P, Mahmoodzadeh H, Satari T (2010) Response of seed germination of safflower to UV-B radiation. Res J Environ Sci 4:70–74. https://doi.org/10.3923/rjes.2010.70.74

Ghodbane S, Lahbib A, Sakly M, Abdelmelek H (2013) Bioeffects of static magnetic fields: oxidative stress, genotoxic effects, and cancer studies. Biomed Res Int 2013:602987. https://doi.org/10.1155/2013/602987

Gicquel M, Taconnat L, Renou JP, Esnault MA, Cabello-Hurtado F (2012) Kinetic transcriptomic approach revealed metabolic pathways and genotoxic-related changes implied in the *Arabidopsis* response to ionising radiations. Plant Sci 195:106–119. https://doi.org/10.1016/j.plantsci.2012.06.015

Golovina EA, Tikhonov AN, Hoekstra FA (1997) An electron paramagnetic resonance spin-probe study of membrane permeability changes with seed ageing. Plant Phys 114:383–389

Guajardo-Flores D, Serna-Guerrero D, Serna-Saldívar SO, Jacobo-Velázquez DA (2014) Effect of germination and UV-C radiation on the accumulation of flavonoids and saponins in black bean seed coats. Cereal Chem 91:276–279. https://doi.org/10.1094/CCHEM-08-13-0172-R

Hamid N, Jawaid F (2011) Influence of seed pre-treatment by UV-A and UV-C radiation on germination and growth of mung beans. Pakistan J Chem 1:164–167. https://doi.org/10.15228/2011.v01.i04.p04

Hegazi AZ, Hamideldin N (2010) The effect of gamma irradiation on enhancement of growth and seed yield of okra [*Abelmoschus esculentus* (L.) Monech] and associated molecular changes. J Hortic For 2:38–51

Heisler GM, Grant RH, Gao W, Slusser JR (2003) Ultraviolet radiation and its impacts on agriculture and forests. Agric For Meteorol 120:3–7. https://doi.org/10.1016/j.agrformet.2003.08.007

Hideg E, Jansen MAK, Strid A (2013) UV-B exposure, ROS, and stress: inseparable companions or loosely linked associates? Trends Plant Sci 18:107–115. https://doi.org/10.1016/j.tplants.2012.09.003

Hollósy F (2002) Effects of ultraviolet radiation on plant cells. Micron 33:179–197. https://doi.org/10.1016/S0968-4328(01)00011-7

Hussain S, Zheng M, Khan F, Khaliq A, Fahad S, Peng S, Huang J, Cui K, Nie L (2015) Benefits of rice seed priming are offset permanently by prolonged storage and the storage conditions. Sci Rep 5:8101. https://doi.org/10.1038/srep08101

Irfaq M, Nawab K (2001) Effect of gamma irradiation on some morphological characteristics of three wheat (*Triticum aestivum* L.) cultivars. J Biol Sci 1:935–937. https://doi.org/10.3923/jbs.2001.935.937

Iuliana C, Caprita R, Giancarla V, Sorina R (2013) Response of barley seedlings to microwaves at 2.45 GHz. Sci Pap Anim Sci Biotechnol 46:185–191

Jakubowski T (2010) The impact of microwave radiations at different frequencies on the weight of seed potato germs and crop of potato tubers. Agric Eng 6:57–64

Javed N, Ashraf M, Akram NA, Al-Qurainy F (2011) Alleviation of adverse effects of drought stress on growth and some potential physiological attributes in maize (*Zea mays* L.) by seed electromagnetic treatment. Photochem Photobiol 87:1354–1362. https://doi.org/10.1111/j.1751-1097.2011.00990.x

Jayasanka SMDH, Asaeda T (2014) The significance of microwaves in the environment and its effect on plants. Environ Rev 22:220–228. https://doi.org/10.1016/j.biortech.2015.02.055

Jayawardena SDL, Peiris R (1988) Food crops breeding in Sri Lanka - achievements and challenges. BIO News 4:22–34

Knox OGG, McHugh MJ, Fountaine JM, Havis ND (2013) Effects of microwaves on fungal pathogens of wheat seed. Crop Prot 50:12–16. https://doi.org/10.1016/j.cropro.2013.03.009

Kotwaliwale N, Singh K, Kalne A, Jha SN, Seth N, Kar A (2014) X-ray imaging methods for internal quality evaluation of agricultural produce. J Food Sci Technol 51:1–15. https://doi.org/10.1007/s13197-011-0485-y

Kovács E, Keresztes Á (2002) Effect of gamma and UV-B/C radiation on plant cells. Micron 33:199–210. https://doi.org/10.1016/S0968-4328(01)00012-9

Kovalchuk I, Molinier J, Yao Y, Arkhipov A, Kovalchuk O (2007) Transcriptome analysis reveals fundamental differences in plant response to acute and chronic exposure to ionizing radiation. Mutat Res 624:101–113. https://doi.org/10.1016/j.mrfmmm.2007.04.009

Krylov A, Tarakanova GA (1960) Magnetotropism of plants and its nature. Plant Phys 7:156–160

Kurdziel M, Dlubacz A, Weselucha-Birczynska A, Filek M, Labanowska M (2015) Stable radicals and biochemical compounds in embryos and endosperm of wheat grains differentiating sensitive and tolerant genotypes-EPR and Raman studies. J Plant Phys 183:95–107. https://doi.org/10.1016/j.jplph.2015.05.018

Labanowska M, Kurdziel M, Filek M (2016) Changes of paramagnetic species in cereal grains upon short-term ozone action as a marker of oxidative stress tolerance. J Plant Physiol 190:54–66. https://doi.org/10.1016/j.jplph.2015.10.011

Leprince O, Hoekstra FA (1998) The responses of cytochrome redox state and energy metabolism to dehydration support a role for cytoplasmic viscosity in desiccation tolerance. Plant Physiol 1181253–1181264. https://doi.org/10.1104/pp.118.4.1253

Leprince O, Vertucci CW, Hendry GAF, Atherton NM (1995) The expression of desiccation-induced damage in orthodox seeds is a function of oxygen and temperature. Physiol Plant 94:233–240. https://doi.org/10.1034/j.1399-3054.1995.940208.x

Luckey TD (1980) Hormesis with ionizing radiation. CRC Press, Boca Raton

Luckey TD (2006) Radiation hormesis: the good, the bad, and the ugly. Dose Response 4:169–190. https://doi.org/10.2203/dose-response.06-102.Luckey

Macovei A, Garg B, Raikwar S, Balestrazzi A, Carbonera D, Buttafava A (2014) Synergistic exposure of rice seeds to different doses of γ-ray and salinity stress resulted in increased antioxidant enzyme activities and gene-specific modulation of TC-NER pathway. Biomed Res Int 170:780–787. https://doi.org/10.1155/2014/676934

Maherchandani N (1975) Effects of gamma radiation on the dormant seed of *Avena fatua* L. Radiat Bot 15:439–443. https://doi.org/10.1016/0033-7560(75)90018-6

Maity JP, Mishra D, Chakraborty A, Saha A, Santra SC, Chanda S (2005) Modulation of some quantitative and qualitative characteristics in rice (*Oryza sativa* L.) and mung (*Phaseolus mungo* L.) by ionizing radiation. Radiat Phys Chem 74:391–394. https://doi.org/10.1016/j.radphyschem.2004.08.005

Majeed A, Ahmad H, Muhammad Z (2010) Variation in chlorophyll contents and grain yield of *Lepidium sativum* L as induced by gamma irradiation. Int J Biol Sci Eng 1:147–151

Marcu D, Damian G, Cosma C, Cristea V (2013) Gamma radiation effects on seed germination, growth and pigment content, and ESR study of induced free radicals in maize (*Zea mays*). J Biol Phys 39:625–634. https://doi.org/10.1007/s10867-013-9322-z

Mattson MP (2008) Hormesis defined. Ageing Res Rev 7:1–7. https://doi.org/10.1016/j.arr.2007.08.007

Minisi FA, El-Mahrouk ME, Rida ME-DF, Nasr MN (2013) Effects of gamma radiation on germination, growth characteristics and morphological variations of *Moluccella laevis* L. Am Eurasian J Agric Environ Sci 13:696–704

Mitchell LM, Cambrosio A (1997) The invisible topography of power: electromagnetic fields, bodies and the environment. Soc Stud Sci 27:221–271. https://doi.org/10.1177/030631297027002002

Mokobia CE, Anomohanran O (2005) The effect of gamma irradiation on the germination and growth of certain Nigerian agricultural crops. J Radiol Prot 25:181–188. https://doi.org/10.1088/0952-4746/25/2/006

Moussa HR (2006) Role of gamma irradiation in regulation of NO3 level in rocket (*Eruca vesicaria* subsp. sativa) plants. Russ J Plant Physiol 53:193–197. https://doi.org/10.1134/S1021443706020075

Ouhibi C, Attia H, Rebah F, Msilini N, Chebbi M, Aarrouf J, Urban L, Lachaal M (2014) Salt stress mitigation by seed priming with UV-C in lettuce plants, growth, antioxidant activity and phenolic compounds. Plant Phys Biochem 83:126–133. https://doi.org/10.1016/j.plaphy.2014.07.019

Paparella S, Araújo SS, Rossi G, Wijayasinghe M, Carbonera D, Balestrazzi A (2015) Seed priming: state of the art and new perspectives. Plant Cell Rep 34:1281–1293. https://doi.org/10.1007/s00299-015-1784-y

Poinapen D, Brown DCW, Beeharry GK (2013) Seed orientation and magnetic field strength have more influence on tomato seed performance than relative humidity and duration of exposure to non-uniform static magnetic fields. J Plant Physiol 170:1251–1258. https://doi.org/10.1016/j.jplph.2013.04.016

Qi W, Zhang L, Wang L, Xu H, Jin Q, Jiao Z (2015) Pretreatment with low-dose gamma irradiation enhances tolerance to the stress of cadmium and lead in *Arabidopsis thaliana* seedlings. Ecotoxicol Environ Saf 115:243–249. https://doi.org/10.1016/j.ecoenv.2015.02.026

Rajendra P, Nayak HS, Rao BB, Subramanyam C, Devendranath D, Gunasekaran B, Aradhya SRS, Bhaskaran A (2005) Effects of power frequency electromagnetic fields on growth of germinating *Vicia faba* L., the broad bean. Electromagn Biol Med 24:39–54. https://doi.org/10.1081/JBC-200055058

Reddy MVB, Raghavan GSV, Kushalappa AC, Paulitz TC (1998) Effect of microwave treatment on quality of wheat seeds infected with Fusarium graminearum. J Agric Eng Res 71:113–117. https://doi.org/10.1006/jaer.1998.0305

Reddy KV, Reshma SR, Jareena S, Nagaraju M (2012) Exposure of greengram seeds (*Vigna radiate* var. radiata) to static magnetic fields: effects on germination and alfa-amylase activity. Res J Seed Sci 5:106–114. https://doi.org/10.3923/rjss.2012.106.114

Sacandè M, Golovina EA, Van Aelst AC, Hoekstra FA (2001) Viability loss of neem (*Azadirachta indica*) seeds associated with membrane phase behaviour. J Exp Bot 52:919–931. https://doi.org/10.1093/jexbot/52.358.919

Sahin H (2014) Effects of microwaves on the germination of weed seeds. J Biosyst Eng 39:304–309. https://doi.org/10.5307/JBE.2014.39.4.304

Sahu ID, McCarrik RM, Lorigan GA (2013) Use of electron paramagnetic resonance to solve biochemical problems. Biochemistry 52:5967–5984. https://doi.org/10.1021/bi400834a

Scialabba A, Tamburello C (2002) Microwave effects on germination and growth of radish (*Raphanus sativus* L.) seedlings. Acta Bot Gall 149:113–123. https://doi.org/10.1080/12538078.2002.10515947

Shaukat S, Farooq M, Siddiqui M, Zaidi S (2013) Effect of enhanced UV-B radiation on germination, seedling growth and biochemical responses of *Vigna mungo* (L.) Hepper. Pak J Bot 45:779–785

Shine MB, Guruprasad KN, Anand A (2011) Enhancement of germination, growth, and photosynthesis in soybean by pre-treatment of seeds with magnetic field. Bioelectromagnetics 32:474–484. https://doi.org/10.1002/bem.20656

Siddiqui A, Dawar S, Javed Zaki M, Hamid N (2011) Role of ultra violet (UV-C) radiation in the control of root infecting fungi on groundnut and mung bean. Pak J Bot 43:2221–2224

Sjodin J (1962) Some observations in X1 and X2 of *Vicia faba* L., after treatment with different mutagens. Hereditas 48:565–586

Smirnov AI, Golovina HA, Yakimchenko OE, Aksyonov SJ, Lebedev YS (1992) In vivo seed investigation by electron paramagnetic resonance spin probe technique. J Plant Phys 140:447–452. https://doi.org/10.1016/S0176-1617(11)80823-0

Soran M-L, Stan M, Niinemets Ü, Copolovici L (2014) Influence of microwave frequency electromagnetic radiation on terpene emission and content in aromatic plants. J Plant Physiol 171:1436–1443. https://doi.org/10.1016/j.jplph.2014.06.013

Steffen-Heins A, Steffens B (2015) EPR spectroscopy and its use in planta-a promising technique to disentangle the origin of specific ROS. Front Environ Sci 3:15. https://doi.org/10.3389/fenvs.2015.00015

Talei D, Valdiani A, Maziah M, Mohsenkhah M (2013) Germination response of MR 219 rice variety to different exposure times and periods of 2450 MHz microwave frequency. Sci World J 2013:408026. https://doi.org/10.1155/2013/408026

Teixeira da Silva JA, Dobránszki J (2015) Magnetic fields: how is plant growth and development impacted? Protoplasma 253:231–248. https://doi.org/10.1007/s00709-015-0820-7

Vashisth A, Nagarajan S (2010) Effect on germination and early growth characteristics in sunflower (*Helianthus annuus*) seeds exposed to static magnetic field. J Plant Physiol 167:149–156. https://doi.org/10.1016/j.jplph.2009.08.011

Ventura L, Donà M, Macovei A, Carbonera D, Buttafava A, Mondoni A, Rossi G, Balestrazzi A (2012) Understanding the molecular pathways associated with seed vigor. Plant Physiol Biochem 60:196–206. https://doi.org/10.1016/j.plaphy.2012.07.031

Vian A, Roux D, Girard S, Bonnet P, Paladian F, Davies E, Ledoigt G (2006) Microwave irradiation affects gene expression in plants. Plant Signal Behav 1:67–70. https://doi.org/10.4161/psb.1.2.2434

Weil JA, Bolton JR (2006) Electron paramagnetic resonance: elementary, theory and practical applications. John Wiley and Sons, Inc, Hoboken

Wolff SA, Coelho LH, Karoliussen I, Jost A-IK (2014) Effects of the extraterrestrial environment on plants: recommendations for future space experiments for the MELiSSA higher plant compartment. Life 4:189–204. https://doi.org/10.3390/life4020189

Zaka R, Chenal C, Misset M (2002) Study of external low irradiation dose effects on induction of chromosome aberrations in *Pisum sativum* root tip meristem. Mutat Res 517:87–99. https://doi.org/10.1016/S1383-5718(02)00056-6

Zeeman P (1897) The effect of magnetisation on the nature of light emitted by a substance. Nature 55:347. https://doi.org/10.1038/055347a0

# Effect of Seed Priming on Seed Dormancy and Vigor

**Muhammad Junaid Rao, Sajjad Hussain, Muhammad Akbar Anjum, Muhammad Saqib, Riaz Ahmad, Muhammad Fasih Khalid, Muhammad Sohail, Muhammad Nafees, Muhammad Arif Ali, Niaz Ahmad, Iqra Zakir, and Shakeel Ahmad**

## Contents



M. J. Rao
Key Laboratory of Horticultural Plant Biology (Ministry of Education), Huazhong Agricultural University, Wuhan, People's Republic of China

S. Hussain (✉) · M. A. Anjum · M. Saqib · R. Ahmad · M. F. Khalid · M. Sohail
Department of Horticulture, Bahauddin Zakariya University, Multan, Punjab, Pakistan
e-mail: sajjad.hussain@bzu.edu.pk

M. Nafees
Department of Horticultural Sciences, University college of Agriculture and Environmental Sciences, The Islamia University Bahawalpur, Bahawalpur, Punjab, Pakistan

M. A. Ali · N. Ahmad
Department of Soil Science, Bahauddin Zakariya University, Multan, Punjab, Pakistan

I. Zakir · S. Ahmad (✉)
Department of Agronomy, Bahauddin Zakariya University, Multan, Punjab, Pakistan
e-mail: shakeelahmad@bzu.edu.pk



M. Hasanuzzaman, V. Fotopoulos (eds.), *Priming and Pretreatment of Seeds and Seedlings*, https://doi.org/10.1007/978-981-13-8625-1_6

**Abstract** Seed dormancy is a prerequisite for preservation and cultivation of crops. The genetic characteristics of seeds and external environmental conditions combinedly influence the seed vigority. Therefore, seed dormancy and vigor can modify the growth, development, and yield capacity of both agronomic and horticultural crops. There are some physiological and biochemical stimulators by which seed dormancy period and intensity are regulated. Seed priming significantly regulates the different types of seed dormancy as well as enhances the vigority. Thus, different seed priming treatments can be potentially used in crop production to increase the uniformity of germination along with better growth and developments. This chapter will discuss about the seed priming-induced regulation of seed dormancy and vigor.

**Keywords** Seed viability · Seed treatment · Abiotic stress · Crop yield · Normal seedling · Germination

# 1 Introduction

Seed priming is a pre-sowing treatment of seed for uniform and rapid germination with reduction in germination time. It is applied in annual as well as in perennial crops. In the last 25 years, seed priming is commonly applied in different vegetable and flower species to increase the rate and uniformity of seedlings. Seed treatment also removes or decreases the level of dormancy in dormant seeds.

# 2 Seed Dormancy and Vigor

Seed is said to be dormant when it is not germinating even though it has the capacity to germinate under favorable environmental conditions (moisture, temperature, light) (Baskin and Baskin 2004).

Seed dormancy should be differentiated from quiescence which is the condition of seed in which seed does not germinate due to unfavorable environmental conditions affecting seed metabolic activities and resulting in temporary suspension of growth (Amen 1963).

Seed dormancy is induced in the seed in the seed development and maturation phase. The environmental conditions (temperature, light, and soil nitrate) faced by the mother plant during seed maturation stage influence the severity of dormancy (He et al. 2014). The level of seed dormancy is more high in freshly harvested seeds and is reduced after a period of dry storage (after ripening) (Née et al. 2017). The term seed vigor is coined for those seeds which determine the seed lot during germination or performance of the seed and seedling emergence and potential level of activity. During seed development, the seed vigor is decided according to two main aspects: environment factors and genetic composition. On the mother plant, the vigor of seed gradually rises with the passage of time and seed maturity. The seed vigor level might be altered by humidity, temperature, harvest stage, mechanical injury during harvesting, soil water, soil nutrition, and so on.

## 2.1 Categories of Seed Dormancy

Seed dormancy is categorized in various ways by different researchers as Harper (1957), Lang (1987), Lang et al. (1985), and Nikolaeva et al. (1985, 1999). However, the seed dormancy comprehensive classification system has been suggested by Baskin and Baskin (2004). According to this system, seed dormancy has five classes.

### 2.1.1 Physiological Dormancy (PD)

Physiological dormancy is the supreme abundant practice found in all major angiosperms and seeds of gymnosperms species. In the temperate seeds, it is the most predominant form of dormancy. Physiological dormancy can be distributed into three distinct groups: deep, non-deep, and intermediate (Baskin and Baskin 2004).

*Deep*: In deep level of physiological dormancy, excised embryos from these seeds either produce abnormal seedlings or will not grow; to break this dormancy, cold or warm stratification is required. However, dormancy condition didn't break by the gibberellic acid treatment (Baskin and Baskin 2004; Baskin and Baskin 2005). Examples are *Acer pseudoplatanus* and *Acer platanoides* (PD deep).

*Intermediate*: In this type of physiological dormancy, excised embryo produces normal seedling; however, seeds need to be cold stratified for 2–3 months to break dormancy, and gibberellic acid treatment stimulates germination in few species (not in all). However, the dry storage has potential to lessen the cold stratification period.

*Non-deep*: It is the utmost predominant level of physiological dormancy (Baskin and Baskin 2004). This level of dormancy can be overcome through gibberellic acid treatment and it also varies from species to species; dormancy can also be broken by dry storage and warm or cold stratification (Finch-Savage and Leubner-Metzger 2006).

### 2.1.2 Morphological Dormancy (MD)

Morphological dormancy exists in the seeds that are under-developed (in size) but differentiated into hypocotyl-radical and cotyledons. These embryos are not dormant (physiologically) but required some time to germinate and grow, e.g., celery (*Apium graveolens*) (Jacobsen and Pressman 1979).

### 2.1.3 Morphophysiological Dormancy (MPD)

In this class of dormancy, seeds have combined form of morphological and physiological dormancy. Normally, seeds contain an immature embryo with dormancy condition. Thus, to break the dormancy condition, they require a pretreatment. One example of morphophysiological dormancy (MPD) seed is *Trollius* (Ranunculaceae), and embryo required longer period for growth/radicle germination as compared with MD embryo (Hepher and Roberts 1985).

### 2.1.4 Physical Dormancy (PY)

Water-impermeable layers around the seed make it dormant and radicle cannot come out, and this type of dormancy is known as physical dormancy. Scarification either chemical or mechanical has the ability to break this dormancy, e.g., *Trigonella* and *Melilotus* (family Fabaceae) (Baskin and Baskin 1998).

### 2.1.5 Combinational Dormancy (PY + PD)

Combinational dormancy is a complex form of dormancy in which seeds are physiologically dormant with impermeable seed layers, e.g., *Geranium* and *Trifolium*. A period of dry storage is needed to remove physiological component followed by cold stratification to make the seed coat permeable to water.

## 2.2 *The Physiological and Biochemical Mechanisms of Dormancy Accumulation and Release*

The seed dormancy stimulation is governed by various series of regulators that respond at different stages. Some of these are discussed here.

### 2.2.1 Seed Maturation Regulators

Embryogenesis and seed maturation are two fundamental phases involved in seed development. In seed maturation phase, the seed dormancy is also prompted at once, some storage compounds are also accumulated, and drying tolerance and the latency of metabolic activity are acquired. Some transcription factors, such as LEAFY COTYLEDON 1 (LEC1), LEC2, FUSCA 3 (FUS3), and ABSCISIC ACID INSENSITIVE 3 (ABI3), play a dominant role in stimulating the seed maturation and transition phase from embryo to seedling. In these four transcription factors, mutation results in abnormal seed growth which leads to mutant phenotypes and decline in seed dormancy. The discrete roles of these transcription factors and their complex interactions were studied previously in detail (Holdsworth et al. 2008).

### 2.2.2 Hormonal Regulation

Natural hormones are considered to be strongly involved in seed dormancy process. Various reviews have been published focusing on the role of hormones such as abscisic acid (ABA) gibberellins, ethylene, brassinosteroids, cytokinins, and auxins in the phenomenon of dormancy as well as germination. Especially, the balance between the levels and respective pathways of abscisic acid (ABA) and gibberellin is vital for promotion of germination and stimulation of both conservation and induction of dormancy (Finkelstein et al. 2008). The regulators of hormones in the release of seed dormancy are briefly described in Fig. 1.

#### 2.2.2.1 Abscisic Acid

Abscisic acid is the most important hormone involved in the seed dormancy induction and maintenance and causing delay in germination (Hilhorst 1995; Koornneef et al. 2002; Nambara and Marion-Poll 2003). ABA synthesis is controlled by the genes involved in genetic expression of hormones, and overexpression of these genes can increase the ABA level causing seed dormancy or delay in germination. In seed development phase, ABA is produced by two processes; embryo itself produces its ABA or it is produced by maternal tissues. However, ABA synthesized only by embryo itself is responsible for the induction of dormancy.

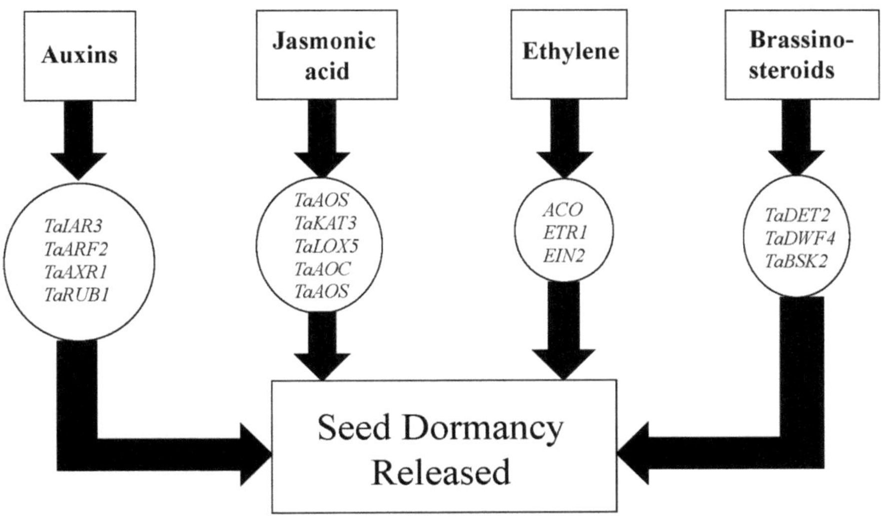


**Fig. 1** Regulators of auxin, jasmonic acid, ethylene, and brassinosteroid pathways in seed dormancy release

#### 2.2.2.2 Gibberellins (GA)

Gibberellins are another group of hormones which play important role in the promotion of germination and dormancy breakdown. GA mainly imparts two functions in germinating embryos. First, it enhances the growth of potential embryos, and second, it is essential to overwhelm the mechanical restriction imposed through outer cover of seed by waning the tissues which surround the radicle. GA biosynthesize and its localization may be affected by the environmental situations such as light and temperature (Yamauchi et al. 2004). The processes involved in dormancy release and germination are complex and these are governed by the interactive mechanisms of ABA and GA with environmental conditions.

#### 2.2.2.3 Ethylene

Ethylene is involved in the promotion of germination and it works antagonistic to ABA. Various studies indicated the role of ethylene in dormancy release in different species (Corbineau et al. 1990). In some species, it releases dormancy, whereas in other it alone is unable to break the dormancy even when it stimulates germination of a particular species in nondormant seeds. The ethylene's prime action might be the expansion of radial cells in the hypocotyl, increased water potential, or rapid seed respiration, and several hypotheses have been proposed describing the role of ethylene in dormancy breakdown and as germination stimulator (Esashi 1991; KeÇpczyński and KeÇpczyńska 1997; Matilla 2000).

#### 2.2.2.4 Brassinosteroids

Brassinosteroids also play positive role in seed germination mechanism. GA and brassinosteroids (BR) interactively regulate the shoot elongation and seedlings' photomorphogenesis through spaciously independent pathways. Various studies indicated the presence of endogenous BR in germinating seeds, and its exogenous application also enhances the germination in certain cereals (Yamaguchi et al. 1987), angiosperms (Takeuchi et al. 1991, 1995), tobacco, and *Arabidopsis* mutants (Steber and McCourt 2001).

#### 2.2.2.5 Cytokinins

Developing seeds may accumulate the cytokinins in liquid endosperm (Mok and Mok 2001). Cytokinins play the role of cell expansion in developing embryos and might have role in embryonic pattern development, in enhancing sink strength, in cereals by the filling of grains in early periods, and in embryogenesis. Cytokinins are famous among several plant species where it alone has the capability to break down the seed dormancy (Cohn and Butera 1982).

### 2.3 *The Influence of Environmental Factors on Seed Dormancy*

Seeds are the key sensors for the different environmental factors. These factors control the dormancy cycles of seeds and give the adequate timing of seed germination and seedling establishment. Temperature, light, water, nitrate, oxygen, smoke, and allelochemicals are considered as key environmental factors influencing dormancy level and its severity either during seed development or in the soil-related seed bank.

Chiang et al. (2009) identified the response of temperature and light on seed dormancy in seed maturation phase. Especially, temperature is a primary factor affecting the level of primary dormancy through its action in regulating the ABA and GA contents. Seed dormancy in the bank of soil is controlled by environmental signals, with soil moisture and temperature which are the key issues (Batlla and Benech-Arnold 2010; Footitt et al. 2011). Level of dormancy and its gene expressions were associated with seasonal alterations in the temperature of soil. Soil temperature regulates dormancy level (deep or non-deep).

Soil temperature affects the expression of genes for ABA and GA synthesis. In winter, seeds' dormancy increased due to low soil temperature and as gene expression for ABA biosynthesis increased. In spring and summer, dormancy then declined (Graeber et al. 2012).

In addition to temperature and humidity, allelochemicals also have an impact on the level of seed dormancy in soil. These chemicals are released from single plant and affect the physiological mechanism in nearby plants. Wild tobacco (*Nicotiana attenuata)* is an example of this phenomenon. It is an annual plant which germinates by receiving smoke signals after forest fire in wild forests. However, four terpenes and ABA leaching from the litter from the dominant vegetation have the capability to encourage dormancy in the seeds of *N. attenuata* (Krock et al. 2002; Preston et al. 2002; Linkies and Leubner-Metzger 2012).

## 3 Effect of Seed Priming on Breaking Seed Dormancy and Improving Seed Vigor

Seed priming effects are directly interrelated to repairing as well as building up nucleic acids, enhancing the protein synthesis, and membrane repairing (Wang et al. 2003). Treated seeds increase the antioxidant enzymes' activities as compared to untreated seeds (Hsu et al. 2003). Seed priming also plays a significant role in improvement of seed germination as well as uniformity of heterogeneously matured seed lots. Priming comprises different categories such as solid matrix priming, hydropriming, and osmopriming for breakage of seed dormancy and to enhance seed vigor in many horticultural as well as agronomic crops (Neamatollahi and Darban 2010). Recently, numerous researchers recommended that seeds priming be considered as a helpful strategy for good germination and healthier seedling growth resulting in higher yields (Naba'ee et al. 2013).

### 3.1 Hydropriming Effects on Seed Dormancy and Vigor

Hydropriming is more useful for the breakage of seed dormancy, improvement of seed germination, and seedlings' emergence which ultimately results in higher yield. The beneficial impacts of seed priming have been elaborated by several researchers in numerous field crops as well as horticultural crops (Ghassemi-Golezani et al. 2010; Naba'ee et al. 2013). Previous research work confirmed that hydro seed priming greatly depends on type of plant species, water potentiality, priming duration, seed vigor, temperature, humidity and storage conditions, and many others. Nowadays, several priming methods have been adopted and successfully used in many field crops. Among them, hydropriming is a very simple, easy, very low cost, less time required, and environmental-friendly type of priming (Ghassemi-Golezani et al. 2008). Hydropriming is found to be more helpful and efficient type of priming in decreasing the possibility of poor stand establishment in an extensive variation of ecological circumstances. Mahmoodi et al. (2011) examined the effect of hydropriming approximately for 18 h and found it to be very superior for improvement of seedlings' vigor as well as seedling emergence of maize crop. These effects of hydropriming enhance the capability of seedling establishment as well as field performance of many cereal crops all over the world.

### 3.2 Osmopriming Effects on Seed Dormancy and Vigor

Osmopriming is famous as one of the most special types of seed priming which can be used to improve the growth and yield of horticultural and field crops (Ghassemi-Golezani et al. 2008; Neamatollahi and Darban 2010). In osmopriming, seeds were soaking in aerated little water prospective solutions which enhance pre-germination accomplishments to remain continue before actual germination. Seed priming with NaCl increases the morpho-physical, biochemical, and molecular changes in plants which are cultivated under high salinity stress. The seeds treated with NaCl during sowing showed significant positive effects on seed germination, plant growth, and development of tomatoes. Tomato seeds were primed or treated with PEG-6000 and were found to have significant improvement in germination rates. Potassium salts were also used for seed priming and contributed in accelerated germination in corn seed. Moreover, it has been concluded that osmopriming is necessary for better seed germination, which resulted in good plant phenology with excellent crop yield (Khan et al. 2008).

### 3.3 Solid Matrix Priming Effects on Seed Dormancy and Vigor

Solid matrix seed priming has diverse applications for okra seeds. Solid matrix seed priming can be successfully used for acceleration of seed germination as well as acts as delivery system for selective fungicides to control insects-pests and soil-borne diseases (Mercado and Fernandez 2002). Okra is a vegetable crop and found to be very more delicate under cool temperature. This priming can be more efficiently used for priming of seeds before sowing which have greater capability to enhance seed germination cultivated under cold soils. Seed priming might be used to provide the highest membrane reliability in the embryo and the emerging seedling decreasing seepage with the membranes. Conclusively, solid matrix priming can be used to attain improved seed germination. Numerous types of constituents can be efficiently used for solid matrix priming. However, optimum circumstances for moisture content and seed priming period might be evaluated for every matrix (Mereddy 2015).

## 4 Conclusion

Nowadays, seed dormancy and vigor are very much important in crop production. There are many types of dormancy, i.e., physical, morphological, morphophysiological, and physiological dormancy, which resist the seed for not germinating under favorable conditions. Seed dormancy occurs due to many factors like environmental factors and hormonal regulations. To overcome this, seed priming is done which helps dormant seed to germinate. Seed priming also plays a significant role in the improvement of seed germination and its vigor as well as uniformity of heterogeneously matured seed lots. Priming comprises different categories such as solid matrix priming, hydropriming, and osmopriming for breakage of seed dormancy and to enhance seed vigor in many horticultural as well as agronomic crops.

## References

Amen RD (1963) The concept of seed dormancy. Am Sci 51(4):408–424
Baskin CC, Baskin JM (1998) Seeds: ecology, biogeography, and, evolution of dormancy and germination. Elsevier, San Diego
Baskin JM, Baskin CC (2004) A classification system for seed dormancy. Seed Sci Res 14(1):1–16
Baskin CC, Baskin JM (2005) Underdeveloped embryos in dwarf seeds and implications for assignment to dormancy class. Seed Sci Res 15(4):357–360
Batlla D, Benech-Arnold RL (2010) Predicting changes in dormancy level in natural seed soil banks. Plant Mol Biol 73(1–2):3–13
Chiang GC, Barua D, Kramer EM, Amasino RM, Donohue K (2009) Major flowering time gene, FLOWERING LOCUS C, regulates seed germination in *Arabidopsis thaliana*. Proc Natl Acad Sci 106(28):11661–11666
Cohn MA, Butera DL (1982) Seed dormancy in red rice (*Oryza sativa*). II. Response to cytokinins. Weed Sci 30(2):200–205
Corbineau F, Bagniol S, Côme D (1990) Sunflower (*Helianthus annuus* L.) seed dormancy and its regulation by ethylene. Isr J Bot 39(4–6):313–325
Esashi Y (1991) Ethylene and seed germination. In: Mattoo AK, Suttle JC (eds) The plant hormone ethylene. CRC Press, Boca Raton, pp 133–157
Finch-Savage WE, Leubner-Metzger G (2006) Seed dormancy and the control of germination. New Phytol 171(3):501–523

Finkelstein R, Reeves W, Ariizumi T, Steber C (2008) Molecular aspects of seed dormancy. Annu Rev Plant Biol 59:387
Footitt S, Douterelo-Soler I, Clay H, Finch-Savage WE (2011) Dormancy cycling in Arabidopsis seeds is controlled by seasonally distinct hormone-signaling pathways. Proc Natl Acad Sci 108(50):20236–20241
Ghassemi-Golezani K, Sheikhzadeh-Mosaddegh P, Valizadeh M (2008) Effects of hydro-priming duration and limited irrigation on field performance of chickpea. Res J Seed Sci 1:34–40
Ghassemi-Golezani K, Chadordooz-Jeddi A, Nasrollahzadeh S, Moghaddam M (2010) Influence of hydro-priming duration on field performance of pinto bean (*Phaseolus vulgaris* L.) cultivars. Afr J Agric Res 5:893–897
Graeber KAI, Nakabayashi K, Miatton E, Leubner-Metzger GERHARD, Soppe WJ (2012) Molecular mechanisms of seed dormancy. Plant Cell Environ 35(10):1769–1786
Harper JL (1957) The ecological significance of dormancy and its importance in weed control. In: Proceedings of the 4th international congress of crop protection, vol 1, pp 415–420, Braunschweig (1960)
He H, de Souza Vidigal D, Snoek LB, Schnabel S, Nijveen H, Hilhorst H, Bentsink L (2014) Interaction between parental environment and genotype affects plant and seed performance in Arabidopsis. J Exp Bot 65(22):6603–6615
Hepher A, Roberts JA (1985) The control of seed germination in *Trollius ledebouri*: the breaking of dormancy. Planta 166(3):314–320
Hilhorst HW (1995) A critical update on seed dormancy. I. Primary dormancy. Seed Sci Res 5(2):61–73
Holdsworth MJ, Bentsink L, Soppe WJ (2008) Molecular networks regulating Arabidopsis seed maturation, after-ripening, dormancy and germination. New Phytol 179(1):33–54
Hsu CC, Chen CL, Chen JJ, Sung JM (2003) Accelerated aging-enhanced lipid peroxidation in bitter gourd seeds and effects of priming and hot water soaking treatments. Sci Hortic 98(3):201–212
Jacobsen JV, Pressman E (1979) A structural study of germination in celery (*Apium graveolens* L.) seed with emphasis on endosperm breakdown. Planta 144(3):241–248
KeÇpczyński J, KeÇpczyńska E (1997) Ethylene in seed dormancy and germination. Physiol Plant 101(4):720–726
Khan A, Khalil SK, Khan AZ, Marwat KB, Afzal A (2008) The role of seed priming in semi-arid area for mung bean phenology and yield. Pak J Bot 40(6):2471–2480
Koornneef M, Bentsink L, Hilhorst H (2002) Seed dormancy and germination. Curr Opin Plant Biol 5(1):33–36
Krock B, Schmidt S, Hertweck C, Baldwin IT (2002) Vegetation-derived abscisic acid and four terpenes enforce dormancy in seeds of the post-fire annual, *Nicotiana attenuata*. Seed Sci Res 12(4):239–252
Lang GA (1987) Endo-, para-and ecodormancy: physiological terminology and classification for dormancy research. Hortic Sci 22:271–277
Lang GA, Early JD, Arroyave NG, Darnell RL, Martin GC, Stutte GW (1985) Dormancy: toward a reduced universal terminology. HortScience 20:809–811
Linkies A, Leubner-Metzger G (2012) Beyond gibberellins and abscisic acid: how ethylene and jasmonates control seed germination. Plant Cell Rep 31(2):253–270
Mahmoodi TM, Ghassemi-Golezani K, Habibi D, Paknezhad F, Ardekani MR (2011) Effect of hydro-priming duration on seedling vigour and field establishment of maize (*Zea mays* L.). Res Crops 12(2):341–345
Matilla AJ (2000) Ethylene in seed formation and germination. Seed Sci Res 10(2):111–126
Mercado MFO, Fernandez PG (2002) Solid matrix priming of soybean seeds. Philipp J Crop Sci 27(2):27–35
Mereddy R (2015) Solid matrix priming improves seedling vigor of okra seeds. Proc Oklahoma Acad Sci 80:33–37
Mok DW, Mok MC (2001) Cytokinin metabolism and action. Annu Rev Plant Biol 52(1):89–118

Naba'ee M, Roshandel P, Khani MA (2013) The effects of plant growth regulators on breaking seed dormancy in *Silybum marianum* L. J Cell Tissue Res 4(1):45–54

Nambara E, Marion-Poll A (2003) ABA action and interactions in seeds. Trends Plant Sci 8(5):213–217

Neamatollahi E, Darban SA (2010) Investigation of hydropriming and osmopriming effects on canola (*Brassica napus* L.) cultivars. Int J Agric Res 5:87–92

Née G, Xiang Y, Soppe WJ (2017) The release of dormancy, a wake-up call for seeds to germinate. Curr Opin Plant Biol 35:8–14

Nikolaeva MG, Rasumova MV, Gladkova VN (1985) Pravocnik po prorascivanij pokojascichsja semjan. In: Danilova MF (ed) Reference book on dormant seed germination. 'Nauka' Publishers, Leningrad (in Russian)

Nikolaeva MG, Lyanguzova IV, Pozdova LM (1999) Biology of seeds. V.L. Komarov Botanical Institute, Russian Academy of Sciences, St. Petersburg (in Russian with English summary and English table of contents)

Preston CA, Betts H, Baldwin IT (2002) Methyl jasmonate as an allelopathic agent: sagebrush inhibits germination of a neighboring tobacco, Nicotiana attenuata. J Chem Ecol 28(11):2343–2369

Steber CM, McCourt P (2001) A role for brassinosteroids in germination in Arabidopsis. Plant Physiol 125(2):763–769

Takeuchi Y, Worsham AD, Awad AE (1991) Effects of brassinolide on conditioning and germination of witchweed (*Striga asiatica*) seeds. In: Brassinosteroids, ACS Symposium series. American Chemical Society, Washington, DC

Takeuchi Y, Omigawa Y, Ogasawara M, Yoneyama K, Konnai M, Worsham AD (1995) Effects of brassinosteroids on conditioning and germination of clover broomrape (*Orobanche minor*) seeds. Plant Growth Regul 16(2):153–160

Wang HY, Chen CL, Sung JM (2003) Both warm water soaking and matriconditioning treatments enhance anti-oxidation of bitter gourd seeds germinated at sub-optimal temperature. Seed Sci Technol 31(1):47–56

Yamaguchi T, Wakizuka T, Hirai K, Fujii S, Fujita A (1987) Stimulation of germination in aged rice seeds by pretreatment with brassinolide. Proc Plant Growth Regul Soc Am 14:26–27

Yamauchi Y, Ogawa M, Kuwahara A, Hanada A, Kamiya Y, Yamaguchi S (2004) Activation of gibberellin biosynthesis and response pathways by low temperature during imbibition of *Arabidopsis thaliana* seeds. Plant Cell 16(2):367–378

# Alterations in Plant Secondary Metabolism by Seed Priming

**Qasim Ali, Sana Ashraf, Muhammad Kamran, Abdul Rehman, and Shakeel Ahmad**

## Contents



**Abstract** Seed technologists are extensively using seed priming with water alone or with some additives to improve germination potential and stress tolerance in seed. Priming is also very important to seed bank managers who require conserving germplasm of important crops ex situ. Taking care of the plant species, physiology, and seed morphology, different priming techniques and methods can be applied. All these treatments activate the pre-germinative seed metabolism. Alteration in the secondary metabolism starts from the beginning of water uptake along with seed-repairing process. The seed repair process includes induction of antioxidant system and activation of DNA repair pathways. Both these mechanisms are essential for the integrity of seed genome, successful germination, and seedling establishment. With

Q. Ali (✉) · A. Rehman
College of Agriculture, Bahauddin Zakariya University, Layyah, Pakistan
e-mail: ali_qasim@bzu.edu.pk

S. Ashraf
Institute of Soil and Environmental Sciences, University of Agriculture, Faisalabad, Pakistan

M. Kamran
Department of Agronomy, University College of Agriculture, University of Sargodha, Sargodha, Pakistan

S. Ahmad
Department of Agronomy, Bahauddin Zakariya University, Multan, Punjab, Pakistan

 
M. Hasanuzzaman, V. Fotopoulos (eds.), *Priming and Pretreatment of Seeds and Seedlings*, https://doi.org/10.1007/978-981-13-8625-1_7

the beginning of the water uptake, reactive oxygen species (ROS) start to accumulate in the seeds that can cause severe damage to seed proteins and lipids. However, to cope up with these hazardous ROS molecules, seeds have strong antioxidant system involving a number of enzymatic and nonenzymatic antioxidants. These antioxidant molecules play an active role in scavenging the injurious oxidative molecules, thus making the seed germination a successful event. This book chapter will provide an overview of priming and its impact on the alteration in production of different secondary metabolites which are produced in response to priming process.

**Keywords** Seed priming · Secondary metabolism · Germination · Antioxidants · Seed vigor

## 1 Introduction

High-quality seeds are becoming dire need of the day to fulfill the increasing demand of the better quality seeds for agricultural market. For better and healthy crop production, fast germination rate is the major requirement because slow germinating seeds are usually susceptible toward soil-born infections and are unable to withstand the adverse environmental conditions (Osburn and Schroth 1989). Seed priming is a well-known treatment to enhance the seed quality and health. Priming seeds germinate at a greater emergence rate, resulting in better crop production. Primed seeds also show the resistance to biotic and abiotic stresses. It also improves the seed vigor which is a complex agronomic character that is directly controlled by a set of genes and a number of environmental factors (Jisha et al. 2013; Rajjou et al. 2012).

Priming can be defined as a method to rehydrate seed with water alone or with some other additives to initiate the metabolic processes that usually activate during the start of the germination, but avoiding the seed transition towards complete germination. Therefore, treatment must be ceased before the occurrence of any loss in desiccation tolerance. A number of positive benefits are attributed to seed priming in the literature. For example, priming reduces the thermo- and photo-dormancy and thus provides a wider range of seed germination temperatures. Moreover, it gives seed good strength to cope with pathogens and weeds (Ellis and Butcher 1988; Hill et al. 2008). Further, it helps to manage water shortage and harvest timings (Hill et al. 2008). Different factors such as priming methods, time of priming, plant species and physiology, seed vigor, and lot determine the benefits of seed priming (Parera and Cantliffe 1990). Priming is usually used to treat vegetable seeds, mainly leek and onion (*Genus Allium*), lettuce (*Lactuca sativa* L.), pepper (*Genus Capsicum*), and tomato (*Solanum lycopersicum* L.) (Di Girolamo and Barbanti 2012; Dearman et al. 1987). Priming is also used to increase the flower quality in flowers seed industry. It is also used to prime the seeds of selected varieties (Di Girolamo and Barbanti 2012; Momin 2013). Priming the cereal seeds at large and commercial scale is much more difficult; however, benefits are reported widely (Murungu et al. 2004).

Priming procedure greatly alters the production and storage of many antioxidants such as ascorbic acid, glutathione, α-tocopherol, etc. Further, priming causes the partial digestion of starch that significantly alters the concentration of reducing sugars, and this increased concentration has direct impact on different chemical reactions going on in the seed such as protein glycation and reaction between reducing sugar and amino group of several proteins (Ventura et al. 2012). Similarly, priming also has significant impact on the alteration of proline content in the seed (Kubala et al. 2015a). There are also many reports regarding the impact of priming on alteration in the concentration of many other metabolites such as leucine, glutamate, fumarate, aspartate, threonate, or pinitol content of seeds of several species (Chen and Arora 2011; Di Girolamo and Barbanti 2012; Krol et al. 2015). Another molecule methionin, which is an important precursor of polyamine and ethylene, is activated during the priming of seeds. As a small aliphatic molecule, polyamines are involved to influence all facets of plant growth and development as well as germination. Ethylene is also known as stress hormone which also plays a significant role in many reactions of seed germination, and it has been reported that seed priming can alter the kinetics of ethylene production from its precursor ACC (Wu et al. 2014). There are also some plant hormones (auxin, gibberellins, cytokinins, ABA, etc.) which are considered to take part in many reactions in seed germination, but their quantitative impact due to priming is still under study.

## 2 Pre-germinative Metabolism and Seed Priming

Beneficial impacts of priming are because of the alteration in the particular metabolic processes that took place in seed when it starts to uptake the water (Bray 1995). On rehydration, many physiology processes are induced, for example, a new synthesis of nucleic acid and protein, activation of DNA, ATP synthesis, cell buildup of phospholipids and sterols, and increased antioxidant mechanism. Recent details about the alteration in secondary metabolism or pre-germinated metabolism explaining the shift from the dormant stage of dry seed to the active state of primed germinating seed have been given at different levels using 'omic' tools, for example, proteomics (Galland et al. 2014; Tan et al. 2013).

Seeds always have to face the adverse environmental conditions, whether on the parent plant during the maturity stage or during the postharvest storage and also during the initial stage of germination. During all these stresses, seeds are subjected to severe oxidative damage of nucleic acid, protein, and lipids (Ventura et al. 2012; Kranner et al. 2010). In response to this, secondary metabolites are produced to protect seeds/seedlings from this damage. Moreover, repairing process of DNA must also be set aside at appropriate level in the embryo to maintain vigor and enhance the potential of germination of seeds. When DNA is conserved after the repair process, cell cycle continuation is resumed in the embryo cells that lead to

DNA replication. However, in case of defective repair process, oxidative injury resulted in the death of cells (Waterworth et al. 2010; Ventura et al. 2012). DNA restoration is the main mechanism of the pre-germinative metabolism initiated when seed starts to uptake water and by imbibition, causing the nonstop buildup of reactive oxygen species. All the important DNA restoration pathways are induced during the initial stage of seed imbibition to conserve integrity of genome (Cordoba-Canero et al. 2014). Transcription-coupled nucleotide excision repair, a sub pathway which explicitly recognizes and eliminates abrasions from the transcribed strands of transcriptionally active genes, has been examined in seed germination in rice by Macovei et al. (2014) who focused on the need for DNA helicases to enable appropriate unwinding of the DNA duplex during transcription. Moreover, microRNAs selecting helicases in connection to DNA impairment response and restoration in rice seeds/sprouts have also been examined under oxidative stress conditions induced by $\gamma$-ray by Macovei and Tuteja (2013), mounting the existing knowledge on the seed response to treatments arbitrated by 'physical rigorization' or ionizing radiation (Grover and Khan 2014). Appropriate DNA restoration depends on the effective ligase-mediated rejoining of fragmented strands and DNA ligase VI has been recognized as the main factor of seed vigor in *Arabidopsis* (Waterworth et al. 2010). In secondary metabolism, production of ROS has an important role in the activation of seed germination; however, the mechanisms involved in the production of signaling molecules have not been completely explained (Diaz-Vivancos et al. 2013). It is observed that the radicals with short half-life will play their role at the site of induction, while the radicals with long half-life are carried to distant locations (Moller et al. 2007).

Many other related molecules of ROS-mediated pathways such as calcium-binding proteins and mitogen-activated protein kinases have been identified in the plants, but their function in seed is still under observation (Foyer and Noctor 2005). Alternatively, it has been found that severe oxidative injury to many biomolecules is caused when they react with ROS. DNA and RNA both are partially damaged due to ROS reactions. However, DNA injury can be repaired by the specific restoration processes, while it is much difficult for RNA due to its severe sensitivity to injury caused by the ROS due to the lack of damage recovery mechanism. Moreover, injury caused by ROS to proteins is both irreversible and reversible (El-Maarouf-Bouteau et al. 2013). Enhanced production of antioxidant enzymes enables the seeds to control the ROS accretion during the water absorption. The detoxifying action of the ROS in the seed is termed as seed antioxidant potential, which shows the ability of seed to cope with the stress situations and increase germination potential (Liu et al. 2007). The expression profile of different genes involving the production of antioxidant enzymes, for example, superoxide dismutase, can help in determining the seed antioxidant potential. Likewise, role of different antioxidant enzymes such as catalase, glutathione, and ascorbate peroxidase was increased at protein level showing the stimulation in the antioxidant defense system. Other useful indicators of the antioxidant response in germinating seeds are metallothionein genes encoding different isoforms of the ROS scavenger protein metallothionein (Leszczyszyn et al. 2013; Chen et al. 2014). Different studies show the overexpression of the metallothione during seed rehydration such as in the sacred lotus, *Silene* spp. and *Arabidopsis* seeds. This overexpression also showed improved resistance to accelerated aging (Jin et al. 2014). Productions of many secondary metabolites are increased during the priming process, which enables the seeds to pass through severe physiological and metabolic alterations. All these make the way for seeds towards improved seed germination and successful seedling growth.

# 3 Alteration in Oxidative Metabolism by Priming

The important part of the seed physiology is to withstand the severe oxidative damage due to increased ROS production. Starting from seed development by maturation on the mother plant and seed germination, a huge alteration in the seed metabolic activity along with seed moisture content takes place. Alteration in the metabolite concentration is mainly because of the enhanced respiratory activities of mitochondria and increased activities of many enzymes such as NADPH oxidases, oxalate oxidases, peroxidases, and β-oxidation pathways. On the other hand, production of ROS causes oxidative damage that directly affects the germination process. Among ROS molecules, $H_2O_2$ plays its role as signaling molecule. However, seed must be equipped with strong ROS detoxifying system which promptly reduces the concentration of damaging molecules. This detoxification is carried out by an efficient system involving enzymatic (i.e., superoxide dismutase, glutathione reductase, monodehydroascorbate reductase, catalase) and nonenzymatic antioxidant molecules (i.e., reduced glutathione, ascorbic acid). Production of ROS at lower concentrations is also recognized to play its part as signaling molecules, such as $H_2O_2$, hydrogen sulfide, and nitric oxide, are responsible for changing dormant to germination phase (Bailly et al. 2008; Wojtyla et al. 2016; Krasuska and Gniazdowska 2012).

Similarly, it has been confirmed by Kubala et al. (2015a) that the process of seed hydration and drying back to original moisture content resulted in the increase of ROS production as well as induction of antioxidant enzyme system. The induction of AsA, GSH, and APX after the osmopriming of spinach seeds was also accompanied by higher production of superoxide dismutase and catalase (Chen and Arora 2011). So, the induction of AsA-GSH cycle during hydropriming of spinach seeds is a clear indication that it can reduce the level of lipid peroxidation components in the imbibed seeds. Furthermore, Kubala et al. (2015a) have also described that induction of antioxidant defensive system is the mere requirement of the establishment of the seedlings happened during the last phases of germination due to the increased production of catalase. They also have confirmed through integrated proteomic and transcriptomic approach that enhanced and successful germination due to priming is closely related to induction of antioxidant system. During the study of seed priming on *Brassica napus*, authors recorded the upregulation of CAT2 and PER21 encoding genes due to osmopriming and also observed the accumulation of GR protein. Similarly, they also have recorded the upregulation of gene (PER 13), which resulted in the accumulation of peroxidase, peroxiredoxin proteins, and DHAR during the germination of primed seeds. Moreover, the authors postulated a correlation between stress tolerance potential of hydro-primed seeds of *B. napus* and induction of antioxidant metabolic activities during the germination process. The increased production of CAT, SOD, and APX is associated with amplified expression of respective genes. Similarly, Nouman et al. (2014) conferred that seed priming of *Moringa* with *Moringa* leaf extract enhances the plant growth even under salt-stressed condition and it is all because of the production of the antioxidant enzymes such as POD, CAT, and SOD. Likewise, increased tolerance to stresses is caused by NaCl and polyethylene glycol due to enhanced activities of POX and SOD during germination of seeds primed with β-amino butyric acid. Also in tobacco varieties, increased tolerance of chilling stress was obtained because of antioxidant enzymes induced after seed priming with putrescine (Nouman et al. 2014). In a research conducted by Islam et al. (2015) on hydro-primed seeds of wheat, increased production in APX, POD, and CAT resulted in the salinity tolerance in the wheat seedlings. Effects of severe dry conditions caused by the application of PEG on seed germination and establishment of seedlings of sorghum were improved by osmopriming which was associated with increased activities of antioxidants including SOD, POD, CAT, and APX (Zhang et al. 2015a, b). Similarly, increased activity of APX along with decreased activity of CAT, POD, and SOD was found to reduce the negative effects caused by the ZnO nanoparticles stress in seeds of rice prime with PEG (Salah et al. 2015). ROS production significantly affects the seed age during storage period. Induction of the ROS deteriorates the quality of the seed. Seed quality restoration from aging can be recovered by the activation of CAT and seed priming plays a significant role in the activation of CAT. Further, sunflower seed aging has been observed to increase the storage of hydrogen peroxide and loss of CAT at gene expression level as well as protein content (Kibinza et al. 2011).

Fascinatingly, the negative outcomes of aging were restored by seed hydropriming, which resulted in the synthesis of CAT by inducing the gene expression and translation of the enzyme.

As a conclusion, priming causes a great alteration in the metabolism of seed which increased the production of many secondary metabolites. These metabolites play significant part in relieving the seeds from the hazardous oxidative stress, restoring the seeds from the aging, and regulating the production of ROS.

## 4 Reserve Mobilization and Priming

Latest transcriptome and proteome researches on cabbage, *Arabidopsis* seeds, and rape seeds osmopriming have shown the fact that germination and priming transformed analogous methods (Kubala et al. 2015a; Gallardo et al. 2001; Soeda et al. 2005). Certainly, it is projected that procedures linked with germination like respiration, metabolism of energy, and initial reserve mobilization process can also take place in the priming (Varier et al. 2010). Speedy and constant germination of rice seed as a result of priming resulted in enhanced α-amylase activity, consequently increasing the concentration of available sugars in primed kernels (Farooq et al. 2006). Sung and Chang (1993) have revealed that primed maize seeds exhibited enhanced enzymatic activity of carbohydrates such as α and β amylases and lipids, for example, isocitrate lyase solubilization. Chickpea seeds primed with mannitol showed enhanced amylase activity, invertases, sucrose synthase, and sucrose phosphate synthase in shoot portion of primed plants. The enhanced amylase concentration in shoots proposed a rapid breakdown of newly synthesized starch present in the shoots of primed plants, resulting in further accessibility of glucose for plant development (Kaur et al. 2002). Excessive concentration of soluble aldolase, protein, and ICL performance has been noticed in halo-primed seeds of pepper as compared to unprimed control (Smith and Cobb 1992). Furthermore, the accumulation of α-glucosidase resulted in higher concentration of globulin degradation stuff in the priming procedure for seeds of sugar beet. Parallel results regarding priming for seeds of sugar beet have been observed by Capron et al. (2000), which depicted enhanced mobilization of 11S-globulin-β-subunit as a result of osmopriming. Gallardo et al. (2001) have noticed elevated level of polypeptides in both hydro- and osmo-primed seeds of *Arabidopsis*. They were recognized as yields of 12S-cruciferin-β-subunit breakdown.

Kubala et al. (2015a, b) have documented that, in the germination of osmo-primed and post-primed *B. napus* seeds, storage of transcript and proteins happened and also enhanced storage of cruciferin CRU1 has been observed. The stimulation of genes related with lipid catabolism was interrelated with the instigation of genes required for lipid transport (Kubala et al. 2015a). Priming also decreases the concentration of oleosins-proteins that cover oil bodies.

During priming, stimulation of respiration and rapid ATP fabrication are main metabolic processes (McDonald 2000) and advanced respiratory action is essential to provide energy required for seed germination. The enhanced ATP level after priming was detected in araucaria, cabbage, eggplant, oat, spinach, and tomato (Mazor et al. 1984; Corbineau et al. 2000). Primed seeds require a high level of fuel to fulfill energy requirement for higher mobilization of reserve. Consequently, it resulted in enhanced energy level and metabolic rate of primed seeds and hence caused improved germination and stress-bearing potential.

## 5 Holistic "omic" Methods of Seed Priming

Identification of the biomarkers for sequenced primed seed is an important goal for plant physiologists. The order of processes involved in $H_2O$ holding priming includes a slight hydration stage in the soaking phase followed by a sufficient or inadequate fast dehydration period. It can be examined that every step has its own configuration of activation/deactivation. There is no link between transcription and translation, causing in few circumstances a restricted correlation between mRNA and protein levels. Certain proteins produced in incubation may be destroyed in dehydration, while the degree of degeneration may be affected by the level of drying. Lastly, protein production may also take place as a consequence of the translation of prolonged mRNA formerly produced throughout seed development (Taylor et al. 1998; Jisha et al. 2013; Boudet et al. 2006).

The benefit of these corresponding universal methods is that they offer a wide-ranging set of data permitting physiologist to get a worldwide opinion of all metabolic factors included in priming. However, the main drawback is that these practices are costly. As a result, specimens are regularly examined after treatment time and the time-reliant evolution of factors is hardly measured. It is quite problematic to reconstruct a kinetic method, but available readings deliver deep information regarding dynamic priming method.

## 6 Alteration in Transcriptomics by Priming

Transcriptomics is the procedure of examination of the transcriptome, the whole set of transcripts of RNA formed via genome under particular conditions by means of high-throughput approaches. Numerous practices are accessible for transcriptomic methods comprising c-DNA amplified fragment length polymorphism, microarray, disclosed sequence tag sequencing, and sequential examination of gene expression, RNA seq, and immense corresponding signature sequencing (Haider and Pal 2013). Buitink et al. (2006) have stated that above 1300 genes may be controlled in the priming. Those genes whose expressions are controlled in priming procedure are normally classified based on the role of the parallel proteins (metabolism, regulation of cell cycle, processing of DNA, regulation of transcription, stress reactions, cellular communications, and transport). But, a regular portion of the recognized genes is yet not explained.

In case of *B. napus*, frequent priming-structured genes are included in gluconeogenesis which is indispensable for triacylglycerol disintegration into minute fragments. Additional significant type of genes included in $H_2O$ holding priming of rape seed encodes for transcription factors and this is particularly the situation after hormopriming with ABA (Patade et al. 2012). In the similar types, it was verified that germination of primed seeds includes particular genes relatively to germination of unprimed seeds (Kubala et al. 2015a). In fresh plants of primed seeds, the appearance of stress-linked genes is frequently speedier as compared to the plants from unprimed seeds, as currently demonstrated for cold-bearing potential (Patade et al. 2012). Priming encouraged the genes involved both in protein production and protein breakdown. Certain transcription features may be downregulated in the soaking and upregulated during drying (Kubala et al. 2015a). Genes included in the production of osmoprotectant, like proline, can be controlled at several stages of priming (Kubala et al. 2015b).

Transcripts of *Brassica oleracea* are plentiful in dry seeds which quickly reduce in the osmopriming, and germination activity is started in the osmopriming. For these researchers, genes manifested in slow drying succeeding the soaking phase are interrelated with the stress tolerance characteristics of primed matter. It can be a particular consideration of cell cycle-regulated genes, enzymes included in carbon metabolism, and factors of the translation mechanism. Acquirement of the desiccation resistance after the soaking phase is a vital characteristic regarding priming. Transcriptomic data revealed that the molecular study related to such feature may be alike with those included in the usual seed dehydration procedure in the maturation. In *Medicago truncatula*, numerous regulatory genes manifested in response to scarcity stress are also upregulated in the seed development. For priming, these genes are required for post-soaking dehydration, while genes controlling cell wall biogenesis, cell cycle, and energy metabolism are suppressed (Buitink et al. (2006). Transcripts stored after PEG-priming action include those encoding for LEA and seed loading proteins or genes regulating seed dormancy, although genes incorporated in cell wall alterations and photosynthesis process are normally suppressed (Maia et al. 2011). Beside the contribution of transcription aspects, it may be of main significance in the upcoming time to examine the epigenetic alterations brought by priming.

## 7 Alteration in Proteomics by Priming

Proteome depicts the complete set of the protein present at a particular time in particular biological sample and proteomic is the comprehensive study that enables to identify and quantify these proteins. Two-dimensional gel electrophoresis and peptide mass fingerprinting are being practiced to separate proteins in different plant systems. To isolate the individual peptide from sequence information, tandem mass spectrometry is being used. Quantitative proteome analysis shows the alteration in protein level among the different biological samples (Wong and Cagney 2010). Another alternative tool is gel-free shotgun proteomic that is used to identify and quantify protein at large scale. This technique was applied in remote past to recognize the important biomarkers of priming-activated salinity tolerance in wheat by concomitantly using the hydro gel nanoparticles and protein fractionation enrichment methods (Fercha et al. 2013). Authors recorded the important alteration in 72 different proteins induced by osmopriming. Most of these proteins played their role in protein synthesis, proteolysis, defense or disease prevention, and metabolism regulation. A complete alteration in the pattern of proteome signature has been observed after priming with ascorbic acid. The proteins identified in the study are involved in antioxidant defense system, DNA repair process, metabolism regulation, and methionine-related metabolism. To estimate the accurate quantity of protein biosynthesis during these steps, an inhibitor of the protein synthesis such as cycloheximide can be added to a priming solution. In some cases, a new protein synthesis is not required by the seed because it depends only on the prestored-translated mRNA accumulated during the maturation of seed (Pace et al. 2012). Hydropriming with cycloheximide induces resistance to Cd stress in

*Trifoliun repens* and *Poa pratensis*, and this tolerance induction was associated with reduction in α and β-amylase (de Lespinay et al. 2010). There are two pathways for the alteration in the concentration of proteins in the seeds: first involves the synthesis of afresh proteins, while the other requires digestion of accumulated proteins through protease induction. It has been observed that priming can influence the activities of proteases and stimulate the expression of protease encoding gene (de Lespinay 2009).

To understand the alteration in the production of secondary metabolites, some techniques are even used to detect protein in the embryo or tissue covering the embryo. Monocots have aleurone layers that are related to hydrolytic enzyme synthesis (Bonsager et al. 2007). However, the separation of the aleurone is much difficult in the monocots grasses having very small seeds such as turf grass, while it is easily practiced in wheat or barley having large sized seeds. It has been recorded that the set of proteins that are directly regulated by priming in the embryo and neighboring regions is much more different. The number and quantity of proteins regulated by priming is much higher in embryo compared to other tissues of seeds. Proteins regulated in embryos are directly associated with regulation metabolism (particularly methionine biosynthesis, amino acids synthesis, etc.), cell growth, and preservation of cell structure and energy supply, while proteins upregulated in nearby seed parts are largely linked to reserves' mobilization or managing of the oxidative stress. On the other hand, only ascorbic acid is found in embryo of barley, while many other proteins taking part in redox reaction were found at the commencement of radicle elongation (Bonsager et al. 2007). Priming is regarded as a stimulating technique and numerous proteomic methods were used to identify the biomarker of seed vigor. A set of 18 accumulating proteins due to osmopriming has been identified in sugar beet seeds by Catusse et al. (2011). Further, these proteins are found to be reduced directly during the aging. Seed vigor seems directly associated with starch and lipid mobilization, production of protein, and methionine cycle. In another study, more than 81 proteins indicated an important alteration in their concentration as compared to same proteins found in seeds with varied vigor (Zhang et al. 2015a, b). In the light of these findings, reduction in seed vigor is a process that requires energy to proceed along with synthesis and degradation of protein. Salicylic acid is considered as a stimulating agent that enhances germination under salt-stressed conditions. Salicylic acid boosted the activation of antioxidant enzymes, enhanced the protein quality, and reactivated the maturation process during the early phases of germination. Likewise, Tanou et al. (2009) confirmed the activation and alteration in many secondary metabolic processes such as MAP kinase signaling, tyrosine nitration, N-nitrosilation, and redox proteomic using the "prime-omic" technique. Hydropriming was also observed to activate drought tolerance in *Medicago truncatula* (Boudet et al. 2006). It has been confirmed by the proteomic analysis that drought tolerance character is associated with the production of proteins from variant groups related to late embryogenesis. Moreover, Fourier transform infrared spectroscopy was used to study the alteration in protein in seeds after priming compared to dry seeds and results showed that the proteins acquire α helices and β-sheets configuration during drying process.

Many other techniques are used to study the protein patterns after priming in plant species whose genome is still not sequenced. An experiment was carried out to analyze the proteome of pearl millet seeds in response to seed priming with β-aminobutyric acid. Results indicated the overexpression of genes encoding proteins related to glucose metabolism and a number of these proteins are directly associated with energy (Anup et al. 2015). It has also been observed that seedlings originated from these primed seeds were more resistant to diseases such as downy mildew. It is also a noteworthy point that many of the proteins were found in the energy-related organelles such as mitochondria and chloroplast as well as in the extracellular spaces.

## 8 Alteration in Metabolomics by Priming

Metabolome can be referred to as the entire set of the metabolites having smaller molecular mass occurring within a plant tissue at a particular time. So metabolome can be recognized as quantitative measurement of the complete set of compounds which are directly participating in the metabolism of specified sample. Plant metabolomic analysis has gained a special importance in functional genomics. To study the complete range of metabolites using a sole analytical method is much more difficult. Therefore, a combination of several techniques and tools (i.e., gas chromatography, HPLC, atmospheric pressure chemical ionization, electrospray ionization assisted with mass spectrometry, secondary mass spectrometry, and capillary electrophoresis) is being used to examine the maximum number and quantity of metabolite. However, the metabolic tools are much harder to study the seed material, having the large proportion of molecules obtained from reverse mobilization. Superfluous soluble sugar present in the graminaea and a lot of lipid molecules present in the oil digestion sample obtained from the oleaginous plant material can hinder the separation of some other compounds with minute fraction but are involved in the improved seed germination in response to priming. In the same way, many of the metabolites are only synthesized or found in the particular tissues or organelle of the cell and undergo some chemical reactions during the extraction process. Gibberellins are an example which play an important part in the smoothing of germination process but are reported to react with phenols or sugars, so this property makes it a hindrance in isolation and/or identification of many other metabolites (Wu et al. 2014).

# 9 Conclusion

It can be concluded that priming has significant impact on secondary metabolites ranging from simple sugars to complex protein molecules. These metabolites have significant role in the germination potential. However, the exact quality- and quantity-based research is needed to fully explore all the features of modification process involving the priming. The study of this metabolite can help to improve the quality of seed and germination rate. Advanced molecular techniques applied to translational research programs in seed science will address the recent prospects of seed industry. This will enhance the seed response to industrial treatments, especially priming technology for the rationale management/production of competitive commercial seed lots. The accessibility of innovative guidelines will also facilitate researches on novel materials to be included in industrial formulations formed to enhance seed quality.

# References

Anup CP, Prasad M, Nataraj S, Mayuri NG, Manbali J, Hussain A, Kukkundoor RK (2015) Proteomic analysis of elicitation of downy mildew disease resistance in pearl millet by seed priming with β-aminobutyric acid and Pseudomonas fluorescens. J Proteome 120:58–74

Bailly C, El-Maarouf-Bouteau H, Corbineau F (2008) From intracellular signaling networks to cell death: the dual role of reactive oxygen species in seed physiology. C R Biol 331:806–814. http://sci-hub.tw/10.1016/j.crvi.2008.07.022

Bonsager BC, Finnie C, Roepstorff P, Svensson B (2007) Spatio-temporal changes in germi- nation and radicle elongation of barley seeds tracked by proteome analysis of dissected embryo, aleurone layer, and endosperm tissues. Proteomics 7:4528–4540

Boudet J, Buitnink J, Hoekstra FA, Rogniaux H, Larré C, Satour P, Leprince O (2006) Com- parative analysis of the heat stable proteome of radicles of Medicago truncatula seeds during germination identifies late embryogenesis abundant proteins associated with desiccation tolerance. Plant Physiol 140:1418–1436

Bray CM (1995) Biochemical processes during the osmopriming of seeds. In: Kigel J, Galili G (eds) Seed development and germination. Marcel Dekker, New York, pp 767–789

Buitink J, Jegger JJ, Guisle I, Wu BL, Wuillème LG, Bars AL, Meur NL, Becker A, Küster H, Leprince O (2006) Transcriptome profiling uncovers metabolic and regulatory processes occurring during the transition from desiccation sensitive to desiccationtolerant stages in Medicago truncatula seeds. Plant J 47:735–750

Capron I, Corbineau F, Dacher F, Job C, Côme D, Job D (2000) Sugarbeet seed priming: effects of priming on germination, solubilization of 11-S globulin and accumulation of LEA proteins. Seed Sci Res 10:243–254. http://sci-hub.tw/10.1017/S0960258500000271

Catusse J, Meinhard J, Job C, Strub JM, Fischer U, Pestova E, Westohoff P, Van Dorselaer A, Job D (2011) Proteomics reveals potential boiomarkers of seed vigor in sugarbeet. Proteomics 11:1569–1580

Chen K, Arora R (2011) Dynamics of the antioxidant system during seed osmopriming, postpriming germination, and seedling establishment in spinach (*Spinacia oleracea*). Plant Sci 180:212–220. http://sci-hub.tw/10.1016/j.plantsci.2010.08.007

Chen C, Letnik I, Hacham Y, Dobrev P, Ben-Daniel BH, Vankova R, Amir R, Miller G (2014) ASCORBATE PEROXIDASE6 protects Arabidopsis desiccating and germinating seeds from

stress and mediates cross talk between reactive oxygen species, abscisic acid, and auxin. Plant Physiol 166:370–383
Corbineau F, Özbingol N, Vineland D, Côme D (2000) Improvement of tomato seed germination by osmopriming as related to energy metabolism. In: Black M, Bradford KJ, Vasquez-Ramos J (eds) Seed biology advances and applications: proceedings of the sixth international workshop on seeds, Merida, Mexico. CABI, Cambridge, pp 467–474. http://sci-hub.tw/10.1079/9780851994048.0467
Cordoba-Canero D, Roldan-Arjona T, Ariza RR (2014) Arabidopsis ZDP DNA 30-phosphatase and ARP endonuclease function in 8-oxoG repair initiated by FPG and OGG1 DNA glycosylases. Plant J 79:824–834
de Lespinay A (2009) Study of seed priming mechanisms of three plant species used in revegetation of industrial sites. PhD thesis, Université Catholique de Louvain, Belgium, p 253
de Lespinay A, Lequeux H, Lambillotte B, Lutts S (2010) Protein synthesis is differentially required for germination in Poa pratensis and Trifolium repens in the absence or in the presence of cadmium. Plant Growth Regul 61:205–214
Dearman J, Brocklehust PA, Drew RLK (1987) Effect of osmotic priming and aging on the germination and emergence of carrot and leek seed. Ann Appl Biol 111:717–722
Di Girolamo G, Barbanti L (2012) Treatment conditions and biochemical processes influencing seed priming effectiveness. Ital J Agron 7:8–18
Diaz-Vivancos P, Barba-Espın G, Hernandez JA (2013) Elucidating hormonal/ROS networks during seed germination: insights and perspectives. Plant Cell Rep 32:1491–1502
Ellis RH, Butcher PD (1988) The effects of priming and ‘natural’ differences in quality amongst onion seed lots on the response of the rate of germination to temperature and the identification of the characteristics under genotypic control. J Exp Bot 39:935–950
El-Maarouf-Bouteau H, Meimoun P, Job C, Job D, Bailly C (2013) Role of protein and mRNA oxidation in seed dormancy and germination. Front Plant Sci 4:77
Farooq M, Basra SMA, Khalid M, Tabassum R, Mahmood T (2006) Nutrient homeostasis, metabolism of reserves, and seedling vigor as affected by seed priming in coarse rice. Can J Bot 84:1196–1202. http://sci-hub.tw/10.1139/b06-088
Fercha A, Capriotti AL, Caruso G, Cavaliere C, Gherroucha H, Samperi R, Stampachiacchierea S, Laganaa A (2013) Gel-free proteomics reveal potential biomarkers of priminginduced salt tolerance in durum wheat. J Proteome 91:486–499. http://sci-hub.tw/10.1016/j.jprot.2013.08.010
Foyer CH, Noctor G (2005) Redox homeostasis and antioxidant signaling: a metabolic interface between stress perception and physiological responses. Plant Cell 17:1866–1875
Galland M, Huguet R, Arc E, Cueff G, Job D, Rajjou L (2014) Dynamic proteomics emphasizes the importance of selective mRNA translation and protein turnover during Arabidopsis seed germination. Mol Cell Proteomics 13:252–268
Gallardo K, Job C, Groot SPC, Puype M, Demol H, Vandekerckhove J, Job D (2001) Proteomic analysis of Arabidopsis seed germination and priming. Plant Physiol 126:835–848. http://sci-hub.tw/10.1104/pp.126.2.835
Grover S, Khan AS (2014) Effect of ionizing radiation on some characteristics of seeds of wheat. Int J Sci Technol Res 3:2277–8616
Haider S, Pal R (2013) Integrated analysis of transcriptomic and proteomic data. Curr Genomics 14:91–110
Hill H, Bradford KJ, Cunningham J, Taylor AG (2008) Primed lettuce seeds exhibit increased sensitivity to moisture during aging. Acta Hortic 782:135–141
Islam F, Yasmeen T, Ali S, Ali B, Farooq MA, Gill RA (2015) Priming-induced antioxidative responses in two wheat cultivars under saline stress. Acta Physiol Plant 37:153. http://sci-hub.tw/10.1007/s11738-015-1897-5
Jin S, Sun D, Wang J, Li Y, Wang X, Liu S (2014) Expression of the rgMT gene, encoding for a rice metallothionein-like protein in Saccharomyces cerevisiae and Arabidopsis thaliana. J Genet 93:709–718

Jisha KC, Vijayakumari K, Puthur JT (2013) Seed priming for abiotic stress tolerance: an overview. Acta Physiol Plant 35:1381–1396. http://sci-hub.tw/10.1007/s11738-012-1186-5
Kaur S, Gupta AK, Kaur N (2002) Effect of osmo- and hydropriming of chickpea seeds on seedling growth and carbohydrate metabolism under water deficit stress. Plant Growth Regul 37:17–22. http://sci-hub.tw/10.1023/A:1020310008830
Kibinza S, Bazin J, Bailly C, Farrant JM, Corbineau F, El-Maarouf-Bouteau H (2011) Catalase is a key enzyme in seed recovery from ageing during priming. Plant Sci 181:309–315. http://sci-hub.tw/10.1016/j.plantsci.2011.06.003
Kranner I, Beckett RP, Minibayeva FV, Seal CE (2010) What is stress? Concepts, definitions and applications in seed science. New Phytol 188:655–673
Krasuska U, Gniazdowska A (2012) Nitric oxide and hydrogen cyanide as regulating factors of enzymatic antioxidant system in germinating apple embryos. Acta Physiol Plant 34:683–692. http://sci-hub.tw/10.1007/s11738-011-0868-8
Krol P, Igielski R, Pollmann S, Kepczynska E (2015) Priming of seeds mith methyljasmonate induced resistance to hemibiotroph Fusarium oxysporum f. sp. Lycopersici in tomato via 12-oxo-phytodienoic acid, salicylic acid, and flavonol accumulation. J Plant Physiol 179:122–132
Kubala S, Garnczarska M, Wojtyla Ł, Clippe A, Kosmala A, Żmieńko A, Lutts S, Quinet M (2015a) Deciphering priming-induced improvement of rapeseed (*Brassica napus* L.) germination through an integrated transcriptomic and proteomic approach. Plant Sci 231:94–113. http://sci-hub.tw/10.1016/j.plantsci.2014.11.008
Kubala S, Wojtyla Ł, Quinet M, Lechowska K, Lutts S, Garnczarska M (2015b) Enhanced expression of the proline synthesis gene P5CSA in relation to seed osmopriming improvement of *Brassica napus* germination under salinity stress. J Plant Physiol 183:1–12. http://sci-hub.tw/10.1016/j.jplph.2015.04.009
Leszczyszyn OI, Imam HT, Blindauer CA (2013) Diversity and distribution of plant metallothioneins: a review of structure, properties and functions. Metallomics 5:1146–1169
Liu X, Xing D, Li L, Zhang L (2007) Rapid deterioration of seed vigour based on the level of superoxide generation during early imbibition. Photochem Photobiol Sci 6:767–774
Macovei A, Tuteja N (2013) Different expression of miRNA targeting helicases in rice in response to low and high dose rate c-ray treatments. Plant Sign Behav 8:e25128
Macovei A, Garg B, Raikwar S, Balestrazzi A, Carbonera D, Buttafava A, Bremont JFJ, Gill SS, Tuteja N (2014) Synergistic exposure of rice seeds to different doses of gamma-ray and salinity stress resulted in increased antioxidant enzyme activities and gene-specific modulation of TC-NER pathway. Biomed Res Int 2014:676934
Maia J, Dekkers BJW, Provart NJ, Ligterink W, Hilhorst HWM (2011) The re-establishement of desiccation tolerance in germinated Arabidopsis thaliana seeds and its associated transcriptome. PLoS One 6:e29123
Mazor L, Perl M, Negbi M (1984) Changes in some ATP-dependent activities in seeds during treatment with polyethylene glycol and during the redrying process. J Exp Bot 35:1119–1127
McDonald MB (2000) Seed Priming. In: Black M, Bewley JD (eds) Seed technology and its biological basis. Sheffield Academic Press, Sheffield, pp 287–325
Moller IM, Jensen PE, Hansson A (2007) Oxidative modifications to cellular components in plants. Ann Rev Plant Biol 58:459–481
Momin KC (2013) Challenges of the flower seed industry. Indian J Appl Res 3:244–245
Murungu FS, Chiduza C, Nyamugafata P, Clark LJ, Whalley WR, Finch-Savage WE (2004) Effects of 'on-farm seed priming' on consecutive daily sowing occasions on the emergence and growth of maize in semi-arid Zimbabwe. Field Crops Res 89:49–57
Nouman W, Basra SMA, Yasmeen A, Gull T, Hussain SB, Zubair M, Gul R (2014) Seed priming improves the emergence potential, growth and antioxidant system of Moringa oleifera under saline conditions. Plant Growth Regul 73:267–278. http://sci-hub.tw/10.1007/s10725-014-9887-y
Osburn RM, Schroth MN (1989) Effect of osmopriming sugar beet deed on exudation and subsequent damping-off caused by Pythium ultimum. Phytopathology 78:1246–1250

Pace R, Benincasa P, Ghanem ME, Quinet M, Lutts S (2012) Germination of uintreated and primed seeds in rapeseed (Brassica napus var oleifera Del.) under salinity and low matric potential. Exp Agric 48:238–251
Parera C, Cantliffe D (1990) Improved stand establishment of sh 2 sweet corn by solid matrix priming. In: Proc Nat Symp Stand Estab Hort Crops, Minneapolis, MN, pp 91–96
Patade VY, Khatri D, Manoj K, Kumari M, Ahmed Z (2012) Cold tolerance in thiourea primed capsicum seedlings is associated with transcript regulation of stress responsive genes. Mol Biol Rep 39:10603–10613
Rajjou L, Duval M, Gallardo K, Catusse J, Bally J, Job C, Job D (2012) Seed germination and vigor. Annu Rev Plant Biol 63:507–533
Salah SM, Yajing G, Dongdong C, Jie L, Aamir N, Qijuan H, Weimin H, Mingyu N, Jin H (2015) Seed priming with polyethylene glycol regulating the physiological and molecular mechanism in rice (*Oryza sativa* L.) under nano-ZnO stress. Sci Rep 5:14278. http://sci-hub.tw/10.1038/srep14278
Smith PT, Cobb BG (1992) Physiological and enzymatic characteristic of primed, re-dried, and germinated pepper seeds (*Capsicum annum* L). Seed Sci Technol 20:503–513
Soeda Y, Konings MCJM, Vorst O, van Houwelingen AMM, Stoopen GM, Maliepaard CA, Kodde J, Bino RJ, Groot SPC, van der Geest AHM (2005) Gene expression programs during Brassica oleracea seed maturation, osmopriming, and germination are indicators of progression of the germination process and the stress tolerance level. Plant Physiol 137:354–368. http://sci-hub.tw/10.1104/pp.104.051664
Sung FJM, Chang YH (1993) Biochemical activities associated with priming of sweetcorn seed to improve vigor. Seed Sci Technol 21:97–105
Tan L, Chen S, Wang T, Dai S (2013) Proteomic insights into seed germination in response to environmental factors. Proteomics 13:1850–1870
Tanou G, Job C, Rajjou L, Arc E, Belghazi M, Diamentidis G, Molasiotis A, Job D (2009) Proteomics reveals the overlapping roles of hydrogen peroxide and nitric oxide in the acclimation of citrus plants to salinity. Plant J 60:795–804
Taylor AG, Allen PS, Bennett MA, Bradford JK, Burris JS, Mishra MK (1998) Seed enhancements. Seed Sci Res 8:245–256. http://sci-hub.tw/10.1017/S0960258500004141
Varier A, Vari AK, Dadlani M (2010) The subcellular basis of seed priming. Curr Sci 99:450–456
Ventura L, Dona M, Macovei A, Carbonera D, Buttafava A, Mondoni A, Rossi G, Balestrazzi A (2012) Understanding the molecular pathways associated with seed vigor. Plant Physiol Biochem 60:196–206
Waterworth WM, Masnavi G, Bhardwaj RM, Jiang Q, Bray CM, West CE (2010) A plant DNA ligase is an important determinant of seed longevity. Plant J 63:848–860
Wojtyla Ł, Lechowska K, Kubala S, Garnczarska M (2016) Different modes of hydrogen peroxide action during seed germination. Front Plant Sci 7:66. http://sci-hub.tw/10.3389/fpls.2016.00066
Wong JW, Cagney G (2010) An overview of label-free quantitation methods in proteomics by mass spectrometry. Methods Mol Biol 604:273–283
Wu X, Li N, Li H, Tang H (2014) An optimized method for NMR-based plant seed metabolomics analysis with maximized polar metabolite extraction efficiency, signal-to-noise ratio, and chemical shift consistency. Analyst 139:1769–1778
Zhang F, Yu J, Johnston CR, Wang Y, Zhu K, Lu F, Zhang Z, Zou J (2015a) Seed priming with polyethylene glycol induces physiological changes in sorghum (*Sorghum bicolor* L. Moench) seedlings under suboptimal soil moisture environments. PLoS One 10:e0140620. http://sci-hub.tw/10.1371/journal.pone.0140620
Zhang H, Wang WQ, Liu SJ, Moller JM, Song SQ (2015b) Proteome analysis of poplar seed vigor. PLoS One 10:e0132509

# Seed Priming-Induced Early Vigor in Crops: An Alternate Strategy for Abiotic Stress Tolerance

**Meenakshi Thakur, Poornima Sharma, and Anjali Anand**

## Contents



**Abstract** Plants experience various stresses during the course of their growth and development. Seed priming involves the induction of a particular physiological stage of growth in the seed, prior to germination, using different approaches like hydro-, osmo-, chemical, hormonal, biological, matrix, and magnetopriming. Seed priming-induced early vigor in crops reduces time of seed emergence, accomplishes uniform emergence, and gives a better crop stand. Hence, seed priming is a smart, potent, and feasible alternative that facilitates plant protection against various stress conditions during seedling establishment. Primed seeds imbibe water rapidly and

M. Thakur
College of Horticulture and Forestry, Neri, Hamirpur, Dr. Y.S. Parmar University of Horticulture and Forestry, Nauni, Solan, Himachal Pradesh, India

P. Sharma · A. Anand (✉)
Division of Plant Physiology, ICAR-Indian Agricultural Research Institute, New Delhi, India



M. Hasanuzzaman, V. Fotopoulos (eds.), *Priming and Pretreatment of Seeds and Seedlings*, https://doi.org/10.1007/978-981-13-8625-1_8

revive the seed metabolism through repair and buildup of nucleic acids and proteins. Various signaling pathways activate in the early stages of seedling growth resulting in faster plant defense responses. It leads to the accumulation of inactive signaling molecules in primed cells which, upon later subjection to stress condition, lead to hyper-activation of the signaling proteins, thereby amplifying signal transduction, thus guiding toward more rapid and/or more intense activation of defense responses. It imparts "stress memory" that activates stress-responsive system in primed seeds, equipping them to be defensive to later stress exposure. The activities of antioxidative enzymes and synthesis of metabolites such as proline and malondialdehyde under various stress conditions are also enhanced in response to priming. The molecular changes during stress in primed seeds include cell division and elongation, plasma membrane fluidity, induction of stress-responsive proteins (heat-shock proteins and late embryogenesis abundant proteins), and changes in transcriptome and proteome. All these changes lead to enhanced and synchronized seed germination, early vigor, and improved plant growth, biomass, plant height, leaf area, root-shoot length, dry weight, and yield. In this chapter, we will discuss the physiological, biochemical, and molecular mechanisms of seed priming under various abiotic stress conditions.

**Keywords** Antioxidant · Drought · Germination · Priming · ROS · Priming · Seed metabolism · Salinity · Temperature stress · Vigor

## 1 Introduction

Seed quality is a measure of viability, seed lot purity, health, and mechanical damage with a more arcane trait of seed vigor (Perry 1980). Plants are challenged by various stress factors of environment such as less available water, salinity, mechanical impedance due to crusting, soil temperature (high or low), insects, pests, and oxygen that inhibit seed germination and seedling emergence (Zhao et al. 2007). Under stressful growing conditions, seeds wait for the favorable environmental conditions and take longer time to germinate. However, the seedbed deteriorates during this waiting period, and the seedlings experience increased stress leading to mortality. The descent of seedbed condition before seedling emergence is affected by temperature and water availability. Seed germination occurs successfully, if the moisture content is sufficient in the seedbed (Trait 1). Once the radicle emerges out of the seed, further growth depends upon the temperature of the seedbed as determined by drying of the soil surface from upward toward downward. Thus, the roots rapidly grow downward (Trait 2) facilitating the maintenance of root tip contact with receding moisture. The soil strength increases due to its continuous drying, and the hypocotyls extend out of the soil before its emergence (Trait 3). Induction of seed vigor in a seed lot suggests that priming treatments offer a strategy for avoidance of stress that ensures rapid germination of seeds and subsequent growth in impeded soils. The seed germinates rapidly and exhibits rapid growth of roots and shoots in the soils of increasing impedance. The physiological changes after seedling emergence result in establishment of a uniform crop stand and increased yield under stress (Fig. 1).

Successful crop establishment under deteriorating seedbed can be attained by enhanced vigor in Traits 1–3 and increased rate of seedling development. Therefore, robust priming treatments are required to withstand changes under different stresses experienced by the seedbed (Finch-Savage et al. 2010). Seed priming techniques involve controlled hydration-dehydration steps that allow metabolic activities during imbibition to proceed but restrict emergence of radicle (Bradford 1986).

Seed reserves in the form of starch, protein, and lipids are deposited in the seed during seed development. Prior to autotrophic growth, the reserves are mobilized to the growing seedling and sustain the seedling during photosynthesis-independent growth. Although in-depth knowledge about seedling emergence under stress is scarce, the effect of seedbed's stress environment on initial seedling growth can be determined. The physical stress in the seedbed decides the seed reserve-dependent germination and seedling development pattern prior to emergence. Germination and priming are essentially similar processes as both exhibit similar altered pattern of expression of genes/proteins (Soeda et al. 2005). A triphasic pattern of water uptake during germination includes phase I, a physical process that involves water uptake and is irrespective of dormant or nondormant, viable or nonviable seeds; phase II, a lag phase that involves little net water uptake where seed moves from dormant to

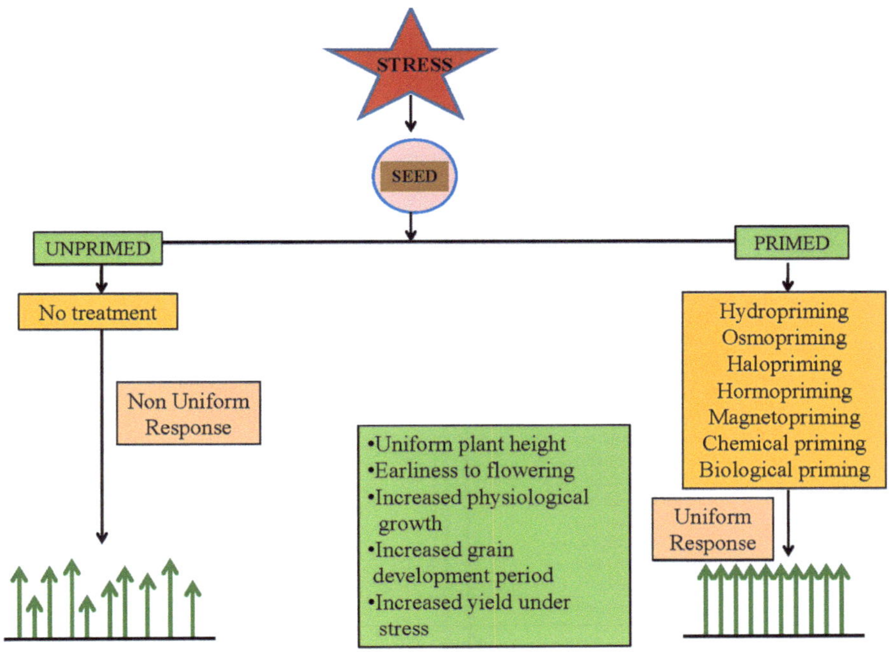


**Fig. 1** Physiological response in plants raised from primed seeds leading to the establishment of a uniform plant population

germinating state; and phase III, a marked increase in water uptake associated with radicle protrusion and growth (Bewley and Black 1994). Similar to phases I and II of germination, priming can also be considered as a special type of preparative germination phase in which phase II is prolonged (Nonogaki et al. 2007). Faster absorption of water enables a "head start" of transition state that possibly improves the germination potential which is reflected by greater germination percentage, speed, and uniformity (Fig. 2) (Myint et al. 2010). Seed treatment with natural and synthetic compounds prior to germination provides the advantage due to priming, which results in the production of signal molecules that have the potential to enhance seed germination and radicle emergence during stress conditions. Thus, priming generates moderate stress during soaking and dehydration (Ashraf and Foolad 2005) that prepare the seeds to tolerate abiotic stresses during seedling establishment (Yacoubi et al. 2011). The germination rate is enhanced as a consequence of revival of seed metabolism following priming treatment (McDonald 2000).

Various approaches are employed for seed priming which include partial shortening of the duration of imbibition (hydropriming); keeping the seeds in solutions with low external water potential (osmopriming); chemical, hormonal, biological, solid matrix and magnetopriming etc. Different priming agents exhibit varied response to crop species and type of stress (Iqbal and Ashraf 2005), although each treatment improves the germination rate, uniformity, and growth of seedlings particularly under stressful conditions (Parera and Cantliffe 1991). DNA damage dur-

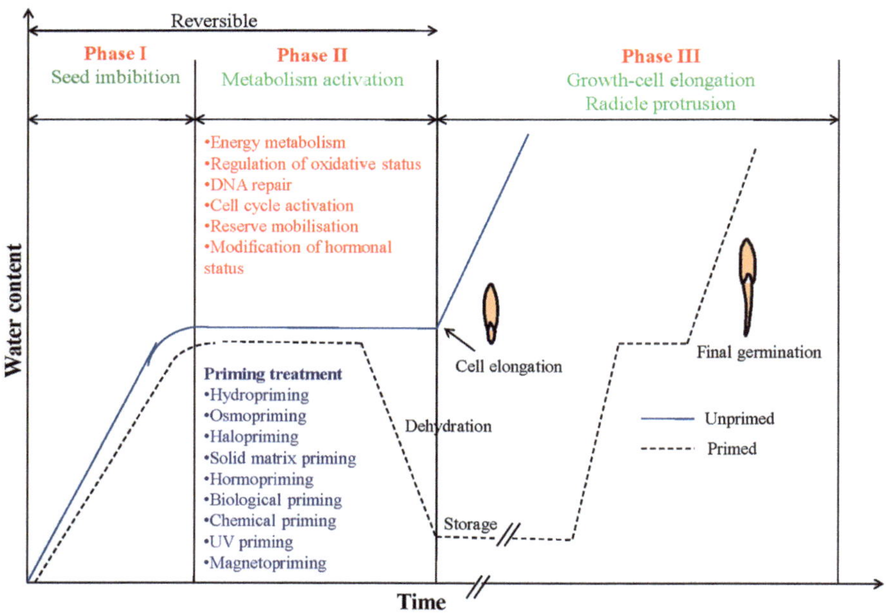


**Fig. 2** Seed hydration curves and germinating phases in unprimed and primed seeds. (Adapted from Lutts et al. 2016)

ing seed aging can be repaired by upregulation of DNA repair machinery that helps in improving seed viability (Thornton et al. 1993). Priming activates the respiratory pathway resulting in production of energy during the seed imbibition (Weitbrecht et al. 2011). In osmopriming, water is absorbed at a slow rate due to reduced water potential of the solution used for priming, giving an advantage to the seed to complete the repair of DNA. Primed seeds exhibit less damage to the cells which is otherwise an inevitable feature during rehydration of the unprimed seed in the germination process (Balestrazzi et al. 2011). Chemical compounds, that act as priming agents, lead to production of signaling molecules or inactive transcription factors, which rapidly act or get activated to ensure protection upon being exposed to stressful condition (Bruce et al. 2007). In addition, such agents have been shown to reduce oxidative stress that is associated with various abiotic stresses (Demidchik 2015), in a similar mode of action in both the seedlings and mature plants. In the following section, we discuss the different types of priming treatments and their effectiveness in alleviating the stressed condition.

## 2 Types of Priming

### 2.1 *Hydropriming*

Hydropriming relies on wetting of the seeds in water followed by re-drying to original moisture content before sowing. This method is not so expensive and environmentally friendly. On-farm priming, which is synonymous to hydropriming, consists of conducting the practice of seed wetting followed by surface drying on farm, prior to sowing (Harris et al. 2005). This technique may be useful for farmers in marginal tropical environment with limited resources. However, the main disadvantage of hydropriming is nonuniform degree of seed hydration due to uncontrolled water uptake and unsynchronized emergence due to difference in concurrent activation of metabolic pathways within seeds. Hence, it becomes highly important to optimize the hydration conditions such as duration of treatment, temperature, and water volume used in hydropriming to ensure uniform seed hydration and synchronized emergence. Drum priming – a commercial variant of hydropriming (Rowse 1996) involves rotation of seeds in drum followed by gradual hydration by addition of water in vapor form. Drum priming is better over conventional hydropriming as it allows controlled seed imbibition. At the cellular level, hydropriming of maize seeds for 12 h changed the protein profiles in the embryos with abundance of two proteins, namely, embryonic protein DC-8 and globulin-1, which could serve as potential candidates for the protein markers for priming effect and seed vigor (Gong et al. 2013). Earlier studies have revealed the association of proteins like actin isoform or cytosolic glyceraldehyde 3-phosphate dehydrogenase, which operate during seed dehydration process, with germination and priming process (Gallardo et al. 2001).

## 2.2 Osmopriming and Halopriming

In osmopriming, seeds are soaked in low water potential solutions such as polyethylene glycol, mannitol, sorbitol, glycerol, or inorganic salts such as $CaCl_2$, KCl, $KNO_3$, $K_3PO_4$, $KH_2PO_4$, $MgSO_4$, and NaCl (Yacoubi et al. 2013). When inorganic salts are used as priming agents, it is often referred to as halopriming. Soaking of the seeds in osmotic solutions of low water potential results in gradual seed imbibition as water enters the seeds slowly, which results in the activation of early phases of germination (Di Girolamo and Barbanti 2012). Also, halopriming activates seed metabolism by production of osmolytes and/or signaling molecules, thus promoting germination. Lara et al. (2014) suggested that priming of tomato seeds with potassium nitrate ($KNO_3$) promoted faster germination as the absorbed nitrate entered the seed embryo metabolism through the nitrate reductase enzyme pathway, resulting in the production of signaling molecule, nitrite/nitric oxide. Sahin et al. (2011) observed that NaCl priming in tomato seeds resulted in the accumulation of osmoregulating defense molecules (e.g., proline and anthocyanin) and antioxidative enzyme system activation.

## 2.3 Solid Matrix Priming

Osmopriming possesses some limitations due to expensive osmotic agents and technical problems with aeration (Paparella et al. 2015). Solid matrix priming (matriconditioning) is an alternative method to osmopriming which involves controlled water uptake by seeds. Solid matrix priming is cost-effective as it involves mixing and incubation of seeds with wet solid water carrier for a specific duration. Afterward, seeds are separated from matrix, washed, and dried again. The use of solid medium permits slow hydration of seeds, thus simulating the natural imbibition process that occurs when seed is sown in the soil (McDonald 2000).

## 2.4 Hormopriming

During hormopriming, plant growth regulators such as auxin, gibberellin, kinetin, polyamines, abscisic acid, ethylene, and salicylic acid are used during seed imbibition, which can have direct impact on seed germination.

## 2.5 Biological Priming

Biological priming involves seed imbibition together with plant growth-promoting bacteria (Callan et al. 1990). This priming method provides protection to seeds against the soil and seed-borne pathogens in addition to enhanced rate and uniformity of germination.

## 2.6 Chemical Priming

In this method of priming, seeds are treated with certain natural or synthetic compounds (nitric oxide, hydrogen peroxide, hydrogen sulfide, melatonin, and polyamines) to improve germination rate and enhance abiotic stress tolerance.

## 2.7 UV Priming

In UV priming, seeds are treated with low doses of UV radiations to improve yield and quality and enhance biotic and abiotic stress tolerance in plants. Dhanya Thomas and Puthur (2017) observed that UV radiation (0.004–4 $Wm^{-2}$) stimulated germination, growth, biomass, and fresh and dry weight of seedlings in various crops such as rice, wheat, maize, cucumber, and cowpea. The activities of ROS scavenging system also increased, resulting in abiotic stress tolerance in plants.

## 2.8 Magnetopriming

Magnetopriming involves treatment of seeds with magnetic field of different strength. It is a dry seed priming technique that does not involve hydration of seeds and allows storage of seeds at room temperature. In germinating soybean seeds, magnetopriming resulted in the stimulation of seedling growth through the production of ROS (Shine et al. 2012). Podlesny et al. (2005) observed that increased level of free radicals in faba bean and pea seeds was accompanied with enhanced germination after treatment with laser light and magnetic field.

# 3 Physiological, Biochemical, and Molecular Aspects of Seed Priming for Abiotic Stress Tolerance

Abiotic stress causes an array of physiological, biochemical, and metabolic changes in plants which causes oxidative stress, modification in metabolism, and eventually poor yield (Xiong and Zhu 2002; Shafi et al. 2009). Seed priming technology has dual benefits during stressed conditions: (1) promotes high vigor favoring enhanced, fast, and uniform emergence and (2) improves yields in plants from primed seeds (Basra et al. 2005; Kaur et al. 2005). According to McDonald (2000), primed seeds attain the capacity for faster imbibition and revival of seed metabolism that includes repair and synthesis of nucleic acids and proteins followed by post-translational processing and targeted proteolysis (Kubala et al. 2015). Chromosomal damage may be repaired during priming (Sivritepe and Dourado, 1995), accompanied by early DNA repair, and increase in RNA and protein synthesis (Manonmani et al. 2014). Lipid peroxidation is reduced (Chiu et al. 2006), and antioxidative enzyme activities increased in primed seeds (Espanany et al. 2016). Seed quality enhancement by seed priming has also been linked to molecular changes during stress which include increased cell division and elongation, fluidity of plasma membrane, induction of heat-shock and late embryogenesis abundant proteins, and changes at transcriptome and proteome level (Varier et al. 2010; Mahajan et al. 2011). The physiological, biochemical, and molecular changes observed by various researchers in primed seeds in response to various stress conditions have been discussed below:

## 3.1 *Salt Stress*

Salinity adversely affects early seed germination and growth of seedling and ultimately alters the metabolic processes in a mature plant (Nawaz et al. 2012). Salt stress lowers down the osmotic potential external to the seed and prevents water uptake; thereby increased $Na^+$ and $Cl^-$ ions negatively affect the germinating seed (Khajeh-Hosseini et al. 2003). However, the first physiological process to be affected under salinity is reduced water uptake by the seed due to low water potential of the soil. This is accompanied by a disturbance in the mobility of nutrients to developing tissues (Ashraf and Wahid 2000), reduced hydrolysis of seed reserve (Ahmad and Bano 1992), and accumulation of osmolytes (Zidan and Elewa 1995). All these physiological perturbations can be mitigated if the seed is induced to complete the seedling establishment phase within a short time period. This is accomplished by pre-sowing priming treatments that embolden the seed to defend the first encounter to stresses. Cayuela et al. (1996) investigated the effect of seed priming with 6 M NaCl solution with respect to growth and physiological responses of tomato plants (*Lycopersicon esculentum* L.). It was observed that primed seeds emerged earlier in comparison to non-treated seeds. Primed seeds also accumulated $Na^+$ and $Cl^-$ ions in the roots. An increased sugar and organic acid content in the leaves of primed seeds was also observed. Positive effects of osmopriming of sweet sorghum seeds for mitigation of adverse response in salt and thermal stress during germination and radicle growth were evident in studies by Patane et al. (2009). Germination under saline media improved in seeds of tomato and asparagus treated with 0.8 MPa PEG 8000 (Pill et al. 1991). Cucumber seeds osmoconditioned with 0.7 M mannitol improved the rate of germination in 200 mM NaCl solutions (Passam and Kakouriotis 1994). Gao et al. (1999) studied aquaporins in osmoconditioned oilseed rapeseed (*Brassica napus*) during germination under both salt and osmotic stresses. They identified two aquaporin genes *Bn-PIP1* belonging to the family of plasma membrane intrinsic protein (PIPs) and *Bn-TIP2* to tonoplast intrinsic protein. They suggested that *Bn-PIP1* was required for hydration and enzymatic breakdown of storage reserves at early stages of germination, whereas *Bn-TIP2* was related to cell growth associated with radicle protrusion.

Magnetopriming of chickpea seeds was effectively used to overcome the adverse effects of salinity at germination and early seedling growth (Thomas et al. 2013). In magnetoprimed wheat seeds, Na/K ratio was less compared to unprimed plants under saline conditions. De Souza et al. (2016) reported 72% germination in osmoprimed *Physalis angulata* L. seeds grown under salt stress of 14 dS $m^{-1}$ EC or higher. The higher biomass in the primed seeds may be due to the upregulation of genes related to ion transport and antioxidant enzymes. The relative expression of *APX*, *GST*, *TXN*, *HAK1*, and *SOS1* was higher in primed seeds which could be correlated with the induction of salt tolerance. ROS production during salt stress induces a signaling cascade that can change the gene expression and protect the primed seed from oxidative stress under high salt concentration.

The metabolic processes required for germination under stress conditions are maintained by induced expression of genes specific to water homeostasis, inorganic ion transport and metabolism, cell wall biogenesis, and signal transduction (Bertorello and Zhu 2009; Peng et al. 2009). Salt tolerance in rice seedlings was mediated through ion homeostasis that helped in increasing anthocyanin and chlorophyll content in seeds primed with spermidine and gibberellic acid (Chunthaburee et al. 2014). The phenolic content and antioxidant capacity were higher in the primed than the control seedlings.

## 3.2 Drought Stress

Water stress may potentially affect seed germination and early seedling growth (Ahmad et al. 2009). Yan (2015) observed that priming of Chinese cabbage (*Brassica rapa*) seeds with 200 mM potassium nitrate and 200 mM urea stimulated germination and early seedling growth under drought stress. Seed priming treatments provided enhanced drought tolerance by modulating the activities of peroxidase, superoxide dismutase, and catalase and levels of soluble sugar and proline. These results suggested that seed priming can be effectively used to increase the germination and early seedling growth of Chinese cabbage under moisture stress conditions.

Chickpea plants grown from magnetically treated seeds and subjected to −0.2 MPa of moisture deficit showed an increase in water utilization, biomass, and radiation use efficiency (Mridha et al. 2016). In maize, magnetopriming helped the seedlings to withstand moisture stress due to improved soil-water relations (Anand et al. 2012). Seed priming with hormonal (methyl jasmonate, salicylic acid) or chemical elicitor (paclobutrazol) resulted in increased total phenolic content, antioxidant activity, and expression of rice drought-responsive (*RD1* and *RD2*) genes (of AP2/ERF family) in drought-tolerant rice genotype under drought stress (Samota et al. 2017). Growth and development of the plants raised from primed seeds were found to be better under control and drought stress conditions compared to that of the plants raised from unprimed seeds under the stress.

Fatemi (2014) primed the seeds of sunflower (*Helianthus annuus* L.) with ascorbic acid and concluded that primed seeds exhibited higher germination percentage, seed stamina index, and growth as compared to control under low water potentials (−2 and − 4 bar) induced by PEG 6000. The exogenously applied ascorbic acid was effective in ameliorating the adverse effects of drought stress. Similar results were also observed in ascorbic acid priming of safflower seeds where the antioxidant system of enzymatic and nonenzymatic compounds was increased to alleviate the drought stress (Rajaji et al. 2012). Primed seeds also exhibited higher α-amylase and/or β-amylase activities, which resulted in an increased breakdown of starch and subsequent buildup of sugar levels. This was followed by higher rate of respiration, viability of seed, speed of germination, and establishment of seedling as compared to non-primed seeds (Zheng et al. 2009; Amooaghaie and Nikzad 2013; Li et al. 2013, 2014a, b). Li et al. (2014a, b) reported that chemical agents such as polyamines upregulate the expression of β-amylase gene under water stress during the initial stages of seed germination.

Soybean plants pretreated with $H_2O_2$ when exposed to drought stress resulted in higher levels of myo-inositol and galactinol. These sugars are considered to be the osmoprotectants and scavengers of reactive oxygen species. Increased gene expression of D-myo-inositol 3-phosphate synthase 2 (GmMIPS2) and galactinol synthase (GolS), the key enzymes for the biosynthesis of the oligosaccharides, is correlated with the accumulation of myo-inositol and galactinol (Ishibashi et al. 2011). The role of DC-8 in the protection of embryo cellular structures during drought stress tolerance to desiccation was implicated by Sghaier-Hammami et al. (2010). The PEG priming in *Arabidopsis* and *Medicago* seeds upregulated AP2/EREBP regulon during germination which act as a component of the hormone, sugar, and redox signaling for abiotic stresses (Buitink et al. 2006; Dietz et al. 2010; Maia et al. 2011).

Salah et al. (2015) highlighted the effect of polyethylene glycol seed priming on the physiological and molecular mechanism of two cultivars of *Oryza sativa* L. under different levels of zinc oxide nanorods. Under nano-ZnO stress, PEG priming resulted in significantly reduced antioxidant enzyme activities as well as malondialdehyde contents. Gene expression analysis suggested that PEG priming downregulated the expression of *APXa*, *APXb*, *CATa*, *CATb*, *CATc*, *SOD1*, *SOD2*, and *SOD3* genes as compared to non-primed seeds under stress. The ultrastructural analysis showed that nano-ZnO stress significantly damaged the leaf mesophyll and root cells in both cultivars. However, seed priming with PEG significantly alleviated the toxic effects of nano-ZnO stress and improved the cell structures of leaf and roots in both cultivars.

Rice plants pretreated with $H_2O_2$ showed higher mRNA levels for $\Delta'$-pyrroline-5-carboxylate synthase (P5CS), involved in the biosynthesis of proline (Uchida et al. 2002).

The DNA repair system was activated in *Artemisia sphaerocephala* and *Artemisia ordosia* during seed priming that could improve seed viability under stressed environment (Huang et al. 2007). In another study, soaking the seeds of *Medicago truncatula* in water increased the level of 8-oxodG, the main oxidative DNA lesion occurring under oxidative stress, while in osmoprimed seeds, the oxidative damage was temporally shifted indicating the protective role of osmotic agent within the first few hours of rehydration (Balestrazzi et al. 2011).

## *3.3 Temperature Stress*

Extreme temperature stress has a detrimental effect on the metabolic activities of crop plants. Seed priming techniques are used by various researchers to combat temperature stress in various crops. Heat stress adversely affects yield of late-sown wheat. Osmoprimed late-sown wheat ameliorated the adverse effect of high temperature by increasing the number of tillers, biological yield, and harvest index (Mustafa et al. 2017). On the other hand, chilling stress is also responsible for limiting the growth and productivity of many crops. Low temperature results in poor and erratic germination as a result of reduced water uptake (Li et al. 2013), membrane injury, less cellular respiration (Xing and Rajashekar 2001), and elevated ROS levels (Nayyar et al. 2005). Low temperature-induced damage to cell organelles results in disruption of key physiological processes. It is manifested by inhibition of cell division and elongation and metabolic imbalance in plant tissues which is eventually translated into disturbed seed morphology (Ruelland et al. 2009; Li et al. 2013).

In rice plants, Hussain et al. (2016) studied the function of various seed priming techniques such as hydropriming, osmopriming, redox priming, chemical priming, and hormonal priming to impart tolerance against chilling. Low temperature affected various physiological processes like delayed germination and poor seedling growth in unprimed rice. The metabolic events associated with seed reserve mobilization and respiration was reduced. Control rice seedlings showed an increased lipid peroxidation and hydrogen peroxide accumulation. Seed priming safeguards the seedlings from oxidative stress by triggering the activities of antioxidant enzyme system like superoxide dismutase, peroxidase, and catalase and the accumulation of glutathione and free proline. Corn seeds primed with melatonin show increased tolerance to chilling stress upon germination (Kołodziejczyk et al. 2016).

To summarize, we have postulated a schematic diagram to depict the physiological, biochemical, and molecular changes in primed seeds in Fig. 3.

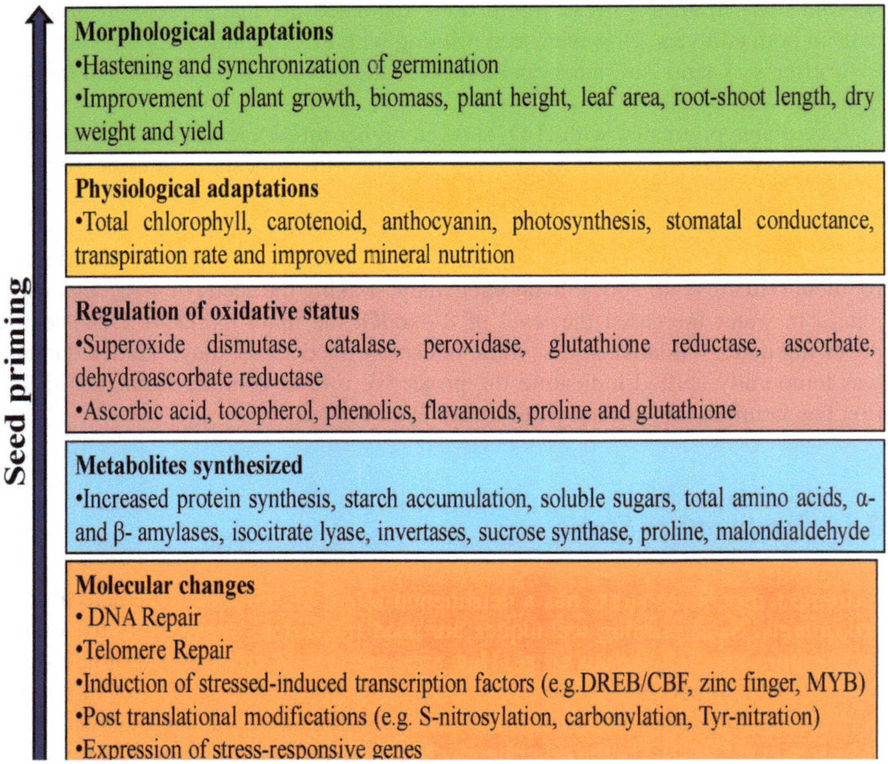


**Fig. 3** Physiological, biochemical, and molecular changes in primed seeds

## 4 The Concept of "Stress Memory" During Seed Priming

Priming is akin to pre-germination stress exposure that can establish a "stress memory" in the seeds (Bruce et al. 2007). Thus, the enhanced tolerance during post-priming germination may be due to the priming-induced "cross-tolerance." Various studies have elucidated the "cross-tolerance" response in various crops, e.g., NaCl priming imparted tolerance to PEG-induced drought stress in sugarcane (*Saccharum officinarum*) and high-temperature tolerance in barley (*Hordeum vulgare*); $H_2O_2$ priming induced salt and heat tolerance in wheat seedlings and maize seeds, respectively; and hydro-, $CaCl_2$, and ABA priming imparted salt and drought stress tolerance in Indian mustard (*Brassica juncea*) (Wahid et al. 2007, 2008; Mei and Song 2008; Patade et al. 2009; Srivastava et al. 2010). Figure 4 represents the stress signaling mechanism mediated by stress memory in primed seeds.

Chen and Arora (2011) summarized the strategies to improve seed stress tolerance with two approaches: (1) ensuring an advanced status of the germination and (2) providing priming-induced "stress imprint." A noticeable response of priming effect was seen into the later plant developmental stages, e.g., soybean seeds coated

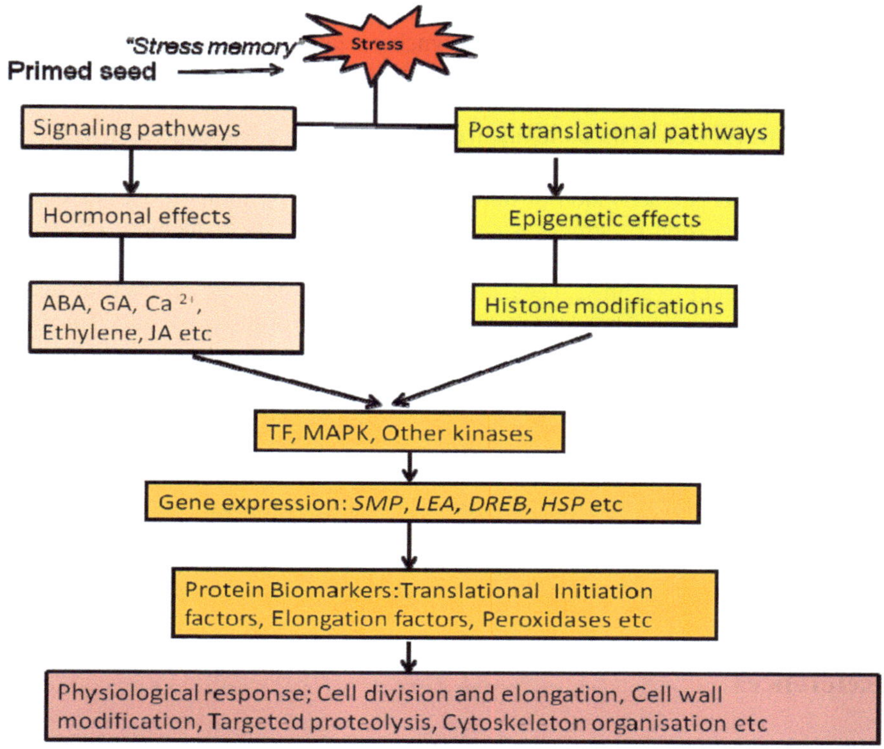


**Fig. 4** Mechanism of stress signaling in primed seeds

with melatonin showed improved tolerance to drought (Wei et al. 2014). Similar results were observed in wheat seeds treated with $H_2O_2$ (He et al. 2009). All these studies reinforce the opinion that seed priming is an efficient method for improvement of stress tolerance in plants through the application of "stress memory," most likely through epigenetic modifications.

## 5 Conclusion

Abiotic stress at the germination stage is caused by extreme temperatures, water excess or limitation, soil crusting, and salinity individually or in combination, leading to poor germination, irregular plant stand, and low yields. The mitigation of the adverse effects of the stresses at the germination stage can improve the chances of a good crop stand which further leads to higher yield from the suboptimal growing environment. Seed priming is a pre-sowing controlled hydration of seed with different techniques followed by dehydration, to increase the speed and uniformity of germination and emergence under stress conditions. It has the advantages of being practical and inexpensive that guarantees its acceptance by the farmers. At the physiological level, it results in hastening and synchronization of germination, leading to uniform plant growth under abiotic stress. Biochemical responses of seed priming include activation of various signaling pathways and cellular responses in the early stages of growth such as the antioxidant defense system, increased protein synthesis, sugar and starch accumulation, etc. Molecular changes in primed seeds in response to stress involve the activation of transcription factors, posttranslational modifications, and expression of stress-responsive genes. Also, stress memory or "stress imprint" that is generated during priming due to abiotic stresses activates the stress-responsive system in primed seeds and makes them to be more resistant to subsequent stress exposure. These mechanisms establish a "priming memory" in seeds which can be employed for stress tolerance at a later exposure stage. In the holistic sense, seed priming can induce early vigor response brought about by activation of the germination metabolism that helps the seed to mitigate stress and establish the crop stand. Upscaling of the priming techniques and studies on maintaining the longevity of primed seeds under storage would help in expanding their application under field conditions. Basic research in seed priming complemented with molecular biology can help in deciphering the mechanisms that are pivotal in explaining the early seedling vigor and improved stand establishment of primed seed.

## References


Ahmad J, Bano M (1992) The effect of sodium chloride on physiology of cotyledons and mobilization of reserved food in *Cicer arietinum*. Pak J Bot 24:40–48

Ahmad S, Ahmad R, Ashraf MY, Ashraf M, Waraich EA (2009) Sunflower (*Helianthus annuus* L.) response to drought stress at germination and seedling growth stages. Pak J Bot 41:647–654

Amooaghaie R, Nikzad K (2013) The role of nitric oxide in priming-induced low-temperature tolerance in two genotypes of tomato. Seed Sci Res 23:123–131

Anand A, Nagarajan S, Verma APS, Joshi DK, Pathak PC, Bhardwaj J (2012) Pre-treatment of seeds with static magnetic field ameliorates soil water stress in seedlings of maize (*Zea mays* L.). Ind J Biochem Biophys 49:63–70

Ashraf M, Foolad MR (2005) Pre-sowing seed treatment-a shotgun approach to improve germination, plant growth, and crop yield under saline and non-saline conditions. Adv Agron 88:223–271

Ashraf M, Wahid S (2000) Time-course changes in organic metabolites and mineral nutrients in germinating maize seeds under salt (NaCl) stress. Seed Sci Technol 28:641–656

Balestrazzi A, Confalonieri M, Macovei A, Carbonera D (2011) Seed imbibition in *Medicago truncatula* Gaertn.: expression profiles of DNA repair genes in relation to PEG-mediated stress. J Plant Physiol 168:706–713

Basra SMA, Farooq M, Tabassum R (2005) Physiological and biochemical aspects of seed vigour enhancement treatments in fine rice (*Oryza sativa* L.). Seed Sci Technol 33:623–628

Bertorello AM, Zhu JK (2009) SIK1/SOS2 networks: decoding sodium signals via calcium-responsive protein kinase pathways. Eur J Appl Physiol 458:613–619

Bewley JD, Black M (1994) Seeds: physiology of development and germination, 2nd edn. Plenum Press, New York

Bradford KJ (1986) Manipulation of seed water relations via osmotic priming to improve germination under stress conditions. Hort Sci 21:1105–1112

Bruce TJA, Matthes MC, Napier JA, Pickett JA (2007) Stressful 'memories' of plants: evidence and possible mechanisms. Plant Sci 173:603–608

Buitink J, Leger JJ, Guisle I, Vu BL, Wuilleme Lamirault G, Bars AL, Meur NL, Becker A, Kuster H, Leprince O (2006) Transcriptome profiling uncovers metabolic and regulatory processes occurring during the transition from desiccation-sensitive to desiccation-tolerant stages in *Medicago truncatula* seeds. Plant J 47:735–750

Callan NW, Marthre DE, Miller JB (1990) Bio-priming seed treatment for biological control of *Pythium ultimum* preemergence damping-off in sh-2 sweet corn. Plant Dis 74:368–372

Cayuela E, Perez-Alfocea K, Caro M, Bolarin MC (1996) Priming of seeds with NaCl induces physiological changes in tomato plants grown under salt stress. Physiol Plant 96:231–236

Chen K, Arora R (2011) Dynamics of the antioxidant system during seed osmopriming, post-priming germination, and seedling establishment in spinach (*Spinacia oleracea*). Plant Sci 180:212–220

Chiu KY, Chuang SJ, Sung JM (2006) Both antioxidation and lipid-carbohydrate conversion enhancements are involved in priming-improved emergence of *Echinacea purpurea* seeds that differ in size. Sci Hortic 108:220–226

Chunthaburee S, Sanitchon J, Pattanagul W, Theerakulpisut P (2014) Alleviation of salt stress in seedlings of black glutinous rice by seed priming with spermidine and gibberellic acid. Notuale Bot Horti Agrobot Cluj-Napoca 42(2):405–413

De Souza MO, Pelacani CR, Willems LAJ, De Castro RD, Hilhorst HWM, Ligterink W (2016) Effect of osmopriming on germination and initial growth of *Physalis angulata* L. under salt stress and on expression of associated genes. Ann Braz Acad Sci 88:503–516

Demidchik V (2015) Mechanisms of oxidative stress in plants: from classical chemistry to cell biology. Environ Exp Bot 109:212–228

Dhanya Thomas TT, Puthur JT (2017) UV radiation priming: a means of amplifying the inherent potential for abiotic stress tolerance in crop plants. Environ Exp Bot 138:57–66

Di Girolamo G, Barbanti L (2012) Treatment conditions and biochemical processes influencing seed priming effectiveness. Ital J Agron 7:8–18

Dietz KJ, Vogel MO, Viehhauser A (2010) AP2/EREBP transcription factors are part of gene regulatory networks and integrate metabolic, hormonal and environmental signals in stress acclimation and retrograde signaling. Protoplasma 245:3–14

Espanany A, Fallah S, Tadayyon A (2016) Seed priming improves seed germination and reduces oxidative stress in black cumin (*Nigella sativa*) in presence of cadmium. Ind Crop Prod 79:195–204

Fatemi SN (2014) Germination and seedling growth in primed seeds of sunflower under water stress. Annu Res Rev Biol 4(23):3459–3469

Finch-Savage WE, Clay HA, Lynn JR, Morris K (2010) Towards a genetic understanding of seed vigor in small-seeded crops using natural variation in *Brassica oleracea*. Plant Sci 179(6):582–589

Gallardo K, Job C, Groot SP, Puype M, Demol H, Vandekerckhove J, Job D (2001) Proteomic analysis of *Arabidopsis* seed germination and priming. Plant Physiol 126:835–848

Gao YP, Young L, Bonham-Smith P, Gusta LV (1999) Characterization and expression of plasma and tonoplast membrane aquaporins in primed seed of *Brassica napus* during germination under stress conditions. Plant Mol Biol 40:635–644

Gong F, Wu X, Wang W (2013) Comparative proteomic identification of embryo proteins associated with hydropriming induced rapid-germination of maize seeds. Plant Omics J 6(5):333–339

Harris D, Breese WA, Kumar Rao JVDK (2005) The improvement of crop yield in marginal environments using 'on-farm' seed priming: nodulation, nitrogen fixation and disease resistance. Aust J Agric Res 56:1211–1218

He L, Gao Z, Li R (2009) Pretreatment of seed with $H_2O_2$ enhances drought tolerance of wheat (*Triticum aestivum* L.) seedlings. Afr J Biotechnol 8:6151–6157

Huang Z, Boubriak I, Osborne DJ, Dong M, Gutterman Y (2007) Possible role of pectin-containing mucilage and dew in repairing embryo DNA of seeds adapted to desert conditions. Ann Bot 101(2):277–283

Hussain S, Khan F, Hussain HA, Nie L (2016) Physiological and biochemical mechanisms of seed priming-induced chilling tolerance in rice cultivars. Front Plant Sci 7:116
Iqbal M, Ashraf M (2005) Changes in growth, photosynthetic capacity and ionic relations in spring wheat (*Triticum aestivum* L.) due to pre-sowing seed treatment with polyamines. Plant Growth Regul 46:19–30
Ishibashi Y, Yamaguchi H, Yuasa T, Iwaya-Inoue M, Arima S, Zheng SH (2011) Hydrogen peroxide spraying alleviates drought stress in soybean plants. J Plant Physiol 168:1562–1567
Kaur S, Gupta AK, Kaur N (2005) Seed priming increases crop yield possibly by modulating enzymes of sucrose metabolism in chickpea. J Agron Crop Sci 19:81–87
Khajeh-Hosseini M, Powell AA, Bingham IJ (2003) The interaction between salinity stress and seed vigour during germination of soyabean seeds. Seed Sci Technol 31:715–725
Kołodziejczyk I, Dzitko K, Szewczyk R, Posmyk MM (2016) Exogenous melatonin improves corn (*Zea mays* L.) embryo proteome in seeds subjected to chilling stress. J Plant Physiol 193:47–56
Kubala S, Garnczarska M, Wojtyla L, Clippe A, Kosmala A, Zmienko A et al (2015) Deciphering priming induced improvement of rapeseed (*Brassica napus* L.) germination through an integrated transcriptomic and proteomic approach. Plant Sci 231:94–113
Lara TS, Lira JMS, Rodrigues AC, Rakocevic M, Alvarenga AA (2014) Potassium nitrate priming affects the activity of nitrate reductase and antioxidant enzymes in tomato germination. J Agric Sci 6(2):72–80
Li X, Jiang H, Liu F, Cai J, Dai T, Cao W et al (2013) Induction of chilling tolerance in wheat during germination by pre-soaking seed with nitric oxide and gibberellin. Plant Growth Regul 71:31–40
Li X, Cai J, Liu F, Dai T, Cao W, Jiang D (2014a) Cold priming drives the sub-cellular antioxidant systems to protect photosynthetic electron transport against subsequent low temperature stress in winter wheat. Plant Physiol Biochem 82:34–43
Li Z, Peng Y, Zhang XQ, Ma X, Huang LK, Yan YH (2014b) Exogenous spermidine improves seed germination of white clover under water stress via involvement in starch metabolism, antioxidant defenses and relevant gene expression. Molecules 19(11):18003–18024
Lutts S, Benincasa P, Wojtyla L, Kubala S, Pace R, Lechowska K, Quinet M, Garnczarska M (2016) Seed priming: new comprehensive approaches for an old empirical technique. In: Araujo S, Balestrazzi A (eds) New challenges in seed biology – basic and translational research driving seed technology. InTechOpen, Rijeka, pp 1–46
Mahajan G, Sarlach RS, Japinder S, Gill MS (2011) Seed priming effects on germination, growth and yield of dry direct-seeded rice. J Crop Improv 25:409–417
Maia J, Dekkers BJW, Provart NJ, Ligterink W, Hilhorst HWM (2011) The re-establishment of desiccation tolerance in germinated *Arabidopsis thaliana* seeds and its associated transcriptome. PLoS One 6:e29123
Manonmani V, Begum MAJ, Jayanthi M (2014) Halo priming of seeds. Res J Seed Sci 7:1–13
McDonald MB (2000) Seed priming. In: Black M, Bewley JD (eds) Seed technology and its biological basis. Sheffield Academic Press, Sheffield, pp 287–325
Mei YQ, Song SQ (2008) Cross-tolerance is associated with temperature and salinity stress during germination of barley seeds. Seed Sci Technol 36:689–698
Mridha N, Chattaraj S, Chakraborty D, Anand A, Aggarwal P, Nagarajan S (2016) Pre-sowing static magnetic field treatment for improving water and radiation use efficiency in chickpea (*Cicer arietinum* L.) under soil moisture stress. Bioelectromagnetics 37:400–408
Mustafa HSB, Mahmood T, Ullah A, Sharif A, Bhatti AN, Muhammad Nadeem M, Ali R (2017) Role of seed priming to enhance growth and development of crop plants against biotic and abiotic stresses. Bull Biol Allied Sci Res 2:1–11
Myint T, Chanprasert W, Srikul S (2010) Effect of seed weight on germination potential of different oil palm (*Elaeis guineensis* Jacq.) crosses. Seed Sci Technol 58:125–135
Nawaz A, Amjad M, Jahangir MM, Khan SM, Cui H, Hu J (2012) Induction of salt tolerance in tomato (*Lycopersicon esculentum* Mill.) seeds through sand priming. Aust J Crop Sci 6(7):1199–1203

Nayyar H, Chander K, Kumar S, Bains T (2005) Glycine betaine mitigates cold stress damage in chickpea. Agron Sustain Dev 25:381–388
Nonogaki H, Chen F, Bradford KJ (2007) Mechanisms and genes involved in germination sensu stricto. In: Bradford KJ, Nonogaki H (eds) Seed development, dormancy and germination. Blackwell, Oxford, pp 264–304
Paparella S, Araujo SS, Rossi G, Wijayasinghe M, Carbonera D, Balestrazzi A (2015) Seed priming: state of the art and new perspectives. Plant Cell Rep 34:1281–1293
Parera CA, Cantliffe DJ (1991) Improved germination and modified imbibitions of shrunken-2 sweet corn by seed disinfection and solid matrix priming. J Am Soc Hortic Sci 116:942–945
Passam HC, Kakouriotis D (1994) The effects of osmoconditioning on the germination, emergence and early plant growth of cucumber under saline conditions. Sci Hortic 57:233–240
Patade VY, Bhargava S, Suprasanna P (2009) Halopriming imparts tolerance to salt and PEG induced drought stress in sugarcane. Agric Ecosyst Environ 134:24–28
Patane C, Cavallaro V, Cosentino SL (2009) Germination and radicle growth in unprimed and primed seeds of sweet sorghum as affected by reduced water potential in NaCl at different temperatures. Ind Crop Prod 30:1–8
Peng Z, Wang MC, Li F, Lv HJ, Li CL, Xia GM (2009) A proteomic study of the response to salinity and drought stress in an introgression strain of bread wheat. Mol Cell Proteomics 8:2676
Perry DA (1980) The concept of seed vigor and its relevance to seed production techniques. In: Hebblethwaite PD (ed) Seed production. Butterworths, London, pp 585–591
Pill WG, Frett JJ, Morneau DC (1991) Germination and seedling emergence of primed tomato and asparagus seeds under adverse conditions. HortScience 26:1160–1162
Podlesny J, Stochmal A, Podlesna A, Misiak LE (2005) Effect of laser light treatment on some biochemical and physiological processes in seeds and seedlings of white lupine and faba bean. Plant Growth Reg 67(3):227–233
Rajaji A, Asli DE, Farzanian M (2012) The effects of seed priming with ascorbic acid on drought tolerance and some morphological and physiological characteristics of safflower (*Carthamus tinctorius* L.). Ann Biol Res 3(8):3984–3989
Rowse HR (1996) Drum priming-a non-osmotic method of priming seeds. Seed Sci Technol 24(2):281–294
Ruelland E, Vaultier MN, Zachowski A, Hurry V (2009) Cold signaling and cold acclimation in plants. Adv Bot Res 49:35–150
Sahin FI, Iseri OD, Haberal M (2011) NaCl priming improves salinity response of tomato (*Lycopersium esculentum* Mill.) at seedling stage. Curr Opin Biotechnol 22:S46–S47
Salah SM, Yajing G, Dongdong C, Jie L, Amir N, Qijuan H, Weimin H, Mingyu N, Jin H (2015) Seed priming with polyethylene glycol regulating the physiological and molecular mechanism in rice (*Oryza sativa* L.) under nano-ZnO stress. Sci Rep 5:14278
Samota MK, Sasi M, Awana M, Yadav OP, Mithra SVA, Tyagi A, Kumar S, Singh A (2017) Elicitor-induced biochemical and molecular manifestations to improve drought tolerance in rice (*Oryza sativa* L.) through seed-priming. Front Plant Sci 8:934
Sghaier-Hammami B, Jorrin Novo J, Gargouri-Bouzid R, Drira N (2010) Abscisic acid and sucrose increase in the protein content in date palm somatic embryos, causing changes in 2-DE profile. Phytochemistry 71:1223–1236
Shafi M, Bakht J, Hassan MJ, Raziuddin M, Zhang G (2009) Effect of cadmium and salinity stresses on growth and antioxidant enzyme activities of wheat (*Triticum aestivum* L.). Bull Environ Contam Toxicol 82:772–776
Shine MB, Guruprasad KN, Anand A (2012) Enhancement of germination, growth, and photosynthesis in soybean by pre-treatment of seeds with magnetic field. Bioelectromagnetics 32(6):474–484
Sivritepe HO, Dourado AM (1995) The effect of priming treatments on the viability and accumulation of chromosomal damage in aged pea seeds. Ann Bot 75:165–171
Soeda Y, Konings MC, Vorst O, van Houwelingen AM et al (2005) Gene expression programs during *Brassica oleracea* seed maturation, osmopriming, and germination are indicators of progression of the germination process and the stress tolerance level. Plant Physiol 137:354–368

Srivastava AK et al (2010) Comparative evaluation of hydro-, chemo-, and hormonal-priming methods for imparting salt and PEG stress tolerance in Indian mustard (*Brassica juncea* L.). Acta Physiol Plant 32:1135–1144

Thomas S, Anand A, Chinnusamy V, Dahuja A, Basu S (2013) Magnetopriming circumvents the effect of salinity stress on germination in chickpea seeds. Acta Physiol Plant 35:3401–3411

Thornton JM, Collins ARS, Powell AA (1993) The effect of aerated hydration on DNA synthesis in embryos of *Brassica oleracea* L. Seed Sci Res 3:195–199

Uchida A, Jagendorf AT, Hibino T, Takabe T et al (2002) Effects of hydrogen peroxide and nitric oxide on both salt and heat stress tolerance in rice. Plant Sci 163:515–523

Varier A, Vari AK, Dadlani M (2010) The subcellular basis of seed priming. Curr Sci 99:450–456

Wahid A, Perveen M, Gelani S, Basra SMA (2007) Pretreatment of seed with $H_2O_2$ improves salt tolerance of wheat seedlings by alleviation of oxidative damage and expression of stress proteins. J Plant Physiol 164:283–294

Wahid A, Sehar S, Perveen M, Galani S et al (2008) Seed pretreatment with hydrogen peroxide improves heat tolerance in maize at germination and seedling growth stages. Seed Sci Technol 36:633–645

Wei W, Li QT, Chu YN, Reiter RJ et al (2014) Melatonin enhances plant growth and abiotic stress tolerance in soybean plants. J Exp Bot 66:695–707

Weitbrecht K, Muller K, Leubner-Metzger G (2011) First off the mark: early seed germination. J Exp Bot 62:3289–3309

Xing W, Rajashekar CB (2001) Glycine betaine involvement in freezing tolerance and water stress is *Arabidopsis thaliana*. Environ Exp Bot 46:21–28

Xiong L, Zhu JK (2002) Molecular and genetic aspects of plant responses to osmotic stress. Plant Cell Environ 25:131–139

Yacoubi R, Job C, Belghazi M, Chaibi W, Job D (2011) Toward characterizing seed vigor in alfalfa through proteomic analysis of germination and priming. J Proteome Res 10:3891–3903

Yacoubi R, Job C, Belghazi M, Chaibi W, Job D (2013) Proteomic analysis of the enhancement of seed vigor in osmoprimed alfalfa seeds germinated under salinity stress. Seed Sci Res 23:99–110

Yan M (2015) Seed priming stimulate germination and early seedling growth of Chinese cabbage under drought stress. S Afr J Bot 99:88–92

Zhao TJ, Liu Y, Yan YB, Feng F, Liu WQ, Zhou HM (2007) Identification of the amino acids crucial for the activities of drought responsive element binding factors (DREBs) of *Brassica napus*. FEBS Lett 581:3044–3050

Zheng C, Jiang D, Liu E, Cao W et al (2009) Exogenous nitric oxide improves seed germination in wheat against mitochondrial oxidative damage induced by high salinity. Environ Exp Bot 67:222–227

Zidan MA, Elewa MA (1995) Effect of salinity on germination, seedling growth and some metabolic changes in four plant species (Umbelliferae). Ind J Plant Physiol 38:57–61

# Seed Priming Alleviates Stress Tolerance in Rice (*Oryza sativa* L.)

**R. K. Sarkar, A. K. Mukherjee, and K. Chakraborty**

## Contents



R. K. Sarkar (✉) · K. Chakraborty
Department of Crop Physiology & Biochemistry, ICAR-National Rice Research Institute, Cuttack, India

A. K. Mukherjee
Department of Crop Protection, ICAR-National Rice Research Institute, Cuttack, India



M. Hasanuzzaman, V. Fotopoulos (eds.), *Priming and Pretreatment of Seeds and Seedlings*, https://doi.org/10.1007/978-981-13-8625-1_9

**Abstract** Seed priming is an age-old practice in agriculture. Seed germination and seedling growth can be improved through seed priming (seed hydration–dehydration–rehydration techniques) and seed coating with different living and nonliving substances. Seed priming is an age-old practice in agriculture dates to 1926 showed that rapid germination and seedling growth due to chlorine water priming. Seed priming break ups the seed cover, decreases inhibitor concentration in hull and endosperm and transforms the seeds into a higher state of activation. Both germination rate and seedling growth turn out to be fast and superior in primed seeds compared to non-primed seeds, while capsulated seed gets some essential ingredients required for fast and uniform germination from the coating substances. It is observed that different priming agents improve tolerance to excess and deficit water, salinity, metal toxicity and temperature and different biotic stresses in rice. Reports show that capsulated seeds with appropriate pelleting agents induce tolerance to abiotic stresses such as germination stage oxygen deficiency, cold, drought and salinity. Seed treatments have greater impact to stabilize rice yields under adverse conditions. In this chapter, physiological and biochemical status of primed seeds are discussed in relation to tolerance to diverse abiotic stresses in rice.

**Keywords** Abiotic stresses · Rice · Seed invigoration · Tolerance · Yield

## Abbreviations

| | |
|---|---|
| ABA | abscisic acid |
| ALA | 5-aminolevulinic acid |
| As | arsenic |
| Bo | boron |
| $CaCl_2$ | calcium chloride |
| CAT | catalase |
| Cd | cadmium |
| CL | chitosan-lignosulphonate polymer |
| Cr | chromium |
| DNA | deoxyribonucleic acid |
| E | eugenol |
| GA | gibberellic acid |
| GPX | glutathione peroxidase |
| GSOD | germination stage oxygen deficiency |
| $H_2O_2$ | hydrogen peroxide |
| Hsp70 | heat shock protein 70 |
| IAA | indole-3-acetic acid |
| KCl | potassium chloride |
| $KNO_3$ | potassium nitrate |
| LEA | late embryogenesis abundant |

| | |
|---|---|
| MDA | malondialdehyde |
| Mn | manganese |
| mRNA | messenger RNA |
| P | phosphorus |
| PAs | polyamines |
| Pb | lead |
| PEG | polyethylene glycol |
| RNA | ribonucleic acid |
| ROS | reactive oxygen species |
| RQ | respiratory quotient |
| SA | salicylic acid |
| Se | selenium |
| SOD | superoxide dismutase |
| SPD | spermidine |
| SPM | spermine |
| Zn | zinc |

## 1 Introduction

Rice, the world's most important cereal, provides food and livelihood in South and Southeast Asian countries. Almost half of the world populace meet their daily energy requirement from it. In Asia approximately 90% of rice is produced and consumed (Khush 2005), so any disturbances that decrease the rice yield bring poverty and food insecurity (Zeigler and Barclay 2008; Sarkar et al. 2018). Rice production is now at verge of fall, even in the areas of assured irrigation due to weather-related consequences (Rang et al. 2011; Brown et al. 2018). Several impeding factors starting from industrialization to housing risk-free premium land is converted from agricultural to non-agricultural land. Rice cultivation is now concentrating more in riskier areas. These areas are now more vulnerable than before by the effect of drought, salinity, submergence, temperature and many more other stresses due to unprecedented occurrence of climate change (Roy and Guha 2017; Brown et al. 2018; Sarkar et al. 2018). Millions of hectares of rice land are affected because of these stresses (Sarkar et al. 2006; Bailey-Serres and Voesenek 2008; Radanielson et al. 2018). In vast South and Southeast Asian countries, rice is mainly cultivated during rainy season (June–November) (Seck et al. 2012; Singh and Sarkar 2014; Smith et al. 2017). So, sustainable rice production during this period is the guarantee of stress-free life during the rest time of the year. During rainy season, among the weather-related main problems are drought, submergence and salinity, whereas during dry season, high and cold temperature and salinity at coastal areas are predominant abiotic stresses (Reddy et al. 2009; Sarkar et al. 2013, 2018; Ray et al. 2016; Radanielson et al. 2018). High temperature stress at flowering and grain-filling stages is not so severe to rainy season compared to dry season in the case of rice. Temperature during the month of October–November (flowering to grain-filling period of rainy season rice) is within the optimal range, whereas the temperatures from April to May (flowering to grain-filling period of dry season rice) sometimes greatly shoot up above the optimal range (33–35 °C). It was reported that even 1-h exposure at temperature of >35 °C at anthesis could create greater sterility in rice (Yoshida 1981; Rang et al. 2011).

Therefore, to get rid of food insecurity, improving productivity of rice under the adverse effect of climate change is crucial. Alleviation of stress-induced damage can secure greater productivity. Among the different techniques employed to alleviate stresses, seed priming is thought to be an easy technique, which can improve productivity under different stressful environments (Abichandani and Ramiah 1951; Bradford 1995; Farooq et al. 2009; Sarkar 2012; Singh et al. 2016). 'Sampietro in 1926 showed that rice seed soaked in chlorine-water produced more rapid germination, with subsequent seedling development as much as 27% better than with untreated seed' (Mikkelsen and Sinah 1961). Abichandani and Ramiah (1951) reported that seed treatment with nutrient solution improved yield in rice. Germination takes place in three phases such as (1) phase I, imbibitional stage; (2) phase II, activation stage; and (3) phase III, germination stage (Fig. 1). Seed priming is the termination of germination process at phase II. Germination process is stopped through dehydration. These seeds preserve the biochemical constituents of activation stage even after dehydration, and therefore, time to germination is shorten in prime seeds compared to non-primed seeds (Sarkar 2012; Sasaki et al. 2015). It is to be looked whether the benefit of early germination is translated in later stages of growth or not. Rice is cultivated either through direct seeding or transplanting. Is seed priming more remunerative to direct seeded rice or transplanted rice? Likely the question arises: does crop raised with the primed seeds tolerate the abiotic stresses better at seedling to early seedling stages or at later period of growths? As seed priming is done with different agents, does the effect of such agents continue up to the stage of maturity or not? These show that getting benefit of seed priming in rice, an in-depth understanding is needed: how and where the primed seeds are to be sown so that it can alleviate the stress tolerance level and in return can deliver greater benefit.

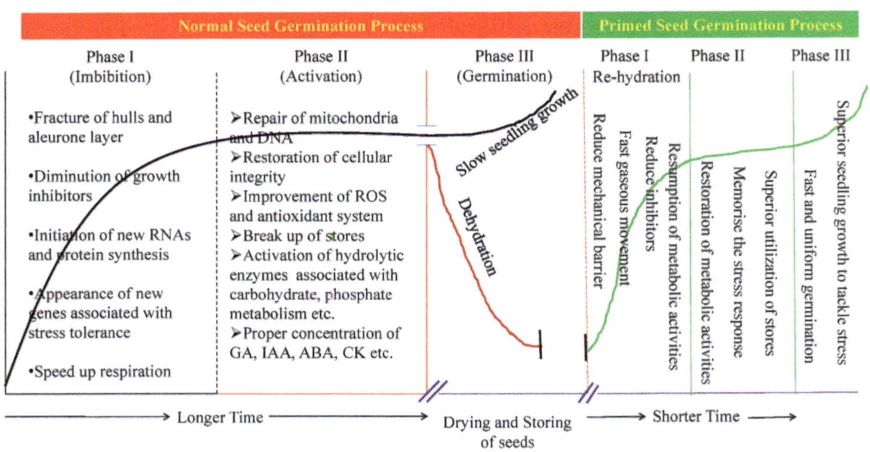


**Fig. 1** Major events associated with seed priming in rice

## 2 Difference Between Seed Invigoration and Seed Priming

Basically, there is no difference between seed invigoration and priming. Any pre-sowing treatment of seeds that improves the vigour or status of the seeds is known as seed invigoration. Seed priming is a part of seed invigoration, which involves control hydration and drying of seeds. Seed invigoration has been done in different ways to improve the germination percentage and plant establishment. To control disease and pest, treatment of seeds with fungicide and pesticide are done. Besides, several agents either living (e.g. *Bacillus subtilis*, *Trichoderma* sp.) or nonliving (e.g. calcium peroxide, iron oxide) and other materials are used for seed coating before sowing. Seed coating or mixing of seeds with different substances is done to improve production under a set of environment, e.g. coating of seeds with calcium peroxide improves plant establishment under oxygen deficiency as calcium peroxide supplies extra oxygen to germinating seeds under hypoxia/anoxia (Vijayan et al. 2018). Likely treating the seed of rice with *Trichoderma harzianum* improves salinity tolerance in rice (Rawat et al. 2012). Basically, coating does not initiate the process of seed germination; however, when seeds are germinated, it alters the plant growth. Biopriming is a term commonly used when biological agents act as coating substances (Rawat et al. 2012; Singh et al. 2016). Due to the popularity of the term ‘priming’, the terminology ‘priming’ is now frequently used even where the technique of controlled hydration-dehydration is not practised.

## 3 Factors Influencing Seed Priming

As the definition goes, seed priming is the initiation of germination process followed by termination of the process. Before root initiation, the hydrated seeds are dried up, and later these dried seeds are used up for cultivation. Besides, soaking of seed is also done with different water-soluble chemicals, extracts of different organic materials and bio-agents. It is thought that presence of such materials in half-germinated seeds may boost up plant growth. So, another aspect of seed priming is to laden up growth-promoting substances into the seeds. Non-dormant seeds started to germinate when the environment is favourable. Temperature plays a key role during imbibition followed by germination. Seedling growth largely depends on seed reserves, and the temperature quotient for growth rate is nearly double (Yoshida 1973). Seed germination rate decreases if the temperature is above or below 30 °C (Uneo and Miyoshi 2005; Tilebeni et al. 2012). Depending on the temperature, time to seed priming differs (Farooq et al. 2006a). It may be 1 day at 25 °C but 4 days at 15 °C; however at optimal temperature of 30 °C, it may be 16–18 h (Farooq et al. 2009; Sarkar 2012). Likely, any seed priming medium that hinders imbibition process can delay the germination process. Timings of seed priming depend on the methods of priming. Limited work has been done on seed hydration with aeration. Farooq et al. (2007) reported that osmopriming under aerated condition resulted in improved germination percentage, stand establishment and yield in rice. Under elevated oxygen level, timing to germination may decrease compared to hypoxia (suboptimal level of oxygen) (Vijayan et al. 2018). All these show that there are no fixed timings for incubation of seeds for priming. Standardization of procedure is important to get the effectiveness of the seed priming technique.

# 4 Seed Priming Methods

Different techniques of seed priming are being described in literature depending on constituent of the priming medium. To alleviate the stress tolerance, simple hydropriming is not always pertinent; therefore methodologies with or without associated agents to counteract the stress effects are important.

## *4.1 Seed Priming Through Hydration-Dehydration Technique*

Hydropriming: Seeds are soaked with water without any additional substances in the solution.

Osmopriming: When solutes that alter the water potential of solution is used as priming agents, it is known as osmopriming. Polyethylene glycol, calcium chloride and sodium chloride are some examples.

Redox priming: When chemicals that change the redox state of the solution is used as priming agent, it is called as redox priming. Hydrogen peroxide ($H_2O_2$) is a good agent for redox priming.

Nutrient/chemical priming: Chemicals that supply the major and minor nutrients such as N, P, K, S, Zn, B, Se, etc. are used as in the case of nutrient/chemical priming.

Organic/hormonal priming: Different organic substances or hormones are used as priming agent, e.g. polyamines, and plant growth regulator such as kinetin, gibberellic acid, auxin, salicylic acid, etc.

Botanical priming: Instead of chemicals or synthetic materials, plant extracts are used as priming agent.

Nanopriming: Priming is done with nanoparticles like biocompatible silver nanoparticles, etc.

Matripriming: Instead of putting the seeds inside the liquid medium, here seeds are mixed with granulated clay particulates or vermiculite medium. Water is slowly absorbed by the seeds.

On-farm seed priming: This technique is popular among the rice farmers. Seeds are soaked overnight (10–12 h) at room temperature ranging from 25 °C to 32 °C. Surface of the soaked seeds is dried up for easy handling before sowing (Harris et al. 1999, 2001). The seeds are sown either on the seedbed or in the field where moisture is optimum.

Basically, each technique of seed priming is not much different from another technique. In literature the different terminologies are used based on the usage of substances as priming agents.

### 4.2 *Seed Invigoration Through Coating with Chemicals/ Bio-agents*

Inorganic coating: Capsulated seeds are made with different chemicals/substances of inorganic origin.

Organic coating (biopriming): Biological agents mainly living organisms are coated with seeds.

## 5 Seed Priming and Rice Seeds

### 5.1 *Hull: Water Uptake and Gaseous Transport*

Imbibition is the first step of seed germination. Decrease in water uptake by germinating seeds slows down the germination process. Alternatively better oxygen availability to germinating embryo enhances the speed of germination and seedling growth (Roberts 1961). Intact rice hull restricts both water and gaseous transports (Toole et al. 1956). Seed germination occurs when adequate amount of water is absorbed, and seed coat becomes soft and elastic (Steinbrecher and Leubner-Metzger 2018). Initially rice hull restricts water absorption, and therefore, water uptake is slowed down at initial period of imbibition. With time, water uptake is increased. The entry of water during imbibition into the cells of dry seeds breaks the hull and aleurone layer (Bewley 1997). To impose seed priming, soaked seeds are again dried up. These seeds are more efficient to absorb water and do not create much restriction in gas movement. Restrict movement of water and gases have long impact on seedling vigour and growth (Parija 1943; Toole et al. 1956; Welbaum et al. 1998). Due to lesser inhibitory action of hull, primed seeds are more capable to germinate fast and to give better seedling (Parija 1943; Toole et al. 1956).

### 5.2 *Growth Inhibitor*

Hull is the protective layer of embryo. Under unfavourable condition, it protects the seed. Rice is grown mostly in rainy season. Frequent rain during seed maturation stage creates favourable environment for seed germination. Several reports have shown that presence of growth inhibitors either in the hull or in the endosperm or in both hull and endosperm inhibits germination and protects the seeds from unwanted germination (Toole et al. 1956; Mishra and Patnaik 1959; Takahashi et al. 1976). Presence of growth inhibitor slows down the germination process (Waheed et al. 2012; Shiratsuchi et al. 2017). Sircar et al. (1955) observed that greater concentration of auxin in endosperm retarded the growth of the embryo at the initial stages of germination. They removed a fraction of endosperm and observed better germination and seedling growth. Later, it was observed that antagonism between ABA and GA controlled the germination (Hayashi 1987; Zhu et al. 2009). Greater concentration of ABA in seeds inhibits germination, whereas GA promotes germination (Hilhorst and Toorop 1997; Bewley and Nonogaki 2017). ABA-deficient mutants have lower water potential and thereby improve the water uptake vis-à-vis germination. So, mutants with lesser ABA content take up water easily and confront less inhibitory action on germination. Once seeds are soaked, in addition of weakening of the hull tissues and loosening of cell wall of the embryo, many growth inhibitors are also leached out from the seeds (Bradford 1995; Hilhorst and Toorop 1997; Bewley and Nonogaki 2017). During rehydration processes, in absence of inhibitory concentrations of inhibitors, the action of growth promoters is accelerated, and germination becomes faster (Bewley and Nonogaki 2017).

## 5.3 *Physiological, Biochemical and Molecular Status at Phases I and II of Seed Germination*

Seed germination is an intricate procedure. Mechanical barrier exerted by seed coat can be overcome by the germinating embryo when seed coat is loosened and weakened. The processes are under control of hormones such as GA, ABA, ethylene, auxin, jasmonates and nitric oxide (Finch-Savage and Leubner-Metzger 2006; Steinbrecher and Leubner-Metzger 2018). ABA which inhibits germination, the level of it is high in matured seeds (Finkelstein and Gibson 2002). During imbibitional process, concentrations of ABA decrease, while concentrations of GA start to increase. Through seed priming, optimal balance of the plant hormones is achieved (Varier et al. 2010). Activity of endo-β-mannanase increases during the phases I and II of germination (Welbaum et al. 1998; Wang et al. 2005). This enzyme weakens cell wall of the endosperm around radical to make its emergence easier (Bewley 1997). GA enhances endo-β-mannanase activity, whereas ABA inhibits its activity (Wang et al. 2005). Seeds of rice contain mannan polymers, which is cleared by the action of endo-β-mannanase that helps in storage mobilization (Dirk et al. 1995; Wang et al. 2005). Several hydrolytic enzymes such as α-amylase, debranching enzymes, α-glucosidase, acid and alkaline phosphatase as well as phytase are overexpressed during phase II of seed germination to utilize the storage (Panda et al. 2017). New protein synthesis and respiration speed up during phase, I whereas during phase II, the biochemical changes become stable. In both stages, new mRNAs and proteins are synthesized to continue the germination process (Das et al. 2015). Greater numbers of mitochondria appear to support the energy requirement during these stages (Bai et al. 2017). Phase II is a critical developmental stage to implement seed priming. Germination is aborted at this phase. Major metabolic events such as mitochondrion repair, restoration of cellular integrity, DNA repair and breakdown of stores and synthesis of new RNAs and proteins take place during phase II. Gene expression and maximal metabolic events depend on the timing of seed imbibition. During phase I, mainly gene associated with stress appeared, whereas during phase II, gene associated with carbohydrate metabolism, protein synthesis and signalling dominates. Cheng et al. (2017) seeing the protein expression pattern of glucose-1-phosphate adenylyltransferase, aminotransferase and prolamin precursor commented that optimum time to stop seed soaking was 24 h at 25 °C after initiation of imbibition. These also showed that during phases I and II of seed germination, a great alteration takes place inside the germinating seed. The changes persist even after abortion of germination process (He and Yang 2013; Cheng et al. 2017; Narsai et al. 2017). So seed priming stimulates germination processes as it takes less time to complete the events of germination (Fig. 1).

## 5.4 *Seed Priming and Storage*

Physiology of primed seeds is different from non-primed seeds. Viability of seeds is lost due to decrease in respiration rate, increase in RQ and malfunction of electron transport chain. Once the germination process is stopped, respiration starts to decline, and several other abnormalities such as activation of DNase, production of free radicals and degradation of protein occur. These lead to loss of seed viability (Rajjou and Debeaujon 2008; Wang et al. 2018). It was observed that in many cases, primed seeds lost their viability during storage (Varier et al. 2010; Hussain et al. 2015). Commercial success with primed seeds depends on storability of it. If the viability of seeds lost within shortest time, then commercial success is not achievable with priming. Hussain et al. (2015) reported that viability of rice seeds declined within 15 days of priming at storage temperature of 25 °C, whereas Guzman and Aquino (2007) did not find any failure of seed viability even at room temperature of 25 °C to 30 °C. They observed that 50% of germinability is lost within 2.4–3.5 months and 7.2–8.5 months of storage in non-primed and hydroprimed seeds, respectively. Seed priming did not deteriorate seed viability as observed by several other workers also (Basu et al. 1974; Basu and Pal 1980). Hussain et al. (2015) reported that storing of primed seeds at 4 °C did not deteriorate seed viability even after 210 days of storage. Wang et al. (2018) observed that primed rice seeds stored under vacuum condition either low or room temperature and at room temperature with low humidity did not lose their viability. It showed that interaction between cultivars and technique of seed priming has great impact in influencing the storability of primed seeds (Basu and Pal 1980; Hussain et al. 2015; Wang et al. 2018). So, seed viability could be extended even in primed seeds through choice of suitable rice cultivars and use of suitable seed priming technique.

# 6 Seed Priming Alleviates Stress Tolerance in Rice

## *6.1 Drought Tolerance*

Seed priming improves germination rate, germination percentage and seedling vigour as we know now. Rice is mainly cultivated as direct seeding condition under upland conditions. Maintaining proper crop stand in upland is a great hurdle due to improper germination and seedling growth (Matsushima and Sakagami 2013). Matsushima and Sakagami (2013) found a decrease of emergence time in japonica cv. Koshihikari due to osmo- and hydropriming compared to non-priming. Root and shoot growth increased >1.3 and > 1.2 times due to osmo- and hydropriming, respectively, compared to non-priming. Goswami et al. (2013) observed superiority of priming over non-priming in cv. Kshitish. Among the different priming methods, priming with 5% PEG-6000 solution found to be more effective compared to hydropriming and priming with NaCl and $KH_2PO_4$ (Goswami et al. 2013). Priming improved both germination and seedling growth under drought stress due to better detoxifying of reactive oxygen species (ROS). The contents of malondialdehyde (MDA), a product of lipid peroxidation, decreased significantly in primed seeds compared to non-primed seeds. Status of glutathione peroxidase (GPX), manganese superoxide dismutase (MnSOD) and heat shock protein 70 (Hsp70) increased due to priming, so primed seeds performed better under drought stress than non-primed seeds. Some cultivars responded better compared to others under seed priming (Basra et al. 2006; Li and Zhang 2012). Hussain et al. (2016a) observed better germination and seedling growth under drought in pigmented rice japonica cv. Heug Jinju Byeo compared to non-pigmented rice cv. Anjoong under hydropriming and osmopriming with $CaCl_2$ (−1.25 MPa). Seedlings grown from primed seeds showed better antioxidant and photosynthetic activities and maintained greater leaf water status and total plant biomass. Zheng et al. (2016) reported that priming with spermidine (0.5 mmol $L^{-1}$) enhanced emergence rate and seedling growth in two indica rice cvs. Huanghuazhan (inbred) and Yangliangyou 6 (hybrid) compared to hydro-, nutrient and osmopriming. Seed priming done either with hormone or nutrient solutions had similar effect in improving drought/osmotic stress tolerance in rice (Li and Zhang 2012; Wojtyla et al. 2016). Positive effect of hormonal priming with phytohormones, viz. ABA, GA and SA, on drought tolerance has been reported by several workers (Basra et al. 2006; Li and Zhang 2012). During dehydration on the cessation of phase II, large amount of LEA protein is formed that supports the seedling growth and imparts tolerance to drought, salinity and cold stresses (Rodriguez-Valentin et al. 2014; Wojtyla et al. 2016). The biochemical changes that occur during seed priming imprint the stress memory in seeds. While encountering drought stress, the 'primary memory' starts to work and activates several genes pertaining to drought stress, and therefore, seed priming modulates the level of signalling molecule vis-à-vis drought tolerance (Bruce et al. 2007; Wojtyla et al. 2016).

## 6.2 Salinity Tolerance

Salinity threat is increasing due to erratic rainfall, excessive irrigation and deposition of salt into the soil, intrusion of seawater and other climatic anomalies day by day (Pereira 2016). Rice is highly susceptible to salt stress at early seedling and flowering stages whereas comparatively tolerant at germination, late vegetative and grain-filling stages (Sarkar et al. 2013; Singh and Sarkar 2014). Rice once germinates if exposed to salt stress dies out quickly. Robust seedlings counterbalance the salt stress better compared to weak seedlings due to salt dilution at tissue level (Munns and Tester 2008; Sarkar et al. 2013). Salinity affects seedling growth through osmotic and ionic stresses. Like drought, salinity also impairs cell membrane structure and function and damages nucleic acid and protein through excessive production of ROS. Seedlings grown from primed seeds are more capable to detoxify the ROS (Ibrahim 2016). Afzal et al. (2012) reported that osmopriming with $CaCl_2$ ($\Psi s = -1.25$ MPa) and KCl ($\Psi s = -1.25$ MPa) and redox priming with $H_2O_2$ (50 mM) each for 36 h improved salt tolerance in two rice cvs. Shaheen Basmati (salt tolerant) and Basmati-2000 (salt sensitive). Germination rate and speed and seedling vigour were more under osmopriming in these two cultivars compared to hydropriming and non-priming conditions (Afzal et al. 2012). In their observation, redox priming was injurious compared to hydropriming and non-priming. Chunthaburee et al. (2014) observed that chemical priming with spermidine (1 mM) was more effective than hormonal priming with $GA_3$ (0.43 mM) in combating salt stress in Niewdam 00621 (salt tolerant) and KKU-LLR-039 (salt sensitive). Seedlings grown from primed seed exhibited better antioxidant activities. Seedlings raised from primed seeds accumulated fewer concentrations of $Na^+$ and $H_2O_2$ and more amounts of $K^+$ and proline in leaves. The beneficial impact of priming was more in salt-susceptible cultivar compared to salt-tolerant cultivar (Chunthaburee et al. 2014). Seed priming with polyamines (PAs), viz. spermine (SPM) and spermidine (SPD), improved the transcriptome profile of key genes governing antioxidant systems, osmolyte and ABA biosynthesis, ion transport, late embryogenesis abundant genes, photosynthetic genes, PA metabolic enzymes and several transcript factors, which were associated with salt tolerance in both salt-susceptible cv. IR64 and salt-tolerant cv. Nonabokra (Paul and Roychoudhury 2017). Polyamines were found to be the best material to be used as seed priming agent to improve salt tolerance in rice (Roychoudhury et al. 2011; Yu et al. 2016).

## 6.3 Temperature Tolerance

Temperature stress either low or high affects rice growth. During germination, chilling stress affects rice growth in those areas where rice is sown during the months of winter or early spring season. The same crop may experience high temperature stress in the months of April to May at the time of flowering to grain-filling stages.

Temperatures below or high of normal during flowering to grain-filling stages impair grain development and thus decrease rice production. Seed priming has great impact to alleviate chilling/cold stress in rice at germination and early vegetative stages. Seed priming with Se (50 mM) and SA (100 mg $L^{-1}$) enhanced seed germination under chilling stress (night temperature 13.4 °C) by 21–26% of the cvs. Huanghuazhan (inbred) and Yangliangyou 6 (hybrid) (Wang et al. 2016). Several other priming methods such as hydropriming, osmopriming with $CaCl_2$, redox priming with $H_2O_2$ and seed coating with Hanyubaomu and Miaoboshi were not so effective compared to Se and SA to combat chilling stress in rice (Wang et al. 2016). Pouramir-Dashtmian et al. (2014) observed better seedling growth at 8 °C and 12 °C in rice cv. Khazar due to SA priming. Nutrient priming with $KNO_3$ was ineffective under chilling stress (Tonel et al. 2013). Seed priming with Se and SA improved the activities of antioxidant enzymes such as superoxide dismutase, peroxidase and catalase and triggered the accumulation of glutathione and free proline and enhanced respiration (Pouramir-Dashtmian et al. 2014; Hussain et al. 2016b; Sheteiwy et al. 2017). Primed seeds with Se and SA were able to counterbalance the negative effects of chilling stress (18 °C) through diminution of oxidative stress, increment of respiration rate, α-amylase activity and activation of storage utilization to supply energy to growing plant tissues and protection of structural and functional ability of protein, cell membrane and nucleic acids (Wang et al. 2016; Hussain et al. 2016b; Sheteiwy et al. 2017). SPD (5 mM) and ALA (8.5 mM) were found to increase chilling tolerance in rice cvs. ZY 6 and QY 1 (Sheteiwy et al. 2017). In addition to antioxidant system and enhanced carbohydrate utilization, priming with these two substances also enhanced the activities of PA biosynthesis enzymes such as arginine decarboxylase, ornithine decarboxylase and S-adenosylmethionine decarboxylase. Relative expression of genes associated with PA biosynthesis was greater under SPD and ALA priming compared to non-priming (Sheteiwy et al. 2017).

## 6.4 Submergence Tolerance

Rice is grown at different ecosystems starting from rainfed lowland flood-prone to aerobic to irrigate ecologies. Under irrigated ecology, transplanting is the main means of rice cultivation, whereas under aerobic condition, direct seeding is the only way. Under rainfed lowland, both direct seeding and transplanting methods are followed. Direct seeding reduces the cost of cultivation under irrigated condition, whereas in addition to reduction of cultivation cost in rainfed lowland, it exerts more robustness in seedlings, enabling them to withstand complete submergence during flooding. Direct seeding has several advantages, yet it needs a paradigm shift in the process of cultivation to realize the full potential of the method (Das et al. 2004; Vijayan et al. 2018). Sowing of seeds on the surface of soil surface invites bird menace. Splattering of seeds occur due to torrential rain. Floating of germinating seedlings under standing water causes impairment of seedling establishment and greater lodging compared to the seeds sown beneath the soil surface. The ultimate consequence is poor plant stand and reduced grain yield. However, the problem is more aggravated if seeds are sown beneath the soil surface, and due to unwanted rain, stagnation of water occurs on the surface of the field. Flooding creates a hypoxic zone beneath the soil surface (Hagiwara et al. 1987). Under such condition though germination occurs in rice, the growth of coleoptile is minimal, while the growth of radical is severely inhibited. Under submergence, seed priming was found effective in plant establishment owing to fast emergence of radical and better coleoptile growth (Das et al. 2004; Ella et al. 2011; Sarkar 2012; Illangakoon et al. 2016). Seed priming with 2% jamun (*Syzygium cumini*) leaf extract or water (hydropriming) greatly accelerated the activities of total amylase and alcohol dehydrogenase in rice (Ella et al. 2011; Sarkar 2012). Non-primed seeds showed poor SOD and CAT activities and greater lipid peroxidation than primed seeds in both tolerant (Khaiyan and Khao Hlan On) and susceptible (FR13A and IR42) rice cultivars (Ella et al. 2011). It was observed that the rice cultivar tolerant to GSOD responded well under priming than cultivar susceptible to GSOD (Illangakoon et al. 2016). So, seed priming is an additional tool to improve crop establishment under stagnation of water over and above the tolerant genotype. Submergence during germination creates oxygen deficiency (Ray et al. 2016). The rice tolerant to GSOD performs well due to better carbohydrate metabolism, shifting of aerobic to anaerobic respiration, counteracting oxidative stress, efficient N metabolism and increased production of ethylene along with co-ordinated action of ABA, GA and G proteins (Vijayan et al. 2018). Genes associated with cell proliferation and development, and of such metabolic events are overexpressed in GSOD tolerant rice compared to non-tolerant rice (Vijayan et al. 2018). As we observed, seed priming enhanced several metabolic activities, which in turn helped rice to tolerate oxygen deficiency at germination stage (Ella et al. 2011; Sarkar 2012; Illangakoon et al. 2016; Hussain et al. 2016c). Hussain et al. (2016c) found greater alteration in transcriptomic profiling of Se and SA primed seeds than non-primed seeds. Genes associated with GSOD tolerance such as carbohydrate and nitrogen metabolism, oxidative stress and cell elongation and development are overexpressed in Se and SA primed seeds compared to non-primed seeds under oxygen deficiency/submergence.

## *6.5 Metal Toxicity*

Growing rice in heavy metal-contaminated areas is one of the most challenging tasks. Heavy metals such as Pb, Cd, Cr and Mn and metalloids such as arsenic (As) affect not only plant productivity but also cause severe toxicity to animals and humans, leading to various diseases and death (Fan et al. 2017). Arsenic toxicity to human is widespread causing several deaths throughout the world (Kalita et al. 2018). Seed priming has been found to be an easy technique to grow rice under metal- and metalloid-contaminated soils (Shinwari et al. 2015; Moulick et al. 2016, 2018). Salicylic acid priming has been found better than hydropriming to counter the chromium (Cr) toxicity in two rice cvs. Basmati 385 and Shaheen Basmati

(Shinwari et al. 2015). Seed priming accelerated seed germination and seedling growth; maintained greater pigment levels; improved status of metabolites, viz. proline and glycinebetaine, total soluble proteins and sugars; slowed down the cell membrane damage; and decreased the absorption of Cr by the seedlings. Seed priming with Se (0.8–1 mg $L^{-1}$) alleviated As toxicity in rice cvs. Kranti, IR36 and IET4094 (Moulick et al. 2016, 2018). Arsenic accumulation in both shoot and root decreased significantly in seedlings grown from Se primed seeds compared to non-primed seeds. Seedlings grown from Se priming seeds were able to protect cell membrane damage from As toxicity and maintained greater concentrations of metabolites such as proline and total phenolics. Salah et al. (2015) found effectiveness of PEG as primed agent to decrease the Zn toxicity in rice cvs. Zhu Liang You 06 and Qian You No. 1. PEG priming protected the cell structures of leaf and roots from nano-zinc oxide stress.

# 7 Seed Coating and Biopriming in Rice

## *7.1 Seed Coating and Nutrient Efficiency*

Seed coatings with micro or major elements have been done to improve the performances of plants under nutrient-deficient soil condition to boost up initial seedling growth and efficient utilization of applied fertilizers. Coating seeds with rock phosphate (0.5 Kg $Kg^{-1}$ seed) improved seedling emergence and growth significantly on low P soil (Rosa et al. 2000). In boron (B)-deficient soil, seed coating using 'Arabic gum' with 1–2 g B $Kg^{-1}$ seed was found appropriate to improve germination percentage and seedling growth in aromatic rice cvs. Super Basmati and Shaheen Basmati (Rehman et al. 2015). Seeds treated with Zn significantly improved the rice yield and Zn content in seeds under Zn-deficient soil (Slaton et al. 2001). Johnson et al. (2005) did not find any benefit of nutrient priming in rice. Seed coating with dolomitic limestone and aluminium silicate did not improve seedling emergence; however, it improved the shoot biomass in rice cvs. IRGA424 and IRGA 422 CL (Tavares et al. 2012).

## *7.2 Seed Coating and Abiotic Stress Tolerance*

Direct seeding is beneficial to farmers in many respects starting from saving of energy to early harvesting (Sarkar et al. 1999; Sarkar and Das 2003; Kumar and Ladha 2011). Like seed priming (please see 6.4 part), seed coating has been found to be effective to induce GSOD tolerance (Yamada 1952; Ota and Nakayama 1970). Coating of seeds with calcium peroxide increases germination and seedling growth under flooded condition. The calcium peroxide-coated seeds create an oxidized zone at the vicinity of the seeds through liberation of oxygen which helps in proper germination and establishment of rice under flooded soil condition due to greater α-amylase and lower alcohol dehydrogenase and pyruvate decarboxylase activities (Ota and Nakayama 1970; Mei et al. 2017). Hara (2017) observed improved seedling emergence under flooded soil in tungstic acid- and ammonium phosphotungstate-coated rice seeds. Yamauchi (2017) develop a technology based on Fe coating for direct seeding of rice. Fe-coated seeds can be stored at room temperature for >1 year and can be handled just like non-coated seeds. The technology is useful not only for water and wet seeding but also for dry seeding of rice. This technology has great potential to replace transplanting and thus is capable to reduce the cost of rice cultivation (Yamauchi 2017). Fenangad and Orge (2015) developed a direct seeding technology based on the materials available locally in the Philippines. Seed coating materials were prepared by mixing cow and buffalo manures, carbonized rice husk, rice bran and vermin compost. After coating the seeds with these mixtures, the seeds were dried up and used later. Seeds coated with these materials escaped bird damage and were able to germinate under submergence.

Besides submergence, seed coating was found to be effective in inducing salt tolerance in rice (Songlin et al. 2002). Salt tolerance of hybrid rice, viz. Shanyou 10 and Eryoupeijiu, significantly alleviated through seed coating with chitosan (1.5%). Chitosan coating accelerated the activities of β-amylase and improved germination percentage, proline content and total soluble sugar, fructose and sucrose contents of germinating seeds. Rice is vulnerable at water shortage. During germination, shortage of water leads to impair germination and seedling growth. Super absorbent polymer either used as a seed coat or applied in the planting grooves improved moisture availability of seeds and thus helped in crop establishment and growth under limited water supply (Rehman et al. 2011; Barros et al. 2017). High-qing et al. (2007) observed that seed coating with different pharmaceuticals improved cold tolerance in cvs. Jinyou 974 and Xiangzhaoxian 31. Seed coating improves the antioxidant status of the growing seedlings under cold stress, protects pigment destruction and cell membrane and induces better germination and seedling growth. So, like seed priming, seed coating with different agents has great impact to induce abiotic stress tolerance in rice.

## 7.3 *Biopriming*

### 7.3.1 Disease and Pests

To combat plant diseases, chemical pesticides are mostly used which are neither safe for the environment nor for human health. In recent times, seed biopriming with different biocontrol agents like *Trichoderma* sp. and *Pseudomonas fluorescence* has been observed to be one promising tool to combat these biotic stresses (David 2008). To do biopriming, attachment of biological agents to the seeds is the most important. Castañeda et al. (2014) developed a technique called 'electrospinning technique'. Through this technique, seeds are capsulated with polymer nano-fibres. The microcapsules on seed surface contain the active ingredients to control the diseases and pests. This technique improves the germination % of seeds as well as phytosanitary level. Seed priming with one organism to diminish the action of another organism has great importance to induce biotic stress tolerance. Palupi et al. (2017) reported that isolates of *Pseudomonas diminuta* A6 and *Bacillus subtilis* 5/B in combination could diminish the effect of *Xoo* infection intensity of the seeds.

### 7.3.2 Abiotic Stress Tolerance

Seed priming with biological organisms along with organic and inorganic chemicals was found to improve status of the seeds to induce abiotic stress tolerance. Seed coating with chitosan-lignosulphonate polymer and eugenol incorporated into chitosan-lignosulphonate polymer was found as good materials to incorporate both organic and inorganic and biological agents (Thobunluepop et al. 2008). A very thin film of it is coated along with biological organism on the surface of the seeds. When these seeds started to germinate, the organism attached to the seeds also grew along with the seeds. Rutin, a synthetic plant hormone at the concentration of 1 picomolar, boosts up the biofilm formation of the bacterium *Bacillus subtilis*. Biopriming of rice seeds with *B. subtilis* and rutin improved root and shoot growth, chlorophyll and carotenoid concentrations in rice seedlings and other beneficial aspects on plant development that in turn boost up the capacity to survive of the rice seedlings under adverse conditions (Singh et al. 2016). Biopriming of rice seeds (cv. Kalanamak 3131) with different salt-tolerant isolates of *Trichoderma harzianum* alleviated salt tolerance (Rawat et al. 2012). The seeds treated with *T. harzianum* exhibited better seedling growth, maintained photosynthetic and antioxidant systems and accumulated greater contents of proline and other metabolites compared to the seedlings grown from non-biopriming seeds.

## 8 Seed Priming Translates to Yield?

Seed priming improves germination and early seedling growth. If rice is cultivated through transplanting method, then germination at field level has no significance as rice seeds are raised with optimum condition when transplanting method is followed. Yet, there is probability that the seedlings grown from primed seeds might be vigorous, which might help in production. Seed priming with $CaCl_2$ improved fertile tiller and grain number per unit area in Basmati rice (Farooq et al. 2007). Seedling emergence and seedling growth at nursery showed highly significant positive association with grain yield (Farooq et al. 2007). Ghodrat et al. (2013) observed that seed priming with phytohormones such as indolebutyric acid (IBA), gibberellic acid (GA) and salicylic acid (SA) improved overall status of the seedlings at nursery. Transplanting of rice with such stout seedlings improved the yield per unit area in rice cv. Sadri. Mondo et al. (2016) observed that though seed priming improved the germination % and seedling growth at germination stage, such benefit did not translate into grain yield. During flowering there was no significant difference between the plants raised through primed seeds or non-primed seeds.

Transplanting is practised mainly under irrigated conditions, whereas direct seeding is primerly practiced under rainfed lowland and upland conditions. Farmers of Southeast Asian countries are shifting from transplanting to direct seeding even under irrigated conditions due to low cost of direct seeding than transplanting (Sarkar et al. 1999; Kumar and Ladha 2011). Though direct seeding is economical, adoption of this technology is poor due to uncertainty in germination and seedling establishment. Seed priming is supposed to tackle such problem. Under Indo-British rainfed farming project, mega on-farm trials were conducted in three states such as Rajasthan, Gujarat and Madhya Pradesh of India (Harris et al. 1999). The trials were conducted in eight villages under farmer management conditions except seed priming. Twelve hundred fifty on-farm trials were conducted with primed seeds. Farmers got higher grain yield by cultivating rice through primed seeds. Seedlings came out from the primed seeds were robust, showed earlier flowering and had better drought tolerance. Incidence of re-sowing was less with primed seeds compared to non-primed seeds. Crop duration reduced while using primed seeds compared to non-primed seeds, and farmers got more time for the next crop. The same types of trials were also conducted in Zimbabwe (Harris et al. 2001). It showed that seed priming had great importance to enhance yield under direct seeded conditions. Several abiotic stresses such as drought, salinity and water stagnation directly affect seed germination and seedling growth at early stage of crop development. In most of the cases, crop stand becomes suboptimal. Seed priming is one of the choices to improve crop stand to stabilize yield under the adversity of abiotic stresses (Farooq et al. 2006b; Zeng and Shi 2008; Rehman et al. 2011, 2015; Mahajan et al. 2011; Sarkar 2012; Binang et al. 2012).

## 9 Conclusion

Seed priming advances seed germination and seedling growth due to the alteration in physiological and biochemical state of the seed (Fig. 1). Genes associated with stress tolerance is overexpressed due to seed priming. The memory of stress persists in later stage of plant growth so that primed seeds are more prudent to overcome the stresses. Different priming agents are used for seed priming, which make the seeds more accustomed to withstanding the abiotic stresses such as drought, submergence, salinity, metal toxicity and temperature and even biotic stresses such as disease and pests. With a lot of benefits, seed priming has one disadvantage. It deteriorates the longevity in certain rice cultivars. Efforts are needed to reduce the disadvantages associated with seed priming. Knowledge pertaining to interaction between suitable priming technique and rice varieties can solve the seed storage problem of primed

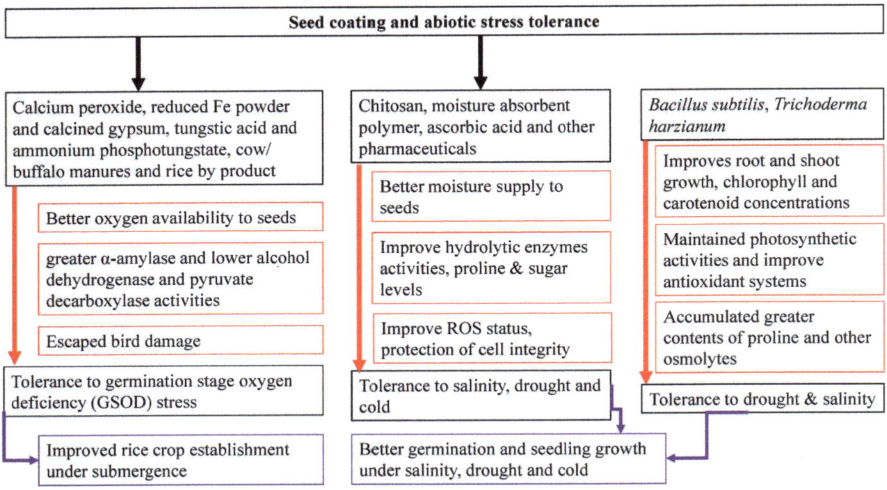


**Fig. 2** Seed coating with different nonliving and living elements increases metabolic activities, supplies greater oxygen to seeds, improves the concentrations of osmolytes and antioxidant status and pigment concentration and thus imparts abiotic stress tolerance in rice

seeds. Like seed priming, seed coating with living and nonliving resources greatly induced tolerance to different abiotic stresses such as GSOD, drought, salinity, cold and nutrient deficiency (Fig. 2). Direct seeded crop gets more benefit from seed priming compared to transplanted crop. Still date, most of the studies are done under laboratory conditions. Field studies under the adversary of abiotic stresses with primed seeds are not so much. Seed priming/coating has great business potential. The main bottleneck is the seed longevity and cost of coating materials. Knowledge on seed longevity and cost of materials and understanding of the nature of the stress can help in business potential and farm productivity. Seed priming is a climate-resilient technology, and to sustain production, seed priming provides an extra arm to tackle the abiotic stresses under the vagaries of climate change.

**Acknowledgement** Authors are grateful to the Indian Council of Agriculture, India, for providing financial support through Emeritus Scientist Scheme to RKS.

## References

Abichandani CT, Ramiah K (1951) Treatment of rice seed in nutrient solution as means of increasing yield. Curr Sci 20:270–271

Afzal I, Butt A, Rehman HU, Basra AMA, Afzal A (2012) Alleviation of salt stress in fine aromatic rice by seed priming. Aust J Crop Sci 6:1401–1407

Bai B, Shi B, Hou N, Cao Y, Meng Y, Bian H, Zhu M, Han N (2017) microRNAs participate in gene expression regulation and phytohormone cross-talk in barley embryo during seed development and germination. BMC Plant Biol 17:150. https://doi.org/10.1186/s12870-017-1095-2

Bailey-Serres J, Voesenek LACJ (2008) Flooding stress: acclimations and genetic diversity. Annu Rev Plant Biol 59:313–339
Basra SMA, Farooq M, Wahid A, Khan MB (2006) Rice seed invigoration by hormonal and vitamin priming. Seed Sci Technol 34:753–758
Basu RN, Pal P (1980) Control of rice seed deterioration by hydration-dehydration pretreatments. Seed Sci Technol 8:151–160
Basu RN, Chattopadhyay K, Pal P (1974) Maintenance of seed viability in rice (*Oryza sativa* L.) and jute (*Corchorus capsularis* L. and *C. olitorius* L.). Indian Agric 18:75–79
Bewley JD (1997) Seed germination and dormancy. Plant Cell 9:1055–1066
Bewley JD, Nonogaki H (2017) Seed maturation and germination. In: Reference module in life sciences, Elsevier. isbn:9780128096338. https://doi.org/10.1016/B978-0-12-809633-8.05092-5
Binang WB, Shiyam JO, Ntia JD (2012) Effect of seed priming method on agronomic performances and cost effectiveness of rainfed dry-seeded Nerica rice. Res J Seed Sci 5:136–143
Bradford KJ (1995) Water relations in seed germination. In: Kigel J, Galili G (eds) Seed development and germination. Arcel Dekker, New York, pp 351–396
Brown PR, Tuan VV, Nhan DK, Dung LC, Ward J (2018) Influence of livelihoods on climate change adaptation for smallholder farmers in the Mekong delta Vietnam. Int J Agric Sustain 16:255–271
Bruce TJA, Matthes MC, Napier J, Pickett JA (2007) Stressful "memories" of plants: evidence and possible mechanisms. Plant Sci 173:603–608
Castañeda LMF, Genro C, Roggia I, Bender SS, Bender RJ, Pereira CN (2014) Innovative rice seed coating (*Oryza sativa*) with polymer nanofibres and microparticles using the electrospinning method. J Res Updat Polym Sci 3:33–39
Cheng J, Wang L, Zeng P, He Y, Zhou R, Zhang H, Wang Z (2017) Identification of genes involved in rice seed priming in the early imbibition stage. Plant Biol 19:61–69
Chunthaburee S, Sanitchon J, Pattanagul W, Theerakulpisut P (2014) Alleviation of salt stress in seedlings of black glutinous rice by seed priming with spermidine and gibberellic acid. Not Bot Horti Agrobot 42:405–413
Das KK, Panda D, Nagaraju M, Sharma SG, Sarkar RK (2004) Antioxidant enzymes and aldehyde releasing capacity of rice cultivars (*Oryza sativa* L.) as determinants of anaerobic seedling establishment capacity. Bulg J Plant Physiol 30:34–44
Das SS, Karmakar P, Nandi AK, Mishra NS (2015) Small RNA mediated regulation of seed germination. Front Plant Sci 6:828. https://doi.org/10.3389/fpls.2015.00828
David BV (2008) Biotechnological approaches in IPM and their impact on environment. J Biopest 1:1–5
de Barros AF, Pimentel LD, Araujo EF, de Macedo LR, Martinez HEP, Batista VAP, da Paixão MQ (2017) Super absorbent polymer application in seeds and planting furrow: it will be a new opportunity for rainfed agriculture. Semina Ciên Agrár Londrina 38:1703–1714
de Guzman LEP, Aquino AL (eds) (2007) Longevity of hydro-primed rice seeds, Philipp J Crop Sci 32:77–88
Dirk LMA, Griffen LA, Downie B, Bewley JD (1995) Multiple isozymes of endo-β-d-mannanase in dry and imbibed seeds. Phytochemistry 40:1045–1056
Ella ES, Dionisio-Sese ML, Ismail AM (2011) Seed pre-treatment in rice reduces damage, enhances carbohydrate mobilization and improves emergence and seedling establishment under flooded conditions. AoB Plants 2011:plr007. https://doi.org/10.1093/aobpla/plr007
Fan Y, Zhu T, Li M, He J, Huang R (2017) Heavy metal contamination in soil and brown rice and human health risk assessment near three mining areas in Central China. J Healthcare Eng 2017:4124302. https://doi.org/10.1155/2017/4124302
Farooq M, Basra SMA, Afzal I, Khaliq A (2006a) Optimization of hydropriming techniques for rice seed invigoration. Seed Sci Technol 34:507–512
Farooq M, Basra SMA, Tabassum R, Afzal I (2006b) Enhancing the performance of direct seeded fine rice by seed priming. Plant Prod Sci 9:446–456
Farooq M, Basra SMA, Khan MB (2007) Seed priming improves growth of nursery seedlings and yield of transplanted rice. Arch Agron Soil Sci 53:315–326

Fenangad DB, Orge RF (2015) Simple seed coating technology for improved seedling establishment in direct-seeded rice. OIDA Int J Sustain Dev 8:35–42

Finch-Savage WE, Leubner-Metzger G (2006) Seed dormancy and the control of germination. New Phytol 171:501–523

Finkelstein R, Gibson SI (2002) ABA and sugar interactions regulating development: "cross-talk" or "voices in a crowd"? Curr Opin Plant Biol 5:26–32

Ghodrat V, Moradshahi A, Rousta MJ, Karampour A (2013) Improving yield and yield components of rice (*Oryza sativa* L.) by indolebutyric acid (IBA), gibberellic acid ($GA_3$) and salicylic acid (SA) pre-sowing seed treatments. Am Eurasian J Agric Environ Sci 13:872–876

Goswami A, Banerjee R, Raha S (2013) Drought resistance in rice seedlings conferred by seed priming. Protoplasma 250:1115–1129

Hagiwara M, Imura M, Mitsuishi S (1987) Oxidation-reduction state in the vicinity of the paddy rice seed sown into flooded soil. Jpn J Crop Sci 56:356–362

Hara Y (2017) Comparison of the effects of seed coating with tungsten and molybdenum compounds on seedling establishment rates of rice, wheat, barley, and soybean under flooded conditions. Plant Prod Sci 20:406–411

Harris D, Joshi A, Khan PA, Gothkar P, Sodhi PS (1999) On-farm seed priming in semi-arid agriculture: development and evaluation in maize, rice and chickpea in India using participatory methods. Exp Agric 35:15–29

Harris D, Pathan AK, Gothkar P, Joshi A, Chivasa W, Nyamudeza P (2001) On-farm seed priming: using participatory methods to revive and refine a key technology. Agric Syst 69:151–164

Hayashi M (1987) Relationship between endogenous germination inhibitors and dormancy in rice seeds. Jpn Agric Res Quart 21:153–161

He D, Yang P (2013) Proteomics of rice seed germination. Front Plant Sci 4:246. https://doi.org/10.3389/fpls.2013.00246

High-qing Z, Ying-bin Z, Guo-chao X, Yuan-fu X (2007) Effect and mechanism of cold tolerant seed-coating agents on the cold tolerance of early indica rice seedlings. Agric Sci China 6:792–801

Hilhorst HWM, Toorop PE (1997) Review on dormancy, germinability, and germination in crop and weed seeds. Adv Agron 61:111–165

Hussain M, Farooq M, Lee D-J (2016a) Evaluating the role of seed priming in improving drought tolerance of pigmented and non-pigmented rice. J Agro Crop Sci 203:269–276

Hussain S, Yin H, Peng S, Khan FA, Khan F, Sameeullah M, Hussain HA, Huang J, Cui K, Nie L (2016b) Comparative transcriptional profiling of primed and non-primed rice seedlings under submergence stress. Front Plant Sci 7:1125. https://doi.org/10.3389/fpls.2016.01125

Hussain S, Khan FA, Hussain HA, Nie L (2016c) Physiological and biochemical mechanisms of seed priming-induced chilling tolerance in tice cultivars. Front Plant Sci 7:116. https://doi.org/10.3389/fpls.2016.00116

Illangakoon TK, Ella ES, Ismail AM, Marambe B, Keerthisena RSK, Bentota AP, Kulatunge S (2016) Impact of variety and seed priming on anaerobic germination-tolerance of rice (*Oryza sativa* L.) varieties in Sri Lanka. Trop Agric Res 28:26–37

Johnson SE, Lauren JG, Welch RM, Duxbury JM (2005) A comparison of the effects of micronutrient seed priming and soil fertilization on the mineral nutrition of chickpea (*Cicer arietinum*), lentil (*Lens culinaris*), rice (*Oryza sativa*) and wheat (*Triticum aestivum*) in Nepal. Exp Agric 41:427–448

Kalita J, Pradhan AK, Shandilya ZM, Tanti B (2018) Arsenic stress responses and tolerance in rice: physiological, cellular and molecular approaches. Rice Sci 25:235–249

Khush GS (2005) What it will take to feed 5.0 billion rice consumers in 2030. Plant Mol Biol 59:1–6

Kumar V, Ladha JK (2011) Direct seeding of rice: recent developments and future research needs. Adv Agron 111:297–413

Li X, Zhang L (2012) SA and PEG-induced priming for water stress tolerance in rice seedling.

In: Zhu E, Sambath S (eds) Information technology and agricultural engineering, vol 134. Springer, Berlin/Heidelberg, pp 881–887
Mahajan G, Sarlach RS, Japinder S, Gill MS (2011) Seed priming effects on germination, growth and yield of dry direct-seeded rice. J Crop Improv 25:409–417
Matsushima K-I, Sakagami J-I (2013) Effects of seed hydropriming on germination and seedling vigor during emergence of rice under different soil moisture conditions. Am J Plant Sci 4:1584–1593
Mei J, Wang W, Peng S, Nie L (2017) Seed pelleting with calcium peroxide improves crop establishment of direct-seeded rice under waterlogging conditions. Sci Rep 7:4878. https://doi.org/10.1038/s41598-017-04966-1
Mikkelsen DS, Sinah MN (1961) Germination inhibition in *Oryza sativa* and control by preplanting soaking treatments. Crop Sci 1:332–335
Mishra G, Patnaik SN (1959) Effect of coumarin on the germination and growth of rice seedlings. Nature 183:989–990
Mondo VHV, Nascente AS, Neves PDCF, Taillebois JE, Oliveira FHS (2016) Seed hydropriming in upland rice improves germination and seed vigor and has no effects on crop cycle and grain yield. Aust J Crop Sci 10:1534–1542
Moulick D, Ghosh D, Santra SC (2016) Evaluation of effectiveness of seed priming with selenium in rice during germination under arsenic stress. Plant Physiol Biochem 109:571–578
Moulick D, Santra SC, Ghosh D (2018) Rice seed priming with Se: a novel approach to mitigate as induced adverse consequences on growth, yield and as load in brown rice. J Hazard Mater 355:187–196
Munns R, Tester M (2008) Mechanisms of salinity tolerance. Annu Rev Plant Biol 59:651–681
Narsai R, Gouil Q, Secco D, Srivastava A, Karpievitch YV, Liew LC, Lister R, Mathew G, Lewsey MG, Whelan J (2017) Extensive transcriptomic and epigenomic remodelling occurs during *Arabidopsis thaliana* germination. Genom Biol 18:172. https://doi.org/10.1186/s13059-017-1302-3
Ota Y, Nakayama M (1970) Effect of seed coating with calcium peroxide on germination under submerged condition in rice plant. Proc Crop Sci Soc Jpn 39:535–536
Palupi T, Ilyas SAS, Machmud M, Widajati E (2017) Effect of seed coating with biological agents on seed quality of rice. Biodiversitas 18:727–732
Panda D, Rao DN, Das KK, Sarkar RK (2017) Role of starch hydrolytic enzymes and phosphatases in relation to under water seedling establishment in rice. Indian J Plant Physiol 22:279–286
Parija P (1943) On the seed treatment and phasic development. Curr Sci 12:88–89
Pereira A (2016) Plant abiotic stress challenges from the changing environment. Front Plant Sci 7:1123. https://doi.org/10.3389/fpls.2016.01123
Pouramir-Dashtmian F, Khajeh-Hosseini M, Esfahani M (2014) Improving chilling tolerance of rice seedling by seed priming with salicylic acid. Arch Agron Crop Sci 60:1291–1302
Radanielson AM, Gaydon DS, Li T, Angeles O, Roth CH (2018) Modeling salinity effect on rice growth and grain yield with ORYZA v3 and APSIM-Oryza. Eur J Agron 100:44–55. https://doi.org/10.1016/j.eja.2018.01.015
Rajjou L, Debeaujon I (2008) Seed longevity: survival and maintenance of high germination ability of dry seeds. C R Biol 331:796–805
Rang ZW, Jagadish SVK, Zhou QM, Craufurd PQ, Heuer S (2011) Effect of high temperature and water stress on pollen germination and spikelet fertility in rice. Environ Exp Bot 70:58–65
Rawat L, Singh Y, Shukla N, Kumar J (2012) Seed biopriming with salinity tolerant isolates of *Trichoderma harzianum* alleviates salt stress in rice: growth, physiological and biochemical characteristics. J Plant Pathol 94:353–365
Ray S, Vijayan J, Sarkar RK (2016) Germination stage oxygen deficiency (GSOD): an emerging stress in the era of changing trends in climate and rice cultivation practice. Front Plant Sci 7:671. https://doi.org/10.3389/fpls.2016.00671
Reddy JN, Sarkar RK, Patnaik SSC, Singh DP, Singh US, Ismail AM, Mackill DJ (2009)

Improvement of rice germplasm for rainfed lowland of eastern India. SABRAO J Breed Genet 41(Special supplement). ISSN:102907073
Rehman HU, Basra SMA, Farooq M (2011) Field appraisal of seed priming to improve the growth, yield, and quality of direct seeded rice. Turk J Agric 35:357–365
Rehman HU, Kamran M, Basra SMA, Afzal I, Farooq M (2015) Influence of seed priming on performance and water productivity of direct seeded rice in alternating wetting and drying. Rice Sci 22:189–196
Roberts GH (1961) Dormancy in rice seeds 1. The distribution of dormancy period. J Exp Bot 12:315–329
Rodriguez-Valentin R, Campos F, Battaglia M, Sol6rzano RM, Rosales MA, Covarrubias AA (2014) Group 6 late embryogenesis abundant (LEA) proteins in monocotyledonous plants: genomic organization and transcript accumulation patterns in response to stress in *Oryza sativa*. Plant Mol Biol Rep 32:198–208
Rosa C, Bell RW, White PF (2000) Phosphorus seed coating and soaking for improving seedling growth of *Oryza sativa* (rice) cv. IR66. Seed Sci Technol 28:201–211
Roy C, Guha I (2017) Economics of climate change in the Indian Sundarbans. Glob Bus Rev 18:493–508
Roychoudhury A, Basu S, Sengupta DN (2011) Amelioration of salinity stress by exogenously applied spermidine or spermine in three varieties of Indica rice differing in their level of salt tolerance. J Plant Physiol 168:317–328
Salah SM, Yajing G, Dongdong C, Jie L, Aamir N, Qijuan H, Weimin H, Mingyu N, Jin H (2015) Seed priming with polyethylene glycol regulating the physiological and molecular mechanism in rice (*Oryza sativa* L.) under nano-ZnO stress. Sci Rep 5:14278. https://doi.org/10.1038/srep14278
Sarkar RK (2012) Seed priming improves agronomic trait performance under flooding and non-flooding conditions in rice with QTL *SUB1*. Rice Sci 19:286–294
Sarkar RK, Das S (2003) Yield of rainfed lowland rice with medium water depth under anaerobic direct seeding and transplanting. Trop Sci 43:192–198
Sarkar RK, Bera SK, De RN (1999) Rice (*Oryza sativa*) cultivars for anaerobic seeding. Indian J Agric Sci 69:73–76
Sarkar RK, Reddy JN, Sharma SG, Ismail AM (2006) Physiological basis of submergence tolerance in rice and implications for crop improvement. Curr Sci 91:899–906
Sarkar RK, Mahata KR, Singh DP (2013) Differential responses of antioxidant system and photosynthetic characteristics in four rice cultivars differing in sensitivity to sodium chloride stress. Acta Physiol Plant 35:2915–2926
Sarkar RK, Chakraborty K, Chattopadhyay K, Ray S, Panda D, Ismail AM (2018) Response of rice to individual and combined stresses of flooding and salinity. In: Hasanuzzaman M, Fujita M, Biswas JK, Nahar K (eds) Advances in rice research for abiotic stress tolerance. Elsevier. https://doi.org/10.1016/B978-0-12-814332-2.00013-7
Sasaki K, Kishitani S, Fumitaka Abe F, Sato T (2015) Promotion of seedling growth of seeds of rice (*Oryza sativa* L. cv. Hitomebore) by treatment with $H_2O_2$ before sowing. Plant Prod Sci 8:509–514
Seck PA, Diagne A, Mohanty S, Wopereis MC (2012) Crops that feed the world 7: Rice. Food Sec 4:7–24
Sheteiwy M, Shena H, Xua J, Guana Y, Songa W, Hu J (2017) Seed polyamines metabolism induced by seed priming with spermidine and 5-aminolevulinic acid for chilling tolerance improvement in rice (*Oryza sativa* L.) seedlings. Environ Exp Bot 137:58–72
Shinwari KI, Jan M, Shah G, Khattak SR, Urehman S, Daud MK, Naeem R, Jamil M (2015) Seed priming with salicylic acid induces tolerance against chromium (VI) toxicity in rice (*Oryza sativa* L.). Pak J Bot 47:161–170
Shiratsuchi H, Ohdaira Y, Yamaguchi H, Fukuda A (2017) Breaking the dormancy of rice seeds with various dormancy levels using steam and high temperature treatments in a steam nursery

cabinet. Plant Prod Sci 20:183–192
Singh DP, Sarkar RK (2014) Distinction and characterization of salinity tolerant and sensitive rice cultivars as probed by the chlorophyll fluorescence characteristics and growth parameters. Funct Plant Biol 41:727–736
Singh A, Gupta R, Pandey R (2016) Rice seed priming with picomolar rutin enhances rhizospheric *Bacillus subtilis* CIM colonization and plant growth. PLoS One 11:e0146013. https://doi.org/10.1371/journal.pone.0146013
Sircar SM, Das TM, Lahiri AN (1955) Germination of rice embryo under water and its relation of growth to endosperm fractions. Nature 175:1046–1047
Slaton NA, Wilson JCE, Ntamatungiro S, Norman RJ, Boothe DL (2001) Evaluation of zinc seed treatments for rice. Agron J 93:152–157
Smith A, Snapp S, Chikowo R, Thorne P, Bekunda M, Glover J (2017) Measuring sustainable intensification in smallholder agroecosystems: a review. Glob Food Sec 12:127–138
Songlin R, Qingzhong X, Bao ZWX (2002) Effects of chitosan coating on seed germination and salt-tolerance of seedling in hybrid rice (*Oryza sativa* L.). Eur PMC 28:803–808
Steinbrecher T, Leubner-Metzger G (2018) Tissue and cellular mechanics of seeds. Curr Opin Genet Dev 51:1–10
Takahashi N, Kato T, Tsunagawa M, Sasaki N, Kitahara Y (1976) Mechanisms of dormancy in rice seeds. II. New growth inhibitors, momilactone –A and –B isolated from the hulls of rice seeds. Jpn J Breed 26:91–9198
Tavares LC, Rufino CDA, Dörr CS, Barros ACSA, Peske ST (2012) Performance of lowland rice seeds coated with dolomitic limestone and aluminum silicate. Revista Brasileira de Sementes 34:202–211
Thobunluepop P, Pawelzik E, Vearasilp S (2008) The perspective effects of various seed coating substances on rice seed variety Khao DAWK Mali 105 storability I: the case study of physiological properties. Pak J Biol Sci 11:2291–2299
Tilebeni HG, Yousefpour H, Farhadi R, Golpayegani A (2012) Germination behavior of rice (*Oriza Sativa* L.) cultivars seeds to difference temperatures. Adv Environ Biol 6:573–577
Tonel FR, Marini P, Bandeira JDM, Silva ACSD, Sampaio NV, Villela FA (2013) Osmotic priming of rice seeds subjected to low temperatures. J Seed Sci 35:231–235
Toole EH, Hendricks SB, Borthwick HA, TOOLE VK (1956) Physiology of seed germination. Annu Rev Plant Physiol 7:299–324
Uneo K, Miyoshi K (2005) Difference of optimum germination temperature of seeds of intact and dehusked japonica rice during seed development. Euphytica 143:271–275
Varier A, Vari AK, Dadlani M (2010) The subcellular basis of seed priming. Curr Sci 99:450–456
Vijayan J, Senapati S, Ray S, Chakraborty K, Molla KA, Basak N, Pradhan B, Yeasmin L, Chattopadhyay K, Sarkar RK (2018) Transcriptomic and physiological studies identify cues for germination stage oxygen deficiency tolerance in rice. Environ Exp Bot 147:234–248
Waheed A, Ahmad H, Abbasi FM (2012) Different treatment of rice seed dormancy breaking, germination of both wild species and cultivated varieties (*Oryza sativa* L.). J. Mater Environ Sci 3:551–560
Wang A, Wang X, Ren Y, Gong X, Bewley JD (2005) Endo-β-mannanase and β-mannosidase activities in rice grains during and following germination, and the influence of gibberellin and abscisic acid. Seed Sci Res 15:219–227
Wang W, Chen Q, Hussain S, Mei J, Dong H, Peng S, Huang J, Cui K, Nie L (2016) Pre-sowing seed treatments in direct-seeded early rice: consequences for emergence, seedling growth and associated metabolic events under chilling stress. Sci Rep 6:19637. https://doi.org/10.1038/srep19637
Wang W, He A, Peng S, Huang J, Cui K, Nie L (2018) The effect of storage condition and duration on the deterioration of primed rice seeds. Front Plant Sci 9:172. https://doi.org/10.3389/fpls.2018.00172
Welbaum GE, Bradford KJ, Yim K-O, Booth DT, Oluoch MO (1998) Biophysical, physiological and biochemical processes regulating seed germination. Seed Sci Res 8:161–172

Yamada N (1952) Calcium peroxide as an oxygen supplier for crop plants. (in Japanese, with English summary). Proc Crop Sci Soc Jpn 21:65–66

Yamauchi M (2017) A review of iron-coating technology to stabilize rice direct seeding onto puddled soil. Agron J 109:739–750

Yoshida S (1973) Effects of temperature on growth of the rice plant (*Oryza sativa* L.) in a controlled environment. Soil Sci Plant Nutr 19:299–310

Yoshida S (1981) Fundamentals of rice crop science. IRRI, Manila

Yu J, Lai Y, Wu X, Wu G, Guo C (2016) Overexpression of OsEm1 encoding a group I. LEA protein confers enhanced drought tolerance in rice. Biochem Biophys Res Comm 478:703–709

Zeigler RS, Barclay A (2008) The relevance of rice. Rice 1:3–10

Zeng D, Shi Y (2008) Preparation and application of a novel environmentally friendly organic seed coating for rice. Am Euras J Agron 1:19–25

Zheng M, Tao Y, Hussain S, Jiang Q, Peng S, Huang J, Cui K, Nie L (2016) Seed priming in dry direct-seeded rice: consequences for emergence, seedling growth and associated metabolic events under drought stress. Plant Growth Regul 78:167–178

Zhu G, Ye N, Zhang J (2009) Glucose-induced delay of seed germination in rice is mediated by the suppression of ABA catabolism rather than an enhancement of ABA biosynthesis. Plant Cell Physiol 50:644–651

## *Further Reading*

Farooq M, Kobayashi N, Wahid A, Ito O, Basra SMA (2009) Strategies for producing more rice with less water. Adv Agron 101:351–388

Hussain S, Zheng M, Khan F, Khaliq A, Fahad S, Peng S, Huang J, Cui K, Nie L (2015) Benefits of rice seed priming are offset permanently by prolonged storage and the storage conditions. Sci Rep 5:8101. https://doi.org/10.1038/srep08101

Ibrahim EA (2016) Seed priming to alleviate salinity stress in germinating seeds. J Plant Physiol 192:38–46

Paul S, Roychoudhury A (2017) Effect of seed priming with spermine/spermidine on transcriptional regulation of stress-responsive genes in salt-stressed seedlings of an aromatic rice cultivar. Plant Gene 11:133–142

Wojtyla L, Lechowska K, Kubala S, Garnczarska M (2016) Molecular processes induced in primed seeds – increasing the potential to stabilize crop yields under drought conditions. J Plant Physiol 203:116–126

# Seed Priming-Mediated Improvement of Plant Morphophysiology Under Salt Stress

**Abdul Rehman, Babar Shahzad, Aman Ullah, Faisal Nadeem, Mohsin Tanveer, Anket Sharma, and Dong Jin Lee**

## Contents



**Abstract** This chapter is describing the adverse effect of the salinity stress on the crop growth and development and how seed priming can alleviate salinity-induced devastating effects on plants. Growth of plant under salt stress is affected negatively due to oversynthesis of reactive oxygen species (ROS), leading to oxidative damage to biomolecule and plant membranes. The water stress and accumulation of toxic ions are the other major effects observed under salt stress. Overproduction of ROS reacts with key cellular molecules and metabolites including proteins, lipids, photosynthetic pigments, and DNA. However, numerous plant species have effective

---

A. Rehman · D. J. Lee
Department of Crop Science and Biotechnology, Dankook University, Chungnam, Republic of Korea

B. Shahzad (✉) · M. Tanveer
Tasmanian Institute of Agriculture, University of Tasmania, Hobart, TAS, Australia
e-mail: Babar.Shahzad@utas.edu.au

A. Ullah · F. Nadeem
Department of Agronomy, Faculty of Agriculture, University of Agriculture, Faisalabad, Pakistan

A. Sharma
Plant Stress Physiology Laboratory, Department of Botanical & Environmental Sciences, Guru Nanak Dev University, Amritsar, India



M. Hasanuzzaman, V. Fotopoulos (eds.), *Priming and Pretreatment of Seeds and Seedlings*, https://doi.org/10.1007/978-981-13-8625-1_10

defense system based on antioxidants that activates once plant undergoes any abiotic stress. Among various antioxidants, nonenzymatic and enzymatic are essential to detoxify ROS and its scavenging. Recently, seed priming has gained popularity as it develops tolerance in plants against salinity during the germination process and seedling development stage. In various types of environmental stresses, the different priming techniques as osmopriming, hydropriming, hormonal priming, nutrient priming, chemical priming, bio-priming, matrix priming, and redox priming are employed. There has been increasing evidence that priming stimulates the cellular defense response that induces tolerance to biotic and abiotic stresses upon exposure in the field.

**Keywords** Seed priming · Salt stress · Reactive oxygen species · Antioxidants · Abiotic stress tolerance

## 1 Introduction

Among various abiotic stresses, salt stress is considered as a major threat to agricultural crops globally. Crop plant experiencing salinity shows poor growth due to toxic ion accumulation and disturbed water relation balance (Rehman et al. 2016). The enhanced accumulation of reactive oxygen species including hydrogen peroxide ($H_2O_2$), singlet oxygen ($^1O_2$), and superoxide anions ($O_2^{\bullet-}$) under salt stress induces oxidative damage to plants (Ashraf 2009; Ali et al. 2017). These species are highly reactive and promptly react with biomolecule substances and essential cellular metabolites such as DNA, proteins, pigments, and lipids (Ashraf 2009). Accumulation of ROS leads to multiple impairments such as deactivation of antioxidative defense system and lipid peroxidation (Tanou et al. 2009; Anjum et al. 2015, 2016a; Shahzad et al. 2018 a, b; Fahad et al. 2019), although $H_2O_2$ production under stressful environment causes oxidative damage. However, this molecule functions as a signaling molecule in a variety of biological processes including activation of antioxidant enzymatic system that helps plant to adjust under stress environment (Qiao and Fan 2008; Tanou et al. 2009; Hernandez et al. 2010; Tanveer and Shabala 2018).

Accumulation of ROS, e.g., malondialdehyde, that appears to be mutagenic product of lipid peroxidation resulting in cell damage and effectively used as marker for cell membrane injury (Riahi and Ehsanpour 2013; Anjum et al. 2016a, b). Conversely, Li et al. (2010) reported that malondialdehyde production depends on stress type, cultivar/crop type, and strength of antioxidative defense mechanism. Effective antioxidative plant defense system including accumulation of enzymatic [catalase (CAT), glutathione reductase (GR), ascorbate peroxidase (APX)] and nonenzymatic (phenolics, proline, flavonoids, tocopherol, ascorbic acid) antioxidants helps in ROS detoxification (Ali and Ashraf 2011; Anjum et al. 2017).

Ascorbic acid (AsA) and phenolics are important physiological nonenzymatic antioxidant (Rice-Evans et al. 1996); AsA has a powerful role to complete and strengthen the antioxidation and plant defense process. Moreover, proline concentration in leaf, AsA, tocopherol, and glutathione significantly contribute to ROS scavenging and are able to maintain cell water potential, i.e., osmotic adjustment (Kishor et al. 2014). Contrarily, a study by Signorelli et al. (2016) reported that proline has no significant ability to scavenge ROS (NO2, $O_2^{\bullet -}$, NO, peroxynitrite) due to its selective antioxidant potential to convert OH as a second defense line. Generally, it has been observed that proline accumulates in plants under salt stress; but salt-sensitive plant accumulates less proline than salt tolerant (Habib et al. 2012; Anjum et al. 2015).

Under stress conditions, cultivars having high yield potential seem to be more stress sensitive leading to their poor growth and performance. To achieve high yield potential, stress tolerance imparts in cultivars following various non-breeding and breeding approaches with particular focus on imperative physiological and biochemical traits. Furthermore, exogenous application of different chemicals (with required nutrient) using different methods is also an attractive option to enhance stress tolerance (Anwar et al. 2012) in crops. Among various methods, seed priming with nutrients is considered as handsome and effective approach (Ibrahim 2016; Rehman et al. 2018).

In many crops, germination and seedling emergence are more sensitive to adverse growing conditions and significantly contribute in uniform crop stand (Ali et al. 2017), resulting in higher yield. Similarly, salt stress negatively affects the germination by disturbing physio-biochemical processes including ionic imbalance, oxidative stress, and osmotic stress. Various studies show that salt stress slows down the germination of seed by reducing water imbibition, disruption of proteins structure, and stored food mobilization (Ibrahim 2016; Shahzad et al. 2019). Conversely, priming of seed helps in early seed germination by accelerating the pre-germination metabolic process (Paparella et al. 2015).

Seed priming stimulates the germination and improves seedling development predominantly through decreasing the time period of lag phase, DNA repair during water imbibition, activation of enzymes and germination metabolites, and osmotic adjustment (Lee and Kim 2000; Farooq et al. 2006; Brocklehurst and Dearman 2008; Hussian et al. 2015). Seed priming includes different methods such as hydropriming, osmopriming, chemical priming, nutrient priming, hormonal priming, and redox priming which are considered promising techniques for various crops and can be adopted under wide stressful environments (Jisha et al. 2013; Paparella et al. 2015). Various reports have confirmed the beneficial impact of seed priming under stress conditions that includes cellular defense response activation that helps crop plant to tolerate under abiotic and biotic stress under field conditions (jisha et al. 2013).

# 2 Priming and Salt Stress Tolerance

## 2.1 Plant Morphology

In salinity stress, the germination is substantially reduced or delayed because of ion toxicity ($Na^+$, $Cl^-$) and obstructed water uptake that induce osmotic stress (Hasegawa et al. 2000; Khaje-Hosseini et al. 2003; Farsiani and Ghobadi 2009). However, under salinity stress, seed priming seems to be an effective technique that helps to remove problem of poor germination. For instance, germination rate and seedling vigor index of salt-stressed maize crop substantially improved with hydropriming (Janmohammadi et al. 2008). A research study conducted by Ashraf and Rauf (2001) found that under salt stress, water soaking of maize seed showed uniform and vigorous germination.

Different seed priming methods have been evolved and widely practiced in various field crops including wheat, chickpea, and cotton (Iqbal and Ashraf 2007; Kaur et al. 2002; Casenave and Toselli 2007a; Ullah et al. 2019) to enhance the uniformity of germination, seedling vigor, and acceleration of vegetative growth resulting in higher grain yield (Farooq et al. 2018a, b, c). Under suboptimal conditions, primed seed showed early germination, healthy seedling, and overall better growth than non-primed seeds (Khalil et al. 2001; Sivritepe et al. 2003). Earlier, several reports are available that support the positive impact of seed priming on the germination, growth, and tolerance against stressful environment in various crops such as maize, wheat, lentil, sugarcane, and cucumber (Foti et al. 2008; Ghiyasi et al. 2008; Ghassemi-Golezani et al. 2008; Patade et al. 2009; Ghassemi-Golezani and Esmaeilpour 2008).

Under salinity stress, higher amount of salt accumulated in spaces between the cells that cause imbalance in water relations (Zhang et al. 2006). Seed priming has also been adopted in numerous field and horticultural crops, to decrease the emergence/germination time, to ensure uniform stand establishment (Farooq et al. 2005; Ashraf and Foolad 2005), and to improve the allometric traits.

Uniform germination, growth, and vigorous crop stand are not achieved in late-planted wheat causing low yield under rice-wheat rotation; thereby, sowing of primed wheat seed helps in early germination with uniform crop stand, ultimately leading to better grain yield (Kant et al. 2006). Lee et al. 1998; Kant et al. 2006 stated that uniform crop stand in response to seed priming occurs due to stimulation in germination metabolism, which accelerates germination rate with uniform seedling. Seed priming ensures the timely breakdown of food reserve through activating germination metabolism which helps seed to make up germination process within less time (Kant et al. 2006; Farooq et al. 2007, 2017).

In addition, sowing of primed seed produced highest tillers in terms of number and fertility. In a comparison of osmopriming and hydropriming, it was found that hydropriming shortens the emergence period and improved the vigor and dry weight of seedling (Ahmadi et al. 2007). Harris (2006) also found that seed priming in wheat (for <12 h) proves to be more beneficial than other techniques as it positively improves germination, decreases emergence time, and has early flowering and maturity phase, resulting in higher grain yield.

Priming of various canola cultivars with different solutions (KCl 2%, KCl 0.5%, $KH_2PO_4$ 0.01%) showed positive relation with germination, stem growth, and root dry biomass and also strengthens plant defense to cope stressful environment (Saeidi et al. 2008).

## 2.2 *Plant Physiology*

Seed imbibition and lag phase are completed during the process of priming, and when these primed seeds are sown, these two phases of water absorption of seed are shortened (Khan et al. 2009). Improved germination of primed seed attributed to inside swelling of seed embryo (Elouaer and Hannachi 2012) that mediates in water uptake. Activation of pre-germination metabolism except radicle protrusion occurs during the process of seed priming (Farooq et al. 2007). During the process of imbibition, certain steps including, decrease in resistance of endosperm, membrane and DNA repairing, embryos become mature and the germination inhibitors are leached (Bewley et al. 2013). Hence, under adverse environments, the seedlings grow/emerge at a faster and vigorous rate and perform better than the non-primed seeds (Sadeghi et al. 2011).

Induction of tolerance against stress conditions in primed germinated seeds may be attributed to manifestation of cross tolerance mechanism which consists of two approaches (Chen and Arora 2013). The first strategy is the activation of germination metabolism such as mobilization of reserve materials, endosperm weakening, embryo expansion, and enhanced energy metabolism (Li et al. 2010; Sun et al. 2010) which cause quiescent dry seed transition into a germinating phase with high germination vigor. The second approach reflects the imposition of abiotic stress environment on seed which inhibits emergence of radicle but accelerates cross tolerance in response to stress. The combination of these approaches establishes "priming memory" within the seed that reactivates later, upon exposure to stress which builds up strong tolerance in germinated seeds (Bruce et al. 2007; Pastor et al. 2013).

The process of osmoregulation in plant starts upon the active uptake of inorganic ions. Alleviation of negative impact of salt stress on seed germination and health seedling development can be achieved by seed priming that increases the accumulation of $Ca^{+2}$ and $K^+$ and reduces accumulation of $Cl^-$ and $Na^+$ in developing seedling (Iqbal et al. 2006; Afzal et al. 2008; Bakht et al. 2011), resulting in more water uptake with low osmotic potential (Ashraf 2009). Potassium is essential for activation of enzymes, for turgor and membrane potential balance, and in osmotic regulation in cells (Cherel 2004). Like potassium, calcium (Ca) also plays very important roles in cell elongation and division, maintains cell wall integrity, regulates the uptake of nutrients across the membrane, and improves uptake of water in plants and alleviates the adverse effect of $Na^+$ during plant growth (Patade et al. 2009; Gobinathan et al. 2009; Summart et al. 2010).

## 2.3 Leaf Gas Exchange

Seed priming with saponin improves the tolerance against salinity in quinoa by stabilizing the stomatal conductance and leaf photosynthetic rate resulting in better gas relations (Yang et al. 2018). Under salinity, abscisic acid (ABA) concentration increases in leaf (Amjad et al. 2014) which may inhibit some $CO_2$ diffusion rate due to increased stomatal closure induced by higher ABA concentration resulting in disrupted stomatal conductance (Liu et al. 2005, 2006). Contrarily, priming with saponin decreases the ABA concentration and improves gas relations, i.e., stomatal conductance (Yang et al. 2018). Another study showed that seed priming with triacontanol improves leaf gas exchange (Sarwar et al. 2017) due to its substantial role in stomatal regulation by upregulating the photosynthetic genes and increases $CO_2$ rate under salt stress (Chen et al. 2002; Perveen et al. 2010). Seed priming with auxin enhances the gas relations in salt-tolerant and non-salt-tolerant cultivars that is attributed to the tryptophan-dependent indole acetic acid involving in the opening of stomata (Merritt et al. 2001) and enhances $CO_2$ assimilation rate under salinity (Iqbal and Ashraf 2013).

## 2.4 Transpiration

Seed priming with tryptophan increased the rate of transpiration in both salt-sensitive and salt-tolerant wheat cultivars (Iqbal and Ashraf 2013). Brassinolide priming in maize improved the transpiration rate by 11% under drought stress (Anjum et al. 2011). This could be due to the improvement in leaf water balance by brassinolide application (Sairam 1994). Various studies in literature show extensive work on brassinosteroid application in improving abiotic stress tolerance in plants (Shahzad et al. 2018a; Sharma et al. 2018; Tanveer et al. 2018a, b). Priming of barley seeds with $CaCl_2$ improved drought tolerance attributed to enhanced transpiration rate without negative effects on the leaf turgor status and better stomatal aperture (Kaczmarek et al. 2017). Priming of *Brassica juncea* (Fariduddin et al. 2003) and safflower (Mohammadi et al. 2017) with salicylic acid helped to improve the stomatal conductance, better water status, and mesophyll conductance resulting in higher transpiration rate. A positive correlation has been established between transpiration and stomatal conductance, with increase in stomata opening rise in transpiration observed. Seed priming helps in vigorous plant growth with well-established root system; thus, enhancement of transpiration rate directly links with better plant water status that may be due to more water uptake by deeper roots (Mohammadi et al. 2011; Abdolahi and Shekari 2013).

## 2.5 Photosynthesis

In mung bean, priming of β-amino butyric acid aided in alleviating several abiotic stresses, i.e., salinity, high temperature, and drought (Jakab et al. 2001; Cohen 2001; Zimmerli et al. 2008) by enhancing mitochondrial activities, photosynthetic pigments, photosynthesis, and chlorophyll *a* contents resulting in better photosynthesis process (Jisha and Puthur 2016). Moreover, β-amino butyric acid priming substantially stimulated the activities of photosystems I and II and increased the transport, absorption, and trapping of electron per photosystem II (Jisha and Puthur 2016). In rice seeds, priming with PEG (Li and Zhang 2012) and polyamine priming in cucumber (Zhang et al. 2009) enhanced the photosynthetic efficiency by increasing the photochemical efficiency of photosystem II. Seed priming helps to increase mitochondrial number and improve the outer membrane integrity of mitochondria (Benamar et al. 2003; Varier et al. 2010). In maize, seed priming with silicon enhanced the photosynthetic efficiency by improving the chlorophyll contents under alkaline stress (Abdel Latef and Tran 2016) that helped plant to stay green for longer time and tolerate under stressful environment. Another study reported that soybean primed seeds with $CaCl_2$, $ZnSO_4$, and gibberellic acid improved the photosynthesis attributed to higher photosynthetic pigments and better integrity of chloroplast, mitochondria, and membranes (Dai et al. 2017). In safflower, salicylic acid application through seed priming enhanced the photosynthetic rate and chlorophyll content index and the number of chlorophyll pigments such as photoreceptor antennas that also has direct positive effect on the photosynthetic device such as improvement in electron transport chain and stimulates the activities of enzymes involving in the photosynthesis, e.g., rubisco (Mohammadi et al. 2017). Furthermore, salicylic acid prevented the degradation of chloroplast and improved the electron transport capacity by photosystem II resulting in stimulation of net photosynthetic rate and overall photosynthesis process (Shakirova et al. 2003; Fariduddin et al. 2003; Khodary 2004) as observed in soybean, corn, and barley.

## 2.6 Antioxidant Activities

Seed priming with comprehensive agents aids plant to tolerate against abiotic stresses by improving the activities of antioxidant enzymes by detoxifying the ROS (reactive oxygen species) (Dai et al. 2017). Improvement in antioxidant capacity improves the potential of plant to mitigate damage induced by ROS. Catalase (CAT) and superoxide dismutase (SOD) are considered as the most effective antioxidant enzymes that provide first-line defense against toxic ROS level (Gill and Tuteja 2010). Catalases with maximum turnover have the ability to convert about six million hydrogen peroxide ($H_2O_2$) molecules to oxygen ($O_2$) and water ($H_2O$) per minute only using one molecule of catalases (Gill and Tuteja 2010). Seed priming with $CaCl_2$ and $ZnSO_4$ enhances antioxidant enzyme activities including CAT and SOD and reduces the malondialdehyde contents and lipid peroxidation (Dai et al. 2017). Priming of seed with silicon plays pivotal role in enhancing the tolerance against alkaline stress by accumulating the osmoprotectants and activating the antioxidant machinery (Abdel Latif and Tran 2016) such as SOD, CAT, and peroxidases leading to oxidative stress mitigation. Moreover, silicon application through seed priming improves the contents of antioxidant phenols under alkaline stress (Abdel Latif and Tran 2016). Another study indicated that β-amino butyric acid alleviated the oxidative stress and lipid peroxidation by reducing the malondialdehyde contents and enhancing the proline contents and activities of SOD and guaiacol peroxidase, chitinase, nitrate reductase, and polyphenol oxidase (Jisha and Puthur 2016) under drought and saline stress. Ascorbic acid and salicylic acid priming stimulate antioxidant enzyme activities such as ascorbic acid, ascorbate peroxidases, POD, and CAT and accumulate the osmoprotectants that help to maintain the plant water status under abiotic stress, e.g., salt stress (Carvalho et al. 2011; Ahmad et al. 2012), and protect plant from oxidative stress. In a study, Carvalho et al. (2011) showed that hormonal priming (methyl jasmonate, salicylic acid, and chloroethylphosphonic acid (CEA)) protected maize seedlings from salt stress damage by activating the antioxidant machinery such as glutathione reductase, ascorbate peroxidase, SOD, POD, and CAT.

## 3 Summary and Future Research Perspectives

Seed priming is an important technique to attain desirable results against several abiotic stresses including salt stress. During the last few years, it has been emerged as a promising approach in inducing stress tolerance due to its involvement in improving overall plant defense against these abiotic stresses. Moreover, it provides a realistic, effective, and smart choice for successful plant protection. Although exact mechanism behind crop improvement is still unknown, however, it has been suggested that seed priming normally helps to regulate plant signaling through activating certain cell signaling pathways and cellular responses. It is therefore needed to further explore the molecular mechanism involving these signaling pathways.

## References

Abdel Latef AA, Tran LSP (2016) Impacts of priming with silicon on the growth and tolerance of maize plants to alkaline stress. Front Plant Sci 7:243

Abdolahi M, Shekari F(2013) Effect of priming by salicylic acid on the vigor and performance of wheat seedlings at different planting dates. Cereal Res 3:17–32. (In Persian with English abstract)

Afzal I, Rauf S, Barsa SMA, Murtaza G (2008) Halopriming improves vigour, metabolism of reserves and ionic contents in wheat seedlings under salt stress. Plant Soil Environ 54:382–388

Ahmad I, Ahmad TKA, Basra SMA, Hasnain Z, Ali A (2012) Effect of seed priming with ascorbic acid, salicylic acid and hydrogen peroxide on emergence, vigor and antioxidant activities of maize. Afr J Biotechnol 11:1127–1137

Ahmadi A, Sio-Se Mardeh A, Poustini K, Esmailpour Jahromi M (2007) Influence of osmo and hydropriming on seed germination and seedling growth in wheat (*Triticum aestivum* L.) cultivars under different moisture and temperature conditions. Pak J Biol Sci 10:4043–4049

Ali Q, Ashraf M (2011) Exogenously applied glycinebetaine enhances seed and seed oil quality of maize (*Zea mays* L.) under water deficit conditions. Environ Exp Bot 71:249–259

Ali Q, Javed MT, Noman A, Haider MZ, Waseem M, Iqbal N, Waseem M, Shah MS, Shahzad F, Perveen R (2017) Assessment of drought tolerance in mung bean cultivars/lines as depicted by the activities of germination enzymes, seedling's antioxidative potential and nutrient acquisition. Arch Agron Soil Sci:84–102

Amjad M, Akhtar J, Anwar-ul-Haq M, Yang AZ, Akhtar SS, Jacobsen SE (2014) Integrating role of ethylene and ABA in tomato plants adaptation to salt stress. Sci Hortic 172:109–116

Anjum SA, Wang LC, Farooq M, Hussain M, Xue LL, Zou CM (2011) Brassinolide application improves the drought tolerance in maize through modulation of enzymatic antioxidants and leaf gas exchange. J Agron Crop Sci 197:177–185

Anjum SA, Tanveer M, Hussain S, Bao M, Wang L, Khan I, Ullah E, Tung SA, Samad RA, Shahzad B (2015) Cadmium toxicity in maize (*Zea mays* L.): consequences on antioxidative systems, reactive oxygen species and cadmium accumulation. Environ Sci Pollut Res 22:17022–17030

Anjum SA, Tanveer M, Hussain S, Shahzad B, Ashraf U, Fahad S, Hassan W, Jan S, Khan I, Saleem MF, Bajwa AA (2016a) Osmoregulation and antioxidant production in maize under combined cadmium and arsenic stress. Environ Sci Pollut Res 23:11864–11875

Anjum SA, Tanveer M, Ashraf U, Hussain S, Shahzad B, Khan I, Wang L (2016b) Effect of progressive drought stress on growth, leaf gas exchange, and antioxidant production in two maize cultivars. Environ Sci Pollut Res 23:17132–17141

Anjum SA, Ashraf U, Tanveer M, Khan I, Hussain S, Shahzad B, Zohaib A, Abbas F, Saleem MF, Ali I, Wang LC (2017) Drought induced changes in growth, osmolyte accumulation and antioxidant metabolism of three maize hybrids. Front Plant Sci 8

Anwar MP, Juraimi AS, Puteh A, Selamat A, Rahman MM, Samedani B (2012) Seed priming influences weed competitiveness and productivity of aerobic rice. Acta Agri Scand. 62(6):499-509

Anwar S, Iqbal M, Raza SH, Iqbal N (2013) Efficacy of seed preconditioning with salicylic and ascorbic acid in increasing vigor of rice (*Oryza sativa* L.) seedling. Pak J Bot 45:157–162

Ashraf M (2009) Biotechnological approach of improving plant salt tolerance using antioxidants as markers. Biotechnol Adv 27:84–93

Ashraf M, Foolad MR (2005) Presowing seed treatment – a shotgun approach to improve germination, plant growth, and crop yield under saline and non-saline conditions. Adv Agron 88:223–265

Ashraf M, Rauf H (2001) Inducing salt tolerance in maize (*Zea mays* L.) through seed priming with chloride salts: growth and ion transport at early growth stages. Acta Physiol Plant 23:407–414

Bakht J, Shafi M, Jamal Y, Sher H (2011) Response of maize (*Zea mays* L.) to seed priming with NaCl and salinity stress. Span J Agric Res 9:252–261

Benamar A, Tallon C, Macherel D (2003) Membrane integrity and oxidative properties of mitochondria isolated from imbibing pea seeds after priming or accelerated ageing. Seed Sci Res 13:35–45

Bewley JD, Bradford KJ, Hilhorst HWM, Nonogaki H (2013) Seeds physiology of development. In: Germination and dormancy, 3rd edn. Springer, New York

Brocklehurst PA, Dearman J (2008) Interaction between seed priming treatments and nine seed lots of carrot, celery and onion II. Seedling emergence and plant growth. Ann Appl Biol 102:583–593

Bruce T, Matthes MC, Napier JA, Pickett JA (2007) Stressful memories of plants: evidence and possible mechanisms. Plant Sci 173:603–608

Carvalho RF, Piotto FA, Schmidt D, Peters LP, Monteiro CC, Azevedo RA (2011) Seed priming with hormones does not alleviate induced oxidative stress in maize seedlings subjected to salt stress. Sci Agric 68:598–602
Casenave EC, Toselli ME (2007a) Hydropriming as a pre-treatment for cotton germination under thermal and water stress conditions. Seed Sci Technol 35:88–98
Casenave EC, Toselli ME (2007b) Hydropriming as a pre-treatment for cotton germination under thermal and water stress conditions. Seed Sci Technol 35:88–98
Chen K, Arora R (2013) Priming memory invokes seed stress-tolerance. Environ Exp Bot 94:33–45
Chen X, Yuan H, Chen R, Zhu L, Du BQ, Weng Q, He G (2002) Isolation and characterization of triacontanol regulated genes in rice (*Oryza sativa* L.): possible role of triacontanol as plant growth stimulator. Plant Cell Physiol 43:869–876
Cherel L (2004) Regulation of K+ channel activities in plants: from physiological to molecular aspects. J Exp Bot 55:337–351
Cohen Y (2001) The BABA story of induced resistance. Phytoparasitica 29:375–378
Dai LY, Zhu HD, Yin KD, Du JD, Zhang YX (2017) Seed priming mitigates the effects of saline-alkali stress in soybean seedlings. Chilean J Agric Res 77:118–125
Elouaer MA, Hannachi C (2012) Seed priming to improve germination and seedling growth of safflower (*Carthamus tinctorius*) under salt stress. Eur Asian J Biosci 6:76–84
Fariduddin Q, Hayat S, Ahmad A (2003) Salicylic acid influences the net photosynthetic rate, carboxylation efficiency, nitrate reductase activity and seed yield in *Brassica juncea*. Photosynthetica 41:281–284
Farooq M, Ullah A, Lee DJ, Alghamdi SS, Siddique KHM (2018a) Desi chickpea genotypes tolerate drought stress better than kabuli types by modulating germination metabolism, trehalose accumulation, and carbon assimilation. Plant Physiol Biochem 126:47–54
Farooq M, Ullah A, Lee DJ, Alghamdi SS (2018b) Terminal drought-priming improves the drought tolerance in desi and Kabuli chickpea. Int J Agric Biol 20:1129–1136
Farooq M, Ullah A, Lee DJ, Alghamdi SS (2018c) Effects of surface drying and re-drying primed seeds on germination and seedling growth of chickpea. Seed Sci Technol 46:211–215
Farooq M, Hussain M, Nawaz A, Lee DJ, Alghamdi SS, Siddique KHM (2017) Seed priming improves chilling tolerance in chickpea by modulating germination metabolism, trehalose accumulation and carbon assimilation. Plant Physiol Biochem 111:274–283
Farooq M, Basra SMA, Hafeez K (2006) Seed invigoration by osmohardening in coarse and fine rice. Seed Sci Technol 34:181–187
Farooq M, Basra SMA, Hafeez K, Ahmad N (2005) Thermal hardening: a new seed vigor enhancement tool in rice. Acta Bot Sin 47:187–193
Farooq M, Basra SMA, Rehman H, Hussain M, Amanat Y (2007) Pre-sowing salicylicate seed treatments improve the germination and early seedling growth in fine rice. Pak J Agric Sci 44:1–8
Farsiani A, Ghobadi ME (2009) Effects of PEG and NaCl stress on two cultivars of corn (*Zea mays* L.) at germination and early seedling stages. World Acad Sci Eng Technol 57:382–385
Foti R, Abureni K, Tigere A, Gotosa J, Gerem J (2008) The efficacy of different seed priming osmotica on the establishment of maize (*Zea mays* L.) caryopses. J Arid Environ 72:1127–1130
Ghassemi-Golezani K, Esmaeilpour B (2008) The effect of salt priming on the performance of differentially matured cucumber (*Cucumis sativus*) seeds. Not Bot Horti Agrobot Cluj-Napoca 36:67–70
Ghassemi-Golezani K, Sheikhzadeh-Mosaddegh P, Valizadeh M (2008) Effects of hydropriming duration and limited irrigation on field performance of chickpea. Res J Seed Sci 1:34–40
Ghiyasi M, Abbasi Seyahjani A, Mehdi T, Reza A, Hojat S (2008) Effect of osmopriming with polyethylene glycol (2008) on germination and seedling growth of wheat (*Triticum aestivum* L.) seeds under salt stress. Res J Biol Sci 3:1249–1251
Gill SS, Tuteja N (2010) Reactive oxygen species and antioxidant machinery in abiotic stress tolerance in crop plants. Plant Physiol Biochem 48:909–930

Gobinathan P, Sankar B, Murali PV, Panneerselvam RN (2009) Effect of calcium chloride on salinity-induced oxidative stress in *Pennisetum typoidies*. Bot Res Int 2:143–148

Habib N, Ashraf M, Ali Q, Perveen R (2012) Response of salt stressed okra (*Abelmoschus esculentus* Moench) plants to foliar-applied glycine betaine and glycine betaine containing sugar beet extract. S Afr J Bot 83:151–158

Harris D (2006) Development and testing of on-farm seed priming. Adv Agron 90:129–278

Hasegawa PM, Bressan RA, Zhu JK, Bohnert HJ (2000) Plant cellular and molecular response to high salinity. Annu Rev Plant Physiol Plant Mol Biol 51:463–499

Hernandez M, Fernandez-Garcia N, Diaz-Vivancos P, Olmos E (2010) A different role for hydrogen peroxide and the antioxidative system under short and long salt stress in *Brassica oleracea* roots. J Exp Bot 61:521–535

Hussain S, Zheng M, Khan F, Khaliq A, Fahad S, Peng S (2015) Benefits of rice seed priming are offset permanently by prolonged storage and the storage conditions. Sci Rep 5:8101

Ibrahim EA (2016) Seed priming to alleviate salinity stress in germinating seeds. J Plant Physiol 192:38–46

Iqbal M, Ashraf M (2013) Salt tolerance and regulation of gas exchange and hormonal homeostasis by auxin-priming in wheat. Pesq Agrop Brasileira 48:1210–1219

Iqbal M, Ashraf M, Jamil A, Ur-Rehman S (2006) Does seed priming induce changes in the levels of some endogenous plant hormones in hexaploid wheat plants under salt stress. J Integr Plant Biol 48:181–189

Iqbal M, Ashraf M (2007) Seed treatment with auxins modulates growth and ion partitioning in salt-stressed wheat plants. J Integr Plant Biol 49:1003–1015

Jakab G, Cottier V, Toquin V, Rigoli G, Zimmerli L, Metraux JP, MauchMani B (2001) D-Aminobutyric acid-induced resistance in plants. Eur J Plant Pathol 107:29–37

Janmohammadi M, Dezfuli PM, Sharifzadeh F (2008) Seed invigoration techniques to improve germination and early growth of inbred line of maize under salinity and drought stress. Gen Appl Plant Physiol 34:215–226

Jisha KC, Puthur JT (2016) Seed priming with BABA (β-amino butyric acid): a cost-effective method of abiotic stress tolerance in *Vigna radiata* (L.) Wilczek. Protoplasma 253:277–289

Jisha KC, Vijayakumari K, Puthur JT (2013) Seed priming for abiotic stress tolerance: an overview. Acta Physiol Plant 35:1381–1396

Kaczmarek M, Fedorowicz-Strońska O, Głowacka K, Waśkiewicz A, Sadowski J (2017) CaCl2 treatment improves drought stress tolerance in barley (*Hordeum vulgare* L.). Acta Physiol Plant 39:41

Kant S, Pahuja SS, Pannu RK (2006) Effect of seed priming on growth and phenology of wheat under late-sown conditions. Trop Sci 44:9–150

Kaur S, Gupta AK, Kaur N (2002) Effect of osmo- and hydropriming of chickpea seeds on seedling growth and carbohydrate metabolism under water deficit stress. Plant Growth Regul 37:17–22

Khaje-Hosseini M, Powell AA, Bingham IJ (2003) The interaction between salinity stress and seed vigour during germination of soybean seeds. Seed Sci Technol 31:715–725

Khalil SK, Mexal JG, Murray LW (2001) Germination of soybean seed primed in aerated solution of polyethylene glycol (8000). J Biol Sci 1:105–107

Khan HA, Ayub CM, Pervez MA, Bilal RM, Shahid MA, Ziaf K (2009) Effect of seed priming with NaCl on salinity tolerance of hot pepper (*Capsicum annuum* L.) at seedling stage. Soil Environ 28:81–87

Khodary SFA (2004) Effect of salicylic acid on the growth, photosynthesis and carbohydrate metabolism in salt stressed maize plants. Int J Agric Biol 6:5–8

Kishor K, Polavarapu B, Sreenivasulu N (2014) Is proline accumulation per se correlated with stress tolerance or is proline homeostasis a more critical issue? Plant Cell Environ 37:300–311

Lee SS, Kim JH, Hong SB, Yun SH, Park EH (1998) Priming effect of rice seeds on seedling establishment under adverse soil conditions. Korean J Crop Sci 43:194–198

Lee SS, Kim JH (2000) Total sugars, α-amylase activity, and germination after priming of normal and aged rice seeds. Korean J Crop Sci. 45(2):108–111

Li X, Zhang L (2012) SA and PEG induced priming for water stress tolerance in rice seedling. In: Zhu E, Sambath S (eds) Information technology and agricultural engineering, AISE, vol 134. Springer, Berlin/Heidelberg, pp 25–87

Li G, Wan S, Zhou J, Yang Z, Qin P (2010) Leaf chlorophyll fluorescence, hyperspectral reflectance, pigments content, malondialdehyde and proline accumulation responses of castor bean (*Ricinus communis* L.) seedlings to salt stress levels. Ind Crop Prod 31:13–19

Liu FL, Jensen CR, Shahanzari A, Andersen MN, Jacobsen SE (2005) ABA regulated stomatal control and photosynthetic water use efficiency of potato (Solanum tuberosum L.) during progressive soil drying. Plant Sci 168:831–836

Liu L, Shahnazari A, Andersen MN, Jacobsen SE, Jensen CR (2006) Physiological responses of potato (Solanum tuberosum L.) to partial root-zone drying: ABA signalling, leaf gas exchange, and water use efficiency. J Exp Bot 57:3727–3735

Merritt F, Kemper A, Tallman G (2001) Inhibitors of ethylene synthesis inhibit auxin-induced stomatal opening in epidermis detached from leaves of *Vicia faba* L. Plant Cell Physiol 42:223–230

Mohammadi L, Shekari F, Saba J, Zangani E (2011) Seed priming by salicylic acid affected vigor and morphological traits of safflower seedlings. Modern Agric Sci 7:63–72

Mohammadi L, Shekari F, Saba J, Zangani E (2017) Effects of priming with salicylic acid on safflower seedlings photosynthesis and related physiological parameters. J Plant Physiol Breed 7:1–13

Paparella S, Araújo SS, Rossi G, Wijayasinghe M, Carbonera D, Balestrazzi A (2015) Seed priming: state of the art and new perspectives. Plant Cell Rep 34:1281–1293

Pastor V, Luna E, Mauch-Mani B, Ton J, Flors V (2013) Primed plants do not forget. Environ Exp Bot 94:46–56

Patade VY, Sujata B, Suprasanna P (2009) Halopriming imparts tolerance to salt and PEG induced drought stress in sugarcane. Agric Ecosyst Environ 134:24–28

Perveen S, Shahbaz M, Ashraf M (2010) Regulation in gas exchange and quantum yield of photosystem II (PSII) salt stress and nonstressed wheat plants raised from seed treated with triacontanol. Pak J Bot 42:3073–3081

Qiao W, Fan LM (2008) Nitric oxide signaling in plant responses to abiotic stresses. J Integr Plant Biol 50:1238–1246

Rehman MZ, Rizwan M, Sabir M, Shahjahan Ali S, Ahmed HR (2016) Comparative effects of different soil conditioners on wheat growth and yield grown in saline-sodic soils. Sains Malaysiana 45:339–346

Rehman A, Farooq M, Naveed M, Nawaz A, Shahzad B (2018) Seed priming of Zn with endophytic bacteria improves the productivity and grain biofortification of bread wheat. Eur J Agron 94:98–107

Sharma A, Kumar V, Kumar R, Shahzad B, Thukral AK, Bhardwaj R (2018) Brassinosteroid-mediated pesticide detoxification in plants: a mini-review. Cogent Food Agric 4(1):1436212

Tanveer M, Shahzad B, Sharma A, Biju S, Bhardwaj R (2018a) 24-Epibrassinolide; an active brassinolide and its role in salt stress tolerance in plants: a review. Plant Physiol Biochem 130:69–79

Tanveer M, Shahzad B, Sharma A, Khan EA (2018b) 24-Epibrassinolide application in plants: an implication for improving drought stress tolerance in plants. Plant Physiol Biochem 135:295–303

Fahad S, Rehman A, Shahzad B, Tanveer M, Saud S, Kamran M, Ihtisham M, Khan SU, Turan V, Ur Rahman MH (2019) Rice responses and tolerance to metal/metalloid toxicity. In: Advances in rice research for abiotic stress tolerance. Woodhead Publishing, pp 299–312

Shahzad B, Fahad S, Tanveer M, Saud S, Khan IA (2019) Plant responses and tolerance to salt stress. In: Approaches for enhancing abiotic stress tolerance in plants. Taylor & Francis, pp 61–77

Riahi M, Ehsanpour AA (2013) Responses of transgenic tobacco (*Nicotiana plumbaginifolia*) over-expressing P5CS gene under in vitro salt stress. Prog Biol Sci 2:76–84

Rice-Evans CA, Miller NJ, Paganga G (1996) Structure-antioxidant activity relationships of flavonoids and phenolic acids. Free Radic Biol Med 20:933–956
Sadeghi H, Khazaei F, Yari L, Sheidaei S (2011) Effect of seed osmopriming on seed germination behaviour and vigor of soybean (*Glycine max* L.). J Agric Biol Sci 6:39–43
Saeidi MR, Abdolghaium A, Hassanzadeh M, Rouhi A, Nikzad P (2008) Investigation of seed priming on some germination aspects of different canola cultivars. J Food Agric Environ 6:188–191
Sairam RK (1994) Effects of homobrassinolide application on plant metabolism and grain yield under irrigated and moisture-stress conditions of two wheat varieties. Plant Growth Regul 14:173–181
Sarwar M, Amjad M, Ayyub CM (2017) Alleviation of salt stress in cucumber (*Cucumis sativus*) through seed priming with triacontanol. Int J Agric Biol 19:771–778
Shahzad B, Tanveer M, Che Z, Rehman A, Cheema SA, Sharma A, Song H, Rehman S, Zhaorong D (2018a) Role of 24-epibrassinolide (EBL) in mediating heavy metal and pesticide induced oxidative stress in plants: a review. Ecotoxicol Environ Saf 147:935–944
Shahzad B, Tanveer M, Rehman A, Cheema SA, Fahad S, Rehman S, Sharma A (2018b) Nickel; whether toxic or essential for plants and environment-a review. Plant Physiol Biochem 132:641–651
Shakirova FM, Sakhabutdinova RA, Bezrukova MV, Fatkhutdinova RA, Fatkhutdinova DR (2003) Changes in the hormonal status of wheat seedlings induced by salicylic acid and salinity. Plant Sci 164:317–322
Sivritepe N, Sivritepe HO, Eris A (2003) The effect of NaCl priming on salt tolerance in melon seedling grown under saline conditions. Sci Hortic 97:229–237
Signorelli S, Imparatta C, Rodríguez-Ruiz M, Borsani O, Corpas FJ, Monza J (2016) In vivo and in vitro approaches demonstrate proline is not directly involved in the protection against superoxide, nitric oxide, nitrogen dioxide and peroxynitrite. Funct Plant Biol 43:870–879
Summart J, Thanonkeo P, Panichajakul S, Prathepha P, McManus MT (2010) Effect of salt stress on growth, inorganic ion and proline accumulation in Thai aromatic rice, Khao Dawk Mali 105, callus culture. Afr J Biotechnol 9:145–152
Sun YY, Sun YJ, Wang MT, Li XY, Guo X, Hu R, Ma J (2010) Effects of seed priming on germination and seedling growth under water stress in rice. Acta Agron Sin 36:1931–1940
Tanou G, Molassiotis A, Diamantidis G (2009) Hydrogen peroxide-and nitric oxide-induced systemic antioxidant prime-like activity under NaCl-stress and stress-free conditions in citrus plants. J Plant Physiol 166:1904–1913
Tanveer M, Shabala, S. (2018). Targeting redox regulatory mechanisms for salinity stress tolerance in crops. In: Salinity responses and tolerance in plants, vol. 1, pp 213–234. Springer, Cham
Ullah A, Farooq M, Hussain M, Ahmad R, Wakeel A (2019) Zinc seed priming improves stand establishment, zinc uptake and early seedling growth of chickpea. J Anim Plant Sci. In press
Varier A, Vari AK, Dadlani M (2010) The sub cellular basis of seed priming. Curr Sci 99:450–456
Yang A, Akhtar SS, Iqbal S, Qi Z, Alandia G, Saddiq MS, Jacobsen SE (2018) Saponin seed priming improves salt tolerance in quinoa. J Agron Crop Sci 204:31–39
Zhang WP, Jiang B, Li WG, Song H, Yu YS, Chen JF (2009) Polyamines enhance chilling tolerance of cucumber (Cucumis sativus L.) through modulating antioxidative system. Sci Hortic 122:200–208
Zhang J, Jia W, Yang J, Ismal AM (2006) Role of ABA integrating plant responses to drought and salt stresses. Field Crop Res 97:111–119
Zimmerli L, Hou BH, Tsai CH, Jakab G, Mauch-Mani B, Somerville S (2008) The xenobiotic beta-aminobutyric acid enhances Arabidopsis thermo-tolerance. Plant J 53:144–156

# Mechanisms of Seed Priming Involved in Salt Stress Amelioration

**Magdi T. Abdelhamid, Raafat R. El-Masry, Darwish S. Darwish, Mazhar M. F. Abdalla, Shinya Oba, Ragab Ragab, Ayman EL Sabagh, Mahmoud H. El Kholy, and Elsayed Omer**

## Contents



M. T. Abdelhamid (✉) · R. R. El-Masry
Botany Department, National Research Centre, Giza, Egypt

D. S. Darwish · M. M. F. Abdalla
Agronomy Department, Faculty of Agriculture, Cairo University, Giza, Egypt

S. Oba
Plant Production Control Laboratory, Faculty of Applied Biological Sciences, Gifu University, Gifu, Japan

R. Ragab
Centre for Ecology and Hydrology (CEH), Wallingford, UK

A. EL Sabagh
Department of Agronomy, Faculty of Agriculture, Kafrelsheikh University, Kafrelsheikh, Egypt

M. H. El Kholy
Soil and Water Research Institute, Agricultural Research Center, Giza, Egypt

E. Omer
Medicinal and Aromatic Plants Research Department, National Research Centre, Giza, Egypt



M. Hasanuzzaman, V. Fotopoulos (eds.), *Priming and Pretreatment of Seeds and Seedlings*, https://doi.org/10.1007/978-981-13-8625-1_11

**Abstract** Reduction of photosynthetic rates in plants under salt pressure is triggered primarily by reducing water potential and/or elevated levels of $Na^+$ and/or $Cl^-$ accumulated in chloroplasts and chlorophyll that are essential for plant health. Much work has been done to create strategies to deal with salinity stress. The use of organic and inorganic compounds, nutrients, and seed priming is among the promising strategies adopted to alleviate the unfavorable consequences of salinity and improve plant yields and yield quality. Seed priming is a commercially used method to improve the germination and vigor of seeds. The purpose of this work is to review the recent literature on plant response to seed priming under salinity stress. A schematic diagram describing the chain reaction following the seed priming was proposed. The diagram illustrates the possible impact of the seed priming on different physiological processes, enhancing tolerance level and final yield. Several mechanisms have been proposed to explain how plants build up many physiological and biochemical adaptations to regulate themselves when using seed priming under saline conditions. The variation on photosynthetic pigment (chlorophyll a, chlorophyll b, carotenoids), photosystem II, and net $CO_2$ photosynthetic rate; ionic balance; $K^+$:$Na^+$ ratio difference; osmolyte accumulation, i.e., proline, glycine betaine, amino acids, and sugars; accumulation of enzymatic like superoxide dismutase, catalase, ascorbate peroxidase, peroxidase, and nonenzymatic antioxidants (i.e., total phenols, flavonoids, ascorbic acid, total carotenoids); membrane stability; and levels of $H_2O_2$ and malondialdehyde (MDA) contents are among the most important physiological and biochemical plant reaction mechanisms for seed priming that allow plants to cope under saline conditions.

**Keywords** Antioxidants · Enzymes · Growth · Ionic balance · Osmolytes · Oxidative stress · Photosynthesis · Photosynthetic pigment · Salinity · Salt · Seed priming

## 1 Introduction

Salinity is one of the most common environmental problems that limits the productivity of plants, particularly in arid and semiarid areas (Ashraf and Harris 2004). Based on the Food and Agriculture Organization (FAO) Soils portal (http://www.fao.org/soils-portal/soil-management/), soil is considered to be saline soil when electrical conductivity (EC) is >0.7 dS $m^{-1}$. Nevertheless salt-affected soils are characterized by a specific electrical conductivity (EC) of extracts of saturated soil-paste above 4 dS $m^{-1}$ (temp at 25 °C) according to the *USDA Agriculture Handbook*

(1954). More recently, Bresler et al. (1982) stated that soils are rated salt affected beginning at an EC level of 2 dS $m^{-1}$, as some vegetable, fruit, and ornamental crops undergo from salinity within the EC of 2–4 dS $m^{-1}$.

Salt-affected soil refers to soils where, with normal plant growth, salts interfere. Depending on salt quantity, salt type, sodium present, and soil alkalinity, salt-affected soils can be divided into saline, saline-sodic, and sodic. There will be different characteristics for each type of salt-affected soil, which will also determine how it can be managed. Based on the FAO Soils portal using the world's FAO/UNESCO soil map (1970–1980), the FAO estimated the total global saline soil area was 397 million hectares and the total sodic soils 434 million ha. However, Nelson and Mareida (2001) estimated approx. 12 million ha of irrigated land may possibly have disappeared from production accordingly of salinization.

Salinity results in many unfavorable effects on the growth of plants due to low osmotic soil potential (osmotic stress), ion effects (salt stress), element imbalances, or a combination of these causes (Ashraf 2004). All these factors cause physiological and biochemical unfavorable effects on plant growth and development (Munns and James 2003). To evaluate plant tolerance to salinity stress, the plant's growth or survival is measured as it integrates many physiological mechanisms that occur within the plant. For plants growing in the saline medium, osmotic balance is essential.

The main strategies for the use of salt-affected soils are use of reclaiming and protective measures to make salt-affected soils suitable for agriculture and utilize salt-affected soils by growing halophytes or salt-tolerant crops/cultivars. The latter approach has been called the "biological approach" (Ashraf 1994) and has significant potential to mitigate the worldwide soil salinity problem. This chapter discuss biological strategies that can increase crop salinity tolerance. Strategies employed to relieve the impacts of salt stress in plants include applying chemical, biological, and physical treatments to seeds, seedlings, or plants prior to exposure to salinity stress. These treatments, mostly used before seed sowing, are anticipated to activate physiological and molecular pathways that allow the seed to react more quickly and/or more strongly after exposure to an environmental stress factor.

Seed priming is a knowledgeable, rational, and low-cost method for fast emergence, increased seedling strength, and improved plant yields under adverse environments (Jisha et al. 2013; Paparella et al. 2015). It is a restricted hydration method that activates metabolic processes before radicular protrusion at the early stage of germination (Hussain et al. 2015).

Under various environmental stresses, different methods of seed priming were utilized that include hydropriming, osmopriming, Chemopriming, nutrient priming, and hormonal priming (Jisha et al. 2013; Paparella et al. 2015). Seed priming promotes germination throughout the process by soaking seeds in solutions with exogenous molecules, i.e., salicylic acid (SA) and acetylsalicylic acid (ASA) (Khan et al. 2009a, b), or metals (Mirshekari et al. 2012), or hormones (Nakaune et al. 2012). Seed priming was proposed to trigger a sequence of physiological developments which increase plant performance grown in salt stress, including the introduction of antioxidant systems (Varier et al. 2010; Eisvand et al. 2010).

The purpose of this chapter is to present a wide overview of plant response under salt stress and the primary treatments in specific organic and inorganic substances and seed priming utilized to seeds/plants with a critical debate of the key accomplishments outlined in this area.

## 2 Plant Response to Salt Stress

A soil is considered saline once the soil solution's electrical conductivity (EC) is approximately 4 dS $m^{-1}$ (corresponding to 40 mM NaCl), creating an osmotic pressure of approximately 0.2 MPa, which reduces most crop yields (Munns and Tester 2008). Salt stress induces ion toxicity, osmotic stress, nutrient deficiency, and oxidative stress on crops, thus restricting the absorption of soil water and eventually decreasing yield output (Awad et al. 2012; Hellal et al. 2012; Abdelhamid et al. 2013b; Rady et al. 2015, 2016b). Figure 1 shows response of soybean crop grown in soils with three salinity intensities, i.e., 1.12, 6.25, and 9.38 dS $m^{-1}$, at Fayoum Governorate, Egypt.

Plant species vary for salt tolerance (Blaylock 1994; Bargaz et al. 2016). Salt-tolerant crops (plants less affected by salinity) can adjust the osmotic impacts of elevated levels of salt internally as compared to salt-sensitive crops. Also, salt-tolerant crops can absorb salt soil water more easily. On the other side, salt-sensitive crops have restricted ability for adjustment and are injured at comparatively small levels of salt. In reaction, in sensitive crop cultivars, reductions in yields caused by salinity look more widespread than in tolerant crop cultivars. Several authors report such a reduction in salinity yield (Blaylock 1994; Abdelhamid et al. 2013a; Dawood et al. 2016).

Global studies have shown that salinity affects germination and growth of seedlings in different crops, such as Bagayoko (2012). The effect of salt stress on plant

1.12 EC salinity 6.25 EC salinity 9.38 EC salinity

**Fig. 1** Soybean crop grown in soils with different salinity levels at Fayoum Governorate, Egypt

can be attributed to the adverse effects of salinity on the cell cycle and differentiation. A general reduction in plant biomass was observed in all plant tissues under salt stress, but it is particularly noticeable in the aerial part (Abdelhamid et al. 2010). Different authors have associated a decrease in plant biomass with a decrease in leaf numbers or leaf abscissions (Dawood et al. 2014b) or a decrease in net assimilation rate and relative growth rate (Rady et al. 2015).

Salinity has several unfavorable effects on plant growth owing to low soil osmotic stress, ionic imbalances, or combined effects of these factors (Ashraf 2004). All these variables affect plant growth and development with adverse physiological and biochemical impacts (Munns and James 2003). The plant's development or survival is evaluated by integrating many physiological processes that happen within the plant to assess the salinity stress tolerance of the plant.

The stress of salinity in the soil significantly reduces nutrient concentrations such as N, K, Ca, and Mg (Sadak et al. 2015; Talaat et al. 2015; Rady et al. 2016b). In addition, salinity reduces plant P uptake owing to precipitation of phosphate ions with Ca ions (Bano and Fatima 2009; Abdelhamid et al. 2010; Bargaz et al. 2016). Some elements have particular toxic effects on plants, such as sodium, chlorine, and boron. Excessive accumulation of $Na^+$ in cell walls can rapidly trigger osmotic stress and cell death (Munns 2002). Furthermore, salinity has damaging impacts on photosynthesis, primarily due to decreased leaf area, chlorophyll content, and stomatal conductance, and to some degree owing to decreased effectiveness of photosystem II (Netondo et al. 2004).

There is no doubt that there has been noteworthy recognition of osmotic adjustment as a significant and efficient mechanism for resisting salinity in plants. Under saline environment, a positive growth relationship with osmotic adjustment was found, primarily because of building up more $K^+$ and glycinebetaine, while a reduction of sugars (Ochiai and Matoh 2001). Correspondingly, Orabi and Abdelhamid (2016) found that salt-tolerant faba bean cultivar "Giza 843" was correlated with a superior capacity of osmotic adjustment by building up proline and ions, e.g., P, $K^+$, $Ca^{2+}$, and $Mg^{2+}$, compared to the salt-sensitive cultivar "Giza 3." Moreover, the salt-tolerant *Phaseolus vulgaris* genotype "RIL115" has been found to have superior osmotic adjustment by building up more soluble sugar, proline, amino acid, and ions, e.g., P, $K^+$, $Ca^{2+}$, $Mg^{2+}$, and $Mn^{2+}$, compared to the salt-sensitive genotype "RIL147" (Bargaz et al. 2016).

The accumulation of soluble carbohydrates in crops has been commonly recorded as a reaction to salinity (Murakeozy et al. 2003; Dawood et al. 2016). Amino acids were reported to accumulate under salinity stress in higher plants (vascular plants) (Mansour 2000). Di Martino et al. (2003) reported that building up amino acids and glycinebetaine caused osmoregulation, considered one of the predominant methods used by spinach crops to withstand salt stress. Several authors reported that in salt-stressed plants, proline accumulates more than other amino acids (Ashraf 1994; Bekheta et al. 2009; Abdelhamid et al. 2013a; Rady et al. 2016b).

Antioxidant metabolism, including antioxidant enzymes and nonenzymatic compounds, plays a critical role in salinity-induced detoxification of reactive oxygen species (ROS). Salinity tolerance is strongly associated with antioxidant enzyme activity, i.e., catalase (CAT), superoxide dismutase (SOD), glutathione peroxidase (GPX), and ascorbate peroxidase (APX), and nonenzymatic antioxidant, i.e., glutathione reductase (GR) (Gupta et al. 2005; Orabi and Abdelhamid 2016). Moreover, Orabi and Abdelhamid (2016) reported that faba bean plant is protected against the damaging impact of salinity primarily by enhancing the activity of enzymes, e.g., superoxide dismutase (SOD), catalase (CAT), peroxidase (POX), and phenylalanine acid lyase (PAL) enzymes.

It is well established in the literature that salinity causes severe decrease in plant yield. Several studies have already determined the fact that crops are adversely affected by this phenomenon, from planting to harvesting and even in storage. Salinity not only adversely affects seed yield and yield components in many plant species such as lupine (Dawood et al. 2016), common bean (Rady et al. 2016a), wheat (Ouda et al. 2015; Rady et al. 2016a), and faba bean (Orabi and Abdelhamid 2016) but also causes a nutritional imbalance in seed quality in terms of decreasing oil percentage, carbohydrates, and concentrations of 17 amino acids in lupine seeds (Dawood et al. 2016) and decreasing protein content in leaves and oat grains, with increasing level of salt in irrigation water (Kumar et al. 2010). In contrast, salinity raised carbohydrate and protein in seed yield of two faba bean cultivars (Orabi and Abdelhamid 2016) and increased protein, phenolic concentration, and lupine seed alkaloids (Dawood et al. 2016).

## 3 Techniques Used to Overcome Salt Stress in Plants

Salinity can be limited by salt leaching from the root rhizosphere, altering practices of farm management, and using salt-tolerant plants. Using new irrigation approaches such as deficit irrigation and using irrigation systems such as drip and sprinkler irrigation to optimize water use efficiently can sustain irrigation practices. Farming systems could also be modified to include better crop rotation in order to integrate perennials in crop rotation with annual crops and legumes within rotations and to adopt a precise agricultural approach. While using these tactics to sustainable agriculture that can alleviate the reduction of yields under the stress of salinity, the application is often limited due to the cost and availability of good-quality water or lack of water resources. Hence, developing an efficient, low-cost, easily adaptable technique to enhance plant performance under salt stress is a challenge. Globally, comprehensive study has been carried out to create strategies to deal with the stress of salinity, for example, the development of salt-tolerant varieties. Using organic and inorganic compounds, nutrients, and seed priming is one of the promising strategies adopted to alleviate the unfavorable outcomes of salinity and improve plant yields and yield quality.

### 3.1 Use of Osmoprotectants, Plant Growth Regulators (PGRs), Vitamins, and Nutrients

Much attention has been given to the approach of using osmolytes, osmoprotectants, plant growth regulators, and plant nutrient supply to overcome plant salt stress and boost crop production under saline conditions. Osmoprotectants or compatible solutes are greatly soluble compounds of low molecular weight, which can accumulate in cells without impairing cellular function at high levels. They safeguard crops against stress by contributing to cellular osmotic adjustment, detoxifying reactive oxygen species (ROS), protecting the structure of membranes, and stabilizing proteins (Ashraf and Foolad 2007). Exogenous use of glycinebetaine (GB) in perennial ryegrass significantly reduces salt stress negative impacts. The suggestion that GB increased salt tolerance in perennial ryegrass was primarily associated with increased SOD, CAT, and APX activity and cell membrane damage alleviation by reducing membrane lipid oxidation and improving ion homeostasis under salt stress (Alasvandyari et al. 2017). Exogenous proline applications alleviated oxidative stress and increased plant growth. Abdelhamid et al. (2013a) reported that proline increased the activity of antioxidant enzymes, superoxide dismutase (SOD), catalase (CAT), and peroxidase (POD), as well as carotenoid, ascorbic and endogenous proline, and P and $K^+$ and $K^+$:$Na^+$ concentrations in salt-affected common bean plants. In addition, proline provided osmoprotection and facilitated the growth of salt-stressed faba bean plants (Dawood et al. 2014b). In order to improve their salt tolerance, a number of natural or synthetic PGRs were exogenously applied to crops. A foliar spray of indole-3-acetic acid at 150 ppm, for example, gave the highest plant growth values (flag leaf area and SPAD values), yield (grain and straw yield), and chemical composition (total crude protein, P and K percent) of wheat grown under salt stress conditions (El-Metwally et al. 2015). Khodary (2004) reported a significant increase in growth and development of maize under saline conditions due to applied 0.1 mM of salicylic acid to maize plants. A number of other organic compounds were also exogenously applied to test their effect on the plant's response to saline conditions. In faba bean, for example, Semida et al. (2014) suggested that antioxidant $\alpha$-tocopherol could activate antioxidants in plants and decrease oxidative damage resulting in physiological modifications in plants grown in saline soil. Furthermore, Dawood et al. (2016) concluded that the damaging impacts of salt stress on lupine plant growth and seed yield nutritional value were to some extent alleviated, mainly at 200 ppm $\alpha$-tocopherol. In soybean, Rady et al. (2015) reported that 100 mg $L^{-1}$ $\alpha$-tocopherol exogenous application increased net assimilation rate, relative growth rate, chlorophyll, proline, free amino acids, soluble sugars, N, P, $K^+$, and $K^+$:$Na^+$ ratios in soybean plants under salt stress. Exogenous use of 200 or 400 mg $l^{-1}$ nicotinamide (vitamin B3) increased the nutrient concentration, sucrose, soluble sugars, free amino acids, lipid peroxidation, oxidative enzymes, photosynthetic pigments, plant growth, and seed yield and improved the quality of the seed yield, i.e., carbohydrates and protein of faba bean salt-stressed plants (Abdelhamid et al. 2013b).

Amino acids are well-known biostimulants that affect plant growth and yield positively, thus mitigating abiotic stress injuries significantly. For example, amino acid mixture on faba bean has been reported to have increased shoot length, leaf number per plant, fresh and dry shoot weight, photosynthetic pigments, total carbohydrates, polysaccharides, DNA, and nucleic acid RNA (Sadak et al. 2015). In addition, amino acid exogenous foliar application at rate of 500 or 1000 or 1500 mg $L^{-1}$ enhanced salt tolerance of faba bean plants by increasing osmotic solutes, phenolic content, IAA, and endogenous polyamine content, while it reduced antioxidant enzymes and lipid peroxidation (Sadak and Abdelhamid 2015).

Salt-induced nutritional disorders can be alleviated by adding mineral nutrients to the growth medium, according to numerous reports. Adding N, for example, has improved growth and yield of sorghum crop (Esmaili et al. 2008). In addition, Rady et al. (2016b) recommended use of $\frac{1}{3}\,H_4NO_3$ at rate of 55 kg N $ha^{-1}$ + cerealine (at rate of 4 kg $ha^{-1}$ bio-fertilizer) + cattle manure (at rate of 10 t $ha^{-1}$) in saline soils to improve wheat growth and nutritional yield quality with reduced heavy metal content. Also, in wheat, El-Lethy et al. (2013) reported important increases in photosynthetic pigments, antioxidant enzyme activity, soluble sugar, starch and total phenols, N, P, $K^+$, and $K^+$:$Na^+$ ratio and wheat grain yield due to soil application of 150 mg $K_2O$ $kg^{-1}$. It was also concluded that 150 mg $K_2O$ $kg^{-1}$ soil reduced the unfavorable effects of NaCl salinity by enhancing photosynthetic pigments, antioxidant enzyme activity, osmoprotectant concentrations, and $K^+$:$Na^+$ ratio, which all were reflected in increasing the growth of common bean plants (Dawood et al. 2014a). Bargaz et al. (2016) found major increase in P, $K^+$, $Ca^{2+}$, and $Mg^{2+}$ in leaves of both common bean recombinant inbred lines (RILs) and noteworthy increase in chlorophyll, carotenoids, soluble sugars, free amino acids and proline, and seed yields in all salinity intensities up to 60 kg $ha^{-1}$ P. Like P, it is known that calcium (Ca) improves the unfavorable effects of salinity in plants, most probably by making possible superior selectivity of $K^+$:$Na^+$ ratios (Hasegawa et al. 2000). Foliar fertilization is generally a useful technique for supplementing plant-deficient nutrients when soil conditions restrict the availability of nutrients to roots (Mengel 2002). Under saline conditions, the absorption of K, $Ca^{2+}$, and N through the roots and their supply to growing shooting regions is significantly impaired (Munns 2005), and thus the plant's optimum concentration of essential nutrients is reduced (Marschner 1995). In this case, foliar fertilization can be used to promote mineral nutrition, suppress physiological disorders, and promote growth and yield of plant (Pervez et al. 2004). Efficiency of foliar applied nutrients depends on their mobility within the plant (Mengel 2002). Of all nutrients, N, K, and Mg are more phloem-mobile, whereas Ca and Fe are least phloem-mobile. Thus, foliar application of the latter two nutrients is not effective (Mengel 2002).

## 3.2 *Seed Priming*

Proof of seed priming benefits has been reported extensively. Seed priming is a water-based method that allows regulated seed rehydration to activate the generally activated metabolic classes during early germination but stops the transition of seed into complete germination. Priming is commonly used to treat vegetable seeds, mainly carrots, leek, onions, tomatoes, peppers, cucumbers, broccoli, lettuce, and okra (Dearman et al. 1987; Mahmoudi et al. 2012; Demirkaya 2014; Azooz et al. 2015; Hassini et al. 2017; Sarwar et al. 2017; Rinez et al. 2018), and field crop seeds, mostly wheat, canola lentil, alfalfa, bean, mung bean, common bean, safflower, barley, and maize (Jabeen and Ahmad 2013; Jamil et al. 2012; Al-Tawaha and Al-Ghzawi 2013; Azooz et al. 2013; Movaghatian and Khorsandi 2014; Younesi and Moradi 2014; Ali et al. 2017; Keshavarza and Moghadam 2017; Imran et al. 2018; Tabassum et al. 2018). Priming helps improve the quality of the product in flower seed industry that is usually utilized in best cultivars of *Petunia* (Momin 2013). Herbs, such as ajwain, dill, wild foxglove, rosemary, and scarlet sage, as well enjoy seed priming (Di Girolamo and Barbanti 2012; Mahdavi and Rahimi 2013; Ghassemi-Golezani and Nikpour-Rashidabad 2017; Masondo et al. 2018). Priming has also been used on *Arabidopsis thaliana*, which is the preferred model system for plant biology research (Jiménez-Arias et al. 2015). In seed conditioning in agriculture, several seed priming techniques have been successfully used to accelerate germination rates, improve seed uniformity, and improve yield and yield quality such as hydropriming, osmopriming, and halopriming (e.g., Jabeen and Ahmad 2013; Mahmoudi et al. 2012; Ouhibi et al. 2014; Maswada and Abd El-Kader 2016; Masondo et al. 2018; Panuccioa et al. 2018; Tabassum et al. 2018). Seed priming also promotes many plants to neutralize environmental stress negative impacts, such as salinity stress (see references in Table 1). A growing research has discussed and provided an outline of seed priming expertise, summarizing the currently available physical, chemical, and biological treatments (see recent reviews, e.g., Paparella et al. 2015; Ibrahim 2016; Mahmood et al. 2016). The following is a brief outline of seed priming methods:

**Table 1** The functions of seed priming in plant under salt stress

| Plant species | Priming agent/concentration/time | Main findings | References |
|---|---|---|---|
| Tomato (*Solanum lycopersicum*) | Menadione sodium bisulfite/1.3 mg $L^{-1}$ | Improved net photosynthesis, regulated stomatal aperture and maintained water balance, and increased relative growth rate | Jiménez-Arias et al. (2019) |
| | NaCl/0.05 M | Increased germination, root length, hypocotyl lengths, fresh weights, proline, and anthocyanin | İşeri et al. (2014) |
| | NaCl/5 M | Increased chlorophyll content, Na, $K^+$, Ca, Mg, and soluble sugars | Demirkaya (2014) |
| Chickpea (*Cicer arietinum*) | Mannitol/2 and 4%/24 h | Increased plant survival and growth | Sarwar et al. (2006) |

(continued)

**Table 1** (continued)

| Plant species | Priming agent/concentration/time | Main findings | References |
|---|---|---|---|
| Pepper (*Capsicum annuum* L.) | Aqueous extracts of *Padina pavonica*/60 g $L^{-1}$ or *Jania rubens*/80 g $L^{-1}$ | Increased proline, total soluble sugar, K, and P, while reduced leakage of electrolytes and MDA content | Rinez et al. (2018) |
| Maize (*Zea mays* L.) | Deionized water/24 h or $ZnSO_4$ $7H_2O$/4 mM/24 h | Increased biomass production, nutrient status, and early seedling development | Imran et al. (2018) |
| | Distilled water or leaf extracts of *Rosmarinus officinalis* or *Artemisia* | Increased germination, photosynthetic pigments, chlorophyll fluorescence, photosynthetic efficiency, and antioxidant system | Panuccioa et al. (2018) |
| | Magnetic fields (MF)/week (15 mT) or strong (150 mT) | Enhanced rate and percentage of germination and improved plant growth | Karimi et al. (2017) |
| | Tap water or salicylic acid/0.5 mM or NaCl/2 dS $m^{-1}$ | Increased CAT, SOD, peroxidase, soluble proteins, K, rate of emergence and colonization %, while $Na^+$ decreased | Kazemi et al. (2018) |
| *Z. mays*, *P. sativum*, and *L. sativus* | Gibberellic acid/0.2 g $L^{-1}$/12 h | Increased percentage of germination, length of shoot and root, and total crop weight, while reduced time for germination | Tsegay and Andargie (2018) |
| Rice (*Oryza sativa*) | Gibberellic acid/0, 50, 100 and 150 ppm | Increased germination, seedling growth, and weight, while decreased toxic ions such as $Na^+$ | Imran et al. (2016) |
| | Validamycin A/30 μM | Increased grain yield, weight of 1000 grains, and percentage of filled grain, while decreased trehalose activity, leading to accumulation of trehalose | Hathout et al. 2014 |
| *Barley* (*Hordeum vulgare*) | $dH_2O$ or $CaCl_2$/1.5% or *Enterobacter* spp. strain FD17 | Improved leaf area, grain yield, harvest index, chlorophyll a, chlorophyll b, total soluble phenolics, proteins, proline, glycine betaine, $K^+$, relative water content, osmotic and pressure potential, and stability of the cell membrane while decreased MDA and $Na^+$ | Tabassum et al. (2018) |

(continued)

**Table 1** (continued)

| Plant species | Priming agent/concentration/time | Main findings | References |
|---|---|---|---|
| Sorghum (*Sorghum bicolor*) | PEG-6000/-0.86 Mpa/48 h | Increased activities of catalase (CAT) and guaiacol peroxidase (POX) | Oliveira et al. (2012) |
| | PEG-6000/-0.86 Mpa/48 h | Increased proline, while reduced $Na^+$ and $Cl^-$ | Oliveira et al. (2011) |
| | PEG-6000/-0.86 Mpa/48 h | Increased root growth | Patanè et al. (2009) |
| Safflower (*Carthamus tinctorius*) | Glycine betaine/0, 10, 30, and 60 mM | Increased enzyme activities (CAT, SOD), cell membrane stability, $K^+$, and protein, while reduced peroxidase, MDA, and $Na^+$:$K^+$ ratio | Alasvandyari et al. (2017) |
| | Chitosan/0, 0.25, 0.50, and 0.75% dissolved in 1% acetic acid/3 h | Increased germination %, while decreased MDA, proline, and activity of CAT and POX | Jabeen and Ahmad (2013) |
| Sunflower (*Helianthus annuus*) | $KNO_3$/-1.0 M Pa/24 h | Increased percentage of germination, radicular length, height of seedling, biomass, leaf number, and $K^+$ | Bajehbaj (2010) |
| Dill (*Anethum graveolens*) | Gibberellic acid/1 mM/4 h or salicylic acid/1 mM/4 h | Increased ratio of $K^+$:$Na^+$, osmolytes (i.e., proline, glycine betaine, soluble sugars), enzymes (i.e., SOD, APX, POX), production of essence, and yield of essential oil | Ghassemi-Golezani and Nikpour-Rashidabad (2017) |
| Cucumber (*Cucumis sativus*) | Triacontanol/25 and 50 μM/12 h | Increased seedling emergence rate, uniformity and early growth, shoot and root length, dry weight seedling, chlorophyll, proline, and exchange gas traits | Sarwar et al. (2017) |
| Broccoli sprouts (*Brassica oleracea*) | KCl/50 mM | Increased water relation traits (i.e., osmotic and water potential and hydraulic root conductivity) and glucosinolate | Hassini et al. (2017) |
| Pea (*Pisum sativum*) | Leaf extracts of *Typha angustifolia*/40 g $L^{-1}$/48 h (seed weight/volume; 1:5 (w/v)) | Increased germination%, osmotic (proline, soluble sugars, $K^+$, P), chlorophyll, carotenoid, secondary metabolites (total phenolic, total flavonoid), while reduced electrolyte leakage and MDA | Ghezal et al. (2016) |

(continued)

**Table 1** (continued)

| Plant species | Priming agent/concentration/time | Main findings | References |
|---|---|---|---|
| Mung bean (*Vigna radiata*) | β-Amino butyric acid/0, 0.5, 1.0, 1.5, 2.0, and 2.5 mM/6 h | Increased fluorescence chlorophyll a, proline, protein, carbohydrate, nitrate reductase, and antioxidant enzymes of POX and SOD, photosynthetic pigments, activity of photosynthetic and mitochondrial while decreased MDA | Jisha and Puthur (2016) |
| | Salicylic acid (SA)/0, 0.00001, 0.001, and 10 mM/6 h | Increased germination %, germination rate, radicle length, plumule length, and vigor index. SA with 0.00001 mM is recommended | Movaghatian and Khorsandi (2014) |
| Okra (*Hibiscus esculentus*) | Jasmonate/50 $\mu$M solution for 10 h | Increased fresh weight, dry weight, water content, chlorophyll, relative water content, osmoprotectants (soluble sugars, soluble proteins, proline), mineral ($K^+$, $Ca^{2+}$, $Mg^{2+}$), $K^+$:$Na^+$ and $Ca^{2+}$:$Na^+$ ratios, membrane stability index, antioxidant enzyme (POD, APX), while reduced $Cl^-$, $Na^+$, $H_2O_2$, and MDA | Azooz et al. (2015) |
| *Arabidopsis* (*Arabidopsis thaliana*) | Menadione sodium bisulfite/2, 20, and 40 mM/2 d | Increased dry and fresh weight, relative rate of growth, and proline | Jiménez-Arias et al. (2015) |
| Broad bean (*Vicia faba* L.) seeds | Salicylic acid/0.2 mM/10 h | Increased growth (i.e., dry root, shoot weight, leaf area), water relations (i.e., plant water content, leaf relative water content), osmotic solutes (i.e., soluble carbohydrate, protein, amino acids), enzymes (i.e., CAT, POD, APX, GR), and ascorbic acid while decreased MDA and EC percentage | Azooz (2009) |
| Alfalfa (*Medicago sativa*) | Gibberellic acid/0, 3, 5, and 8 mM/24 h | Increased germination rate, dry weight, enzyme activity (SOD, CAT, POX, APX), and ascorbate, while reduced membrane permeability (electrolyte leakage) and MDA | Younesi and Moradi (2014) |

(continued)

**Table 1** (continued)

| Plant species | Priming agent/concentration/time | Main findings | References |
|---|---|---|---|
| Lettuce (*Lactuca sativa*) | $dH_2O$/2 h or $KNO_3$/0.5%/2 h Gibberellic acid/3 mM/12 h | Increased dry weight of leaves, number of leaves, area of leaves, leaf water, chlorophyll, carotenoids, enzymes (CAT, GPX), ascorbate, while decreased leakage of electrolytes | Mahmoudi et al. (2012) |
| Chinese cabbage (*B. oleracea capitata*) | $dH_2O$/10 h or gibberellic acid/100, 150, and 200 mg $L^{-1}$/10 h | Increased seed water uptake, germination rate, leaf area, chlorophyll, relative water content, fresh and dry weights, membrane stability, while decreased electrolyte leakage | Jamil et al. (2012) |
| Melon (*Cucumis melo*) | NaCl/18 dS $m^{-1}$/3 d | Increased stomatal conductance, chlorophyll, fruit set, $K^+$ leaf and stem, Ca leaf and stem, while reduced $Na^+$ leaf and stem | Sivritepe et al. (2005) |
| Sugarcane (*Saccharum officinarum*) | NaCl/100 mM | Increased germination %, rate of germination, and shoot length | Patade et al. (2009) |
| Wheat (*Triticum aestivum*) | Sodium nitroprusside (a nitric oxide donor)/0, 0.1, and 0.2 mM/12 h | Increased shoot and root biomass, shoot length, 100 grain weight, grain yield, antioxidant enzymes (i.e., SOD, POD, CAT), ascorbic acid, proline, and total phenolics, while deceased $H_2O_2$ and MDA | Ali et al. (2017) |
| | $dH_2O$ or $CaCl_2$/1.5% /12 h (seed to solution ratio 1:5 (w/v) | Increased area of the leaf, water relations, proline, glycine betaine, and yield of grain while decreased MDA and $Na^+$ | Tabassum et al. (2017) |
| | $dH_2O$ or KCl/100 mM or $CaCl_2$/100 mM | Increased proline, phenolic compounds, antioxidant enzyme activities (CAT, POD, APX), while reduced $H_2O_2$ and MDA | Islam et al. (2015) |
| | Chilling/3 °C/2 weeks | Increased net photosynthetic rate, growth, yield greater, root, $K^+$, and $Ca^{2+}$, while reduced $Na^+$ in the shoots | Iqbal and Ashraf (2010) |
| | $CaCl_2$/50 mM/12 h or NaCl/50 mM/12 h or $CaSO_4$/50 mM/12 h | Increased germination, length of root, fresh and dry weights, and $K^+$. The most effective priming agents were $CaSO_4$ and $CaCl_2$ | Afzal et al. (2008) |

### 3.2.1 Hydropriming

In hydropriming, seeds are soaked in water with or without aeration at optimum temperature (generally on or after 5–20 °C) and regularly for 6–24 h. This method is especially helpful in rural regions where crop production is damaged by unfavorable climatic environments, e.g., drought, and is required to enhance the efficiency of water use by decreasing chemical exposure (McDonald 2000). Hydropriming has long been known for its advantages, although it is now used less frequently than other methods. Harris (1992) suggested hydropriming as an inexpensive practice, explained for several crops in developing countries as "on-farm seed priming."

### 3.2.2 Osmopriming

Osmopriming is known as osmotic priming or osmotic conditioning as well. It is a common pre-sowing practice that uses low-water-potential osmotic solutions to help manage water uptake. Osmopriming's main goal is to limit the oxidative damage caused by ROS by preventing water from entering. As a result, the osmotic agent's water potential is a critical factor (Heydecker and Coolbear 1977; Taylor et al. 1998). In osmopriming, seeds are soaked in osmotic agent solutions for a certain time period.

This technique mainly utilizes sugars, polyethylene glycol (PEG), glycerol, sorbitol, or mannitol, followed by air drying pre-sowing.

### 3.2.3 Solid Matrix Priming

The priming of solid matrix is also known as "matriconditioning." It has been developed as a replacement to avoid high osmopriming costs, which require big volumes of osmotic solution and costly temperature management and aeration systems (Paparella et al. 2015). There are a number of natural materials that were used for solid priming as matrices, e.g., charcoal, sawdust, coal, calcined kaolin, and vermiculite, and marketable materials, e.g., Agro-Lig®. Solid matrix priming is best performed in an air circulation sealed container while avoiding too much evaporation.

### 3.2.4 Biopriming

Biopriming is a method using a combination of useful or bioactive particles. Is it well acknowledged that the mixture of plants with particular fungi or bacteria yields very appropriate outcomes as these microorganisms are capable of creating endophytic connections with the plant, leading to enhanced plant development and manufacturing of phytohormones and enhanced biotic/abiotic stress resistance (Waller et al. 2005). *Bacillus* spp., *Enterobacter* spp., *Pseudomonas* spp., and *Trichoderma* spp. are the most frequently used biopriming species (Niranjan Raj et al. 2004). In addition, biopriming is accomplished by adding secondary metabolites to the priming mixture and phytohormones (Hamayun et al. 2010).

### 3.2.5 Chemopriming

By addition of conventional decontaminators like sodium hypochlorite (NaOCl), hydrochloric acid (HCl), natural substances, and agrochemicals (e.g., herbicides, insecticides, fungicides, fumigants) to the priming solution for microbial pollution prevention, Chemopriming is accomplished (Parera and Cantliffe 1990). Natural compounds with a wide range of antimicrobial activity, including organic acids, essential oils, crude extracts of plants, and milk products, are mainly used for seed disinfection in organic farming (Van der Wolf et al. 2008). Additional environmentally acceptable means such as those made from insecticidal and fungicidal microencapsulated plant extracts are being tested (Gaidau et al. 2014).

### 3.2.6 Thermopriming

Thermopriming is a technique that seeds subject to different temperatures before sowing. This practice has been widely demonstrated to enhance germination effectiveness under negative environmental circumstances, decreasing seed germination thermo-inhibition (Huang et al. 2002). Thermopriming is accomplished by presowing seed at different temperatures.

## 4 Mechanisms Induced by Seed Priming to Increase Plant Tolerance to Salt Stress

Seed priming has been shown to increase germination percentage, shoot and root length, shoot to root ratio, shoot to root weight, germination percentage, germination time, radicular length, radicular surface area, average radicle and radicle volume diameter, germination stress tolerance, seedling vigor, shoot and root length stress, fresh and root weight, and germination speed (Patade et al. 2009; Bajehbaj 2010; Movaghatian and Khorsandi 2014; Ghezal et al. 2016; Al-Tabbal 2017; Nimac et al. 2018); increased photosynthetic pigments in terms of chlorophyll, carotenoids, and chlorophyll fluorescence (Anwar et al. 2011; Demirkaya 2014; Azooz et al. 2015; Ghezal et al. 2016; Jisha and Puthur 2016; Abdel Latef et al. 2017; Panuccioa et al. 2018; Tabassum et al. 2018); increased plant water content (relative water content, osmotic and water potential, and root hydraulic conductivity) (Azooz 2009; Jamil et al. 2012; Ouhibi et al. 2014; Hassini et al. 2017; Tabassum et al. 2017; Yang et al. 2017; Jiménez-Arias et al. 2019); improved leaf photosynthetic rate and stomatal conductance (Iqbal and Ashraf 2013; Jisha and Puthur 2016; Sarwar et al. 2017; Yang et al. 2017; Panuccioa et al. 2018; Jiménez-Arias et al.

2019); increased osmoprotectants (soluble sugars, soluble proteins, and proline) (Oliveira et al. 2011; Jabeen and Ahmad 2013; Azooz et al. 2015; Abdel Latef et al. 2017; Ghassemi-Golezani and Nikpour-Rashidabad 2017); increased enzymatic antioxidant activity, i.e., catalase (CAT), peroxidase (POX), superoxide dismutase (SOD), ascorbate peroxidase (APX), and glutathione reductase (GR) (Azooz 2009; Jabeen and Ahmad 2013; Oliveira et al. 2012; Abdel Latef et al. 2017; Alasvandyari et al. 2017; Mekawy et al. 2018); increased nutrient uptake, i.e., $K^+$, $Ca^{2+}$, and $Mg^{2+}$ (Afzal et al. 2008; Iqbal and Ashraf 2010; Anwar et al. 2011; Demirkaya 2014; Azooz et al. 2015); increase plant growth (relative growth rate, unit leaf rate, leaf area ratio, specific leaf area) (Maswada and Abd El-Kader 2016); enhanced organic solutes, i.e., soluble sugar, soluble protein, free amino acids, proline, and betaine (Anwar et al. 2011; Oliveira et al. 2011; Azooz et al. 2015; Jiménez-Arias et al. 2015; Ghezal et al. 2016; Jisha and Puthur 2016; Alasvandyari et al. 2017; Ghassemi-Golezani and Nikpour-Rashidabad 2017; Sarwar et al. 2017; Keshavarza and Moghadam 2017; Tabassum et al. 2017); reduced leakage of electrolytes and $H_2O_2$ content and MDA content (Jisha and Puthur 2014; Alasvandyari et al. 2017; Mekawy et al. 2018; Rinez et al. 2018; Tabassum et al. 2018); and consequently increased plant yield (biomass, grain yield, yield attributes) (Iqbal et al. 2006; Iqbal and Ashraf 2010, 2013; Mohammadi et al. 2012; Ghezal et al. 2016) and accordingly is capable to boost plant performance and tolerance under salinity stress in several crops such as okra (Azooz et al. 2015), wheat (Iqbal and Ashraf 2010; Maswada and Abd El-Kader 2016), quinoa (Yang et al. 2017), and pepper (Rinez et al. 2018).

Table 1 shows the functions of seed priming in plant under salt stress. In Table 1, there are many plant species that were tested for seed priming under salt stress, namely, tomato (*Solanum lycopersicum*), sea fennel (*Crithmum maritimum* L.), quinoa (*Chenopodium quinoa* Willd.), chickpea (*Cicer arietinum*), pepper (*Capsicum annuum* L.), maize (*Zea mays* L.), rice (*Oryza sativa* L.), *barley* (*Hordeum vulgare*), wild foxglove (*Ceratotheca triloba*), sorghum (*Sorghum bicolor*), safflower (*Carthamus tinctorius* L.), sunflower (*Helianthus annuus* L.), lupine (*Lupinus termis*), dill (*Anethum graveolens* L.), cucumber (*Cucumis sativus*), broccoli sprouts (*Brassica oleracea*), common bean (*Phaseolus vulgaris*), pea (*Pisum sativum* L.), mung bean (*Vigna radiata* L.), okra (*Hibiscus esculentus*), *Arabidopsis* (*Arabidopsis thaliana*), faba bean (*Vicia faba* L.), alfalfa (*Medicago sativa*), lettuce (*Lactuca sativa*), lentil (*Lens culinaris*), Chinese cabbage (*B. oleracea capitata*), canola (*Brassica napus*), melon (*Cucumis melo*), sugarcane (*Saccharum officinarum* L.), ajwain (*Carum copticum*), and wheat (*Triticum aestivum* L.). These plant species were tested for seed priming techniques using many materials with different concentrations, namely, distilled water ($dH_2O$), menadione sodium bisulfite ($C_{11}H_9NaO_5S$), gibberellic acid ($C_{19}H_{22}O_6$), saponin ($C_{36}H_{58}O_9$), zinc sulfate ($ZnSO_4$), salicylic acid ($C_7H_6O_3$), apigenin ($C_{15}H_{10}O_5$), validamycin A ($C_{20}H_{35}NO_{13}$), phloroglucinol ($C_6H_6O_3$), polyethylene glycol 6000 (*PEG-6000*) ($HO(C_2H_4O)_nH$), glycine betaine ($C_5H_{11}NO_2$), chitosan ($C_{56}H_{103}N_9O_{39}$), acetic acid ($C_2H_4O_2$ or $CH_3COOH$), potassium nitrate ($KNO_3$), zinc oxide (ZnO nanoparticles), triacontanol ($C_{30}H_{62}O$), vitamin B12 ($C_{63}H_{88}CoN_{14}O_{14}P$), zinc sulfate ($ZnSO_4$), β-amino butyric acid ($C_4H_9NO_2$), jasmonic acid ($C_{12}H_{18}O_3$), ascorbic acid (vitamin C) ($C_6H_8O_6$ or $HC_6H_7O_6$), nicotinamide (vitamin B3) ($C_6H_6N_2O$), sodium nitroprusside ($Na_2[Fe(CN)_5(NO)]$), kinetin ($C_{10}H_9N_5O$), indole acetic acid ($C_{10}H_9NO_2$), indolebutyric acid ($C_{12}H_{13}NO_2$), tryptophan ($C_{11}H_{12}N_2O_2$), calcium sulfate ($CaSO_4$), calcium chloride ($CaCl_2$), potassium chloride (KCl), sodium chloride (NaCl), putrescine ($C_4H_{12}N_2$), spermidine ($C_7H_{19}N_3$), spermine ($C_{10}H_{26}N_4$), Zn-glutamine ($C_5H_{10}N_2O_3$), Zn-glycine ($C_4H_8N_2O_4Zn$), Zn-arginine ($C_6H_{14}N_4O_2$), and Zn-histidine ($C_6H_9N_3O_2Zn$), UV-C irradiation, *Enterobacter* spp. strain FD17, magnetic fields, aqueous extracts of *Padina pavonica* or *Jania rubens* or *Typha angustifolia* or *Rosmarinus officinalis* or *Artemisia*, commercial seaweed extract, and smoke-compound karrikinolide.

It has also been previously documented that plants are constructing many physiological and biochemical adaptations to adjust themselves under saline environments. With regard to using seed priming to alleviate the salt stress effects on plants, we proposed a schematic diagram describing the chain reaction following the seed priming (Fig. 2). The diagram illustrates the possible impact of the seed priming on different physiological processes, enhancing tolerance level and final yield.

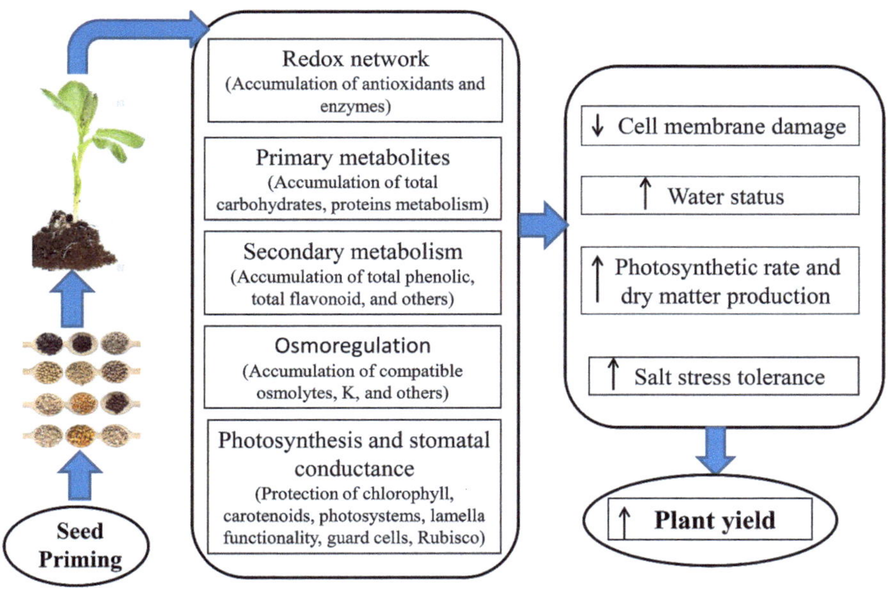


**Fig. 2** Schematic diagram describing the chain reaction following the seed priming. The diagram illustrates the possible impact of the seed priming on different physiological processes, enhancing tolerance level and final yield

### 4.1 *Induced Salt Tolerance by Seed Priming Through Plant Physiological Response*

Photosynthesis is the biochemical pathway through which plants transform solar energy into the chemical energy needed for growth. Under salt stress, reduced photosynthetic rates in plants are primarily due to reduced water potential and/or greater of $Na^+$ and/or $Cl^-$ in chloroplasts that slow down chlorophyll (Iqbal and Ashraf 2007; Khayyat et al. 2014). Reducing the content of chlorophyll is common under salt stress and has been used in many reports as a sensitive indicator for cell metabolism (Chutipaijit et al. 2011). Reductions in chlorophyll a, chlorophyll b, carotenoids, and photosynthetic pigments have been recorded in faba bean following 44-day use of diluted seawater with 3.13 or 6.25 dS $m^{-1}$, with reduced total pigments by 6.7% and 15.6%, respectively, using both seawater treatments compared to control crops (Dawood et al. 2014b). In another study in soybean plants that are exposed to diluted salinity of seawater, chlorophyll a, chlorophyll b, total carotenoids, and total concentrations of pigments gradually decreased with increased levels of salt stress (Rady et al. 2015). Membrane deterioration is considered one of several reasons which might affect negatively on the chlorophyll under stress (Mane et al. 2010). Moreover, increased chlorophyll a and chlorophyll b content was reported by Tabassum et al. (2018), which was correlated with increased cell membrane stability.

Photosystem II (PS II), when subjected to salt stress, is a fragile component of the photosynthetic structure (Allakhverdiev et al. 2000). The efficiency of PS II, the electron transportation chain (ETC), and the net rate of $CO_2$ assimilation have been reported to be reduced under salt stress (Piotr and Grazyna 2005). Moreover, changes in *Scenedesmus obliquus*' photosynthetic attributes resulting in reduced biomass accumulation were reported (Demetriou et al. 2007). There was no significant difference in salt tolerant-cultivar *Vicia faba* cv. Giza 843 a susceptible one Giza 4 under the saline condition in Fv/Fm (used to estimate PS II's potential efficiency through darkly adapted measurements), while Giza 843 had a significantly higher PI (photosynthetic performance index) (Semida et al. 2014). Salt stress decreased the development of citrus by decreasing the net photosynthetic rate, stomatal behavior, PS II performance, and photosynthetic efficiency (Lòpez-Climent et al. 2008). It has been reported that salinity stress affects barley growth by altering chlorophyll fluorescence (PS II) and developing complex oxygen functions (Kalaji et al. 2011). In addition, salt stress decreased *Brassica* development by influencing PS II, electron transport rates, and D1 protein (Mittal et al. 2012). Moreover, under salt stress, there are other reasons that reduce net $CO_2$ assimilation rates, such as cell membrane dehydration, which decreases carbon dioxide permeability, salt toxicity, increased senescence, variations in cytoplasmic structure-induced enzyme activity, and negative reaction to decreased sink activity (Iyengar and Reddy 1996).

As discussed above, a plant under salt stress faces a reduction in photosynthetic rates due to a decreasing in water potential accompanied/or not with too much concentration of $Na^+$ and/or $Cl^-$ in the chloroplasts and chlorophyll. Therefore, any means that could increase plant water potential and exclude $Na^+$ and $Cl^-$ may increase chlorophyll content and improve photosynthetic performance of PS II and consequently result in enhancing net $CO_2$ photosynthetic rate and finally boost plant yield. In this regard, several authors reported that seed priming materials (e.g., saponin; $dH_2O$; $CaCl_2$; zinc nanoparticles (ZNPs); vitamin B12; NaCl; indole acetic acid; indolebutyric acid; tryptophan) increased photosynthetic pigment (chlorophyll *a*, chlorophyll *b*, carotenoids), photosystem II, and net $CO_2$ photosynthetic rate in several crops, e.g., quinoa, *barley*, lupine, common bean, mung bean, and wheat (Jisha and Puthur 2014; Abdel Latef et al. 2017; Keshavarza and Moghadam 2017; Yang et al. 2017; Tabassum et al. 2018). More details about the materials used for priming and their concentrations are shown in Table 1. A schematic diagram describing the chain reaction following the seed priming is shown in Fig. 2. The diagram illustrates the possible impact of the seed priming on different physiological processes, enhancing tolerance level and final yield.

It was reported that improved photosynthetic pigment (chlorophyll *a*, chlorophyll *b*, carotenoids), photosystem II, and net $CO_2$ assimilation rate by seed priming application ($CaCl_2$; ZNPs; vitamin B12; aqueous leaf extracts of *Typha angustifolia*; β-amino butyric acid; NaCl; jasmonate sodium nitroprusside) were associated with decreased malondialdehyde and hydrogen peroxide in plants under salt stress (Jisha and Puthur 2014, 2016; Azooz et al. 2015; Ghezal et al. 2016; Maswada and Abd El-Kader 2016; Abdel Latef et al. 2017; Keshavarza and Moghadam 2017; Tabassum et al. 2018). Various studies (e.g., Sivritepe et al. 2005; Iqbal and Ashraf 2010; Azooz et al. 2013, 2015; Maswada and Abd El-Kader 2016; Ghezal et al. 2016; Tabassum et al. 2018) reported that enhanced photosynthetic pigment (chlorophyll *a*, chlorophyll *b*, carotenoids), photosystem II, and net $CO_2$ photosynthetic rate by seed priming application ($CaCl_2$; chilling at 3 °C; NaCl; sodium nitroprusside; diluted seawater with 9.00 dS $m^{-1}$; ascorbic acid; nicotinamide; jasmonate; aqueous leaf extracts of *Typha angustifolia*) were associated with increased contents of nutrients (P, $K^+$, $Ca^{2+}$, $Mg^{2+}$, $Ca^{2+}$, Zn), raised $Ca^{2+}$:$Na^+$ and $K^+$:$Na^+$ ratios, and reduced $Na^+$ and $Cl^-$ in plants under salt stress. Enhanced photosynthetic pigment (chlorophyll *a*, chlorophyll *b*, carotenoids), photosystem II, and net $CO_2$ photosynthetic rate by seed priming application ($CaCl_2$; NaCl; ZNPs; jasmonate; ascorbic acid; nicotinamide; EC 9.00 dS $m^{-1}$; sodium nitroprusside) were correlated with increased organic solute content (proline and glycinebetaine, free amino acids, soluble sugar, soluble protein) under salt stress (Anwar et al. 2011; Azooz et al. 2013, 2015; Jisha and Puthur 2014; Maswada and Abd El-Kader 2016; Abdel Latef et al. 2017; Tabassum et al. 2018). Increased photosynthetic pigment (chlorophyll *a*, chlorophyll *b*, carotenoids), photosystem II, and net $CO_2$ photosynthetic rate by seed priming application (ZNPs; vitamin B12; aqueous leaf extracts of *Typha angustifolia*; jasmonate; $KNO_3$; $dH_2O$; gibberellic acid) were associated with increased enzymatic antioxidants (SOD, CAT, POD, and APX, POX) and nonenzymatic antioxidants (total phenols, flavonoids, ascorbic acid, total carotenoids) in salt-stressed plants (Mahmoudi et al. 2012; Azooz et al. 2015; Ghezal et al. 2016; Abdel Latef et al. 2017; Keshavarza and Moghadam 2017).

### 4.2 Induced Salt Tolerance by Seed Priming Through Ionic Balance

When plants are stressed by salt, the absorption of essential mineral elements is interrupted. The increase in concentration of $Na^+$ in the root rhizosphere directly affects negatively on the absorption of many necessary elements, such as $K^+$, while $Na^+$ shows opposite correlation with $K^+$ (Kohler et al. 2009). In most plant species, salt tolerance was correlated with only tiny quantities of $Na^+$ and $Cl^-$ in the shoots (Greenway and Munns 1980), while $Cl^-$ under saline conditions is the most prevalent anion. Noble and Rogers (1992) proposed shoot $Cl^-$ levels as selection standard for salt tolerance. For instance, in four *Vicia faba* cultivars, Abdelhamid et al. (2010) recorded a partial ion exclusion mechanism, stating that partial toxic ion exclusion was suggested as the main selection parameter for faba bean salt tolerance. They reported that faba bean salt-tolerant cultivars (Giza 429, Giza 843, Misr 1) had higher N, P, $K^+$, $Ca^{2+}$, and $Mg^{2+}$ than susceptible cultivars (Giza 3), which showed lesser $K^+$, $Ca^{2+}$, and $Mg^{2+}$ while higher $Na^+$ and $Cl^-$. Ions' uptake in plants can be exploited as a marker of salt tolerance since they are genetically controlled but also influenced by the environment (Chaubey and Senadhira 1994). The *Vicia faba* salt-tolerant cultivar Giza 429 has been confirmed by Semida et al. (2014) that it had higher levels of N, P, $K^+$, and $Ca^{2+}$, whereas susceptible salt cultivar Giza 40 had lower levels of N, P, $K^+$, and $Ca^{2+}$ but higher levels of $Na^+$. Furthermore, under saline conditions, high $Na^+$:$K^+$ and $Na^+$:$Ca^{2+}$ ratios occur in the soil due to excessive amounts of exchangeable $Na^+$. Thus, plants grown in such environments have elevated $Na^+$ levels, while $K^+$ and $Ca^{2+}$ take-up is decreased. For the integrity and functioning of cell membranes, rational quantities of both K+ and $Ca^{2+}$ are requisite (Wenxue et al. 2003). For several major metabolic processes in plant cells, optimal $K^+$ is essential (Tomar and Agarwal 2013). A good amount of $K^+$:$Na^+$ and $Ca^{2+}$:$Na^+$ ratios are also considered beneficial in maintaining ion balance in plant leaves and demonstrating salt tolerance in plant (Abdelhamid et al. 2010; Semida et al. 2014). Semida et al. (2014) confirmed that *Vicia faba* cv. Giza 429 salt-tolerant cultivar had higher $K^+$:$Na^+$ and $Ca^{2+}$:$Na^+$ ratio, while susceptible salt cultivar Giza 40 had lower ratios. In addition, Bargaz et al. (2016) reported that *Phaseolus vulgaris* recombinant inbred lines (RIL) RIL115, exhibited more salt tolerance with greater uptake of P, $K^+$, $Ca^{2+}$, $Mg^{2+}$, and $Mn^{2+}$ and higher uptake of $K^+$:$Na^+$ than susceptible salt tolerance of RIL147. However, contradictory results for different plant species were reported, where Ashraf et al. (1994) questioned ion exclusion mechanism validation. Moreover, salt-tolerant and salt-sensitive lines were found to vary a little in $Na^+$ and $Cl^-$ shoot of the species *Atylosia* (Subbarao et al. 1990).

There are several reports examining how seed priming is affecting plant growth and mechanisms involved to help a plant to be more tolerant under salt stress condition. As discussed above that a plant under salt stress with greater elemental contents, i.e., $K^+$ and $Ca^{2+}$, and lower $Na^+$ could be considered salt tolerant. In this regard, several authors reported that seed priming (namely, NaCl, $CaCl_2$, gibberellic acid, PEG-6000, SA, glycinebetaine, $ZnSO_4$, aqueous leaf extracts of *Typha angustifolia*, jasmonate, ascorbic acid, nicotinamide, UV-C irradiation, $CaCl_2$, $dH_2O$, seawater with 9.00 dS $m^{-1}$, sodium nitroprusside, and $CaSO_4$) improved $K^+$, $Ca^{2+}$, and the ratios of $K^+$:$Na^+$ and $Ca^{2+}$:$Na^+$, while it reduced $Na^+$ in wheat (*Triticum aestivum*) (Afzal et al. 2008; Maswada and Abd El-Kader 2016; Tabassum et al. 2017), barley (*Hordeum vulgare*) (Anwar et al. 2011; Tabassum et al. 2018), rice (*Oryza sativa* L.) (Imran et al. 2016), sorghum (*Sorghum bicolor*) (Oliveira et al. 2011), dill (*Anethum graveolens* L.) (Ghassemi-Golezani and Nikpour-Rashidabad 2017), safflower (*Carthamus tinctorius*) (Alasvandyari et al. 2017), common bean (*Phaseolus vulgaris*) (Gulmezoglu et al. 2016), pea (*Pisum sativum* L.) (Ghezal et al. 2016), okra (*Hibiscus esculentus*) (Azooz et al. 2015), faba bean (*Vicia faba* L.) (Azooz et al. 2013), lettuce (*Lactuca sativa*) (Ouhibi et al. 2014), and melon (*Cucumis melo*) (Sivritepe et al. 2005). More details about the materials used for priming and their concentrations are shown in Table 1.

Seed priming has been shown to be capable of promoting plant growth under salinity stress through increased nutrient uptake of $K^+$, $Ca^{2+}$, and $Mg^{2+}$ while lower uptake of $Na^+$ and $Cl^-$ which could be considered as one of the seed priming approaches to mitigate salt stress (Afzal et al. 2008; Iqbal and Ashraf 2010; Anwar et al. 2011; Demirkaya 2014; Azooz et al. 2015). This could make seed priming useful to alleviate the plant's salinity stress. Seed priming also participates with important function in regulating the ion transportation and membrane proteins, which manage plant ion homeostasis. For example, Yang et al. (2017) reported little $Na^+$ and elevated $K^+$ in quinoa plants grown under 400 mM NaCl stress primed with concentrations of 0.5, 2, 5, 10, 15, 25, and 35% saponin. In rice, 24-h seed priming with 10 ppm of apigenin reduced $Na^+$ concentration in salt-stressed plants and helped maintain a superior $K^+$:$Na^+$ ratio in all plant parts compared to control plants, possibly by adjusting the appearance of confident central $Na^+$ transporter encoding genes (Mekawy et al. 2018). Iqbal and Ashraf (2010) compared two salt-sensitive and salt-tolerant *Triticum aestivum* cultivars after priming their seeds with chilling at 3 °C for 2 weeks and distilled water hydropriming for 12 h at 22 ± 3 °C. They assigned the increased growth and yield to enhanced concentration of $K^+$ and $Ca^{2+}$ in the roots and decreased $Na^+$ in wheat cultivars (Iqbal and Ashraf 2010). A schematic diagram describing the chain reaction following the seed priming is shown in Fig. 2. The diagram illustrates the possible impact of the seed priming on different physiological processes, enhancing tolerance level and final yield.

## *4.3 Induced Salt Tolerance by Seed Priming Through Osmolyte Organic Solutes*

In order to combat stress, plant metabolism is altered in many respects by integrating compatible solute manufacturing to set proteins and cellular structures and/or retain cell turgor through osmotic adjustment and polymerization metabolism to eradicate ROS surpluses and resume the cellular redox equilibrium (Janska et al. 2010; Krasensky and Jonak 2012). Organic plant osmolytes are compounds of low molecular weight, methylated tertiary N and amino acids, and other metabolites of low molecular weight (Chen and Jiang 2010). Osmolytes have unique responses that can defend plant cells other than osmotically (Yancey 2005). Plant performance was also enhanced by the use of exogenous osmolytes under salt stress. A well-recognized adaptive mechanism in salt-stressed plants is the accumulation of osmolytes consisting of proline, glycine betaine, and sugar (Parida and Das 2005; Ashraf and Foolad 2007).

In many crop plants, e.g., safflower (*Carthamus tinctorius*), wheat (*Triticum aestivum*), sorghum (*Sorghum bicolor*), faba bean (*Vicia faba*), lupine (*Lupinus termis*), and common bean (*Phaseolus vulgaris*), and in reaction to salt stress, soluble sugars, free amino acids, proline, and glycine betaine build up (El-Lethy et al. 2013; Semida et al. 2014; Bargaz et al. 2016; Orabi and Abdelhamid 2016; Alasvandyari et al. 2017). The increment in accumulation of these organic osmolytes corresponded with a rise in the rate of salinity (Bargaz et al. 2016). Tolerant genotypes accumulated in some species more proline than susceptible one in reaction to salt stress, e.g., common bean salt-tolerant genotype RIL115 accumulated significant higher proline than the salt-susceptible RIL147 genotype (Bargaz et al. 2016), and in faba bean, the salt-tolerant cultivar Giza 843 had higher proline more than susceptible salt tolerance Giza 3 cultivar (Orabi and Abdelhamid 2016), while in wheat, salt-tolerant cultivar Sakha 93 accumulated significantly more soluble sugars than the salt-susceptible Gemiza 9 cultivar (El-Lethy et al. 2013). In addition, genotype 115 salt-tolerant faba beans appear to be associated with its capacity to accumulate more osmotic solutes than susceptible one (Azooz 2009). Proline therefore prevents crops from salt stress injury by keeping osmoregulation. Furthermore, the roles of proline as an osmolyte or osmoprotective compatible and its antioxidant characteristics to decrease ROS have become a strong plant safeguard against negative abiotic stress, including salt stress (Matysik et al. 2002).

Therefore, as discussed above, a plant under salt stress with greater compatible solute, e.g., glycine betaine, amino acids, soluble sugars, and proline, could be considered salt tolerant. In this regard, several authors reported that seed priming techniques with different materials, e.g., $dH_2O$, $CaCl_2$, PEG-6000, glycinebetaine, ZnO nanoparticles, gibberellic acid, triacontanol, vitamin B12, aqueous leaf extracts of *Typha angustifolia*, β-amino butyric acid, NaCl, jasmonate, menadione sodium bisulfite, soluble sugars, salicylic acid, glycine betaine, amino acids, and proline, in many crops i.e., in *barley* (Anwar et al. 2011), sorghum (Oliveira et al. 2011), safflower (Alasvandyari et al. 2017), lupine (Ghassemi-Golezani and Nikpour-Rashidabad 2017), cucumber (Sarwar et al. 2017), common bean (Keshavarza and Moghadam 2017), pea (Ghezal et al. 2016), mung bean (Jisha and Puthur 2016), okra (Azooz et al. 2015), *Arabidopsis* (Jiménez-Arias et al. 2015), and wheat (Tabassum et al. 2017). More details about the materials used for priming and their concentrations are shown in Table 1.

On the other hand, exogenous application of compatible solute might not only raise their levels but also protects components of cells, thus raising cellular osmotic pressure. Exogenous amino acids enhanced free amino acids, proline, and soluble sugars in salt-stressed faba bean (Sadak and Abdelhamid 2015). Furthermore, exogenous 5.0 mM proline applications increased proline in common bean plants in saline soil (Abdelhamid et al. 2013a). Exogenous use of glycinebetaine has been reported to improve $CO_2$ assimilation, decrease stomatal conductance, and improve the efficiency of PS II in maize salt stress plants (Yang et al. 2005). Glycinebetaine functions an important job in protecting plant against salt stress (Ashraf and Foolad 2007) through protein and RuBisCO stabilization (Mäkelä et al. 2000), photosynthetic device protection (Cha-Um and Kirdmanee 2010), osmotic adjustment (Gadallah 1999), and ROS reduction (Ashraf and Foolad 2007). A schematic diagram describing the chain reaction following the seed priming is shown in Fig. 2. The diagram illustrates the effects of seed priming on different physiological processes, enhancing tolerance level and final yield.

Seed priming in 0, 10, 30, and 60 mM glycinebetaine increased salt tolerance in safflower, mainly linked with enhanced CAT and SOD enzyme activity, and protects cell membranes as a consequence of decreased lipid peroxidation and enhanced ion homeostasis (Alasvandyari et al. 2017). Tabassum et al. (2017) reported increased leaf proline and glycinebetaine, and this increase was positively correlated with increases in leaf area, water content, and grain yield while negatively associated with decreases in MDA and $Na^+$ content when wheat primed with $dH_2O$ and $CaCl_2$. The increased activity of POD and APX antioxidant enzymes with greater amount of osmoprotectants (soluble sugars, soluble proteins, and proline) resulted in protecting okra plants from oxidative injury induced by NaCl stress owing to 50 μM jasmonate seed priming (Azooz et al. 2015).

## *4.4 Induced Salt Tolerance by Seed Priming Through Antioxidant Defense Response*

Salinity causes cell-level oxidative damage in plants. The oxidative burst findings since the molecule of oxygen ($O_2$) occurs under stress circumstances as an electron acceptor, leading in the building up of reactive oxygen species (ROS) in subcellular compartments, especially in mitochondria and chloroplast. ROS contains superoxide radical ($O_2^{\bullet-}$), singlet oxygen ($^1O_2$), hydroxyl radical ($^{\bullet}OH$), and hydrogen peroxide ($H_2O_2$). These are oxidizing compounds that can cause injury to DNA, proteins, and lipids (Quiles and López 2004).

Most plant antioxidant studies show that plant production of antioxidants increases in response to salinity to counteract high levels of salt-induced ROS in cells. However, plants with elevated activities of antioxidants can scavenge/detoxify ROS and thus lead to higher salt tolerance (Wise and Naylor 1987; Garratt et al. 2002). Some of the antioxidant enzymes are directly engaged in salt-induced ROS detoxification, i.e., glutathione reductase (GR), superoxide dismutase (SOD), catalase (CAT), glutathione peroxidase (GPX), and ascorbate peroxidase (APX). Ascorbate, glutathione, and flavonoid are the main nonenzymatic antioxidants engaged in ROS detoxification by salinity (Van Oosten et al. 2013; Begara-Morales et al. 2014).

Significant differences in antioxidant expression for both enzymatic and nonenzymatic are apparent at interspecific or intraspecific concentrations in reaction to salt stress and may rely on germplasms, species, geographic distribution, climate variables, and seasons. For example, plants vary in antioxidant activity under salt environment, and this difference can be assigned to genotypic diversity and the degree of stomata closure that alters the $CO_2$ fixation rate and avoids photoinhibition (Munns and Tester 2008). While determining if guaiacol peroxidase (POX), the SOD and CAT operate an important function in protecting and recovering cowpea from oxidative salt stress (Cavalcanti et al. 2004). It was reported, however, that the principal antioxidant mechanism, which includes CAT, SOD, and POX, in cowpea mature leaves did not interfere with salt-stressed plants' survival (Cavalcanti et al. 2004).

Many studies confirm the positive association between elevated antioxidant activity and the salt tolerance grade in diverse plant species, even though in some other instances, this is not always right. For instance, the activity of PPO, POX, SOD, CAT, and APX antioxidant enzymes improved with higher concentrations of faba bean salt-stressed plants (Sadak and Abdelhamid 2015). Orabi and Abdelhamid (2016) recorded enhanced activity of antioxidant enzymes POX, CAT, SOD, and PAL in two faba bean cultivars under salt stress, and the greater activity was observed in Giza 843 salt-tolerant cultivar compared to Giza 3 susceptible one.

As discussed above that a plant under salt stress with greater antioxidants, both enzymatic and nonenzymatic could be considered a tolerant one. In this regard, several authors reported that seed priming techniques (e.g., vitamin B12; glycinebetaine; jasmonate; sodium nitroprusside; KCl; $CaCl_2$; jasmonate; glycine betaine; ZNPs; chilling at 3 °C) increased antioxidants in many crops, e.g., wheat, broad bean, lupine, and safflower (Azooz 2009; Islam et al. 2015; Abdel Latef et al. 2017; Alasvandyari et al. 2017). More details about the materials used for priming and their concentrations are shown in Table 1. A schematic diagram describing the chain reaction following the seed priming is shown in Fig. 2. The diagram illustrates the effects of seed priming on different physiological functions, enhancing tolerance level and increased final yield.

The PEG-6000 solution used for seed priming with an osmotic potential of −0.86 Mpa increased CAT and POX activity, and these enzymes can protect sorghum seedlings from oxidative damage in nutrient solutions due to accelerated seed aging and salinity (Oliveira et al. 2012). In latest research by Mekawy et al. (2018), when rice seeds were submerged in a 10 ppm apigenin solution, apigenin pretreatment was connected with the induction of the antioxidant rice protection mechanism by increasing antioxidant CAT and APX activity in the roots, in addition to enhanced accumulation of nonenzymatic antioxidant carotenoids and flavonoids in the shoots, so apigenin pretreatment may reduce the damaging impacts of salinity on rice seedlings by probably triggering induction of antioxidant defense mechanism. Furthermore, Alasvandyari et al. (2017) found that priming with 60 mM glycinebetaine increased salt tolerance in the safflower, mainly owing to enhanced activity of CAT and SOD and decreased cell membrane damage due to decreased lipid peroxidation and enhanced ion homeostasis. Seed priming with ZNPs generally stimulate stressed plant growth, accompanied by increased ascorbic acid and Zn, plus enhanced activity of SOD, CAT, POD, and APX enzymes. Priming with ZNPs in stressed crops led to a decrease in MDA and $Na^+$ content. These results therefore indicate that seed priming with ZNPs, especially 60 mg $L^{-1}$ ZnO, is an effective method that can be utilized to enhance salt tolerance of lupine plants (Abdel Latef et al. 2017). Improved SOD, APX, and POX activity through seed priming (i.e., vitamin B12; glycine betaine; jasmonate; sodium nitroprusside; KCl; $CaCl_2$) reduced levels of $H_2O_2$ and MDA in plants under salt stress (Azooz et al. 2015; Islam et al. 2015; Ali et al. 2017; Alasvandyari et al. 2017; Keshavarza and Moghadam 2017). Seed priming with jasmonate, glycinebetaine, ZNPs, and 3 °C chilling increased activity of SOD, APX, and POX which was associated with increased nutrient content of $K^+$, $Ca^{2+}$, $Mg^{2+}$, $Ca^{2+}$, and Zn in salt-stressed plants (Iqbal and Ashraf 2010; Azooz et al. 2015; Abdel Latef et al. 2017; Alasvandyari et al. 2017). Enhanced SOD, APX, and POX activities were associated with increased organic solutes of proline, free amino acids, soluble sugar, and soluble protein, in salt-stressed plants by seed priming application, i.e., vitamin B12, ZNPs, gibberellins, and salicylic acid (Abdel Latef et al. 2017; Ghassemi-Golezani and Nikpour-Rashidabad 2017; Keshavarza and Moghadam 2017). Moreover, Azooz (2009) reported that *Vicia faba* genotype 115 had a higher accumulation of CAT, POD, APX, and GR antioxidant enzyme activity than genotype 125, when grown in salt with 0 or 140 mM NaCl and seed priming with 0.2 mM salicylic acid; thus they concluded that the salt tolerance genotype 115 appears to be associated with increased antioxidant enzyme activity.

## 5 Concluding Remarks and Future Perspectives

In conclusion, salt stress is a condition shown in Fig. 1 where too many salts in the soil solution bring about plant growth inhibition or plant death. Salt stress presents a growing threat to plant farming. In order to regulate themselves under saline conditions, plants build up many physiological and biochemical adaptations. Seed priming is a technique that might advance seed performance under salt stress. Seed priming could build up various tolerance mechanisms in seeds against salt stress, for example, variation in photosynthetic pigment, osmotic adjustment, ionic balance, and ion exclusion and antioxidant protection system. A schematic diagram describing the chain reaction following the seed priming is shown in Fig. 2. The diagram illustrates the possible impact of the seed priming on different physiological processes, enhancing tolerance level and final yield. The up-to-date knowledge on seed priming management used in a variety of crops under saline condition is summarized in Table 1. The use of Chemopriming to boost plant tolerance against salt stress is highly promising. Because only limited useful information is available for a large number of plant species, further research is needed to study appropriate substances with precise concentration of priming agents that can guarantee successful seed germination and growth of plant seedling in saline environments. Future research should also focus on molecular, physiological, and metabolic changes caused by salt stress priming mediators. Moreover, further research in the directions proposed in this chapter would be a great extent support for establishment of this know-how in plant management under salt stress in the near future.

**Acknowledgments** The first author, Dr. Magdi Abdelhamid, would like to express his gratitude to his colleagues at the Department of Botany, National Research Center, Giza, Egypt, for their important contribution to this study in some phases.

## References

Abdel Latef AA, Abu Alhmad MF, Abdelfattah KE (2017) The possible roles of priming with ZnO nanoparticles in mitigation of salinity stress in lupine (*Lupinus termis*) plants. J Plant Growth Regul 36(1):60–70

Abdelhamid MT, Shokr M, Bekheta MA (2010) Growth, root characteristics, and leaf nutrients accumulation of four faba bean (*Vicia faba* L.) cultivars differing in their broomrape tolerance and the soil properties in relation to salinity. Commun Soil Sci Plant Anal 41(22):2713–2728

Abdelhamid MT, Rady M, Osman A, Abdalla M (2013a) Exogenous application of proline alleviates salt-induced oxidative stress in *Phaseolus vulgaris* L. plants. J Hortic Sci Biotechnol 88:439–446

Abdelhamid MT, Sadak MSH, Schmidhalter U, El-Saady A (2013b) Interactive effects of salinity stress and nicotinamide on physiological and biochemical parameters of faba bean plant. Acta Biol Colomb 18:499–510

Afzal I, Rauf S, Basra SMA, Murtaza G (2008) Halopriming improve vigor, metabolism of reserves and ionic contents in wheat seedling under salt stress. Plant Soil Environ 54:382–388

Al-Tabbal JASM (2017) Germination and physiological traits to ascertain the ability of hormonal priming to improve salinity tolerance in Sorghum bicolor. J Agron 16:138–146

Al-Tawaha ARM, Al-Ghzawi ALA (2013) Effect of chitosan coating on seed germination and salt tolerance of lentil (*Lens culinaris* L.). Res Crops 14(2):489–491

Alasvandyari F, Mahdavi B, Hosseini SM (2017) Glycine betaine affects the antioxidant system and ion accumulation and reduces salinity-induced damage in safflower seedlings. Arch Biol Sci 69(1):139–147

Ali Q, Daud MK, Haider MZ, Ali S, Aslam N, Noman A, Iqbal N, Shahzad F, Rizwan M, Deeba F, Ali I, Jin ZS (2017) Seed priming by sodium nitroprusside improves salt tolerance in wheat (*Triticum aestivum* L.) by enhancing physiological and biochemical parameters. Plant Physiol Biochem 119:50–58

Allakhverdiev SI, Sakamoto A, Nishiyama Y, Murata N (2000) Inactivation of photosystems I and II in response to osmotic stress in *Synechococcus*. Contribution of water channels. Plant Physiol 122:1201–1208

Anwar S, Shafi M, Bakht J, Mohammad TJ, Yousaf H (2011) Effect of salinity and seed priming on growth and biochemical parameters of different barley genotypes. Afr J Biotechnol 10:15278–15286

Ashraf M (1994) Breeding for salinity tolerance in plants. Crit Rev Plant Sci 13:17–42

Ashraf M (2004) Some important physiological selection criteria fort salt tolerance in plants. Flora 199:361–376

Ashraf M, Foolad MR (2007) Roles of glycine betaine and proline in improving plant abiotic stress resistance. Environ Exp Bot 59:207–216

Ashraf M, Harris PJC (2004) Potential biochemical indicators of salinity tolerance in plants. Plant Sci 166:3–16

Ashraf M, Zafar ZU, Tufail M (1994) Intra-specific variation for salt tolerance in a potential oil-seed crop, brown mustard (*Brassica juncea* (L.) Czern & Coss.). Arch Agron Crop Sci 38:449–458

Awad N, Turky A, Abdelhamid M, Attia M (2012) Ameliorate of environmental salt stress on the growth of *Zea mays* L. plants by exopolysaccharides producing bacteria. J Appl Sci Res 8:2033–2044

Azooz MM (2009) Salt stress mitigation by seed priming with salicylic acid in two faba bean genotypes differing in salt tolerance. Int J Agric Biol 11:343–350
Azooz MM, Alzahrani AM, Yuoussef MM (2013) The potential role of seed priming with ascorbic acid and nicotinamide and their interactions to enhance salt tolerance in broad bean (*Vicia faba* L.). Aust J Crop Sci 7:291–2100
Azooz MM, Metwally A, Abou-Elhamd MF (2015) Jasmonate-induced tolerance of Hassawi okra seedlings to salinity in brackish water. Acta Physiol Plant 37:77
Bagayoko M (2012) Soil salinity alkalinity effects on germination and seedling growth of vegetable crops in the Office du Niger zone. J Res Environ Sci Toxicol 1(12):328–337
Bajehbaj AA (2010) The effects of NaCl priming on salt tolerance in sunflower germination and seedling grown under salinity conditions. Afr J Biotechnol 9:1764–1770
Bano A, Fatima M (2009) Salt tolerance in *Zea mays* (L.) following inoculation with *Rhizobium* and *Pseudomonas*. Biol Fertil Soils 45:405–413
Bargaz A, Nassar RMA, Rady MM, Gaballah MS, Thompson SM, Brestic M, Schmidhalter U, Abdelhamid MT (2016) Improved salinity tolerance by phosphorus fertilizer in two *Phaseolus vulgaris* recombinant inbred lines contrasting in their phosphorus deficiency sensitivity. J Agron Crop Sci 202:497–507
Begara-Morales JC, Sanchez-Calvo B, Chaki M, Valderrama R, Mata-Pérez C, López-Jaramillo J, Padilla MN, Carreras A, Corpas FJ, Barroso JB (2014) Dual regulation of cytosolic ascorbate peroxidase (APX) by tyrosine nitration and S-nitrosylation. J Exp Bot 65:527–538
Bekheta MA, Abdelhamid MT, El-Morsi AA (2009) Physiological response of *Vicia faba* to prohexadione–calcium under saline conditions. Planta Daninha 27:769–779
Blaylock AD (1994) Soil salinity, salt tolerance, and growth potential of horticultural and landscape plants. Cooperative Extension Service, University of Wyoming, Department of Plant, Soil, and Insect Sciences, College of Agriculture, Laramie
Bresler E, McNeal BL, Carter DL (1982) Saline and sodics soils. Advanced series in agricultural series, 1st edn. Springer, Berlin, 236p
Cavalcanti FR, Oliveira JTA, Martins-Miranda AS, Viégas RA, Silveira JAG (2004) Superoxide dismutase, catalase and peroxidase activities do not confer protection against oxidative damage in salt-stressed cowpea leaves. New Phytol 163:563–571
Cha-um S, Kirdmanee C (2010) Effect of glycinebetaine on proline, water use, and photosynthetic efficiencies, and growth of rice seedlings under salt stress. Turk J Agric For 34:517–527
Chaubey CN, Senadhira D (1994) Conventional plant breeding for tolerance to problem soils. In: Yeo AR, Flowers TJ (eds) Soil mineral stresses, approaches to crop improvement. Springer, Berlin, pp 37–60
Chen H, Jiang JG (2010) Osmotic adjustment and plant adaptation to environmental changes related to drought and salinity. Environ Rev 18:309–319
Chutipaijit S, Cha-um S, Sompornpailin K (2011) High contents of proline and anthocyanin increase protective response to salinity in *Oryza sativa* L. spp. indica. Aust J Crop Sci 5:1191–1198
Dawood MG, Abdelhamid MT, Schmidhalter U (2014a) Potassium fertiliser enhances the salt-tolerance of common bean (*Phaseolus vulgaris* L.). J Hortic Sci Biotechnol 89:185–192
Dawood MG, Taie HAA, Nassar RMA, Abdelhamid MT, Schmidhalter U (2014b) The changes induced in the physiological, biochemical and anatomical structure of *Vicia faba* by the exogenous application of proline under seawater stress. S Afr J Bot 93:54–63
Dawood MG, El-Metwally IM, Abdelhamid MT (2016) Physiological response of lupine and associated weeds grown at salt-affected soil to α-tocopherol and hoeing treatments. Gesunde Pflanzen 68:117–127
Dearman J, Brocklehust PA, Drew RLK (1987) Effect of osmotic priming and aging on the germination and emergence of carrot and leek seed. Ann Appl Biol 111:717–722
Demetriou G, Neonaki C, Navakoudis E, Kotzabasis K (2007) Salt stress impact on the molecular structure and function of the photosynthetic apparatus – the protective role of polyamines. Biochim Biophys Acta 1767:272–280

Demirkaya M (2014) Improvement in tolerance to salt stress during tomato cultivation. Turk J Biol 38:1–7

Di Girolamo G, Barbanti L (2012) Treatment conditions and biochemical processes influencing seed priming effectiveness. Ital J Agric 7:e25

Di Martino C, Delfine S, Pizzuto R, Loreto F, Fuggi A (2003) Free amino acids and glycine betaine in leaf osmoregulation of spinach responding to increasing salt stress. New Phytol 158(3):455–463

Eisvand HR, Tavakkol-Afshari R, Sharifzadeh F, Maddah Arefi H, Hesamzadeh Hejazi SM (2010) Effects of hormonal priming and drought stress on activity and isozyme profiles of antioxidant enzymes in deteriorated seed of tall wheatgrass (Agropyron elongatum host). Seed Sci Technol 38:280–297

El-Lethy SR, Abdelhamid MT, Reda F (2013) Effect of potassium application on wheat (*Triticum aestivum* L.) cultivars grown under salinity stress. World Appl Sci J 26:840–850

El-Metwally IM, Ali OAM, Abdelhamid MT (2015) Response of wheat (Triticum aestivum L.) and associated grassy weeds grown in salt-affected soil to effects of graminicides and indole acetic acid. Agriculture (Poľnohospodárstvo) 61(1):1–11

Esmaili E, Kapourchal SA, Malakouti MJ, Homaee M (2008) Interactive effect of salinity and two nitrogen fertilizers on growth and composition of sorghum. Plant Soil Environ 54:537–546

Gadallah MAA (1999) Effects of kinetin on growth, grain yield and some mineral elements in wheat plants growing under excess salinity and oxygen deficiency. Plant Growth Regul 27:63–74

Gaidau C, Niculescu M, Stepan E, Epure D-G, Gidea M (2014) New mixes based on collagen extracts with bioactive properties, for treatment of seeds in sustainable agriculture. Curr Pharm Biotechnol 14:1389–2010

Garratt LC, Janagoudar BS, Lowe KC, Anthony P, Power JB, Davey MR (2002) Salinity tolerance and antioxidant status in cotton cultures. Free Radic Biol Med 33:502–511

Ghassemi-Golezani K, Nikpour-Rashidabad N (2017) Seed pretreatment and salt tolerance of dill: Osmolyte accumulation, antioxidant enzymes activities and essence production. Biocatal Agric Biotechnol 12:30–35

Ghezal N, Rinez I, Sbai H, Saad I, Farooq M, Rinez A, Zribi I, Haouala R (2016) Improvement of Pisum sativum salt stress tolerance by bio-priming their seeds using *Typha angustifolia* leaves aqueous extract. S Afr J Bot 105:240–250

Greenway H, Munns R (1980) Mechanisms of salt tolerance in nonhalophytes. Annu Rev Plant Physiol 31:49–90

Gulmezoglu N, Aydogan C, Turhan E (2016) Physiological, biochemical and mineral dimensions of green bean genotypes depending on Zn priming and salinity. Legum Res 39(5):713–721

Gupta KJ, Stoimenova M, Kaiser WM (2005) In higher plants, only root mitochondria, but not leaf mitochondria reduce nitrite to NO, in vitro and in situ. J Exp Bot 56:2601–2609

Hamayun M, Khan SA, Khan AL, Tang DS, Hussain J, Ahmad B, Anwar Y, Lee IJ (2010) Growth promotion of cucumber by pure cultures of gibberellin-producing Phoma sp. GAH7. World J Microbiol Biotechnol 26:889–894

Harris D (1992) Staying in control of rainfed crops. In: Proceedings of the first annual scientific conference of the SADCC/ODA land and water management programme, Gaborone, Botswana, pp 257–262

Hasegawa PM, Bressan RA, Zhu JK, Bohnert HJ (2000) Plant cellular and molecular responses to high salinity. Annu Rev Plant Phys 51:463–499

Hassini I, Martinez-Ballesta MC, Boughanmi N, Moreno DA, Carvajal M (2017) Improvement of broccoli sprouts (*Brassica oleracea* L. var. italica) growth and quality by KCl seed priming and methyl jasmonate under salinity stress. Sci Hortic 226:141–151

Hathout TA, El-Khallal SM, Abdelgawad ZA, Said EM, Al Mokadem AZ (2014) Enhancing rice salt stress tolerance by priming with validamycin A. Int J Bot 10:1–12

Hellal FA, Abdelhameid MT, Abo-Basha DM, Zewainy RM (2012) Alleviation of the adverse effects of soil salinity stress by foliar application of silicon on faba bean (*Vicia faba* L.). J Appl Sci Res 8:4428–4433

Heydecker W, Coolbear P (1977) Seed treatments for improved performance-survey and attempted prognosis. Seed Sci Technol 5:353–425

Huang YM, Wang HH, Chen KH (2002) Application of seed priming treatments in spinach (*Spinacia oleracea* L.) production. J Chin Soc Hortic Sci 48:117–123

Hussain S, Zheng M, Khan F, Khaliq A, Fahad S, Peng S (2015) Benefits of rice seed priming are offset permanently by prolonged storage and the storage conditions. Sci Rep 5:8101

Ibrahim EA (2016) Seed priming to alleviate salinity stress in germinating seeds. J Plant Physiol 192:38–46

Imran QM, Kamran M, Rehman SU, Ghafoor A, Falak N, Kim KM et al (2016) GA mediated OsZAT-12 expression improves salt resistance of rice. Int J Agric Biol 18:330–336

Imran M, Boelt B, Muhling K-H (2018) Zinc seed priming improves salt resistance in maize. J Agro Crop Sci 204:390–399

Iqbal M, Ashraf M (2007) Seed preconditioning modulates growth, ionic relations, and photosynthetic capacity in adult plants of hexaploid wheat under salt stress. J Plant Nutr 30(3): 381–396. https://doi.org/10.1080/01904160601171330

Iqbal M, Ashraf M (2010) Changes in hormonal balance: a possible mechanism of pre-sowing chilling-induced salt tolerance in spring wheat. J Agron Crop Sci 196:440–454

Iqbal M, Ashraf M (2013) Salt tolerance and regulation of gas exchange and hormonal homeostasis by auxin-priming in wheat. Pesqui Agropecu Bras 48:1210–1219

Iqbal M, Ashraf M, Rehman S, Rha ES (2006) Does polyamine seed pretreatment modulate growth and levels of some plant growth regulators in hexaploid wheat (*Triticum aestivum* L.) plants under salt stress? Bot Stud 47:239–250

İşeri ÖD, Sahin FI, Haberal M (2014) Sodium chloride priming improves salinity response of tomato at seedling stage. J Plant Nutr 37:374–392

Islam F, Yasmeen T, Ali S, Ali B, Farooq MA, Gill RA (2015) Priming-induced antioxidative responses in two wheat cultivars under saline stress. Acta Physiol Plant 37:1–12

Iyengar ERR, Reddy MP (1996) Photosynthesis in high salt-tolerant plants. In: Pesserkali M (ed) Hand book of photosynthesis. Marshal Dekar, Baten Rose, pp 56–65. 08247 97086

Jabeen N, Ahmad R (2013) The activity of antioxidant enzymes in response to salt stress in safflower (*Carthamus tinctorius* L.) and sunflower (*Helianthus annuus* L.) seedlings raised from seed treated with chitosan. J Sci Food Agric 93:1699–1705

Jamil M, Ashraf M, Rha E (2012) Alleviation of salt stress using gibberellic acid in Chinese cabbage. Acta Agron Hung 60:345–355

Janska A, Marsik P, Zelenkova S, Ovesna J (2010) Cold stress and acclimation: what is important for metabolic adjustment? Plant Biol 12:395–405

Jiménez-Arias D, Pérez JA, Luis JC, Martín-Rodríguez V, Valdés-González F, Borges AA (2015) Treating seeds in menadione sodium bisulphite primes salt tolerance in Arabidopsis by inducing an earlier plant adaptation. Environ Exp Bot 109:23–30

Jiménez-Arias D, García-Machado FJ, Morales-Sierra S, Suárez E, Pérez JA, Luis JC, Garrido-Orduña C, Herrera AJ, Valdés F, Sandalio LM, Borges AA (2019) Menadione sodium bisulphite (MSB): beyond seed-soaking. Root pretreatment with MSB primes salt stress tolerance in tomato plants. Environ Exp Bot 157:161–170

Jisha KC, Puthur JT (2014) Halopriming of seeds imparts tolerance to NaCl and PEG induced stress in *Vigna radiata* (L.) Wilczek varieties. Physiol Mol Biol Plants 20:303–312

Jisha KC, Puthur JT (2016) Seed priming with BABA (β-amino butyric acid): a cost-effective method of abiotic stress tolerance in *Vigna radiata* (L.) Wilczek. Protoplasma 253(2):277–289

Jisha KC, Vijayakumari K, Puthur JT (2013) Seed priming for abiotic stress tolerance: an overview. Acta Physiol Plant 35:1381–1396

Kalaji MH, Govindjee BK, Kościelniak J, Żuk-Gołaszewska K (2011) Effects of salt stress on photosystem II efficiency and CO2 assimilation of two Syrian barley landraces. Environ Exp Bot 73:64–72

Karimi S, Eshghi S, Karimi S, Hasan-nezhadian S (2017) Inducing salt tolerance in sweet corn by magnetic priming. Acta Agric Slove 109:89–102

Kazemi JS, Aboutalebian MA, Hamzei J, Meskarbashee M (2018) Mycorrhiza and seed priming effect to improve the balance of sodium and potassium and some changes in antioxidants in the leaves of maize (*Zea mays* L.) under soil salinity. J Agron 17:18–27

Keshavarza H, Moghadam RSG (2017) Seed priming with cobalamin (vitamin B12) provides significant protection against salinity stress in the common bean. Rhizosphere 3:143–149

Khan HA, Ayub CM, Pervez MA, Bilal RM, Shahid MA, Ziaf K (2009a) Effect of seed priming with NaCl on salinity tolerance of hot pepper (*Capsicum annuum* L.) at seedling stage. Soil Environ 28:81–87

Khan HA, Pervez MA, Ayub CM, Ziaf K, Bilal RM, Shahid MA, Akhtar N (2009b) Hormonal priming alleviates salt stress in hot pepper (*Capsicum annuum* L.). Soil Environ 28:130–135

Khayyat M, Tehranifar A, Davarynejad GH, Sayyari-Zahan MH (2014) Vegetative growth, compatible solute accumulation, ion partitioning and chlorophyll fluorescence of 'Malas-e-Saveh' and 'Shishe-Kab' pomegranates in response to salinity stress. Photosynthetica 52:301–312

Khodary SEA (2004) Effect of salicylic acid on the growth, photosynthesis and carbohydrate metabolism in salt-stressed maize plants. Int J Agric Biol 6:5–8

Kohler J, Hernández JA, Caravaca F, Roldán A (2009) Induction of antioxidant enzymes is involved in the greater effectiveness of a PGPR versus AM fungi with respect to increasing the tolerance of lettuce to severe salt stress. Environ Exp Bot 65:245–252

Krasensky J, Jonak C (2012) Drought, salt, and temperature stress-induced metabolic rearrangements and regulatory networks. J Exp Bot 63:1593–1608

Kumar A, Agarwal S, Kumar P, Singh A (2010) Effects of salinity on leaf grain protein in some genotypes of oat (*Avena sativa* L.). Res Sci Technol 2:85–87

Lòpez-Climent MF, Arbona V, Pérez-Clemente RM, Gòmez-Cadenas A (2008) Relationship between salt tolerance and photosynthetic machinery performance in citrus. Environ Exp Bot 62:176–184

Mahdavi B, Rahimi A (2013) Seed priming with chitosan improves the germination and growth performance of ajowan (*Carum copticum*) under salt stress. Eur Asia J Biosci 7:69–76

Mahmood A, Turgay OC, Farooq M, Hayat R (2016) Seed biopriming with plant growth promoting rhizobacteria: a review. FEMS Microbiology Ecology 92:fiw112. https://doi.org/10.1093/femsec/fiw112

Mahmoudi H, Ben Massoud R, Baatour O, Tarchoune I, Ben Salah I, Nasri N, Abidi W, Kaddour R, Hannoufa A, Lachaal M, Ouerghi Z (2012) Influence of different seed priming methods for improving salt stress tolerance in lettuce plants. J Plant Nutr 35:1910–1922

Mäkelä P, Kärkkäinen J, Somersalo S (2000) Effect of glycine-betaine on chloroplast ultrastructure, chlorophyll and protein content, and RuBPCO activities in tomato grown under drought or salinity. Biol Plant 43:471–475

Mane AV, Karadge BA, Samant JS (2010) Salinity induced changes in photosynthetic pigments and polyphenols of *Cymbopogon Nardus* (L.) Rendle. J Chem Pharm Res 2:338–347

Mansour MMF (2000) Nitrogen containing compounds and adaptation of plants to salinity stress. Biol Plant 43:491–500

Marschner H (1995) Mineral nutrition of higher plants. Academic, London

Masondo NA, Kulkarni MG, Finnie JF, Van Staden J (2018) Influence of biostimulants-seed-priming on *Ceratotheca triloba* germination and seedling growth under low temperatures, low osmotic potential and salinity stress. Ecotoxicol Environ Saf 147:43–48

Maswada HF, Abd El-Kader NIK (2016) Redox halopriming: a promising strategy for inducing salt tolerance in bread wheat. J Agron Crop Sci 202:37–50

Matysik J, Bhalu AB, Mohanty P (2002) Molecular mechanisms of quenching of reactive oxygen species by proline under stress in plants. Curr Sci 82:525–532

McDonald MB (2000) Seed priming. In: Black M, Bewley JD (eds) Seed technology and its biological basis. Sheffield Academic Press, Sheffield, pp 287–325

Mekawy AMM, Abdelaziz MN, Ueda A (2018) Apigenin pretreatment enhances growth and salinity tolerance of rice seedlings. Plant Physiol Biochem 2018(130):94–104

Mengel K (2002) Alternative or complementary role of foliar supply in mineral nutrition. Acta Hortic 594:3348

Mirshekari B, Baser S, Allahyari S, Hamendanlu N (2012) 'On-farm' seed priming with Zn plus Mn is an effective way to improve germination and yield of marigold. Afr J Microbiol Res 6:5796–5800

Mittal S, Kumari N, Sharma V (2012) Differential response of salt stress on *Brassica juncea*: photosynthetic performance, pigment, proline, D1 and antioxidant enzymes. Plant Physiol Biochem 54:17–26

Mohammadi SK, Shekari F, Fotovat R, Darudi A (2012) Effect of laser priming on canola yield and its components under salt stress. Int Agrophys 26:45–51

Momin KC (2013) Challenges of the flower seed industry. Indian J Appl Res 3:244–245

Movaghatian A, Khorsandi F (2014) Salicylic acid effects on germination of mungbean (*Vigna radiata* L.) under salinity stress. Adv Environ Biol 8:566–570

Munns R (2002) Comparative physiology of salt and water stress. Plant Cell Environ 25:239–250

Munns R (2005) Genes and salt tolerance: bringing them together. New Phytol 167:645–663

Munns R, James RA (2003) Screening methods for salinity tolerance: a case study with tetraploid wheat. Plant Soil 253:201–218

Munns R, Tester M (2008) Mechanisms of salinity tolerance. Annu Rev Plant Biol 2008(59):651–681

Murakeozy EP, Nagy Z, Duhaze C, Bouchereau A, Tuba Z (2003) Seasonal changes in the levels of compatible osmolytes in three halophytic species of inland saline vegetation in Hungary. J Plant Physiol 160:395–401

Nakaune M, Hanada A, Yin Y, Matsukura C, Yamaguchi S, Ezura H (2012) Molecular and physiological dissection of enhanced seed germination using short-term low-concentration salt seed priming in tomato. Plant Physiol Biochem 52:28–37

Nelson M, Mareida M (2001) Environmental impacts of the CGIAR: an assessment, in Doc. No. SDR/TAC:IAR/01/11 presented to the mid-term meeting, 21–25 May, Durban, South Africa

Netondo GW, Onyango JC, Beck E (2004) Sorghum and salinity: II. Gas exchange and chlorophyll fluorescence of sorghum under salt stress. Crop Sci 44:806–811

Nimac A, Lazarević B, Petek M, Vidak M, Šatović Z, Carović-Stanko K (2018) Effects of salinity and seed priming on germination of sea fennel (*Crithmum maritimum* L.). Agric Conspec Sci 83(2):181–185

Niranjan Raj S, Shetty NP, Shetty HS (2004) Seed bio-priming with Pseudomonas fluorescens isolates enhances growth of pearl millet plants and induces resistance against downy mildew. Int J Pest Manag 50:41–48

Noble CL, Rogers ME (1992) Arguments for the use of physiological criteria for improving the salt tolerance in crops. Plant Soil 146:99–107

Ochiai K, Matoh T (2001) Mechanism of salt tolerance in the grass species, *Anneurolepidium chinense*—I. Growth response to salinity and osmotic adjustment. Soil Sci Plant Nutr 47:579–585

Oliveira ABD, Alencar NLM, Prisco JT, Gomes-Filho E (2011) Accumulation of organic and inorganic solutes in NaCl-stressed sorghum seedlings from aged and primed seeds. Sci Agric 68(6):632–637

Oliveira AB, Gomes-Filho E, Enéas-Filho J, Prisco JT, Alencar NLMA (2012) Seed priming effects on growth, lipid peroxidation, and activity of ROS scavenging enzymes in NaCl-stressed sorghum seedlings from aged seeds. J Plant Interact 7:151–159

Orabi SA, Abdelhamid MT (2016) Protective role of α-tocopherol on two *Vicia faba* cultivars against seawater-induced lipid peroxidation by enhancing capacity of anti-oxidative system. J Saudi Soc Agric Sci 15:145–154

Ouda S, Noreldin T, Mounzer O, Abdelhamid MT (2015) CropSyst model for wheat irrigation water management with fresh and poor quality water. J Water Land Dev 27:41–50

Ouhibi C, Attia H, Rebah F, Msilini N, Chebbi M, Aarrouf J, Urban L, Lachaal M (2014) Salt stress mitigation by seed priming with UV-C in lettuce plants: growth, antioxidant activity and phenolic compounds. Plant Physiol Biochnol 83:126–133

Panuccioa MR, Chaabanib S, Roulab R, Muscoloa A (2018) Bio-priming mitigates detrimental effects of salinity on maize improving antioxidant defense and preserving photosynthetic efficiency. Plant Physiol Biochem 132(2018):465–474

Paparella S, Arau JSS, Rossi G, Wijayasinghe M, Carbonera D, Balestrazzi A (2015) Seed priming: state of the art and new perspectives. Plant Cell Rep 34:1281–1293
Parera C, Cantliffe D (1990) Improved stand establishment of sh2 sweet corn by solid matrix priming. Proceedings of the national symposium on stand establishment horticultural crops, Minneapolis, MN, USA, pp 91–96
Parida AK, Das AB (2005) Salt tolerance and salinity effects on plants: a review. Ecotoxicol Environ Saf 60:324–349
Patade VY, Sujata B, Suprasanna P (2009) Halopriming imparts tolerance to salt and PEG induced drought stress in sugarcane. Agric Ecosyst Environ 134:24–28
Patanè C, Cavallaro V, Cosentino SL (2009) Germination and radicle growth in unprimed and primed seeds of sweet sorghum as affected by reduced water potential in NaCl at different temperatures. Ind Crop Prod 30:1–8
Pervez H, Ashraf M, Makhdum MI (2004) Influence of potassium rates and sources on seed cotton yield and yield components of some elite cotton cultivars. J Plant Nutr 27:1295–1317
Piotr S, Grazyna K (2005) Antioxidant defense in the leaves of $C_3$ and $C_4$ plants under salinity stress. Physiol Plant 125:31–40
Quiles MJ, López NI (2004) Photoinhibition of photosystems I and II induced by exposure to high light intensity during oat plant growth effects on the chloroplast NADH dehydrogenase complex. Plant Sci 166:815–823
Rady MM, Sadak MS, El-Lethy SR, Abdelhamid EM, Abdelhamid MT (2015) Exogenous α-tocopherol has a beneficial effect on *Glycine max* (L.) plants irrigated with diluted sea water. J Hortic Sci Biotechnol 90:195–202
Rady MM, Semida WM, Hemida KA, Abdelhamid MT (2016a) The effect of compost on growth and yield of *Phaseolus vulgaris* plants grown under saline soil. Int J Recycl Org Waste Agric 5:311–321
Rady MM, Mounzer OH, Alarcón JJ, Abdelhamid MT, Howladar SM (2016b) Growth, heavy metal status and yield of salt-stressed wheat (Triticum aestivum L.) plants as affected by the integrated application of bio-, organic and inorganic nitrogen-fertilizers. J Appl Bot Food Qual 89:21–28
Rinez I, Ghezal N, Rinez A, Farooq M, Dbara S, Saad I, Haouala R (2018) Improving salt tolerance in pepper by bio-priming with Padina pavonica and Jania rubens aqueous extracts. Int J Agric Biol 20:513–523
Sadak MS, Abdelhamid MT (2015) Influence of amino acids mixture application on some biochemical aspects, antioxidant enzymes and endogenous polyamines of Vicia faba plant grown under seawater salinity stress. Gesunde Pflanzen 67:119–129
Sadak MS, Abdelhamid MT, Schmidhalter U (2015) Effect of foliar application of amino acids on plant yield and physiological parameters in bean plants irrigated with seawater. Acta Biolo Colomb 20:141–152
Sarwar N, Yousaf S, Jamil FF (2006) Induction of salt tolerance in chickpea by using simple and safe chemicals. Pak J Bot 38:325–329
Sarwar M, Amjad M, Ayyub CM (2017) Alleviation of salt stress in cucumber (*Cucumis sativus* L.) through seed priming with triacontanol. Int J Agric Biol 19:771–778
Semida WM, Taha RS, Abdelhamid MT, Rady MM (2014) Foliar-applied α-tocopherol enhances salt-tolerance in *Vicia faba* L. plants grown under saline conditions. S Afr J Bot 95:24–31
Sivritepe HO, Sivritepe N, Eris A, Turhan E (2005) The effects of NaCl pre-treatments on salt tolerance of melons grown under long-term salinity. Sci Hortic 106:568–581
Subbarao GV, Johansen C, Jana MK, Kuma Rao JVDK (1990) Physiological basis of difference in salinity tolerance of pigeon pea and its related wild species. J Plant Physiol 137:64–71
Tabassum T, Farooq M, Ahmad R, Zohaib A, Wahid A (2017) Seed priming and transgenerational drought memory improves tolerance against salt stress in bread wheat. Plant Physiol Biochem 118:362–369
Tabassum T, Ahmad R, Farooq M, Basra SMA (2018) Improving salt tolerance in barley by Osmopriming and biopriming. Int J Agric Biol 20:2455–2464

Talaat NB, Ghoniem AE, Abdelhamid MT, Shawky BT (2015) Effective microorganisms improve growth performance, alter nutrients acquisition and induce compatible solutes accumulation in common bean (*Phaseolus vulgaris* L.) plants subjected to salinity stress. Plant Growth Regul 75:281–295

Taylor AG, Allen PS, Bennet MA, Bradford KJ, Burris JS, Misra MK (1998) Seed enhancements. Seed Sci Res 8:245–256

Tomar NS, Agarwal RM (2013) Influence of treatment of Jatropha curcas L. leachates and potassium on growth and phytochemical constituents of wheat (Triticum aestivum L.). Am J Plant Sci 4:1134–1150

Tsegay BA, Andargie M (2018) Seed priming with gibberellic acid (GA3) alleviates salinity induced inhibition of germination and seedling growth of *Zea mays* L., *Pisum sativum* Var. abyssinicum A. Braun *and Lathyrus sativus* L. J Crop Sci Biotechnol 21(3):261–267

USDA Agriculture Handbook (1954) Diagnosis and improvement of saline and alkaline soils: USDA agriculture handbook. No. 60. United States Department of Agriculture, Washington, DC, 160pp

Van der Wolf JM, Birnbaum Y, van der Zouwen PS, Grrot SPC (2008) Disinfection of vegetable seed by treatment with essential oils, organic acids and plant extracts. Seed Sci Technol 36:76–88

Van Oosten MJ, Sharkhuu A, Batelli G et al (2013) The Arabidopsis thaliana mutant air implicates SOS3 in the regulation of anthocyanins under salt stress. Plant Mol Biol 83:405–415

Varier A, Vari AK, Dadlani M (2010) The subcellular basis of seed priming. Curr Sci India 99:451–456

Waller F, Achatz B, Baltruschat H, Fodor J, Becker K, Fischer M, Heier T, Huckelhoven R, Neumann C, Von-Wettstein D (2005) The endophytic fungus piriformis indica reprograms barley to salt-stress tolerance, disease resistance, and higher yield. Proc Natl Acad Sci U S A 102:13386–13391

Wenxue W, Bilsborrow PE, Hooley P, Fincham DA, Lombi E, Forster BP (2003) Salinity induced differences in growth, ion distribution and partitioning in barley between the cultivar Maythorpe and its derived mutant golden promise. Plant Soil 250:183–191

Wise RR, Naylor AW (1987) Chilling-enhanced photooxidation: evidence for the role of singlet oxygen and endogenous antioxidants. Plant Physiol 83:278–282

Yancey PH (2005) Organic osmolytes as compatible, metabolic and counteracting cytoprotectants in high osmolarity and other stresses. J Exp Biol 208:2819–2830

Yang X, Liang Z, Lu C (2005) Genetic engineering of the biosynthesis of glycinebetaine enhances photosynthesis against high temperature stress in transgenic tobacco plants. Plant Physiol 138:2299–2309

Yang A, Akhtar SS, Iqbal S, Qi Z, Alandia G, Saddiq MS, Jacobsen S-E (2017) Saponin seed priming improves salt tolerance in quinoa. J Agron Crop Sci 2018(204):31–39

Younesi O, Moradi A (2014) Effect of priming of seeds of *Medicago sativa* 'bami' with gibberellic acid on germination, seedlings growth and antioxidant enzymes activity under salinity stress. J Hortic Res 22:167–174

# Seed Priming and Seedling Pre-treatment Induced Tolerance to Drought and Salt Stress: Recent Advances

**Smita Sahoo, Pankaj Borgohain, Bedabrata Saha, Debojyoti Moulick, Bhaben Tanti, and Sanjib Kumar Panda**

## Contents



**Abstract** Plants are sessile entities and hence have to face the environmental havocs without any chance of avoiding it. Abiotic stress like drought and salinity encumbers plant growth and developmental activities rendering drastic drop in crop yields. Though plants have evolved stress response mechanisms, many a times it doesn't suffice, and the plant succumbs to stress intensity and duration. But sometimes priming and pretreatment with exogenous agents (selenium, silicon, zinc, copper, etc.) enhance the inherent tolerance capability of plants to some extent. These priming and pretreatment technologies (using sodium nitroprusside, hydrogen peroxide, etc.) are cost-effective and user-friendly for stress alleviation in various plants. This chapter centre rounds these ameliorating agents and the mechanisms involved in enhancing the tolerance capability. Various reports suggested different techniques and treatments in combinations while using variety of agents. This chapter aims to compare and summarise the technologies used. Though we mainly represent the mitigations reported in the past decade or so, this chapter is divided into two sections, with a glimpse of various inorganic and organic amendments used to alleviate salinity and drought stresses in wide range of plants.

The authors Smita Sahoo, Pankaj Borgohain, Bedabrata Saha and Debojyoti Moulick have contributed equally to this chapter.

S. Sahoo · P. Borgohain · B. Saha · D. Moulick (✉) · S. K. Panda (✉)
Plant Molecular Biotechnology Laboratory, Department of Life Science and Bioinformatics, Assam University, Silchar, Assam, India

B. Tanti
Department of Botany, Gauhati University, Guwahati, Assam, India



M. Hasanuzzaman, V. Fotopoulos (eds.), *Priming and Pretreatment of Seeds and Seedlings*, https://doi.org/10.1007/978-981-13-8625-1_12

**Keywords** Abiotic stress · Cross tolerance · Osmotic stress · Oxidative stress · Hydropriming

# 1 Introduction

Being sessile in nature, plant has to develop its own strategies to deal/cope with different environmental stresses (Barkla et al. 2013). Environmental stresses like high temperature, low temperature, salinity, drought and heavy metals have a great detrimental effect on crop growth and productivity worldwide (Cheeseman 2016). Among all the abiotic factors, drought and salinity are the most potent deterrents inflicting damage to 50% of world arable lands. Although conventional breeding and genetic manipulation through genetic engineering have great potential to enhance the tolerance capacity of crop plants against various abiotic stresses, seed priming or prestress exposure is emerging rapidly as an attractive alternative strategy in crop stress management and plant stress physiology to enhance plant's stress tolerance capacity under abiotic stress in the subsequent stages of development (Moshelion and Altman 2015; Savvides et al. 2016). Now both seed priming and pretreatment have emerged as an effective and practical method to enhance different abiotic stress tolerance/avoidance capacity (Bayat and Sepehri 2012; Khan et al. 2012; Hussain et al. 2016). Unlike conventional breeding and genetic engineering which are expensive, complicated and time-consuming approaches or even unacceptable in many countries, seed priming is more simple, of low cost, user-friendly and an effective method to enhance seed germination, seedling growth and crop production under adverse condition. In this method, seeds are pre-soaked with different ions, hormones, chemicals, organic compounds and antioxidants before planting (Ahmad et al. 2012; Moulick et al. 2016). Under multiple abiotic conditions, primed seeds have more stress tolerance capacity and better germination ability compared to non-primed seeds in different crop species (Zhuo et al. 2009).

## *1.1 Seed Priming*

Increased seed quality is the need of the hour and has an increased demand in agricultural seed markets. Seed priming has emerged as less arduous approach in this direction. Seed priming technologies are said to induce 'pre-germinative metabolism' which includes activation of DNA repair pathways and antioxidant metabolism. Based on the agents used, seed priming has been characterised into hydropriming, biopriming, osmopriming, solid matrix priming, chemopriming and thermopriming. Hydropriming is the most primitive of all the priming technologies available which involves soaking of seeds under optimal temperature for partial hydration (McDonald 2000; Chen et al. 2012; Moulick et al. 2017). Here the main critical point is to maintain optimum temperature and humidity to avoid radicle protrusion. Although during this seed priming process, germination is not allowed through the controlled hydration technique, all other pre-germination metabolic activities required for embryonic and radicle growth are initiated (Jisha et al. 2013; Chen and Arora 2013; Paparella et al. 2015). So, as a result of this priming process, primed seeds have a better seed germination rate and uniform seedling growth compared to non-primed seeds (Chen and Arora 2013; Hakeem 2015). Biopriming involves the use of microorganisms or biologically active molecules as priming agents (Niranjan Raj et al. 2004). These biologically active molecules often involve plant secondary metabolites and hormones. In case of osmopriming, seeds are treated with an osmoticum at low water potential to facilitate better control of water uptake (Zhou et al. 2013). Solid matrix priming is an adaptation of osmopriming developed to reduce the huge cost of osmopriming. Instead of liquid here solid osmoticum is used like coal, sawdust, charcoal, etc. (Paparella et al. 2015). Chemopriming involves priming with chemical agents like salicylic acid, gibberellic acid, selenium, zinc sulphate, copper sulphate, sodium selenite, etc. (Hamayun et al. 2010; Moulick et al. 2018a, b, c). Thermopriming involves temperature treatment which has been reported to improve germination rate and percentage under adverse environmental conditions (Paparella et al. 2015). Besides this system of classification, different people have used different terminologies for classification based on the agents used like hormonal priming (priming with plant hormones) and redox priming (priming with ROS molecules).

Hameed et al. (2010) reported that priming the wheat seeds with ions, hormones, organic compounds and antioxidants increases the seed performance capacity against salt stress, whereas priming tomato cultivars with polyamines enhances the germination and seedling growth (Afzal et al. 2009). Application of salicylic acid (hormonal priming) and paclobutrazol in maize plants also decreased the impact of drought stress on growth and yield (Bayat and Sepehri 2012). Combination of 5-aminolevulinic acid (ALA) and drought treatment enhanced antioxidant enzyme activity and decreased ROS activity in cucumber under drought stress (Li et al. 2011). Application of ALA in low concentration alleviated the physiological changed in *Brassica napus* under salinity stress (Naeem et al. 2012). Seed priming of *Salicornia utahensis* with growth regulators like thiourea, kinetin, fusicoccin and ethephon improved salt tolerance capacity (Gul and Khan 2003). Under drought stress *Agropyron elongatum* seed priming with abscisic acid (ABA) and gibberellin (GA) induced CAT and SOD activities (Eisvand et al. 2010).

Under drought stress, priming wheat seeds with sodium silicate and ascorbic acid enhanced germination and early seedling growth (Hameed et al. 2013; Ahmed et al. 2016; Farooq et al. 2012; Jafar et al. 2012). Barley seed priming with polyethylene glycol at four osmotic potential levels (0, −7, −10 and − 14 MPa) increased germination rate and seedling growth under drought stress at 0, −3, −6 and − 9 MPa (Rouhollah Amini 2013). In this process, stored proteins are solubilised and lipid peroxidation activity is reduced. Antioxidant activities and osmolyte accumulation are increased through better metabolic processes (Jafar et al. 2012; Delavari et al. 2010). Similar results were also reported in rice, rapeseeds and basil under drought stress (Goswami et al. 2013; Zheng et al. 2016; Kubala et al. 2015; Farahani and Maroufi 2011; Srivastava et al. 2010; Zheng et al. 2016). In soybean and sunflower, seed priming with $KNO_3$ (osmopriming) increased seed germination rate, seedling growth and proline accumulation under salinity stress (Ahmad et al. 2012; Bajehbaj 2010). Application of exogenous spermidine ameliorated salt stress in *indica* rice varieties (Roychoudhury et al. 2011). Under drought condition, spermidine increased seed germination of white clove (Li et al. 2014) with an enhanced antioxidant enzyme activity and lower accumulation of ROS molecule and malondialdehyde (MDA) content.

Seed priming with $H_2O_2$ or hydrogen peroxide (redox priming) showed better performance than priming with mannitol in *Cakile maritima* which was evident from lower accumulation of oxidative stress biomarkers and higher accumulation of antioxidant enzyme activity (Ellouzi et al. 2017). In redox priming method, thiourea, a thiol compound, helped in maintaining the integrity as well as function of mitochondria in *Brassica juncea* seeds, exposed to drought stress (Srivastava et al. 2009).

Selenium priming (chemical priming) of wheat seed was found effective in seedling growth and biochemical changes under water limitation condition (Nawaz et al. 2013). In maize, seed priming with zinc sulphate ($ZnSO_4$) and copper sulphate ($CuSO_4$) (chemical priming) significantly increased the *Zea mays* L. caryopses under drought stress (Foti et al. 2008). Triticale seed priming with monopotassium phosphate ($KH_2PO_4$) increased seed germination percentage and seedling growth under both osmotic stresses (salt and drought) (Yagmur and Kaydan 2008). Another important chemical primer, paclobutrazol, has a significant role in periwinkle (*Catharanthus roseus*) in terms of antioxidant system under salinity stress (Jaleel et al. 2007). Application of choline to rice seeds increased salinity tolerance capacity in terms of photosynthetic ability, growth and accumulation of glycinebetaine (Su et al. 2006; Cha-um et al. 2006).

Melatonin priming of faba bean seeds increased salt tolerance capacity in plants incorporated to salt stress (Dawood and EL-Awadi 2015). Green gram seed priming with β- aminobutyric acid reduced MDA content and increased antioxidant defence mechanism (Jisha and Puthur 2016). Seed priming with exogenous spermidine and gibberellic acid alleviated salt stress in tomato and black glutinous rice seedlings (Hu et al. 2012; Saleethong et al. 2013; Chunthaburee et al. 2014). Under drought stress exogenous spermidine application alleviated drought stress in white clove through stimulation of antioxidant defence mechanism. Exogenous application of melatonin enhances plants ability to cope with different abiotic stresses like drought, high temperature, radiation, chemical stress and salinity (Li et al. 2012a, b; Wei et al. 2014; Zhang et al. 2015). Whereas, exogenous application of spermine in soybean seeds alleviated drought stress by increasing antioxidant activity and decreasing lipid peroxidation level (Radhakrishnan and Lee 2013). Application of exogenous nitric oxide in terms of sodium nitroprusside (SNP as nitric oxide donor) improved seed germination in wheat under salinity stress (Zheng et al. 2009). Alfalfa seed priming by hydrogen sulphide (sodium hydrosulphide, as $H_2S$ donor) increases salt tolerance by reducing oxidative damage and enhancing seed germination percentage as well as antioxidant metabolism (Wang et al. 2012). Seed priming mitigation by ultraviolet light (UV-C) of range 200–280 nm decreased salt stress effect in lettuce seeds in terms of plant growth, accumulation of phenolic compounds and antioxidant properties (Ouhibi et al. 2014).

## *1.2 Seedling Pretreatment*

Initial treatment before onset of stress pre-induces various physiological, biochemical and molecular changes for better adaptation to stress. A lot of studies showed that initial treatment or priming with appropriate levels of chemicals (organic/inorganic) enhances tolerance by modulating various physiological processes such as photosynthesis and by modulating multiple stress-responsive pathways such as the reactive oxygen species (ROS) and methylglyoxal (MG) detoxification pathways. Wheat (Hasanuzzaman et al. 2011) and rice (Corpas et al. 2011) plants showed enhanced tolerance towards salinity stress after giving sodium nitroprusside pretreatment for 24 h through enhanced induction of antioxidative defence. Similar observation was observed when roots of orange plants (Tanou et al. 2009), maize (de Azevedo Neto et al. 2005) and rice (Uchida et al. 2002) were pretreated with $H_2O_2$. Initial exposure of $H_2O_2$ in seedlings stage enhances tolerance towards salinity and drought through the modulation of various physiological processes (de Azevedo Neto et al. 2005; Chao et al. 2009; Liu et al. 2010). Seven-day-old mung bean seedlings (Hossain et al. 2011) and barley (Cuin and Shabala 2005) pretreated with proline or betaine showed an increase in GSH (reduced glutathione) and other similar metabolite contents along with antioxidative enzyme activities; potassium efflux hence decreased oxidative stress damage posed by salinity stress in a synergistic fashion.

Pretreatment with salicylic acid increases the tolerance in wheat (Kang et al. 2013), barley (El-Tayeb 2005) and bean and tomato plants (Senaratna et al. 2000). Spermine treatment in *Cucumis sativus* also alleviates oxidative stress induced by salt stress, rendered with improvement in photochemical efficiency of PSII (Shu et al. 2013). Silicon (Si), the second abundant element in soil, has been found to alleviate salinity stress after pretreatment in *Sorghum* (Liu et al. 2015), barley (Liang et al. 2006), rice (Gong et al. 2006), wheat (Saqib et al. 2008) and cucumber (Zhu et al. 2004). Alleviation of oxidative damage modulated by L-arginine was observed after pretreatment in sunflower (Nejadalimoradi et al. 2014) and rice (Kakkar et al. 2000). Polyamine pretreatment has often been applied to ameliorate the detrimental effects of salinity in maize (Pandolfi et al. 2010). Mitigation of NaCl-induced $K^+$ flux was also observed in barley roots after pretreatment with amino acids (Cuin and Shabala 2007). Methyl jasmonate and jasmonic acid are collectively known as jasmonates which are involved in diverse developmental processes such as root growth, fruit ripening, fertility and senescence (Creelman and Mulpuri 2002). The role of methyl jasmonate in mitigating NaCl- induced salinity stress on soybean was observed along with enhancement in endogenous level of

ABA and proline level (Yoon et al. 2009). Uses of phytohormones (indole-3-acetic acid, IAA; gibberellic acid $GA_3$; and kinetin Kin) in restoring the metabolic alteration caused by oxidative stress generation due to salt stress were investigated in *Vigna radiata.* The mitigation effects under salinity conditions were also observed when seedlings of *Malus hupehensis* were pretreated with melatonin, a low molecular weight molecule similar to indole acetic acid in functions (Li et al. 2012a, b).

Amelioration of drought-induced damages is also minimised by pretreatment with foliar application of glycine betaine (GB) in wheat plants (Ma et al. 2006). It mainly protects the PSII complex, which means GB enhance the photoinhibition tolerance of PSII. Improvement of yield under stress conditions is another prime goal for researchers since a long time. Foliar application of boron as a pretreatment in wheat enhanced yield grown under drought conditions (Abdel-Motagally and El-Zohri 2016). The selenium pretreatment boosted the metabolite and antioxidative defence system under drought stress conditions in rapeseed seedlings (Hasanuzzaman et al. 2011). Exogenously pretreated salicylic acid alleviated drought stress in *Nigella sativa* (Kabiri et al. 2014), maize (Saruhan et al. 2012) and wheat (Horváth et al. 2007). Brassinosteroids have been used to increase the tolerance capacity in *Chorispora bungeana* and tomato to drought stress. Malondialdehyde content, membrane permeability and proline content were less increased in EBR-pretreated plants under drought stress (Li et al. 2012a, b; Damghan 2009). The foliar application of 5-aminolevulinic acid (ALA) in cucumber leaves before onset of drought stress confers lower damage induced by oxidative stress by altering antioxidative enzyme activities (Li et al. 2011).

## 2 Future Prospects

Salinity and drought induces adverse effects on various field crops and hence are the most sought after research topic fetching attention of experts from around the globe. Besides inhibiting growth and reducing yield, these two abiotic stresses have perhaps the most sound impact on economy too. Several mitigation options have been investigated and reported in this direction. Among the mitigation options tried so far, breeding of tolerant varieties (conventional), generation of transgenic varieties, water management processes (alternating drying and wilting), soil health monitoring and adequate management as well as various agronomic practices are also tried side by side. Now, keeping an eye in future and remaining in vigil to climate change scenario, a comprehensive or integrated stress management system should be chosen. In such integrated crop management system, a continuous monitoring and adequate remedies like seed priming/pre-treatment should be prescribed according to regional preferences. Besides these, local, regional and international funding or governing body(s) should try to merge the gaps that exists among the various research groups or institutions working in a similar or closely related areas, and bringing them under the same roof will definitely smoothen the way ahead.

## References


Abdel-Motagally FM, El-Zohri M (2016) Improvement of wheat yield grown under drought stress by boron foliar application at different growth stages. J Saudi Society Agric Sci. https://doi.org/10.1016/j.jssas.2016.03.005

Afzal I, Munir F, Ayub CM, Basra SM, Hameed A, Nawaz A (2009) Changes in antioxidant enzymes, germination capacity and vigour of tomato seeds in response of priming with polyamines. Seed Sci Technol 37(3):765–770

Ahmad I, Ahmad TKA, Basra SM, Hasnain Z, Ali A (2012) Effect of seed priming with ascorbic acid, salicylic acid and hydrogen peroxide on emergence, vigor and antioxidant activities of maize. Afr J Biotechnol 11(5):1127–1137

Ahmed M, Qadeer U, Ahmed ZI, Hassan FU (2016) Improvement of wheat (*Triticum aestivum*) drought tolerance by seed priming with silicon. Arch Agron Soil Sci 62(3):299–315

Amini R (2013) Drought stress tolerance of barley (*Hordeum vulgare* L.) affected by priming with PEG. Intl J Farm Alli Sci 2(20):803–808

Bajehbaj AA (2010) The effects of NaCl priming on salt tolerance in sunflower germination and seedling grown under salinity conditions. Afr J Biotechnol 9(12)

Barkla BJ, Vera-Estrella R, Pantoja O (2013) Progress and challenges for abiotic stress proteomics of crop plants. Proteomics 13(12–13):1801–1815

Bayat S, Sepehri A (2012) Paclobutrazol and salicylic acid application ameliorates the negative effect of water stress on growth and yield of maize plants. Int J Res Agric Sci 8:127–139

Chao YY, Hsu YT, Kao CH (2009) Involvement of glutathione in heat shock–and hydrogen peroxide–induced cadmium tolerance of rice (*Oryza sativa* L.) seedlings. Plant Soil 318(1–2):37

Cha-um S, Supaibulwatana K, Kirdmanee C (2006) Water relations, photosynthetic ability and growth of Thai rice (*Oryza sativa* L. ssp. *Indica* cv. KDML 105) to salt stress by application of exogenous glycinebetaine and choline. J Agron Crop Sci 192:25–36

Cheeseman J (2016) Food security in the face of salinity, drought, climate change, and population growth. In: Halophytes for food security in dry lands. Academic, Oxford, pp 111–123

Chen K, Arora R (2013) Priming memory invokes seed stress-tolerance. Environ Exp Bot 94:33–45

Chen K, Fessehaie A, Arora R (2012) Dehydrin metabolism is altered during seed osmopriming and subsequent germination under chilling and desiccation in *Spinacia oleracea* L. cv. Bloomsdale: possible role in stress tolerance. Plant Sci 183:27–36

Corpas FJ, Leterrier M, Valderrama R, Airaki M, Chaki M, Palma JM, Barroso JB (2011) Nitric oxide imbalance provokes a nitrosative response in plants under abiotic stress. Plant Sci 181(5):604–611

Creelman RA, Mulpuri R (2002) The oxylipin pathway in Arabidopsis. The Arabidopsis book/American Society of Plant Biologists 1

Cuin TA, Shabala S (2005) Exogenously supplied compatible solutes rapidly ameliorate NaCl-induced potassium efflux from barley roots. Plant Cell Physiol 46(12):1924–1933

Cuin TA, Shabala S (2007) Amino acids regulate salinity-induced potassium efflux in barley root epidermis. Planta225(3):753

Damghan IR (2009) Exogenous application of brassinosteroid alleviates drought-induced oxidative stress in *Lycopersicon esculentum* L. Gen Appl Plant Physiol 35(1–2):22–34

Dawood MG, El-Awadi ME (2015) Alleviation of salinity stress on *Vicia faba* L. plants via seed priming with melatonin. Acta Biolo Colomb 20(2):223–235

de Azevedo Neto AD, Prisco JT, Enéas-Filho J, Medeiros JVR, Gomes-Filho E (2005) Hydrogen peroxide pre-treatment induces salt-stress acclimation in maize plants. J Plant Physiol 162(10):1114–1122

Delavari PM, Baghizadeh A, Enteshari SH, Kalantari KM, Yazdanpanah A, Mousavi EA (2010) The effects of salicylic acid on some of biochemical and morphological characteristic of *Ocimum basilicucm* under salinity stress. Aus J Basic Appl Sci 4:483–4845

Eisvand HR, Tavakkol-Afshari R, Sharifzadeh F, Maddah Arefi H, Hesamzadeh Hejazi SM (2010) Effects of hormonal priming and drought stress on activity and isozyme profiles of antioxidant enzymes in deteriorated seed of tall wheatgrass (*Agropyron elongatum* host). Seed Sci Technol 38:280–297

Ellouzi H, Sghayar S, Abdelly C (2017) $H_2O_2$ seed priming improves tolerance to salinity; drought and their combined effect more than mannitol in *Cakile maritima* when compared to *Eutrema salsugineum*. J Plant Physiol 210:38–50

El-Tayeb MA (2005) Response of barley grains to the interactive e. ect of salinity and salicylic acid. Plant Growth Regul 45(3):215–224

Farahani HA, Maroufi K (2011) Effect of hydropriming on seedling vigour in basil (*Ocimum basilicum* L.) under salinity conditions. Adv Environ Biol 5:828–833

Farooq M, Irfan M, Aziz T, Ahmad I, Cheema SA (2012) Seed priming with ascorbic acid improves drought resistance of wheat. J Agron Crop Sci 199:12–22

Foti R, Abureni K, Tigere A, Gotosa J, Gere J (2008) The efficacy of different seed priming osmotica on the establishment of maize (*Zea mays* L.) caryopses. J Arid Environ 72:1127–1130

Gong HJ, Randall DP, Flowers TJ (2006) Silicon deposition in the root reduces sodium uptake in rice (*Oryza sativa* L.) seedlings by reducing bypass flow. Plant Cell Environ 29(10):1970–1979

Goswami A, Banerjee R, Raha S (2013) Drought resistance in rice seedlings conferred by seed priming. Protoplasma 250(5):1115–1129

Gul BI, Khan MA (2003) Effect of growth regulators and osmotic in alleviating salinity effects on the germination of *Salicornia utahensis*. Pak J Bot 35:877–886

Hakeem A (2015) Study on seed osmopriming for alleviation of stresses during postgerminative seedling growth in wheat. Nanjing Agricultural University, Jiangsu, China Ph.D. thesis

Hamayun M, Khan SA, Khan AL, Tang DS, Hussain J, Ahmad B, Anwar Y, Lee IJ (2010) Growth promotion of cucumber by pure cultures of gibberellin-producing Phoma sp. GAH7. World J Microbiol Biotechnol 26:889–894

Hameed A, Afzal I, Iqbal N (2010) Seed priming and salinity induced variations in wheat (*Triticum aestivum* L.) leaf protein profile. Seed Sci Technol 38(1):236–241

Hameed A, Sheikh MA, Jamil A, Basra SM (2013) Seed priming with sodium silicate enhances seed germination and seedling growth in wheat (*Triticum aestivum* L.) under water deficit stress induced by polyethylene glycol. Pak J Life Soc Sci 11(1):19–24

Hasanuzzaman M, Hossain MA, Fujita M (2011) Nitric oxide modulates antioxidant defense and the methylglyoxal detoxification system and reduces salinity-induced damage of wheat seedlings. Plant Biotechnol Rep 5(4):353

Horváth E, Pál M, Szalai G, Páldi E, Janda T (2007) Exogenous 4-hydroxybenzoic acid and salicylic acid modulate the effect of short-term drought and freezing stress on wheat plants. Biol Plant 51(3):480–487

Hossain MA, Hasanuzzaman M, Fujita M (2011) Coordinate induction of antioxidant defense and glyoxalase system by exogenous proline and glycinebetaine is correlated with salt tolerance in mung bean. Front Agric China 5(1):1–14

Hu X, Zhang Y, Shi Y, Zhang Z, Zou Z, Zhang H, Zhao J (2012) Effect of exogenous spermidine on polyamine content and metabolism in tomato exposed to salinity–alkalinity mixed stress. Plant Physiol Biochem 57:200–209

Hussain S, Hanqi Y, Deng S, Khan F (2016) Comparative transcriptional profiling of primed and non-primed rice seedlings under submergence stress. Front Plant Sci 7:1125. https://doi.org/10.3389/fpls.2016.01125

Jafar MZ, Farooq M, Cheema MA, Afzal I, Basra SMA, Wahid MA, Aziz T, Shahid M (2012) Improving the performance of wheat by seed priming under saline conditions. J Agron Crop Sci 198:38–45

Jaleel CA, Manivannan P, Gopi R, Panneerselvam R (2007) Responses of antioxidant defense system of Catharanthus roseus (L.) G. Don. To paclobutrazol treatment under salinity. Acta Physiol Plant 29:205–209

Jisha KC, Puthur JT (2016) Seed priming with BABA (β-amino butyric acid): a cost-effective method of abiotic stress tolerance in *Vigna radiata* (L.) Wilczek. Protoplasma 253(2):277–289

Jisha KC, Vijayakumari K, Puthur JT (2013) Seed priming for abiotic stress tolerance: an overview. Acta Physiol Plant 35:1381–1396. https://doi.org/10.1007/s11738–012–1186-5

Kabiri R, Nasibi F, Farahbakhsh H (2014) Effect of exogenous salicylic acid on some physiological parameters and alleviation of drought stress in *Nigella sativa* plant under hydroponic culture. Plant Prot Sci 50(1)

Kakkar RK, Bhaduri S, Rai VK, Kumar S (2000) Amelioration of NaCl stress by arginine in rice seedlings: changes in endogenous polyamines. Biol Plant 43(3):419–422

Kang GZ, Li GZ, Liu GQ, Xu W, Peng XQ, Wang CY, Zhu YJ, Guo TC (2013) Exogenous salicylic acid enhances wheat drought tolerance by influence on the expression of genes related to ascorbate-glutathione cycle. Biol Plant 57(4):718–724

Khan SU, Bano A, Din JU, Gurmani AR (2012) Abscisic acid and salicylic acid seed treatment as potent inducer of drought tolerance in wheat. Pak J Bot 44:43–49

Kubala S, Wojtyla Ł, Quinet M, Lechowska K, Lutts S, Garnczarska M (2015) Enhanced expression of the proline synthesis gene P5CSA in relation to seed osmopriming improvement of *Brassica napus* germination under salinity stress. Plant Sci 183:1–12

Li DM, Zhang J, Sun WJ, Li Q, Dai AH, Bai JG (2011) 5-Aminolevulinic acid pretreatment mitigates drought stress of cucumber leaves through altering antioxidant enzyme activity. Sci. Hortic 130(4):820–828

Li C, Wang P, Wei Z, Liang D, Liu C, Yin L, Jia D, Fu M, Ma F (2012a) The mitigation effects of exogenous melatonin on salinity-induced stress in *Malus hupehensis*. J Pineal Res 53:298–306

Li YH, Liu YJ, Xu XL, Jin M, An LZ, Zhang H (2012b) Effect of 24-epibrassinolide on drought stress-induced changes in *Chorispora bungeana*. Biol Plant 56(1):192–196

Li Z, Peng Y, Zhang XQ, Ma X, Huang LK, Yan YH (2014) Exogenous spermidine improves seed germination of white clover under water stress via involvement in starch metabolism, antioxidant defenses and relevant gene expression. Molecules 19(11):18003–18024

Liang Y, Zhang W, Chen Q, Liu Y, Ding R (2006) Effect of exogenous silicon (Si) on H+-ATPase activity, phospholipids and fluidity of plasma membrane in leaves of salt-stressed barley (*Hordeum vulgare* L.). Environ Exp Bot 57(3):212–219

Liu ZJ, Guo YK, Bai JG (2010) Exogenous hydrogen peroxide changes antioxidant enzyme activity and protects ultrastructure in leaves of two cucumber ecotypes under osmotic stress. J Plant Growth Regul 29(2):171–183

Liu P, Yin L, Wang S, Zhang M, Deng X, Zhang S, Tanaka K (2015) Enhanced root hydraulic conductance by aquaporin regulation accounts for silicon alleviated salt-induced osmotic stress in *Sorghum bicolor* L. Environ Exp Bot 111:42–51

Ma QQ, Wang W, Li YH, Li DQ, Zou Q (2006) Alleviation of photoinhibition in drought-stressed wheat (*Triticum aestivum*) by foliar-applied glycinebetaine. J Plant Physiol 163(2):165–175

McDonald MB (2000) Seed priming. In: Black M, Bewley JD (eds) Seed technology and its biological basis. Sheffield Academic Press, Sheffield, pp 287–325

Moshelion M, Altman A (2015) Current challenges and future perspectives of plant and agricultural biotechnology. Trends Biotechnol 33:337–342

Moulick D, Ghosh D, Santra SC (2016) Evaluation of effectiveness of seed priming with selenium in rice during germination under arsenic stress. Plant Physiol Biochem 109:571–578

Moulick D, Santra SC, Ghosh D (2017) Seed priming with se alleviate as induced phytotoxicity during germination and seedling growth by restricting as translocation in rice (*Oryza sativa* L. cv IET-4094). Ecotoxicol Environ Saf 145:449–456

Moulick D, Santra SC, Ghosh D (2018a) Effect of selenium induced seed priming on arsenic accumulation in rice plant and subsequent transmission in human food chain. Ecotoxicol Environ Saf 152:67–77

Moulick D, Santra SC, Ghosh D (2018b) Rice seed priming with se: a novel approach to mitigate as induced adverse consequences on growth, yield and as load in brown rice. J Hazard Mater 355:187–196

Moulick D, Santra SC, Ghosh D (2018c) Seed priming with se mitigates as-induced phytotoxicity in rice seedlings by enhancing essential micronutrient uptake and translocation and reducing as translocation. Environ Sci Pollut Res. https://doi.org/10.1007/s11356-018-2711-x

Naeem MS, Warusawitharana H, Liu H, Liu D, Ahmad R, Waraich EA, Xu L, Zhou W (2012) 5-Aminolevulinic acid alleviates the salinity-induced changes in *Brassica napus* as revealed by the ultrastructural study of chloroplast. Plant Physiol Biochem 57:84–92

Nawaz F, Ashraf MY, Ahmad R, Waraich EA (2013) Selenium (se) seed priming induced growth and biochemical changes in wheat under water deficit conditions. Biol Trace Elem Res 151(2):284–293

Nejadalimoradi HA, Nasibi FA, Kalantari KM, Zanganeh RO (2014) Effect of seed priming with L-arginine and sodium nitroprusside on some physiological parameters and antioxidant enzymes of sunflower plants exposed to salt stress. Agric Commun 2:23–30

Niranjan Raj S, Shetty NP, Shetty HS (2004) Seed bio-priming with *Pseudomonas fluorescens* isolates enhances growth of pearl millet plants and induces resistance against downy mildew. Int J Pest Manag 50:41–48

Ouhibi C, Attia H, Rebah F, Msilini N, Chebbi M, Aarrouf J, Urban L, Lachaal M (2014) Salt stress mitigation by seed priming with UV-C in lettuce plants: growth, antioxidant activity and phenolic compounds. Plant Physiol Biochem 83:126–133

Pandolfi C, Pottosin I, Cuin T, Mancuso S, Shabala S (2010) Specificity of polyamine effects on NaCl-induced ion flux kinetics and salt stress amelioration in plants. Plant Cell Physiol 51(3):422–434

Paparella S, Araújo SS, Rossi G, Wijayasinghe M, Carbonera D, Balestrazzi A (2015) Seed priming: state of the art and new perspectives. Plant Cell Rep 34(8):1281–1293

Radhakrishnan R, Lee IJ (2013) Ameliorative effects of spermine against osmotic stress through antioxidants and abscisic acid changes in soybean pods and seeds. Acta Physiol Plant 35:263–269

Roychoudhury A, Basu S, Sengupta DN (2011) Amelioration of salinity stress by exogenously applied spermidine or spermine in three varieties of *indica* rice differing in their level of salt tolerance. J. Plant Physiol 168:317–328

Saleethong P, Sanitchon J, Kong-Ngern K, Theerakulpisut P (2013) Effects of exogenous spermidine (spd) on yield, yield-related parameters and mineral composition of rice (*Oryza sativa* L. ssp. "*indica*") grains under salt stress. Aust J Crop Sci 7:1293–1301

Saqib M, Zörb C, Schubert S (2008) Silicon-mediated improvement in the salt resistance of wheat (*Triticum aestivum*) results from increased sodium exclusion and resistance to oxidative stress. Funct Plant Biol 35(7):633–639

Saruhan N, Saglam A, Kadioglu A (2012) Salicylic acid pretreatment induces drought tolerance and delays leaf rolling by inducing antioxidant systems in maize genotypes. Acta Physiol Plant 34(1):97–106

Savvides A, Ali S, Tester M, Fotopoulos V (2016) Chemical priming of plants against multiple abiotic stresses: mission possible? Trends Plant Sci 21:329–340

Senaratna T, Touchell D, Bunn E, Dixon K (2000) Acetyl salicylic acid (aspirin) and salicylic acid induce multiple stress tolerance in bean and tomato plants. Plant Growth Regul 30(2):157–161

Shu S, Yuan LY, Guo SR, Sun J, Yuan YH (2013) Effects of exogenous spermine on chlorophyll fluorescence, antioxidant system and ultrastructure of chloroplasts in *Cucumis sativus* L. under salt stress. Plant Physiol Biochem 63:209–216

Srivastava AK, Ramaswamy NK, Mukopadhyaya R, Chiramal Jincy MG, D'Souza SF (2009) Thiourea modulates the expression and activity profile of mtATPase under salinity stress in seeds of *Brassica juncea* L. Ann Bot 103:403–410

Srivastava AK, Lokhande VH, Patade VY, Suprasanna P, Sjahril R, D'Souza SF (2010) Comparative evaluation of hydro-, chemo-, and hormonal priming methods for imparting salt and PEG stress tolerance in Indian mustard (*Brassica juncea* L.). Acta Physiol Plant 32:1135–1144

Su J, Hirji R, Zhang L, He C, Selvaraj G, Wu R (2006) Evaluation of the stress-inducible production of choline oxidase in transgenic rice as a strategy for producing the stress-protectant glycine betaine. J Exp Bot 57:1129–1135

Tanou G, Molassiotis A, Diamantidis G (2009) Hydrogen peroxide-and nitric oxide-induced systemic antioxidant prime-like activity under NaCl-stress and stress-free conditions in citrus plants. J Plant Physiol 166(17):1904–1913

Uchida A, Jagendorf AT, Hibino T, Takabe T, Takabe T (2002) Effects of hydrogen peroxide and nitric oxide on both salt and heat stress tolerance in rice. Plant Sci 163(3):515–523

Wang Y, Li L, Cui W, Xu S, Shen W, Wang R (2012) Hydrogen sulfide enhances alfalfa (*Medicago sativa*) tolerance against salinity during seed germination by nitric oxide pathway. Plant Soil 351(1–2):107–119

Wei W, Li QT, Chu YN, Reiter RJ, Yu XM, Zhu DH, Zhang WK, Ma B, Lin Q, Zhang JS, Chen SY (2014) Melatonin enhances plant growth and abiotic stress tolerance in soybean plants. J Exp Bot 66:695–707

Yagmur M, Kaydan D (2008) Alleviation of osmotic strength of water and salt in germination and seedling growth of triticale with seed priming treatments. Afr J Biotechnol 7:2156–2162

Yoon JY, Hamayun M, Lee SK, Lee IJ (2009) Methyl jasmonate alleviated salinity stress in soybean. J Crop Sci Biotechnol 12(2):63–68

Zhang N, Sun Q, Zhang H, Cao Y, Weeda S, Ren S, Guo YD (2015) Roles of melatonin in abiotic stress resistance in plants. J Exp Bot 66:647–656

Zheng C, Jiang D, Liu F, Dai T, Liu W, Jing Q, Cao W (2009) Exogenous nitric oxide improves seed germination in wheat against mitochondrial oxidative damage induced by high salinity. Environ Exp Bot 67:222–227

Zheng M, Tao Y, Hussain S, Jiang Q, Peng S, Huang J, Cui K, Nie L (2016) Seed priming in dry direct-seeded rice: consequences for emergence, seedling growth and associated metabolic events under drought stress. Plant Growth Regul 78(2):167–178

Zhu Z, Wei G, Li J, Qian Q, Yu J (2004) Silicon alleviates salt stress and increases antioxidant enzymes activity in leaves of salt-stressed cucumber (*Cucumis sativus* L.). Plant Sci 167(3):527–533

Zhuo J, Wang W, Lu Y, Sen W, Wang X (2009) Osmopriming-regulated changes of plasma membrane composition and function were inhibited by phenylarsine oxide in soybean seeds. J Integr Plant Biol 51:858–867

# Seed Priming Toward Enhanced Chilling Tolerance in Field Crops: An Overview

**Hafiz Athar Hussain, Sadam Hussain, Shakeel Ahmad Anjum, and Saddam Hussain**

## Contents



**Abstract** Abiotic stresses are the most important limiting factors for crop productivity worldwide in the wake of climatic change. Among the various kinds of abiotic stresses, chilling is one of the important components that limit the growth and final productivity of field crops. Chilling-induced adversities in plant growth and yield mainly occur due to physical and biochemical damages, physiological alterations, and molecular disruptions. Several management approaches are being tested in reducing the detrimental impacts of chilling stress. Seed priming can be a good approach to overcome the negative effects of the chilling stress in different crops. Primed seeds show increased germination rates and better seedling establishment which result in high level of chilling stress tolerance and vigorous plant growth.

H. A. Hussain
Institute of Environment and Sustainable Development in Agriculture, Chinese Academy of Agricultural Sciences, Beijing, China

S. Hussain · S. A. Anjum · S. Hussain (✉)
Department of Agronomy, University of Agriculture Faisalabad, Faisalabad, Punjab, Pakistan



M. Hasanuzzaman, V. Fotopoulos (eds.), *Priming and Pretreatment of Seeds and Seedlings*, https://doi.org/10.1007/978-981-13-8625-1_13

This chapter provides an overview of physiological, morphological, and biochemical responses of crops to chilling stress and highlights the role of seed priming in augmenting chilling tolerance in crop plants.

**Keywords** Seed priming · Chilling tolerance · Crop plants · Morphophysiological growth

## Abbreviations

| | |
|---|---|
| ABA | Abscisic acid |
| APX | Ascorbate peroxidase |
| AQP | Aquaporin |
| But | Butenolide |
| CAT | Catalase |
| Ct | Chitosan |
| GA3 | Gibberellic acid |
| $H_2O_2$ | Hydrogen peroxide |
| $KNO_3$ | Potassium nitrate |
| LEAs | Late embryogenesis abundant proteins |
| PAs | Polyamines |
| PB | Plant biostimulants |
| PEG | Polyethylene glycol |
| POD | Peroxidase |
| Put | Putrescine |
| ROS | Reactive oxygen species |
| SMP | Solid matrix priming |
| SOD | Superoxide dismutase |

## 1 Introduction

Crop plants often face various types of abiotic stresses during their life cycle under natural environments which pose detrimental effects on crop growth and productivity (Hussain et al. 2016a, 2018a). Among various abiotic stresses, suboptimal temperature is the major limiting factor for the crop growth, development, and productivity. In temperate climates, chilling temperatures lead to a significant reduction or complete failure of crop due to either direct damage or delayed maturation (Lukatkin et al. 2012). During early growth phase, chilling stress inhibits seed germination and disrupts plant growth and developmental processes (Allen and Ort 2001), while during reproductive phase of plants, low-temperature exposure delays heading, causes pollen sterility, and ultimately leads to a significant loss in crop yields (Suzuki et al. 2008). Previously, several researchers have demonstrated the detrimental influence of chilling on growth and development of rice (*Oryza sativa*) (Wang et al. 2016; Hussain et al. 2016a, b), cotton (*Gossypium hirsutum* L.) (Zhao et al. 2012), potato (*Solanum tuberosum*) (Svensson et al. 2002), tomato (*Solanum lycopersicum* L.) (Starck et al. 2000), sugarcane (*Saccharum officinarum*) (Thakur et al. 2010; Zhu et al. 2013), and muskmelons (*Cucumis melo*) (Wang et al. 2004).

Due to its thermophilic nature, chilling stress induces shoot water deficit and reduces the water uptake by the plant roots (Stewart et al. 1990). It also weakens the kinetics of several metabolic and physiological processes in crop plants (Ruelland et al. 2009). Yan et al. (2012) observed that the low temperatures may reduce the rates of nutrient uptake that, in turn, limit the plant growth. Induction of the oxidative stress is also an indicator for the harmful impacts of chilling on crop plants (Hussain et al. 2016a). A large amount of ROS is produced in plant cell under chilling stress conditions which trigger lipid peroxidation rate in cellular membranes (Gill and Tuteja 2010; Chen et al. 2015). However, plant antioxidative defense system plays a pivot role in ameliorating the negative effects of chilling-induced ROS production (Wang et al. 2016; Hussain et al. 2016a, b). Various researchers have reported that higher antioxidant activities under chilling conditions were concomitant with stress tolerance in maize (Zhang et al. 1995; Hodges et al. 1997; Prasad 1997; Taka 2004), rice (Kang and Saltveit 2002a; Huang and Guo 2005; Hussain et al. 2016a, b), tobacco (Deng et al. 2012), and cucumber (Kang and Saltveit 2002b) cultivars.

In recent years, seed priming has appeared as an effective, easy, and adoptive approach to improve the seedling emergence, vigor, and stress tolerance capability of many field crops (Jisha et al. 2013; Hussain et al. 2015; Hussain et al. 2018a, b). A plenty of studies have advocated the beneficial effects of various seed priming methods (such as hydropriming, hormonal priming, osmopriming/osmoconditioning, chemical priming, nutri-priming, and redox priming) under chilling stress in different field crops (Jisha et al. 2013; Paparella et al. 2015; Hussain et al. 2016a, b, 2018a). Primed seeds showed vigorous head start and enhanced tolerance to abiotic stresses principally due to osmotic adjustment, well-developed embryo, efficient energy metabolism, quicker cellular defense responses, and enhanced enzyme activation (Jisha et al. 2013). Moreover, seed priming-evoked enhancements in plant antioxidative defense system have also been well-documented (Khaliq et al. 2015; Zheng et al. 2016; Hussain et al. 2016a, b). Under chilling stress, seed emergence and plant growth were significantly enhanced by combined priming with hydrogen peroxide ($H_2O_2$) and salicylic acid (SA), which was closely related with the suppression of ROS accumulation and regulation of hormonal metabolism, energy supply, and accumulation of metabolites (Li et al. 2017). Xu et al. (2011) stated that seed priming treatments enhanced tobacco (*Nicotiana tabacum* L.) germination and subsequent growth under chilling stress owing to activation of plant antioxidant defense system. Likewise, seed priming triggered the speed and rate of germination in maize and improved seedling growth under chilling conditions (Guan et al. 2009). Elkoca et al. (2007) suggested the hydro- and osmopriming in chickpea for better germination and strong growth of seedling under chilling stress. This chapter provides an overview of the impacts of chilling stress on physiological, morphological, and biochemical responses in field crops and discusses the role of seed priming in enhancing chilling tolerance in crop plants.

# 2 Responses of Field Crops to Chilling Stress

Crop plants need an optimum temperatures limit for the proper growth and better development. Chilling stress inflicts detrimental effects on plant growth; however, such effects may vary with the severity of stress and the specific growth stage of crops. Effects of chilling on morphological growth, yield and yield attributes, and physiological and biochemical responses in different crops are discussed below.

## 2.1 *Morphological Growth and Yield*

Every crop has an optimum temperature range for proper growth and development; nevertheless, suboptimal temperature badly distresses the normal growth and associated metabolisms in plants (Hasanuzzaman et al. 2012; Kumar et al. 2013; Khan et al. 2017). Previously, a plenty of studies have indicated that chilling adversely affected the uniformity and rate of germination, reduced the seedling vigor, and delayed the development stages of crops, which resulted in severe reductions in seed yields (Kang and Saltveit 2002a; Cruz and Milach 2004; Oliver et al. 2007; Cheng et al. 2007; Ruelland et al. 2009, Hussain et al. 2018a). Moreover, leaf growth as well as shoot development strongly depends on the optimum temperature; suboptimal temperature slows down the leaf initiation rate resulting in reduced leaf number (Warrington and Kanemasu 1983). Low temperature may decrease the rate of both cell elongation and division; therefore, smaller leaf area under low temperature fails to support the requisite growth of plant (Ben-Haj-Salah and Tardieu 1995). During reproductive phase, chilling induces the pollen sterility, flower abscission, distortion of pollen tube, ovule abortion, and poor fruit set, which eventually decreases seed yield (Thakur et al. 2010). Chilling stress affects the growth of the root system, root branching, and its dry weights, finally reducing the root surface area for uptake of the water and nutrients (Cutforth et al. 1986; Richner et al. 1996; Hussain et al. 2018a). Kaspar and Bland (1992) stated that the root length is more susceptible to chilling stress than root dry weight. Furthermore, aboveground growth of the plants is also inhibited by the insufficient root development under chilling stress (Richner et al. 1996; Hussain et al., 2016a, b). Farooq et al. (2009) stated that less root branching, lower hydraulic conductance, thicker root axis, and reduced root length under chilling stress lead to the significant reductions in plant nutrient uptake. In a series of laboratory and field experiments, it was found that chilling stress (18 °C or below) adversely affected the growth of newly established seedlings and upset the associated metabolic and physiological processes in rice (*Oryza sativa* L.) (Wang et al. 2016; Hussain et al. 2016a, b). In temperate growing areas, cold stress is responsible for reasonable yield reduction (30–40%) in rice because of panicle distortion and spikelet sterility (Andaya and Mackill 2003). In the grain crops, low temperature during vegetative stage causes growth inhibition, chlorosis, wilting, and necrosis (Janowiak et al. 2002), while during reproductive phase, it induces pollen sterility, flower abscission, abortion of ovule, and poor fruit set and hence reduces the final grain yield (Thakur et al. 2010; Hussain et al. 2018a). Clarke and Siddique (2004) and Nayyar et al. (2005) reported that chilling stress during reproductive stage of chickpea caused flower abortion and poor pod development. In wheat (*Triticum aestivum* L.), low-temperature exposure at jointing and booting stages significantly declined the number of productive tillers and spikes (Thakur et al. 2010; Li et al. 2015). Conclusively, chilling stress harmfully affects the growth and yield and yield attributes of field crops as the outcome of stress-induced alterations in metabolic and physiological activities and corresponding adverse consequences for plant reproductive organs.

## 2.2 Physiological and Biochemical Responses

Under chilling stress, photosynthesis is one of the major biochemical functions which in turn hinders the plant growth and limits the final productivity (Khan et al. 2017). The exposure of low temperature decreased the $CO_2$ assimilation, photosynthesis, chlorophyll contents, and transpiration rates because of the reduced ATP synthase activity, which further limited the Rubisco regeneration and restricted the photophosphorylation (Allen and Ort 2001; Yordanova and Popova 2007). In plants, important physiological functions such as photosynthesis and respiration are more susceptible to chilling stress (Yadav 2010; Hussain et al. 2016a). Cold temperatures decrease the capacity and efficacy of photosynthetic apparatus because of changes in gases exchange attributes, poor chloroplast development, and reduced chlorophyll synthesis (Hussain et al. 2018a). Low temperature declines the chlorophyll fluorescence and enhances the flux reaction that regulates oxidative damage in plants (Khan et al. 2017). Moreover, it severely disrupts the electron transport in thylakoid and restricts the stomatal conductivity (Allen and Ort 2001). The higher ROS accumulation under chilling stress is highly toxic to metabolic and cellular functions in plants because of damage to proteins, DNA, lipids, carbohydrates, and cellular membranes (Gill and Tuteja 2010). Under excessive ROS generation, the perturbation of mitochondrial and chloroplastic metabolism results in reduced respiration and poor energy supply to growing tissues of plants (Taylor et al. 2002). Cold temperature negatively affected the activities of enzymes and altered the membrane properties leading to tissue necrosis (Chinnusamy et al. 2007). In soil, temperature may affect the physicochemical and biological processes which influence nutrient availability and in turn affect the uptake of nutrients by plants (Hussain and Maqsood 2011). Moreover, low temperature slows down the rate of carbon dioxide fixation and restricts the supply of $NADP^+$ (Wise 1995). Low temperature reduced the water content in shoot of maize by significant reductions in the uptake of water by roots instead of transpiration rate (Janowiak and Markowski 1994). Aroca et al. (2003) stated that cold-induced reductions in vapor pressure deficit between atmosphere and leaf surface reduced water uptake by the roots and transpiration rate. Generally, the cold-sensitive plants exhibit drought stress symptoms under chilling, which mainly occur due to poor hydraulic conductivity of roots followed by loss of turgor pressure and reductions in leaf water potential (Aroca et al. 2003).

# 3 Seed Priming: A Viable Option to Enhance Chilling Tolerance in Crop Plants

Seed priming could be a crucial and facile technique in increasing seed germination, enhancing vigor, and improving stand establishment of crop plants under various abiotic stresses (Zhang et al. 2006; Casenave and Toselli 2007; Paparella et al. 2015; Wang et al. 2016; Zheng et al. 2016; Li et al. 2017; Hussain et al. 2018a, b). Seed priming is a pre-sowing treatment which encompasses the control hydration of seed for commencement and regulation of pre-germinative metabolic processes (Hussain et al. 2015) such as protein synthesis, as well as synthesis and repair of DNA and mitochondria (Paparella et al. 2015). Several studies have reported that seedlings emerged from the primed seeds showed greater vigor and performed well under chilling stress (Jisha et al. 2013; Wang et al. 2016; Hussain et al. 2016a, b). Priming treatments enhance the shoot and rooting frequency of seedling, improve vigor index, and ultimately increase crop yields (Paparella et al. 2015; Hussain et al. 2018a). Currently, different priming treatments have been used in different plant species such as hydropriming (using distilled water), hormonal priming, chemical priming, osmopriming (using osmotic solutions), redox priming, solid matrix priming, and biological priming (Jisha et al. 2013; Hussain et al. 2015; Paparella et al. 2015; Wang et al. 2016). These techniques are being used to accelerate the seed germination, increase vigor and uniformity of seedling establishment, enhance vegetative growth, and improve the economic yield in different field crops such as wheat (Iqbal and Ashraf 2007; Jisha et al. 2013), maize (Foti et al. 2008), rice (Khaliq et al. 2015; Wang et al. 2016), cotton (Casenave and Toselli 2007), and chickpea (Kaur et al. 2002). However, the efficacy of seed priming varies with priming agent (compounds), crop species, and severity of stress (Jisha et al. 2013).

## *3.1 Hydropriming*

Hydropriming has been proposed as a safe, simple, and inexpensive technique for improving the capacity of seed for osmotic adjustment, faster germination, better stand establishment, vigorous vegetative growth, and higher crop productivity under adverse environmental conditions (Harris et al. 1999; McDonald 1999; Kaur et al. 2002; Elkoca et al. 2007; Jisha et al. 2013). In hydropriming, the seeds are soaked in the sterilized distilled water kept at optimal temperature (25 °C), and duration of priming is adjusted by controlling imbibition of seed during germination (Kaya et al. 2006; Jisha et al. 2013). In hydropriming, the beneficial fact of priming is the augmentation of physiological and biochemical events in seed even when the germination is postponed by negligible matric potential and lower osmotic potential of seed (Jisha et al. 2013). In addition, the decreased viscosity of protoplasm of hydroprimed seeds leads to greater permeability of water and nutrients (Thomas et al. 2000; Jisha et al. 2013). In hydropriming, better germination and improved seedling growth is associated with greater uptake of water by the primed seeds (Yagmur and Kaydan 2008). Abebe and Modi (2009) recommended hydropriming as a safe, simple, and economical method for vigorous seed germination and better seedling establishment in different grain crops. Hydropriming was observed to be the most effective technique for mustard and barley seeds (Ajouri et al. 2004; Srivastava et al. 2010). In semiarid tropics, on-farm priming technique has been shown to improve the germination, better stand establishment, and higher grain yield in maize (Jisha et al. 2013). Hydropriming decreased the ceiling temperature and optimal temperature for seed germination and played a role in decreasing germination time and increasing final emergence in maize (Finch-Savage et al. 2004). Posmyk and Janas (2007) reported that hydropriming along with exogenously applied proline can be used as an effective strategy for enhancing germination, stand establishment, and seedling growth of *Vigna radiata* L. (mung bean) under chilling stress. In rice, effectiveness of hydropriming in enhancing chilling tolerance was also reported by Wang et al. (2016) and Hussain et al. (2016a, b).

## 3.2 *Osmopriming*

Osmopriming is the soaking of seeds in well-aerated solutions with low water potential in order to control the extent and rate of imbibitions (Bradford 1986; Chen and Arora 2011). Osmopriming is bit like a prolonged early seed imbibition that leads to a steady progression of several pre-germinative metabolic processes. For the preparation of solutions with low water potential, a number of chemicals are being used. For instance, polyethylene glycol (PEG) is usually used because of its nontoxic behavior and large molecular size, which reduce the water potential without entering in the seeds during soaking period (Thomas et al. 2000; Chen and Arora 2011; Zheng et al. 2016). Many other types of the chemicals such as $KH_2PO_4$, $KNO_3$, $K_3PO_4$, $KH_2PO_4$, KCl, $CaCl_2$, NaCl, $MgSO_4$, and mannitol are also used for this purpose (Farooq et al. 2009; Jisha et al. 2013; Zheng et al. 2016). Osmopriming is technically, economically, and methodologically more efficient as compared to hydropriming, because osmotic primed seeds are produced more easily, and offers better stand establishment and ultimately higher yield (Jett et al. 1996; Foti et al. 2008; Moradi and Younesi 2009; Jisha et al. 2013). Increasing evidence suggest that osmopriming enhanced the tolerance of germinating seeds against different abiotic stresses including chilling stress (Iqbal and Ashraf 2007; Elkoca et al. 2007; Farooq et al. 2009; Chen et al. 2010; Chen and Arora 2011). Better maintenance of the plasma membrane structure in osmoprimed seeds was noted as compared to hydropriming, which ultimately resulted in better germination owing to longer and controlled hydration of seeds (Jett et al. 1996; Jisha et al. 2013). Seed priming with PEG enhanced the germination potential of the seeds, strengthening the antioxidant system, and ultimately resulted in greater stress tolerance in the germinating seeds (Chen and Arora 2011; Hussain et al. 2015; Zheng et al. 2016). Osmopriming with PEG (−0.5 MPa) for 24 h resulted in better and faster germination of chickpea (*Cicer arietinum* L.) under low temperature (Elkoca et al. 2007). Moreover, increased number of roots, greater root length, and higher fresh and dry weights were recorded in primed seeds than nontreated seeds (Elkoca et al. 2007). Pre-sowing seed treatments changed the pattern of calcium and nitrogen concentrations in seeds and newly developed seedling, which were associated with enhanced activities of α-amylase (Farooq et al. 2009). Osmopriming with $KNO_3$ effectively improved the uniformity and rate of seedling and seedling vigor index in sorghum (*Sorghum bicolor*) and sunflower varieties (Kaya et al. 2006; Moradi and Younesi 2009).

Polyamines (PAs) priming was effective in enhancing the tolerance against low-temperature stress in sessile plants (He et al. 2002; Nayyar 2005; Cuevas et al. 2008; Alcazar et al. 2010; Kovacs et al. 2010; Xu et al. 2011), preventing chlorophyll loss (Besford et al. 1993), and inhibiting the activities of protease and RNase (Altman 2006). Among the different types of polyamines, spermine (Spm), putrescine (Put), and spermidine (Spd) are more associated with abiotic stress tolerance in plants (Martínez-Tellez et al. 2002; He et al. 2002; Xu et al. 2011; Zheng et al. 2016). Put accumulation could alleviate the chilling injury in different plant species, such as chickpea (*Cicer arietinum*) (Nayyar 2005) and maize (*Zea mays* L.) (Gao et al. 2009) by the enhanced activities antioxidants such as SOD, POD, CAT, and APX (Xu et al. 2011; Khaliq et al. 2015).

## 3.3 Chemical Priming

Chemical seed priming is an efficient and rapidly emerging field in modern stress management (Xu et al. 2011; Savvides et al. 2016). Plants treated with different chemical agents (natural and synthetic compounds) prior to stress events showed greater tolerance when exposed to stress conditions (Savvides et al. 2016). Chemical seed priming may also be reported as a cost-effective methodology with low cost-benefit ratio (CBR) if applied in crop stress management (Savvides et al. 2016). A number of chemicals compounds were used to bring about priming in different field crops under different abiotic stress conditions such as ethanol, butenolide (But), paclobutrazol, selenium, Put, and chitosan (Shao et al. 2005; Jain et al. 2006; Foti et al. 2008; Guan et al. 2009; Hasanuzzaman et al. 2010; Xu et al. 2011; Demir et al. 2012; Khaliq et al. 2015; Hussain et al. 2016a, b).

Butenolide (But) has been shown to improve the emergence and seedling growth of *Capsicum annum* L. (sweet pepper) (Demir et al. 2012). Jain et al. (2006) recommended the beneficial aspects of priming for reducing the effects of cold stress in tomato seeds. Chitosan is a large cationic polysaccharide, also used as priming agents in different plants. In maize, seed priming using chitosan (0.50%) enhanced the speed of germination, increased seedling growth, and furnished better stand establishment under low-temperature stress conditions (Guan et al. 2009). In groundnut (*Arachis hypogaea* L.), chitosan priming increased the activities of lipases and levels of gibberellic acid and indole acetic acid (Zhou et al. 2002; Rakesh et al. 2017). Xu et al. (2011) demonstrated that, in tobacco, Put priming can enhance the cold tolerance during early germination by the regulation of antioxidant defense system in plant cells. Hussain et al. (2016a, b) while studying on rice reported that selenium priming effectively increased the chilling tolerance by enhancing the respiration rate, regulating starch metabolism, decreasing oxidative damage, and augmenting antioxidative defense system.

## 3.4 Hormonal Priming

Better seed performance can also be achieved by the inclusion of plant growth hormones and regulators as pre-sowing seed treatments (Lee et al. 1998; Hamza and Suggars 2001). Abscisic acid (ABA) is a plant hormone widely reported to be involved in response to different abiotic stresses such as cold and osmotic stress (Devinar et al. 2013; Fahad et al. 2015). In *Brassica juncea* (Indian mustard), ABA priming was effective in increasing the germination rate and subsequent growth as compared to non-ABA-treated seeds (Srivastava et al. 2010). In *Brassica napus*, seed priming with ABA showed the earlier and faster germination with respect to nontreated seeds during low-temperature (8 °C) conditions (Gao et al. 2002). Various other studies on rice and wheat have revealed the efficacy of different seed priming compounds such as $GA_3$, SA, ascorbic acid (AA), and kinetin in enhancing the tolerance against chilling stress (Singh and Usha 2003; Khan et al. 2011; Hussain et al. 2016a, b). Sakhabutdinova et al. (2003) documented that hormonal priming in wheat may also maintain the levels of cytokinin and auxin in plant tissues, which may regulate the rate of cell division.

## 3.5 Nutri-priming

Seed priming with limiting nutrients is a very useful, indispensable, and cost-effective technique to improve early plant growth in various field crops (Ros et al. 2000; Jisha et al. 2013). Nutri-priming combines the beneficial effects of priming with improved essential nutrients supply to the plants (Al-Mudaris and Jutzi 1999; Arif et al. 2007), and increasing evidence also suggest that plant mineral nutrient status plays a crucial role in augmenting plant tolerance against various abiotic stresses (Marschner 1995). Seed priming with Zn increased final seed yield in wheat and chickpea (Yilmaz et al. 1998; Arif et al. 2007; Rehman et al. 2018), while in barley, priming with 10 mM Zn and 50 mM P enhanced the germination as well as seedling growth (Ajouri et al. 2004). In maize (*Z. mays*), seed priming with 1%

$ZnSO_4$ improved the plant vegetative growth and increased the final grain yield and grain-Zn contents (Harris et al. 2007). Likewise, Foti et al. (2008) stated that seed priming with $ZnSO_4$ and $CuSO_4$ in maize significantly increased the seed emergence and final caryopses germination by 29 and 43%, respectively. Seed treatment with $KH_2PO_4$ showed great potential to improve the germination percentage, seedling emergence, and plant growth of winter wheat (Ghana and Schillinger 2003). Recently, Imran et al. (2013) discussed that in maize seeds, nutrient priming with Zn + Mn and Fe significantly enhanced the concentrations of these nutrient elements (Fe, Zn, and Mn) in seed. Moreover, significant increases in root length and shoot and root fresh biomass (60–70%) were recorded under the influence of Fe and Zn + Mn seed treatments as compared with hydropriming (Imran et al. 2013).

## *3.6 Redox Priming*

Redox state of cell is a key indicator regulating the major processes in plant growth, development, and tolerance against abiotic stresses (Miao et al. 2006; Miller et al. 2009). In reply to any external stimuli, a plant adjusts its redox state; however, the extent of such change is dependent on stimulus nature and dose and the time to which the tissues of the plants are exposed to these stimuli (Miller et al. 2009; Suzuki et al. 2012). In wheat and mustard, pre-sowing seed treatments and foliar spray with thiol compounds and thiourea (bioregulatory molecule) could upsurge the stress tolerance of the seedlings and increase the crop yield (Sahu et al. 2005). In plants, ROS and RNS (such as $H_2O_2$and NO) are involved in plant stress responses (including cold responses) (Neill et al. 2003; Hancock et al. 2011). Seed priming with $H_2O_2$ (1.5 mM) was recorded to improve the tolerance of *Capsicum annum* against chilling stress (Yadav et al. 2011). In *B. juncea*, combination of $H_2O_2$ with 24-epibrassinolide (24-EBL) enhanced the cold tolerance by regulating the activities of different enzymes of plant antioxidative defense system like CAT, SOD, and APX (Manish et al. 2010). In pearl millet (*Pennisetum glaucum* L.), faster seed germination and higher seedling vigor because of seed treatment with NO donors (2-nitroso-1-naphthol, sodium nitroprusside, and nitroso-*R*-salt) was observed by Manjunatha et al. (2008). $H_2O_2$ could play a role as signaling molecule in the start of germination process and is involved in specific changes at the transcriptomic, hormonal, and proteomic levels (Barba-Espın et al. 2012; Macovei and Tuteja 2013). Glutathione (low molecular mass antioxidant) is a major compound in plant antioxidant system and contributes in cellular redox signaling systems that affect plant growth and defense (Maughan and Foyer 2006). Moreover, in sunflower (*Helianthus annuus* L.), priming with antioxidant substances solutions like tocopherol, AA, and glutathione increased the seed vigor under chilling stress (Draganić and Lekić 2012).

### 3.7 Solid Matrix Priming

Solid matrix priming (SMP) encompasses the use of solid and/or semisolid medium with a small amount of water (Taylor et al. 1998). This method exploits the physical and chemical characters of solid material to limit the uptake of the water by seeds (Jisha et al. 2013). During this technique, water is slowly available to seeds leading to controlled and slow imbibition and allowing the activation of repair mechanisms. Nowadays, a number of solid matrices such as Agro-lig, expanded calcined clay, sodium polypropionate gel, exfoliated vermiculite, synthetic calcium silicate, and bituminous soft coal are commonly used (Kubik et al. 1988; Jisha et al. 2013). Some locally available materials like compost, press mud, volcanic cinder, sawdust, gunny bag, and charcoal are also utilized as solid matrices (Lorenzo 1991; Afzal et al. 2002). SMP was effective in improvement in seed germination percentage in invigorating seeds of soybean (*Glycine max*) (Mercado and Fernandez 2002). The SMP combined with *Trichoderma viride* (or Captan) was effective for increasing germination, seedling development, and productivity of okra under low-temperature stress (Pandita et al. 2010).

### 3.8 Biological Priming

Seed priming using beneficial microorganisms is a key component of agricultural practices (Timmusk et al. 2014). Use of beneficial microorganisms (e.g., *P. chlororaphis* MA342, *C. rosea* IK726d11, *P. fluorescens* CHA0, *C. rosea* IK726d11) as priming agent may ensure better stand establishment, because seed-applied microorganisms may establish in root zone of plant and promote the growth and plant health (Bennett and Whipps 2008). Biological priming comprises seed coating with biocontrol agent (e.g., *Pseudomonas fluorescens*) followed by hydration under warm temperature on moist germination blotters, in moist vermiculite, or in plastic bag, and seed is removed prior to radicle emergence (Callan et al. 1990). In the rhizosphere of rice plant, different types of fungi such as mycorrhizal were found to increase the accumulation of number of proteins and transcripts and play an essential role in plant defense system (Barea et al. 1993; Sharmila et al. 2000; Pozo and Azcón-Aguilar 2007). In sunflower, biological priming with *Pseudomonas fluorescens* (UTPf86 and UTPf76) enhanced the ability of seedlings to grow uniformly and provided vigorous stand establishment (Moeinzadeh et al. 2010). The use of biostimulants as organic material to counter the effects of abiotic stresses (such as low temperature) is well-recognized (Sharma et al. 2014; Bulgari et al. 2014), and enhanced growth, development, and crop quality by the use of biostimulants were attributed to increase the uptake of nutrients, enhanced nutrient efficacy, and greater stress tolerance. Biostimulants consisting of growth-promoting elements include a variety of hormones, humic substances, manures, and/or seaweed extracts (Hamza and Suggars 2001).

# 4 Mechanisms of Seed Priming-Induced Chilling Tolerance

Seed priming regulates several physiological, molecular, biochemical processes in plants which contribute to enhanced chilling tolerance (Wojtyla et al. 2016). The beneficial effects of seed priming in augmenting stress tolerance and improving crop performance in terms of rapid and uniform germination, vigorous seedling establishment, higher plant growth, and enhanced yield have been well described in several species (Chen et al. 2012a, b; Patade et al. 2012; Jisha et al. 2013; Wojtyla et al. 2016). Primed seeds attain the potential to imbibe more rapidly and regulate the seed metabolisms, thus improving the rate of germination (McDonald 2000). In primed seeds, the increases in seed vigor and germination rate occur mainly due to better reserve mobilization, repair mechanisms, activation of enzymes, and increased synthesis of RNA and DNA (Nascimento and West 1998; Arif et al. 2008). Primed seeds also show enhanced activities of amylases (α-amylase and β-amylase) and thus higher sugar accumulation and starch degradation, resulting in higher respiration, germination rate, seed vigor, seedling establishment, and vegetative growth than non-primed seeds (Li et al. 2013; Amooaghaie and Nikzad 2013; Zheng et al. 2009, 2010; Li et al. 2014; Hussain et al. 2016a, b; Wang et al. 2016; Hussain et al. 2018a). Under low temperature, plants from primed seeds recorded higher tissue level of soluble carbohydrates, soluble proteins, and free proline than plants from non-primed seeds (Turk et al. 2014; Liu et al. 2011; Hussain et al. 2016a, b). Seed priming also persuades the expression of some genes/proteins which are essential for cell division. In *Arabidopsis* seeds, both osmopriming (−0.75 MPa PEG 6000) and hydropriming led to the accumulation of α tubulin and β tubulin (tubulin subunits) (Gallardo et al. 2001; Kubala et al. 2015).

The ectopic expression of some aquaporin genes (AQPs, e.g., SoPIP2;1) was enhanced in primed seeds that also revealed greater tolerance against chilling stress (Chen et al. 2013). The transmembrane water transport through overexpression of AQPs enhanced the capability of seeds to absorb, transport, and distribute water within plant tissues. During imbibition in primed seeds, AQPs-induced increase in transport and supply of water for the growing embryo lead to enhanced rate of germination as well as stress resistance (Chen et al. 2013; Wojtyla et al. 2016). Different late embryogenesis abundant proteins (hydrophilic proteins (LEAs) also accumulate in response to cell/tissue dehydration under temperature stresses (Dure et al. 1989; Fahad et al. 2015). Numerous studies have indicated better performance of the plants under cold stress owing to upregulation of LEAs in different species (Steponkus et al. 1998; Verslues et al. 2006; Hundertmark and Hincha 2008; Nakayama et al. 2008; Boucher et al. 2010; van Buer et al. 2016). The LEA proteins also constitute some cold-responsive proteins, such as cold responsive-15A (COR15A; At2g42540), zinc finger of *Arabidopsis thaliana*-10 (ZAT10; At1g27730), phenylammonium lyase (PAL1; At2g37040), and chalcone synthase (CHS; At5g13930) (Wilhelm and Thomashow 1993; Baker et al. 1994; Leyva et al. 1995; Steponkus et al. 1998; Rossel et al. 2007; Beckers et al. 2009; Candat et al. 2014; Sethi et al. 2014; van Buer et al. 2016). COR15A (cold regulated 15-A;

At2g42540) encodes a cold-inducible chloroplast protein (Zarka et al. 2003; van Buer et al. 2016), which protects the inner wrapping of chloroplasts from the damages of low temperature (Steponkus et al. 1998). The ZAT10 (also called STZ) controls the activation of some genes known to be involved in antioxidant protection (Gadjev et al. 2006; Mittler et al. 2006; Rossel et al. 2007; Blomster et al. 2011) and the adjustment of mesophyll development to environmental stimuli (Munekage et al. 2015). In *Brassica oleracea*, seed priming regulated the expression of RAB18 (responsive to ABA 18; AT5G66400) and Em6 (AT2G40170), which were important in chilling stress tolerance (Soeda et al. 2005). Likewise, dehydrins which are also identified as Group II LEAs (Bies-Ethève et al. 2008; Candat et al. 2014) were reported to be linked with improved chilling stress tolerance in primed seeds (Wechsberg et al. 1994; Sarhan et al. 1997).

Different antioxidants have been reported to play a pivotal role in enhancing chilling tolerance in primed seedlings (Bolkhina et al. 2003). These enzymes (e.g., POD, APX, SOD, and CAT) may protect the cellular organelles and membranes against the harmful effects of ROS (Bowler et al. 1992; Posmyk et al. 2001; Chen et al. 2016; Khan et al. 2018). All seed priming treatments regulated the activities of CAT, SOD, and POD under normal and stressful conditions (Goswami et al. 2013; Zheng et al. 2016). Increased activities of SOD, CAT, and GR were noted in osmoprimed soybean seeds under chilling stress as compared to non-primed seeds, indicating activation of antioxidant defense mechanism in osmoprimed seeds (Posmyk et al. 2001). Hussain et al. (2016a, b) concluded that better tolerance of primed rice seedlings against chilling stress was concomitant with the higher activities/levels of enzymatic as well as nonenzymatic antioxidants compared with non-primed seedlings.

## 5 Conclusion

Among various environmental extremes, chilling stress is a major limiting factor for the crop growth, development, and productivity. During early growth phase, chilling stress inhibits seed germination and disrupts plant growth and developmental processes, while during reproductive phase of plants, it causes pollen sterility, delays flowering, and ultimately leads to significant losses in crop yields. Seed priming has appeared as an effective, simple, and adoptive approach to mitigate the detrimental effects of chilling stress in many field crops. Seed priming is a pre-sowing treatment which encompasses the control hydration of seed for commencement and regulation of pre-germinative metabolic processes while preventing radicle emergence. A plenty of studies have suggested the beneficial effects of various seed priming techniques such as hydropriming, hormonal priming, osmopriming/osmo-conditioning, chemical priming, nutri-priming, and redox priming under chilling stress. Primed seeds show speedy emergence and uniform crop establishment by providing a vigorous head start under chilling stress compared with non-primed seeds. Moreover, plants which emerged from primed seeds possess greater stress tolerance because of modifications in several physiological, molecular, biochemical processes in plants. The available studies suggest that positive effects of seed priming on germination and seedling growth are mainly associated with higher starch metabolism and better reserve mobilization, increased synthesis of RNA and DNA, enhanced respiration rates, higher synthesis and accumulation of metabolites, maintenance of membrane stability, and higher activities/levels of antioxidants. Seed priming also regulates the expression of some cold-responsive genes/proteins including COR15A, ZAT10, RAB18, DHNs, and AQPs which induce the low-temperature tolerance of crop plants. However, further research at proteomic and transcriptomic levels is needed to explore the in-depth mechanisms of seed priming-evoked chilling tolerance. Efforts should also be made to design suitable and novel priming agents for different field crops in order to meet the challenges of environmental extremes under natural settings.

## References

Abebe AT, Modi AT (2009) Hydropriming in dry bean (*Phaseolus vulgaris* L.). Res J Seed Sci 2:23–31

Afzal I, Basra SM, Ahmad N, Cheema MA, Warraich EA, Khaliq A (2002) Effect of priming and growth regulator treatments on emergence and seedling growth of hybrid maize (*Zea mays* L.). Int J Agric Biol 4:303–306

Ajouri H, Asgedom H, Becker M (2004) Seed priming enhances germination and seedling growth of barley under conditions of P and Zn deficiency. J Plant Nutr Soil Sci 167:630–636

Alcazar R, Altabella T, Marco F, Bortolotti C, Reymond M, Koncz C, Carrasco P, Tiburcio A (2010) Polyamines: molecules with regulatory functions in plant abiotic stress tolerance. Planta 231:1237–1249

Allen DJ, Ort DR (2001) Impacts of chilling temperatures on photosynthesis in warm-climate plants. Trends Plant Sci 6:36–42

Al-Mudaris MA, Jutzi SC (1999) The influence of fertilizer-based seed priming treatment on emergence and seedling growth of *Sorghum bicolor* and *Pennisetum glaucum* in pot trials under greenhouse conditions. J Agron Crop Sci 182:135–141

Altman A (2006) Polyamines and wounded storage tissues–inhibition of RNase activity and solute leakage. Physiol Plant 54:194–198

Amooaghaie R, Nikzad K (2013) The role of nitric oxide in priming-induced low-temperature tolerance in two genotypes of tomato. Seed Sci Res 23:123–131

Andaya VC, Mackill DJ (2003) Mapping of QTLs associated with cold tolerance during the vegetative stage in rice. J Exp Bot 54:257–2585

Arif M, Waqas M, Nawab K, Shahid M (2007) Effect of seed priming in Zn solutions on chickpea and wheat. Afr Crop Sci Conf Proc 8:237–240

Arif M, Jan MT, Marwat KB, Khan MA (2008) Seed priming improves emergence and yield of soybean. Pak J Bot 40:1169–1177

Aroca R, Vernieri P, Irigoyen JJ, Sancher-Diaz M, Tognoni F, Pardossi A (2003) Involvement of abscisic acid in leaf and root of maize (*Zea mays* L.) in avoiding chilling-induced water stress. Plant Sci 165:671–679

Baker SS, Wilhelm KS, Thomashow MF (1994) The 5′-region of Arabidopsis thaliana Cor15a has cis-acting elements that confer cold-regulated, drought-regulated and ABA-regulated gene-expression. Plant Mol Biol 24:701–713

Barba-Espın G, Hernandez JA, Diaz-Vivancos P (2012) Role of $H_2O_2$ in pea seed germination. Plant Signal Behav 7:193–195
Barea JM, Salamanca CP, Herrera MA (1993) Inoculation of woody legumes with selected arbuscular mycorrhizal fungi and rhizobia to recover desertified mediterranean ecosystem. Appl Environ Microbiol 59:129–133
Beckers GJM, Jaskiewicz M, Liu YD (2009) Underwood WR, He SY, Zhang SQ, Conrath U. mitogen-activated protein kinases 3 and 6 are required for full priming of stress responses in *Arabidopsis thaliana*. Plant Cell 21:944–953
Ben-Haj-Salah H, Tardieu F (1995) Temperature affects expansion rate of maize leaves without change in spatial distribution of cell length. Plant Physiol 109:861–870
Bennett AJ, Whipps JM (2008) Dual application of beneficial microorganisms to seed during drum priming. Appl Soil Ecol 38:83–89
Besford RT, Richardson CM, Campos JL, Tiburcio AF (1993) Effect of polyamines on stabilization of molecular complexes in thylakoid membranes of osmotically stressed oat leaves. Planta 189:201–206
Bies-Ethève N, Gaubier-Comella P, Debures A, Lasserre E, Jobet E, Raynal M, et al. (2008) Inventory, evolution and expression profiling diversity of the LEA (late embryogenesis abundant) protein gene family in *Arabidopsis thaliana*. Plant Mol Biol 67:107–124
Blomster T, Salojarvi J, Sipari N, Brosche M, Ahlfors R, Keinanen M, Overmyer K, Kangasjarvi J (2011) Apoplastic reactive oxygen species transiently decrease auxin signaling and cause stress-induced morphogenic response in Arabidopsis. Plant Physiol 157:1866–1883
Bolkhina O, Virolainen E, Fagerstedt K (2003) Antioxidants, oxidative damage and oxygen deprivation stress: a review. Ann Bot 91:179–194
Boucher V, Buitink J, Lin X, Boudet J, Hoekstra FA, Hundertmark M, Renard D, Leprince O (2010) MtPM25 is an atypical hydrophobic late embryogenesis-abundant protein that dissociates cold and desiccation-aggregated proteins. Plant Cell Environ 33:418–430
Bowler C, Van Montagu M, Inze D (1992) Superoxide dismutase and stress tolerance. Annu Rev Plant Physiol Plant Mol Biol 43:83–116
Bradford KJ (1986) Manipulation of seed water relations via osmotic priming to improve germination under stress conditions. Hortic Sci 21:1005–1112
Bulgari R, Cocetta G, Trivellini A, Vernieri P, Ferrante A (2014) Biostimulants and crop responses: a review. Biol Agric Hortic 31:1–17
Callan NW, Mathre DE, Miller JB (1990) Bio-priming seed treatment for biological control of Pythium ultimum pre emergence damping off in sh2 sweet corn. Plant Dis 74:368–372
Candat A, Paszkiewicz G, Neveu M, Gautier R, Logan DC, Avelange-Macherel MH, Macherel D (2014) The ubiquitous distribution of late embryogenesis abundant proteins across cell compartments in Arabidopsis offers tailored protection against abiotic stress. Plant Cell:114
Casenave EC, Toselli ME (2007) Hydropriming as a pre-treatment for cotton germination under thermal and water stress conditions. Seed Sci Technol 35:88–98
Chen K, Arora R (2011) Dynamics of the antioxidant system during seed osmopriming, post-priming germination, and seedling establishment in spinach (*Spinacia oleracea*). Plant Sci 180:212–220
Chen K, Arora R, Arora U (2010) Osmopriming of spinach (*Spinacia oleracea* L. cv. Bloomsdale) seeds and germination performance under temperature and water stress. Seed Sci Technol 38:45–57
Chen H, Chu P, Zhou Y, Li Y, Liu J, Ding Y et al (2012a) Overexpression of AtOGG1, a DNA glycosylase/AP lyase, enhances seed longevity and abiotic stress tolerance in Arabidopsis. J Exp Bot 63:4107–4121
Chen K, Fessehaie A, Arora R (2012b) Dehydrin metabolism is altered during seed osmopriming and subsequent germination under chilling and desiccation in *Spinacia oleracea* L. cv. Bloomsdale: possible role in stress tolerance. Plant Sci 183:27–36
Chen K, Fessehaie A, Arora R (2013) Aquaporin expression during seed osmopriming and post-priming germination in spinach. Biol Plant 57:193–198

Chen W, Guo C, Hussain S, Zhu B, Deng F, Xue Y, et al. (2015) Role of xylo-oligosaccharides in protection against salinity-induced adversities in Chinese cabbage. Environ Sci Pollut Res 23:1254–1264

Chen W, Guo C, Hussain S, Zhu B, Deng F, Xue Y, Geng M, Wu L (2016) Role of xylo-oligosaccharides in protection against salinity-induced adversities in Chinese cabbage. Environ Sci Pollut Res 23:1254–1264

Cheng C, Yun KY, Ressom HW, Mohanty B, Bajic VB, Jia Y et al (2007) An early response regulatory cluster induced by low temperature and hydrogen peroxide in seedlings of chilling-tolerant japonica rice. BMC Genomics 8:175

Chinnusamy V, Zhu J, Zhu JK (2007) Cold stress regulation of gene expression in plants. Trends Plant Sci 12:444. https://doi.org/10.1016/j.tplants.2007.07.00

Clarke HJ, Siddique KHM (2004) Response of chickpea genotypes to low temperature stress during reproductive development. Field Crop Res 90:323–334

Cruz R, Milach S (2004) Cold tolerance at the germination stage of rice: methods of evaluation and characterization of genotypes. Sci Agric 61:1–8

Cuevas JC, Lopez-Cobollo R, Alcazar R, Zarza X, Koncz C, Altabella T, Salinas J, Tiburcio AF, Ferrando A (2008) Putrescine is involved in Arabidopsis freezing tolerance and cold acclimation by regulating abscisic acid levels in response to low temperature. Plant Physiol 148:1094–1105

Cutforth HW, Shaykewich CF, Cho CM (1986) Effect of soil water and temperature on corn (*Zea mays* L.) root growth during emergence. Can J Soil Sci 66:51–58

Demir I, Ozuaydın I, Yasar F, Staden JV (2012) Effect of smoke derived butenolide priming treatment on pepper and salvia seeds in relation to transplant quality and catalase activity. S Afr J Bot 78:83–87

Deng B, Du W, Liu C, Sun W, Tian S, Dong H (2012) Antioxidant response to drought, cold and nutrient stress in two ploidy levels of tobacco plants: low resource requirement confers polytolerance in polyploids? Plant Growth Regul 66:37–47

Devinar G, Llanes A, Masciarelli O, Luna V (2013) Different relative humidity conditions combined with chloride and sulfate salinity treatments modify abscisic acid and salicylic acid levels in the halophyte *Prosopis strombulifera*. Plant Growth Regul 70:247–256

Draganić I, Lekić S (2012) Seed priming with antioxidants improves sunflower seed germination and seedling growth under unfavorable germination conditions. Turk J Agric For 36:421–428

Dure L, Crouch M, Harada JJ, Ho T, Mundy J, Quatrano RS, Thomas TL, Sung ZR (1989) Common amino acid sequence domains among the LEA proteins of higher plants. Plant Mol Biol 12:475–486

Elkoca E, Haliloglu K, Esitken A, Ercisli S (2007) Hydro- and osmopriming improve chickpea germination. Acta Agric Scand Sect B Soil Plant Sci 57:193–200

Fahad S, Nie L, Chen Y, Wu C, Xiong D, Saud S, Hongyan L, Cui K, Huang L (2015) Crop plant hormones and environmental stress. Sustain Agric Rev 15:371–400

Farooq M, Aziz T, Wahid A, Lee DJ, Siddique KHM (2009) Chilling tolerance in maize: agronomic and physiological approaches. Crop Pasture Sci 60:501–516

Finch-Savage WE, Dent KC, Clark LJ (2004) Soak conditions and temperature following sowing influence the response of maize (*Zea mays* L.) seeds to on-farm priming (pre-sowing seed soak). Field Crop Res 90:361–374

Foti R, Aburenia K, Tigerea A, Gotosab J, Gerec J (2008) The efficacy of different seed priming osmotica on the establishment of maize (*Zea mays* L.) caryopses. J Arid Environ 72:1127–1130

Gadjev I, Vanderauwera S, Gechev TS, Laloi C, Minkov IN, Shulaev V, Apel K, Inze D, Mittler R, Van Breusegem F (2006) Transcriptomic footprints disclose specificity of reactive oxygen species signaling in Arabidopsis. Plant Physiol 141:436–445

Gallardo K, Job C, Groot SPC, Puype M, Demol H, Vandekerckhove J, Job D (2001) Proteomic analysis of Arabidopsis seed germination and priming. Plant Physiol 126:835–848

Gao YP, Bonham-Smith PC, Gusta LV (2002) The role of peroxiredoxin antioxidant and calmodulin in ABA-primed seeds of Brassica napus exposed to abiotic stresses during germination. J Plant Physiol 159:951–958

Gao CH, Hu J, Zhang S (2009) Association of polyamines in governing the chilling sensitivity of maize genotypes. Plant Growth Regul 57:31–38

Ghana SG, Schillinger WF (2003) Seed priming winter wheat for germination, emergence, and yield. Crop Sci 43:2135–2141

Gill SS, Tuteja N (2010) Reactive oxygen species and antioxidant machinery in abiotic stress tolerance in crop plants. Plant Physiol Biochem 48:909–930

Goswami A, Banerjee R, Raha S (2013) Drought resistance in rice seedlings conferred by seed priming: role of the anti-oxidant defense mechanisms. Protoplasma 250:1115–1129

Guan YJ, Hu J, Wang XJ, Shao CX (2009) Seed priming with chitosan improves maize germination and seedling growth in relation to physiological changes under low temperature stress. J Zhejiang Univ Sci B 10:427–433

Hamza B, Suggars A (2001) Biostimulants: myths and realities. TurfGrass Trends 8:6–10

Hancock JT, Neill SJ, Wilson ID (2011) Nitric oxide and ABA in the control of plant function. Plant Sci 181:555–559

Harris D, Joshi A, Khan PA, Gothkar P et al (1999) On-farm seed priming in semiarid agriculture: development and evaluation in maize, rice and chickpea in India using participatory methods. Exp Agric 35:15–29

Harris D, Rashid A, Miraj G, Arif M, Shah H (2007) 'On-farm' seed priming with zinc sulphate solution – a cost-effective way to increase the maize yields of resource-poor farmers. Field Crop Res 102:119–127

Hasanuzzaman M, Anwar HM, Fujita M (2010) Selenium in higher plants: physiological role, antioxidant metabolism and abiotic stress tolerance. J Plant Sci 5:354–375

Hasanuzzaman M, Hossain MA, Fujita M (2012) Exogenous selenium pretreatment protects rapeseed seedlings from cadmium-induced oxidative stress by upregulating antioxidant defense and methylglyoxal detoxification systems. Biol Trace Elem Res 149:248–261

He L, Nada K, Tachibana S (2002) Effects of spermidine pretreatment through the roots on growth and photosynthesis of chilled cucumber plants (*Cucumis sativus* L.). J Jpn Soc Hort Sci 71:490–498

Hodges DM, Andrews CJ, Johnson DA, Hamilton RI (1997) Antioxidant enzyme responses to chilling stress in differentially sensitive inbred maize lines. J Exp Bot 48:1105–1113

Huang M, Guo Z (2005) Responses of antioxidative system to chilling stress in two rice cultivars differing in sensitivity. Biol Plant 9:81–84

Hundertmark M, Hincha DK (2008) LEA (late embryogenesis abundant) proteins and their encoding genes in *Arabidopsis thaliana*. BMC Genomics 9:118. https://doi.org/10.1186/1471-2164-9-118

Hussain A, Maqsood MA (2011) Root zone temperature influences nutrient accumulation and use in maize. Pak J Bot 43:1551–1556

Hussain S, Zheng M, Khan F, Khaliq A, Fahad S, Peng S et al (2015) Benefits of rice seed priming are offset permanently by prolonged storage and the storage conditions. Sci Rep 5:8101. https://doi.org/10.1038/srep08101

Hussain S, Khan F, Hussain HA, Nie L (2016a) Physiological and biochemical mechanisms of seed priming-induced chilling tolerance in rice cultivars. Front Plant Sci 7:116. https://doi.org/10.3389/fpls.2016.00116

Hussain S, Khan F, Cao W, Wu L, Geng M (2016b) Seed priming alters the production and detoxification of reactive oxygen intermediates in rice seedlings grown under sub-optimal temperature and nutrient supply. Front Plant Sci 7:439

Hussain HA, Hussain S, Khaliq A, Ashraf U, Anjum SA, Men S, Wang L (2018a) Chilling and drought stresses in crop plants: implications, cross talk, and potential management opportunities. Front Plant Sci 9:393. https://doi.org/10.3389/fpls.2018.00393

Hussain S, Khaliq A, Tanveer M, Matloob A, Hussain HA (2018b) Aspirin priming circumvents the salinity-induced effects on wheat emergence and seedling growth by regulating starch metabolism and antioxidant enzyme activities. Acta Physiol Plant 40(4):68

Imran M, Mahmood A, Römheld V, Neumann G (2013) Nutrient seed priming improves seedling development of maize exposed to low root zone temperatures during early growth. Eur J Agron 49:141–148

Iqbal M, Ashraf M (2007) Seed preconditioning modulates growth, ionic relations, and photosynthetic capacity in adult plants of hexaploid wheat under salt stress. J Plant Nutr 30:381–396

Jain N, Kulkarni MG, Van Staden J (2006) Abutenolide isolated from smoke can overcome the detrimental effect of extreme temperatures during tomato seed germination. Plant Growth Regul 49:263–267
Janowiak F, Markowski A (1994) Changes in leaf water relations and injuries in maize seedlings induced by different chilling conditions. J Agron Crop Sci 172:19–28
Janowiak F, Maas B, Dörffling K (2002) Importance of abscisic acid for chilling tolerance of maize seedlings. J Plant Physiol 159:635–643
Jett LW, Welbaum GE, Morse RD (1996) Effect of matric and osmotic priming treatments on broccoli seed germination. J Am Soc Hortic Sci 121:423–429
Jisha KC, Vijayakumari K, Puthur JT (2013) Seed priming for abiotic stress tolerance: an overview. Acta Physiol Plant 35:1381–1396
Kaspar TC, Bland WL (1992) Soil temperature and root growth. Soil Sci 154:290–299
Kang HM, Saltveit ME (2002a) Chilling tolerance of maize, cucumber and rice seedling leaves and roots are differentially affected by salicylic acid. Physiol Plant 115(4):571–576
Kang HM, Saltveit ME (2002b) Reduced chilling tolerance in elongating cucumber seedling radicles is related to their reduced antioxidant enzyme and DPPH-radical scavenging activity. Physiol Plant 115(2):244–250
Kaur S, Gupta AK, Kaur N (2002) Effect of osmo- and hydropriming of chickpea on seedling growth and carbohydrate metabolism under water deficit stress. Plant Growth Regul 37:17–22
Kaya MD, Okcu G, Atak M, Cikili Y, Kolsarici O (2006) Seed treatments to overcome salt and drought stress during germination in sunflower (*Helianthus annuus* L.). Eur J Agron 24:291–295
Khaliq A, Aslam F, Matloob A, Hussain S, Geng M, Wahid A et al (2015) Seed priming with selenium: consequences for emergence, seedling growth, and biochemical attributes of rice. Biol Trace Elem Res 166:236–244
Khan MB, Gurchani MA, Hussain M, Freed S, Mahmood K (2011) Wheat seed enhancement by vitamin and hormonal priming. Pak J Bot 43:1495–1499
Khan TA, Fariduddin Q, Yusuf M (2017) Low-temperature stress: is phytohormones application a remedy? Environ Sci Pollut Res 24:21574–21590
Khan F, Hussain S, Tanveer M, Khan S, Hussain HA, Iqbal B, Geng M (2018) Coordinated effects of lead toxicity and nutrient deprivation on growth, oxidative status, and elemental composition of primed and non-primed rice seedlings. Environ Sci Pollut Res 25:21185. https://doi.org/10.1007/s11356-018-2262-1
Kovacs Z, Simon-Sarkadi L, Szucs A, Kocsy G (2010) Differential effects of cold, osmotic stress and abscisic acid on polyamine accumulation in wheat. Amino Acids 38:623–631
Kubala S, Garnczarska M, Wojtyla Ł, Clippe A, Kosmala A, Zmienko A et al (2015) Deciphering priming-induced improvement of rapeseed (Brassica napus L.) germination through an integrated transcriptomic and proteomic approach. Plant Sci 231:94–113
Kubik KK, Eastin JA, Eskridge KM (1988) Solid matrix priming of tomato and pepper. In: Proceedings of the international conference on stand establishment for horticultural crops. Lancaster, Pennsylvania, pp 86–96
Kumar RR, Goswami S, Singh K, Rai GK, Rai RD (2013) Modulation of redox signal transduction in plant system through induction of free radical/ROS scavenging redox-sensitive enzymes and metabolites. Aus J Crop Sci 7:1744–1751
Lee SS, Kim JH, Hong SB, Yuu SH, Park EH (1998) Priming effect of rice seeds on seedling establishment under adverse soil conditions. Korean J Crop Sci 43:194–198
Leyva A, Jarillo JA, Salinas J, Martinezzapater JM (1995) Low-temperature induces the accumulation of phenylalanine ammonia-lyase and chalcone synthase messenger-RNAs of Arabidopsis thaliana in a light-dependent manner. Plant Physiol 108:39–46
Li X, Jiang H, Liu F, Cai J, Dai T, Cao W, Jiang D (2013) Induction of chilling tolerance in wheat during germination by pre-soaking seed with nitric oxide and gibberellin. Plant Growth Regul 71:31–40
Li Z, Peng Y, Zhang XQ, Ma X, Huang LK, Yan YH (2014) Exogenous spermidine improves seed germination of white clover under water stress via involvement in starch metabolism, antioxidant defenses and relevant gene expression. Molecules 19:18003–18024

Li XN, Pu HC, Liu FL, Zhou Q, Cai J, Dai TB et al (2015) Winter wheat photosynthesis and grain yield responses to spring freeze. Agron J 107:1002–1010

Li Z, Xu J, Gao Y, Wang C, Guo G, Luo Y, Huang Y, Hu W, Sheteiwy MS, Guan Y, Hu J (2017) The synergistic priming effect of exogenous salicylic acid and $H_2O_2$ on chilling tolerance enhancement during maize (*Zea mays* L.) seed germination. Front Plant Sci 8:1153. https://doi.org/10.3389/fpls.2017.01153

Liu X, Wang L, Liu L, Guo Y, Ren H (2011) Alleviating effect of exogenous nitric oxide in cucumber seedling against chilling stress. Afr J Biotechnol 10:4380–4386

Lorenzo MCB (1991) Seed invigoration of soybean and corn through solid matrix priming. BS, Philippines

Lukatkin AS, Brazaitytė A, Bobinas Č, Duchovskis P (2012) Chilling injury in chilling-sensitive plants: a review. Zemdirbyste-Agriculture 99:111–124

Macovei A, Tuteja N (2013) Different expression of miRNAs targeting helicases in rice in response to low and high dose rate γ-ray treatments. Plant Signal Behav 8:25128

Manish K, Geetika S, Renu B, Sandeep K, Gaganodeep J (2010) Effect of exogenous H2O2 on antioxidant enzymes of Brassica juncea L. seedlings in relation to 24-epibrassinolide under chilling stress. Indian J Biochem Biophys 47:378–382

Manjunatha G, Raj SN, Shetty NP, Shetty HS (2008) Nitric oxide donor seed priming enhances defense responses and induces resistance against pearl millet downy mildew disease. Pestic Biochem Physiol 91:1–11

Marschner H (1995) Mineral nutrition of higher plants, 2nd edn. Academic, San Diego

Martínez-Tellez MA, Ramos-Clamont MG, Gardea AA, Vargas-Arispuro I (2002) Effect of infiltrated polyamines on polygalacturonase activity and injury response in zucchini squash (*Cucurbita pepo* L.). Biochem Biophys Res Commun 295:98–101

Maughan S, Foyer CH (2006) Engineering and genetic approaches to modulating the glutathione network in plants. Physiol Plant 126:382–397

McDonald MB (1999) Seed deterioration: physiology, repair and assessment. Seed Sci Technol 27:177–237

McDonald MB (2000) Seed priming. In: Black M, Bewley JD (eds) Seed technology and its biological basis. Sheffield Academic Press, Sheffield, pp 287–325

Mercado MFO, Fernandez PG (2002) Solid matrix priming of soybean seeds. Philipp J Crop Sci 27:27–35

Miao Y, LvD WP, Wang XC, Chen J, Miao C, Song CP (2006) An Arabidopsis glutathione peroxidase functions as both a redox transducer and a scavenger in abscisic acid and drought stress responses. Plant Cell 18:2749–2766

Miller G, Suzuki N, Ciftci-Yilmaz S, Mittler R (2009) Reactive oxygen species homeostasis and signaling during drought and salinity stresses. Plant Cell Environ 33:453–467

Mittler R, Kim Y, Song LH, Coutu J, Coutu A, Ciftci-Yilmaz S, Lee H, Stevenson B, Zhu JK (2006) Gain- and loss-of-function mutations in Zat10 enhance the tolerance of plants to abiotic stress. FEBS Lett 580:6537–6542

Moeinzadeh A, Sharif-Zadeh F, Ahmadzadeh M, Heidari Tajabadi F (2010) Biopriming of sunflower (*Helianthus annuus* L.) seeds with Pseudomonas fluorescens for improvement of seed invigoration and seedling growth. Aust J Crop Sci 4:564–570

Moradi A, Younesi O (2009) Effects of osmo- and hydropriming on seed parameters of grain sorghum (*Sorghum bicolor* L.). Aust J Basic Appl Sci 3:1696–1700

Munekage YN, Inoue S, Yoneda Y, Yokota A (2015) Distinct palisade tissue development processes promoted by leaf autonomous signalling and long-distance signalling in A rabidopsis thaliana. Plant, Cell Environ 38(6):1116–116

Nakayama K, Okawa K, Kakizaki T, Inaba T (2008) Evaluation of the protective activities of a late embryogenesis abundant (LEA) related protein, Cor15am, during various stresses in vitro. Biosci Biotechnol Biochem 72:1642–1645

Nascimento WM, West SH (1998) Priming and seed orientation affect emergence and seed coat adherence and seedling development of muskmelon transplants. Hortic Sci 33:847–848

Nayyar H (2005) Putrescine increases floral retention, pod set and seed yield in cold stressed chickpea. J Agron Crop Sci 191:340–345
Nayyar H, Bains T, Kumar S (2005) Low temperature induced floral abortion in chickpea: relationship to abscisic acid and cryoprotectants in reproductive organs. Environ Exp Bot 53:39–48
Neill SJ, Desikan R, Hancock JT (2003) Nitric oxide signalling in plants. New Phytol 159:11–35
Oliver SN, Dennis ES, Dolferus R (2007) ABA regulates apoplastic sugar transport and is a potential signal for cold-induced pollen sterility in rice. Plant Cell Physiol 48:1319–1330
Pandita VK, Anand A, Nagarajan S, Seth R, Sinha SN (2010) Solid matrix priming improves seed emergence and crop performance in okra. Seed Sci Technol 38:665–674
Paparella S, Araújo SS, Rossi G, Wijayasinghe M, Carbonera D, Balestrazzi A (2015) Seed priming: state of the art and new perspectives. Plant Cell Rep 34:1281–1293
Patade VY, Khatri D, Manoj K, Kumari M, Ahmed Z (2012) Cold tolerance in thiourea primed capsicum seedlings is associated with transcript regulation of stress responsive genes. Mol Biol Rep 39:10603–10613
Posmyk MM, Janas KM (2007) Effects of seed hydropriming in presence of exogenous proline on chilling injury limitation in *Vigna radiata* L. seedlings. Acta Physiol Plant 29:509–517
Posmyk MM, Corbineau F, Vinel D, Bailly C, Côme D (2001) Osmoconditioning reduces physiological and biochemical damage induced by chilling in soybean seeds. Physiol Plant 111:473–482
Pozo MJ, Azcón-Aguilar C (2007) Unraveling mycorrhiza-induced resistance. Curr Opin Plant Biol 10:393–398
Prasad TK (1997) Role of catalase in inducing chilling tolerance in pre emergence maize seedlings. Plant Physiol 114:1369–1376
Rakesh P, Prasad RD, Devi GU, Bhat BN (2017) Effect of biopolymers and synthetic seed coating polymers on castor and groundnut seed. Int J Pure Appl Biosci 5(4):2043–2048
Rehman A, Farooq M, Naveed M, Nawaz A, Shahzad B (2018) Seed priming of Zn with endophytic bacteria improves the productivity and grain biofortification of bread wheat. Eur J Agron 94:98–107
Richner W, Soldati A, Stamp P (1996) Shoot-to-root relations in field-grown maize seedlings. Agron J J88:56–61
Ros CR, Belland W, White PF (2000) Phosphorous seed coating and soaking for improving seedling growth of Oryza sativa (rice) cv IR66. Seed Sci Technol 28:391–401
Rossel JB, Wilson PB, Hussain D, Woo NS, Gordon MJ, Mewett OP, Howell KA, Whelan J, Kazan K, Pogson BJ (2007) Systemic and intracellular responses to photooxidative stress in Arabidopsis. Plant Cell 19:4091–4110
Ruelland E, Vaultier MN, Zachowski A, Hurry V (2009) Cold signalling and cold acclimation in plants. Adv Bot Res 49:35–150
Sahu MP, Kumawat SM, D'Souza SF, Ramaswamy NK, Singh G (2005) Sulphydryl bioregulator technology for increasing mustard production. Res Bull RAUBARC:1–52
Sakhabutdinova AR, Fatkhutdinova DR, Beazrukova MV, Shakirova FM (2003) Salicylic acid prevents the damaging action of stress factor of wheat plants. Bulg J Plant Physiol Spec Issue:314–319
Sarhan F, Oullet F, Vazquez-Tello A (1997) The wheat wcs 120 gene family. A useful model to understand the molecular genetics of freezing tolerance in cereals. Physiol Plant 101:439–445
Savvides A, Ali S, Tester M, Fotopoulos V (2016) Chemical priming of plants against multiple abiotic stresses: mission possible? Trends Plant Sci 21:329–340
Sethi V, Raghuram B, Sinha AK, Chattopadhyay S (2014) A mitogen-activated protein kinase cascade module, MKK3-MPK6 and MYC2, is involved in blue light-mediated seedling development in Arabidopsis. Plant Cell 26:3343–3357
Shao CX, Hu J, Song WJ, Hu WM (2005) Effects of seed priming with chitosan solutions of different acidity on seed germination and physiological characteristics of maize seedling. J Zhejiang Univ Agric Life Sci 31:705–708

Sharma HSS, Fleming C, Selby C, Rao JR, Martin T (2014) Plant biostimulants: a review on the processing of macroalgae and use of extracts for crop management to reduce abiotic and biotic stresses. J Appl Phycol 26:465–490
Sharmila P, Puthur JT, Pardha Saradhi P (2000) Vesicular arbuscular mycorrhizal fungi improves establishment of micropropagated plants. In: Mukerji KG, Chamola BP, Singh J (eds) Mycorrhizal biology. Kluwer Academic/Plenum, New York, pp 236–250
Singh B, Usha K (2003) Salicylic acid induced physiological and biochemical changes in wheat seedlings under water stress. Plant Growth Regul 39:137–141
Soeda Y, Konings MC, Vorst O, van Houwelingen AM, Stoopen GM, Maliepaard CA et al (2005) Gene expression programs during *Brassica oleracea* seed maturation, osmopriming, and germination are indicators of progression of the germination process and the stress tolerance level. Plant Physiol 137:354–368
Srivastava AK, Lokhande VH, Patade VY, Suprasanna P, Sjahril R, D'Souza SF (2010) Comparative evaluation of hydro-, chemo-, and hormonal priming methods for imparting salt and PEG stress tolerance in Indian mustard (*Brassica juncea* L.). Acta Physiol Plant 32:1135–1144
Starck Z, Niemyska B, Bogdan J, Tawalbeh RNA (2000) RNA response of tomato plants to chilling stress in association with nutrient or phosphorus starvation. Plant Soil 226:99–106
Steponkus PL, Uemura M, Joseph RA, Gilmour SJ, Thomashow MF (1998) Mode of action of the COR15a gene on the freezing tolerance of Arabidopsis thaliana. Proc Natl Acad Sci U S A 95:14570–14575
Stewart CR, Martin BA, Reding L, Cerwick S (1990) Seedling growth, mitochondrial characteristics, and alternative respiratory capacity of corn genotypes differing in cold tolerance. Plant Physiol 92:761–766
Suzuki K, Nagasuga K, Okada M (2008) The chilling injury induced by high root temperature in the leaves of rice seedlings. Plant Cell Physiol 49:433–442
Suzuki N, Koussevitzky SHAI, Mittler RON, Miller GAD (2012) ROS and redox signalling in the response of plants to abiotic stress. Plant Cell Environ 35(2):259–270
Svensson J, Ismail A, Palva ET, Close TJ (2002) Dehydrins. In: Storey KB, Storey JM (eds) Sensing, signalling and cell adaptation. Elsevier Science, Amsterdam, pp 155–171
Taka T (2004) The relationship of antioxidant enzymes and some physiological parameters in maize during chilling. Plant Soil Environ 50:27–32
Taylor AG, Allen PS, Bennett MA, Bradford KJ, Burris JS, Misra MK (1998) Seed enhancements. Seed Sci Res 8(2):245–256
Taylor NL, Day DA, Millar AH (2002) Environmental stress causes oxidative damage to plant mitochondria leading to inhibition of glycine decarboxylase. J Biol Chem 277:42663–42668
Thakur P, Kumar S, Malik JA, Berger JD, Nayyar H (2010) An overview: cold stress effects on reproductive development in grain crops. Environ Exp Bot 67:429–443
Thomas UC, Varughese K, Thomas A, Sadanandan S (2000) Seed priming–for increased vigour, viability and productivity of upland rice. Leisa India 4:14
Timmusk S, Abd El-Daim IA, Copolovici L, Tanilas T, Kännaste A, Behers L et al (2014) Drought-tolerance of wheat improved by rhizosphere bacteria from harsh environments: enhanced biomass production and reduced emissions of stress volatiles. PLoS One 9:96086
Turk H, Erdal S, Genisel M, Atici O, Demir Y, Yanmis D (2014) The regulatory effect of melatonin on physiological, biochemical and molecular parameters in cold-stressed wheat seedlings. Plant Growth Regul 74:139–152
van Buer J, Cvetkovic J, Baier M (2016) Cold regulation of plastid ascorbate peroxidases serves as a priming hub controlling ROS signaling in *Arabidopsis thaliana*. BMC Plant Biol 16(1):163
Verslues PE, Agarwal M, Katiyar-Agarwal S, Zhu J, Zhu JK (2006) Methods and concepts in quantifying resistance to drought, salt and freezing, abiotic stresses that affect plant water status. Plant J 45:523–539
Wang B, Vinocur O, Altman AS (2004) Role of plant heat-shock proteins and molecular chaperones in the abiotic stress response. Trends Plant Sci 9:244–252

Wang W, Chen Q, Hussain S, Mei J, Dong H, Peng S, Huang L, Cui K, Nie L (2016) Pre-sowing seed treatments in direct-seeded early rice: consequences for emergence, seedling growth and associated metabolic events under chilling stress. Sci Rep 6. https://doi.org/10.1038/srep19637

Warrington IJ, Kanemasu ET (1983) Corn growth response to temperature and photoperiod I. seedling emergence, tassel initiation, and anthesis. Agron J 75:749–754

Wechsberg GE, Probert RJ, Bray CM (1994) The relationship between 'dehydrinlike' protein sand seed longevity in Ranunculus sceleratus L. J Exp Bot 45:1027–1030

Wilhelm KS, Thomashow MF (1993) Arabidopsis thaliana Cor15b, an apparent homolog of Cor15a, is strongly responsive to cold and ABA, but not drought. Plant Mol Biol 23:1073–1077

Wise RR (1995) Chilling-enhanced photo oxidation: the production, action and study of re active oxygen species produced during chilling in the light. Photosynth Res 45:7997

Wojtyla Ł, Lechowska K, Kubala S, Garnczarska M (2016) Molecular processes induced in primed seeds—increasing the potential to stabilize crop yields under drought conditions. J Plant Physiol 203:116–126

Xu S, Hu J, Li Y, Ma W, Zheng Y, Zhu S (2011) Chilling tolerance in Nicotiana tabacum induced by seed priming with putrescine. Plant Growth Regul 63:279–290

Yadav SK (2010) Cold stress tolerance mechanisms in plants. Agron Sustain Dev 30:515–527

Yadav PV, Kumari M, Ahmed Z (2011) Seed priming mediated germination improvement and tolerance to subsequent exposure to cold and salt stress in capsicum. Res J Seed Sci 4:125–136

Yagmur M, Kaydan D (2008) Alleviation of osmotic strength of water and salt in germination and seedling growth of triticale with seed priming treatments. Afr J Biotechnol 7:2156–2162

Yan Q, Duan Z, Jingdong M, Xun L, Fei D (2012) Effects of root-zone temperature and N, P, and K supplies on nutrient uptake of cucumber (*Cucumis sativus* L.) seedlings in hydroponics. Soil Sci Plant Nutr 58:707–717

Yilmaz AH, Ekiz I, Gültekin B, Torun H, Barut K, Cakmak I (1998) Effect of seed zinc content on grain yield and zinc concentration of wheat grown in zinc-deficient calcareous soils. J Plant Nutr 21:2257–2264

Yordanova R, Popova L (2007) Effect of exogenous treatment with salicylic acid on photosynthetic activity and antioxidant capacity of chilled wheat plants. Gen Appl Plant Physiol 33:155–170

Zarka DG, Vogel JT, Cook D, Thomashow MF (2003) Cold induction of Arabidopsis CBF genes involves multiple ICE (inducer of CBF expression) promoter elements and a cold-regulatory circuit that is desensitized by low temperature. Plant Physiol 133(2):910–918

Zhang J, Cui S, Li J, Wei J, Kirkham MB (1995) Protoplasmic factors, antioxidant responses, and chilling resistance in maize. Plant Physiol Biochem 33:567–575

Zhang S, Hu J, Liu N, Zhu Z (2006) Presowing seed hydration treatment enhances the cold tolerance of direct-sown rice. Seed Sci Technol 34(3):593–601

Zhang H, Tan ZQ, HuLY WSH, Luo JP, Jones RL (2010) Hydrogen sulfide alleviates aluminum toxicity in germinating wheat seedlings. J Integr Plant Biol 52(6):556–567

Zhao J, Li S, Jiang T, Liu Z, Zhang W (2012) Chilling stress- the key predisposing factor for causing Alternaria alternata infection and leading to cotton (*Gossypium hirsutum* L.) leaf senescence. PLoS One 7:36126

Zheng C, Jiang D, Liu F, Dai T, Liu W, Jing Q, Cao W (2009) Exogenous nitric oxide improves seed germination in wheat against mitochondrial oxidative damage induced by high salinity. Environ Exp Bot 67:222–227

Zheng M, Tao Y, Hussain S, Jiang Q, Peng S, Huang J et al (2016) Seed priming in dry direct-seeded rice: consequences for emergence, seedling growth and associated metabolic events under drought stress. Plant Growth Regul 78:167–178

Zhou YG, Yang YD, Qi YG, Zhang ZM, Wang XJ, Hu XJ (2002) Effects of chitosan on some physiological activity in germinating seed of peanut. J Peanut Sci 31:22–25

Zhu JJ, Li YR, Liao JX (2013) Involvement of anthocyanins in the resistance to chilling-induced oxidative stress in *Saccharum officinarum* L. leaves. Plant Physiol Biochem 73:427–433

# Seed Priming and Metal/Metalloid Stress Tolerance in Plants

**Abdul Saboor, Ghulam Mustafa, Muhammad Arshad, Muhammad Ahmad, Sajjad Hussain, Niaz Ahmed, Shakeel Ahmad, Muhammad Shahid, and Muhammad Arif Ali**

## Contents



---

A. Saboor · G. Mustafa · N. Ahmed · M. A. Ali (✉)
Department of Soil Science, Bahauddin Zakariya University, Multan, Punjab, Pakistan
e-mail: arif1056@bzu.edu.pk

M. Arshad
Institute of Environmental Sciences and Engineering, School of Civil and Environmental Engineering, National University of Sciences and Technology, Islamabad, Pakistan

M. Ahmad
Center for Health & Bioresources, Austrian Institute of Technology, Vienna, Austria

S. Hussain
Department of Horticulture, Bahauddin Zakariya University, Multan, Punjab, Pakistan

S. Ahmad (✉)
Department of Agronomy, Bahauddin Zakariya University, Multan, Punjab, Pakistan
e-mail: shakeelahmad@bzu.edu.pk

M. Shahid
Department of Environmental Sciences, COMSATS University Islamabad, Vehari Campus, Pakistan



M. Hasanuzzaman, V. Fotopoulos (eds.), *Priming and Pretreatment of Seeds and Seedlings*, https://doi.org/10.1007/978-981-13-8625-1_14

**Abstract** Metal contamination, as a consequence of anthropogenic activities, poses threat to plants and their ecosystems. Plants growing on metal-contaminated soils show severe aberration in their metabolism leading to stunted plant growth and low biomass production. Recently, seed priming has enabled us to improve germination and yield on such heavy metals toxic soils by different mechanisms. Seed priming is conferring tolerance against abiotic besides biotic stresses. Different seed priming procedures like hydropriming, chemo-priming, osmopriming, redox priming, nutrio-priming, hormo-priming, and biopriming are being used under metal/ metalloid stress conditions. Priming of seeds is generally done with natural or synthetic compounds before sowing of seeds. It is the simplest, easiest, and cost-effective technique for improving seed vigor and robust seedling emergence ultimately adding increments to the economy of country with high returns. Seed priming treatment effectively alleviates the negative impact of metal stress through different mechanisms. In this chapter seed priming is discussed in context of their potential for alleviation of different metal stresses. Furthermore, different mechanisms of seed priming are elaborated.

**Keywords** Seed priming · Metal stress · Metal tolerance · Metalloid

## 1 Introduction

Plants are grown on open fields; therefore they have the maximum chances of getting exposed to numerous stresses with a slight change in weather required for their growth and yield and other most important vicinity in which they grow (Shakoor et al. 2011). Abiotic stress has devastating effects on plant, such as drought being the most common one and predicted to be worse with the shortage of available water resources in 2050 (Harrison et al. 2014; Shahzad et al. 2016). Salinity is another major stress constraint in the world because no continent is free from saline soils and excessive evaporation and less water for irrigation further exacerbated the situation (Ali et al. 2017; Shahzad et al. 2017); others include heat and cold stress (Zhao et al. 2007; Bunty et al. 2018; Eremina et al. 2016). Abiotic stresses cause osmotic disturbance in plant cells which create a disruption at different levels including damaged photosynthetic apparatus and ionic channels leading to changes in phenotypic appearance and ultimately end with total dry plant (Xiong and Zhu 2002). Abiotic stresses like drought salinity, heavy metals, temperature extremes (low and high), mineral toxicities including metals and metalloids impedes plant growth and interferes with the normal growth of plants (Barakat 2011). Crops are grown in open fields; therefore they suffer from slight deviation with the changes in environment (Mittler 2006). Plants thrive well under abiotic stress conditions by adjusting to the existing conditions as being immobile to approach toward favorable growing conditions (Verslues et al. 2006). When plants face stress, they give signals to the other parts to adjust to the stress conditions (Xiong and Zhu 2001), in response to these stress stimuli resulting in regulation of physiological and biochemical activities in plants against stress (Bacha et al. 2017). Volatile molecules play an important role against stress by signaling (Vickers et al. 2009). However their concentration varies depending upon the severity of stress and time in which the plant is going to be under stress (from hours to days, it varies according to environment) (Vickers et al. 2009). Heavy metal stress regulates numerous ion channels and pumps to inhibit their uptake and removal from the xylem channel (Thakur et al. 2016). Crop improvement for heavy metal stress varies from direct selection-based conventional breeding to modern molecular breeding (Ashrafzadeh and Leung 2015). Currently contemporary genomic approaches have added value for developing stress-tolerant cultivars (Hu et al. 2016; Ashraf et al. 2018). Genetic engineering and genome editing techniques have paved path and broaden our horizon by targeted insertion of genes into desired cultivars (Gust et al. 2010). These all modern approaches have significantly increased yield of crops against stress (Mickelbart et al. 2015). Besides their merits transgenic crops have also some demerits that may cause pleiotropic effect and gene silencing (Kola et al. 2015). Considering the harmful effects of transgenic approaches, seed priming emerges as a safe and easy technique for growing crops on heavy metal-polluted soils (Galhaut et al. 2014). Being the simplest technique, it is easy to carry either in lab or on farm (Hussain et al. 2018a, b). Furthermore it has the potential to cope with adverse soil environment (Tabassum et al. 2017). It provides nutrients to seeds at the initial germinating stage (Iqbal et al. 2017). Keeping in view the current problems of abiotic stress, it is the need of the hour to find efficient, environmentally safe, and cost-effective mechanisms. So we have tried to compile some possible mechanism of seed priming for better crop growth as discussed below.

## 2 Types of Heavy Metals and Their Stress Level

Soil is a reservoir of various elements consisting of essential and nonessential plant elements attached on mineral sites. Among the heavy metals, some metals are needed by crops in small quantity, whereas some metals are not required by plants; even their small amount become toxic.

## 3 Beneficial Heavy Metals

Among heavy metals, cobalt, copper, nickel, iron, zinc, molybdenum, and manganese are essential heavy metals. These heavy metals are needed by crops in small amount for completing their development besides life cycle (Alloway 2013). These metal ions affected the enzymatic activity and cellular metabolism. These metals have a crucial role in synthesis of protein, nucleic acid, and a part of chlorophyll.

Cobalt complex is an active part of vitamin B12, which is required by both animals and humans. It is involved in water utilization and reduction of transpiration rate in plants (Dal Corso et al. 2014). It also helps in photosynthesis process (Gad 2005). Nickel is an important component of urease enzyme. Copper (Cu) is an essential micronutrient which takes part in photosynthetic electron transport and is a cofactor of superoxide dismutase and ascorbate oxidase enzymes (Yruela 2005). However, slightly high concentration of Cu will become phytotoxic and inhibit many functions in plants. Cu toxicity results in the production of reactive oxygen species, for example, $O_2^{\bullet-}$, $H_2O_2$, and malondialdehyde; those enhance the oxidation rate of proteins (Zaho et al. 2010). Several enzymes contain zinc (Zn) as their essential part, like carbonic anhydrase, alcohol dehydrogenase, and superoxide dismutase besides RNA polymerase. Zinc takes part in many oxidation reactions and is essential to maintain integrity of ribosome. It also plays a key role in transcription factors as it is a part of RNA polymerase. Manganese is involved in splitting of water in photosystem II. It is a part of enzyme malic dehydrogenase, oxalosuccinic decarboxylase, and superoxide dismutase. Iron is involved in photosynthesis, respiration and nitrogen fixation process, and chloroplast development besides biosynthesis. It is also a main component of heme protein and iron-sulfur proteins (Moller et al. 2007). Molybdenum (Mo) enhances the concentration of chlorophyll in plants (McGarth et al. 2010).

## 4 Non-beneficial Heavy Metals

Selenium, mercury, chromium, and lead are considered as non-beneficial heavy metals as they are not required by plants. Their little quantity can damage plant growth and reduce yield (Richmond and Sussman 2003).

## 5 Causes of Metal/Metalloid Toxicity Stress on Plants

Zn (zinc) is an indispensable micronutrient needed for crop growth. Due to anthropogenic activities, smelting, and industrial activities, its concentration in soil is increasing day by day. Zn concentrations range in soil 100–300 mg $kg^{-1}$ (Kupper and Elisa 2016; Warne et al. 2008). Several plant metabolic functions retarded by high soil Zn cause stunted growth and leaf senescence and decreased shoot and root growth (Choi et al. 1996; Rout and Das 2009; Fontes and Cox 1998). Younger leaves become chlorotic and can spread to older leaves if exposed to high Zn concentration in soil for long time. Zn toxicity also caused the deficiency of $Fe^{+2}$ and $Mn^{+2}$ in plant leaves which is due to the Zn hindrance for the translocation of micronutrient from root to shoot (Ebbs and Kochian 1997). Zn toxicity causes purplish-red color in leaves that is attributed to deficiency of phosphorus (Mourato et al. 2015; Lee et al. 1996).

The role of Cu in plants is well recognized, and it is considered as an indispensable micronutrient for crop growth (Thomas et al. 1998) and has a significant role in

$CO_2$ assimilation plus ATP synthesis. Cu is an important constituent of several proteins of photosynthetic system besides cytochrome oxidase of respiratory electron transport chain (Demirevska-kepova et al. 2004). Cu occurrence in ecosystems is increasing due to industrial and mining activities. Various anthropogenic activities like mining besides smelting of Cu-containing ores are also adding Cu to soil. Huge amount of waste rocks besides tailings are generated by mining activities which are deposited at the surface. Cytotoxic role is that it induces stress besides injury to crops that eventually cause leaf chlorosis and stunted growth (Lewis et al. 2001). Plants suffer from oxidative stress and generate ROS under high Cu concentrations (Stadtman and Oliver 1991). Due to oxidative stress, metabolic pathways are disturbed and macromolecules are damaged (Hegedus et al. 2001).

Heavy addition of mercury (Hg) into arable lands has contaminated whole food chain by Hg. Hg is unique owing to its occurrence in various forms, e.g., HgS, $Hg^{2+}$, $Hg^0$, and methyl-Hg. However, $Hg^{2+}$ is the dominant form in agricultural soil (Clarkson and Magos 2006; Han et al. 2006). Hg rests in solid phase due to adsorption to the sulfides, organic matter, and clay content. There are increasing evidences that $Hg^{2+}$ accumulate in higher besides aquatic plants readily (Isaksson et al. 2007; Kamal et al. 2004; Israr et al. 2006). Higher concentration of $Hg^{2+}$ is highly toxic to plant cells which caused visible tissue injuries besides physiological malfunctioning in plants (Zhou et al. 2007). For example, $Hg^{2+}$ can cause stomatal disclosure by binding with water channel proteins; thus water flow in plants is physically obstructed (Zhang and Tyerman 1999). $Hg^{2+}$ toxicity stimulates ROS generation inducing oxidative stress and interferes the mitochondrial activity. As a result, biomembrane lipids besides cellular metabolism in crops are disrupted (Messer et al. 2005; Israr and Sahi 2008; Cargnelutti et al. 2006).

Industrial revolution has dramatically polluted biosphere with toxic metals (Swaminathan 2003). Chromium (Cr) is a heavy metal, and severe environmental contamination in soil and sediments besides groundwater is caused by high Cr levels (Shanker et al. 2004). One of key consumers of water is tanning industry. High quantity of Cr (1.07–7.80 mg $L^{-1}$) has been found in the waste water as discharge of tanning industry. It has been estimated that freshwater bodies contain 3550 t Cr by the anthropogenic discharge of Cr worldwide (Nriagu 1990). The World Health Organization (WHO), Environmental Protection Agency (EPA), and International Agency for Research on Cancer (IARC) have established Cr (VI) as a powerful epithelial irritant, very toxic, and a proven human carcinogen. It has been studied in many pants about the toxicity of Cr. Nutrient imbalance, root injury, chlorosis in fresh leaves, and wilting of tops besides inhibition of plant growth are caused by the excessive concentration of Cr (Chatterjee and Chatterjee 2000; Sharma et al. 2013). It has also been described in terrestrial plants about the chlorophyll biosynthesis inhibition (Vazques et al. 1987). Around, 40% inhibition of growth was shown by barley saplings grown in 100 μM Cr (Skeffington et al. 1976). Growth of stem, leaves, and roots and alterations in the germination process are caused by the toxic effect of Cr on growth of plants. Plants productivity besides total biomass production is affected by exposure to higher level of Cr (Shanker et al. 2004). Plant physiological processes like water relations and photosynthesis are also deleteriously affected by Cr. Metabolites or its ability to generate ROS and direct effect on enzymes describes the metabolic alterations by the exposure of Cr to plants (Shanker et al. 2003).

Lead is distributed worldwide, the most abundant toxic element of the pedosphere. Lead toxicity results from ejection of municipal sewage sludge, mining besides smelting activities, paints having Pb, and gasoline besides explosives. It imparts harmful impacts on morphology, growth, and photosynthetic activities of plants. Toxic levels of Pb lead to retardation of enzymatic activities and imbalance in water regulation and disturb membrane permeability besides mineral nutrition (Sharma and Dubey 2005). At cellular level, Pb reacts with sulfhydryl groups; thus it inhibits activity of enzymes. High Pb concentration increasing the production of reactive oxygen species in plants causes oxidative stress that disturbs plants functions (Reddy et al. 2005).

Arsenate (As) and phosphate (P) are competitor of each other for their uptake by plants because they are transported through the same transporter into the plasma lemma (Bolan et al. 2013). Numerous plant species have been developed for arsenic tolerance and have explored mechanism to survive at toxic arsenic stress level (Chandra et al. 2018; Anawar et al. 2018; Bianucci et al. 2018). One of the mechanisms observed is the suppression of phosphate that resultantly uptakes more arsenic by plants (Kalita et al. 2018). Currently many genes regulating As tolerance have been reported (Chen et al. 2018). Instead of these modern approaches for its tolerance, plants still accumulate arsenic toxicity nominal rate as compared to arsenic free soils (Wu et al. 2016). When plants accumulate arsenic, it has further transformation mechanism to convert it into less toxic form to protect the plants photosynthetic machinery (Mirza et al. 2014). The possible form of its conversion includes arsenate to arsenite or monomethylarsinic acid (MMA) (Meharg and Hartley-Whitaker 2002).

Cobalt (Co) is a natural mineral element existing in our soils in different forms (Lange et al. 2016). In earlier times cobalt was given less importance as a toxic element, and there were no control measures for it (Barrio-Parra et al. 2018). It has become the part of soil toxicity from different normal daily life usage materials such as cobalt alloys, domestic heating system, and batteries and also from the waste of different industries (De Miguel et al. 1999; Kim et al. 2015).

Cobalt is toxic to plants even with a minute quantity of t, and its uptake disrupts plant's growth and development (López-Moreno et al. 2016). Its harmful effects have been reported in different crops on their growth and yield (Li et al. 2009). It also interferes the uptake of essential mineral elements necessary for their potent growth (López-Moreno et al. 2016). Likely it damages the chlorophyll contents, transpiration which results in stomatal closure and creates pool of reactive oxygen species (Panda et al. 2016).

Nickel (Ni) is also a naturally found trace metal ion in earth crusts (Rinklebe and Shaheen 2017). However, its concentration is accumulating in many areas through human activities like mining and smelting, waste from industries, and application of fertilizers besides pesticides (Wang et al. 2018). Nickel concentration in normal soils is up to 1000 mg $kg^{-1}$, while its concentration is higher in contaminated soils

(200–26,000 mg $kg^{-1}$) which is 20-fold higher than normal soil. Plants uptake nickel in the form of $Ni^{2+}$, and its excessive uptake creates physiological changes, and its phenotypic symptoms are chlorosis and necrosis (Torres et al. 2016). It also triggers oxidative stress like other heavy metal stresses in plants resulting in high malondialdehyde (MDA) production (Younis et al. 2015). Biochemical changes elucidated those MDA level increases in sensitive cultivars compared to tolerant one which further aggravates the situation for plants by disrupting potassium and calcium channels and limiting water uptake (Soares et al. 2016). Decrease level of water uptake is linked with symptoms of $Ni^{2+}$ toxicity (Khaliq et al. 2016).

## 6 Factors Affecting Metal and Metalloid Toxicity

Metals are natural components of the soil; however their concentration varies in different soils depending upon mineral fraction (Gadd 2008). Metals also exist in organic fraction in a bound form, and decomposition of organic fraction causes their mobility. Factors affecting the mobility of Cr, Se, Co, Pb, As, Ni, and Cu include pH, humic besides fulvic acids, and root exudates besides nutrients (Violante et al. 2010). Heavy metals remain in soil with a residence time of thousands of years impacting negative influence on soil microflora and ground cover (Pehlivan et al. 2009). Mining of metals is considered as a double sword because at one side it gives benefits to earn by selling them and their usage in different sectors (Csavina et al. 2012), while on the other side, it is really very harmful causing contamination to the soil and surrounding environments (Holden and Jacobson 2008). It may be spilled to the surrounding soil or may emit toxic dust that will settle too far from its mining and cause long spread toxicity especially to the crops grown by near to that place (Johnson and Hallberg 2005). The fate of these contaminants ends up in soil as deposit and results in phytotoxicity (Pruvot et al. 2006).

Every soil contains heavy metal from less harmful range to the toxic levels depending upon the parent material from which the soil was formed (Alloway 2013). Soils formed from ultramafic rocks parent materials also hold notable amount of heavy metals specifically Ni, Cr, and Co (Ross et al. 2007). Decomposed ironstones such as iron oxide sediments create high arsenic metal accumulation (Robinson et al. 2007). Heavy metals spillage to river or any water body causes suspension, and when flood comes, it overflows to surrounding land ultimately causing metal toxicity (Bai et al. 2012). Livestock manures contain toxic elements depending upon their feed; mostly they contain Cu, Zn, As, and Cd (Sager 2007; Toor et al. 2007). It can be accumulated at the top surface of soil due to long-term usage (Adeli et al. 2007; He et al. 2009); thereby it is reported to be the source for creating metal/metalloid toxicity (Qian et al. 2003; Wang et al. 2013). General practice of managing sewage sludge is also harmful source for agricultural soils. Globally the sewage sludge is managed by dumping and landfills posing threats of metalloids contaminations (Walter et al. 2006). Another option that exists for its degradation is pyrolysis of sewage sludge which emits residual waste of heavy metals (Shinogi et al. 2003, Hwang et al. 2007). It pollutes the soil by Pb, Cd, Ni, Cr, and Hg, if it contains waste from industrial sewage (Przewrocki et al. 2004; Dai et al. 2006; Singh and Agrawal 2007).

Inorganic fertilizers are also abundant source of heavy metal load into agricultural soils globally (Savci 2012). Excessive fertilizer application gathers metals in soils (Huang and Jin 2008). Some fertilizer contains high concentration of cadmium which ultimately ended up their journey in soil deposits (Brigden et al. 2002). Phosphatic fertilizer application increases Cd and other toxic elements such as F, Hg, and Pb (Raven et al. 1998).

Insecticides and weedicides applied to control pest and weed also cause toxic levels to our soil. These include huge amount of Pb, Cu, and phenyl mercuric chloride (Robinson et al. 2007). Different building material used on agricultural land such as fences and other materials corrosion leads to toxicity of heavy metals (Alloway 2013). Military exercise sites create contamination in the form of bullets, shells, bombs, damaged vehicles, and leakage of fuel and burning of something on or near agricultural lands (Schleich et al. 2010).

## 7 Seed Priming

Seed priming is a simple technique of treating seeds for better germination and improved seedling emergence especially to cope with harsh soil environment (Farooq et al. 2006a). It includes soaking of seed in appropriate solution that triggers the metabolic activities of seed and drying of seed for sowing (Farooq et al. 2005). Imbibition of seed includes partial hydration of seed to initiate germination process but prevents radicle protrusion (Ashraf and Foolad 2005); however faster and synchronized seed germination can be achieved with seed priming (Farooq et al. 2008). It has direct influence on farmer's livelihood by providing increased rate of seed germination and crop development within a less period of time giving higher yield returns instead of complete crop failure which includes several factors under specific circumstances and in specific regions (Bennet et al. 1992).

Currently various types of contemporary priming techniques have been practiced from farm to industry level for better crops (Harris et al. 2007; Fig. 1). Hydropriming is the simplest environment-friendly technique which employs soaking of seeds in sterilized water at suitable temperature (Farooq et al. 2006b). It invigorates the seed and provides sturdy crop establishment especially in tropical regions (Nawaz et al. 2013). Osmopriming is the incubation of seed generally with a salt and polyethylene glycol solution for controlling imbibition and hampers radicle protrusion (Mouradi et al. 2016). It protects the seed from low external potential of water to have synchronized germination (Chen and Arora 2011). Osmopriming is reported to be a precise technique as compared to hydropriming because it produced quicker results (Moradi and Younesi 2009). Chemical and biopriming are specifically used

**Fig. 1** Types of seed priming

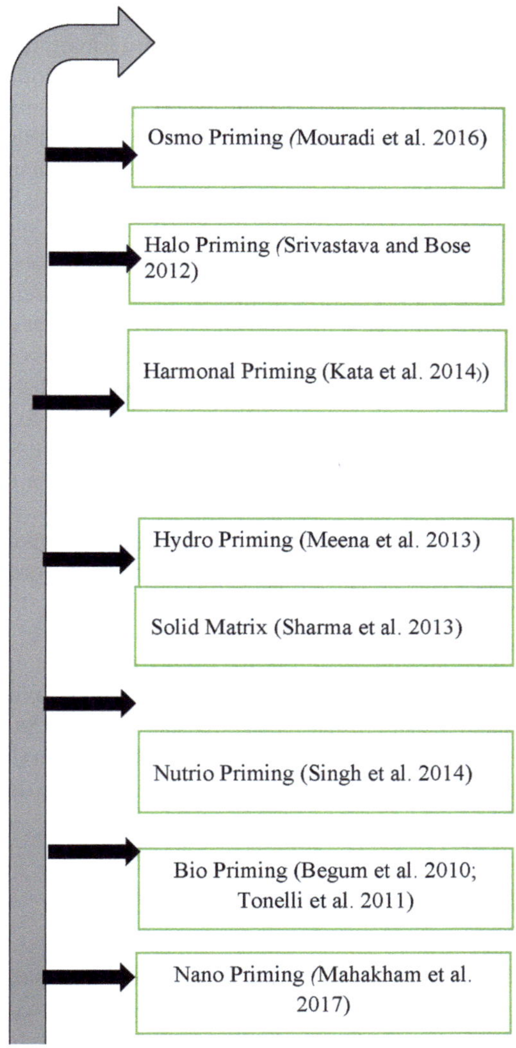


for providing nutrition to seeds and protecting them from any disease attacks (Begum et al. 2010; Tonelli et al. 2011).

Seedling state of plant growth is the most sensitive stage which sometimes can face water limitations and ultimately leads to uneven crop establishment and ultimately poor crop yield (Demir et al. 2006). Inorder to overcome the above-mentioned issues, seed priming has been proven to have evident effects during unfavorable conditions for crop (Parera and Cantliffe 1994). The profitable benefits of seed priming have taken the attention of seed industry to produce seed notably for adverse conditions (Basra et al. 2005). However appropriate priming agent is a prerequisite for formation of seed to grow vigorously under existing adverse field environment (Job et al. 2000).

Seed priming has been used against abiotic stress since the discovery of this technique (Heydecker et al. 1975). Immediately after its success for seed improvement, it has been tested in numerous horticultural and field crops against stress tolerance; therefore it is considered as chief element of plant immunity system offering protection against environmental stresses (Ton et al. 2009). Drought is the condition with no or limited water availability for seed germination (Hussain et al. 2018a, b). Primed seed has been reported to produce robust root and shoot emergence under drought stress (Iqbal and Ashraf 2013). Seed priming with potassium chloride increased seed emergence under drought stress (Eivazi 2012). The phenomenon is primed seeds absorb moisture to start metabolic activity (Kazemi and Eskandari 2012). It also played a role in mitigating the harmful effects of salt stress conditions during radicle growth (Patane et al. 2009). Saline soils are more problematic for seedling emergence (Munns and Tester 2008). Saline soil contains high salt concentration especially in upper layer of soil which is seed zone. Priming technique can be used in such salinity zone where evapotranspiration is high resulting in high water loss (Jisha et al. 2013). Priming has obvious positive effects that enable glycophyte to adapt to saline conditions overcoming salinity problems (Chen 2011; Maiti and Pramanik 2013).

Priming is also associated with the repair of nucleic acid and increase in synthesis of protein; therefore it exhibits higher water uptake and requirements (Oliveria et al. 2010; McDonald 2000). Enhanced antioxidative enzyme activity has also been observed in primed seeds (Hussain et al. 2018a, b), which in turn reduces lipid peroxidation (Chiu et al. 2006). It enhances antioxidant activity and works as a repairer of membrane under abiotic stress (Ibrahim 2016).

## 8 Factors Affecting Seed Priming

The efficiency of priming is influenced by several factors such as temperature, oxygen availability, plant species, priming methods, duration of priming, and seed quality. Optimal conditions for priming can vary from species to species and even for different cultivars of a species. For example, when two different cultivars of *Capsicum annuum* were primed with polyethylene glycol and $KNO_3$, one of the cultivars were more responsive to priming in terms of mean times of germination and mean time to emergence (Bradford et al. 1990). The duration of priming treatment can also influence the success of priming. However, the duration of priming is widely varied from plants to plants, type of priming, temperature, and other factors. For example, in sunflower 8 h of salt priming is sufficient to observe its significant effect, while for that of ornamental species, 2 weeks are required to achieve the similar effect. In *Zea mays* 36 h of osmopriming resulted in 15% more seed germination as compared to 12-h treatment (Dezfuli et al. 2008). Prolonged exposure of seeds to priming solutions can lead to irreversible damage to seeds and hence poor germination of seeds. For example, in *Z. mays* 48 h of osmopriming reduced the seed germination by 4% (Dezfuli et al. 2008). Although this reduction was not significant as compared to 36-h treatment, however, it indicates that further increase in duration of priming might lead to poor germination of maize seeds.

Temperature is another crucial parameter which can largely affect the success of seed priming. This is because metabolic activities in plants are mainly temperature dependent and higher or lower temperature can affect these metabolic processes. In general, 15–30 °C have been shown to be optimal for different priming treatments (Bradford 1986). However, temperature below 10 °C was found to increase the mean germination time of seeds (Black and Bewley 2000). When wheat seeds were primed with different solutions at 20 °C, they showed improved germination, and seedlings were more vigorous compared to other temperatures tested in the experiment (Yari et al. 2010). Although seed priming is mainly employed to improve seed quality, however, the initial seed quality of seeds prior to priming is a prerequisite to observe benefits of seed priming. Seeds which are devoid of pathogens and are vigorous could be more responsive to priming treatments.

## 9 Seed Priming in Relation to Metal/Metalloid Toxicity

Metal toxicity negatively interferes with all stages of plant growth from germination to seed setting, ultimately reducing the yield of economically important crops. Seeds of various plant species planted in soil saturated with different heavy metals had reduced germination, less number of seedlings, lower root or shoot length, and reduced dry biomass. Metals when in excess affect different physiological and biochemical processes of seeds and thereby negatively affect the seed health. For example, cadmium toxicity has been shown to delay the seed germination via disturbing the balance in cotyledon/embryo of total carbohydrates, amino acids, and minerals. Metals are essential for plant growth within a limited range, but when they increase in concentration, it causes toxicity and hampers plant growth (Aihemaiti et al. 2018). Anthropogenic activities expose more heavy metal toxicity to soil including sewage dumping; automobiles and industries are primary factors for causing toxicity to agricultural lands in their vicinity (Bitton and Morel 2018; Ayturan et al. 2018; Fig. 2).

Seed has prime importance in agriculture because almost every crop is grown from seed; however sometimes it faces soil toxicity which restricts seedling emergence and eventually poor plant growth (Bisen et al. 2015). Plants grown on toxic soils deviate from the normal growth activities as it causes changes in their physiological and biochemical mechanisms (Ali et al. 2014). Zinc, a major micronutrient, plays crucial biological functions. Zinc plays an important role for growth, but its nano-zinc oxide (ZnO) toxicity causes ZnO stress which impedes germination percentage and germination index of seeds (Singh et al. 2018; Chen et al. 2015). ZnO stress can be overcome by exploiting seed priming by polyethylene glycol (PEG); it hastens seed germination besides reduced germination time (Salah

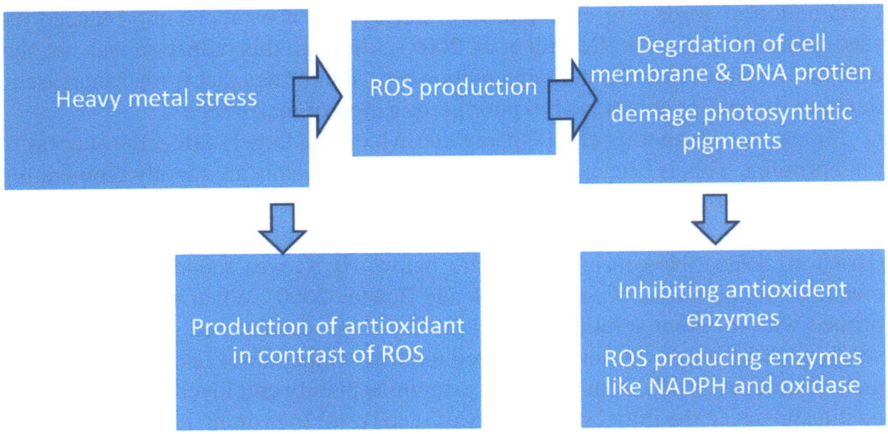


**Fig. 2** A representation of heavy metal stress and ROS production (Shabala 2017)

et al. 2015). Priming improves germination by stimulating metabolic activities of seed providing sturdy root and shoot development under ZnO stress conditions (Sheteiwy et al. 2016).

Cadmium toxicity is most abundant in soil which interferes with the normal growth of seedling and inhibits its growth (Modlitbova et al. 2018). Priming improves vitality of seed (Afzal et al. 2013). Numerous types of priming materials have been reported to improve seed germination under cadmium toxicity, but seed priming with salicylic acid boosted the germination percentage (Espanany et al. 2016). Salicylic acid is a hormone which is reported to have a role in regulating physiological changes against stress (Fariduddin et al. 2018). It was implicated to stimulate seed germination; therefore it was used as priming agent against cadmium toxicity (Lu et al. 2018). Presoaking of seeds with salicylic acid maintains optimum nutrition of mineral; furthermore it enhances absorption of other essential nutrients under cadmium toxicity (Belkhadi et al. 2010; Gu et al. 2018). Salicylic primed seeds depicted improvement in morphological and biochemical characteristics of plants against cadmium toxicity (Moussa and El-Gamal 2010; Matewally et al. 2003; Choudhry and Panda 2004). Salicylic acid priming benefits are of broad spectrum which include potent root development to protection of chlorophyll machinery of plant in cadmium toxic conditions (Panda and Patra 2007). Hormonal priming such as with auxin, gibberellin, and abscisic acid also played a role in improving germination of seed with cadmium stress (Sneideris et al. 2015). Cadmium causes significant reduction in relative water content; however indole acetic acid primed seed reported to lessen adversarial effects of cadmium on seed and throughout the course of plant growth (Agami and Mohamed 2013).

Arsenic is a heavy metal which is distributed extensively in soil by humans due to its use as a pesticide, herbicide, and additive to animal feed (Stevens et al. 2018). Arsenic exposure to plants results in formation of reactive oxygen species which eventually leads to premature cell and death of plant (Farooq et al. 2016). Until now various options have been executed to alleviate the toxic effect of arsenic on plants, and among all those options, seed priming is the only easy-to-use application (Khan and Gupta 2018). Selenium primed seeds assist in protecting the plants by reduction in proline contents and soluble proteins and enhances the chlorophyll contents (Moulick et al. 2017). It is speculated that selenium priming has quenching ability of reactive oxygen species harmful effects which in turn speeds up the growth (Moulick et al. 2018). The other perception is that priming of selenium creates sufficient selenium pool around seed coat which is later supplied to roots and shoots to minimize the arsenic toxicity (Moulick et al. 2017).

Industrialization has worsened the agricultural soil and environmental conditions; lead (Pb) is also one of the by-products of industry accumulating deeply in soil (Alloway 2013). Lead enters plant through roots and starts disruption in normal functioning of plants as other heavy metals, i.e., reduced germination, water relations, and physiological process (Amari et al. 2017). Seed priming by $KNO_3$ meaningfully inhibits harmful effects of lead and lowers the translocation from roots to above part of plants which contrarily increased the catalase and peroxidase besides superoxide dismutase in roots (Nawaz et al. 2017).

Rice seed priming by different concentrations of selenium (Se) resulted in higher seed germination besides seedling development under arsenic (As) stress (Moulick et al. 2016). The Se priming did not only reduce the contact of rice root with As but also reduced the translocation of absorbed As to shoots (Moulick et al. 2017). Interestingly, the adult plants of primed seeded rice in As-infested soil were taller in height and had higher number of tillers and biomass (Moulick et al. 2018), suggesting that seed priming is not only effective against As stress at earlier stages, but the priming effect is likely to remain viable throughout the plant growth cycle. Black cumin seeds treated with either $KNO_3$ or KCl led to higher rate and earlier germination of seeds as compared to untreated plants under cadmium (Cd) stress. Furthermore, the seedlings of primed seeds had higher overall length and dry weight (Espanany et al. 2016). Similarly, Nawaz et al. (2017) showed that when maize seeds are pre-treated with $KNO_3$, they are better able to tolerate lead (Pb) stress and this tolerability was linked to higher activities of antioxidant enzymes such as catalases, peroxidases, and superoxide dismutase. Kumar et al. (2016) indicated that when wheat seeds are treated by variable concentrations of $Ca(NO_3)_2$, they tend to germinate faster and can alleviate the effect of mercuric chloride ($HgCl_2$). Improvement in germination rate of different crop seeds using chemicals under metal stress might be due to shift in the production of different oxides. For example, $KNO_3$ treatment leads to higher production of nitric oxide (NO) in germinating seeds perhaps due to catabolization of nitrites and nitrates, hence improving the availability of these nutrients in germinating seeds. Moreover, salt treatment also contributes to endosperms degeneration and higher activities of different proteases, amylases, and lipases, which might contribute to higher germination of seeds under metal stress conditions.

## 10 Seed Priming with Naturally Occurring Metabolites of Plants Under Metal Stress

Another category of seed priming agents is naturally occurring secondary metabolites of plants such as plant hormones. Plant synthesizes these metabolites endogenously; however, their exogenous application has been proven effective in alleviation of different stresses (Savvides et al. 2016). The evidence of their role in seed priming and amelioration of metal toxicity are also accumulating (Rakshit and Singh 2018). Maize seeds pre-treated with different concentrations of salicylic acid (SA) resulted in improved seedling growth and reduced accumulation of chromium (Cr) when compared with non-treated seeds (Singh et al. 2016). The declined accumulation of Cr in seedlings was correlated with higher activities of ROS scavenging enzymes (Singh et al. 2016) suggesting that SA treatment may act as an early signal to protect the germinating seedlings from Cr stress through fine-tuning of metabolic machinery involved in Cr uptake. Salicylic acid has priming potential to ameliorate the toxic effects of other metals and it has also been evaluated. For example, when *Vigna radiata* seeds were treated with SA and spiked with toxic levels of cadmium (Cd), the germinating seedlings performed better in all growth parameters measured in the study. Likewise, the tolerance was positively correlated with higher activities of antioxidant enzymes and lower concentration of $H_2O_2$. Similar effects were observed for gibberellic acid (GA) treatment; however, GA effect was significantly more pronounced than SA (Hassan and Mansoor 2017).

## 11 Mechanism of Seed Priming

Heavy metal pollution occurs naturally by forest fires due to sudden increase in temperature, volcanic eruptions, or surface mineralization. Anthropogenic activities and poor management further trigger the formation of metal-polluted environments ranging from different chemical applications in agriculture to boost the yield of crop, pesticide applications to control pest and disease and mining of new metals, poor management of industrial waste effluents, and removal of toxic gases to clean the air to lastly our heavy automobiles usage. All these abovementioned factors are responsible for polluting our planet day by day with the burgeoning population in the world (El-Ramady et al. 2015; He et al. 2017; Luoma 2017).

When we talk about heavy metals, it is classified into further two groups, i.e., essential and nonessential metal ions (Mohammad et al. 2015). The first group includes essential micronutrients Cu, Co, Fe, Mn, Ni, and Zn that work differently ranging from electron transfer to redox reaction and metabolic procedures (Tahervand and Jalali 2017). The second group covers nonessential metals As, Cd, Cr, Hg, and Pb, with their slight increase in concentration they cause toxicity to soil and plants (Dago et al. 2014). Another group in which we can divide is metals related with bioactivity on the basis of physicochemical properties, i.e., redox (Cr, Cu, Fe,

Mn) and non-redox metals (Al, Cd, Hg, Ni, Zn) (Panda et al. 2016). The first group of metal members creates oxidative damage to plants after their uptake via Fenton reaction (Liochev 2018), which leads in the production of reactive oxygen species (ROS) in crops and causes detrimental effects by disruption of homeostasis in cell, regulation of gene expression, production of hydrogen peroxide ($H_2O_2$), damage to the photosynthetic machinery, and eventually plant death (Adrees et al. 2015). Non essential metals group plays an indirect role by depleting glutathione and sulfhydryl group binding (Urbancsok et al. 2018) and hampering activity of antioxidant enzymes against ROS triggering oxidative stress in plants (Bielen et al. 2013).

Chromium is reported to be the toxic element for plant growth and also for humans by their incorporation into the food chain (Lukina et al. 2016). It creates environmental pollution in two forms, i.e., hexavalent $Cr^{6+}$ and trivalent $Cr^{3+}$. Plant does not uptake chromium metal directly or absorbs via plant roots (Shahid et al. 2017); it makes their way to plants through specific ion carriers (Ali et al. 2015a, b). Fate of chromium in plants is mostly roots after making their way to plants rather than being accumulated in other parts of plants (Stambulska et al. 2018), because it is an immobile metal ion; thereby it piles up in the roots (Batool et al. 2017). Chromium effects are visible during the early growth stages of plant and thereby reduce stem development (Ali et al. 2015a, b). Moreover chromium stress limits the cell division in roots which ultimately results in short root system (Stambulska et al. 2018). Chromium uptake by root interferes with the normal functioning ranging from water uptake to limiting the essential nutrients uptake necessary for plant throughout the course of growth (Srivastava and Jain 2011). However plants evolved different mechanisms to cope with heavy metal stress (Emamverdian et al. 2015). The first defense mechanism that plants developed with the course of time toward heavy metal toxicity is physical barriers, i.e., activating trichome tissues, formation of thick cuticular membrane, and beneficial association of mycorrhizae, besides cell wall activation immediately after signaling metal stress (Trinh et al. 2014; Hristozkova et al. 2017; Shahid et al. 2017). Root hairs (trichome) block their movement further upward by working as a storage organ, and also secretion of metabolites detoxify metal ions (Oksanen 2018). Seed priming maintains the balance of ROS scavengers to cope with ROS produced under stress (Hussain et al. 2017). It decreases the production of malondialdehyde and hydrogen peroxide and increases proline concentration (Hussain et al. 2016a, b; Hossain et al. 2015). Thus, all ROS scavenging enzymes work together one after one for alleviating oxidative stress; the final enzyme that ends in forming $H_2O$ is glutathione reductase which is also triggered by seed priming (Khaliq et al. 2015).

## 12 Conclusion

Seed priming is an efficient and environment-friendly method to diminish the harmful effects of heavy metal in plants. Seed priming is an easy to perform, effective, and efficient option for better crop growth. Seed priming firstly protects from harsh conditions during the initial growth stages of seed. It further induces cellular responses and signal stimulus against metal stress. However precise processes of seed priming under stress still need to be explored. It also accumulates in active signaling proteins in primed seeds. These inactive proteins become activated after sensing stress. It is reported from different studies revealing the beneficial effects of seed priming that alleviate harmful effects of metal toxicity. Currently numerous priming studies have been done, and most of them have achieved positive results, but there are still some gaps needed to be filled. As a conclusion we can say that either it regulates changes in crops at diverse growth stages or not. Seed priming agents and their different concentrations respond toward different crops variedly. It is also the need of the hour to unravel molecular and genetic facets of plant species against heavy metal stress in relation to seed priming.

## References


Adeli A, Sistani KR, Tewolde H, Rowe DE (2007) Broiler litter application effects on selected trace elements under conventional and no-till systems. Soil Sci 172:349–365

Adrees M, Ali S, Rizwan M, Zia-ur-Rehman M, Ibrahim M, Abbas F, Farid M, Qayyum MF, Irshad MK (2015) Mechanisms of silicon-mediated alleviation of heavy metal toxicity in plants: a review. Ecotoxicol Environ Saf 119:186–197

Afzal I, Basra SMA, Cheema MA, Farooq M, Jafar MZ, Shahid M, Yasmeen A (2013) Seed priming: a shotgun approach for alleviation of salt stress in wheat. Inter J Agri Biol 15(6):1199–1203

Agami RA, Mohamed GF (2013) Exogenous treatment with indole-3-acetic acid and salicylic acid alleviates cadmium toxicity in wheat seedlings. Ecotoxicol Environ Saf 94:164–171

Aihemaiti A, Jiang J, Li DA, Liu N, Yang M, Meng Y, Zou Q (2018) The interactions of metal concentrations and soil properties on toxic metal accumulation of native plants in vanadium mining area. J Environ Manag 222:216–226

Ali B, Qian P, Jin R, Ali S, Khan M, Aziz R, Tian T, Zhou W (2014) Physiological and ultra-structural changes in Brassica napus seedlings induced by cadmium stress. Biol Plant 58:131–138

Ali N, Masood S, Mukhtar T, Kamran MA, Rafique M, Munis MFH, Chaudhary HJ (2015a) Differential effects of cadmium and chromium on growth, photosynthetic activity, and metal uptake of Linum usitatissimum in association with Glomus intraradices. Environ Monit Assess 187(6):311

Ali S, Bharwana SA, Rizwan M, Farid M, Kanwal S, Ali Q, Ibrahim M, Gill RA, Khan MD (2015b) Fulvic acid mediates chromium (Cr) tolerance in wheat (Triticum aestivum L.) through lowering of Cr uptake and improved antioxidant defense system. Environ Sci Pollut Res 22(14):10601–10609

Ali S, Rizwan M, Qayyum MF, Ok YS, Ibrahim M, Riaz M, Arif MS, Hafeez F, Al-Wabel MI, Shahzad AN (2017) Biochar soil amendment on alleviation of drought and salt stress in plants: a critical review. Environ Sci Pollut Res 24(14):12700–12712

Alloway BJ (2013) Sources of heavy metals and metalloids in soils. In: Heavy metals in soils. Springer, Dordrecht, pp 11–50

Amari T, Ghnaya T, Abdelly C (2017) Nickel, cadmium and lead phytotoxicity and potential of halophytic plants in heavy metal extraction. South Afric J Bot 111:99–110

Anawar HM, Rengel Z, Damon P, Tibbett M (2018) Arsenic-phosphorus interactions in the soil-plant-microbe system: dynamics of uptake, suppression and toxicity to plants. Environ Pollut 233:1003–1012

Ashraf M, Foolad MR (2005) Pre-sowing seed treatment-a shotgun approach to improve germination growth and crop yield under saline and none-saline conditions. Adv Agron 88:223–271

Ashraf MA, Akbar A, Askari SH, Iqbal M, Rasheed R, Hussain I (2018) Recent advances in abiotic stress tolerance of plants through chemical priming: an overview. In: Advances in seed priming. Springer, Singapore, pp 51–79

Ashrafzadeh S, Leung DM (2015) In vitro breeding of heavy metal-resistant plants: a review. Hortic Environ Biotechnol 56(2):131–136

Ayturan ZC, Kunt F, Dursun S (2018) Metal bioaccumulation/toxicity test for metal industry wastewaters. Int J Environ Pollut Environ Model 1(1):1–5

Bacha H, Tekaya M, Drine S, Guasmi F, Touil L, Enneb H, Triki T, Cheour F, Ferchichi A (2017) Impact of salt stress on morpho-physiological and biochemical parameters of Solanum lycopersicum cv. Microtom leaves. S Afr J Bot 108:364–369

Bai J, Xiao R, Zhang K, Gao H (2012) Arsenic and heavy metal pollution in wetland soils from tidal freshwater and salt marshes before and after the flow-sediment regulation regime in the Yellow River Delta, China. J Hydrol 450–451:244–253

Barakat MA (2011) New trends in removing heavy metals from industrial wastewater. Arab J Chem 4(4):361–377

Barrio-Parra F, Elío J, De Miguel E, García-González JE, Izquierdo M, Álvarez R (2018) Environmental risk assessment of cobalt and manganese from industrial sources in an estuarine system. Environ Geochem Health 40:737–748

Basra SMA, Farooq M, Tabassum R (2005) Physiological and biochemical aspects of seed vigour enhancement treatments in fine rice (Oryza sativa L.). Seed Sci Technol 33:623–628

Batool R, Tabassum T, Ali M (2017) Evaluation of Cr (VI) remediation potential of Eichornia sp in conjunction with chromium-resistant bacterial strains. Trop J Pharm Res 16(5):1005–1011

Begum MM, Sariah M, Puteh AB, Abidin MZ, Rahman MA, Siddiqui Y (2010) Field performance of bio-primed seeds to suppress Colletotrichum truncatum causing damping off and seedling stand of soybean. Biol Control 53:18–23

Belkhadi A, Hediji H, Abbes Z, Nouairi I, Barhoumi Z, Zarrouk M, Chaïbi W, Djebali W (2010) Effects of exogenous salicylic acid pre-treatment on cadmium toxicity and leaf lipid content in Linum usitatissimum L. Ecotoxicol Environ Saf 73(5):1004–1011

Bennett MA, Fritz VA, Callan NW (1992) Impact of seed treatments on crop stand establishment. Hortic Technol 2(3):345–349

Bianucci E, Godoy A, Furlan A, Peralta JM, Hernández LE, Carpena-Ruiz RO, Castro S (2018) Arsenic toxicity in soybean alleviated by a symbiotic species of Bradyrhizobium. Symbiosis 74(3):167–176

Bielen A, Remans T, Vangronsveld J, Cuypers A (2013) The influence of metal stress on the availability and redox state of ascorbate, and possible interference with its cellular functions. Inter J Mol Sci 14(3):6382–6413

Bisen K, Keswani C, Mishra S, Saxena A, Rakshit A, Singh HB (2015) Unrealized potential of seed biopriming for versatile agriculture. In: Nutrient use efficiency: from basics to advances. Springer, New Delhi, pp 193–206

Bitton G, Morel JL (2018) Microbial enzyme assays for the detection of heavy metal toxicity. In: Microscale testing in aquatic toxicology. CRC Press, Boca Raton, pp 143–152

Black M, Bewley JD (2000) Seed technology and its biological basis. CRC Press, Boca Raton

Bolan N, Mahimairaja S, Kunhikrishnan A, Choppala G (2013) Phosphorus–arsenic interactions in variable-charge soils in relation to arsenic mobility and bioavailability. Sci Total Environ 463:1154–1162

Bradford KJ (1986) Manipulation of seed water relations via osmotic priming to improve germination under stress conditions. Hort Sci 21(5):1105–1112

Bradford KJ, Steiner JJ, Trawatha SE (1990) Seed priming influence on germination and emergence of pepper seed lots. Crop Sci 30(3):718–721

Brigden K, Stringer R, Santillo D (2002). Heavy metal and radionuclide contamination of fertilizer products and phosphogypsum waste produced by The Lebanese Chemical Company, Greenpeace Research Laboratories, Department of Biological Sciences, University of Exeter, Exeter EX4 4PS, UK

Bunty S, Singh VK, Sheoran RK, Thakral NK, Anu N (2018) Heat stress in plants. Ann Biol 34(1):70–78

Cargnelutti D, Tabaldi LA, Spanevello RM, Jucoski GO, Battisti V, Redin M, Linares CEB, Dressler VL, Flores MM, Nicoloso FT, Morsch VM, Schetinger MRC (2006) Mercury toxicity induces oxidative stress in growing cucumber seedlings. Chemosphere 65:999–1106

Chandra S, Saha R, Pal P (2018) Assessment of arsenic toxicity and tolerance characteristics of bean plants (Phaseolus vulgaris) exposed to different species of arsenic. J Plant Nutr 41(3):340–347

Chatterjee J, Chatterjee C (2000) Phytotoxicity of cobalt, chromium and copper in cauliflower. Environ Pollut 109:69–74

Chen K (2011). Antioxidants and dehydrin metabolism associated with osmopriming-enhanced stress tolerance of germinating spinach (Spinacia oleracea L. cv. Bloomsdale) seeds

Chen K, Arora R (2011) Dynamics of the antioxidant system during seed osmopriming, post-priming germination, and seedling establishment in spinach (*Spinacia oleracea*). Plant Sci 180(2):212–220

Chen J, Liu X, Wang C, Yin SS, Li XL, Hu WJ, Simon M, Shen ZJ, Xiao Q, Chu CC, Peng XX (2015) Nitric oxide ameliorates zinc oxide nanoparticles-induced phytotoxicity in rice seedlings. J Hazard Mater 297:173–182

Chen Y, Cao Y, Rathinasabapathi B, Ma L (2018) Novel genes of Hyperaccumulator ferns in arsenic tolerance, uptake, and metabolism: implications for crop improvement. In: Current advances in fern research. Springer, Cham, pp 361–379

Chiu KY, Chuang SJ, Sung JM (2006) Both anti-oxidation and lipid-carbohydrate conversion enhancements are involved in priming-improved emergence of Echinacea purpurea seeds that differ in size. Sci Hortic 108:220–226

Choi JM, Pak CH, Lee CW (1996) Micronutrient toxicity in French marigold. J Plant Nutr 19:901–916

Choudhury S, Panda SK (2004) Role of salicylic acid in regulating cadmium induced oxidative stress in Oryza sativa L. roots. Bulg J Plant Physiol 30(3–4):95–110

Clarkson TW, Magos L (2006) The toxicology of mercury and its chemical compounds. Crit Rev Toxicol 36(8):609–662

Csavina J, Field J, Taylor MP, Gao S, Landazuri A, Betterton EA, Sáez AE (2012) A review on the importance of metals and metalloids in atmospheric dust and aerosol from mining operations. Sci Total Environ 433:58–73

Dago À, González I, Ariño C, Díaz-Cruz J. M, Esteban M (2014). Chemometrics applied to the analysis of induced phytochelatins in Hordeum vulgare plants stressed with various toxic non-essential metals and metalloids. Talanta, 118, pp. 201–209

Dai JY, Ling C, Zhao JF, Na MA (2006) Characteristics of sewage sludge and distribution of heavy metal in plants with amendment of sewage sludge. J Environ Sci 18(6):1094–1100

Dal Corso G, Manara A, Piasentin S, Furini A (2014) Nutrient metal elements in plants. Metallomics 6(10):1770–1788

De Miguel ME, Llamas JF, Chacón E, Mazadiego LF (1999) Sources and pathways of trace elements in urban environments: a multi-elemental qualitative approach. Sci Total Environ 235(1–3):355–357

Demir AO, Goksoy AT, Buyukcangaz H, Turan ZM, Koksal ES (2006) Deficit irrigation of sunflower (Helianthus annuus L.) in a sub-humid climate. Irrig Sci 24:279–289

Demirevska-Kepova K, Simova-Stoilova L, Stoyanova Z, Holzer R, Feller U (2004) Biochemical changes in barley plants after excessive supply of copper and manganese. Environ Exp Bot 52:253–266

Dezfuli PM, Sharif-Zadeh F, Janmohammadi M (2008) Influence of priming techniques on seed germination behavior of maize inbred lines (Zea mays L.). J Agric Biol Sci 3(3):22–25

Ebbs SD, Kochian LV (1997) Toxicity of zinc and copper to Brassica species: implications for phytoremediation. J Environ Qual 26:776–781

Eivazi A (2012) Induction of drought tolerance with seed priming in wheat cultivars (Triticum aestivum L.). Acta Agric Slov 99(1):21–29

El-Ramady H, Abdalla N, Alshaal T, Domokos-Szabolcsy E, Elhawat N, Prokisch J, Shams MS (2015) Selenium in soils under climate change, implication for human health. Envron Chem Lett 13(1):1–19

Emamverdian A, Ding Y, Mokhberdoran F, Xie Y (2015) Heavy metal stress and some mechanisms of plant defense response. Sci World J 2015:1–18

Eremina M, Rozhon W, Poppenberger B (2016) Hormonal control of cold stress responses in plants. Cell Mol Life Sci 73(4):797–810

Espanany A, Fallah S, Tadayyon A (2016) Seed priming improves seed germination and reduces oxidative stress in black cumin (Nigella sativa) in presence of cadmium. Ind Crop Prod 79:195–204

Fariduddin Q, Khan TA, Yusuf M, Aafaqee ST Khalil RRAE (2018) Ameliorative role of salicylic acid and spermidine in the presence of excess salt in Lycopersicon esculentum. Photosynthetica, pp 1–13

Farooq M, Basra SMA, Saleem BA, Nafees M, Chishti SA (2005) Enhancement of tomato seed germination and seedling vigor by osmopriming. Pak J Agric Sci 42:36–41

Farooq M, Basra SMA, Afzal I, Khaliq A (2006a) Optimization of hydropriming techniques for rice seed invigoration. Seed Sci Technol 34(2):507–512

Farooq M, Basra SMA, Tabassum R, Afzal I (2006b) Enhancing the performance of direct seeded fine rice by seed priming. Plant Prod Sci 9:446–456

Farooq M, Basra SMA, Rehman H, Saleem BA (2008) Seed priming enhances the performance of late sown wheat (Triticum aestivum L.) by improving chilling tolerance. J Agron Crop Sci 194(1):55–60

Farooq MA, Gill RA, Islam F, Ali B, Liu H, Xu J, Zhou W (2016) Methyl jasmonate regulates antioxidant defense and suppresses arsenic uptake in *Brassica napus* L. Front in Plant Sci 7:468

Fontes RLS, Cox FR (1998) Zinc toxicity in soybean grown at high iron concentration in nutrient solution. J Plant Nutr 21:1723–1730

Gad N (2005) Interactive effect of salinity and cobalt on tomato plants. II. Some physiological parameters as affected by cobalt and salinity. Res J Agric Biol Sci 1(3):270–276

Gadd JM (2008) Transformation and mobilization of metals, metalloids, and radionuclides by microorganisms. In: Violante A, Huang PM, Gadd GM (eds) Biophysico-chemical processes of metals and metalloids in soil environments, Wiley-Jupac Series, Vol 1. Wiley, Hoboken, pp 53–96

Galhaut L, Lespinay D, Walker A, Bernal, Correal MP, Lutts ES (2014) Seed priming of Trifolium repens L. improved germination and early seedling growth on heavy metal-contaminated soil. Water Air Soil Pollut 225(4):1905

Gu CS, Yang YH, Shao YF, Wu KW, Liu ZL (2018) The effects of exogenous salicylic acid on alleviating cadmium toxicity in Nymphaea tetragona Georgi. S Afr J Bot 114:267–271

Gust AA, Brunner F, Nürnberger T (2010) Biotechnological concepts for improving plant innate immunity. Curr Opin Biotechnol 21(2):204–210

Han FX, Su Y, Monts DL, Waggoner AC, Plodinec JM (2006) Binding distribution, and plant uptake of mercury in a soil from Oak Ridge, Tennessee, USA. Sci Total Environ 368:753–768

Harris D, Rashid A, Miraj G, Arif M, Shah H (2007) 'On-farm' seed priming with zinc sulphate solution. A cost-effective way to increase the maize yields of resource-poor farmers. Field Crop Res 102(2):119–127

Harrison MT, Tardieu F, Dong Z, Messina CD, Hammer GL (2014) Characterizing drought stress and trait influence on maize yield under current and future conditions. Glob Chang Biol 20(3):867–878

Hassan M, Mansoor S (2017) Priming seeds with phytohormones alleviates cadmium toxicity in mung bean (Vigna radiata L. Wilczek) seedlings. Pak J Bot 49(6):2071–2078

He ZQ, Endale DM, Schomberg HH, Jenkins MB (2009) Total phosphorus, zinc, copper, and manganese concentrations in cecil soil through 10 years of poultry litter application. Soil Sci 174:687–695

He K, Sun Z, Hu Y, Zeng X, Yu Z, Cheng H (2017) Comparison of soil heavy metal pollution caused by e-waste recycling activities and traditional industrial operations. Environ Sci Pollut Res 24(10):9387–9398

Hegedus A, Erdei S, Horvath G (2001) Comparative studies of H2O2 detoxifying enzymes in green and greening barley seedlings under cadmium stress. Plant Sci 160:1085–1109

Heydecker W, Higgins J, Turner YJ (1975) Invigoration of seeds. Seed Sci Technol 3:881–888
Holden WN, Jacobson RD (2008) Civil society opposition to nonferrous metals mining in Guatemala. Volunt Int J Volunt Nonprofit Org 19(4):325–350
Hossain MA, Bhattacharjee S, Armin SM, Qian P, Xin W, Li HY, Burritt DJ, Fujita M, Tran LSP (2015) Hydrogen peroxide priming modulates abiotic oxidative stress tolerance insights from ROS detoxification and scavenging. Front Plant Sci 6:420
Hristozkova M, Geneva M, Stancheva I, Iliev I, Azcón-Aguilar C (2017) Symbiotic association between golden berry (Physalis peruviana) and arbuscular mycorrhizal fungi in heavy metal-contaminated soil. J Plant Prot Res 57(2):173–184
Hu T, Zhu S, Tan L, Qi W, He S, Wang G (2016) Overexpression of OsLEA4 enhances drought, high salt and heavy metal stress tolerance in transgenic rice (Oryza sativa L.). Environ Exp Bot 123:68–77
Huang SW, Jin JY (2008) Status of heavy metals in agricultural soils as affected by different patterns of land use. Environ Monit Assess 139(1–3):317–327
Hussain S, Khan F, Cao W, Wu L, Geng M (2016a) Seed priming alters the production and detoxification of reactive oxygen intermediates in rice seedlings grown under sub-optimal temperature and nutrient supply. Front Plant Sci 7:439
Hussain S, Khan F, Hussain HA, Nie L (2016b) Physiological and biochemical mechanisms of seed priming-induced chilling tolerance in rice cultivars. Front Plant Sci 7:116
Hussain M, Farooq M, Nawaz A, Al-Sadi AM, Solaiman ZM, Alghamdi SS, Siddique KH (2017) Biochar for crop production: potential benefits and risks. J Soils Sediment 17(3):685–716
Hussain M, Farooq M, Sattar A, Ijaz M, Sher A, Ul-Allah S (2018a) Mitigating the adverse effects of drought stress through seed priming and seed quality on wheat (Triticum aestivum L.) productivity. Pak J Agric Sci 55(2):313
Hussain S, Khaliq A, Tanveer M, Matloob A, Hussain HA (2018b) Aspirin priming circumvents the salinity-induced effects on wheat emergence and seedling growth by regulating starch metabolism and antioxidant enzyme activities. Acta Physiol Plant 40(4):68
Hwang IH, Ouchi Y, Matsuto T (2007) Characteristics of leachate from pyrolysis residue of sewage sludge. Chemosphere 68(10):1913–1919
Ibrahim EA (2016) Seed priming to alleviate salinity stress in germinating seeds. J Plant Physiol 192:38–46
Iqbal M, Ashraf M (2013) Gibberellic acid mediated induction of salt tolerance in wheat plants: growth, ionic partitioning, photosynthesis, yield and hormonal homeostasis. Environ Exp Bot 86:76–85
Iqbal S, Farooq M, Cheema SA, Afzal I (2017) Boron seed priming improves the seedling emergence, growth, grain yield and grain biofortification of bread wheat. Int J Agric Biol 19(1)
Isaksson R, Balogh SJ, Farris MA (2007) Accumulation of mercury by the aquatic plant Lemna minor. Int J Environ Stud 64(2):189–194
Israr M, Sahi SV (2008) Promising role of plant hormones in translocation of lead in *Sesbania drummondii* shoots. Environ Poll 153(1):29–36
Israr M, Sahi S, Datta R, Sarkar D (2006) Bioaccumulation and physiological effects of mercury in Sesbania drummondii. Chemosphere 65:591–598
Jisha KC, Vijayakumari K, Puthur JT (2013) Seed priming for abiotic stress tolerance: an overview. Acta Physiol Plant 35(5):1381–1396
Job D, Capron I, Job C, Dacher F, Corbineau F, Come D (2000) Identification of germination-specific protein markers and their use in seed priming technology. In: Black M, Bradford KJ, Vazguez-Ramos J (eds) Seed biology: advances and applications. CAB International, Wallingford, pp 449–459
Johnson DB, Hallberg KB (2005) Acid mine drainage remediation options: a review. Sci Total Environ 338(1–2):3–14
Kalita J, Pradhan AK, Shandilya ZM, Tanti B (2018) Arsenic stress responses and tolerance in rice: physio cellular molecular approaches. Rice Sci 25(5):235–249
Kamal M, Ghalya AE, Mahmouda N, Cote R (2004) Phytoaccumulation of heavy metals by aquatic plants. Environ Int 29:1029–1039

Kazemi K, Eskandari H (2012) Does priming improve seed performance under salt and drought stress? J Basic Appl Sci Res 2(4):3503–3507
Khaliq A, Aslam F, Matloob A, Hussain S, Geng M, Wahid A, Rehman H (2015) Seed priming with selenium: consequences for emergence, seedling growth, and biochemical attributes of rice. Biol Trace Elem Res 166(2):236–244
Khaliq A, Ali S, Hameed A, Farooq MA, Farid M, Shakoor MB, Mahmood K, Ishaque W, Rizwan M (2016) Silicon alleviates nickel toxicity in cotton seedlings through enhancing growth, photosynthesis, and suppressing Ni uptake and oxidative stress. Arch Agron Soil Sci 62(5):633–647
Khan E, Gupta M (2018) Arsenic–silicon priming of rice (*Oryza sativa* L.) seeds influence mineral nutrient uptake and biochemical responses through modulation of Lsi-1, Lsi-2, Lsi-6 and nutrient transporter genes. Sci Rep 8(1):10301
Kim JH, Jung JH, Phee YG (2015) Exposure assessment of airborne cobalt in manufacturing industries. J Korean Soc Occup Environ Hyg 25(2):166–173
Kola VSR, Renuka P, Madhav MS, Mangrauthia SK (2015) Key enzymes and proteins of crop insects as candidate for RNAi based gene silencing. Front Physiol 6:119
Kumar MP, Bhawna MS, Bose B (2016) Hydro and halo priming: influenced germination responses in wheat Var-HUW-468 under heavy metal stress. Acta Physiol Plant 38(9):199. https://doi.org/10.1007/s11738-016-2226-3
Kupper H, Elisa A (2016) Mechanisms of metal toxicity in plants. Metallomics 8(3):269–285
Lange B, Pourret O, Meerts P, Jitaru P, Cancès B, Grison C, Faucon MP (2016) Copper and cobalt mobility in soil and accumulation in a metallophyte as influenced by experimental manipulation of soil chemical factors. Chemosphere 146:75–84
Lee CW, Choi JM, Pak CH (1996) Micronutrient toxicity in seed geranium (Pelargonium 9 hortorum Baley). J Am Soc Hortic Sci 121:77–82
Lewis S, Donkin ME, Depledge MH (2001) Hsp 70 expression in Enteromorpha intestinalis (Chlorophyta) exposed to environmental stressors. Aquat Toxicol 51:277–291
Li HF, Gray C, Mico C, Zhao FJ, McGrath SP (2009) Phytotoxicity and bioavailability of cobalt to plants in a range of soils. Chemosphere 75(7):979–986
Liochev SI (2018) The mechanism of "Fenton-like" reactions and their importance for biological systems. A biologist's view. Met Ions Biol Syst 36:1–39
López-Moreno ML, Avilés LL, Pérez NG, Irizarry BÁ, Perales O, Cedeno-Mattei Y, Román F (2016) Effect of cobalt ferrite ($CoFe_2O_4$) nanoparticles on the growth and development of Lycopersicon lycopersicum (tomato plants). Sci Total Environ 550:45–52
Lu Q, Zhang T, Zhang W, Su C, Yang Y, Hu D, Xu Q (2018) Alleviation of cadmium toxicity in Lemna minor by exogenous salicylic acid. Ecotoxicol Environ Saf 147:500–508
Lukina AO, Boutin C, Rowland O, Carpenter DJ (2016) Evaluating trivalent chromium toxicity on wild terrestrial and wetland plants. Chemosphere 162:355–364
Luoma SN (2017) Processes affecting metal concentrations in estuarine and coastal marine sediments. In: Heavy metals in the marine environment. CRC Press, Boca Raton, pp 51–66
Maiti R, Pramanik K (2013) Vegetable seed priming: a low cost, simple and powerful techniques for farmers' livelihood. Int J Bio-Resource Stress Manag 4:475–481
Matewally A, Finkemeir I, Georgi M, Dietz K-J (2003) Salicylic acid alleviates cadmium toxicity in barley seedlings. Plant Physiol 132:272–281
McDonald MB (2000) Seed priming. In: Black M, Bewley JD (eds) Seed technology and its biological basis. Sheffield Academic Press, Sheffield
McGrath SP, Micó C, Curdy R, Zhao FJ (2010) Predicting molybdenum toxicity to higher plants: influence of soil properties. Environ Pollut 158:3095–3102
Meharg AA, Hartley-Whitaker J (2002) Arsenic uptake and metabolism in arsenic resistant and nonresistant plant species. New Phytol 154(1):29–43
Messer RL, Lockwood PE, Tseng WY, Edwards K, Shaw M, Caughman GB, Lewis JB, Wataha JC (2005) Mercury (II) alters mitochondrial activity of monocytes at sublethal doses via oxidative stress mechanisms. J Biomed Mater Res B 75:257–263
Mickelbart MV, Hasegawa PM, Bailey-Serres J (2015) Genetic mechanisms of abiotic stress tolerance that translate to crop yield stability. Nat Rev Genet 16(4):237

Mirza N, Mahmood Q, Maroof Shah M, Pervez A, Sultan S (2014) Plants as useful vectors to reduce environmental toxic arsenic content. Sci World J 2014:1
Mittler R (2006) Abiotic stress, the field environment and stress combination. Trends Plant Sci 11(1):15–19
Modlitbova P, Pořízka P, Novotný K, Drbohlavova J, Chamradova I, Farka Z, Zlámalová-Gargošová H, Romih T, Kaiser J (2018) Short-term assessment of cadmium toxicity and uptake from different types of Cd-based Quantum Dots in the model plant Allium cepa L. Ecotoxicol Environ Saf 153:23–31
Mohammad J, Khan S, Shah MT, Islam-ud-din AA (2015) Essential and nonessential metal concentrations in morel mushroom (Morchella esculenta) in Dir-Kohistan, Pakistan. Pak J Bot 47(SI):133–138
Møller IM, Jensen PE, Hansson A (2007) Oxidative modifications to cellular components in plants. Annu Rev Plant Biol 58:459–481
Moradi A, Younesi O (2009) Effects of osmo-and hydro-priming on seed parameters of grain sorghum (*Sorghum bicolor* L.). Aus J Basic Applied Sci 3(3):1696–1700
Moulick D, Ghosh D, Santra SC (2016) Evaluation of effectiveness of seed priming with selenium in rice during germination under arsenic stress. Plant Physiol Biochem 109:571–578
Moulick D, Santra SC, Ghosh D (2017) Seed priming with Se alleviate As induced phytotoxicity during germination and seedling growth by restricting As translocation in rice (Oryza sativa L c. v. IET-4094). Ecotoxicol Environ Saf 145:449–456. https://doi.org/10.1016/j.ecoenv.2017.07.060
Moulick D, Santra SC, Ghosh D (2018) Rice seed priming with Se: A novel approach to mitigate as induced adverse consequences on growth, yield and as load in brown rice. J Hazard Mater 355:187–196. https://doi.org/10.1016/j.jhazmat.2018.05.017
Mouradi M, Bouizgaren A, Farissi M, Makoudi B, Kabbadj A, Very AA, Sentenac H, Qaddoury A, Ghoulam C (2016) Osmopriming improves seeds germination, growth, antioxidant responses and membrane stability during early stage of Moroccan alfalfa populations under water deficit. Chil J Agric Res 76(3):265–272
Mourato M, Moreira I, Leitão I, Pinto F, Sales J, Martins L (2015) Effect of heavy metals in plants of the genus Brassica. Int J Mol Sci 16(8):17975–17998
Moussa HR, El-Gamal SM (2010) Effect of salicylic acid pretreatment on cadmium toxicity in wheat. Biol Plant 54(2):315–320
Munns R, Tester M (2008) Mechanisms of salinity tolerance. Annu Rev Plant Biol 59:651–681
Nawaz J, Hussain M, Jabbar A, Nadeem GA, Sajid M, Subtain MU, Shabbir I (2013) Seed priming a technique. Inter J Agric Crop Sci 6(20):1373
Nawaz F, Naeem M, Akram A, Ashraf MY, Ahmad KS, Zulfiqar B (2017) Seed priming with KNO3 mediates biochemical processes to inhibit lead toxicity in maize (Zea mays L.). J Sci Food Agric 97(14):4780–4789. https://doi.org/10.1002/jsfa.8347
Nriagu JO (1990) A global assessment of natural sources of atmospheric trace metals. Nature 338:47–49
Oksanen E (2018) Trichomes form an important first line of defence against adverse environment new evidence for ozone stress mitigation. Plant Cell Environ 41:1497
Oliveira AB, Prisco JT, Enéas-Filho J, Gomes-Filho E (2010) Salinity effects on germination and establishment of sorghum seedlings from artificially aged and primed seeds. J New Seeds 11:399–411
Panda SK, Patra HK (2007) Effect of salicylic acid potentiates cadmium-induced oxidative damage in Oryza sativa L. leaves. Acta Physiol Plant 29(6):567–575
Panda SK, Choudhury S, Patra HK (2016) Heavy-metal-induced oxidative stress in plants: physiological and molecular perspectives. In: Abiotic stress response in plants. Wiley, Hoboken
Parera CA, Cantliffe DJ (1994) Presowing seed priming. Hortic Rev 16:109–114
Patane C, Cavallaro V, Cosentino SL (2009) Germination and radicle growth in unprimed and primed seeds of sweet sorghum as affected by reduced water potential in NaCl at different temperatures. Ind Crop Prod 30:1–8
Pehlivan E, Özkan AM, Dinç S, Parlayici Ş (2009) Adsorption of $Cu^{2+}$ and $Pb^{2+}$ ion on dolomite powder. J Hazard Mater 167(1–3):1044–1049

Pruvot C, Douay F, Hervé F, Waterlot C (2006) Heavy metals in soil, crops and grass as a source of human exposure in the former mining areas (6 pp). J Soils Sediments 6(4):215–220
Przewrocki P, Kulczycka J, Wzorek Z, Kowalski Z, Gorazda K, Jodko M (2004) Risk analysis of sewage sludge-Poland and EU comparative approach. Pol J Environ Stud 13(2):237–244
Qian P, Schoenau JJ, Wu T, Mooleki SP (2003) Copper and zinc amounts and distribution in soil as influenced by application of animal manure in east-Central Saskatchewan. Can J Soil Sci 83(2):197–202
Rakshit A, Singh HB (eds) (2018) Advances in seed priming. Singapore: springer Singapore. Springer, Imprint
Raven PH, Berg LR, Johnson GB (1998) Environment, 2nd edn. Saunders College Publishing, New York
Reddy AM, Kumar SG, Jyotsnakumari G, Thimmanayak S, Sudhakar C (2005) Lead induced changes in antioxidant metabolism of horsegram (Macrotyloma uniflorum (Lam.) Verdc.) and bengalgram (Cicer arietinum L.). Chemosphere 60:97–104
Richmond KE, Sussman M (2003) Got silicon? The non-essential beneficial plant nutrient. Curr Opin Plant Biol 6(3):268–272
Rinklebe J, Shaheen SM (2017) Redox chemistry of nickel in soils and sediments: a review. Chemosphere 179:265–278
Robinson GR, Larkins P, Boughton CJ, Bradley WR, Sibrell PL (2007) Assessment of contamination from arsenic pesticide use on orchards in the Great Valley Region, Virginia and West Virginia. J Environ Qual 36:654–663
Ross SM, Wood MD, Copplestone D, Warriner M, Crook P (2007) UK soil and herbage pollutant survey, Environmental concentrations of heavy metals in UK soil and herbage (Report No. 7). Environment Agency, Bristol
Rout GR, Das P (2009) Effect of metal toxicity on plant growth and metabolism: I. Zinc. In: Sustainable agriculture. Springer, Dordrecht, pp 873–884
Sager M (2007) Trace and nutrient elements in manure, dung and compost samples in Austria. Soil Biol Biochem 39:1383–1390
Salah SM, Yajing G, Dongdong C, Jie L, Aamir N, Qijuan H, Weimin H, Mingyu N, Jin H (2015) Seed priming with polyethylene glycol regulating the physiological and molecular mechanism in rice (Oryza sativa L.) under nano-ZnO stress. Sci Rep 5:14278
Savci S (2012) An agricultural pollutant: chemical fertilizer. Int J Environ Sci Dev 3(1):73
Savvides A, Ali S, Tester M, Fotopoulos V (2016) Chemical priming of plants against multiple abiotic stresses: mission possible? Trends Plant Sci 21(4):329–340. https://doi.org/10.1016/j.tplants.2015.11.003
Schleich CE, Beltrame MO, Antenucci CD (2010) Heavy metals accumulation in the subterranean rodent Ctenomys talarum (Rodentia: Ctenomyidae) from areas with different risk of contamination. Folia Zool 59(2):108–114
Shabala Sergey, ed (2017). Plant stress physiol. Boston CABI.
Shahid M, Shamshad S, Rafiq M, Khalid S, Bibi I, Niazi NK, Dumat C, Rashid MI (2017) Chromium speciation, bioavailability, uptake, toxicity and detoxification in soil-plant system: a review. Chemosphere 178:513–533
Shahzad S, Chaudhry UK, Anwar B, Saboor A, Yousaf MF, Saeed F, Yaqoob S (2016) Drought stress effect on morphological and physiological characteristics of different varieties of annual verbena (Verbena hybrid). J Biol Environ Sci 9:32–46
Shahzad S, Khan MY, Zahir ZA, Asghar HN, Chaudhry UK (2017) Comparative effectiveness of different carriers to improve the efficacy of bacterial consortium for enhancing wheat production under salt affected field conditions. Pak J Bot 49(4):1523–1530
Shakoor U, Saboor A, Ali I, Mohsin AQ (2011) Impact of climate change on agriculture: empirical evidence from arid region. Pak J Agric Sci 48(4):327–333
Shanker AK, Djanaguiraman M, Pathmanabhan G, Sudhagar R, Avudainayagam S (2003) Uptake and phytoaccumulation of chromium by selected tree species. In: Proceedings of the international conference on water and environment held in Bhopal, India
Shanker AK, Djanaguiraman M, Pathmanabhan G, Sudhagar R, Avudainayagam S (2004) Uptake and phytoaccumulation of chromium by selected tree species. India

Sharma P, Dubey RS (2005) Lead toxicity in plants. Braz J Plant Physiol 17:35–52
Sharma AD, Rathore SVS, Srinivasan K, Tayagi R (2013) Comparison of various seed priming method for seed germination, seedling vigour and fruit yield in okra (*Abelmoschus esculentus* L. Moench). Sci Hortic 165:75–81
Sheteiwy MS, Fu Y, Hu Q, Nawaz A, Guan Y, Li Z, Huang Y, Hu J (2016) Seed priming with polyethylene glycol induces antioxidative defense and metabolic regulation of rice under nano-ZnO stress. Environ Sci Pollut Res 23(19):19989–20002
Shinogi Y, Yoshida H, Koizumi T, Yamaoka M, Saito T (2003) Basic characteristics of low-temperature carbon products from waste sludge. Adv Environ Res 7(3):661–665
Singh RP, Agrawal M (2007) Effects of sewage sludge amendment on heavy metal accumulation and consequent responses of Beta vulgaris plants. Chemosphere 67(11):2229–2240
Singh VP, Kumar J, Singh M, Singh S, Prasad SM, Dwivedi R, Singh MPVVB (2016) Role of salicylic acid-seed priming in the regulation of chromium (VI) and UV-B toxicity in maize seedlings. Plant Growth Regul 78(1):79–91. https://doi.org/10.1007/s10725-015-0076-4
Singh A, Singh NB, Afzal S, Singh T, Hussain I (2018) Zinc oxide nanoparticles: a review of their biological synthesis, antimicrobial activity, uptake, translocation and biotransformation in plants. J Mater Sci 53(1):185–201
Skeffington RA, Shewry PR, Peterson PJ (1976) Chromium uptake and transport in barley seedlings (Hordeum vulgare L.). Planta 132(3):209–214
Sneideris LC, Gavassi MA, Campos ML, D'Amico-Damiao V, Carvalho RF (2015) Effects of hormonal priming on seed germination of pigeon pea under cadmium stress. An Acad Bras Cienc 87(3):1847–1852
Soares C, de Sousa A, Pinto A, Azenha M, Teixeira J, Azevedo RA, Fidalgo F (2016) Effect of 24-epibrassinolide on ROS content, antioxidant system, lipid peroxidation and Ni uptake in Solanum nigrum L. under Ni stress. Environ Exp Bot 122:115–125
Srivastava S, Jain R (2011) In-situ monitoring of chromium cytotoxicity in sugarcane. J Environ Biol 32(6):759–763
Stadtman ER, Oliver CN (1991) Metal-catalyzed oxidation of proteins. Physiological consequences. J Biol Chem 266:2005–2008
Stambulska UY, Bayliak MM, Lushchak VI (2018, 2018) Chromium (VI) toxicity in legume plants: modulation effects of rhizobial symbiosis. Biomed Res Int
Stevens B, Betts A, Miller B, Scheckel K, Anderson R, Bradham K, Basta N (2018) Arsenic speciation of contaminated soils/solid wastes and relative oral bioavailability in swine and mice. Soil Syst 2(2):27
Swaminathan R (2003) Magnesium metabolism and its disorders. The Clinical Biochemist Reviews 24(2):47
Tabassum T, Farooq M, Ahmad R, Zohaib A, Wahid A (2017) Seed priming and transgenerational drought memory improves tolerance against salt stress in bread wheat. Plant Physiol Biochem 118:362–369
Tahervand S, Jalali M (2017) Sorption and desorption of potentially toxic metals (Cd, Cu, Ni and Zn) by soil amended with bentonite, calcite and zeolite as a function of pH. J Geochem Explor 181:148–159
Thakur S, Singh L, Wahid ABZ, Siddiqui MF, Atnaw SM, Din MFM (2016) Plant-driven removal of heavy metals from soil: uptake, translocation, tolerance mechanism, challenges, and future perspectives. Environ Monit Assess 188(4):206
Thomas F, Malick C, Endreszl EC, Davies KS (1998) Distinct responses to copper stress in the halophyte, Mesembryanthemum crystallinum. Physiol Plant 102:360–368
Ton J, Ent S, Hulten MHA, Pozo M, Oosten V, Loon LC, Mauch-Mani B, Turlings TCJ, Pieterse CMJ (2009) Priming as a mechanism behind induced resistance against pathogens, insects and abiotic stress. Int Org Biol Control 44:3–13
Tonelli ML, Furlan A, Taurian T, Castro S, Fabra A (2011) Peanut priming induced by biocontrol agents. Physiol Mol Plant 75:100–105
Toor GS, Haggard BE, Donoghue AM (2007) Water extractable trace elements in poultry litters and granulated products. J Appl Poult Res 16:351–360

Torres GN, Camargos SL, OLDS W, KDB M, Scaramuzza WL, Pereira M (2016) Growth and micronutrient concentration in maize plants under nickel and lime applications. Revista Caatinga 29(4):796–804

Trinh NN, Huang TL, Chi WC, Fu SF, Chen CC, Huang HJ (2014) Chromium stress response effect on signal transduction and expression of signaling genes in rice. Physiol Plant 150(2):205–224

Urbancsok J, Bones AM, Kissen R (2018) Arabidopsis mutants impaired in glutathione biosynthesis exhibit higher sensitivity towards the glucosinolate hydrolysis product allyl-isothiocyanate. Sci Rep 8(1):9809

Vazques MD, Poschenrieder C, Barcelo J (1987) Chromium (VI) induced structural changes in bush bean plants. Ann Bot 59:427–438

Verslues PE, Agarwal M, Katiyar Agarwal S, Zhu J, Zhu JK (2006) Methods and concepts in quantifying resistance to drought, salt and freezing, abiotic stresses that affect plant water status. Plant J 45(4):523–539

Vickers CE, Gershenzon J, Lerdau MT, Loreto F (2009) A unified mechanism of action for volatile isoprenoids in plant abiotic stress. Nat Chem Biol 5(5):283

Violante A, Cozzolino V, Perelomov L, Caporale AG, Pigna M (2010) Mobility and bioavailability of heavy metals and metalloids in soil environments. J Soil Sci Plant Nutr 10(3):268–292

Walter I, Martinez F, Cala V (2006) Heavy metal speciation and phytotoxic effects of three representative sewage sludges for agricultural uses. Environ Pollut 139(3):507–514

Wang F, Zhao L, Shen Y, Meng H, Xiang X, Cheng H, Luo Y (2013) Analysis of heavy metal contents and source tracing in organic fertilizer from livestock manure in North China. Trans Chin Soc Agric Eng 29(19):202–208

Wang X, Wei D, Ma Y, McLaughlin MJ (2018) Soil ecological criteria for nickel as a function of soil properties. Environ Sci Pollut Res 25(3):2137–2146

Warne MS, Heemsbergen D, Stevens D, McLaughlin M, Cozens G, Whatmuff M, Broos K, Barry G, Bell M, Nash D, Pritchard D, Penney N (2008) Modeling the toxicity of copper and zinc salts to wheat in 14 soils. Environ Toxicol Chem 27:786–792

Wu C, Zou Q, Xue SG, Pan WS, Yue X, Hartley W, Huang L, Mo JY (2016) Effect of silicate on arsenic fractionation in soils and its accumulation in rice plants. Chemosphere 165:478–486

Xiong L, Zhu JK (2001) Abiotic stress signal transduction in plants: molecular and genetic perspectives. Physiol Plant 112(2):152–166

Xiong L, Zhu JK (2002) Molecular and genetic aspects of plant responses to osmotic stress. Plant Cell Environ 25:131–139

Yari L, Aghaalikani M, Khazaei F (2010) Effect of seed priming duration and temperature on seed germination behavior of bread wheat (Triticum aestivum L.). ARPN J Agric Biol Sci 5(1):1–6

Younis U, Qayyum MF, Shah MHR, Danish S, Shahzad AN, Malik SA, Mahmood S (2015) Growth, survival, and heavy metal (Cd and Ni) uptake of spinach (Spinacia oleracea) and fenugreek (Trigonella corniculata) in a biochar-amended sewage-irrigated contaminated soil. J Plant Nutr Soil Sci 178(2):209–217

Yruela I (2005) Copper in plants: acquisition transport and interactions. Funct Plant Boil 36:409–430

Zhang WH, Tyerman SD (1999) Inhibition of water channels by HgCl2 in intact wheat root cells. Plant Physiol 120:849–857

Zhao TJ, Liu Y, Yan YB, Feng F, Liu WQ, Zhou HM (2007) Identification of the amino acids crucial for the activities of drought responsive element binding factors (DREBs) of *Brassica napus*. FEBS Lett 581(16):3044–3050

Zhao Z, Xi M, Jiang G, Liu X, Baia Z, Huang Y (2010) Effects of IDSA, EDDS and EDTA on heavy metals accumulation in hydroponically grown maize (Zea mays L.). J Hazard Mater 181:455–459

Zhou ZS, Huang SQ, Guo K, Mehta SK, Zhang PC, Yang ZM (2007) Metabolic adaptations to mercury-induced oxidative stress in roots of Medicago sativa L. J Inorg Biochem 101:1–9

# Advances in Heavy Metal-Induced Stress Alleviation with Respect to Exogenous Amendments in Crop Plants

**Bedabrata Saha, Bhaben Chowardhara, Saradia Kar, Sanjenbam Sanjibia Devi, Jay Prakash Awasthi, Debojyoti Moulick, Bhaben Tanti, and Sanjib Kumar Panda**

## Contents



**Abstract** Heavy metal contamination of soil due to rapid industrialization and urbanization is an exponentially increasing menace, rendering hectares of arable land barren. Adding to the problem is the global climate change which sums up to the challenge of providing food security and meeting the global sustainability goals. Heavy metals like Cd, Pb, Hg, and As (non essential) and Fe, Mn, Zn, Cr, Cu, and Ni (essential) when present in abnormal amounts in the environment pose a serious threat to growth and yield of crop plants. Various stress factors including heavy metals result in exaggerated synthesis of reactive oxygen species (ROS) causing immense metabolic imbalance in plants. Plants possess both enzymatic and non-enzymatic processes to reduce the load of oxidative stress, but many times, it doesn't suffice. The scope of this chapter centers around the recent strives made in the field of heavy metal-induced heavy metal stress mitigation in crop plants. This chapter specially focuses on exogenous amendments (organic, inorganic, and microbial) to induce stress adaptation in crop plants.

B. Saha · B. Chowardhara · S. Kar · S. S. Devi · J. P. Awasthi
D. Moulick (✉) · S. K. Panda (✉)
Plant Molecular Biotechnology Laboratory, Department of Life Science and Bioinformatics, Assam University, Silchar, Assam, India

B. Tanti
Department of Botany, Gauhati University, Guwahati, Assam, India

 
M. Hasanuzzaman, V. Fotopoulos (eds.), *Priming and Pretreatment of Seeds and Seedlings*, https://doi.org/10.1007/978-981-13-8625-1_15

**Keywords** Toxic metal · Abiotic stress · Cross tolerance · Phytohormones · Biostimulant

## 1 Introduction: Heavy Metals and Oxidative Stress

Any element bearing density above 5.0 $gcm^{-3}$ is conceived as heavy metals. Though around 60% of the elements present in nature are considered as heavy metals, not all of them are important biologically (Saxena and Shekhawat 2013). Heavy metals like Pb (lead), Sb (antimony), Cd (cadmium), Hg (mercury), U (uranium), As (arsenic), and Ag (gold) are not of any functional importance to plants and are toxic, whereas, Fe (iron), Zn (zinc), Cr (chromium), V (vanadium), Co (cobalt), Cu (copper), and Ni (nickel), though are required by plants as micronutrients, or trace elements become toxic when in excess. In many parts of the world, soil is acidic ($pH < 5$); this leads to accumulation of noxious forms ($Al^{3+}$, $Fe^{2+}$, $Mn^{2+}$, etc.) of Earth's crust abundant of heavy metals (Awasthi et al. 2017; Kar and Panda 2018). Anthropogenic activities, like industrialization, urbanization, rampant use of fertilizers, etc., are the main source of heavy metal contamination of the environment leading to phytotoxicity. It was found that some heavy metals can interfere with native soil microbial population as well as interacting or influencing other closely associated floras including crops (Pena-Fernandez et al. 2014; Mandal et al. 2014). Iron toxicity especially directly brings up the ROS levels in the plants through its involvement in Fenton reaction. Iron, in the form of ferrous ion, catalyzes the reaction of the biochemical pathway by forming free radicals like $OH^•$. Also, $O_2$ produced as by-product proceeds further in the formation of more ROS molecules.

$$Fe^{3+} + O_2 \rightarrow Fe^{2+} + O_2$$

$$Fe^{2+} + H_2O_2 \rightarrow Fe^{3+} + OH^• + OH^•$$

$$O_2^{•-} + H_2O_2 \rightarrow O_2 + OH^• + OH^•$$

Plants respond to heavy metal stress showing two types of responses: accumulation and exclusion. Accumulation includes active uptake of metals and subsequent detoxification (through formation of complexes) in aboveground tissues (especially dumped in vacuoles), whereas exclusion involves maintenance of homeostatic level of metal in tissues through exudation, irrespective of heavy metal load in the rhizosphere. Plants also exudate organic acids into the rhizosphere to form complexes with heavy metals so that the latter is not taken up by root cells (Li et al. 2017). Besides the direct responses of heavy metals, it also induces an oxidative burst, disturbing the ROS homeostasis and thus causing oxidative stress. ROS molecules, viz., hydrogen peroxide ($H_2O_2$), superoxide ($O_2^{•-}$), hydroxyl radical ($OH^•$), and singlet oxygen ($^1O_2$), react with cellular biomolecules and cause immense damage to housekeeping and cellular metabolic activities (Singh et al. 2017). Activation of antioxidant defense system under Al stress in many species has been well documented like in tea (Mukhopadyay et al. 2012; Morita et al. 2011), jatropha plant (Ou-Yang et al., 2014), gmelina plant (Dudhane et al. 2012), and rice plant (Ribeiro et al. 2012). Disrupted cellular metabolic activity leads to emergences of visible physiological symptoms like decreased root length, biomass, seed germination, and pigment synthesis in plants.

To evade from this oxidative stress, plants have evolved an intricate cascade of antioxidative response system, both enzymatic and nonenzymatic. These antioxidative activities bring ameliorating effect during heavy metal stress on plants. Some of the important ROS scavenging enzymes under excess iron stress are superoxide dismutase (SOD), ascorbate peroxidase (APX), catalase (CAT), glutathione peroxidase (GPX), monodehydroascorbate reductase (MDHAR), dehydroascorbate reductase (DHAR), and glutathione reductase (GR) (You and Chan 2015). Superoxide dismutases are metaloenzymes that catalyze the dismutation of oxygen radicals. The enzyme is responsible for bringing plant stress tolerance. This first line of defense works for detoxifying ROS under various compartments of plant cells. Cu, Zn, and Fe are the main metals, which form the active prosthetic group for SODs. Under heavy metal stress, high activity of these SODs has been reported. Overexpressing Cu/Zn SOD and Fe-SOD in model plants like tobacco and *Arabidopsis thaliana* results in multiple stress tolerance. A recent report by Verma and Pandey (2016) showed increased activity of SOD with high iron supply in wheat. Similar responses were observed in rice and tobacco under high iron treatment (Stein et al. 2009). Catalase on the other hand directly breaks down $H_2O_2$ into $H_2O$ and $O_2$. With increasing concentration of iron in the medium, the CAT activity was also found to be elevated in crop plants (Li et al. 2011).

The main $H_2O_2$ detoxifying system is the ascorbate-glutathione (GSH) cycle. Ascorbic acid, an antioxidant for the plants, acts as an electron donor to $H_2O_2$ to convert it into water and molecular oxygen. This reaction is catalyzed by APX. The APX activity generally increases along with CAT and SOD. However, their response under heavy metal stress has received mixed reports. In case of excess iron in beans, the generation of APX was induced. Similarly, suppressed APX level in tobacco made the plants more susceptible to excess iron. Glutathione peroxidase on the other hand utilizes reduced glutathione to reduce $H_2O_2$ and delivers similar results as APX. It was reported that silencing of APX genes induced upregulation of other peroxidases. In rice, GPX genes that were found with induced expression of APX genes were silenced (Caverzan et al. 2012). Glutathione reductase, on the other hand, is involved in maintaining the GSH pool of the plants under oxidative stress. GSH is a nonenzymatic indicator under various environmental stresses. Saini et al. (2017) have reported increasing GR activity in rice with gradual increase in iron supply. Some of the plants have even developed high level of heavy metal tolerance to specific metals through evolution and selection from nearby non-metalliferous populations. These unique plants came to be known as metallophytes, viz., *Silene vulgaris*, *Silene dioica*, etc.

Adding to the menace of heavy metal contamination is the global exponential rise in human population, which calls for immediate action to secure the feed for the new mouths. Attenuation of heavy metal stress will go a long way in this regard. Keeping this in mind, a lot of work has been executed to ameliorate heavy metal toxicity especially heavy metal-induced oxidative stress. This review caters to the need of summarizing the works on the topic.

## 2 Methods for Alleviation of Oxidative Stress

Diverse methodologies had been adopted to reduce the exacerbation of oxidative stress induced by different stressors, ranging from conventional breeding to plant genetic modification and also including beforehand priming/hardening of plants (Savvides et al. 2016). Priming with the use of exogenous elements is a rapidly emerging field in crop stress physiology as genetic amendments are currently not acceptable in many countries of the world (Antoniou et al. 2016). Then also endogenous amendments (change in genetic makeup) form an important part in oxidative stress alleviation as it aims to provide for permanent solutions, which are genetically heritable. Here the whole range of amendments that have been used to attenuate heavy metal-induced oxidative stress will be discussed.

Exogenous amendments, have got important role, in sensitization of plants through application of chemicals and biological agents for intense and rapid mobilization of defense responses against stressors. A number of exogenous amendment strategies are employed by researchers to attenuate oxidative stress cropping out of environmental heavy metal pollution. Generally, plants employ a range of mechanisms to avoid or tolerate the heavy metal stress, but many times, when stress is severe and prolonged, these defense responses are overwhelmed. In such cases some amount of external exposure to certain signaling or functional molecules helps stimulate the plants to give a more organized response and hence help survive (Mostofa and Fujita 2013). These stress amending molecules can basically be characterized into three subgroups.

### *2.1 Organic Amendments*

A lot of organic chemicals had been used to reduce the anthropogenic activities that derived heavy metal-induced oxidative stress. Most of these are of biological origin (salicylic acid, brassinosteroids, jasmonates, glycinebetaine, glutathione, etc.), whereas few are not (EDTA, sodium nitroprusside, etc.). As most of these are of biological origin, they function as signaling molecules/stimulants or functional molecules to enhance the plant stress response. Salicylic acid, a phenolic compound, acts by boosting the systemic signaling response to heavy metal stress in plants. Besides this, as an endogenous hormone, it regulates various physiological and developmental procedures (Mostofa and Fujita 2013; Panda and Patra 2007). Brassinosteroids, similar to salicylic acid, are phytohormones responsible for the increase of antioxidant potential, photosynthesis, and detoxification in response to heavy metal stress (Ahammed et al. 2012). It has been reported to attenuate oxidative stress induced by heavy metals stress in chickpea (Hasan et al. 2008), tomato (Hasan et al. 2011), and rice (Sharma et al. 2016). Organic acids (citrate, malate, oxalate, etc.) have long been used as part of their defense response to chelate heavy metals, even when they are applied exogenously (Kambhampati 2013; Cobbett 2000; Blaylock et al. 1997). Not only they chelate heavy metals but also boost up the oxidative defense system in *Solanum nigrum* (Gao et al. 2010), *Boehmeria nivea* (Li et al. 2014), and *Brassica juncea* (Mahmud et al. 2018). EDTA has also been successfully tried in crop plants (Shahid et al. 2014; Habiba et al. 2015). Exogenous application of osmolytes like glycinebetaine, mannitol, and proline has been reported to attenuate oxidative stress caused by heavy metals like Cr, Cd, and Hg in wheat (Ali et al. 2015; Adrees et al. 2015), cotton (Farooq et al. 2016), and rice (Wang et al. 2009), respectively.

Polyamines are another group of ameliorating agents responsible for detoxification of heavy metals. They are low aliphatic amines that are omnipresent in all living entities. Putrescine, spermidine, and spermine are the commonly used polyamines in plants, which are reported to alleviate Cd stress in *Potamogeton crispus* (Yang et al. 2010), *Triticum aestivum* (Benavides et al. 2018; Tajti et al. 2018; Rady and Hemida 2015), *Helianthus annuus* (Benavides et al. 2018), *Oryza sativa* (Pal et al. 2017), *Vigna radiata* (Nahar et al. 2016), etc.

Biochar also attenuates heavy metal toxicity when applied exogenously. It is a porous, carbonaceous product obtained from pyrolysis of organic matter (Dennehy et al. 2017; Feng and Zhu, 2017; Lin et al. 2017). Biochar has high cation exchange capability and high expanse and alkaline in nature. An application of biochar in soils is that it adsorbs/precipitates toxic compounds and enhances the water holding capacity of soils. Many plants were studied to alleviate heavy metal toxicity using biochar like *Oryza sativa* (Zheng et al. 2012; Rizwan et al. 2018; Li et al. 2018a, b), *Brassica chinensis* L. (Bashir et al. 2018a, b), *Triticum aestivum* L (Rehman et al. 2018), *Ipomoea aquatica* (Bashir et al. 2018a, b), *Zea mays* L. (Namgay et al. 2010), *Brassica napus* L. (Houben et al. 2013), etc.

Plant hormones (auxins, gibberellins, cytokinins, abscisic acid, ethylene, jasmonic acid, nitric oxide, brassinosteroids) are major plant growth regulators, which besides playing significant role for plant development also help in stress amendments when applied exogenously (Davies 2010; Asgher et al. 2014; Masood et al. 2012; Piotrowska et al. 2009; Khan et al. 2015) and hence are extensively studied in different plants under heavy metal stress, like auxin in *Arabidopsis thaliana* (Vitti et al. 2013), gibberellic acid in *Parthenium hysterophorus* (Hadi et al. 2014), cytokinins in *Solanum melongena* (Singh and Prasad 2014), ethylene in *Solanum lycopersicum* and *Brassica juncea* (Masood et al. 2012; Asgher et al. 2014), salicylic acid in *Phaseolus aureus* and *Vicia sativa* (Zhang et al. 2011a, b), *Glycine max* (Noriega et al. 2012), jasmonic acid in *Arabidopsis thaliana* and *Glycine max* (Keramat et al.

2009, Piotrowska et al. 2009), and brassinosteroids in *Solanum lycopersicum* (Ahammed et al. 2013).

Plant growth hormones directly or indirectly takes part in intricate pathways resolving the iron stress for plants. Auxin is one of the growth hormones involved in such pathways, which signals for the expression of genes involved in iron homeostasis. It induces genes like *FIT* and *FROs*. FRO is responsible for reduction of $Fe^{3+}$ in the rhizosphere. Exogenous application of auxin stimulated expression of these particular genes (Hindt and Guerinot, 2012). Hence, under Fe-deficient condition, application of auxin can regulate Fe uptake normally. Nitric oxide (NO) is another signaling molecule, which can be implicated in iron uptake. Its function is similar to auxin. Its application in tomato and maize induced expression of genes like FRO, IRT1, and FER (Graziano and Lamattina 2005). These genes are responsible for reducing and transporting iron within the plants. Cytokinin, on the other hand, has been reported as a negative regulator for Fe uptake. Exogenous application of cytokinin resulted into repression of IRT1, FRO2, and FIT genes (Hindt and Guerinot, 2012). Therefore, application of cytokinin can result as another regulating factor for plants against iron toxicity. 24-Epibrassinolide (EBL) is included in brassinosteroids (BRs) which is an endogenous plant growth regulator (PGR). Reports also suggest the use of epibrassinolide (EBL) as exogenous heavy metal stress retardant. EBL boosts antioxidant and non-antioxidant enzymatic system and also influences osmoregulation. EBL has been used in *Raphanus sativus* (Ramakrishna and Rao 2012), *Vigna radiata* (Mir et al. 2015), *Solanum melongena* (Wu et al. 2016), *Solanum lycopersicum* cv. Hezuo903 (Li et al. 2016a, b), and *Brassica juncea* (Kohli et al. 2018).

Abscisic acid (ABA) is a phytohormone necessitated in seed germination; bud dormancy, regulation of organ size; and stomatal closure. It also regulates the plant behavioral responses under biotic and abiotic conditions (Wang et al. 2013; Cao et al. 2014). ABA helps to activate antioxidant defense system that alleviates the toxic effects of heavy metals in plants (Wang et al. 2014). Under stress condition, the levels of ABA increase in plants that stimulates signaling cascades downstream of phytohormones such as salicylic acid (SA) and auxin (IAA), which may also alleviate heavy metal toxicity (Shi et al. 2015). Exogenous application of ABA has been reported to dilute transpiration, contributing to reduced zinc translocation from soil to the aerial part of plants (Disante et al. 2014; Shi et al. 2015).

Salicylic acid, a natural signaling molecule, plays a significant part in determining numerous physiological activities and plant resistance to biotic and abiotic stresses (Wang et al. 2013). It also ameliorates the noxious effects of heavy metals by modulating the antioxidant systems, intake of heavy metals, and maintenance of membrane integrity when exogenously applied. The exogenous application of salicylic acid has been reported in *Brassica juncea* (Kohli et al. 2018), *Phaseolus vulgaris* (Saidi et al. 2013), *Lolium perenne* (Wang et al. 2013), *Glycine max* (Noriega et al. 2012), *Triticum aestivum* (Shakirova et al. 2016), *Linum usitatissimum* (Belkadhi et al. 2014), *Zea mays* (Szalai et al. 2013; Zanganeh et al. 2018), and *Oryza sativa* (Pandey et al. 2013a, b). Salicylic acid has been found to be better ameliorator of Al toxicity when compared to Mg and Ca (Pandey et al. 2013a, b).

5-Aminolevulinic acid (ALA) reported to alleviate abiotic stress like salinity and heavy metal stress. It ameliorates the toxicity effect of Cd in plants such as *Brassica napus* (Ali et al. 2013).

Hemin is also known as ferroprotoporphyrin IX, which is a compound derived from heme and is a powerful inducer of heme oxygenase 1 (HO-1) by upregulating its mRNA abundance and protein level. These responses eventually assure hemin to exert protective effect on stress tolerance (Chen et al. 2017, 2018b).

## 2.2 Inorganic Amendments

Silicon (Si) forms the backbone of inorganic exogenous amendment for heavy metal-induced oxidative stress. It is also a beneficial element with regard to growth, development, yield, and immunity against disease in diverse plant species (Ma et al. 2015; Dorneles et al. 2016). And so, Si has been studied widely, from heavy metal accumulator plants to non-accumulator plants leading to identification of numerous strategies, through which Si mitigates the heavy metal-derived toxicity in plants – lesser uptake of heavy metals, boosting the antioxidant machinery, less damage to photosynthetic apparatus, and chelation of heavy metals (Liu et al. 2013). In case of Al, it forms complexes called aluminosilicates in the cell wall and apoplasm and thereby providing substantial obstacle to penetrate inside, thus protecting the cell from damage (Prabagar et al. 2011). Addition of Si to the medium with Fe-treated rice varieties showed reduction of Fe plaque by 37%–40% (You-Qiang et al. 2012). Exogenous Si helps alleviate heavy metal toxicity with increase in phenolic compounds and antioxidants (Shahnaz et al. 2011; Fleck et al. 2015). Findings of Detmann et al. (2012) disclosed that supplementation of Si ensues in significant modulation of primary metabolism and promotes amino acid remobilization in rice. There is evidence that suggests that in the presence of Si, a noteworthy reduction in arsenic (As) uptake and translocation is observed in rice (Ma and Yamaji 2006; Raab et al. 2007; Bogdan and Schenk 2008). Findings of Matsumoto et al. (2015) and Fleck et al. (2013) indicate that the application of silica gel at 10.0 g $Kg^{-1}$ as soil supplement can effectively reduce As content in aboveground biomass. Silicon as an amendment to metal stress has also been reported in *Gossypium hirsutum* L. (Anwaar et al. 2015), *Zea mays* (de Sousa Paula et al. 2015), *Arundinaria pygmaea* (Emamverdian et al. 2018), *Oryza sativa* L. (Gu et al. 2012; Song et al. 2014), *M. x paradisiaca* (Li et al. 2012), and *Triticum aestivum* (Tripathi et al. 2016).

Similar to Si, selenium (Se) attenuates heavy metal toxicity using diverse procedures which include compartmentalization of heavy metals into cell walls and vacuoles, increasing overall antioxidant activity, restriction of heavy metal transportation in plants, and homogenous distribution of metals in the leaf (Pereira et al. 2015; Zaheer et al. 2018; Jan et al. 2018; Yingang et al. 2018; Wu et al. 2018). In the past few years, in vitro studies were carried out to investigate the consequences of priming rice seeds with Se on germination and seedling growth and cultivation of primed rice seeds in As-contaminated environment, as an alternative mitigation option.

Findings of Moulick et al. (2016, 2017) suggest that priming rice seeds with Se before sowing can promote germination and seedling growth in both soil-based and soilless conditions by facilitating modulation in As-induced redox imbalance and minimizing its uptake. Cultivation of Se primed rice seedlings in arsenic-contaminated pot soil indicated significantly reduced arsenic translocation into the aboveground portion (Moulick et al. 2018a, b). The supplemented Se also promoted intake and translocation of other essential minerals such as Fe, Mn, Zn, etc. (Moulick et al. 2018c).

Sulfur (S) is a key ameliorating agent for reducing the effect of heavy metals in plants. It is one of the essential micronutrients that regulate photosynthesis under normal and stress condition (Nazar et al. 2015). Sulfur forms the basic structural component for certain amino acids (cysteine and methionine), antioxidants (reduced and oxidized GSH), coenzymes, vitamins, secondary metabolites, phytochelatins (PCs), and lipids (Khan et al. 2014). Adequate S availability in *Hordeum vulgare* (Astolfi et al. 2011), *Oryza sativa* (Fan et al. 2010), and *Brassica juncea* (Asgher et al. 2014) led to detoxification of Cd-stimulated oxidative stress through increased production of GSH. Sulfur plays key role in minimizing As accumulation and subsequent translocation in plants (Muñoz-Bertomeu et al. 2009; Zhang et al. 2011a, b; Dixit et al. 2015). Authors like Finnegan and Chen (2012) found that adequate S content (supplemented) becomes essential for plants to survive in As-contaminated environment. Song et al. (2010) observed that upon exposure to heavy metals, plants start producing ligands and PCs generally facilitated by adequate S supply. The sulfhydryl of GSH and PCs facilitates the conversion of arsenate to arsenite and finally modulates the localization into the vacuoles. Supplementation with S has been found to promote the As concentration in roots by stimulating thiolic ligand synthesis and subsequent increase in As complexation in roots, thus minimizing As translocation from roots to shoots (Duan et al. 2011; Dixit et al. 2016).

Calcium is essential macromolecule for plant growth and development. It has role to play as secondary messenger in intracellular signal transduction. Plasma membrane is the first target of heavy metal toxicity in plant cells, which in turn serves as stress perception leading to alteration of Ca signal transduction. Exogenous application of Ca ameliorates heavy metal toxicity by enhancing stability of plasma membrane and mending Ca signaling (Farzadfar et al. 2013). Calcium is known to ameliorate heavy metal stress in *Sesamum indicum* (Abd-Allah et al. 2017), *Matricaria chamomilla* (Farzadfar et al. 2013), *Oryza sativa* (Srivastava et al. 2015), *Vicia faba* (Siddiqui et al. 2012), *Lens culinaris* (Talukdar, 2012), *Cicer arietinum* (Ahmad et al. 2016), and *Arabidopsis thaliana* (Li et al. 2016a, b).

To combat Al toxicity in plants, there is frequent use of Ca amendments like lime, gypsum, or phosphor-gypsum, and the effect of Ca that is ameliorating the Al toxicity has been reported by many authors in various crops especially in acidic soil. Meriño-Gergichevich et al. (2010) reviewed the interaction between the $Al^{3+}$ and $Ca^{2+}$ in acidic soil. Calcium-based amendments lead to formation of less toxic compounds and aluminum hydroxyl sulfate or aluminum sulfate complexes by replacement of $Ca^{2+}$ with $Al^{3+}$. Gypsum amendment has been found to be helpful in diluting Al toxicity in both tolerant and medium cultivars of blueberries. This is an alternative way of liming to ease the long-term Al toxicity in Al-sensitive cultivars besides that this provides a plenty source of nutrients like Ca and S (Reyes-Diaz et al. 2011). Aluminum and Fe show antagonistic effect on each other. At excess Al in *Camellia sinensis*, iron (Fe) shows an ameliorating effect on its stressful effects. In return, exposure to Al under excess Fe treatment reduced its toxic effect on lateral roots and young leaves. Citrate is a common chelating agent for both Al and Fe for their uptake in plants. Therefore, during the metal uptake, competition occurs between Al and Fe (Hajiboland et al. 2013). When exogenously applied, Fe facilitates the accumulation of Fe oxide or promotes the formation of Fe plaque in rice roots and thus reduces As uptake as well as increase the coprecipitation of Fe and As. Due to the formation of a barrier, i.e., Fe plaque, which consists of mostly siderite (5%), goethite (32%), and ferrihydrite (63%), the availability of As gets reduced on the root surface; as a result, the subsequent translocation also decreased (Liu et al. 2004a, b; Nath et al. 2014).

Pretreatment with P ameliorates Al toxicity in rice (Nakagawa et al. 2003), and it is due to the formation of aluminum phosphates in roots. Lu et al. (2010) carried out a vivid field study regarding the status of P in rice shoots across the rice fields of China and found that P content can effectively reduce As content in grain. Neupane and Donahoe (2013) found that co-application of Ca along with P can stimulate the formation of Ca-P complex that can effectively decrease As content in rice grain. According to the opinion of Lee et al. (2016), the antagonistic behavior existing among P and As that finally restricts As mobility and uptake by rice plant, is due to structural and behavioral similarity, which leads to competition among P and As for absorption, which reduces the translocation of As from the root to the shoot.

Boron (B) is an essential micronutrient for plants; less B causes deficiency, while excess causes toxicity. There is a very thin line of B concentration between deficiency and toxicity. When in appropriate amount, B also helps in alleviation of heavy metal toxicity. It acts by encouraging the root alkalization in transition zone via polar auxin transport as reported by Li et al. (2018a, b). Boron boosts up the transport of polar auxin by a transporter gene called auxin efflux transporter PIN2. This contributes to the downregulation of plasma membrane $H^+$-ATPase and leads to increase in pH which results in decrease of Al accumulation in the targeted root zone. The study elucidated an important mechanism for the function of exogenous boron in attenuation of Al deposition and toxicity in plants.

Nitric oxide (NO) is an ameliorating agent which is widely used in plants under heavy metals toxicity in many. Generally, nitric oxide (NO) is a gaseous free radical, and it governs diverse physiological and biochemical activities in plants under stressful and non-stressful environment. Nitric oxide is well-known biological messenger in plant tissues, which governs many growth and development processes. Several reports have found that exogenous application of NO in the form of sodium nitroprusside (SNP) increases plant tolerance against oxidative stress elicited by heavy metals particularly Hg and Cd (Chen et al. 2015; Liu et al. 2015). NO as a stress retardant has been elucidated in *Oryza sativa* (Chen et al. 2015), *Triticum aestivum* (Tripathi et al. 2017), and *Carthamus tinctorius* (Namdjoyan et al. 2017).

Hydrogen sulfide ($H_2S$) is also known to alleviate the heavy metal toxicity in plants from Cu, Al, Cd, and Zn (Chen et al. 2013; Liu et al. 2016; Mostofa et al. 2015; Zhang et al. 2008). $H_2S$ helps to reduce MDA, $H_2O_2$, and electron leakage along with promotion of plant growth chlorophyll content, net photosynthetic rate, and ultrastructural changes. $H_2S$ has been reported to attenuate metal stress through exogenous application in *Brassica oleracea* (Chen et al. 2018a, b), *Brassica napus* (Ali et al. 2014a; Ali et al. 2014b; Ali et al. 2014c), *Sesamum indicum* (Amooaghaie and Enteshari 2017), and *Zea mays* (Zanganeh et al. 2018).

## 2.3 Microbial Amendments

Soil microorganisms are one of the essential parts of soil ecosystem and act by transforming and degrading a wide range of pollutants, ultimately maintaining the soil productivity and ecological functions. Besides the inanimate amendments as discussed above, soil microorganisms have also got important role to play in alleviation of heavy metal toxicity in plants.

Arbuscular mycorrhizal fungi (AMF) are widely available soil microorganisms that can establish mutual symbiosis with almost 80% of terrestrial plant families. AMF can be tolerant under harsh conditions including legacy stressor. AMF also have important part to play in metal tolerance, through sequestration, by mediating interaction between metals and plant roots. Generally, AMF develop glomalin, an alkaline-soluble glycoprotein extracted from the soil, which interacts with heavy metals to form metal-chelated compound (Saini et al. 2017). AMF have been reported to ameliorate zinc toxicity in *Trigonella foenum-graecum* (Saini et al. 2017), *Cajanus cajan* L. Mill. (Garg and Singh 2018), *Sorghum bicolor* L. (Wang et al. 2018a, b), *Solanum lycopersicum* L. cv. Micro-Tom (Ibiang et al. 2018), and *Zea mays* (Wang et al. 2016).

Under aerobic condition, there is large amount of insoluble form of $Fe^{3+}$, which is not a suitable form for uptake. In a situation like this, most of the bacteria are responsible for the production of phytosiderophore, which act as the chelators. The siderophores are the small molecular iron-binding protein (Vegan et al. 2016). In both dicotyledonous (e.g., sunflower) and monocotyledonous (e.g., rice), the presence of such microorganisms has been reported. Few of the bacterial strains reported for the same are *Burkholderia*, *Enterobacter*, *Klebsiella*, *Citrobacter*, etc. (Souza et al. 2015). Therefore, these plant growth-promoting bacteria contribute in nutrition sustenance of crops. Under low pH, it has also been reported that coprecipitation of trace metals occur with the hydrous oxides of Fe and Mn (Charlatchka and Cambier, 2000). Based on these results, some of the Fe (III) reducing bacteria have been reported, which can tolerate other heavy metal contaminations in the soil. This phenomenon can easily be utilized for the plants, which faces trace metal (Cd, Cu, Ni, and Zn) toxicity.

Soil microorganisms (*Bacillus* sp., *Rhodococcus* sp., *Halobacterium* sp.) detoxify arsenic species through sorption at their extracellular surface, which have uronic acids, proteins, and amino sugars with high hydrogen-bonding potential (Huang 2014; Williams et al. 2013; Yang et al. 2012). Whereas the arbuscular mycorrhizal fungi protect the crop plants from As-induced adverse effects by downregulating the mRNA expression of OsLsi1 and OsLsi2, as a result, translocation of arsenite also gets decreased in the aerial part (Chen et al. 2012).

## References

Abd_Allah EF, Hashem A, Alqarawi AA, Wirth S, Egamberdieva D (2017) Calcium application enhances growth and alleviates the damaging effects induced by Cd stress in sesame (*Sesamum indicum* L.). J Plant Interact 12(1):237–243

Adrees M, Ali S, Iqbal M, Bharwana SA, Siddiqi Z, Farid M, Rizwan M (2015) Mannitol alleviates chromium toxicity in wheat plants in relation to growth, yield, stimulation of anti-oxidative enzymes, oxidative stress and Cr uptake in sand and soil media. Ecotoxicol Environ Saf 122:1–8

Ahammed GJ, Choudhary SP, Chen S, Xia X, Shi K, Zhou Y, Yu J (2012) Role of brassinosteroids in alleviation of phenanthrene–cadmium co-contamination-induced photosynthetic inhibition and oxidative stress in tomato. J Exp Bot 64(1):199–213

Ahammed GJ, Ruan YP, Zhou J, Xia XJ, Shi K, Zhou YH, Yu JQ (2013) Brassinosteroid alleviates polychlorinated biphenyls-induced oxidative stress by enhancing antioxidant enzymes activity in tomato. Chemosphere 90(11):2645–2653

Ahmad I, Akhtar MJ, Asghar HN, Ghafoor U, Shahid M (2016) Differential effects of plant growth-promoting rhizobacteria on maize growth and cadmium uptake. J Plant Growth Regul 35(2):303–315

Al Mahmud J, Hasanuzzaman M, Nahar K, Bhuyan MB, Fujita M (2018) Insights into citric acid-induced cadmium tolerance and phytoremediation in Brassica juncea L.: coordinated functions of metal chelation, antioxidant defense and glyoxalase systems. Ecotoxicol Environ Saf 147:990–1001

Ali B, Wang B, Ali S, Ghani MA, Hayat MT, Yang C, Zhou WJ (2013) 5-Aminolevulinic acid ameliorates the growth, photosynthetic gas exchange capacity, and ultrastructural changes under cadmium stress in *Brassica napus* L. J Plant Growth Regul 32(3):604–614

Ali B, Mwamba TM, Gill RA, Yang C, Ali S, Daud MK, Zhou W (2014a) Improvement of element uptake and antioxidative defense in *Brassica napus* under lead stress by application of hydrogen sulfide. Plant Growth Regul 74(3):261–273

Ali B, Song WJ, Hu WZ, Luo XN, Gill RA, Wang J, Zhou WJ (2014b) Hydrogen sulfide alleviates lead-induced photosynthetic and ultrastructural changes in oilseed rape. Ecotoxicol Environ Saf 102:25–33

Ali B, Xu X, Gill RA, Yang S, Ali S, Tahir M, Zhou W (2014c) Promotive role of 5-aminolevulinic acid on mineral nutrients and antioxidative defense system under lead toxicity in Brassica napus. Ind Crop Prod 52:617–626

Ali S, Chaudhary A, Rizwan M, Anwar HT, Adrees M, Farid M, Anjum SA (2015) Alleviation of chromium toxicity by glycinebetaine is related to elevated antioxidant enzymes and suppressed chromium uptake and oxidative stress in wheat (Triticum aestivum L.). Environ Sci Pollut Res 22(14):10669–10678

Amooaghaie R, Enteshari S (2017) Role of two-sided crosstalk between NO and H2S on improvement of mineral homeostasis and antioxidative defense in Sesamum indicum under lead stress. Ecotoxicol Environ Saf 139:210–218

Antoniou C, Savvides A, Christou A, Fotopoulos V (2016) Unravelling chemical priming machinery in plants: the role of reactive oxygen–nitrogen–sulfur species in abiotic stress tolerance enhancement. Curr Opin Plant Biol 33:101–107

Anwaar SA, Ali S, Ali S, Ishaque W, Farid M, Farooq MA, Najeeb U, Abbas F, Sharif M (2015) Silicon (Si) alleviates cotton (*Gossypium hirsutum* L.) from zinc (Zn) toxicity stress by limiting Zn uptake and oxidative damage. Environ Sci Pollut Res 22(5):3441–3450

Asgher M, Khan NA, Khan MI, Fatma M, Masood A (2014) Ethylene production is associated with alleviation of cadmium-induced oxidative stress by sulfur in mustard types differing in ethylene sensitivity. Ecotoxicol Environ Saf 106:54–61

Astolfi S, Zuchi S, Neumann G, Cesco S, di Toppi LS, Pinton R (2011) Response of barley plants to Fe deficiency and Cd contamination as affected by S starvation. J Exp Bot 63(3):1241–1250

Awasthi JP, Saha B, Regon P, Sahoo S, Chowra U, Pradhan A, Roy A, Panda SK (2017) Morpho-physiological analysis of tolerance to aluminum toxicity in rice varieties of North East India. PLoS One 12(4):e0176357

Bashir S, Hussain Q, Shaaban M, Hu H (2018a) Efficiency and surface characterization of different plant derived biochar for cadmium (Cd) mobility, bioaccessibility and bioavailability to Chinese cabbage in highly contaminated soil. Chemosphere 211:632–639

Bashir S, Zhu J, Fu Q, Hu H (2018b) Cadmium mobility, uptake and anti-oxidative response of water spinach (Ipomoea aquatic) under rice straw biochar, zeolite and rock phosphate as amendments. Chemosphere 194:579–587

Belkadhi A, De Haro A, Soengas P, Obregon S, Cartea ME, Chaibi W, Djebali W (2014) Salicylic acid increases tolerance to oxidative stress induced by hydrogen peroxide accumulation in leaves of cadmium-exposed flax (Linum usitatissimum L.). J Plant Interact 9(1):647–654

Benavides MP, Groppa MD, Recalde L, Verstraeten SV (2018) Effects of polyamines on cadmium- and copper-mediated alterations in wheat (Triticum aestivum L) and sunflower (Helianthus annuus L) seedling membrane fluidity. Arch Biochem Biophys 654:27–39

Blaylock MJ, Salt DE, Dushenkov S, Zakharova O, Gussman C, Kapulnik Y, Ensley BD, Raskin I (1997) Enhanced accumulation of Pb in Indian mustard by soil-applied chelating agents. Environ Sci Technol 31(3):860–865

Bogdan K, Schenk MK (2008) Arsenic in rice (Oryza sativa L.) related to dynamics of arsenic and silicic acid in paddy soils. Environ Sci Technol 42:7885–7890

Cao X, Jia J, Zhang C, Li H, Liu T, Jiang X, Polle A, Peng C, Luo ZB (2014) Anatomical, physiological and transcriptional responses of two contrasting poplar genotypes to drought and re-watering. Physiol Plant 151(4):480–494

Caverzan A, Passaia G, Rosa SB, Ribeiro CW, Lazzarotto F, Margis-Pinheiro M (2012) Plant responses to stresses: role of ascorbate peroxidase in the antioxidant protection. Genet Mol Biol 35(4):1011–1019

Charlatchka R, Cambier P (2000) Influence of reducing conditions on solubility of trace metals in contaminated soils. Water Air Soil Pollut 118(1–2):143–168

Chen X, Li H, Chan WF, Wu C, Wu F, Wu S, Wong MH (2012) Arsenite transporters expression in rice (Oryza sativa L.) associated with arbuscular mycorrhizal fungi (AMF) colonization under different levels of arsenite stress. Chemosphere 89(10):1248–1254

Chen J, Wang W-H, Wu F-H, You C-Y, Liu T-W, Dong X-J, He J-X, Zheng H-L (2013) Hydrogen sulfide alleviates aluminum toxicity in barley seedlings. Plant Soil 362(1–2):301–318

Chen J, Liu X, Wang C, Yin SS, Li XL, Hu WJ, Simon M, Shen ZJ, Xiao Q, Chu CC, Peng XX (2015) Nitric oxide ameliorates zinc oxide nanoparticles-induced phytotoxicity in rice seedlings. J Hazard Mater 297:173–182

Chen Q, Zhang X, Liu Y, Wei J, Shen W, Shen Z, Cui J (2017) Hemin-mediated alleviation of zinc, lead and chromium toxicity is associated with elevated photosynthesis, antioxidative capacity; suppressed metal uptake and oxidative stress in rice seedlings. Plant Growth Regul 81(2):253–264

Chen Q, Gong C, Ju X, Zhu Z, Shen W, Shen Z, Cui J (2018a) Hemin through the heme oxygenase 1/ferrous iron, carbon monoxide system involved in zinc tolerance in *Oryza sativa* L. J Plant Growth Regul 1–11

Chen Z, Yang B, Hao Z, Zhu J, Zhang Y, Xu T (2018b) Exogenous hydrogen sulfide ameliorates seed germination and seedling growth of cauliflower under lead stress and its antioxidant role. J Plant Growth Regul 37(1):5–15

Cobbett CS (2000) Phytochelatins and their roles in heavy metal detoxification. Plant Physiol 123(3):825–832

Davies PJ (2010) The plant hormones: their nature, occurrence, and functions. In: Plant hormones. Springer, Dordrecht, pp 1–15

Dennehy C, Lawlor PG, Jiang Y, Gardiner GE, Xie S, Nghiem LD, Zhan X (2017) Greenhouse gas emissions from different pig manure management techniques: a critical analysis. Front Environ Sci Eng 11:11

Detmann KC, Araújo WL, Martins SC, Sanglard LM, Reis JV, Detmann E, Rodrigues FÁ, Nunes-Nesi A, Fernie AR, DaMatta FM (2012) Silicon nutrition increases grain yield, which, in turn, exerts a feed-forward stimulation of photosynthetic rates via enhanced mesophyll conductance and alters primary metabolism in rice. New Phytol 1196(3):752–762

Disante KB, Cortina J, Vilagrosa A, Fuentes D, Hernández EI, Ljung K (2014) Alleviation of Zn toxicity by low water availability. Physiol Plant 150(3):412–424

Dixit G, Singh AP, Kumar A, Singh PK, Kumar S, Dwivedi S, Trivedi PK, Pandey V, Norton GJ, Dhankher OP, Tripathi RD (2015) Sulfur mediated reduction of arsenic toxicity involves efficient thiol metabolism and the antioxidant defense system in rice. J Hazard Mater 298:241–251

Dixit G, Singh AP, Kumar A, Mishra S, Dwivedi S, Kumar S, Trivedi PK, Pandey V, Tripathi RD (2016) Reduced arsenic accumulation in rice (*Oryza sativa* L.) shoot involves sulfur mediated improved thiol metabolism, antioxidant system and altered arsenic transporters. Plant Physiol. Biochemist 99:86–96

Dorneles AO, Pereira AS, Rossato LV, Possebom G, Sasso VM, Bernardy K, Sandri RD, Nicoloso FT, Ferreira PA, Tabaldi LA (2016) Silicon reduces aluminum content in tissues and ameliorates its toxic effects on potato plant growth. Ciência Rural 46(3):506–512

Duan GL, Hu Y, Liu WJ, Kneer R, Zhao FJ, Zhu YG (2011) Evidence for a role of phytochelatins in regulating arsenic accumulation in rice grain. Environ Exp Bot 71:416–421

Dudhane M, Borde M, Jite PK (2012) Effect of aluminium toxicity on growth responses and antioxidant activities in *Gmelina arborea* Roxb. Inoculated with AM fungi. Inter J Phytoremed 14(7):643–655

Emamverdian A, Ding Y, Xie Y, Sangari S (2018) Silicon mechanisms to ameliorate heavy metal stress in plants. Bio Med Res Inter 2018:1–10

Fan JL, Hu ZY, Ziadi N, Xia X (2010) Excessive sulfur supply reduces cadmium accumulation in brown rice (*Oryza sativa* L.). Environ Pollut 158(2):409–415

Farooq MA, Ali S, Hameed A, Bharwana SA, Rizwan M, Ishaque W, Farid M, Mahmood K, Iqbal Z (2016) Cadmium stress in cotton seedlings: physiological, photosynthesis and oxidative damages alleviated by glycinebetaine. South Afr J Bot 104:61–68

Farzadfar S, Zarinkamar F, Modarres-Sanavy SA, Hojati M (2013) Exogenously applied calcium alleviates cadmium toxicity in *Matricaria chamomilla* L. plants. Environ Sci Pollut Res 20(3):1413–1422

Feng Z, Zhu L (2017) Sorption of phenanthrene to biochar modifyed by base. Front Environ Sci Eng 12:1

Finnegan PM, Chen W (2012) Arsenic toxicity: the effects on plant metabolism. Front Physiol 3:1–18

Fleck AT, Mattusch J, Schenk MK (2013) Silicon decreases the arsenic level in rice grain by limiting arsenite transport. J Plant Nutr Soil Sci 176:785–794

Fleck AT, Schulze S, Hinrichs M, Specht A, Waßmann F, Schreiber L, Schenk MK (2015) Silicon promotes exodermal Casparian band formation in Si-accumulating and Si-excluding species by forming phenol complexes. PLoS One 10(9):e0138555

Gao Y, Miao C, Mao L, Zhou P, Jin Z, Shi W (2010) Improvement of phytoextraction and antioxidative defense in *Solanum nigrum* L. under cadmium stress by application of cadmium-resistant strain and citric acid. J Hazard Mater 181(1–3):771–777

Garg N, Singh S (2018) Arbuscular mycorrhiza Rhizophagus irregularis and silicon modulate growth, proline biosynthesis and yield in *Cajanus cajan* L. Millsp.(pigeonpea) genotypes under cadmium and zinc stress. J Plant Growth Regul 37(1):46–63
Graziano M, Lamattina L (2005) Nitic oxide and Iron in plants, an emerging and converging story. Trends Plant Sci 10:4–8
Gu HH, Zhan SS, Wang SZ, Tang YT, Chaney RL, Fang XH, Cai XD, Qiu RL (2012) Silicon-mediated amelioration of zinc toxicity in rice (*Oryza sativa* L.) seedlings. Plant Soil 350(1–2):193–204
Habiba U, Ali S, Farid M, Shakoor MB, Rizwan M, Ibrahim M, Abbasi GH, Hayat T, Ali B (2015) EDTA enhanced plant growth, antioxidant defense system, and phytoextraction of copper by *Brassica napus* L. Environ Sci Pollut Res 22(2):1534–1544
Hadi F, Ali N, Ahmad A (2014) Enhanced phytoremediation of cadmium-contaminated soil by *Parthenium hysterophorus* plant: effect of gibberellic acid (GA3) and synthetic chelator, alone and in combinations. Biorem J 18(1):46–55
Hajiboland R, Barcelo J, Poschenrieder C, Tolra R (2013) Amelioration of iron toxicity: a mechanism for aluminum –induced growth stimulation in tea plants. J Inorg Biochem 128:183–187
Hasan SA, Hayat S, Ali B, Ahmad A (2008) 28-Homobrassinolide protects chickpea (Cicer arietinum) from cadmium toxicity by stimulating antioxidants. Environ Pollut 151(1):60–66
Hasan SA, Hayat S, Ahmad A (2011) Brassinosteroids protect photosynthetic machinery against the cadmium induced oxidative stress in two tomato cultivars. Chemosphere 84(10):1446–1451
Hindt MN, Guerinot ML (2012) Getting a sense for signals: regulation of the plant iron deficiency response. Biochim Biophys Acta (BBA)-Mol Cell Res 1823(9):1521–1530
Houben D, Evrard L, Sonnet P (2013) Beneficial effects of biochar application to contaminated soils on the bioavailability of Cd, Pb and Zn and the biomass production of rapeseed (Brassica napus L.). Biomass Bioenergy 57:196–204
Huang JH (2014) Impact of microorganisms on arsenic biogeochemistry: a review. Water Air Soil Pollut 1848:2–25
Ibiang YB, Innami H, Sakamoto K (2018) Effect of excess zinc and arbuscular mycorrhizal fungus on bioproduction and trace element nutrition of tomato (*Solanum lycopersicum* L. cv. Micro-Tom). Soil Sci Plant Nut 1–10
Jan S, Alyemeni MN, Wijaya L, Alam P, Siddique KH, Ahmad P (2018) Interactive effect of 24-epibrassinolide and silicon alleviates cadmium stress via the modulation of antioxidant defense and glyoxalase systems and macronutrient content in Pisum sativum L. seedlings. BMC Plant Biol 18(1):146
Kambhampati MS (2013) EDTA enhanced phytoremediation of copper contaminated soils using chickpea (Cicer aeritinum L.). Bullet Environ Contam tox 91(3):310–313
Kar S, Panda SK (2018) Iron homeostasis in Rice: deficit and excess. Proc Nat Acad Sci India Sect B Biol Sci:1–9
Keramat B, Kalantari KM, Arvin MJ (2009) Effects of methyl jasmonate in regulating cadmium induced oxidative stress in soybean plant (*Glycine max* L.). Afr J Microbiol Res 3:240–244
Khan NA, Khan MI, Asgher M, Fatma M, Masood A, Syeed S (2014) Salinity tolerance in plants: revisiting the role of sulfur metabolites. J Plant Biochem Physiol 2(120):2
Khan MI, Nazir F, Asgher M, Per TS, Khan NA (2015) Selenium and sulfur influence ethylene formation and alleviate cadmium-induced oxidative stress by improving proline and glutathione production in wheat. J Plant Physiol 173:9–18
Kohli SK, Handa N, Sharma A, Gautam V, Arora S, Bhardwaj R, Alyemeni MN, Wijaya L, Ahmad P (2018) Combined effect of 24-epibrassinolide and salicylic acid mitigates lead (Pb) toxicity by modulating various metabolites in Brassica juncea L. seedlings. Protoplasma 255(1):11–24
Lee CH, Wu CH, Syu CH, Jiang PY, Huang CC, Lee DY (2016) Effects of phosphorous application on arsenic toxicity to and uptake by rice seedlings in As-contaminated paddy soils. Geoderma 270:60–67
Li X, Zhang L, Li Y (2011) Preconditioning alters antioxidative enzyme responses in rice seedlings to water stress. Procedia Environ Sci 11:1346–1351

Li L, Zheng C, Fu Y, Wu D, Yang X, Shen H (2012) Silicate-mediated alleviation of Pb toxicity in banana grown in Pb-contaminated soil. Biol Trace Elem Res 145(1):101–108
Li H, Liu Y, Zeng G, Zhou L, Wang X, Wang Y, Wang C, Hu X, Xu W (2014) Enhanced efficiency of cadmium removal by *Boehmeria nivea* (L.) Gaud. In the presence of exogenous citric and oxalic acids. J Environ Sci 26(12):2508–2516
Li M, Ahammed GJ, Li C, Bao X, Yu J, Huang C, Yin H, Zhou J (2016a) Brassinosteroid ameliorates zinc oxide nanoparticles-induced oxidative stress by improving antioxidant potential and redox homeostasis in tomato seedling. Front Plant Sci 7:615
Li P, Zhao C, Zhang Y, Wang X, Wang X, Wang J, Wang F, Bi Y (2016b) Calcium alleviates cadmium-induced inhibition on root growth by maintaining auxin homeostasis in Arabidopsis seedlings. Protoplasma 253(1):185–200
Li N, Meng H, Xing H, Liang L, Zhao X, Luo K (2017) Genome-wide analysis of MATE transporters and molecular characterization of aluminum resistance in Populus. J Exp Bot 68(20):5669–5683
Li H, Yu Y, Chen Y, Li Y, Wang M, Wang G (2018a) Biochar reduced soil extractable Cd but increased its accumulation in rice (*Oryza sativa* L.) cultivated on contaminated soils. J Soils Sed 1–10
Li X, Li YL, Mai J, Tao L, Qu M, Liu JY, Shen RF, Xu GL, Feng YM, Xiao HD, Wu LS (2018b) B alleviates Al toxicity by promoting alkalization in root transition zone via polar auxin transport. Plant physiol pp-00188
Lin Q, Xu X, Wang L, Chen Q, Fang J, Shen X, Lou L, Tian G (2017) The speciation, leachability and bioaccessibility of Cu and Zn in animal manure-derived biochar: effect of feedstock and pyrolysis temperature. Front Environ Sci Eng 11:1–12
Liu WJ, Zhu YG, Smith FA, Smith SE (2004a) Do iron plaque and genotypes affect arsenate uptake and translocation by rice seedlings (Oryza sativa L.) grown in solution culture? J Exp Bot 55:1707–1713
Liu ZJ, Boles E, Rosen BP (2004b) Arsenic trioxide uptake by hexosepermeases in Saccharomyces cerevisiae. J Biol Chem 279:17312–17318
Liu J, Zhang H, Zhang Y, Chai T (2013) Silicon attenuates cadmium toxicity in Solanum nigrum L. by reducing cadmium uptake and oxidative stress. Plant Physiol Biochem 68:1–7
Liu S-L, Yang R-J, Ma M-D, Dan F, Zhao Y, Jiang P, Wang M-H (2015) Effects of exogenous NO on the growth, mineral nutrient content, antioxidant system, and ATPase activities of *Trifolium repens* L. plants under cadmium stress. Acta Physiol Plant 37(1)
Liu X, Chen J, Wang GH, Wang WH, Shen ZJ, Luo MR, Gao GF, Simon M, Ghoto K, Zheng HL (2016) Hydrogen sulfide alleviates zinc toxicity by reducing zinc uptake and regulating genes expression of antioxidative enzymes and metallothioneins in roots of the cadmium/zinc hyperaccumulator *Solanum nigrum* L. Plant Soil 400(1–2):177–192
Lu Y, Dong F, Deacon C, Chen HJ, Raab A, Meharg AA (2010) Arsenic accumulation and phosphorus status in two rice (*Oryza sativa* L.) cultivars surveyed from fields in South China. Environ Pollut 158(5):1536–1541
Ma JF, Yamaji N (2006) Silicon uptake and accumulation in higher plants. Trends Plant Sci 11:392–397
Ma J, Cai H, He C, Zhang W, Wang L (2015) A hemicellulose-bound form of silicon inhibits cadmium ion uptake in rice (*Oryza sativa*) cells. New Phytol 206(3):1063–1074
Mandal A, Purakayastha TJ, Patra AK (2014) Phytoextraction of arsenic contaminated soil by Chinese brake fern (*Pteris vittata*): effect on soil microbiological activities. Biol Fertil Soils 50(8):1247–1252
Masood A, Iqbal N, Khan NA (2012) Role of ethylene in alleviation of cadmium-induced photosynthetic capacity inhibition by sulphur in mustard. Plant Cell Environ 35(3):524–533
Matsumoto S, Kasuga J, Taiki N, Makino T, Arao T (2015) Inhibition of arsenic accumulation in Japanese rice by the application of iron and silicate materials. Catena 135:328–335

Meriño-Gergichevich C, Alberdi M, Ivanov AG, Reyes-Díaz M (2010) Al 3+-Ca2+ interaction in plants growing in acid soils: al-phytotoxicity response to calcareous amendments. J Soil Sci Plant Nutr 10(3):217–243

Mir BA, Khan TA, Fariduddin Q (2015) 24-epibrassinolide and spermidine modulate photosynthesis and antioxidant systems in *Vigna radiata* under salt and zinc stress. Int J Adv Res 3:592–608

Morita A, Yanagisawa O, Maeda S, Takatsu S, Ikka T (2011) Tea plant (*Camellia sinensis* L.) roots secrete oxalic acid and caffeine into medium containing aluminum. Soil Sci Plant Nutr 57(6):796–802

Mostofa MG, Fujita M (2013) Salicylic acid alleviates copper toxicity in rice (Oryza sativa L.) seedlings by up-regulating antioxidative and glyoxalase systems. Ecotoxicology 22(6):959–973

Mostofa MG, Rahman A, Ansary MM, Watanabe A, Fujita M, Tran LS (2015) Hydrogen sulfide modulates cadmium-induced physiological and biochemical responses to alleviate cadmium toxicity in rice. Sci Rep 5:14078

Moulick D, Ghosh D, Santra SC (2016) Evaluation of effectiveness of seed priming with selenium in rice during germination under arsenic stress. Plant Physiol Biochem 109:571–578

Moulick D, Santra SC, Ghosh D (2017) Seed priming with Se alleviate As induced phytotoxicity during germination and seedling growth by restricting As translocation in rice (*Oryza sativa* L. cv IET-4094). Ecotoxicol Environ Saf 145:449–456

Moulick D, Santra SC, Ghosh D (2018a) Effect of selenium induced seed priming on arsenic accumulation in rice plant and subsequent transmission in human food chain. Ecotoxicol Environ Saf 152:67–77

Moulick D, Santra SC, Ghosh D (2018b) Rice seed priming with se: a novel approach to mitigate as induced adverse consequences on growth, yield and as load in brown rice. J Hazard Mater 355:187–196

Moulick D, Santra SC, Ghosh D (2018c) Seed priming with se mitigates as-induced phytotoxicity in rice seedlings by enhancing essential micronutrient uptake and translocation and reducing as translocation. Environ Sci Pollut Res. https://doi.org/10.1007/s11356-018-2711-x

Mukhopadyay M, Bantawa P, Das A, Sarkar B, Bera B, Ghosh P, Mondal TK (2012) Changes of growth, photosynthesis and alteration of leaf antioxidative defence system of tea [Camellia sinensis (L.) O. Kuntze] seedlings under aluminum stress. Biometals 25(6):1141–1154

Munoz-Bertomeu J, Cascales-Miñana B, Mulet JM, Baroja-Fernández E, Pozueta-Romero J, Kuhn JM, Segura J, Ros R (2009) Plastidial glyceraldehyde-3-phosphate dehydrogenase deficiency leads to altered root development and affects the sugar and amino acid balance in Arabidopsis. Plant Physiol 151(2):541–558

Nahar K, Hasanuzzaman M, Alam MM, Rahman A, Suzuki T, Fujita M (2016) Polyamine and nitric oxide crosstalk: antagonistic effects on cadmium toxicity in mung bean plants through upregulating the metal detoxification, antioxidant defense and methylglyoxal detoxification systems. Ecotoxicol Environ Saf 126:245–255

Nakagawa T, Mori S, Yoshimura E (2003) Amelioration of aluminum toxicity by pretreatment with phosphate in aluminum-tolerant rice cultivar. J Plant Nutr 26(3):619–628

Namdjoyan S, Kermanian H, Soorki AA, Tabatabaei SM, Elyasi N (2017) Interactive effects of salicylic acid and nitric oxide in alleviating zinc toxicity of Safflower (*Carthamus tinctorius* L.). Ecotoxicology 26(6):752–761

Namgay T, Singh B, Singh BP (2010) Influence of biochar application to soil on the availability of As, Cd, Cu, Pb, and Zn to maize (*Zea mays* L.). Soil Res 48(7):638–647

Nath S, Panda P, Mishra S, Dey M, Choudhury S, Sahoo L, Panda SK (2014) Arsenic stress in rice: redox consequences and regulation by iron. Plant Physiol Biochem 80:203–210

Nazar R, Iqbal N, Syeed S, Khan NA (2015) Salicylic acid alleviates decreases in photosynthesis under salt stress by enhancing nitrogen and sulfur assimilation and antioxidant metabolism differentially in two mungbean cultivars. J Plant Physiol 168(8):807–815

Neupane G, Donahoe RJ (2013) Calcium-phosphate treatment of contaminated soil for arsenic immobilization. Appl Geochem 28:145–154

Noriega G, Caggiano E, Lecube ML, Santa Cruz D, Batlle A, Tomaro M, Balestrasse KB (2012) The role of salicylic acid in the prevention of oxidative stress elicited by cadmium in soybean plants. Biometals 25(6):1155–1165

Ou-yang C, Gao S, Mei LJ, Chung TW, Tang L, Wang SH, Chen F (2014) Effects of aluminum toxicity on the growth and antioxidant status in *Jatropha curcas* seedlings. J Med Plant Res 8(3):178–185

Pál M, Csávás G, Szalai G, Oláh T, Khalil R, Yordanova R, Gell G, Birinyi Z, Németh E, Janda T (2017) Polyamines may influence phytochelatin synthesis during Cd stress in rice. J Hazard Mater 340:272–280

Panda SK, Patra HK (2007) Effect of salicylic acid potentiates cadmium-induced oxidative damage in *Oryza sativa* L. leaves. Acta Physiol Plant 29(6):567–575

Pandey P, Srivastava RK, Dubey RS (2013a) Salicylic acid alleviates aluminum toxicity in rice seedlings better than magnesium and calcium by reducing aluminum uptake, suppressing oxidative damage and increasing antioxidative defense. Ecotoxicology 22(4):656–670

Pandey S, Ghosh PK, Ghosh S, De TK, Maiti TK (2013b) Role of heavy metal resistant Ochrobactrum sp. and Bacillus spp. strains in bioremediation of a rice cultivar and their PGPR like activities. J Microbiol 51(1):11–17

Pena-Fernandez A, Gonzalez-Munoz MJ, Lobo-Bedmar MC (2014) Establishing the importance of human health risk assessment for ena metals and metalloids in urban environments. Environ Int 72:176–185

Pereira SI, Barbosa L, Castro PM (2015) Rhizobacteria isolated from a metal-polluted area enhance plant growth in zinc and cadmium-contaminated soil. Int J Environ Sci Technol 12(7):2127–2142

Piotrowska A, Bajguz A, Godlewska-Żyłkiewicz B, Czerpak R, Kamińska M (2009) Jasmonic acid as modulator of lead toxicity in aquatic plant *Wolffia arrhiza* (*Lemnaceae*). Environ Exp Bot 66:507–513

Prabagar S, Hodson MJ, Evans DE (2011) Silicon amelioration of aluminium toxicity and cell death in suspension cultures of Norway spruce (*Picea abies* (L.) Karst.). Environ Exp Bot 70(2–3):266–276

Raab A, Williams PN, Meharg A, Feldmann J (2007) Uptake and translocation of inorganic and methylated arsenic species by plants. Environ Chem 4:197–203

Rady MM, Hemida KA (2015) Modulation of cadmium toxicity and enhancing cadmium-tolerance in wheat seedlings by exogenous application of polyamines. Ecotoxicol Environ Saf 119:178–185

Ramakrishna B, Rao SS (2012) 24-Epibrassinolide alleviated zinc-induced oxidative stress in radish (Raphanus sativus L.) seedlings by enhancing antioxidative system. Plant Growth Regul 68(2):249–259

Rehman u, Zia M, Rizwan M, Hussain A, Saqib M, Ali S, Sohail MI, Shafiq M, Hafeez F (2018) Alleviation of cadmium (Cd) toxicity and minimizing its uptake in wheat (*Triticum aestivum*) by using organic carbon sources in Cd-spiked soil. Environ Pollut 241:557–565

Reyes-Díaz M, Meriño-Gergichevich C, Alarcón E, Alberdi M, Horst WJ (2011) Calcium sulfate ameliorates the effect of aluminum toxicity differentially in genotypes of highbush blueberry (*Vaccinium corymbosum* L.). J Soil Sci Plant Nutr 11(4):59–78

Ribeiro C, Cambraia J, Peixoto PH, Fonseca Júnior ÉM (2012) Antioxidant system response induced by aluminum in two rice cultivars. Braz J Plant Physiol 24(2):107–116

Rizwan M, Ali S, Abbas T, Adrees M, Zia-ur-Rehman M, Ibrahim M, Abbas F, Qayyum MF, Nawaz R (2018) Residual effects of biochar on growth, photosynthesis and cadmium uptake in rice (*Oryza sativa* L.) under Cd stress with different water conditions. J Environ Manag 206:676–683

Saidi I, Ayouni M, Dhieb A, Chtourou Y, Chaïbi W, Djebali W (2013) Oxidative damages induced by short-term exposure to cadmium in bean plants: protective role of salicylic acid. South Afr J Bot 85:32–38

Saini R, Saini HS, Dahiya A (2017) Iron treatment enhances the levels of reduced glutathione, oxidized glutathione and glutathione reductase activity in Rice (*Oryza sativa* L.). J Pharmaco Phytochem 6(4):1321–1328

Savvides A, Ali S, Tester M, Fotopoulos V (2016) Chemical priming of plants against multiple abiotic stresses: mission possible? Trends Plant Sci 21(4):329–340

Saxena I, Shekhawat GS (2013) Nitric oxide (NO) in alleviation of heavy metal induced phytotoxicity and its role in protein nitration. Nitric Oxide 32:13–20

Shahid M, Dumat C, Pourrut B, Silvestre J, Laplanche C, Pinelli E (2014) Influence of EDTA and citric acid on lead-induced oxidative stress to *Vicia faba* roots. J Soils Sediments 14(4):835–843

Shahnaz G, Shekoofeh E, Kourosh D, Moohamadbagher B (2011) Interactive effects of silicon and aluminum on the malondialdehyde (MDA), proline, protein and phenolic compounds in *Borago officinalis* L. J Med Plant Res 5(24):5818–5827

Shakirova FM, Allagulova CR, Maslennikova DR, Klyuchnikova EO, Avalbaev AM, Bezrukova MV (2016) Salicylic acid-induced protection against cadmium toxicity in wheat plants. Environ Exp Bot 122:19–28

Sharma P, Kumar A, Bhardwaj R (2016) Plant steroidal hormone epibrassinolide regulate–heavy metal stress tolerance in *Oryza sativa* L. by modulating antioxidant defense expression. Environ Exp Bot 122:1–9

SHI WG, Li H, LIU TX, Polle A, PENG CH, LUO ZB (2015) Exogenous abscisic acid alleviates zinc uptake and accumulation in P opulus× canescens exposed to excess zinc. Plant Cell Environ 38(1):207–223

Siddiqui MH, Al-Whaibi MH, Sakran AM, Basalah MO, Ali HM (2012) Effect of calcium and potassium on antioxidant system of *Vicia faba* L. under cadmium stress. Int J Mol Sci 13(6):6604–6619

Singh S, Prasad SM (2014) Growth, photosynthesis and oxidative responses of *Solanum melongena* L. seedlings to cadmium stress: mechanism of toxicity amelioration by kinetin. Scien Hort 176:1–10

Singh S, Tripathi DK, Singh S, Sharma S, Dubey NK, Chauhan DK, Vaculík M (2017) Toxicity of aluminium on various levels of plant cells and organism: a review. Environ Exp Bot 137:177–193

Song WY, Park J, Mendoza-Cózatl DG, Suter-Grotemeyer M, Shim D, Hörtensteiner S, Geisler M, Weder B, Rea PA, Rentsch D (2010) Arsenic tolerance in Arabidopsis is mediated by two ABCC-type phytochelatin transporters. Proc Natl Acad Sci U S A 107:21187–21192

Song A, Li P, Fan F, Li Z, Liang Y (2014) The effect of silicon on photosynthesis and expression of its relevant genes in rice (*Oryza sativa* L.) under high-zinc stress. PLoS One 9(11):e113782

Souza R, Ambrosini A, Passaglia LMP (2015) Plant growth promoting bacteria as inoculants in agricultural soils. Genet Mol Biol 38(4):401–419

Srivastava RK, Pandey P, Rajpoot R, Rani A, Gautam A, Dubey RS (2015) Exogenous application of calcium and silica alleviates cadmium toxicity by suppressing oxidative damage in rice seedlings. Protoplasma 252(4):959–975

Stein RJ, Ricachenevsky FK, Fett JP (2009) Differential regulation of the two rice ferritin genes (OsFER1 and OsFER2). Plant Sci 177:563–569

Szalai G, Krantev A, Yordanova R, Popova LP, Janda T (2013) Influence of salicylic acid on phytochelatin synthesis in Zea mays during Cd stress. Turk J Bot 37(4):708–714

Tajti J, Janda T, Majláth I, Szalai G, Pál M (2018) Comparative study on the effects of putrescine and spermidine pre-treatment on cadmium stress in wheat. Ecotoxicol Environ Saf 148:546–554

Talukdar D (2012) Exogenous calcium alleviates the impact of cadmium-induced oxidative stress in *Lens culinaris* Medic. Seedlings through modulation of antioxidant enzyme activities. J Crop Sci Biotechnol 15(4):325–334

Tripathi DK, Singh VP, Prasad SM, Dubey NK, Chauhan DK, Rai AK (2016) LIB spectroscopic and biochemical analysis to characterize lead toxicity alleviative nature of silicon in wheat (*Triticum aestivum* L.) seedlings. J Photochem Photobiol B. Biologicals 154:89–98

Tripathi DK, Mishra RK, Singh S, Singh S, Singh VP, Singh PK et al (2017) Nitric oxide ameliorates zinc oxide nanoparticles phytotoxicity in wheat seedlings: implication of the ascorbate-glutathione cycle. Front Plant Sci 8:1

Vegan P, Abdullah R, Khadiran T, Ismail H, Boyce AN (2016) Role of plant growth promoting Rhizobacteria in agricultural sustainability – a review. Molecules 21(573):1–17

Verma L, Pandey N (2016) Effect of iron stress on oxidative metabolism in wheat plants (*Triticum aestivum* L). Int J Appl Pure Sci Agri 2(12):24–32

Vitti A, Nuzzaci M, Scopa A, Tataranni G, Remans T, Vangronsveld J, Sofo A (2013) Auxin and Cytokinin metabolism and root morphological modifications in *Arabidopsis thaliana* seedlings infected with cucumber mosaic virus (CMV) or exposed to cadmium. Int J Mol Sci 14(4):6889–6902

Wang F, Zeng B, Sun Z, Zhu C (2009) Relationship between proline and Hg2+–induced oxidative stress in a tolerant rice mutant. Arch Environ Contam Toxicol 56(4):723

Wang Q, Liang X, Dong Y, Xu L, Zhang X, Kong J, Liu S (2013) Effects of exogenous salicylic acid and nitric oxide on physiological characteristics of perennial ryegrass under cadmium stress. J Plant Growth Regul 32(4):721–731

Wang J, Chen J, Pan K (2014) Effect of exogenous abscisic acid on the level of antioxidants in Atractylodes macrocephala Koidz under lead stress. Environ Sci Pollut Res 20(3):1441–1449

Wang F, Liu X, Shi Z, Tong R, Adams CA, Shi X (2016) Arbuscular mycorrhizae alleviate negative effects of zinc oxide nanoparticle and zinc accumulation in maize plants–a soil microcosm experiment. Chemosphere 147:88–97

Wang F, Jing X, Adams CA, Shi Z, Sun Y (2018a) Decreased ZnO nanoparticle phytotoxicity to maize by arbuscular mycorrhizal fungus and organic phosphorus. Environ Sci Pollut Res 25:1–12

Wang F, Adams CA, Shi Z, Sun Y (2018b) Combined effects of ZnO NPs and Cd on sweet sorghum as influenced by an arbuscular mycorrhizal fungus. Chemosphere 209:421–429

Williams GP, Gnanadesigan M, Ravikumar S (2013) Biosorption and biokinetic properties of solar Saltern Halobacterial strains for managing Zn2+, As2+ and Cd2+ metals. Geomicrobiol J 30:497–500

Wu XX, Chen JL, Xu S, Zhu ZW, Zha DS (2016) Exogenous 24-epibrassinolide alleviates zinc-induced toxicity in eggplant (*Solanum melongena* L.) seedlings by regulating the glutathione-ascorbate-dependent detoxification pathway. J Hortic Sci Biotechnol 91(4):412–420

Wu Z, Xu S, Shi H, Zhao P, Liu X, Li F, Deng T, Du R, Wang X, Wang F (2018) Comparison of foliar silicon and selenium on cadmium absorption, compartmentation, translocation and the antioxidant system in Chinese flowering cabbage. Ecotoxicol Environ Saf 166:157–164

Yang H, Shi G, Wang H, Xu Q (2010) Involvement of polyamines in adaptation of *Potamogeton crispus* L. to cadmium stress. Aquat Toxicol 100(3):282–288

Yang T, Chen ML, Liu LH, Wang JH, Dasgupta PK (2012) Iron (III) modification of Bacillus subtilis membranes provides record sorption capacity for arsenic and endows unusual selectivity for As(V). Environ Sci Technol 46:2251–2256

Yingang LU, Jun MA, Ying TENG, Junyu HE, Christie P, Lingjia ZHU, Shiping DENG (2018) Effect of silicon on growth, physiology, and cadmium translocation of tobacco (*Nicotiana tabacum* L.) in cadmium-contaminated soil. Pedosphere 28(4):680–689

You J, Chan Z (2015) ROS regulation during abiotic stress response s in crop plants. Front Plant Sci 6:1–15

You-Qiang FU, Hong SHEN, Dao-Ming WU, Kun-Zheng CAI (2012) Silicon-mediated amelioration of Fe2+ toxicity in Rice (*Oryza sativa* L.) roots. Pedosphere 22(6):795–802

Zaheer MM, Yasin NA, Ahmad SR, Khan WU, Ahmad A, Ali A, Rehman SU (2018) Amelioration of cadmium stress in gladiolus (Gladiolus grandiflora L.) by application of potassium and silicon. J Plant Nutr 41(4):461–476

Zanganeh R, Jamei R, Rahmani F (2018) Impacts of seed priming with salicylic acid and sodium hydrosulfide on possible metabolic pathway of two amino acids in maize plant under lead stress. Mol Biol Res Comm 7:83–88

Zhang H, Hu LY, Hu KD, He YD, Wang SH, Luo JP (2008) Hydrogen sulfide promotes wheat seed germination and alleviates oxidative damage against copper stress. J Integr Plant Biol 50(12):1518–1529
Zhang J, Zhao QZ, Duan GL, Huang YC (2011a) Influence of Sulphur on arsenic accumulation and metabolism in rice seedlings. Environ Exp Bot 72:34–40
Zhang W, Gao S, Zhou X, Chellappan P, Chen Z, Zhou X, Zhang X, Fromuth N, Coutino G, Coffey M, Jin H (2011b) Bacteria-responsive microRNAs regulate plant innate immunity by modulating plant hormone networks. Plant Mol Biol 75(1–2):93–105
Zheng R-L, Cai C, Liang J-H, Huang Q, Chen Z, Huang Y-Z, Arp HPH, Sun G-X (2012) The effects of biochars from rice residue on the formation of iron plaque and the accumulation of Cd, Zn, Pb, As in rice (*Oryza sativa* L.) seedlings. Chemosphere 89(7):856–862

# Seed Priming for Disease Resistance in Plants

**Ghulam Mustafa, Sajid Masood, Niaz Ahmed, Abdul Saboor, Shakeel Ahmad, Sajjad Hussain, Muhammad Bilal, and Muhammad Arif Ali**

## Contents




G. Mustafa · N. Ahmed · A. Saboor · M. A. Ali (✉)
Department of Soil Science, Bahauddin Zakariya University, Multan, Punjab, Pakistan
e-mail: arif1056@bzu.edu.pk

S. Masood
State Key Laboratory of Soil and Sustainable Agriculture, Institute of Soil Science, Chinese Academy of Sciences, Nanjing, China

S. Ahmad (✉)
Department of Agronomy, Bahauddin Zakariya University, Multan, Punjab, Pakistan
e-mail: shakeelahmad@bzu.edu.pk

S. Hussain
Department of Horticulture, Bahauddin Zakariya University, Multan, Punjab, Pakistan

M. Bilal
Department of Environmental Sciences, COMSATS University Islamabad, Abbottabad Campus, Pakistan




M. Hasanuzzaman, V. Fotopoulos (eds.), *Priming and Pretreatment of Seeds and Seedlings*, https://doi.org/10.1007/978-981-13-8625-1_16

**Abstract** Crop production is limited by various kinds of diseases throughout the world. The increasing climatic and environmental changes further add to the intensity and severity of diseases in plants. This has created the need for better understanding of various diseases, their mode of spread, level and mechanism of damage, and causes and risk factors in plants. At the same time, there is a need of explanation of plant responses to diseases and their levels and mechanism of tolerance to diseases. The common disease-causing agents are fungi, bacteria, viruses, and nematodes. And the major crop diseases are *powdery* mildew, *downy* mildew, *Fusarium* wilt, and *Fusarium* root rot that severely damaged and are very common among main crop species of important families especially Poaceae, Cucurbitaceae, Solanaceae, and Fabaceae. However, understanding plant's natural adaptation and tolerance to diseases by improving the constitutive and inducible factors has progressed prominently. With the passage of time, the demands of better quality food with enhanced supplies are increasing. There is a need to introduce potential synthetic methods that trigger priming of plant tolerance to diseases. Studies have shown that fungal diseases are the most widespread among them and cause great losses in most of the crop species. Keeping in view the current situation, a lot of practices have been adopted for inducing disease tolerance in plants. However, the traditional crop production practices are not up to mark. Plant biologists are in search of innovative techniques for suitable crop improvements or ways to alleviate disease stresses for achieving the crop production goals. Currently, seed priming has gained a substantial attention due to its crucial role in plant growth and development. Furthermore, the effects of seed priming in disease tolerance in plants are evident since many years. The review of studies has described that different seed priming methods can be used for enhancing disease tolerance in plants. Biopriming and hormonal priming are the most appropriate and effective one. Study of mechanisms of disease tolerance shows that mostly the formation of signaling molecules e.g. salicylic acid (SA) and jasmonic acid (JA) are involved, however, β-aminobutyric acid and azelaic acid have also been reported to be involved in disease tolerance.

Their involvement in defense-related gene expression and in regulation of disease-induced reactive oxygen species (ROS) is clearly noted. However, salicylic acid has diverse roles in physiological changes of lignification and suberization and biochemical functions. Generally, the enzymes involved in seed priming-induced disease tolerance are phenol oxidase, tyrosine ammonia lyase and phenylalanine ammonia lyase, and antioxidant enzymes, superoxide dismutase, catalase, and peroxidase. Moreover, glucanase, peroxidase, and chitinase are also involved in disease tolerance in plants. However, their nature and amounts depend upon the disease type, plant species, and seed priming method. The main proteins participating in disease tolerance in plants are NPR1, NPR3, and NPR4 that act as salicylic acid receptors; however NPR1 is a bona fide salicylic acid-binding agent. The pathways in the inducing disease tolerance in plants are mainly the salicylic acid-dependent pathways. However, salicylic acid-independent pathway or combinations of both have also been found. In this chapter, we review current progress in the seed priming research in broader context of disease tolerance in plants and discuss its protective mechanisms of plants toward disease tolerance.

**Keywords** Seed priming · Disease tolerance · Plants · Growth · Salicylic acid · Jasmonic acid

## Abbreviations

| | |
|---|---|
| 2DE | Two-dimensional electrophoresis |
| ABA | Abscisic acid |
| alfAFP | Alfalfa antifungal peptide |
| AVRFOM2 | Avirulence fusarium oxysporum melonis |
| AZA | Azelaic acid |
| BS | *Bacillus subtilis* |
| cDNA | Complementary DNA |
| CF | Culture filtrate |
| CFU | Colony-forming units |
| C-PTIO | 2-4-Carboxyphenyl-4,4,5,5 tetrazoline-1-oxyl-3-oxide potassium salt |
| CS | Conidial suspension |
| FA | Fusaric acid |
| GSH | Glutathione |
| HRGPs | Hydroxyproline-rich glycoproteins |
| ICS | Isochorismate synthase |
| LAR | Local acquired resistance |
| LCF | Lyophilized culture filtrate |
| MeJA | Methyl jasmonic acid |
| MS | Mass spectroscopy |
| MS | Methyl salicylate |

| | |
|---|---|
| NADP-ME | Nicotine amide diamine penta acetic acid-maleic enzyme |
| NHP | N-Hydroxypipecolic acid |
| NO | Nitric oxide |
| $N_2O$ | Nitrous oxide |
| OMWW | Olive mill waste water |
| PAL | Phenylalanine ammonia lyase |
| PEG | Polyethylene glycol |
| pf1 | *Pseudomonas fluorescens* 1 |
| PGPF | Plant growth-promoting fungus |
| PGPRS | Plant growth-promoting rhizobacteria |
| POX | Peroxidase |
| PR | Pathogenesis-related |
| RCM | Raw cow milk |
| RNA | Ribonucleic acid |
| ROS | Reactive oxygen species |
| SA | Salicylic acid |
| SAR | Systematically induced resistance |
| SNP | Sodium nitroprusside |
| TGA | TGA factors that bind specially to variants of the palindrome TGACGTCA |
| TMV | Tobacco mosaic virus |

## 1 Introduction

According to estimates, the world population will increase by one third that will need 70% more food for ensuring food security (Noble and Ruaysoongnern 2010). This goal will be accomplished only if the efforts are done for improving agricultural production from existing land resources. Moreover, uninterrupted climate changes unfavorably affect the crops from various stresses and diseases. The disease can be defined as the structural or functional disorder of particular system of plants. Diseases are mainly caused by various agents that become active as they find a suitable environment and damage crops (Medda et al. 2015). The commonly known diseases are rust, wilt, blast, blight, canker, decay, and root diseases which represent the widely found diseases in main crop species causing a major reduction in yield. The diseases also deteriorate the quality of the remaining harvested produce (Medda et al. 2015). Conventionally, pesticides are used for controlling disease attacks that incur a lot of expenses. Additionally, the use of pesticides adds to environmental pollution and exerts toxic effects on human health. Although all kinds of diseases have detrimental effects on plants, the fungal diseases are the most commonly prevailing diseases. Their damage to plants in most of plant species is very fast.

Under the current scenario, one of the major aspects of disease management is to understand the mechanism of disease spread, the tolerance levels of plants, and the mechanisms involved. The studies show that plants differ in nature with respect to their responses to diseases. Plants can be classified as discriminative susceptible to diseases, resistant to diseases, or tolerant to diseases. While other are resistant to one type of diseases and tolerant to the other types of diseases, plants may also be moderately resistant and delay the occurrence of diseases. The plant mechanisms to defend against pathogens include resistance, i.e., the ability of the host to limit pathogen multiplication or the growth and the infection of pathogen is inhibited by the plant (Agrawal et al. 1999), or resistance means that a plant completely immunizes itself from a particular disease stress (Horns and Hood 2012). Briefly, the classification of disease resistance can be explained by (a) exclusion, (b) escape, (c) host-parasitic interactions, and (d) endurance or tolerance of a given level of disease (Caldwell 1934).

Tolerance in literature has been described as an intermediate level of an observable resistance or an endurance somewhat between full susceptibility and immunity; however, using both names is confusing. Tolerance is a phenomenon in which severe disease is endured by plants without significantly losing yield or quality (Caldwell 1934). Plants can only endure the tolerance when the yield loss is not directly related to disease severity. The product or yield is one aspect of plant and classical expression of disease is another aspect. In other words, the disease tolerance can be described as the cultivars equivalent in diseases but different loss of yield or the loss of quality. Conversely, when more susceptible cultivars are sown and they have different levels of disease but have similar reduction in yield or quality. Tolerance is the ability of the host to minimize the consequences of infection on its fitness regardless of the level of severity of pathogen attack (Råberg 2014). The proof of tolerance can be established by the claim that a tolerant cultivar is at least as vulnerable as the less tolerant one. Higher or better response of a visibly tolerant cultivar could result from the resistance which is significantly low to be easily recognized. Short-term and/or long-term benefits may be provided by the resistance and tolerance. However, different epidemiological concerns are involved and thus the different evolutionary behaviors are shown by the plants. The global challenge from infectious diseases has prompted the progression of varied host tolerance. Enhanced susceptibility of plants to certain diseases or evolution of pests results in economic losses due to diseases. Compared to disease resistance, plant tolerance to diseases is the ability of the plant to restrain, tolerate, or recover from the insect's attacks under the situations that otherwise typically cause huge amount of damage to other plants of the same species. Tolerance in terms of agricultural production means that despite stress from a disease, the production levels will remain above the economic threshold or yields and are greater than controls. Tolerance can overcome diseases in susceptible plants (i.e., neither tolerant nor resistant) with higher cost than resistance. Tolerance provides comparable direct benefits to the host, because tolerant hosts are under higher disease burden as compared with susceptible competitors. Under such scenario, with high cost of inputs and evolution of pests, there is dire need to induce tolerance in plants against diseases for high plant production (Horns and Hood 2012).

With the passage of time, researchers have adopted various practices for inducing disease tolerance in plants. They are comprised of modern methods like genetic engineering, polyploidy breeding, mutation breeding, and conventional methods of selection and hybridization. However, the genetically inherited characteristics can also influence the ultimate level of damage which can be caused by a particular. From the research, it is clear that the introduction of transgenes into breeding programs is not very simple and easy (Flowers et al. 1997). For this reason, such as pleiotropy and gene silencing, it looks very difficult to keep on track the process of plant breeding with the same precision with which the selection of the gene was started. Some of the examples are briefly described as follows: expression of the alfAFP peptide in the transgenic potato crop provides high resistance in the greenhouse against disease *Verticillium* wilt. More importantly, the resistance is also maintained under field conditions (Gao et al. 2000). Various efforts have been done to produce the transgenic plants that can survive in various kinds of stresses. In this regard, genetic engineering carries the potential, for which one can call it considerably fast and predictable in its outcome (Gust et al. 2010). Thus, the difference between the tolerant and susceptible cultivars is that both show signs and symptoms, but the tolerant cultivar shows less damage due to the infection. Tolerance is the ability of a host plant to survive and give better yields at the level of infection that can cause economic loss to other varieties of the same host species (Rahman et al. 2017).

In response to any external stimuli, it shows many changes at transcriptional, physiological, metabolic, and epigenetic levels. The phase or the stage of plant is called the priming phase. The plant, depending upon the level of challenge, efficiently shows a quick and strong defense response. This stage is called the primed state after experiencing the challenge and results in higher resistance and/or stress tolerance. The priming effect can be pronounced that it can persist throughout the life cycle of plant. It can also be transmitted to the next generations that may be described as immunological memory. Priming tolerance to diseases and/or direct tolerance to diseases is a major opportunity to achieve the targets of crop production. While in the aforementioned methods of inducing tolerance to diseases, the large demands of energy and man power have limited the use of conservative techniques. In addition to this major disadvantage, these methods are also expensive and burdensome. The biosafety regulations and restrictions check the introduction of transgenic crops into the field. Stimuli from the beneficial microbes, arthropods or pathogens, chemicals, and abiotic signals, pathogens can act as warning signals triggering the establishment of priming. Multiplication of pathogen or disease is termed as infection. For example tolerance means that some lines with 65–100% rust showing 44.5% yield reduction whereas yields of others with the same amount of rust were down only 9.5%.

The review of previous studies has created the need of understanding the most economical and feasible technique, seed priming. Seed priming is nowadays being preferred for inducing disease tolerance in plants. The induction of a particular physiological state or biochemical reactions in plants is possible by the treatment of seeds before germination with natural and synthetic compounds. The primed state of a plant is that when a plant is able to better activate defense responses faster (Beckers and Conrath 2007). Seed priming method involved in disease tolerance is drum priming, in which precise amount of water is used to get hydrated seeds. The limitations are that water should be enough for imbibition. There is at least wastage of resources in drum priming as compared with matric priming or osmopriming ensuring enhancement of performance of seed. Drum priming has four phases. Calibration is the process of calibrating an appropriate amount of water required for seed hydration. In this priming method, appropriate temperature of sterilized distilled water is maintained, seeds are immersed in water, and duration of hydropriming is determined by controlling the seed imbibition during germination (Kaya et al. 2006). Hydropriming of maize seeds with plant extracts is effective against toxigenic fungi and improves seed quality parameters (Ayaz et al. 2014). Solid matrix priming or matri conditioning is a type of priming in which the priming agent is a solid matrix instead of an osmotic solution.

Redox priming**:** the cellular redox state is very important and plays a significant role in key processes of plant growth and development. This is also known to induce the disease tolerance in plants. The redox state of plants is modified in response to any external stimuli, and the extent of change is dependent on the nature of stimuli. The redox state changes depend upon the dose, time, and nature of stimuli itself and the ability of plants to tolerate diseases (Jisha et al. 2013). The seed vigor, seed germination, and seedling vigor are improved with UV rays, gamma rays, gamma X-rays, and magneto-priming; however there are concerns of knowledge of exposure time duration and doses of these radiations (Aruajo et al. 2016). The enhancing of microorganisms by priming the seed with beneficial microorganisms is also known as biopriming. The use of plant growth regulators and hormones during priming and other presowing treatments has also known to affect the growth and yield of various crops.

Seed priming enables the plants to have quick cellular responses against diseases. Plants tolerate against diseases and ultimately the plants' growth and yield are enhanced (Jisha et al. 2013). Characteristics produced during the seed priming are transferred to the next generation (Lal et al. 2018). However, there is a need to identify the suitable techniques of seed priming that enhance and ensure tolerance levels of plants against diseases. The objective of this chapter is to enhance our understanding of particular diseases, their mechanisms, and the sustainable, economical, and efficient seed priming methods involved in disease tolerance in plants. Furthermore, we discuss progress till now to understand the functions of seed priming and the responses of plants as a result of diseases, in the developments of plants and plant productions. Moreover, seed priming can induce changes in biochemical and physiological mechanisms which are involved in disease tolerance and signal transduction toward tolerance to diseases.

## 2 Disease and Causes/Factors of Diseases in Plants

The crop yield and quality are extremely threatened by diseases. The main disease-causing agents are fungi, bacteria, viruses, and nematodes. However, some insects also carry fungi, bacteria, viruses, and nematodes. However, indirectly, the insects can also be carriers of fungi, bacteria, and virus that cause diseases (Medda et al. 2015). Diseases are major factor in decreasing crop yields. However, the extent of damage depends upon crop species, pathogen intensity, environmental conditions, etc. The following fungal diseases are very common: downy mildew, powdery mildew, *Fusarium* wilt, *Fusarium* root rot, etc. especially in families Poaceae, Cucurbitaceae, Solanaceae, and Fabaceae. The abiotic factors like temperature, humidity, and radiations support the proliferation of diseases. These diseases can spread in plants only if the plants are susceptible to these diseases.

## 3 Seed Priming and Disease Tolerance in Plants

Seed priming has emerged as a sustainable strategy in diseases and stress management without heavily affecting the fitness of plants (Fig. 1). The method of seed priming can be described as follows: the seeds are dried after seed priming and the seed germination is rapid when they are re-imbibed under normal or stress conditions (Table 1). When diseases attack on plants raised from primed seeds, the plants exhibit instant cellular responses against diseases. The tolerance of plants against diseases that are grown from primed seed is through various cellular and metabolic pathways. Several signaling networks are also involved (Van et al. 2006). Seed priming produces appropriate developmental and physiological processes in plants that enable to adapt to their environment. There is evidence of production of abscisic acid, gibberellic acid, and ethylene by plant growth promoting fungi (Chanclud et al. 2016). Seed priming has been found to improve seed germination, photosynthesis, seedling vigor, cell wall lignification, and suberization.

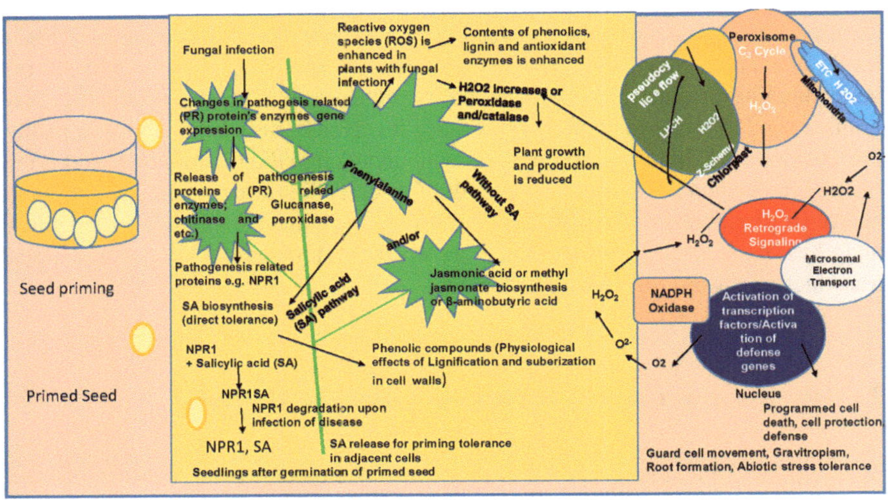


**Fig. 1** Series of steps during disease tolerance in plants by seed priming adopted from (Hossain et al. 2015; Caarls et al. 2015; Vlot et al. 2009)

**Table 1** Seed priming methods used for disease tolerance in plants

| Disease | Plants species | Seed priming | References |
|---|---|---|---|
| *Pythium ultimum* | *Zea mays* | Biopriming | Callan et al. (1990) |
| *Verticillium* wilt | *Brassica napus* | Biopriming | Muller and Berg (2008) |
| *Downy* mildew | *Pennisetum glaucum* | Biopriming | Raj et al. (2004) |
| *Powdery* mildew disease | Tomato | Jasmonic acid or β-aminobutyric acid | Worrall et al. (2012) |
| Defense gene expression against diseases | *Arabidopsis* | Salicylic acid | Wildermuth et al. (2001) |
| *Rice sheath* blight | Rice | Powdered form of *Pseudomonas fluorescens* pf1 | Vidhyasekaran and Muthamilan (1999) |
| *Ragi blast* disease | Ragi | *Pseudomonas fluorescens* isolate (JUP 121) | Patil (2016) |
| *Fusarium solani* | Potato (*Solanum tuberosum*) | Olive oil mill waste (OMW), olive pomace, coffee grounds, and phosphogypsum (0, 10, and 30%) | Samet et al. (2018) |
| *Downy* mildew | Pearl millet (*Pennisetum glaucum*) cultivar HB3 | Priming with trehalose sugar 200 mM | Govinda et al. (2016) |
| *Downy* mildew | Rice | *Bacillus subtilis* and rutin | Singh et al. (2016) |
| *Early blight* of tomato | Tomato | PGPRS priming | Babu et al. (2015) |
| *Pythium ultimum* | ? | PGPRS | Ayaz et al. (2014) |
| *Faba bean root rot* disease | Faba bean | Calcium chloride or benzoic acid | Abdel-Kader et al. (2015) |
| *Downy* mildew | Pearl millet | Plant growth-promoting fungus *P. oxalicum* | Murali and Amurthesh (2015) |
| *Downy* mildew | Pearl millet | Raw cow milk and five enzymes | Sudisha et al. (2011) |
| *Downy* mildew | Pearl millet | ß-aminobutyric acid and *Pseudomonas Fluorescence* | Anup et al. (2015) |
| *Downy* mildew | Pearl millet | Glycine betaine (30 mg $mL^{-1}$) | Lavanya and Amruthesh (2017) |
| *Downy* mildew | Pearl millet | Nitric oxide (NO) donors, 2 nitroso-1-naphthol, nitroso-R-salt and sodium nitroprusside (SNP) | Govind et al. (2016) |
| *Bacterial blight* disease | Rice | Methyl salicylate (an analogous of SA) | Kalaivani et al. (2016) |

These characteristics assist the plants to resist and/or tolerate diseases and to improve plant growth and yield. The plants from primed seeds are able to tolerate the diseases at minimal associated costs. Seed priming promotes direct activation of defense and/or priming of defense (Van et al. 2006). Many of the mechanisms of disease tolerance are poorly understood. Generally, it is known that it includes the expression of increased expression of transcription factors and signaling proteins (Pieterse et al. 2009).

## 3.1 Seed Priming Methods

### 3.1.1 Physical Method

Types of seed priming that involves physical methods are called physical seed priming methods. Out of the physical seed priming methods, hydropriming is a quite easy, economical, and harmless technique. It increases the ability of seeds toward better seedling establishment, disease tolerance, and ultimately plant production (Kaur and Kapoor 2002).

1. Hydration: The appropriate time period during which the seed is properly hydrated or attains a desired moisture level.
2. Incubation: Depending upon plant species, the seed is incubated for a certain period of time, e.g., 14 days.
3. Drying: The added water is removed so that seed reaches to its original water content. Thermo-priming: In this technique, seed is undergone through the process of different temperatures.

Along with other stages of plant growth, seed's germination and health are also affected by low or high temperature. The behavior of seed changes with the temperature, moisture, and physical method of seed priming adopted.

### 3.1.2 Chemical Method

In this method of seed priming, priming with chemical(s) is involved instead of physical method of seed priming. In nutrient priming, solutions of different nutrients are prepared. These nutrient solutions of different concentrations serve different purposes when used in seed priming. There are a lot of evidences of their contribution in disease tolerance in plants (Shah et al. 2012).

### 3.1.3 Biological Method

In this method of seed priming, the biological approaches are used for disease tolerance in plants. Biopriming is multipurpose and induces tolerance in plants via different mechanisms; i.e., the plants produce phytohormones and stimulate immunity, and various genes are expressed to different levels that enhance plant growth parameters. Various other processes such as antibiosis, mycoparasitism, induced phenolic production, activation of antioxidant enzymes, and systematic defense activation in host plants are also triggered (Arora et al. 2013). Only few studies have been conducted on seed priming and disease tolerance. However, a considerable research has been conducted on seed priming and disease resistance. Studies have shown the acceptable results of seed priming on the defense (resistance and tolerance) against diseases in plants.

# 4 Common Diseases in Important Plant Species and Seed Priming for Inducing Disease Tolerance in Plants

Common diseases, their general mode, and mechanism of damage in economically important crop members are described in a comprehensive way. Common diseases, their mechanism of spread, and suitable seed priming for inducing disease tolerance are shown in Table 1.

## *4.1* Downy *Mildew*

*Downy* mildew is very common in most of the economically important crops. Its causal organism is airborne and causes severe damage in important crops. These losses are expected to increase in the future due to changing environmental conditions. The downy mildew of pearl millet (family Gramineae) is also known as green ear disease and is the most important factor reducing yield. *Downy* mildew is spread by an obligate biograph *Sclerospora graminicola*. This disease can cause yield reduction up to 8%. Symptoms of *downy* mildew-affected plants are; formation of powder material deposits on the axial surface of leaves, reduced transpiration, chlorosis, and growth retardation. The inflorescence appears like leaves not like leaf true flowers. Under stress conditions, the pathogens are preserved in the form of oomycetes in soil. As favorable temperature persists that normally ranges from 22 to 25% and optimal relative humidity approaches to 95–100%, the dormant oomycetes are transformed into zoospore form, infecting the plant roots. This disease is predominant in young seedlings of pearl millet and has also been found at later stages. Significant outcomes of seed priming on disease endurance in pearl millet plants are evident from previous studies; however their mechanisms of action are different. The 2DE-MS/MS technique was used to study the proteins, such as priming of seeds of pearl millet with β-aminobutyric acid and *Pseudomonas fluorescens*. This approach further explained the disease tolerance mechanism induced by seed priming with β-aminobutyric acid and *Pseudomonas fluorescens*. The analysis reveals that about 63 proteins are stored that are linked with the defense by manipulation of energy, metabolism, and stress levels. Further this study elaborates that the disease influences the protein dynamics of crop (Anup et al. 2015). Seed priming with chitosan and *Pseudomonas fluorescens* is effective; however the mechanism of inducing tolerance is different. The study shows that this was due to buildup of the levels of hydroxyproline-rich glycoproteins (HRGPs) in the cell walls of coleoptiles after 9 h of seed priming. Maximum HRGPs were found in seedlings. The enzymes peroxidase and hydrogen peroxide play a role in cross-linking HRGP. The chitosan treatment resulted in tissue-specific accumulation of HRGP in the regions around vascular bundles (Sujeeth et al. 2010). Seed priming with nitric oxide donors (nitroso-R-salt, 2-nitroso-1-naphthol, and sodium nitroprusside (SNP)) induces tolerance against *downy* mildew. This increases the reactive oxygen species $H_2O_2$. Priming of seed with aqueous SNP or without polyethylene glycol (PEG) was the best to enable the tolerance against downy mildew in field experiments and in greenhouse. However, seed priming did not show any adverse effects on plant or pathogen. It took a minimum of 3 days after elicitor treatment or subsequent pathogen inoculation to induce resistance (Manjunatha et al. 2008a, b). Raw cow milk (RCM) and five different amino acids enabled the pearl millet to tolerate against *downy* mildew by modulation of physiological and biochemical responses in vitro and in the field. Additionally, they also improved the seed germination and seedling vigor to a great considerable level. It was noted that tolerance against *downy* mildew was 56% with L-proline, 67% with L-isoleucine, and 69% with L-phenylalanine than control under greenhouse conditions. The tolerance against *downy* mildew by RCM is a dose-dependent matter. Among RCM treatments, 10% dilution of RCM enabled the plants to tolerate downy mildew up to 35%. In addition, the growth and yield were also increased over control. In field trials, the effects were comparable, and seeds primed with RCM showed better vegetative and growth parameters. Seed priming also improved reproductive parameters and yield as compared with those primed with distilled water (control) (Sudisha et al. 2011). The plant growth-promoting fungus (PGPF) *Penicillium oxalicum* was screened from the rhizosphere of pearl millet. Seeds of susceptible variety of pearl millet were primed with PGPF inducing (CS, CF, and LCF), all the inducers in the presence of downy mildew enhanced plant growth. CS-primed seeds produced germination of 91% and seedling vigor up to 1427, better than LCF and LC. Temporal studies showed that it took the plants a minimum of 3 days to induce tolerance against *downy* mildew that was maintained thereafter. The disease tolerance levels varied under different growing conditions. The inducers produced significant tolerance against *downy* mildew under greenhouse (58%) and field conditions (62%). The enzymes peroxidase and chitinase have significant contributions (Murali and Amruthesh 2015). Three days of seed priming of pearl millet seed with glycinebetaines at concentration of 30 mg $mL^{-1}$ solution up to 6 h significantly enhanced the seed germination and seedling vigor by enhancing the protection against *downy* mildew (Lavanya and Amruthesh 2017). Seed priming with nitric oxide, a scavenger 2-4-carboxyphenyl-4,4,5,5 tetrazoline-1-oxyl-3-oxide potassium salt (cPTIO) is relatively enhanced the susceptibility of plants to pathogens. Conversely, expression of prime defense reactions like hypersensitive response, lignin deposition, and defense enzyme, e.g., phenylalanine ammonia lyase (PAL–EC 4.3.1.5), was enhanced by NO-donor treatments (Manjunatha et al. 2008a, b). *Downy* mildew of cucumber affects the leaf transpiration (Oerke et al. 2006). Review of studies shows that changes in the transpiration rate and metabolic activities of cucumber (family Cucurbitaceae) leaves during *downy* mildew attack were similar to those members of family Gramineae (Oerke et al. 2006). Variable effects of different seed priming treatments under different growing systems have been described. Priming of seeds with trehalose at 200 mM for 9 h protects the plants from *downy* mildew up to 70.25%. However, only 67.25% has *downy* mildew disease protection under field conditions. The severity of disease was 32.5% greater than those plants in control (Govind et al. 2016).

### *4.2 Powdery Mildew*

Seed priming with JA and BABA enables the plants to tolerate against powdery mildew (Baccelli and Mauch-Mani 2016). Seed priming of cucumber with acibenzolar-S-methyl (ASM) induced resistance in cucumber against powdery mildew (Ramasay et al. 2015).

### *4.3* **Fusarium** *Wilt*

*Fusarium* wilt is found in most of the commonly and economically important crops. It spreads with causal organisms of *Fusarium* species, a fungus that is soil based. It has devastating effects on the crops reducing their yield. It spreads during rainy season or excessive wetness in soil over longer time or favorable conditions for its propagation. Symptoms of *Fusarium* wilt in plants are almost similar with minor differences. Many factors contribute in disease spread, e.g., pathogen intensity in soil, prevalent weather, nutrients, and plant's susceptibility to diseases. The main symptom of *Fusarium* wilt is that vines lose turgor pressure during attack. During minor attacks, vines may recover during the evening, and during severe attacks, vines may sooner or later wilt forever. During *Fusarium* wilt spread by *Fusarium oxysporum* infection in cucumber (family Cucurbitaceae), the hyphae block the xylem and phloem in stem and root attacks that reduce hydraulic conductance. Moreover, there is an increase in the number of nonstomatal pores that enhance the transpiration and increase the water loss, leading to plant death. The stomatal conductance of gases and transpiration rate (E) of infected plants are higher than healthy plant. Furthermore, there is positive correlation between leaf membrane injury and exchange of gas. The study of mechanism of spread shows that fusaric acid (FA) accumulation is the common reason for nonstomatal injury in leaves and nonstomatal water loss in plants infected with *Fusarium* wilt in most of the members of the studied families. During *Fusarium* wilt of cucumber attack, the cells of plants are damaged by fungi and uncontrolled loss of water starts (Wang et al. 2015).

Soil infestation with *Fusarium solani* results in diseased cucumber plants and yield loss in range of 17–54%. Seed priming with bacterial strains was found effective, and five strains of bacteria reduced this disease up to 10%, enhancing cucumber yield up to 9–32% (Adesemoye and Egamberdieva 2013). *Vascular* wilt fungi (*Verticillium dahliae*) secrete proteins, e.g., cysteine-rich proteins, necrosis-inducing proteins, and enzymes, during host colonization. Some of them are required for pathogenicity as reviewed by de Sain and Rep (2015).

In contrary to the *Fusarium* wilt in cucumber, the symptoms of *Fusarium* wilt in watermelon that came from the same family (family Cucurbitaceae) vary greatly. During the start of disease attack, plant leaves appear as dull or gray green and later on start losing the turgor pressure, resulting in wilting of plant leaves. After the wilting of leaves, they become yellow and finally necrosis occurs. The leaves wilt from lower to upper younger foliage of plants. When plants start to vine, the initial symptoms may start; only one runner may wilt and the rest of the plant may look apparently unaffected. When the severe pathogen attack or when the host is susceptible, the entire plant may wilt and die. The surviving plants have considerably low yields. Under high inoculum presence, even the seedlings are damped off as they emerge from soil. The polulation density of pathogen in inoculum is function of disease severity i.e. 166 colony forming units (CFUs) $g^{-1}$ soil are necessary to cause wilt and 367 CFUs are enough to cause 50% wilt (ID50). The minimum amount of inoculum necessary to cause wilt was 166 CFU/g soil and that 367 CFU $g^{-1}$ was enough to cause wilt in 50% of the plants (ID50). Studies have shown that 73% of the fields have 100-1200 CFUs; this heterogeneity of soil shows that disease severity varies and so the different tolerance levels are required in plants (Egel and Martyn 2007).

*Fusarium* root and stem rot are very common and severe diseases in melon. During the comparison of susceptible and resistant successions of the melon to root and stem rot, the microscopic evaluation showed the presence of fungal growth in the intercellular spaces of the root and stem as well as in the xylem (Cohen et al. 2015). Seed priming with β-aminobutyric acid is known to protect against 80 pathogens including bacterial, fungal, viral, and nematic diseases and 40 plant species (Cohen et al. 2016). In contrast to cucumber, the mechanism of *Fusarium* wilt spread in tomato is different. However, fusaric acid production is common in all kinds of *Fusarium* wilt *and* in all plant species. In *Fusarium* wilt of tomato, the fungus produces three kinds of acids, dehydrofusaric acid, fusaric acid, and lycomarasmin, as compared with only fusaric acid in cucumber. In tomato *Fusarium* wilt, the fusaric acid affects differently as compared with its effects in cucumber. Production of fusaric acid in *Fusarium* wilt of tomato (Solanaceae) results in the reduction in photosynthesis, change in the color of leaves (from green to greenish yellow), flaccid plant cell, and cell death. Fusaric acid results in alteration in many proteins (Singh et al. 2017). Pathogen *Botrytis cinerea* based diseases result necrosis in plants. Jasmonic acid (JA) and/or β-aminobutyric acid treatment of tomato seeds increases the tolerance against these diseases. Priming effects are long lasting and do not adversely affect plant growth and development (Worrall et al. 2012). Tomato seed priming with 0.1 mM methyl jasmonic acid (MeJA), seedlings adapt tolerance against *Fusarium* wilt caused by *Fusarium oxysporium*. The seedlings tolerated for 4 weeks against *Fusarium oxysporum* f. sp. *lycopersici*. Methyl jasmonate seed priming with 0.1 mM can be used in practice. The mechanism behind tolerance against *Fusarium* wilt is that MeJA exogenous application enhances the phenolic compounds, e.g., salicylic acid (SA), kaempferol, and quercetin (Król et al. 2015).

Potatoes (*Solanum tuberosum*), an important member of family Poaceae cultivated in a greenhouse, were used to screen compost tea-suppressive ability. Moreover, they were treated with compost teas by mixing olive oil mill waste water (OMWW), olive pomace, coffee grounds, and and were also treated by mixing with phosphogypsum at the rate of 0, 10, and 30%. They repressed the *Fusarium solani* growth, improving the performance of plants even after disease attack. The antagonistic effects of the various treatments were linked with a noticeable rise of the antioxidant enzymes and PR (pathogenesis-related) protein expression that led to reduction in disease severity (Samet et al. 2018).

## 4.4 Fusarium Root Rot

*Fusarium* root rot is very common among crops. Its spread is increasing day by day causing economic losses to crops. It has damaging effects on crops and deteriorates the crop stand. Even at its initial stage, the yield is significantly reduced. In the leguminous plants, seed priming triggers disease tolerance in plants. $CaSO_4$ and $CaCl_2$ enabled the bean varieties to tolerate against *Fusarium oxysporum* (Leyva et al. 2016). Seed priming with *Trichoderma* decreased the cowpea root rot pathogens and enhanced growth as compared to control (Hamed et al. 2015). Pathogenesis-related protein gene expression was decreased in the leaves and increased in roots of maize and other cereals (family Poaceae) by seed priming with *Azospirillum brasilense* strains Ab-V5 and Ab-V6. They produced acids like salicylic acid, indole acetic acid, and indole lactic acid, resulting in improvement of the plant growth (Fukami et al. 2016).

## 4.5 Different Diseases and Seed Priming for Their Tolerance in Plants

Seed priming with *Pseudomonas fluorescens* pf1 preserved in powdered form enhances the bacterial population in rhizosphere. Plants tolerate to rice sheath blight and yield is enhanced as compared with control (Vidhyasekaran and Muthamilan 1999). In the seed priming with *Bacillus subtilis* (BS) and one picomolar solution of rutin, a bioflavonoid enhances the biofilm establishment of bacterium. Additionally, in photosynthetic pigments, chlorophyll and carotenoid are enhanced, and root and shoot lengths are increased. Moreover, increase in phenolic and flavonoid was detected in the leaves. Fluorescent microscopic images show that BS plus rutin enhances callose deposition in the leaves. The amendment of rutin with *Bacillus subtilis* further has added advantage of reducing the formation of reactive oxygen species (ROS) in BS plus rutin-treated rice plants. This further explains that rutin scavenges free radicals and better antioxidant potential. Rutin attracts the this bacteria and plant is protected from diseases by its colonizing in plants rhizosphere in rice (Singh et al. 2016). Seed priming with thiamine produces stronger and more rapid pathogenesis-related (PR) gene expression and enhances the protein kinase C activity in rice (*Oryza sativa*) of family Poaceae, *Arabidopsis* (*Arabidopsis thaliana*), and vegetable crops. Sorghum seed (family Poaceae) priming with *Eclipta alba* extract for 6 h increases yield via two types of mechanisms: hydropriming and an additional effect caused by any of the unknown molecules of plant source (Zida et al. 2015).

Seed priming with the formulations of lignite-based powder, talc-based powder, wettable powder, lignite/fly ash powder, bentonite paste, poly ethylene glycol (PEG), and a water dispersible tablet having a strain of *Bacillus subtilis* AUB-1 is inhibitory to damping off disease. Furthermore, an active rhizosphere colonization was observed in tomato plants (Jayaraj et al. 2005). Seed priming with 0.5 mM salicylic acid (SA) enables the brinjal plants to tolerate against *Verticillium* wilt (family Poaceae) than control when tested under greenhouse conditions (Mahesh et al. 2017). Bacterial communities colonize cucumber (family Cucurbitaceae) seed sown in compost medium after 8 h followed by a 3-h treatment of either rifampicin at 500 ppm or water when dislodged from seed surfaces and subject to RNA extraction and reverse transcription to cDNA (Chen et al. 2012). In *Arabidopsis thaliana* (family Brassicaceae), application of SA to seed suffices to establish SAR, resulting to tolerance against diseases (Wildermuth et al. 2001).

Priming of vegetable seed by biocontrol organisms enables the vegetables to tolerate against damping off disease in seedlings (Jensen et al. 2016). Seed priming of different seeds at various doses with 1–5 mM jasmonic acid does not show any change in final germination; however germination delays by 5 days. It is observed that in seed priming using jasmonic acid at the conc. of 10 mM, the germination is inhibited significantly. Moreover, seed priming with 3 mM jasmonic acid reduces the primary root growth relative to controls. This shows that optimum dose matters (Wasternack 2007). Seed priming with chitosan, a biopolymer, plants can tolerate the fungal and bacterial diseases (Katiyar et al. 2015).

## 5 Pathway and/or Mechanism of Disease Tolerance in Plants by Seed Priming

The primed defense is mostly connected with increase of immunity. Endogenous and exogenous signaling molecules induce plant immune responses. Amino acids and their metabolites affect the plant's immune system. The molecular foundations of priming tolerance against diseases are not well explained. Efforts have been made possible to find molecular basis of defense priming against diseases (Conrath 2011). After the occurrence of diseases, the plants tolerate the diseases due to biochemical changes, i.e., release of different molecules to signaling network and different enzymes that participate in the formation of molecules and pathogenesis-related (PR) protein and enzymes and pathogenesis-related (PR) proteins. Mechanisms are different for different seed priming methods or plant species. The physiological changes of lignification and suberization during or after the disease attack enable the plants to tolerate diseases. Study has shown that phenolic compounds are involved in lignification of adjacent cells during disease attack to enable adjacent cells to prime tolerance against diseases (Saltveit 2017). The general description of tolerance diseases in plants is described as follows. Seed priming induces tolerance in plants by various ways; i) in most of the plants by releasing salicylic acid a signaling molecule in the sequence of events, ultimately gene expression is trigerred, ii) in few plants by releasing jasmonic acid or methyl jasmonates iii) by producing and activating antioxidant enzymes glucanase, catalase and peroxidase etc. iv) in some of plants species β-aminobutyric acid is released that induces tolerance against many diseases. These mechanisms together confer tolerance against diseases in plants.

Review of studies has shown that generally molecules like salicylic acid and/or jasmonic acid, azelaic acid, or β-aminobutyric acid are found to involve in tolerance against diseases. However, in plant system, many sources work together to produce signaling molecules like hormones, proteins, and enzymes that depend upon plant species, plant organs, environmental conditions, and signaling pathway in plant. Different pathways are involved in different plant species for their disease tolerance against diseases. Pathways involved in seed priming induced common disease tolerance in plants which are shown in Table 2.

## 5.1 *Salicylic Acid-Dependent Pathway and/or Mechanism Disease Tolerance in Plants*

Salicylic acid (SA) acts as vital hormone in plant immunity. The most commonly produced molecule is SA that has important role in disease tolerance in plants as a result of seed priming. Generally, salicylic acid (SA) facilitates plant defenses

**Table 2** Pathways and signaling molecules involved in inducing disease tolerance in plants

| Disease | Plant species | Seed priming treatment or seed priming material | Signaling molecule/ pathway | References |
|---|---|---|---|---|
| *Orobanche cumana* | Sunflower | 1 mM salicylic acid | Salicylic acid | Yang et al. (2016) |
| Nematodes | Tomato | Priming with salicylic acid and jasmonic acid | Salicylic acid | Biere and Goverse (2016) |
| Potato virus X (SPCP1 strain) | Tomato | Salicylic acid | Salicylic acid | Cueto-Ginzo et al. (2016) |
| Different diseases | Tobacco, tomato, and *Arabidopsis* | Salicylic acid | Salicylic acid | Yang et al. (2015) |
| *Fusarium* wilt | Plants | 0.1 mm methyl jasmonate | Jasmonic acid | Krol et al. (2015) |

against pathogens and is released during the pathogen attacks or prior to infection in both infected and unaffected leaves as a result of pathogen attack. Salicylic acid is prerequisite for pathogenesis-related gene expression and the production of compounds related to both local and systemic acquired resistance (LAR and SAR) in plants. Under disease tolerance conditions, plants maintain high salicylic acid levels enough to bind pathogenesis-related proteins. The salicylic acid dependent pathway for disease spread and disease tolerance in different crops induced by various seed priming methods is explained by the infection of *Orobanche cumana* in sunflower (*Helianthus annuus*), which reduces the endogenous SA level, and the expression of SA-related genes including *pal* and *chs* is decreased. This also results in reduction of plant biomass. Conversely, the release of reactive oxygen species (ROS) is increased, which enhances the contents of phenolics, lignin, and activities of antioxidant enzymes. Seeds of sunflower variety (TK0409) are primed with 1 mM salicylic acid and treated with *O. cumana* for a month. These treatments increase the sunflower biomass in the form of fresh weight, dry weight, and plant height by 13%, 26%, and 10%, respectively. Salicylic acid reduces the biomass and number of established *O. cumana*. The inhibition production of ascorbate peroxidase and catalase by exogenous application of salicylic acid due to rise in production of hydrogen peroxide contents to 14% is the reason of increase in biomass (Yang et al. 2016).

In another study, *Trichoderma* produces SA- and JA-dependent tolerance in roots of tomato prior to disease spread. As a result of priming of reaction of hormones, the nematode invasion is flexible and depends on parasitism stage. The enhancement of endogenous level of salicylic acid depends upon the dose of exogenous salicylic acid applied (Biere and Goverse 2016). Salicylic acid effects on the *Potato virus X* (SPCP1 strain)-diseased tomato plants were evaluated by analyzing photosynthesis process variation and carbohydrate producing proteins (Cueto-Ginzo et al. 2016). Rutin is a member of flavonoid family, and seed priming of tobacco, tomato, and *Arabidopsis* with rutin enables them to tolerate against various diseases by salicylic acid signaling pathway (Yang et al. 2015). Salicylic acid enables the plants to modify their cell wall against diseases (Napoleão et al. 2017). Salicylic acid induces lignification in cell wall and alters the cell wall to reduce the entry or damage by pathogens (Sattler and Funnell-Harris 2013).

### 5.1.1 Salicylic Acid Synthesis/Production in Plants

Many different molecules participate in the biosynthesis of salicylic acid. The type and source depend on the plant species, plant organs, environmental conditions, and signal pathway involved. Many groups reviewed the mechanism of synthesis and production of salicylic acid that is discussed as follows. Salicylic acid can be synthesized from phenylalanine in plants (Leon et al. 1995); however, SA could also be synthesized when phenylalanine pathway is inhibited (Maunch-Mani et al. 1996). The feeding experiments showed that the specific activity of radiolabeled salicylic acid was lower than the expectations (Cqouz et al. 1998). Some bacterial strains, e.g., *Pseudomonas aeruginosa*, produce SA. They use isochorismate synthase (ICS) and pyruvate lyase enzymes (Serino et al. 1995). The gene linked with *Arabidopsis* disease tolerance as a result of mutationis is (SID2), this results in the production of SA using chorismate via ICS, LAR and SAR responses. This is possible only if SA is made by this pathway (Manohar et al. 2015).

### 5.1.2 Signaling Mechanism of Salicylic Acid

To play a signaling function, a molecule should have required characteristics that facilitate its uninterrupted influence on second messengers. Simple structure, small dimensions, and high diffusivity are clearly noted in salicylic acid. It is evident from different experiments that salicylic acid is an endogenous signal in plants that mediate tolerance to different diseases.

Salicylic acid regulates several biochemical processes, e.g., it binds to pathogenesis-related proteins, inhibits the production of ascorbate peroxidase and catalase, increases the production of hydrogen peroxide contents, and elicits stress protein under disease attack or disease stress conditions. Although the mechanisms by which this is carried out is completely unraveled. However, it is evident from different sources that under disease stressed conditions, salicylic acid interacts with reactive oxygen species (ROS) and glutathione (GSH) (Biere and Goverse 2016).

### 5.1.3 Interaction Between Salicylic Acid and Other Signaling Molecules

Salicylic acid is produced in plants against disease stress. Under the disease stress, salicylic acid acts as signals and ameliorates the disease stress by binding with pathogenesis-related proteins. Salicylic acid binds to and alters the activity of pathogenesis proteins. This represents the shift from the pattern that hormone mediates one or few receptors (Klessig 2017). The best effects of SA are found during the initial stages of the pathogenesis cycle. Salicylic acid increases proteins related to disease tolerance (Cueto-Ginzo et al. 2016). Salicylic acid enhances Pip-43 initiated actions to various levels in the distal tissue of systematically acquired resistance (SAR) processing plants (Bernsdorff et al. 2015). The sunflower seed treatments with 1 mM raise the hydrogen peroxide levels to 14% that is caused by increase of hydrogen peroxide level via minimizing levels of ascorbate peroxidase and catalase decrease (Yang et al. 2016). Salicylic acid partially counterbalances the decrease in photosynthetic rate during disease by improving mesophyll conductance via photosystem II upgradation in disease-tolerant varieties (Cueto-Ginzo et al. 2016).

Evidences of various proteins are found using biochemical approaches and these proteins bind salicylic acid. Mutant screens revealed that protein (NPR1) is a main regulator of salicylic acid-based responses. Proteins (NPR3 and NPR4) are the actual salicylic acid receptors. NPR3 and NPR4 attach to salicylic acid to control NPR1 steadiness. Many different kinds of roles of SA concludes that plants may have abundant SA receptors. During pathogen challenge, the SA gradient generated at the infection site is sensed by NPR3 and NPR4. As a result, NPR1 is transforms to different molecules at the infection site and bound SA is released on effector-based cell decay and defense, whereas NPR1 is stored in surrounding cells to support cell existence and SA-based resistance by binding SA with proteinsfor future release when needed for priming defense (Seyfferth and Tsuda 2014).

### 5.1.4 Role of Salicylic Acid in Gene Expression and Transcriptional Regulation

Seed priming efficient for priming disease tolerance in plants enables the plants to produce salicylic acid in plants. Generally salicylic acid regulates the pathogenesis-related genes in different plant species; however mechanism of regulation is diverse in nature in various plant species. The salicylic acid is involved in the localization and/or activity of transcription factors and co-regulators. Thioredoxins and glutaredoxins mediate the redox changes induced by salicylic acid that modify the transcription regulators. These regulators take part in the downregulation of jasmonic acid-triggered genes, NPR1, and transcription factors (TGA) which control their appropriate site or DNA attachment potential. Additionally, salicylic acid can produce negative regulators and can modify histones at DNA level (Caarls et al. 2015). The AVRFOM2 gene's presence in melon helps melon to escape from *Fusarium* wilt and to control plants to tolerate the *Fusarium* wilt introduced by cultivar forms containing this gene (Schmidt et al. 2016).

## *5.2 Salicylic Acid-Independent Pathway and/or Mechanism of Disease Tolerance in Plants*

Molecules other than salicylic acid are jasmonates or jasmonic acids that are involved but less frequent in the disease tolerance in plants. Methyl jasmonate is able to modify plant's cell wall (Napoleao et al. 2017). The molecule β-aminobutyric acid is a natural product produced in plants against disease infection or disease attack (Thevenet et al. 2016). Seed priming with methyl jasmonate at 0.1 mM induced resistance against *Fusarium* wilt for 4 weeks (Krol et al. 2015). $H_2O_2$ acts as a signal transduction pathway as well as in signaling molecule when maintained in proper balance between production and scavenging. However, its excess amounts reduce the photosynthesis and may lead to cell death as reviewed by Hossain et al. (2015). The signal transduction pathway by chitosan priming that stimulates its responses is hydrogen peroxide and nitric oxide (Pichyangkura and Chadchawan 2015).

### 5.2.1 Jasmonates/Jasmonic Acid Pathway

Jasmonic acid pathway is main in several plant species for tolerance against diseases induced by seed priming in plants (Fu et al. 2015), jasmonic acid pathway is involved in disease tolerance in plants however less frequently than salicylic acid does (Saltveit 2017).

### 5.2.2 Jasmonate Biosynthesis/Production

Evidences from previous studies elaborate the mechanism of synthesis and functions of jasmonic acid. Studies explain that enzymes that participate in synthesis of jasmonic acid are in crystal form and specified on substrate basis. Fatty acid b-oxidation enzymes and 4-Cl-like CoA ligases are well known (Wasternack 2007). Two abscisic acid-responsive lipid ligase genes are involved in JA biosynthesis in *Arabidopsis thaliana* (Wang et al. 2018). Acyl-CoA N-acyltransferase has a role in the biogenesis of jasmonic acid in cotton (Fu et al. 2015).

### 5.2.3 Jasmonate Signaling and/or Interaction with Other Molecules

Esterified oxylipin products in *Arabidopsis* describe roles in defense mechanisms. Jasmonic acid metabolites show specific functions. Transcription factors performing in JA signaling have been described on cross-talk in signaling pathways (Wasternack 2007). Jasmonates have been found to interact with other molecules like salicylic acid and ethylene in different ways depending upon plant species involved. Study showed that seed priming with methyl jasmonate results in accumulation of phenolic compounds like salicylic acid, kaempferol, and quercetin detected in leaf extracts of seedlings. The mechanism behind the disease resistance is the upregulation of phenylalanine ammonia lyase (PAL5) and benzoic acid/salicylic acid carboxyl methyltransferase (BSMT). The downregulation of isochorismate synthase (ICS) gene shows that phenylalanine ammonia lyase (PAL) and not the isochorismate synthase (ICS) gene is the primary root of SA synthesis in tomato. However, elevated levels of SA are accompanied with decrease in jasmonic acid, the precursor of MeJA, and increase of 12-oxo-phytodienoic acid (OPDA), the precursor of JA. So methyl jasmonate can be commercialized (Krol et al. 2015). $H_2O_2$, a small molecule, has long life that permits them to cross the cellular membranes, to different cellular partitions aiding signaling tasks, including retrograde signaling that has its role in signal transduction (Maruta et al. 2012). Although there are many sources of its production, the prime importance is oxidation of glycolate in the peroxisome in photosynthetic carbon oxidation cycle (Miller et al. 2010), and jasmonic acid accumulation also results in priming tolerance against diseases (Mehari et al. 2015).

### 5.2.4 AZA Biosynthesis/Production, AZA Signaling, and AZA Interaction with Other Molecules

Defense priming during *Arabidopsis* SAR to *P. syringae* pv. maculicola is due to production of AZA, released upon localized bacterial infection (Jung et al. 2009). Azelaic acid seemed to be mobile and transported throughout the plant and has roles both in SAR and LAR. However, there is in need of further assessment of its roles (Maldonado et al. 2002).

## *5.3 Salicylic Acid-Dependent Cum Salicylic Acid-Independent Pathway Cumulatively Involves in Disease Tolerance in Plants*

The protein dynamics have prominent role in the tolerance mechanism of pearl millet tolerance to downy mildew (Anup et al. 2015). The *Pseudomonas fluorescens* isolate (JUP 121) priming of Ragi seeds had the best disease endurance of Ragi blast disease as compared to control. This was due to the presence of pathogenesis-related protein enzymes (B-1,3 glucanase and chitinase) (Patil 2016). The nonprotein amino acid pipecolic acid (Pip) contributes in plant systemic acquired resistance (SAR) (Stewart et al. 2018). Flavin-dependent monooxygenase 1 changes to Pip to N-hydroxypipecolic acid (NHP), an important metabolic controller of SAR with or without salicylic acid-dependent pathway in *Arabidopsis* (Lin et al. 2013), β-aminobutyric acid (Cohen 2002), and Pip/FMO1-signaling molecule (Bernsdorff et al. 2015). Different kinds of role of SA in *Arabidopsis* during attack of *Alternaria brassicicola* where it induces susceptibility for this fungus. While in other crops, it induces tolerance or resistance, similar to SA by folic acid pathway (Wittek et al. 2015). Jasmonic acid (JA) and salicylic acid (SA) pathways in tolerance in tomato plants against *Fusarium* wilt have been noted. Their upregulation decreases the disease incidence in plants. The JA-deficient mutant-type def1 showed higher *Fusarium* wilt disease level than wild type (Jogaiah et al. 2018).

# 6 Role of Enzymes in Disease Tolerance in Plants by Seed Priming

The upregulation or downregulation of specific enzymes play a critical role in maintaining disease tolerance, e.g., during the disease tolerance in pearl millet after seed priming with different inducers, conidial suspension (CS), cultural filtrate

(CF), and lyophilized culture filtrate (LCF) of plant growth-promoting fungus (PGPF) was due to peroxidase (POX) activity. When the seeds were treated with CS, the chitinase activity was higher in CS-treated seedlings and was reduced in later stages than control (Murali and Amruthesh 2015). Similarly, seed priming with nitric oxide (NO) donors and sodium nitroprusside (SNP) showed that aqueous SNP seed treatment with or without polythene glycol (PEG) priming both under controlled and field conditions performed better than control treatments. The mechanistic evaluation of tolerance of pearl millet against downy mildew showed that lignin deposition, hypersensitive response, and phenylalanine ammonia lyase (PAL, EC) were enhanced by NO-donor treatment plants (Manjunatha et al. 2008a, b). In the tolerance mechanisms, various enzymes were involved such as phenylalanine ammonia lyase, polyphenol oxidase, tyrosine ammonia lyase, and antioxidant enzymes, viz., activities of superoxide dismutase, catalase, and peroxidase for disease tolerance in plants (Katiyar et al. 2015). Seed priming with chitosan induces several enzymes in plants like glucanase and chitinase. It also induces many enzymes in reactive oxygen species (ROS) scavenging system such as superoxide dismutase (SOD), catalase, and peroxidase (Pichyangkura and Chadchawan 2015). In watermelon crop infected with *Fusarium oxysporum*, NADP-dependent malic enzyme (NADP-ME) had higher relative abundance compared to artificially treated plants. This has role in carbon fixation activity in C4 photosynthesis. In addition, it also participates in monolignol biosynthesis, in flavonoid biosynthesis, and of activated oxygen species (Casati et al. 1999).

The mechanism of tolerance to disease was due to the higher expression of PR (pathogenesis-related protein)-related defense enzymes (B-1,3-glucanase and chitinase). The SA alters the plant physiology and produces these enzymes: 2.5- and 2-fold increased expression of IAA27, MPK1, GPX, chitinase, and β-1,3-glucanase that are related to disease tolerance and plant growth promoting in brinjal (Mahesh et al. 2017). The PGPRS phosphorus solubilizing bacterial strain TN_Vel-35 isolated from rhizosphere induced a strong tolerance against the early blight disease in tomato caused by the *Alternaria solani*. The induced accumulation of antioxidant peroxidase (PO) and polyphenol peroxidase (PPO) upregulation was the cause of control. Furthermore, they enhanced disease tolerance by increasing IAA, nutrient uptake, and chlorophyll contents in tomato (Babu et al. 2015). In the field condition treatment with 200 mM trehalose for 9 h, 67.2% disease protection against downy mildew was recorded, and it reduced the disease severity to 32.7% compared to untreated control treatments which showed 90% of disease severity. Further, trehalose increased levels of polyphenol oxidase, phenylalanine ammonia lyase, and peroxidase, which are indicators of both biotic and abiotic stress responses (Govind et al. 2016). The enzymes involved in the development of resistance were peroxidase and chitinase (Murali and Amruthesh 2015).

# 7 Role of Proteins in Disease Tolerance in Plants by Seed Priming

Pathogenesis-related (PR) proteins are produced in plants when plants are under pathogen attack. These are induced as part of systemic acquired resistance. Upon infections genes are deactivated in plants that produce PR proteins that may be antimicrobial or antifungal depending upon infection-causing agent. Furthermore, the experiments on functional analysis of differentially stored proteins unveil the resistance mechanism of elicitor primed plant against pathogen *S. graminicola* (Anup et al. 2015). Results of the following study show the presence of involvement of proteins involved in disease tolerance, e.g., different mechanisms are involved that mediate reduction in downy mildew using seed priming with β-aminobutyric acid and *P. fluorescens*. The hierarchical clustering analysis shows that proteins are either up- or downregulated. The overrepresentation of proteins related to glucose suggests that this seed priming makes possible the protection against diseases without compromising the normal growth and development. This study formed the basis and is in need of future investigation to uncover the priming of resistance mechanism of this disease using information of differentially accumulated proteins and their functional analysis (Chandra et al. 2015).

# 8 Merits and Demerits of Seed Priming

Seed priming improves the seed germination, emergence, and seedling establishment of many plant species under controlled conditions (Basra et al. 2005). It also repairs damage of aged seeds (Butler et al. 2009) or seeds exposed to abiotic stresses (Yang et al. 2016). Out of various strategies adopted for inducing disease tolerance in plants, seed priming is easy, is economical with low risk involved, and has sustainable technique to minimize the effects of environmental conditions (Ashraf and Foolad 2005). However, seed priming has demerits like losing seed, seedling vigor, and in some cases poor or delayed germination. Induced tolerance against diseases using UV radiations is found to be beneficial against various diseases; however a precautious state developing in low levels of UV may have impairing effect on crop species for encouraging variations with regard to disease stress tolerance and enhanced crop quality and yield (Dhanya and Puthur 2017).

# 9 Future Prospective

It is needed to optimize suitable seed priming techniques for different crop plants to mitigate the challenges raised due to newly introduced diseases and their levels of threat. Moreover, it is required to investigate how different pathways work together with each other in plants at a particular time and in a particular organ or tissue. For example, regarding biosynthesis of salicylic acid and jasmonic acid, future studies should focus on how salicylic acid (SA) or jasmonic acid (JA) is synthesized, in a particular tissue or organ. Finally, the question arises how the SA and JA are sensed by the cells, and such perceptions could conceivably involve direct interaction of SA and JA with different cellular proteins such as enzymes or transcription factors or ion channels. SA- and JA-sensitive enzymes could include signaling enzymes such as protein kinase and phosphatase enzyme activities (Neill et al. 2002a).

Future work will answer to the questions on particular roles of JA and products from SA based on seed priming, on JA and oxylipin perception, on MAP kinase activity in JA signaling, and on the variety of actions of JA in plant-plant and plant-microbe interactions, and, finally, cell- and organ-specific functions of jasmonates in plant development are needed to be analyzed (Wasternack 2007).

## 10 Conclusion

Studies have shown the prominent role of seed priming in the disease tolerance and understanding of the physiological and biochemical changes. The role of seed priming under disease stress and disease tolerance in plants is understood. In this way seed priming has been found to have a vital role in plant growth and development that help disease tolerance in plants. Seed priming has been found to play a significant role in direct tolerance or priming tolerance against diseases. Further, seed priming is a smart, effective, and realistic option for acceptable plant protection strategy without disturbing seed germination and vigor as well as plant performance. Seed priming has the capacity to repair some of the damages caused by seed erosion which in turn results in increased vigor of primed seeds (Arif et al. 2008). Rapidly increasing evidences indicate that seed priming with salicylic acid or biopriming is involved in several physiological processes; however mechanisms vary among different plant species to enable plants to tolerate diseases. Seed priming mobilizes the food and nutrients in the seeds and hence improves the seed germination and vigor; moreover activation and resynthesis of some critical enzymes according to Nascimento and West (1998) are very important.

The type of seed priming is specific for specific disease tolerance in plants. Study of mechanisms shows that seed priming can activate many signaling pathways in the early stages of growth and results in faster plant defense responses. However, the biochemical or molecular basis of some pathways remained unclear. There is also a need to understand how these known pathways cooperate with each other. Future studies are needed on how these particular priming methods affect disease tolerance in different organs or tissues and which pathway is involved. There is need to fully characterize enzymes and a need to know where these are located in particular cells or tissues. Such perception will enable to know which seed priming affects the various cellular proteins, i.e., transcription factors, ion channels, and proteins.

## References


Abdel-Kader MM, Hammam MMA, El-Mougy NS, Abd-Elgawad MMM (2015) Pesticide alternatives for controlling root rot and root knot of cucumber under plastic house conditions. Int J Eng Innovat Technol 11(4):25–31

Adesemoye A, Egamberdieva D (2013) Beneficial effects of plant growth-promoting rhizobacteria on improved crop production: prospects for developing economies. In: Bacteria in agrobiology: crop productivity. Springer, New York, pp 45–63

Agrawal AA, Laforsch C, Tollrian R (1999) Transgenerational induction of defences in animals and plants. Nature 401(6748):60

Arif M, Jan MT, Marwat KB, Khan MA (2008) Seed priming improves emergence and yield of soybean. Pak J Bot 40:1169–1177

Araújo SDS, Paparella S, Dondi D, Bentivoglio B, Carbonera D, Balestrazzi A (2016) Physical methods for seed invigoration: advantages and challenges in seed technology. Front Plant Sci 7

Anup CP, Melvin P, Shilpa N, Gandhi MN, Jadhav M, Ali H, Kini KR (2015) Proteomic analysis of elicitation of *downy* mildew disease resistance in pearl millet by seed priming with β-aminobutyric acid and *Pseudomonas fluorescens*. J Proteome 120:58–74

Arora NK, Tewari S, Singh R (2013) Multifaceted plant-associated microbes and their mechanisms diminish the concept of direct and indirect PGPRs. In: Plant microbe symbiosis: fundamentals and advances. Springer, New Delhi, pp 411–449

Ashraf M, Foolad M (2005) Pre-sowing seed treatment—a shotgun approach to improve germination, plant growth, and crop yield under saline and non-saline conditions. Adv Agron 88:223–271

Ayaz M, Junaid M, Ahmed J, Ullah F, Sadiq A, Ahmad S, Imran M (2014) Phenolic contents, antioxidant and anticholinesterase potentials of crude extract, subsequent fractions and crude saponins from *Polygonum hydropiper* L. BMC Complement Altern Med 141:145

Babu AN, Jogaiah S, Si I, Nagaraj AK, Tran LS (2015) Improvement of growth, fruit weight and early blight disease protection of tomato plants by rhizosphere bacteria is correlated with their beneficial traits and induced biosynthesis of antioxidant peroxidase and polyphenol oxidase. Plant Sci 231:62–73

Baccelli I, Mauch-Mani B (2016) Beta-aminobutyric acid priming of plant defense: the role of ABA and other hormones. Plant Mol Biol 91(6):703–711

Basra SMA, Farooq M, Tabassam R, Ahmad N (2005) Physiological and biochemical aspects of pre-sowing seed treatments in fine rice (*Oryza sativa* L.). Seed Sci and Technol 33:623–628

Beckers GJ, Conrath U (2007) Priming for stress resistance: from the lab to the field. Curr Opin Plant Biol 10(4):425–431

Bernsdorff F, Döring AC, Gruner K, Schuck S, Bräutigam A, Zeier J (2015) Pipecolic acid orchestrates plant systemic acquired resistance and defense priming via salicylic acid-dependent and independent pathways. The Plant Cell TPC2015-00496-RA

Biere A, Goverse A (2016) Plant-mediated systemic interactions between pathogens, parasitic nematodes, and herbivores above-and belowground. Annu Rev Phytopathol 54:499–527

Butler LH, Hay FR, Ellis RH, Smith RD, Murray TB (2009) Priming and re-drying improve the survival of mature seeds of *Digitalis purpurea during storage*. Ann Bot 103(8):1261–1270

Caarls L, Pieterse CM, Van Wees S (2015) How salicylic acid takes transcriptional control over jasmonic acid signaling. Front Plant Sci 6:170

Caldwell J (1934) The physiology of virus diseases in plants: V. The movement of the virus agent in tobacco and tomato. Ann Appl Biol 21(2):191–205

Callan NW (1990) Bio-priming seed treatment for biological control of *Pythium ultimum* preemergence damping-off in sh2 wweet corn. Plant Dis 74(5):368

Casati P, Drincovich MF, Edwards GE, Andreo CS (1999) Malate metabolism by NADP-malic enzyme in plant defense. Photosynth Res 61 (2):99–105

Chanclud E, Kisiala A, Emery NR, Chalvon V, Ducasse A, Romiti-Michel C, Morel JB (2016) Cytokinin production by the *rice blast* fungus is a pivotal requirement for full virulence. PLoS Pathog 12 (2):1005457

Chen MH, Jack AL, McGuire IC, Nelson EB (2012) Seed-colonizing bacterial communities associated with the suppression of *Pythium* seedling disease in a municipal biosolids compost. Phytopathology 102(5):478–489

Cohen Y, Vaknin M, Mauch-Mani B (2016) BABA-induced resistance: milestones along a 55-year journey. Phytoparasitica 44(4):513–538

Cohen YR (2002) β-aminobutyric acid-induced resistance against plant pathogens. Plant Dis 86(5):448–457

Cohen R, Orgil G, Burger Y, Saar U, Elkabetz M, Tadmor Y, Freeman S (2015) Differences in the responses of melon accessions to *Fusarium* root and *stem* rot and their colonization by *Fusarium oxysporum* f. sp. *Radicis-cucumerinum*. Plant Pathol 64 (3):655–663

Conrath U (2011) Molecular aspects of defence priming. Trends in Plant Sci 16 (10):524–531

Cueto-Ginzo AI, Serrano L, Bostock RM, Ferrio JP, Rodríguez R, Arcal L, Medina V (2016) Salicylic acid mitigates physiological and proteomic changes induced by the SPCP1 strain of potato virus X in tomato plants. Physiol Mol Plant Pathol 93:1–11

de Sain M, Rep M (2015) The role of pathogen-secreted proteins in fungal *vascular wilt* diseases. Int J Mol Sci 10:23970–23993

Dhanya TTT, Puthur JT (2017) UV radiation priming: a means of amplifying the inherent potential for abiotic stress tolerance in crop plants. Environ Exp Bot 138:57–66

Egel D, Martyn R (2007) *Fusarium* wilt of watermelon and other cucurbits. The Plant Health Instructor DOI: 10.1094: PHI-I-2007-0122-01

Flowers TJ, Garcia A, Koyama M, Yeo AR (1997) Breeding for salt tolerance in crop plants—the role of molecular biology. Acta Physiol Plant 19(4):427–433

Fu W, Shen Y, Hao J, Wu J, Ke L, Wu C, Huang K, Luo B, Xu M, Cheng X (2015) Acyl-CoA N-acyltransferase influences fertility by regulating lipid metabolism and jasmonic acid biogenesis in cotton. Sci Rep 5:11790

Fukami J, Nogueira MA, Araujo RS, Hungria M (2016) Accessing inoculation methods of maize and wheat with *Azospirillum brasilense*. AMB Express 6(1):3

Gao AG, Hakimi SM, Mittanck CA, Wu Y, Woerner BM, Stark DM, Rommens CM (2000) Fungal pathogen protection in potato by expression of a plant defensin peptide. Nat Biotechnol 18(12):1307

Govind SR, Jogaiah S, Abdelrahman M, Shetty HS, Tran LSP (2016) Exogenous trehalose treatment enhances the activities of defense-related enzymes and triggers resistance against *downy* mildew disease of pearl millet. Front Plant Sci 7:1593

Gust AA, Brunner F, Nürnberger T (2010) Biotechnological concepts for improving plant innate immunity. Curr Opin Biotechnol 21(2):204–210

Hamed ER, Awad HM, Ghazi EA, El-Gamal NG, Shehata HS (2015) *Trichoderma asperellum* isolated from salinity soil using rice straw waste as biocontrol agent for cowpea plant pathogens. J Appl Pharmaceutical Sci 5:091-098

Horns F, Hood ME (2012) The evolution of disease resistance and tolerance in spatially structured populations. Ecol Evol 2(7):1705–1711

Hossain MA, Bhattacharjee S, Armin SM, Qian P, Xin W, Li HY, Tran LSP (2015) Hydrogen peroxide priming modulates abiotic oxidative stress tolerance: insights from ROS detoxification and scavenging. Front Plant Sci 6:420

Jayaraj J, Radhakrishnan N, Kannan R, Sakthivel K, Suganya D, Venkatesan S, Velazhahan R (2005) Development of new formulations of *Bacillus subtilis* for management of tomato damping-off caused by *Pythium aphanidermatum*. Biocontrol Sci Tech 15(1):55–65

Jensen DF, Karlsson M, Sarrocco S, Vannacci G (2016) Biological control using microorganisms as an alternative to disease resistance. In: Plant pathogen resistance biotechnology, Wiley Blackwell, Hoboken, pp 341–363

Jisha K, Vijayakumari K, Puthur JT (2013) Seed priming for abiotic stress tolerance: an overview. Acta Physiol Plant 35(5):1381–1396

Jogaiah S, Abdelrahman M, Tran LSP, Ito SI (2018) Different mechanisms of *Trichoderma virens*-mediated resistance in tomato against Fusarium wilt involve the jasmonic and salicylic acid pathways. Mol Plant Pathol 19(4):870–882

Jung H, Tschaplinski TJ, Wang L, Glazebrook J, Greenberg JT (2009) Priming in systemic plant immunity. Sci 324(5923):89–91

Kalaivani K, Kalaiselvi MM, Senthil-Nathan S (2016) Effect of methyl salicylate (MeSA), an elicitor on growth, physiology and pathology of resistant and susceptible rice varieties. Sci Rep 6(1)

Katiyar D, Hemantaranjan A, Singh B (2015) Chitosan as a promising natural compound to enhance potential physiological responses in plant: a review. Indian J Plant Physiol 20(1):1–9

Kaur C, Kapoor HC (2002) Anti-oxidant activity and total phenolic content of some Asian vegetables. Int J Food Sci Technol 37(2):153–161

Kaya MD, Okçu G, Atak M, Cıkılı Y, Kolsarıcı Ö (2006) Seed treatments to overcome salt and drought stress during germination in sunflower (*Helianthus annuus* L.). Eur J Agron 24(4):291–295

Klessig DF (2017) How does the multifaceted plant hormone salicylic acid combat disease in plants and are similar mechanisms utilized in humans. BMC Biol 15(1):23

Król P, Igielski R, Pollmann S, Kępczyńska E (2015) Priming of seeds with methyl jasmonate induced resistance to hemi-biotroph *Fusarium oxysporum* f. sp. lycopersici in tomato via 12-oxo-phytodienoic acid, salicylic acid, and flavonol accumulation. J Plant Physiol 179:122–132

Lal SK, Kumar S, Sheri V, Mehta S, Varakumar P, Ram B, Reddy MK (2018) Seed priming: an emerging technology to impart abiotic stress tolerance in crop plants In: Advances in seed priming. Springer, Cham, pp 41–50

Lavanya NS, Amruthesh NK (2017) Glycine betaine mediated disease resistance against *Sclerospora graminicola* in pearl millet. J App Biol Biotech 5:045–051

Leon J, Shulaev V, Yalpani N, Lawton MA, Raskin I (1995) Benzoic acid 2-hydroxylase, a soluble oxygenase from tobacco, catalyzes salicylic acid biosynthesis. Proc Nat Acad Sci 92:10413–10417

Leyva M, Rebaza L, Ruiz J (2016) Pre-conditioning treatments in bean seeds for inducing tolerance to *Fusarium oxysporum*. Fitosanidad 20(2):77–84

Lin W, Ma X, Shan L, He P (2013) Big roles of small kinases: the complex functions of receptor-like cytoplasmic kinases in plant immunity and development. J Integr Plant Biol 55(12):1188–1197

Mahesh H, Murali M, Pal MAC, Melvin P, Sharada M (2017) Salicylic acid seed priming instigates defense mechanism by inducing PR-proteins in *Solanum melongena* L. upon infection with *Verticillium dahliae* Kleb. Plant Physiol Biochem 117:12–23

Maldonado AM, Doerner P, Dixon RA, Lamb CJ, Cameron RK (2002) A putative lipid transfer protein involved in systemic resistance signalling in Arabidopsis. Nature 419(6905):399–403

Manjunatha G, Raj SN, Shetty NP, Shetty HS (2008a) Nitric oxide donor seed priming enhances defense responses and induces resistance against pearl millet *downy* mildew disease. Pestic Biochem Physiol 91(1):1–11

Manjunatha G, Roopa K, Prashanth GN, Shekar Shetty H (2008b) Chitosan enhances disease resistance in pearl millet against *downy* mildew caused by *Sclerospora graminicola* and defence-related enzyme activation. Pest Manag Sci 64(12):1250–1257

Manohar M, Tian M, Moreau M, Park SW, Choi HW, Fei Z, von Dahl CC (2015) Identification of multiple salicylic acid-binding proteins using two high throughput screens. Front Plant Sci 5:777

Maruta T, Noshi M, Tanouchi A, Tamoi M, Yabuta Y, Yoshimura K, Ishikawa T, Shigeoka S (2012) $H_2O_2$-triggered retrograde signaling from chloroplasts to nucleus plays specific role in response to stress. J Biol Chem 287: 11717-11729

Mauch-Mani B, Slusarenko AJ (1996) Production of salicylic acid precursors is a major function of phenylalanine ammonia-lyase in the resistance of arabidopsis to *Peronospora parasitica*. Plant Cell 8(2):203

Medda S, Hajra A, Dey U, Bose P, Mondal NK (2015) Biosynthesis of silver nanoparticles from *Aloe vera* leaf extract and antifungal activity against Rhizopus sp. and Aspergillus sp. Appl Nanosci 5(7):875–880

Mehari ZH, Elad Y, Rav-David D, Graber ER, Meller Harel YM (2015) Induced systemic resistance in tomato (*Solanum lycopersicum*) against Botrytis cinerea by biochar amendment involves jasmonic acid signaling. Plant Soil 39(1–2):31–44

Miller G, Suzuki N, Ciftci-Yilmaz S, Mittler R (2010) Reactive oxygen species homeostasis and signalling during drought and salinity stresses. Plant Cell Environ 33(4):453–467

Müller H, Berg G (2008) Impact of formulation procedures on the effect of the biocontrol agent *Serratia plymuthica* HRO-C48 on Verticillium wilt in oilseed rape. BioControl 53(6):905–916

Murali M, Amruthesh KN (2015) Plant growth-promoting fungus *Penicillium oxalicum* enhances plant growth and induces resistance in pearl millet against *downy* mildew disease. J Phytopathol 163(9):743–754

Nascimento WM, West SH (1998) Priming and seed orientation affect seedcoat adherence and seedling development of muskmelon transplants. HortSci 33(5):847–848

Napoleão TA, Soares G, Vital CE, Bastos C, Castro R, Loureiro ME, Giordano A (2017) Methyl jasmonate and salicylic acid are able to modify cell wall but only salicylic acid alters biomass digestibility in the model grass *Brachypodium distachyon*. Plant Sci 263:46–54

Neill SJ (2002) Hydrogen peroxide and nitric oxide as signalling molecules in plants. J Exp Bot 53(372):1237–1247

Noble AD, Ruaysoongnern S (2010) The nature of sustainable agriculture. In: Soil microbiology and sustainable crop production. Springer, pp 1–25

Oerke E, Steiner U, Dehne H, Lindenthal M (2006) Thermal imaging of cucumber leaves affected by *downy* mildew and environmental conditions. J Exp Bot 57(9):2121–2132

Patil SV, Jayamohan NS, Kumudini BS (2016) Strategic assessment of multiple plant growth promotion traits for shortlisting of fluorescent *Pseudomonas* spp. and seed priming against *ragi blast* disease. Plant Growth Regul 80(1):47–58

Pichyangkura R, Chadchawan S (2015) Biostimulant activity of chitosan in horticulture. Sci Hortic 196:49–65

Pieterse CM, Leon-Reyes A, Van der Ent S, Van Wees SC (2009) Networking by small-molecule hormones in plant immunity. Nat Chem Biol 5(5):308

Råberg L (2014) How to live with the enemy: understanding tolerance to parasites. PLoS Biol 12(11):e1001989

Rahman MM, Rahman MA, Miah MG, Saha SR, Karim M, Mostofa MG (2017) Mechanistic insight into salt tolerance of Acacia auriculiformis: the importance of ion selectivity, Osmoprotection, tissue tolerance, and Na+ exclusion. Front Plant Sci 8:155

Raj SN, Shetty NP, Shetty H (2004) Seed bio-priming with Pseudomonas fluorescens isolates enhances growth of pearl millet plants and induces resistance against downy mildew. Int J Pest Manag 50:41–48

Saltveit ME (2017) Synthesis and metabolism of phenolic compounds. Fruit Veg Phytochem 2:115

Samet M, Charfeddine M, Kamoun L, Nouri-Ellouze O, Gargouri-Bouzid R (2018) Effect of compost tea containing phosphogypsum on potato plant growth and protection against *Fusarium solani* infection. Environ Sci Pollut Res 25:1–17

Thevenet D, Pastor V, Baccelli I, Balmer A, Vallat A, Neier R, Glauser G, Mauch-Mani B (2017) The priming molecule -aminobutyric acid is naturally present in plants and is induced by stress. New Phytol 213(2):552–559

Sattler SE, Deanna L. Funnell-Harris (2013) Modifying lignin to improve bioenergy feedstocks: strengthening the barrier against pathogens?. Front Plant Sci 4

Schmidt SM, Lukasiewicz J, Farrer R, Van Dam P, Bertoldo C, Rep M (2016) Comparative genomics of *Fusarium oxysporum* f. sp. melonis reveals the secreted protein recognized by the Fom-2 resistance gene in melon. New Phytol 209(1):307–318

Serino L, Reimmann C, Baur H, Beyeler M, Visca P, Haas D (1995) Structural genes for salicylate biosynthesis from chorismate in *Pseudomonas aeruginosa*. Mol Gen Genet 249:217–228

Seyfferth C, Tsuda K (2014) Salicylic acid signal transduction: the initiation of biosynthesis, perception and transcriptional reprogramming. Front Plant Sci 5:697

Shah H, Jalwat T, Arif M, Miraj G (2012) Seed priming improves early seedling growth and nutrient uptake in mungbean. J Plant Nutr 35(6):805–816

Singh A, Gupta R, Pandey R (2016) Rice seed priming with picomolar rutin enhances rhizospheric *Bacillus subtilis* CIM colonization and plant growth. PLoS One 11(1):e0146013

Singh VK, Singh HB, Upadhyay RS (2017) Role of fusaric acid in the development of '*Fusarium* wilt'symptoms in tomato: physiological, biochemical and proteomic perspectives. Plant Physiol Biochem 118:320–332

Stewart EL, Croll D, Lendenmann MH, Sanchez-Vallet A, Hartmann FE, Palma-Guerrero J, Mcdonald BA (2018) Quantitative trait locus mapping reveals complex genetic architecture of quantitative virulence in the wheat pathogen *Zymoseptoria tritici*. Mol Plant Pathol 19 (1):201–216

Sudisha J, Kumar A, Amruthesh KN, Niranjana SR, Shetty HS (2011a) Elicitation of resistance and defense related enzymes by raw cow milk and amino acids in pearl millet against downy mildew disease caused by *Sclerospora graminicola*. Crop protect 30:794–801

Sudisha J, Kumar A, Amruthesh KN, Niranjana SR, Shetty HS (2011b) Elicitation of resistance and defense related enzymes by raw cow milk and amino acids in pearl millet against *downy* mildew disease caused by *Sclerospora graminicola*. Crop Prot 30(7):794–801

Sujeeth N, Deepak S, Shailasree S, Kini RK, Shetty SH, Hille J (2010) Hydroxyproline-rich glycoproteins accumulate in pearl millet after seed treatment with elicitors of defense responses against *Sclerospora graminicola*. Physiol Mol Plant Pathol 74(3–4):230–237

Van Hulten M, Pelser M, Van Loon L, Pieterse CM, Ton J (2006) Costs and benefits of priming for defense in Arabidopsis. Proc Natl Acad Sci 103 (14):5602–5607

Vidhyasekaran P, Muthamilan M (1999) Evaluation of a powder formulation of *Pseudomonas fluorescens* Pf1 for control of *rice sheath* blight. Biocontrol Sci Tech 9(1):67–74

Wang K, Guo Q, Froehlich JE, Hersh HL, Zienkiewicz A, Howe GA, Benning C (2018) Two abscisic acid-responsive plastid lipase genes involved in jasmonic acid biosynthesis in *Arabidopsis thaliana*. Plant Cell 30(5):1006–1022

Wang M, Sun Y, Sun G, Liu X, Zhai L, Shen Q, Guo S (2015) Water balance altered in cucumber plants infected with *Fusarium oxysporum* f. sp. cucumerinum. Sci Rep 5:7722

Wasternack C (2007) Jasmonates: an update on biosynthesis, signal transduction and action in plant stress response, growth and development. Ann Bot 100(4):681–697

Wildermuth MC, Dewdney J, Wu G, Ausubel FM (2001) Isochorismate synthase is required to synthesize salicylic acid for plant defence. Nature 414(6863):562

Wittek F, Kanawati B, Wenig M, Hoffmann T, Franz-Oberdorf K, Schwab W, Vlot AC (2015) Folic acid induces salicylic acid-dependent immunity in a rabidopsis and enhances susceptibility to a *lternaria brassicicola*. Mol Plant Pathol 16(6):616–622

Worrall D, Holroyd GH, Moore JP, Glowacz M, Croft P, Taylor JE, Roberts MR (2012) Treating seeds with activators of plant defence generates long-lasting priming of resistance to pests and pathogens. New Phytol 193(3):770–778

Yang W, Xu X, Li Y, Wang Y, Li M, Wang Y, Ding X, Chu Z, Lam HM (2016) Rutin-mediated priming of plant resistance to three bacterial pathogens initiating the early SA signal pathway. PLOS ONE 11(1):e0146910

Yang C, Hu L, Ali B, Islam F, Bai Q, Yun X, Zhou W (2016) Seed treatment with salicylic acid invokes defence mechanism of *Helianthus annuus* against *Orobanche cumana*. Ann Appl Biol 169(3):408–422

Zida P, Neya B, Soalla W, Prakash H, Niranjana S, Udayashankar A, Andresen M (2015) Increased feasibility of treating sorghum seeds with *Eclipta alba* extract by lowering concentration of plant extract and soaking time of seeds. Int J Trop Agric 33:2391–2400

# Seed Pretreatment as a Means to Achieve Pathogen Control

**Muhammad Taimoor Shakeel, Rashida Parveen, Idrees Haider, Muhammad Arshad, Shakeel Ahmad, Niaz Ahmad, Sajjad Hussain, Muhammad Riaz, and Muhammad Arif Ali**

## Contents



**Abstract** Seed treatment is the first and basic technique being practiced for seed protection since long. Chemicals and biological agents are applied to protect seeds from microbial pathogens and insect pests. The application of seed treatment is being evolved to more environment-friendly compositions as the time passes. Purposes of seed treatment are to reduce the economic losses to the crop due to

---

M. T. Shakeel · R. Parveen
Department of Plant Pathology, Bahauddin Zakariya University, Multan, Punjab, Pakistan

I. Haider · N. Ahmad · M. A. Ali (✉)
Department of Soil Science, Bahauddin Zakariya University, Multan, Punjab, Pakistan
e-mail: arif1056@bzu.edu.pk

M. Arshad
Institute of Environmental Sciences and Engineering, School of Civil and Environmental Engineering, National University of Sciences and Technology, Islamabad, Pakistan

S. Ahmad (✉)
Department of Agronomy, Bahauddin Zakariya University, Multan, Punjab, Pakistan
e-mail: shakeelahmad@bzu.edu.pk

S. Hussain
Department of Horticulture, Bahauddin Zakariya University, Multan, Punjab, Pakistan

M. Riaz
Department of Environmental Sciences and Engineering, Government College University Faisalabad, Faisalabad, Pakistan

 
M. Hasanuzzaman, V. Fotopoulos (eds.), *Priming and Pretreatment of Seeds and Seedlings*, https://doi.org/10.1007/978-981-13-8625-1_17

pathogens and avoid development of resistant pathogen races and reduction of pesticide residues in the product. The purpose of the seed treatment is to avoid the pathogen attack and protection of seeds and seedlings. Application of chemicals at the early stage has an advantage of effectiveness of low dosage as the pathogen is very vulnerable because the seedborne stage is the weakest part of the life cycle of microbes. On the other hand, the disadvantages include the possibility of accidental poisoning, limited dose capacity, shorter shelf life of the treated seeds, and exposure of the worker to the chemicals when treating larger quantities of seeds. Major fungal diseases including bunts, water molds, rots, and damping-off and bacterial diseases including rots and wilting are commonly controlled by applying chemicals. A few viral diseases are transmitted by the soilborne fungal pathogens that may result into a major problem at the later stages and must be controlled at early crop stage by application of seed treatment of their fungal vectors. Soilborne insects like rootworms, maggots, and wireworms are to be controlled at the sowing stage. To control these pathogens, several biological agents like *Bacillus subtilis*, *Streptomyces griseoviridis*, and *Trichoderma harzianum* are also used along with the synthetic chemicals. The chemical applications are done using carboxin, difenoconazole, fludioxonil, imazalil, mefenoxam, tebuconazole, thiabendazole, and triadimenol. The insecticides commonly used for the seed treatment are chlorpyrifos, imidacloprid, permethrin, and thiamethoxam. Several additives including colors, antifoaming agents, lubricants, carriers, and micronutrients are also used to enhance the efficiency of the seed treatment. Application of biological and chemical treatments at the early stage of the crop is playing a helpful role in maintaining the healthy crop and harvesting an improved yield.

**Keywords** Seed priming · Seed treatment · Fungicides · Insecticides · Seed diseases

# 1 Introduction

Procedures of seed treatment are in practice for thousands of years by the farmers to save their crops from pests and pathogens. At the beginning, the seed treatment was practiced excluding fungal and bacterial pathogens to improve the crop productivity. Seed treatment started with application of lime solutions and brine which got evolved to the application of formulated products. Hit and trial methods kept on altering the formulations and applications of different mixtures and solutions over time, but in the early 1800s, the application of copper sulfate was found to be more efficient than lime solutions for the control of bunt. Later in 1920, dusting of copper carbonate proved to be even more effective than copper sulfate.

The seed treatment products kept on evolving with time, and mercury-based chemicals appeared to be effective against several seedborne biological issues. Although this protective application was very effective, it was still carrying a risk of unintentional poisoning, due to which it was banned in 1970. After the World War

II, several organic fungicides were introduced in the 1980s. The process of evolution in seed treatment methods came up with biological control in 1990.

Seed treatment is defined as the application of seeds with chemical or biological material to avoid or control the pathogens or pests. Seed-treating chemicals might include insecticides, fungicides, or bactericides. Seed treatment chemicals can be applied both to the true seeds and propagative material such as bulbs, tubers, or corms. On the other hand, application of growth regulators, nitrogen-fixing bacteria, and micronutrients to the seeds is not considered to be seed treatment because they are not meant to reduce the pathogen inoculum or pest control.

Seed treatment is essentially applied to the seed to obtain a uniform and well-grown seed. Apart from this, the continuous application of the seed treatments over years has compromised the survival of soilborne pathogens, and the problem of smut or bunt is under control in wheat and barley crops. Also, it has been found effective in reducing the root rots. Although the application of chemicals to the seeds is beneficial, it also poses to be a threat for the workers in case of accidental exposure, and sometimes they contaminate the food by unnoticed mixing of treated seed with feed or food products.

## 2 Purpose of Seed Treatment

Purposes of seed treatment include protection of seeds and seedlings, control of seedborne pathogens, and control of pests. Seedborne disease-causing pathogens may work as infestants or may be present deep inside the seeds, hidden in the cracks of the cervices. These pathogens may be important to be controlled in three possible conditions. Firstly, these pathogens cannot survive in the soil or crop residues and need the soilborne phase to survive the fallow period between the crops. Secondly, these pathogens may survive in the soil or crop residues and behave as soilborne pathogen which adds to its capacity to cause disease. Thirdly, these soilborne pathogens may invade new soils by getting transported with the seed shipments.

Seed treatment might be efficient in controlling the seed-infesting or seed-infecting pathogens. The method to be implemented for the control of the disease is decided by the nature of the pathogen whether it is residing on or inside of the seed.

Soilborne pathogens and pests are threat to the seeds and seedlings, which may cause heavy damage to the crops. Especially, the young seedlings are not able to withstand the stress because of their tender nature and lack of food availability to recover from the damage. Apart from pathogen and pest attack, the stress may include heavy rains, cool soils, soil compaction, hard crust of soils, and postemergence herbicide application. Under the prevalence of unfavorable conditions in the soil environment, the pathogen with less virulence may turn out to be devastating and cause heavy loss to the crop.

Application of the non-systemic seed treatment chemicals can protect the seeds and seedlings from damage by making up a protective barrier on the surface of the seed and avoiding the chewing insects and soilborne pathogens. Systemic seed treatment chemical can help avoiding the damage to the roots by root rotting fungi as well as the aboveground parts of the seedlings from sucking type of pests and the aerial pathogens. The duration of the seed treatment efficacy stands for 11–14 days which is even less if the environmental conditions are warm and humid. Considering the effect of these seed treatment chemicals is not long-lasting but can help in delaying the infection and hence reducing the yield losses.

Usually, seed treatments are not the only way to control soilborne pathogenic diseases. Considering the cost-effectiveness, other methods of disease control like use of certified seeds, crop rotation, heat treatment, resistant varieties, and observing the planting date can be added to the seed treatment to devise an effective control strategy because implementation of a single control measure may not be sufficient to avoid the disease.

## 3 Advantages and Disadvantages of Seed Treatment

There are several advantages of seed treatment, which include the vulnerability of the pathogen as seedborne stage of the plant pathogens is the weakest phase of many plant pathogens during their life cycles that makes seed treatment an effective control measure for the disease. Due to vulnerability of the pathogen to the seed treatment chemical, the low-dose application is cheap and enough to control the disease without developing any unwanted impact on the environment. Pathogens as well as plants are also vulnerable to the diseases during seedling stages, and seed treatment is essential to be executed at this stage. Another advantage is precision targeting of the disease source as the chemical is applied directly to the seed and very little amount of the chemical is lost on untargeted sites.

Disadvantages of the seed treatment with chemicals include the accidental poisoning of animals and birds that may pick/eat the treated seeds as feed from the environment. Low-dose application often results into short-term protection of the plants from the disease. Another disadvantage is low shelf life of the seeds after treatment with the pesticides. Phytotoxicity is also a problem as the higher concentration of the seed-treating chemical is possibly not appropriate for the young seedlings and their tender tissues. If the application rate of the chemicals is not calculated and controlled, the crop may result in low germination percentage or stunting in some cases. Other than plants, human may also become affected by the seed treatment chemicals as the workers are exposed to higher doses of the chemical while treating a greater lot of the seeds.

## 4 Imperative Seed Diseases and Pests

Seed treatment chemical is specific to a group of pathogens or pests. To devise an effective and appropriate control strategy, it is vital to understand the life cycle and behavior of the pathogen or pest. Understanding of their behavior, time of attack, and mode of action will help in defining an appropriate defense strategy, type of seed treatment chemical, and correct time and dose of the application. Pests or the pathogens are not sometimes controlled properly because pesticides with required mode of action are not applied which don't have enough systemic activity. The failure to obtain desired results may also be attributed to the lack of overlapping of the peak periods of pest or pathogen damage and the protection. Economic constraints also hinder the process as sometimes the higher doses of the control chemicals are too expensive to be applied to larger seed quantities. In such situations, the seed treatment is better to be added up with cultural control or use of resistant varieties.

One of the largest and destructive groups of plant pathogens that affect the crop at seedling stage is fungi. Most fungi are decomposers and derive their nutrients from dead organic matter both from plant and animal sources. Some plant pathogenic fungi are not able to survive in the soil for longer period of times, while most stem and root rot fungi are able to stay viable in soils for longer duration even when crop rotation is practiced starving them by not sowing their host crops for a reasonable duration. Seed treatment techniques can be implemented to overcome several seedborne, soilborne, and foliar diseases that appear on the crops during the early stages (Maude 1996).

Very common fungal diseases include bunts, foliar disease like powdery mildew and rusts, water molds like *Pythium* (Hendrix and Campbell 1973) and *Phytophthora*, root rotting fungi like *Fusarium*, and damping-off fungi. Additionally, some bacteria are also pathogenic to crop plants during the early crop stages, but unlike the fungi they cannot move on their own; rather they depend on wind, seed, insects, farm equipment, and other ways that can facilitate them to come in contact with the host plants. They can get access to the host plant only via wounds or natural openings and cause local or systemic symptoms. Bacteria cause several symptoms like leaf spots, soft rots, leaf blights, and wilting, and some of them can be controlled by seed treatments. In beans, the most common diseases caused by bacteria are bacterial blights, halo blights, and brown leaf spots. The spot diseases initiate with small angular, light green, and water-soaked spots, which become dry and brown colored with narrow yellow margins. Halo blight lesions are encircled by a relatively wider pale green to yellow colored tissue. Pod lesions start with the same pattern as followed by leaf lesions, but as they mature, they produce sunken, irregularly shaped reddish brown blotches (Garu et al. 2004). Stewart's bacterial wilt is a common problem of sweet corn with long yellowish streaking symptoms developed parallel to the veins of the infected leaves (Pataky et al. 2000). These symptoms appear to be similar to the symptoms of drought, nutritional deficiency, or insect pest attack.

Plant viruses are spread by the insect vectors from infected to healthy plants. The most common vectors of the viruses are aphids, whiteflies, and leafhoppers. Although the viruses cannot be controlled by application of any pesticide or seed treatment chemical, their mode of transmission can be interrupted by controlling the insect vectors. Vector control is effective in the case of viruses that are transmitted in persistent manner because they cannot be transmitted instantly like those which are transmitted in nonpersistent manner. The viruses that are transmitted in persistent manner need to be accumulated in the vector and can be transmitted to the host plants for several days once they are acquired from a diseased plant. For successful transmission of the virus, the vector needs plentiful time to feed on the healthy plant that provides an opportunity for the application of the insecticide and control the vector. On the other hand, the viruses transmitted by the vector in nonpersistent manner require less feeding time for both acquisition and transmission of the virus making the application of the insecticides ineffective to control the spread of the disease.

In addition to the pathogens, insect pests can be controlled by the seed treatment applications. Under low rootworm pressure, the application of seed treatment chemicals is reliable, but under high pest pressure, the seed treatment must be aided with the soil treatment. Maggots, beetles, wireworms, and grubs are also important pests of the crops that cause damage to the underground parts of the plants. Root feeding may cause nutrient deficiency symptoms on the foliar parts of the plants. Wireworms are common pests of the vegetables, small grains, and cereal crops. Larvae of the wireworm cause damage to the crop by boring into the germinating seed or into the base of the seedling (Van Herk and Vernon 2007). The treatment is beneficial only if applied before the damage. Light pest pressure can be managed by the application of seed treatment, while high pressure can be controlled only by soil application or in severe cases replanting of the crop.

Aphids are one of the major sucking insect pests of several crops. Aphids are commonly green in color but may possess any color. Nymphs are like the adults but lesser in size. The adult aphids have needlelike mouthparts and suck the juice from the host body. They can cause damage to the plant both directly and indirectly (Dedryver et al. 2010). Direct damage is caused by the sucking of sap and injecting the plant toxin that is present in the saliva of the insect. Indirect damage to the crop is caused by transmission of viral diseases caused by *Barley yellow dwarf virus*, *Cucumber mosaic virus*, and *Watermelon mosaic virus* in several host plant species. Several seed treatment chemicals are found to be effective against the aphids and hence reducing the spread of the viruses that are vectored by this group of insects.

## 5 Seed Treatment Products

Seed treatment chemicals are categorized based on different attributes like active ingredients (a.i.), additives, doses, and methods of application. Different ingredients have different limitations of application that's why the seed treatment products have more than one active ingredients. The level of control of the pathogen depends on the strength and dose of the active ingredients because some pathogens require a higher dose of the chemical for the control as compared to others (Halmer 2000).

Evaluation of the seed treatment method depends on several factors like unavailability of the effective control activity of the active ingredient, less systemic activity than required, limited movement of the active ingredient with the developing root system, inability to extend the effective period so that the peak activity time and the damage period of the pathogen overlap, and disease is controlled. Another important factor is the effective dose of the chemical is expensive or is phytotoxic that limits the application of the active ingredient to control the problem. Active ingredients of the seed applications are categorized as systemic and contact. Systemic chemicals penetrate the seeds and move in the stem and reach the leaf tissues, while the contact treatment only protects the outer surface of the seed.

The seed-treating chemicals are divided into bactericides, fungicides, and insecticides based on the pathogens against which they are used. Commonly used bactericide is streptomycin, which is a broad-spectrum chemical to control halo blight pathogen of beans, soft rot, and blackleg of potato.

Fungicides are used to control the fungal seed pathogens. Few examples of the fungicides applied for the control of fungal problems include captan (a.i., ethyl mercaptan, broad-spectrum, non-systemic fungicide to control several seed decaying and rotting fungi such as *Aspergillus*, *Fusarium*, and *Penicillium*), carboxin (systemic fungicide used against smuts, seed rots, damping-off, and seedling blights and commonly used with other fungicides and insecticides), difenoconazole (systemic fungicide used against loose smut, bunt, root rots, powdery mildew, and rusts), fludioxonil (broad-spectrum, non-systemic fungicide used against decay and damping-off fungi including *Rhizoctonia*, *Fusarium*, *Aspergillus*, and *Penicillium*), imazalil (systemic fungicide, used against common and dryland root rot), mefenoxam (narrow-spectrum fungicide, systemic fungicide used against *Pythium*, *Phytophthora*, and powdery mildews), PCNB (pentachloronitrobenzene, non-systemic fungicide, used against *Rhizoctonia*, *Fusarium*, and common bunts), tebuconazole (broad-spectrum, systemic fungicide, used against some fall-season root rots and some foliar diseases), thiabendazole (broad-spectrum, systemic fungicide used against common bunt, seed decay, damping), thiram (broad-spectrum, non-systemic fungicide used against bulb, tuber decay, damping-off, and common bunts), and triadimenol (broad-spectrum, systemic fungicide used against root rots and foliar diseases).

Biological agents are also effective component of the seed treatment strategy and are composed of dormant microorganisms. Under the favorable conditions, the inocula of the biological control agents germinate, grow, and colonize the outer surface of the seed and prove protection layer against the soilborne pathogenic fungi. Biocontrol agents provide protection by releasing antibiotic agents or by competing with the pathogenic microorganisms. Biocontrol agents belong to both bacteria and fungi. Bacterial organisms include *Bacillus subtilis* and *Streptomyces griseoviridis*, and fungal agents include *Trichoderma harzianum* (Lifshitz et al. 1986; Turner and Backman 1991; Mastouri et al. 2010; Burges 2012).

Important insecticides used to control insect pests include chlorpyrifos and diazinon (non-systemic insecticide used against seed-corn maggots and beetles); imidacloprid (systemic insecticide used against aphids, bugs, beetles, leafhopper, thrips, white grubs and wireworms, and whiteflies); lindane, tefluthrin, and permethrin (non-systemic insecticide used against soilborne insects such as wireworms and maggots); and thiamethoxam (systemic insecticide effective against several chewing and sucking types of insects such as Colorado potato beetle, seed-corn maggot, Hessian fly, leafhoppers, thrips, and aphids) (Schmeer et al. 1990, Pike et al. 1993).

## 6 Formulation and Additives

Seed-treating chemicals are commonly formulated as dry flowable (DF), flowable (F), flowable seed treatment (FS), liquid (L), liquid suspension (LS), or wettable powder (WP). Different formulations have different advantages and disadvantages. For example, some formulations don't mix well in the tank, while others readily settle down and make aggregates. On the other hand, some formulations are available in water-soluble form and are relatively safer as they reduce the chance of exposure to the workers.

In addition to active ingredients, the seed treatment chemicals are added with a number of additives like dyes, carriers, binders and stickers, antifoam agents, lubricants (Beilfuss and Gradtke 2001), and micronutrients. Colorants and dyes improve the appearance of the chemical and help in maintaining the uniformity and standardization of the application to the seed lots. Addition of color to the treated seed also helps in detection of treated seeds with the feed or food grains. Carriers and binder are usually added to the seeds to develop the adherence of the chemical to the seed and avoiding the dusting off or cutting off the ingredient. Addition of antifoaming agents in the chemical formulations for seed treatments diminish the troublesome of the foam generation during application.

## 7 Storage of Treated Seeds

Pesticide-treated seed must be stored in a ventilated, clean, and dry place and should never be stored in bulk or close to the edible grains (Ashraf and Foolad 2005). The treated seeds should be stored tightly in woven bags after proper drying since undesired moisture content may deteriorate the seeds quickly. After packaging the seed must be labeled with type of seed and chemicals used to treat the lot. If the seed is stored for longer period, the germination capacity of the lot must be assessed before sale or sowing. If the seed is not good enough for further use, it should be disposed of properly, and the label of the storage should be added with appropriate information regarding the dispose-of methods (Balešević-Tubić et al. 2010).

## References

Ashraf M, Foolad MR (2005) Pre-sowing seed treatment a shotgun approach to improve germination, plant growth, and crop yield under saline and non-saline conditions. Advagro 88:223–271

Balešević-Tubić S, Tatić M, Đorđević V, Nikolić Z, Đukić V (2010) Seed viability of oil crops depending on storage conditions. Helia 33:153–160

Beilfuss W, Gradtke R (2001) U.S. Patent Application No. 09/734,646

Burges HD (2012) Formulation of microbial biopesticides: beneficial microorganisms, nematodes and seed treatments. Springer Science & Business Media, Dordrecht

Dedryver CA, Le Ralec A, Fabre F (2010) The conflicting relationships between aphids and men: a review of aphid damage and control strategies. C R Biol 333:539–553

Grau CR, Dorrance AE, Bond J, Russin JS (2004) Fungal diseases. Soybeans: improvement, production, and uses (soybeans improve) pp 679–763

Halmer P (2000) Commercial seed treatment technology. In: Seed technology and its biological basis. Sheffield Academic Press, Sheffield, pp 257–286

Hendrix FF, Campbell WA (1973) Pythiums as plant pathogens. Annu Rev Phytopathol 11:77–98

Lifshitz R, Windham MT, Baker R (1986) Mechanism of biological control of preemergence damping-off of pea by seed treatment with *Trichoderma* spp. Phytopathology 76:720–725

Mastouri F, Björkman T, Harman GE (2010) Seed treatment with Trichoderma harzianum alleviates biotic, abiotic, and physiological stresses in germinating seeds and seedlings. Phytopathology 100:1213–1221

Maude RB (1996) Seedborne diseases and their control: principles and practice. CAB international, Wallingford

Pataky JK, Michener PM, Freeman ND, Weinzierl RA, Teyker RH (2000) Control of Stewart's wilt in sweet corn with seed treatment insecticides. Plant Dis 84:1104–1108

Pike KS, Reed GL, Graf GT, Allison D (1993) Compatibility of imidacloprid with fungicides as a seed-treatment control of Russian wheat aphid (Homoptera: Aphididae) and effect on germination, growth, and yield of wheat and barley. J Eco Ento 86:586–593

Schmeer HE, Bluett DJ, Meredith R, Heatherington PJ (1990) Field evaluation of imidacloprid as an insecticidal seed treatment in sugar beet and cereals with particular reference to virus vector control. In: Brighton crop protection conference, pests and diseases, vol 1. British Crop Protection Council, pp. 29–36

Turner JT, Backman PA (1991) Factors relating to peanut yield increases after seed treatment with *Bacillus subtilis*. Plant Dis 75:347–353

Van Herk WG, Vernon RS (2007) Soil bioassay for studying behavioral responses of wireworms (*Coleoptera: Elateridae*) to insecticide-treated wheat seed. Environ Entomol 36:1441–1449

# Hydropriming for Plant Growth and Stress Tolerance

**Muhammad Fasih Khalid, Sajjad Hussain, Muhammad Akbar Anjum, Shaghef Ejaz, Maqsood Ahmad, Mehwish Jan, Sumaira Zafar, Iqra Zakir, Muhammad Arif Ali, Niaz Ahmad, Muhammad Junaid Rao, and Shakeel Ahmad**

## Contents



**Abstract** The technique of seed priming has potential in increasing seed longevity during the storage, reducing the time of seedling emergence, improving germination as well as increasing the crop yield, which is reported in various crops. Among the different seed priming techniques, hydropriming is one of the mostly used seed

M. F. Khalid · S. Hussain (✉) · M. A. Anjum · S. Ejaz · M. Jan · S. Zafar
Department of Horticulture, Bahauddin Zakariya University, Multan, Punjab, Pakistan
e-mail: sajjad.hussain@bzu.edu.pk

M. Ahmad
Mirpur University of Science and Technology (MUST), Mirpur, Pakistan

I. Zakir · S. Ahmad (✉)
Department of Agronomy, Bahauddin Zakariya University, Multan, Punjab, Pakistan
e-mail: shakeelahmad@bzu.edu.pk

M. A. Ali · N. Ahmad
Department of Soil Science, Bahauddin Zakariya University, Multan, Punjab, Pakistan

M. J. Rao
Key Laboratory of Horticultural Plant Biology (Ministry of Education), Huazhong Agricultural University, Wuhan, People's Republic of China



M. Hasanuzzaman, V. Fotopoulos (eds.), *Priming and Pretreatment of Seeds and Seedlings*, https://doi.org/10.1007/978-981-13-8625-1_18

priming techniques to start the germination of seeds without the emergence of radicle. Recent reports suggest that hydropriming of seed has the potential to upregulate plants' tolerance to multiple abiotic stresses by enhancing seed germination, seedling growth and development, modulating physiological activities and antioxidant response, as well as expression of genes and proteins. But the changes occurred during hydropriming need to be further addressed by plant scientists using molecular and genomic approaches to cope with abiotic stresses. Therefore, in this chapter, we reviewed the recent knowledge on seed hydropriming focusing abiotic stress tolerance in various plant developmental stages along with the antioxidant metabolism and molecular responses of crops.

**Keywords** Abiotic stress · Antioxidant defence system · Germination · Priming · Physiological response

## 1 Introduction

Delay in seed germination and nonuniformity is a major problem. To establish a healthy crop, unified and rapid germination of seeds is very necessary. Crop yield and quality are directly affected by the growing period from time of sowing to establishment of crop. This growing time of a crop is very important to get better yield. Difference in germination time of seeds is a major factor for low yield of a crop. After uniform and prompt emergence, unified and fast growth of seeds is very essential to get good yield of a crop. The technique of seed priming has potential to cope with the poor seed emergence in crops, which have small seeds. By the technique of seed priming, germination rate of seeds and uniformity in crop establishment is increased. Seed priming only regulates the early process of germination in seeds, but the final phase of radicle emergence is not affected. In the process of seed priming, seeds are soaked in water or either in osmotic solution or in mixture of water and solid medium in a precise ratio. After this process, seeds are dried before the emergence of radical. Previous studies reveal that priming of seed is used to hasten time of seedling emergence, to increase seed longevity during the storage, to improve germination, and to increase the yield. Favorable effects of seed priming are shown in crops such as barley, sugar beet, lentil, chickpea, and grass fox. Beneficial effects of priming are related with various biochemical, molecular, and cellular metabolisms including synthesis of proteins and ribonucleic acid (Bray et al. 1989; Davison and Bray 1991; Dell'Aquila and Bewley 1989).

## 2 Hydropriming

Among different seed priming techniques, hydropriming is one of the mostly used techniques. Seeds are soaked in a solution called priming agent followed by drying in this technique. This technique is used to start the germination of seeds without the emergence of radical (Afzal et al. 2005). Seeds are quickly reached to a specific level of moisture in hydropriming with continuous supply of oxygen that will increase the number of enzymes associated with the production of energy and metabolites associated with the process of germination. The technique is used to get uniformity in germination and to increase the speed of germination for the improvement of final stand. If seeds are contaminated with the pathogens, then hydropriming should be used with complete care. If seeds are affected with fungus, then the growth of fungus will be increased by hydropriming that will cause stand growth of plants. To control the adverse effects of pathogens, biopriming was developed.

# 3 Hydropriming-Mediated Stress Tolerance

## *3.1 Germination*

Seed germination is very much sensitive to abiotic stresses. When exposed to uneven environment, it delays germination. When seed germinates in stress condition, it cannot withstand the field environment and has less vigor and decrease in productivity. Seed germination is utmost salt-sensitive plant growth stage (Goumi et al. 2014) and is severely interrupted by increasing salinity (Bouda and Haddioui 2011). Under salinity, seed germination and plant growth are improved by seed priming (Passam and Kakouriotis 1994). The major purpose of seed priming is to maximize emergence rate; however, postemergence growth are also enhanced by these. Primed seeds give earlier, uniform and occasionally maximum germination and seedling emergence (Bradford 1986). Exposure to soaking the seed technology has been tested in various crops to aid the seed germination depending on the situation and even promote plant tolerance in the event of stress (Idris and Aslam 1975). Priming treatments have proven to promote seed vigor that is effective to achieve fast and uniform seed germination of various vegetable species (Taylor et al. 1998) and many field crops including rice (Lee et al. 1998). Hydropriming is a unique, easy, low-priced, and nature-friendly priming technique due to the use of simple water, and the farmers of developing countries adopt this technique (Mabhaudhi and Modi 2011). Earlier study indicated hydropriming (for 24 h) in wheat showed more production (Kahlon et al. 1992). Hydropriming is an effective method, as suggested by Harris (1992) with various valuable impacts on different field crops, i.e., chickpea, soybean, maize, sunflower, and rice (Ashraf and Foolad 2005; Kaya et al. 2006).

Hydropriming has also been used profitably in sunflower, wheat, cotton, and chickpea (Kaur et al. 2002). Furthermore, hydropriming elevated emergence and growth under salinity (Kaur et al. 2002; Kaya et al. 2006). Hydropriming is a starter procedure for germination without emergence of the radicle that involves soaking of seeds in water followed by drying (Ashraf and Rauf 2001). Roy and Srivastava (2000) proposed that the negative effect of salinity on seed germination can be reduced by seed priming. Generally, seed germination and seedling emergence are enhanced by hydropriming. Hydropriming permits the seeds to rapidly attain good moisture with a persistent oxygen supply, thus resulting in increase of germination process/metabolites which associated with energy production. Generally, hydropriming has been established as profitable and is currently being explored. Hydropriming brings about certain physiological modifications in organic compounds, sugar content, and cumulated ions within the seed, root, and lastly in the leaves of plant leading to high germination rate and great resistance to unpleasant conditions (Alvarado et al. 1987). Chojnowski and Come's (1997) work revealed that the increase in ATP production, the respiration activities, protein synthesis, and induced RNA activity in the treated seeds promoted emergence rate of sunflower seed.

Hydropriming is utilized to enhance the germination and uniformity of seedlings along with improvement in final stand. Hydropriming also deal with attention as seeds can be contaminated with pathogens. In such case, during hydropriming, growth of fungus can be intensified causing plant diseases and restricted development. Hydropriming enhanced the seedling emergence along with growth in *Phaseolus vulgaris* (Kazem et al. 2010). However, in contrast, hydropriming failed to improve germination in common Kentucky blue grass seeds (Pill and Necker 2001), but Basra et al. (2002) found that wheat seeds' treatment with hydropriming (48 h) resulted in the maximum activation followed by 24 h. The favorable priming effect has been related with several biochemical, cellular, and molecular processes containing RNA and protein synthesis (Dell'Aquila and Bewley 1989). By hydropriming, use of chemicals can be reduced or eliminated, and it also avoids disposing of undesirable materials that are hazardous for the atmosphere (McDonald 2000). Hydropriming specifically confirms speedy and uniform emergence along with maximum normal seedling percentage (Singh 1995; Shivankar et al. 2003). Scientists observed that antioxidative enzymes (peroxidase (POD), catalase (CAT), and superoxide dismutase (SOD)) and solute (proline and soluble sugar) contents are significant indicators throughout the process of seed priming and emergence (Wattanakulpakin et al. 2012). Germination metabolites and osmotic adjustment are vital factors to improve seed execution throughout the treatments (Haghpanah et al. 2009). Hydroprimed maize seeds (cv. Zhengdan 958) germinated quicker as compared to non-primed. There was marked change in comparative proteomic analysis between the primed and non-primed seeds, due to identification of eight protein spots (Gong et al. 2013). Hydropriming improved onion seed germination effectively, especially when the seeds were hydrated for 96 h. In cotton, soaking in distilled water (hydropriming) boosted up germination under saline environments as compared to any other seed pretreatment (Shannon and Francois 1977). Moreover, according to Ahmadian et al. (2009), hydropriming under salt stress conditions significantly improved both emergence rate and mean emergence time with maximum normal germination. Results are confirming the investigations in mustard, cauliflower, and brassica. Under osmotic stress, hydropriming promoted fresh weight of seedling. In addition, it resulted in enhanced root growth along with the highest root length. Cayuela et al. (1996) reported that plants from primed seeds have maximum salinity tolerance due to high osmotic adjustment ability. Roots produced from seeds (primed) show high level of sodium and chloride ions, while leaves of primed seeds have more contents of sugars and organic acids as compared with non-primed (Li et al. 2011). Studies showed that hydropriming increases emergence and hasty germination of seedling of pyrethrum in saline and nonsaline conditions, which agreed with results, e.g., in potato seeds, triticale, and sunflower (Demir and Ermis 2003, Yagmur and Kaydan 2008). In hydropriming, the positive effect showed that mobilization of enzymes in tissues of embryo occurred which results in early germination of seeds and production of compounds (i.e., proteins, sugars, and amino acids) in storage organs (Ashraf and Foolad 2005). Practically, due to less expenditure, hydroprimed seeds can be suggested for both salt-affected and non-affected soils to increase probability and rapidity of germination. According to Demir and Mavi (2008), reduced water loss and variances in movement of storage reserves and proteins can cause undesirable effects on germination. In conclusion, the study suggested that pyrethrum seed can be germinated best in the dark by hydropriming treatment. By several seed priming techniques, the negative effect of salt stress on emergence can be relieved (Roy and Srivastava 2000). Akter et al. (2018) proposed that 48 h of hydropriming is the best technique for maize which decreases the salinity stress and had highly significant impact on better germination and superior performance of various seedling attributes. Shukla et al. (2018) investigate that priming is a mechanism that controls germination capacity of seeds by altering the penetrability properties of membranes and enzyme. Rouhi et al. (2011) suggested that hydropriming maximized germination ratios, emergence index, and seed vigor while minimized mean emergence time, the time to get 50% emergence, and energy of emergence. Under salinity and drought, hydropriming resulted in alleviation of harmful effects of both stresses on emergence and growth of the seedling (Kaya et al. 2006). Sung and Chiu (1995) observed that hydropriming hastened mean germination time (MGT) without changing uptake water amount in watermelon. Amini (2011) suggested that hydropriming of alfalfa was an effective method to increase the salt tolerance ability by improving seed emergence and seedling growth under highly saline state. Priming treatment, under saline condition, considerably improved the CAT, POD, SOD activities, and proline content however reduced in malondialdehyde (MDA) accumulation and electrolyte leakage. Hydropriming has proven to be an effective method as compared to halopriming as expressed by the final germination percentage (FGP), root shoot length ratio, and their dry weight (Basra et al. 2006). By priming technique, radicle emergence speeds up through higher enzymatic activities, ATP production, and protein synthesis that cause increase in root length (Parera and Cantliffe 1992). Primed seed results in significantly higher shoot length than that of not primed seeds.

### 3.1.1 Plant Growth and Development

Khazaei et al. (2016) suggested that, by seed priming, positive remarkable effect can be observed on the length of plumule, green area, fresh and dry weight of plumule, fresh and dry weight of radicle, radicle length, and volume along with seedling vigor at various levels of salinity stress in barley, while time to get 50%

germination reduces by the seed treatment. Especially, effective and considerable influence of seed priming on barley was detected under the 200 mM salt stress. Priming causes increment in cell division level within the apical meristem of seedling roots that ultimately results in an increase in plant growth and development (Sakhabutdinova et al. 2003). Hydropriming can be a beneficial tool for production of carrot to increase quality of seed and seedling. Eisvand et al. (2011) work on carrot cultivars (Nantes and Forto) suggested that in both, vigor index, rate of emergence, and length of root and shoot of hydropriming were higher than hormonal priming. Sanchez et al. (2001) found an increase in the root length of pepper and cucumber by hydropriming influence. Salinity stress has adverse effects on growth, chlorophyll content (chlorophyll b reduction was more than chlorophyll a), and mineral composition ($K^+$ and $Ca^{2+}$) in all coriander genotypes. By seed priming, the negative effect of salinity diminished in all genotypes and treated plants exhibited satisfactory reaction to salt stress when compared with untreated or non-primed plants (Meriem et al. 2014). These results are in line with the outcome of many researches on various plant species, e.g., wheat (Mehta et al. 1979), canola (Mohammadi 2009), maize (Bakht et al. 2011), *Sorghum bicolor* (Kadiri and Hussaini 1999), *Vicia faba* (Salam et al. 1999), tomato (Mirabi and Hasanabadi 2012), watermelon (Armin et al. 2010), melon (Sivritepe et al. 2003), *Ziziphus spina-christi* (Takhti and Shekafandeh 2012), and safflower (Elouaer et al. 2012). By primed seed, soluble sugar content and chlorophyll content also amended in leaves of all the coriander cultivars under normal and saline conditions (Meriem et al. 2014).

Seed priming upsurges the action of membrane-bounded enzyme and the free radical scavenging enzymes, i.e., POD, SOD and CAT, to enhance the viability and vitality of plant to survive better under salt-affected soils (Shafi et al. 2009; Chang and Sung 1998). Work on muskmelon, cucumber, and amaranth (Nascimento and West 1999) revealed that seed priming technique amended growth of seedling by reducing viscosity of seed coat particularly in drastic conditions. This betterment is brought by improving synthesis of DNA and protein that has great effect in maximizing the stability of cell membrane in embryo. Phospholipids also have a role in embryo cell membrane as these can enhance the cell membrane permeability and resistance (Bradford 1995). Harris et al. (1999) showed that seed priming (overnight seeds soaking in water) distinctly enhanced quick vigor, and establishment of upland chickpea, maize, and rice results in quicker growth and development with earlier flowering followed by maturity and advanced yields. These priming effects are affiliated with the nucleic acids restoring and building up, repairing of membranes as well as improved proteins synthesis (McDonald 2000). Priming technique has also vital role in enhancement of the antioxidative enzyme's activities in primed seeds (Hsu et al. 2003). Due to availability of ready food throughout emergence (Farooq et al. 2006), primed seeds perform better to accomplish the germination process in a short duration while coping with external stresses (Farooq et al. 2007; Kant et al. 2006). Seedlings (7-day-old) taken from hydroprimed seeds indicated three- to fourfold extra growth in respect of shoot and root length than seedlings acquired from non-treated seeds.

The activities of invertases (alkaline and acid), amylase, sucrose phosphate synthase, and sucrose synthase (SS) were greater in primed seedling shoots. In primed seedling roots, growth in the activities of acid and alkaline invertases and SS was also detected. Cotyledons of primed seedlings also indicated twofold increment in sucrose phosphate synthase-specific activity. The higher activity of amylase was observed in shoots of hydroprimed seedlings that enhanced the speedy hydrolysis of transitory starch of the shoot. This hydrolysis leads to more glucose availability for growth of shoot, and this was confirmed by the presence of low level of starch in primed seedling shoots (Kaur et al. 2002). Pradhan et al. (2015) find that hydropriming considerably amended germination, growth of seedling, and seedling vigor under stress and nonstress environments. Similarly, the work of Janmohammadi et al. (2008) indicates hydropriming offers suitability to seed vigor under both saline and drought conditions.

### 3.1.2 Physiological Responses

Seed priming treatments had significant effects on germination and stress tolerance during which the physiological mechanisms inside the seeds are altered and these changes are characterized by the changes in the level of important indicator solutes such as proline, MDA, CAT, soluble sugars, SOD, and POD. Various studies indicated the effect of unfavorable environment on the seed emergence and stress resistance in different crops as in legumes, wheat (*Triticum aestivum* L.), onion (*Allium cepa* L.), and rice. Yuan-Yuan et al. (2010) reported the influence of drought on primed (hydropriming) and non-primed seeds of rice cultivars. Results concluded that for levels of soluble sugars, MDA were reduced in primed seeds, whereas hydropriming elevated the activities of CAT, SOD, and POD and also enhanced glucose metabolism in stress seeds among all cultivars than control treatment.

### 3.1.3 Antioxidative Response

Hydropriming significantly affects the antioxidative machinery of the plant under stress condition. When plant goes under stress condition, it produced different active oxygen species (AOS) and reactive oxygen species (ROS), which decline in plant health and its yield. To cope with this effect, different techniques used to overcome this issue among, which priming is key part for annual and some of perennial crops, fruit crops are not much affected by this technique in antioxidative manner. Hydropriming plays a vital part in activation of antioxidative enzymes, which cope with the effect of AOS and ROS in plant cell.

Hydropriming mediates stress tolerance by increasing antioxidant enzymatic activities. Superoxide dismutase, catalase, peroxidase, and ascorbate peroxidase are enzymes that can cope with the negative effect of ROS like lipid peroxidation

**Table 1** List of biochemical responses of different plant species affected by hydropriming under stress

| Biochemical | Activity level | Crop | References |
|---|---|---|---|
| Superoxide dismutase | Increase | Alfalfa | Amooaghaie (2011) |
| | | Cucumber | Huang et al. (2006) |
| | | Mung bean | Umair et al. (2012) |
| Peroxidase | Increase | Alfalfa | Amooaghaie (2011) |
| | | Mung bean | Umair et al. (2012) |
| | | Tomato | Araby and Hegazi (2004) |
| Catalase | Increase | Alfalfa | Amooaghaie (2011) |
| | | Cucumber | Huang et al. (2006) |
| | | Mung bean | Umair et al. (2012) |
| | | Tomato | Araby and Hegazi (2004) |
| MDA | Decrease | Cucumber | Huang et al. (2006) |
| | | Alfalfa | Amooaghaie (2011) |
| | | Tomato | Araby and Hegazi (2004) |
| Polyphenol oxidase | Increase | Mung bean | Umair et al. (2012) |
| Ascorbate peroxidase | Increase | Cucumber | Huang et al. (2006) |

mainly measured by malonyldialdehyde. Amooaghaie (2011) observed that the activity of superoxide dismutase increased in alfalfa under salt stress condition when treated with hydropriming. The seed, which are treated with hydropriming, showed more activity of superoxide dismutase and tolerates more under stress condition as compared with non-treated plants. Similar findings were also noted by Huang et al. (2006) and Umair et al. (2012) in cucumber and mung bean.

Peroxidase was also observed higher in the plants grown after the treatment of hydropriming. Alfalfa (Amooaghaie 2011) and mung bean (Umair et al. 2012) and tomato (Araby and Hegazi 2004) observed the increase in the peroxidase contents, which cause in the reduction of ROS. Catalase activity was also observed more in the plant of alfalfa, cucumber, mung bean, and tomato when treated with hydropriming. Polyphenol oxidase and ascorbate peroxidase were also observed more in the plants of mung bean (Umair et al. 2012) and cucumber (Huang et al. 2006) when treated with hydropriming, which means, hydropriming effects increase the stress tolerance by increasing antioxidant enzymes. Cucumber, alfalfa, and tomato when exposed to hydropriming showed decrease or less amount of lipid peroxidation measured by malonyldialdehyde (Table 1).

### 3.1.4 Molecular Responses

Hydropriming approximately for 12 h enhances the seed germination, growth and development and uncertainly enhances alterations in protein profiles of maize embryo. Eight proteins are greatly different in richness between treated and untreated seeds expressly embryonic protein DC-8 as well as globulin-1. Two proteins such as DC-8 and globulin-1 may possibly be the candidates for the molecular markers for priming effect as well as seed vigor. However, more research work is needed to explore the occupation of these well-known embryo proteins in hydropriming in numerous other maize lines all over the world.

# 4 Conclusion

Abiotic factors are one of the major problem in crop production. This phenomenon is also led by climate change. Many studies are taken in this aspect to overcome the effect of abiotic stresses. Priming is used to overcome dormancy and also enhances seed vigor. Priming also acts as mediator in stress tolerance. Among the different seed priming techniques, hydropriming is one of the mostly used. In hydropriming, seeds are soaked in a solution called priming agent followed by drying. Hydropriming helps in stress tolerance during germination, plant growth and development, physiological response, and activation of antioxidative enzymes, which cope with the ROS produced under stress. Further, genetic analysis and molecular marker studies are required to assess the role of hydropriming during stress tolerance. Moreover, there is an urgent need to develop and identify functional genetic approaches to evaluate the changes that occurred during hydropriming.

# References

Afzal I, Basra SMA, Iqbal A (2005) The effects of seed soaking with plant growth regulators on seedling vigour of wheat under salinity stress. J Stress Physiol Biochem 1:6–14

Ahmadian A, Neamatollahi E, Fathi D, Sohani A (2009) Hydropriming and osmopriming effects on cumin (*Cuminum cyminum* L.) seeds germination. Int J Appl Agric Res 4(3):273–282

Akter L, Fakir OA, Alam MK, Islam MU, Chakraborti P, Alam MJ, Rashid MH, Begum M, Kader MA (2018) Amelioration of salinity stress in maize seed germination and seedling growth attributes through seed priming. Open J Soil Sci 8:137–146

Alvarado AD, Bradford KJ, Hewitt JD (1987) Osmotic priming of tomato seed: effect on germination, field emergence, seedling growth and fruit yield. J Am Soc Hortic Sci 112:427–432

Amini R (2011) Effects of osmopriming on drought stress tolerance of perennial rye (*Secale montanum Guss.*) during germination. J Food Agric Environ 9(34):305–308

Amooaghaie R (2011) The effect of hydro and osmopriming on alfalfa seed germination and antioxidant defenses under salt stress. Afr J Biotechnol 10:6269–6275

Araby EL, Hegazi AZ (2004) Response of tomato seeds to hydro and osmo-priming, and possible relations of some antioxidant enzymes and endogenous polyamine fractions. Egypt J Biol 6:81–93

Armin M, Asgharipour M, Razavi-Omrani M (2010) The effect of seed priming on germination and seedling growth of watermelon (*Citrullus lanatus*). Adv Environ Biol 4(3):501–505

Ashraf M, Foolad MR (2005) Pre-sowing seed treatment a shotgun approach to improve germination, plant growth, and crop yield under saline and non-saline conditions. Adv Agron 88:223–271

Ashraf M, Rauf H (2001) Inducing salt tolerance in maize *Zea mays* (L.) through seed priming with chloride salts: growth and ion transport at early growth stages. Acta Physiol Plant 23:407–414
Bakht J, Shafi M, Jamal Y (2011) Response of maize (*Zea mays* L.) to seed priming with NaCl and salinity stresses. Span J Agric Res 9(1):252–261
Basra SMA, Zia MN, Mahmood T, Afzal I, Khaliq A (2002) Comparison of different invigoration techniques in wheat seeds. Pak J Arid Agric 2:11–16
Basra SMA, Afzal I, Anwar S, Anwar-ul-haq M, Shafq M, Majeed K (2006) Alleviation of salinity stress by seed invigoration techniques in wheat (*Triticum aestivum* L.). Seed Technol 28:36–46
Bouda S, Haddioui A (2011) Effect of salt stress on the germination of some species of the genus atriplex. Nat Technol 5:72–79
Bradford KJ (1986) Manipulation of seed water relations via osmotic priming to improve germination under stress conditions. Hortic Sci 21(5):1105–1112
Bradford KJ (1995) Water relations in seed germination. In: Kigel J, Galili G (eds) Seed development and germination. Marcel Dekker, New York, pp 351–396
Bray CM, Davison PA, Ashraf M, Taylor RM (1989) Biochemical changes during priming of leek seeds. Ann Bot 63:185–193
Cayuela E, Perez-Alfocea F, Caro M, Bolarin MC (1996) Priming of seeds with NaCI induces physiological changes in tomato plants grown under salt stress. Plant Physiol 96:231–236
Chang S, Sung M (1998) Deteriorative changes in primed sweet corn seeds during storage. Seed Sci Technol 26:613–626
Chojnowski FC, Come D (1997) Physiological and biochemical changes induced in sunflower seeds by osmopriming and subsequent drying storage and aging. Seed Sci Res 7:323–331
Davison PA, Bray CM (1991) Protein synthesis during osmo-priming of leek (*Allium porrum* L.) seeds. Seed Sci Res 1:29–35
Dell'Aquila A, Bewley JD (1989) Protein synthesis in the axes of polyethylene glycol treated pea seeds and during subsequent germination. J Exp Bot 40:1001–1007
Demir I, Ermis S (2003) Effect of controlled hydration treatment on germination and seedling growth under salt stress during development in tomato seeds. Eur J Hortic Sci 68:53–58
Demir I, Mavi K (2008) Effect of salt and osmotic stresses on the germination of pepper seeds of different maturation stages. Braz Arch Biol Technol 51:897–902
Eisvand HR, Shahrosvand S, Zahedi B, Heidari S, Afrougheh S (2011) Effects of hydro-priming and hormonal priming by gibberellin and salicylic acid on seed and seedling quality of carrot (Daucus carota var. sativus). Iran J Plant Physiol 1(4):233–239
Elouaer MA, Zhani K, Ben Fredj M, Hannachi C (2012) Seed priming for better growth and yield of safflower (*Carthamus tinctorius*) under saline condition. J Stress Physiol Biochem 8(3):135–143
Farooq M, Basra SMA, Wahid A (2006) Priming of field-sown rice seed enhances germination, seedling establishment, allometry and yield. Plant Growth Regul 49:285–294
Farooq M, Basra SMA, Ahmad N (2007) Improving the performance of transplanted rice by seed priming. Plant Growth Regul 51:129–137
Gong F, Wu X, Wang W (2013) Comparative proteomic identification of embryo proteins associated with hydropriming induced rapid-germination of maize seeds. Plant Omics 6(5):333
Goumi YEL, Fakiri M, Lamsaouri O, Benchekroun M (2014) Salt stress effect on seed germination and some physiological traits in three Moroccan barley (*Hordeum vulgare* L.) cultivars. J Mater Environ Sci 5(2):625–632
Haghpanah A, Younesi O, Moradi A (2009) The effect of priming on seedling emergence of differentially matured sorghum (*Sorghum bicolor* L.) seeds. J Appl Sci Res 5:729–732
Harris D (1992) The effects of manure, genotype, seed priming, depth and date of sowing on the emergence and early growth of *Sorghum bicolor* (L.) Moench in semi-arid Botswana. Soil Tillage Res 40:73–88
Harris D, Joshi A, Khan PA, Gothkar P, Sodhi PS (1999) On-farm seed priming in semi-arid agriculture: development and evaluation in maize, rice and chickpea in India using participatory methods. Exp Agric 35:15–29

Hsu CC, Chen CL, Chen JJ, Sung JM (2003) Accelerated aging-enhanced lipid peroxidation in bitter gourd seeds and effects of priming and hot water soaking treatments. Sci Hortic 98:201–212

Huang R, Sukprakarn S, Phavaphutanon L, Juntakool S, Chaikul C (2006) Changes in antioxidant enzyme activity, lipid peroxidation and seedling growth of cucumber seed induced by hydropriming and electric field treatments. Kasetsart J (Nat Sci) 40:825–834

Idris M, Aslam M (1975) The effect of soaking and drying seeds before planting on the germination and growth of Tritium vulgar, under normal and saline conditions. Can J Bot 53:1320–1332

Janmohammadi M, Dezfuli PM, Sharifzadeh F (2008) Seed invigoration techniques to improve germination and early growth of inbred line of maize under salinity and drought stress. Gen Appl Plant Physiol 34:215–226

Kadiri M, Hussaini MA (1999) Effect of hardening pretreatments on vegetative growth, enzyme activities and yield of *Pennisetum americanum* and *Sorghum bicolor*. Glob J Pure Appl Sci 5:179–183

Kahlon PS, Dhaliwal HS, Sharma SK, Randawa AS (1992) Effect of pre-sowing seed soaking on yield of wheat (*Triticum aestivum*) under late sown irrigated conditions. Indian J Agric Sci 62:276–277

Kant S, Pahuja SS, Pannu RK (2006) Effect of seed priming on growth and phenology of wheat under late-sown conditions. Trop Sci 44:9–15

Kaur S, Gupta AK, Kaur N (2002) Effect of osmo- and hydropriming of chickpea seeds on seedling growth and carbohydrate metabolism under water deficit stress. Plant Growth Regul 37(1):17–22

Kaya MD, Okçu G, Atak M, Çıkılı Y, Kolsarıcı O (2006) Seed treatments to overcome salt and drought stress during germination in sunflower (*Helianthus annuus* L.). Eur J Agron 24:291–295

Kazem GG, Afsaneh C, Safar N, Mohammad M (2010) Effect of hydropriming duration on seedling vigour and grain yield of pinto bean (*Phaseolus vulgaris* L.) cultivars. Notulae Botanicae Horti Agrobotanici Cluj-Napoca 38:109–113

Khazaei HR, Nezami A, Saadatian B, Pishe OA, Pordel R (2016) Effect of seed priming on seedling growth of barely (*Hordeum vulgare* L.), under salinity stress in phytogel. Environ Stress Crop Sci 9(1):87–97

Lee SS, Kim JH, Hong SB, Yun SH, Park EH (1998) Priming effect of rice seeds on seedling establishment under adverse soil conditions. Kor J Crop Sci 43:194–198

Li J, Yin LY, Jongsma MA, Wang CY (2011) Effects of light, hydropriming and abiotic stress on seed germination, and shoot and root growth of pyrethrum (*Tanacetum cinerariifolium*). Ind Crop Prod 34(3):1543–1549

Mabhaudhi T, Modi AT (2011) Can hydro-priming improve germination speed, vigour and emergence of maize landraces under water stress? J Agric Sci Technol B 1:20–28

McDonald MB (2000) Seed priming. In: Black M, Bewley JD (eds) Seed technology and its biological basis. Sheffield Academic Press, Sheffield, pp 287–325

Mehta PC, Puntamkar SS, Seth SP (1979) Effect of presoaking of seeds in different salts with varying concentration on the germination and yield of wheat grown on salinized soil. New Agric 6:73–76

Meriem BF, Kaouther Z, Chérif H, Tijani M, André B (2014) Effect of priming on growth, biochemical parameters and mineral composition of different cultivars of coriander (*Coriandrum sativum* L.) under salt stress. J Stress Physiol Biochem 10(3):84–109

Mirabi E, Hasanabadi M (2012) Effect of seed priming on some characteristic of seedling and seed vigor of tomato (*lycopersicun esculentum*). J Adv Lab Res Biol 3:237–240

Mohammadi GR (2009) The influence of NaCl priming on seed germination and seedling growth of canola (*Brassica napus* L.) under salinity conditions. Am Eurasian J Agric Environ Sci 5(5):696–700

Nascimento WM, West SH (1999) Muskmelon transplant production in response to seed priming. Hortic Technol 9:53–55

Parera CA, Cantliffe DJ (1992) Enhanced emergence and seedling vigor in shrunken-2 sweet corn by seed disinfection and solid matrix priming. J Am Soc Hortic Sci 117:400–403

Passam HC, Kakouriotis D (1994) The effects of osmoconditioning on the germination, emergence and early plant growth of cucumber under saline conditions. Sci Hortic 57(3):233–240

Pill WG, Necker AD (2001) The effects of seed treatments on germination and establishment of Kentucky bluegrass (*Poa pratensis* L.). Seed Sci Technol 29:65–72

Pradhan N, Prakash P, Manimurugan C, Tiwari SK, Sharma RP, Singh PM (2015) Screening of tomato genotypes using osmopriming with PEG 6000 under salinity stress. Res Environ Life Sci 8(2):245–250

Rouhi HR, Aboutalebian MA, Sharif-zadeh F (2011) Effects of hydro and osmo priming on drought stress tolerance during germination in four grass species. Int J Agric Sci 1(2):107–114

Roy NK, Srivastava AK (2000) Adverse effect of salt stress conditions on chlorophyll content in wheat (*Triticum aestivum* L.) leaves and its amelioration through pre-soaking treatments. Indian J Agric Sci 70:777–778

Sakhabutdinova AR, Fatkhutdinova DR, Bezrukova MV, Shakirova FM (2003) Salicylic acid prevents the damaging action of stress factors on wheat plants. Bulg J Plant Physiol:314–319

Salam A, Hollington PA, Gorham J, Wyn Jones RG, Gliddon C (1999) Physiological genetics of salt tolerance in wheat (*Triticum aestivum* L.) performance of wheat varieties, inbred lines and reciprocal F1 hybrids under saline conditions. Crop Sci 183(3):145–152

Sanchez JA, Munoz BC, Fresneda J (2001) Combined effects of hardening hydration dehydration and heat shock treatments on the germination of tomato, pepper and cucumber. Seed Sci Technol 29:691–697

Shafi M, Bakht J, Raziuddin ZG (2009) Effect of cadmium and salinity stresses on growth and antioxidant enzymes activity of wheat genotypes. Bull Environ Contam Toxicol 82(6):772–776

Shannon MC, Francois LE (1977) Influence of seed pretreatments on salt tolerance of cotton during germination. Agron J 69:619–622

Shivankar RS, Deore DB, Zode NG (2003) Effect of pre-sowing seed treatment on establishment and seed yield of sunflower. J Oilseeds Res 20:299–300

Shukla N, Kuntal H, Shanker A, Sharma S (2018) Hydro-priming methods for initiation of metabolic process and synchronization of germination in mung bean (*Vigna Radiata* L.) seeds. J Crop Sci Biotechnol 21(2):137–146

Singh BG (1995) Effect of hydration-dehydration seed treatments on vigour and yield of sunflower. Indian J Plant Physiol 38:66–68

Sivritepe N, Sivritepe HO, Eris A (2003) The effect of NaCl priming on salt tolerance in melon seedling grown under saline condition. Sci Hortic 97:229–237

Sung JM, Chiu KY (1995) Hydration effects on seedling emergence strength of watermelon seed differing in ploidy. Plant Sci 110:21–26

Takhti S, Shekafandeh A (2012) Effect of different seed priming on germination rate and seedling growth of Ziziphus Spina-Christi. Adv Environ Biol 6(1):159–164

Taylor AG, Allen PS, Bennett MA, Bradford JK, Burris JS, Misra MK (1998) Seed enhancements. Seed Sci Res 8:245–256

Umair A, Safdar A, Treen MJ, Ali I, Tareen MN (2012) Effects of seed priming on the antioxidant enzymes activity of mungbean (*Vigna radiata*) seedlings. Pak J Nutr 11:140–144

Wattanakulpakin P, Photchanachai S, Ratanakhanokchai K, Kyu KL, Ritthichai P, Miyagawa S (2012) Hydropriming effects on carbohydrate metabolism, antioxidant enzyme activity and seed vigor of maize (*Zea mays* L.). Afr J Biotechnol 11:3537–3547

Yagmur M, Kaydan D (2008) Alleviation of osmotic stress of water and salt in germination and seedling growth of triticale with seed priming treatments. Afr J Biotechnol 7:2156–2162

Yuan-Yuan SUN, Yong-Jian SUN, Ming-Tian WANG, Xu-Yi LI, Xiang GUO, Rong HU, Jun MA, (2010) Effects of seed priming on germination and seedling growth under water stress in rice. Acta Agronomica Sinica 36(11):1931–1940

# Chemical Priming for Multiple Stress Tolerance

**Muhammad Arslan Ashraf, Rizwan Rasheed, Iqbal Hussain, Muhammad Iqbal, Muhammad Riaz, and Muhammad Saleem Arif**

## Contents



**Abstract** Plant growth and productivity is severely limited due to abiotic stresses, which have become more detrimental due to recent climate change. A number of strategies are being exercised in different parts of the world to combat yield reductions in crops due to abiotic stresses including temperature, drought, salinity, UV radiation, trace metal stress, etc. However, some of these strategies referred to as conventional breeding are time-consuming and labor-intensive, while plant genetic modification is another strategy, which is not acceptable in several countries of the world. In contrast, chemical priming appears as an alternative approach that prepares plants to better tolerate future abiotic and biotic stresses. Chemical priming has emerged as potential field in crop stress management strategies. Plants' stress tolerance response can be modified by treating plants with chemical before the occurrence of stress events. Chemical priming imparts significant impact on plant growth, physiology, biochemistry, and molecular mechanism. Until now, the underlying molecular mechanisms mediating plant growth under the influence of chemical priming agents are not well elucidated. In this chapter, we reviewed the functions of emerging chemical priming agents that include polyamines, β-aminobutyric acid,

M. A. Ashraf (✉) · R. Rasheed · I. Hussain · M. Iqbal
Department of Botany, Government College University Faisalabad, Faisalabad, Pakistan
e-mail: marslanashraf@gcuf.edu.pk

M. Riaz · M. S. Arif
Department of Environmental Sciences and Engineering, Government College University Faisalabad, Faisalabad, Pakistan



M. Hasanuzzaman, V. Fotopoulos (eds.), *Priming and Pretreatment of Seeds and Seedlings*, https://doi.org/10.1007/978-981-13-8625-1_19

gamma-aminobutyric acid, melatonin, menadione sodium bisulfite, sodium nitroprusside, sodium hydrosulfide, selenium, and salicylic acid.

**Keywords** Priming agents · Seed soaking · Hormonal priming · Antioxidant enzymes · Reactive oxygen species · Signaling molecules

## Abbreviations

| | |
|---|---|
| APX | ascorbate peroxidase |
| CAT | catalase |
| GR | glutathione reductase |
| $H_2O$ | hydrogen peroxide |
| $H_2S$ | hydrogen sulfide |
| LOX | lipoxygenase |
| MDA | malondialdehyde |
| NaHS | sodium hydrosulfide |
| NO | nitric oxide |
| PEG | polyethylene glycol |
| POD | peroxidase |
| PSII | photosystem II |
| Put | putrescine |
| ROS | reactive oxygen species |
| RWC | relative water contents |
| S | sulfur |
| SNP | sodium nitroprusside |
| SOD | superoxide dismutase |
| Spd | spermidine |
| Spm | spermine |
| UV | ultraviolet |

## 1 Introduction

Plant growth and production is significantly hampered by various environmental constraints such as heavy metal toxicity, UV radiation, excess light, freezing, cold, heat, flooding, drought, and salinity. Global food security is being threatened by the detrimental effects of abiotic stresses on plant growth and productivity, which are further exacerbated by anthropogenic contributions such as urbanization and industrialization (Nagajyoti et al. 2010) and climate change (David and Christopher 2007; Reddy 2015; Savvides et al. 2016). Plants under field conditions face a wide range of abiotic stresses that may occur simultaneously or sequentially (Savvides et al. 2016). Combined abiotic stresses act synergistically to suppress plant growth. Plant exposure to a combination of heat and heavy metals (Zhao et al. 2009), heat and drought (Perdomo et al. 2015), or heat and salinity (Silva et al. 2013) is more detrimental for crop yield compared to the scenario where plants face these stresses individually (Savvides et al. 2016). Therefore, plant scientists are now focusing their research on studying the plant responses toward multiple abiotic restraints (Suzuki et al. 2014).

A number of approaches have been used worldwide to increase crop productivity under multiple abiotic stresses. However, some strategies such as conventional breeding are time consuming, while others such as genetic engineering are not accepted across different countries of the world (Savvides et al. 2016). Alternatively, plants can be made to better tolerate future abiotic and biotic stresses by means of chemical priming. Plant priming also referred to as hardening or sensitization may be induced as a result of initial exposure of plants to environmental stresses that function as reminders for plants to get exposed to the same environmental hazard in the future (Filippou et al. 2013). This makes plants to enter the primed state where the provocation of defensive processes is faster than unprimed plants (Sani et al. 2013). Intriguingly, plants can also achieve primed state through chemical priming, where plants are exposed to priming agents such as synthetic or natural chemical compounds. Chemical priming offers efficient ways for crop stress management so as to counteract the adverse effects of stresses on plant growth and development (Sani et al. 2013; Tables 1, 2, 3 and 4).

## 2 Promising Chemical Priming Agents

Chemical priming agents include but not limited to RONSS (oxygen–nitrogen–sulfur species), melatonin, polyamines, menadione sodium bisulfite, proline, glycine betaine, amino acids, nutrients, and naturally occurring metabolites (Christou et al. 2014a; Savvides et al. 2016).

## 3 Reactive Oxygen, Nitrogen, and Sulfur Species (RONSS)

RONSS such as $H_2O_2$ (hydrogen peroxide), NO (nitric oxide), and $H_2S$ (hydrogen sulfide) play an important part in the abiotic stress acclimation of plants due to their direct impact on gene regulation and signal transduction (Molassiotis and Fotopoulos 2011; Fotopoulos et al. 2015). Over the last few years, there is growing evidence on the application of these molecules or their donors such as sodium hydrosulfide (NaHS) for $H_2S$ and sodium nitroprusside (SNP) for NO at low concentration (1–1000 mM) for the induction of abiotic stress tolerance in crop plants. RONSS molecules mediate important biological functions during primed state resulting in enhanced tolerance to abiotic stress (Antoniou et al. 2016).

**Table 1** Some reports on seed priming-induced mediation in plant salinity tolerance

| Priming agent | Concentration | Stress type | Plant species | Plant responses | References |
|---|---|---|---|---|---|
| Proline | 4 and 8 mM | Salinity | Wheat | Proline priming increased growth, chlorophyll contents, antioxidant properties, and K alongside the significant decline in toxic Na in plants under salinity | Shafiq et al. (2018) |
| Apigenin | 10 mg $L^{-1}$ | Salinity | Rice | Apigenin priming enhanced dry mass accumulation and shoot elongation under both salinity-stressed and non-stressed conditions. Pretreatment with apigenin significantly reduced the accumulation of Na and maintained K/Na in different plant organs under salinity stress. Lipid peroxidation due to ROS was minimal because of better oxidative defense in plants pretreated with apigenin | Mekawy et al. (2018) |
| $ZnSO_4 \cdot 7H_2O$ | 4 mM | Salinity | Maize | Salinity decreased nutrient and water uptake in plants. However, seed priming with Zn markedly improved growth, nutrient uptake, and growth. Plants pretreated with Zn significantly decreased the accumulation of toxic Na that led to salinity tolerance in seedlings | Imran et al. (2018) |
| Aspirin | 125, 250, 375, and 500 mg $L^{-1}$ | Salinity | Wheat | Priming with aspirin increased germination, starch metabolism, and seedling growth. Salinity decreased plant growth, photosynthetic pigments with concomitant increase in phenolics, and lipid peroxidation. However, aspirin priming significantly enhanced the activities of SOD, POD, and CAT that counteracted ROS-induced oxidative damage. Aspirin priming improved oxidative defense system that improved starch metabolism and decreased lipid peroxidation | Hussain et al. (2018) |
| Benzyl aminopurine (BAP), sorghum water extract (SWE) | 5 mg $L^{-1}$ BAP, 5% SWE | Salinity | Wheat | Salinity stress significantly decreased plant growth, chlorophyll, and K contents. Priming resulted in marked increase in K contents, chlorophyll, α-amylase activity, soluble proteins, sugars, and total phenolics alongside the decrease in tissue Na levels. BAP + SWE appeared as the most effective treatment in the alleviation of salinity stress | Bajwa et al. (2018) |

| | | | | | |
|---|---|---|---|---|---|
| Saponin | 0%, 0.5%, 2%, 5%, 10%, 15%, 25%, and 35% | Salinity | Quinoa | Saponin seed priming improved germination, plant growth, photosynthesis, stomatal conductance, plant water relations, yield attributes, and K contents along with significant decrease in toxic Na | Yang et al. (2018a) |
| Polyethylene glycol (PEG) | PEG (−0.5 to −4 MPa); hydropriming (0 MPa) | Salinity | *Amaranthus caudatus* and *Chenopodium quinoa* | Priming treatments increased germination index, germination percentage, and mean germination time in *Amaranthus caudatus* and *Chenopodium quinoa* under salinity stress | Moreno et al. (2018) |
| Phenylalanine | 0.0%, 0.8%, and 1.0% | Salinity | Maize | Priming treatment effectively enhanced antioxidant properties that decreased oxidative damage in salinity stressed plants | Zahra et al. (2018) |
| *Padina pavonica* and *Jania rubens* | Biopriming (60 g $L^{-1}$) | Salinity | Pepper | Salinity ameliorative effect of biopriming was evaluated in pepper. Biopriming was done with the extracts of *Padina pavonica* and *Jania rubens*. The results indicated that salinity caused significant decline in mineral composition of pepper, respiration, and integrity of membranes. Priming treatments significantly increased secondary metabolite accumulation, Na, soluble sugars, and proline in salinity stressed pepper. Increase in salinity tolerance due to priming treatments was attributed to decrease in oxidative damage manifested by decline in lipid peroxidation, electrolyte leakage, secondary metabolite accumulation alongside the stimulated metabolic activity. | Rinez et al. (2018) |
| Salicylic acid, gibberellic acid | 1 mM | Salinity | *Anethum graveolens* L. | Seed priming decreased toxic Na and increased K in plants subjected to salinity. Besides seed priming-induced improvement in salinity tolerance was attributed to greater K/Na, osmolyte accumulation, and activities of antioxidant enzymes | Ghassemi-Golezani and Nikpour-Rashidabad (2017) |
| SA, fish flour | 0.05, 0.1, 2.5, and 5.0 mM | Salinity | Wheat | Seedling pretreatment with SA and fish flour significantly enhanced phenolics, flavonoids, peroxidase activities, and phenylalanine ammonia lyase under salinity | Karadağ and Yücel (2017) |

(continued)

**Table 1** (continued)

| Priming agent | Concentration | Stress type | Plant species | Plant responses | References |
|---|---|---|---|---|---|
| Smoke | Smoke dilution (2:1 v/v) | Salinity | Rice | Smoke priming markedly increased plant fresh and dry masses, protected plants from ion toxicity, and thereby increased productivity under salinity | Malook et al. (2017) |
| $ZnSO_4 \cdot 7H_2O$, $CaCl_2 \cdot 2H_2O$, betaine hydrochloride, gibberellic acid ($GA_3$) | Betaine hydrochloride and 30 mg $L^{-1}$ GA3, 0.025% $ZnSO_4 \cdot 7H_2O$, 52.5 mM $L^{-1}$ $CaCl_2 \cdot 2H_2O$, 0.5 g $L^{-1}$ | Saline-alkali | Soybean | Seed priming increased stress tolerance in terms of better growth, photosynthetic pigments, greater accumulation of soluble sugars, proline contents, and activities of antioxidant enzymes alongside the decrease in lipid peroxidation | Dai et al. (2017) |
| *Trichoderma lixii* ID11D | 1:1 (v/v) | Salinity | Maize | Seed biopriming significantly improved salinity tolerance in maize reflected in the form of higher photochemical quenching, quantum yield of PSII, maximum quantum efficiency of PSII, and electron transport rate (ETR). Besides, priming treatments decreased lipid peroxidation, $H_2O_2$ contents. Increased relative water contents, proline, chlorophyll, and carotenoids were also recorded in salinity stressed maize seedlings | Pehlivan et al. (2017) |
| ZnO nanoparticles | 20, 40, and 60 mg $L^{-1}$ | Salinity | Lupine | Seed priming markedly increased plant growth accompanied by increase in endogenous levels of ascorbic acids, total phenolics, organic solutes, and photosynthetic pigments as well as stimulated activities of antioxidant enzymes (APX, POD, CAT, and SOD). Priming treatments also caused decrement in Na contents and lipid peroxidation that significantly contributed to salinity tolerance in lupine | Abdel Latef et al. (2017) |
| Silver nanoparticles | 0, 2, 5, and 10 mM | Salinity | Wheat | Salinity markedly decreased shoot fresh and dry masses, chlorophyll contents alongside the enhanced activities of antioxidant enzymes. Pretreated plants also accumulated greater levels of $H_2O_2$, MDA, soluble sugars, and proline. Chlorophyll contents were also higher in plants raised from seed priming | Mohamed et al. (2017) |

| | | | | | |
|---|---|---|---|---|---|
| Triacontanol | 25 and 50 μM | Salinity | Cucumber | Priming significantly enhanced seedling uniformity, emergence rate, and growth in salinity stressed cucumber. Besides, priming treatments improved chlorophyll contents and proline in seedlings under salinity | Sarwar et al. (2017) |
| NaCl and $MgCl_2$ | 500 mM | Ionic stress (NaCl) and osmotic stress (PEG) | *Lolium perenne* L. | Priming treatments significantly decreased mean emergence rate, uptake of Cl and Na besides improving water potential during seed germination | Dianati Tilaki and Behtari (2017) |
| Glycinebetaine | 0, 10, 30, and 60 mM | Salinity | Safflower | Priming treatments significantly increased germination and seedling growth under salinity stress. Proline, total soluble sugars, and lipid peroxidation were minimal in plants exposed to salinity. The activities of antioxidant enzymes (SOD, POD) were higher in pretreated plants. Primed plants also manifested decrease in Na contents with concomitant increase in K contents. It was concluded that priming enhanced salinity tolerance reflected in the form of limited lipid peroxidation and improved ion homeostasis. | Alasvandyari et al. (2017) |

Abbreviations: *MDA* malondialdehyde, *Na* sodium, *K* potassium, *Cl* chloride, *SOD* superoxide dismutase, *POD* peroxidase, *CAT* catalase, *APX* ascorbate peroxidase, $H_2O_2$ hydrogen peroxide, *PSII* photosystem II, *Zn* zinc

**Table 2** Some reports on seed priming mediated drought tolerance in plants

| Priming agent | Concentration | Stress type | Plant species | Plant responses | References |
|---|---|---|---|---|---|
| Methyl jasmonate | 2.5 and 5 mM | Osmotic stress | Rice | Osmotic stress caused significant decrease in germination and seedling vigor, lengths, and fresh and dry masses, chlorophyll, photosynthesis, and PSII efficiency. Priming with 2.5 mM markedly improved photosynthesis, transpiration rate, stomatal conductance, intercellular $CO_2$ levels, PSII integrity, and chlorophyll contents. Priming increased relative water content, water use efficiency, and water potential in rice under osmotic stress | Sheteiwy et al. (2018) |
| Fullerol | 10 and 100 mg $L^{-1}$ | Water deficit | Rapeseed | Water deficit conditions significantly decreased germination, plant growth, and photosynthesis while increased ABA levels, ROS accumulation, and endogenous levels of nonenzymatic antioxidant compounds. Priming markedly increased germination, plant growth, and oxidative defense system which in turn decreased lipid peroxidation and improved plant stress tolerance | Xiong et al. (2018) |
| Melatonin | 0, 1, 10, 100, 300, 500, or 1000 μM | Osmotic stress | Wheat | Priming significantly increased seed germination, primary root number and coleoptile length, and fresh and dry masses in wheat seedlings under PEG-induced osmotic stress. The endogenous levels of ROS were significantly lower as a result of stimulated antioxidant enzyme activities as a result of melatonin priming under osmotic stress | Cui et al. (2018) |
| Sugar beet extract | 0%, 10%, 20%, 30%, 40%, and 50% | Drought | Wheat | Seed priming with SBE significantly ameliorated the negative effects of drought stress in wheat. SBE priming increased germination, photosynthetic pigments, antioxidant properties, and nutrient homeostasis that led to enhanced drought tolerance in wheat | Noman et al. (2018b) |

| | | | | | |
|---|---|---|---|---|---|
| Zn-Lys | 0, 1.5, 3, 4.5, or 6 mg $kg^{-1}$ | Drought | Radish | Drought resulted in marked decline in growth, physiological, biochemical, and yield attributes in radish plants. Seed priming with Zn-Lys improved germination, growth, pigments, root growth, nutrient uptake, activities of antioxidant enzymes with concomitant decrease in lipid peroxidation and ROS generation. Ascorbic acid, flavonoids, carbohydrates, protein, and carotenoids were greater in pretreated plants. Nutrient contents such as Zn, P, Fe, Mg, Ca, and K markedly increased in Zn-Lys pretreated plants | Noman et al. (2018a) |
| Polyethylene glycol (PEG) | PEG (−0.5 to −4 MPa); hydropriming (0 MPa) | Osmotic stress | *Amaranthus caudatus* and *Chenopodium quinoa* | Priming treatments increased germination index, germination percentage, and mean germination time in *Amaranthus caudatus* and *Chenopodium quinoa* under salinity stress | Moreno et al. (2018) |
| $CaCl_2$, hydropriming, biopriming | $CaCl_2$ 1.5%, hydropriming (0 MPa), osmopriming, | Drought | Barley | Osmopriming (1.5% $CaCl_2$), hydropriming, and biopriming (Enterobacter sp. strain FD17) effect mediated drought tolerance in barley. Drought significantly decreased grain nutrient contents, water relations, chlorophyll contents, yield, and plant growth alongside marked increase in lipid peroxidation and accumulation of osmolytes. Priming treatments improved grain B, Zn, and Mn, plant water relation attributes, secondary metabolites accumulation, and cell membrane integrity with concomitant decrease in lipid peroxidation. The improvement in drought tolerance was greater in the case of biopriming followed by osmopriming and hydropriming. | Tabassum et al. (2018) |
| Osmopriming | PEG (−0.9 MPa) | Drought | Wheat | Osmopriming treatments increased germination under drought. There existed a significant decrease in leaf water relations, photosynthetic pigments, quantum yield, photosynthesis, and relative water content. Priming treatments significantly enhanced photosynthetic pigments, photosynthesis, and PS II. Oxidative damage was minimal in plants pretreated with osmopriming that also enhanced antioxidant properties | Abid et al. (2018) |

(continued)

**Table 2** (continued)

| Priming agent | Concentration | Stress type | Plant species | Plant responses | References |
|---|---|---|---|---|---|
| $CaCl_2$ | Osmopriming ($\psi$s-1.25 MPa), hydropriming (0.0 MPa) | Drought | Rice | Seed priming influence on drought tolerance in pigmented and nonpigmented rice was examined under drought. Plants subjected to drought stress exhibited significant decline in plant height, leaf area more in pigmented rice, while non-pigmented rice had decrease in dry weight and root length. The endogenous levels of flavonoids, polyphenols besides greater photosynthesis was evident in pigmented rice. Seed priming effectively improved antioxidant properties, flavonoids, polyphenols, and plant growth under drought stress. It was seen that seed priming-mediated drought tolerance was more in pigmented rice compared with non-pigmented rice. | Hussain et al. (2017) |
| Salicylic acid (SA), methyl jasmonate (MJ), paclobutrazol (PB) | 100 μM | Drought | Rice | Seed priming significantly enhanced antioxidant properties and decreased lipid peroxidation under drought | Samota et al. (2017) |
| Mannitol, $H_2O_2$ | Mannitol (2%), $H_2O_2$ (120 μM) | Drought | *Eutrema salsugineum* | Priming treatments significantly circumvented drought effects on plant growth and enhanced antioxidant defense system in plants. Priming with $H_2O_2$ was more beneficial compared with mannitol in improving drought tolerance reflected in the form of limited oxidative injuries and greater endogenous levels of proline, glutathione, and ascorbic acid | Ellouzi et al. (2017) |
| Jasmonic acid and SA | SA (10 mM) and jasmonic acid (100 μM) | Drought | Wheat | Seed priming with jasmonate exhibited better drought ameliorative effects compared with SA in wheat. Priming treatments significantly increased germination, proline, soluble sugars, and water potential in drought stressed plants | Ilyas et al. (2017) |

| | | | | | |
|---|---|---|---|---|---|
| NaCl and $MgCl_2$ | 500 mM | Ionic stress (NaCl) and osmotic stress (PEG) | *Lolium perenne* L. | Priming treatments significantly decreased mean emergence rate and uptake of Cl and Na besides improving water potential during seed germination | Dianati Tilaki and Behtari (2017) |
| $CaCl_2$ | 5, 50, and 500 mM | Drought | Barley | Osmopriming enhanced drought tolerance in terms of germination, photosynthesis, transpiration rate, relative water contents, and chlorophyll contents | Kaczmarek et al. (2016) |
| PEG | PEG (−0.6 MPa) | Drought | Alfalfa | Priming treatments increased nodulation, growth, PSII efficiency, nutrient uptake, N contents, and decreased stomatal conductance in plant under drought | Mouradi et al. (2016) |
| Beta-amino butyric acid | 0, 0.5, 1.0, 1.5, 2.0, and 2.5 mM | Drought | Rice | Priming treatments significantly enhanced growth, photosynthetic pigments, proline, nitrate reductase activity, and activities of antioxidant enzymes and improved chlorophyll a fluorescence-related parameters. Seed priming decreased lipid peroxidation which led to enhanced tolerance to salinity and drought stress | Jisha and Puthur (2016) |

Abbreviations: PEG, polyethylene glycol; $CaCl_2$, calcium chloride; NaCl, sodium chloride; $MgCl_2$, magnesium chloride; SA, salicylic acid; N, nitrogen; Na, sodium; Cl, chloride

**Table 3** Some reports on seed priming mediated tolerance in plants against heavy metal stress

| Priming agent | Concentration | Stress type | Plant species | Plant responses | References |
|---|---|---|---|---|---|
| Distilled water, selenium (Se), salicylic acid (SA) | 60 μM se, 100 mg $L^{-1}$ SA | Lead (Pb) | Rice | Rice plants exposed to 1 mM $PbCl_2$ stress were grown under different nutrition levels, i.e., sufficient nutrient supply, K deprivation, P deprivation, N deprivation. Pb stress significantly decreased growth, increased ROS generation, and lipid peroxidation in P- or N-deprived seedlings. However, sufficient nutrient supply or K-deprived seedlings did not show any growth reduction under Pb toxicity. Priming treatments significantly alleviated the inhibitory effects of Pb in rice under various nutrient treatments. The oxidative damage was also minimal in pretreated plants under Pb toxicity. Priming plants accumulated essential nutrients that might have prevented the uptake of Pb | Khan et al. (2018) |
| Castasterone, citric acid | 0, 0.01, 1, and 100 nM castasterone; 0 and 0.6 mM citric acid | Cadmium (cd) | Mustard | Cd-induced significant decrease in photosynthetic pigments, altered total carbohydrates levels, and induced stomatal closure were attributed to elevated levels of ROS. Priming with castasterone along with citric acid ameliorated Cd toxicity reflected in the form of minimal ROS levels and enhanced photosynthetic pigments | Kaur et al. (2018b) |
| Salicylic acid, gibberellic acid | Salicylic acid (50 μM), gibberellic acid (100 μM) | Cd | Mung bean | Seed priming treatments markedly increased plant growth, antioxidant enzyme activities besides the significant decrease in lipid peroxidation and $H_2O_2$ in plants exposed to Cd stress | Hassan and Mansoor (2017) |
| $KNO_3$ | 0.0% and 0.5% | Pb | Maize | Seed pretreatments markedly improved germination percentage and key physiological indices under Pb toxicity. Seed priming significantly increased photosynthetic pigments in Pb-stressed maize plants. Besides, aerial buildup of Pb was lower along with significant increase in fee amino acid accumulation and activities of antioxidant enzymes in pretreated plants | Nawaz et al. (2017) |
| Smoke (*Cymbopogon jwarancusa*) | Smoke dilutions (1:500 and 1:1000) | Pb | Rice | The increase in the endogenous levels of metabolites (glycinebetain, proline, total soluble protein, proline, $H_2O_2$, MDA) alongside the stimulated activities of antioxidants was greater in non-primed seeds compared with primed seeds under Pb stress | Akhtar et al. (2017) |

Abbreviations: $PbCl_2$, lead chloride; P, phosphorous; K, potassium; N, nitrogen; ROS, reactive oxygen species; MDA, malondialdehyde; $H_2O_2$, hydrogen peroxide

**Table 4** Some reports on seed priming mediated tolerance in plants against temperature stress

| Priming agent | Concentration | Stress type | Plant species | Plant responses | References |
|---|---|---|---|---|---|
| Salicylic acid | $10^{-4}$ M | Heat | Wheat | Heat stress caused significant reduction in growth, chlorophyll, and grain yield alongside the increase in soluble sugars, proteins, and proline. Salicylic acid priming resulted in marked increase in chlorophyll, endogenous levels of sugars, proteins, proline, and grain yield | Munir and Shabbir (2018) |
| $H_2O_2$ | 80 and 120 mM | Heat | *Bombax ceiba* | Hydrogen peroxide seed pretreatment effect on germination, seedling growth, and key physiological attributes was studied in *Bombax ceiba* under heat shock treatment. Heat stress caused significant decline in germination percentage and photosynthesis. However, seed priming with $H_2O_2$ improved seed germination, chlorophyll, photosynthesis, and activity of superoxide dismutase under heat stress. The accumulation of osmoprotectants, namely, proline and soluble sugars was not significant in both primed and unprimed seeds under heat stress. | Zheng et al. (2018) |
| 28-homobrassinolide (28-HBL) | 0, $10^{-6}$, $10^{-9}$, $10^{-12}$ M | Temperature and salinity | Mustard | Temperature and salt stress resulted in a marked decrease in plant growth and increased $H_2O_2$ endogenous levels. However, priming with 28-HBL significantly enhanced the activities of antioxidant enzymes (SOD, CAT, POD, APX, GR, DHAR, MDHAR) that counteracted the ROS-induced oxidative damage and improved plant growth under combined abiotic stresses | Kaur et al. (2018a) |
| Salicylic acid, $CaCl_2$, sorgaab, and moringa leaf extract (MLE) | S 50 mg $L^{-1}$, $CaCl_2$ 2.2%, sorgaab, and MLE (30 times diluted) | Temperature | Wheat | Seeds were primed with moringa leaf extract, sorgaab, salicylic acid, and $CaCl_2$ to improve high temperature stress in wheat. Priming treatments significantly enhanced germination and seedling establishment. However, seeds primed with SA exhibited better germination. Priming increased chlorophyll contents, grain yield, leaf area, and other yield-related attributes. Priming with $CaCl_2$ resulted in significant increase in phenolics, glycinebetaine and relative water content, while priming with MLE exhibited better cell membrane thermo stability. SA priming significantly enhanced total soluble sugars and proline under high temperature stress. | Mahboob et al. (2018) |

(continued)

**Table 4** (continued)

| Priming agent | Concentration | Stress type | Plant species | Plant responses | References |
|---|---|---|---|---|---|
| $CaCl_2$ (osmopriming), distilled water (hydropriming) | Osmopriming (ψs −1.25 MPa) and hydropriming (0.0 MPa) | Chilling | Chickpea | Seed priming alleviated the adverse effects of chilling stress by improving seedling growth, establishment, soluble phenolics, K, N, proline, membrane integrity, oxidative defense system, sugar metabolism, α-amylase activity, photosynthesis, and water relations | Farooq et al. (2017) |
| SA, $H_2O_2$, and ascorbic acid | 20 mg $L^{-1}$ | Heat | Maize | Seed priming improved grain yield, morphological, physiological, and biochemical attributes in maize under heat stress. Higher activities of antioxidant enzymes (SOD, POD, and CAT) as a result of seed priming. Relative water contents, membrane integrity, chlorophyll contents, yield, and yield quality were also higher in pretreated plants exposed to heat stress | Ahmad et al. (2017) |

Abbreviations: *SA* salicylic acid, *K* potassium, *N* nitrogen, *$CaCl_2$* calcium chloride, *SOD* superoxide dismutase, *APX* ascorbate peroxidase, *POD* peroxidase, *CAT* catalase, *GR* glutathione reductase, *MDHAR* monodehydroascorbate reductase, *DHAR* Dehydroascorbate reductase

Priming with different NO improves plant growth and production under environment constraints. Ali et al. (2017) studied the influence of SNP priming on salinity tolerance of wheat cultivars (Galaxy-13, Millat-11, Punjab-11, and Sahar-06). Seeds were soaked in different levels of SNP (0, 0.1, and 0.2 mM). Salinity significantly increased malondialdehyde (MDA), hydrogen peroxide ($H_2O_2$), ascorbic acid, and proline. They recorded significant reduction in growth and grain yield due to salinity stress. Seed priming with SNP ameliorated oxidative damage of salinity and increased growth and grain yield. Seed priming with (0.1 mM) increased germination percentage, seedling vigor, and early establishment of seedlings under salinity. Furthermore, SNP priming enhanced the activities of antioxidant enzymes [superoxide dismutase (SOD), peroxidase (POD), and catalase (CAT)] alongside the concentrations of ascorbic acid, phenolics, and proline. Increase in growth and grain yield was ascribed to upregulation of antioxidant defense system due to SNP priming. In a study by Gadelha et al. (2017), nitric oxide (NO) priming significantly enhanced salinity tolerance of *Jatropha curcas* at the seedling stage. Plants raised from NO priming showed lower accumulation of toxic Cl and Na compared to those of unprimed plants under salinity. NO-priming effectively protected plants from oxidative damage in the form of membrane damage and $H_2O_2$ accumulation. They reported that lower oxidative damage in NO-primed plants was the result of powerful antioxidant system. It was concluded that NO-primed plants showed efficient provoked antioxidant system, lower accumulation of reactive oxygen species (ROS) and toxic ions. Likewise, SNP priming increased polyethylene glycol (PEG)-induced drought tolerance in *Arachis hypogaea* L. Different concentrations of SNP (0, 50, 100 and 15 μ) and PEG (0, −0.4, −0.6, and − 0.8 MPa) were used in the experiment. Priming resulted in significant increase in the activities of SOD, POD, and ascorbate peroxidase (APX), soluble sugars, and proteins that reduced the endogenous levels of $H_2O_2$ and MDA. Seed priming with SNP also increased germination percentage. It was inferred from the results that SNP priming significantly decreased oxidative damage that might have improved seedling vigor and germination (Sepehri and Rouhi 2016). Similarly, priming with 100 μM SNP restored redox status and increased antioxidant defense system in *Arabidopsis* plants under cadmium stress (Méndez et al. 2016). In a study, SNP (0.1 mM) seed priming improved grain yield and plant biomass in rice under salinity. Priming resulted in significant increase in the activities of antioxidant enzymes (SOD, POD, CAT) and the endogenous levels of ascorbic acid and proline. Primed plants had better antioxidant system that decreased oxidative damage in terms of MDA and $H_2O_2$. Better yield in salinity stressed rice was attributed to the provoked antioxidant system (Habib et al. 2016).

Seed priming with SNP significantly increased germination percentage, vigor index, and germination index. Priming also reduced mean germination time when compared with plants from unprimed seeds. The endogenous levels of nonreducing sugars and proteins were higher in primed plants. The activities of enzymes, namely, SOD, POD, α- and β-naphthyl acetate esterase, and α-amylase, were greater in primed plants. In conclusion, SNP priming increased oxidative defense and mobilization of stored carbohydrate, proteins, and lipids (Hameed et al. 2015). Christou et al. (2014b) reported NO and $H_2O_2$ as pivotal signaling molecules mediating plant development under stress conditions. Strawberry plants were grown hydroponically to investigate the functions of SNP (100 μM) and $H_2O_2$ (10 mM) under saline conditions. They reported systemic ameliorative effect of applied reactive species to cellular injuries incurred by salt stress. In a study by Manjunatha et al. (2008), the efficiency of different NO donors, namely, SNP, 2-Nitroso-1-naphthol, and Nitroso-R-salt, was studied to appraise their protective effects in pearl millet against downy mildew disease. SNP seed priming with or without PEG provoked significant host resistance against downy mildew. Likewise, the protective role of SNP pretreatment was studied in barley under arsenic stress. Plants subjected to arsenic stress exhibited decrease in photosynthetic pigments, biomass, and germination percentage, while increase in proline and MDA levels was evident. There was initial increase in the activities of CAT and POD, which decreased drastically at higher concentration of arsenic. Pretreatment with NO significantly enhanced oxidative defense of plants that might have increased plant growth under arsenic stress (Shukla et al. 2015). Kaur and Kaur (2018) reported the effect of NO priming on the antioxidant defense system in two maize cultivars (tolerant cv. CML 32 and susceptible cv. LM 11) with contrasting tolerance to heat stress. Seed priming with SNP significantly increased seedling growth and provoked a number of defense mechanisms. The activities of antioxidant enzymes, namely, SOD, POD, and CAT, were higher in CML 32 that showed strong detoxification of ROS. Furthermore, the degree of oxidative damage in the form of $H_2O_2$ and MDA accumulation was markedly mitigated as a result of seed priming with SNP (75 μM). The oxidative defense mechanism was stronger in cv. CML 32 than that in cv. LM 11. Bibi et al. (2017) studied the effect of SNP priming on the induction of chilling stress in wheat. They reported significant increase in plant growth and relative water content (RWC) of wheat plants as a result of SNP priming. Higher stress tolerance of RONSS-primed plants is the result of the involvement of these molecules in signal transduction (Molassiotis and Fotopoulos 2011). Literature showed the upregulation of stress-related genes as a result of exogenous application of RONSS molecules (Tanou et al. 2012). In a study, wheat seeds were primed with 0.06 mM SNP and measured a number of germination-related parameters. The results of the study showed that SNP priming significantly mitigated salt-induced decrease in seed imbibition rate, vigor index, germination index, and germination percentage. SNP priming also markedly raised beta-amylase activity, whereas the activity of alpha-amylase remained unaffected under salinity. Besides the influence of SNP priming on the expression of germination-related enzymes, SNP priming also decreased the uptake of Na with concomitant increase in K and thereby enhanced K/Na ratio in salinity-stressed plants. It was examined that increase in germination percentage is largely due to elevated activity of beta-amylase in response to SNP priming (Duan et al. 2007). Nejadalimoradi et al. (2014) studied the priming effects of two different NO donors – arginine and SNP – in the mitigation of salinity stress. There was significant growth suppression in salinity-stressed plants, while the ones primed with NO donors showed little decrease in growth. Salinity decreased the activities of antioxidant enzymes, namely, POD, CAT, and APX, while increase in lipoxygenase activity was recorded in leaves and roots. Plants primed with SNP showed increase in all antioxidant enzymes, while arginine priming resulted in enhanced APX and CAT activities. The authors concluded that salinity alleviation effects of SNP and arginine are comparable in terms of antioxidant defense system. In another study by Fan and Liu (2012), the dehydration and drought ameliorative potential of seed priming with SNP (nitric oxide donor) in *Poncirus trifoliata* was studied. Authors also did seed priming with L-Name (nitric oxide synthase inhibitor N G-nitro-l-arginine-methyl ester) to validate the positive influence of SNP seed priming. The priming results were obtained from plants grown in pots or hydroponic culture. It was observed that plants with SNP seed priming from hydroponic culture exhibited smaller stomatal apertures, enhanced antioxidant enzyme activities, and lower ROS accumulation, electrolyte leakage, and water loss than those of control plants. Besides, potted plants pretreated with SNP displayed higher photosynthesis, minimal stomatal conductance, higher chlorophyll contents, alongside the limited electrolyte leakage compared with control plants. The plants pretreated with inhibitor (L-Name) largely reversed the positive alterations in key physiochemical indices induced by SNP priming. The results of the present study displayed increased tolerance to drought and dehydration in plants pretreated with SNP. Conversely, plants pretreated with L-Name made plants sensitive to those stresses. It was concluded that changes in antioxidant enzyme activities and stomatal responses contributed to enhanced stress tolerance in plants pretreated with SNP. Farooq et al. (2009a) reported the ameliorative effects of SNP priming in rice under drought stress. Seeds were primed with SNP (100 and 150 μM), and later the same concentration was also used as foliar spray. Drought was maintained at 50% field capacity (FC). Two controls were used in the experiment, namely, unprimed plant and NO foliar application. Rice plants suffered significant growth reductions under drought; however, exogenous application of NO as seed priming or foliar applications circumvented the adverse effects of drought on plant growth. NO applications resulted in higher water potential, provoked antioxidant system, lower membrane damage, and improved photosynthetic capacity. Authors concluded that foliar application of NO was better in the induction of drought tolerance in rice than that of seed priming. Similarly, the seed soaking with NO or $H_2O_2$ (<10 μM) improved quantum yield for photosystem II compared with plants not pretreated with these signaling molecules. Pretreated plants also exhibited enhanced tolerance to salinity and heat stresses. The results displayed higher activities of ROS scavenging enzymes alongside the increased transcript levels for stress-related genes such as small heat shock protein 26, $\Delta'$-pyrroline-5-carboxylate synthase, and sucrose-phosphate synthase. It was concluded that the enhanced tolerance to both stresses in rice was due to the potential of exogenously applied molecules to function as signaling molecules (Uchida et al. 2002). Likewise, the cross talk between NO and $H_2S$ was reported in maize under heat stress. Seed soaking in SNP decreased lipid peroxidation and electrolyte leakage and increased the survival percentage of maize plants under heat stress. It was also observed that plants pretreated with SNP showed higher accumulation of $H_2S$ as evident from the enhance expression of L-cysteine desulfhydrase. Likewise, pretreatment with $H_2S$ donors GYY4137 and NaHS also increased endogenous $H_2S$ levels. Furthermore, $H_2S$ pretreatment increased survival percentage by circumventing electrolyte leakage.

Intriguingly, SNP-mediated heat tolerance was enhanced by the application of $H_2S$ donors. However, this enhancement in heat tolerance was eliminated in plants treated with inhibitors for $H_2S$ synthesis. It was suggested that $H_2S$ functioned as a downstream signal molecule mediating the NO-induced heat tolerance in maize seedlings (Li et al. 2013b).

The influence of seed presoaking with SNP was studied in wheat under chilling stress. Pretreatment with SNP resulted in increase in germination index, seed germination rate, and length and weight of radical and coleoptile. However, SNP priming decreased the weight of germinated seeds and mean germination time under chilling stress. Seed respiration rate alongside the higher starch degradation reflected as improved amylase activity was observed as a result of seed priming with SNP. Besides, SNP priming provoked efficient antioxidant system that in turn decreased ROS generation and lipid peroxidation. Seed priming with SNP enhanced seedling growth under chilling stress (Li et al. 2013a). In another study, the effect of SNP priming was examined on low temperature tolerance, seedling vigor, and seed germination in two tomato cultivars, namely, Cherry and Falcato, exposed to different temperatures (10, 15, 20, and 25 °C). Treatments in the experiment were broken up as osmopriming (polyethylene glycol 6000) and seed soaking in SNP for 24 h at 25 °C. Reduced alpha-amylase activity, seedling growth, and seed germination was observed in both tomato cultivars at suboptimal temperatures. However, cultivar Falcato showed more sensitivity to low temperature than that of cv. Cherry. Seed pretreatment with SNP in Falcato increased sugar content, $\alpha$- and $\beta$-amylase, seedling growth, germination index, and germination capacity under suboptimal temperatures. It was observed that application of methylene blue (NO-scavenger) diminished the positive effects of NO priming on seed germination (Amooaghaie and Nikzad 2013). Similarly, the mitigative effects of NO pretreatment on light-induced oxidative damage were investigated in two tall fescue cultivars, namely, Arid3 and Houndog5. The adverse effects of high light were more prominent in the leaves of cultivar Houndog5. However, the pretreatment of leaves of both cultivars with SNP exhibited decrease in lipid peroxidation, ROS generation, and electrolyte leakage. The antioxidant enzyme activities (SOD, CAT, APX, GR) were significantly higher in leaves pretreated with SNP. Besides, SNP-treated leaves had decline in lipoxygenase (LOX) activity. The abovementioned responses due to SNP treatment could be eliminated when NO-scavenger cPTIO (2-(4-carboxy-2-phenyl)-4,4,5,5-tetramethylimidazoline-1-oxyl-3-oxide) was applied. Cultivar Arid3 showed maximal rise in the values of nitric oxide synthase activity and NO-production when exposed to high light stress, whereas a little increase was evident in the high light-sensitive cultivar. Increase in oxidative damage due to high light was observed when NOS inhibitor LNNA (N$\omega$-nitro-l-arginine) was given. The intensity of oxidative damage was higher in Arid3. The results of present investigation showed more NO release driven by enhanced NOS activity under high light stress. It was concluded that NO functioned as signaling molecule, increasing antioxidant enzyme activities. The damages due to high light stress were significantly lower in leaves with better oxidative defense system (Xu et al. 2010).

In another study, the interactive effects of NO and $H_2S$ were studied on *Medicago sativa* L. subjected to cadmium toxicity. Pretreatment with SNP and NaHS ameliorated the inhibitory effects of cadmium. Plants from the primed seeds ameliorated Cd-induced inhibition of seedling growth alongside significant decline in lipid peroxidation. Transcript levels and activities of antioxidant enzymes (SOD, POD, APX) showed differential alteration that alleviated the oxidative damage in pretreated plants. The positive impact of $H_2S$ was diminished in plants treated with NO-scavenger cPTIO. It was further reported that plants pretreated with NaHS exhibited enhanced production of NO under stress or nonstress conditions. It was inferred that NO mediated the NaHS which triggered mitigation of Cd stress in alfalfa plants. Authors declared the existence of a cross talk between NO and $H_2S$ (Li et al. 2012). Hydrogen peroxide ($H_2O_2$) functions as an important stress signal and when applied as seed priming induced significant metabolic changes. In this context, wheat seeds were soaked in $H_2O_2$ (1, 40, 80, and 120 μM). Plants treated with $H_2O_2$ (120 μM) increased germination percentage and mean germination time under 150 mM NaCl salinity. Plants from $H_2O_2$-treated seeds accumulated lower endogenous levels for $H_2O_2$ that indicated the presence of stimulated antioxidant system. Exogenous $H_2O_2$ significantly increased photosynthetic capacity alongside higher stomatal conductance. Pretreated plants possessed improved leaf water relations showing enhanced turgor. Plants raised from $H_2O_2$-treated seeds despite higher endogenous levels of $Na^+$ and $Cl^-$ displayed higher $K^+$, $Ca^{2+}$, $PO_4^{3-}$, and $NO_3^-$ under salinity stress. Seed priming with $H_2O_2$ also improved membrane integrity. Two heat stable proteins with 32 and 52 kDa were expressed in plants treated with $H_2O_2$. Seed priming stimulated antioxidant system that protected wheat seedlings from ion-induced oxidative damage. It was concluded that pretreatment of plants-enhanced tolerance to salinity was attributed to the expression of proteins alongside the improved physiological indices (Wahid et al. 2007). In another study, the drought ameliorative influence of $H_2O_2$ seed priming was studied in maize. Seeds were soaked in different concentrations of $H_2O_2$ (20, 40, 60, 80, 100, 120, and 140 mM). Germination rate was higher for 140 mM priming. Plants suffered significant decrease in chlorophyll contents along with increase in total soluble proteins, lipid peroxidation, $H_2O_2$ accumulation, and ascorbic acid under water deficit conditions. Drought decreased $Ca^{2+}$ and $K^+$, while increased P and $Fe^{2+}$ in maize plants. Activities of antioxidant enzymes also increased in drought stressed plants. Pretreatment of plants with 140 mM $H_2O_2$ decreased chlorophyll degradation and lipid peroxidation and increased the levels of ascorbic acid and activities of antioxidant enzymes under water deficit conditions. Authors concluded that enhanced drought tolerance in maize plants was the result of better antioxidant system due to $H_2O_2$ priming (Ashraf et al. 2015). In a study, the influence of exogenously supplemented $H_2O_2$ was examined in *Panax ginseng* under salinity stress. Pretreatment of plants with 100 μM $H_2O_2$ induced physiological tolerance in ginseng against salinity stress. Treatment with $H_2O_2$ showed enhanced activities of APX, CAT, and POD along with the significant decrease in ROS generation and lipid peroxidation. Pretreated plants also showed higher photosynthetic pigments. It was concluded that $H_2O_2$ pretreatment-induced salinity tolerance in ginseng plants was attributed to better oxidative defense system (Sathiyaraj et al. 2014). The role of exogenously applied $H_2O_2$ was investigated in wheat plants under drought stress. Pretreated plants displayed enhanced germination rate (56%) under PEG-induced water scarcity. The accumulation of $H_2O_2$ was minimal in plants treated with $H_2O_2$ that indicated the presence of stimulated antioxidant system. Pretreated plants also showed higher leaf area, dry weight, and net photosynthesis. Besides, $H_2O_2$ treatment also increased proline accumulation and water use efficiency. Exogenously applied $H_2O_2$ decreased lipid peroxidation that markedly enhanced the integrity of membranes. The expression of APX and CAT was higher in treated plants. The results of the present research indicated the functions of $H_2O_2$ as signal molecule that stimulated the antioxidant system in seeds, which protected the seedlings from oxidative damage (He and Gao 2009). The positive role of $H_2O_2$ was studied in cucumber under heat stress. It was seen that pretreated plants showed chloroplast ultrastructure protection due to enhanced antioxidant system. Heat stress increased lipid peroxidation and accumulation of ROS that damaged chloroplasts in leaf cells (71.4%). However, when $H_2O_2$ was applied under heat stress, a significant decrease in lipid peroxidation and ROS generation protected chloroplast ultrastructure. Pretreated plants exhibited enhanced activities of antioxidant enzymes under heat stress. Authors suggested that $H_2O_2$ pretreatment enhanced antioxidant system which in turn decreased lipid peroxidation and hence protected chloroplast ultrastructure (Gao et al. 2010).

# 4 Chemical Priming with Melatonin

Melatonin functions as essential biological hormone mediating plant responses to abiotic constraints. In a study, the influence of added melatonin on photosynthetic attributes and antioxidant system was investigated. Salinity significantly decreased photosystem II (PSII) efficiency, chlorophyll contents, and plant growth along with marked increase in the production of ROS. It was found that melatonin application circumvented ROS-induced oxidative damage and increased photosynthetic efficiency in maize. Melatonin treatment stimulated antioxidant system that effectively scavenged ROS. Immunoblotting displayed decrease in 34 kDa PS II reaction center with rise in subunit S protein in PSII. It was concluded that melatonin activated antioxidant system and mediated PSII proteins (Chen et al. 2018). The positive role of melatonin priming is reported in tomatoes under low temperature stress. Plants pretreated with 100 μM melatonin concentration alleviated the adverse effects of low temperature on photosynthetic efficiency in terms of minimal damage to photosynthetic apparatus, increase in quantum yield of photosystems, increase in electron transport, and protection to thylakoid membrane (Yang et al. 2018b). Melatonin also protected tomato plants from the damages of simulated acid rain (SAR). Higher concentration of bioactive compounds was evident in SAR-stressed plants. Besides the higher antioxidant potential of stressed plants, SAR induced substantial decrease in tomato yield. Plants pretreated with melatonin showed significant improvement in fruit quality, yield attributes, and antioxidant compounds that effectively protected plants from oxidative damage due to ROS. The results of the study showed that melatonin application mediated antioxidant system and thereby improved yield (Debnath et al. 2018). Melatonin also plays an essential role in the acquisition and assimilation of sulfur (S). In this context, tomato plants grown under low S conditions showed stunted growth due to decrease in photosynthesis, chlorophyll content, and biomass production. Plants facing S-deficiency also suffered oxidative damage that led to modifications in cell structure and significant damage to nucleic acids. However, plants treated with exogenous melatonin showed diminution in cellular ROS levels, which mitigated damages to cellular ultrastructures and macromolecules. Besides, melatonin enhanced the expression of genes related to enhanced uptake and assimilation of S. Melatonin mediated 2-cysteine peroxiredoxin and biosynthesis of sulfur compounds. It was concluded that melatonin significantly diminished S deficiency-induced adverse effects by improving redox homeostasis, metabolism, S uptake, and utilization (Hasan et al. 2018). Melatonin pretreatment also enhanced salinity tolerance in wheat seedlings. Pretreated seedlings showed better growth reflected as higher shoot dry weight, chlorophyll content, photosynthesis, indole acetic acid content, and efficiency of photosystems. The accumulation of $H_2O_2$ was also minimal in melatonin pretreated seedlings under salinity. The transcript levels of TaSNAT mediating melatonin biosynthesis were also higher in seedlings treated with melatonin. Melatonin also increased polyamines biosynthesis through enhancing precursor flow from methionine and arginine. Salinity-induced degradation of polyamines was also minimal in seedlings treated with melatonin. It was concluded that melatonin enhanced salinity tolerance through mediating polyamine biosynthesis in wheat seedlings (Ke et al. 2018). Melatonin has also been shown to play a crucial part in the formation of leaf cuticle under water-limited conditions (Ding et al. 2018). Melatonin enhanced drought tolerance in oat seedlings. Oat seedlings treated with 100 μM melatonin showed significant decrease in ROS accumulation. Activities of antioxidant enzymes (SOD, CAT, POD, APX) increased in oat seedlings treated with melatonin under PEG-induced drought stress (Gao et al. 2018). Melatonin pretreatment (5 and 20 μM) significantly enhanced germination in *Stevia rebaudiana* Bertoni. Higher concentration of melatonin (100 and 500 μM) inhibited germination. Melatonin also increased seedling growth and leaf numbers. Seedlings also had greater pigment concentration due to melatonin application (500 μM). However, SOD activity decreased markedly in pretreated seedlings. Seedlings treated with 500 μM melatonin significantly enhanced the activities of antioxidant enzymes (POD, CAT) and the levels of phenolics and sugars (Simlat et al. 2018). The influence of melatonin application was investigated on senescence in kiwifruits during natural aging. Melatonin pretreatment decreased $H_2O_2$ accumulation and membrane damage through the rise of antioxidant enzyme activities (POD, SOD, CAT), which in turn effectively delayed senescence in kiwifruit. Pretreated plants also displayed minimal chlorophyll degradation. The endogenous levels of ascorbic acid, flavonoids, soluble proteins, and sugars were higher in plants treated with melatonin (Liang et al. 2018).

Melatonin also promoted Cd tolerance in wheat seedlings. Melatonin-treated seedlings showed increased plant height, root growth, and biomass accumulation. The enhanced accumulation of $H_2O_2$ due to Cd toxicity was also counteracted by melatonin. Melatonin treatment increased glutathione (GSH) contents under Cd stress. Activities of antioxidant enzymes, namely, APX and SOD, increased in treated seedlings; however, no increase in CAT and POD activities was recorded as a result of melatonin application. It was concluded that melatonin pretreatment improved the scavenging of $H_2O_2$ by stimulated antioxidant system under Cd stress (Ni et al. 2018). In another study, the drought alleviative effects of melatonin pre-treatment in rapeseed (*Brassica napus* L.) seedlings were investigated. Melatonin pretreatment increased seedling growth; activities of APX, POD, and CAT; and accumulation of soluble proteins and sugars. Besides, the accumulation of $H_2O_2$ was also minimal in melatonin pretreated seedlings. Melatonin-induced drought tolerance was associated with improved osmotic adjustment and antioxidant system (Li et al. 2018a). The role of supplemented melatonin in *Cyphomandra betacea* seedlings subjected to Cd toxicity was evaluated. The application of melatonin was done in nutrient solution and soil. The results displayed that seedling growth increased with lower melatonin dose (50 μM), while significant decline in seedling growth was evident with 100–200 μM in both soil and nutrient culture experiment. Likewise, lower dose of melatonin decreased photosynthetic pigment degradation alongside the stimulated activities of antioxidant enzymes. Lower dose of melatonin also enhanced Cd accumulation in different parts of seedlings. It was suggested that lower dose of melatonin enhanced seedling growth under Cd toxicity (Lin et al. 2018). Melatonin also alleviated the salinity damages in rapeseed (*Brassica napus* L.). Melatonin (30 μM) increased growth and activities of APX, CAT, and POD and decreased the accumulation of $H_2O_2$ in plants subjected to salinity. Plants treated with melatonin also displayed enhanced buildup of proteins, sugars, and proline. The results also suggested the inhibitory effects of melatonin when applied at higher concentration. Lower melatonin-induced increase in growth was attributed to better $H_2O_2$ scavenging by antioxidant enzymes and protection from osmotic stress by the accumulation of osmoregulatory substances like soluble glucan, proline, and protein. Melatonin application also enhanced root development and enhanced rapeseed biomass production under salinity (Zeng et al. 2018). Melatonin alleviated oxidative damages in apple leaves subjected to methyl viologen (MV). The decrease in chlorophyll and photosystem efficiency was circumvented by melatonin application. Melatonin pretreated leaves exhibited decreased lipid peroxidation by enhancing activities of antioxidant enzymes (Wei et al. 2018). The cold stress significantly inhibited growth and development in *Camellia sinensis* L. Melatonin pretreatment enhanced cold stress tolerance by decreasing ROS accumulation and lipid peroxidation alongside the enhanced chlorophyll content, photosynthesis, and PSII efficiency. Melatonin pretreatment also displayed better antioxidant properties and redox homeostasis. This study highlighted the key role of melatonin in circumventing the inhibitory effects of frost damage in plants (Li et al. 2018b). In another study, the functions of melatonin were investigated in watermelon subjected to vanadium (V) stress. Pretreatment of watermelon seedlings with melatonin (0.1 μM) increased chlorophyll content, photosynthesis, and seedling growth under 50 mg $L^{-1}$ V stress. The transport of V was significantly lower in melatonin pretreated seedlings. Furthermore, activities of antioxidant enzymes (SOD, CAT) markedly increased, which resulted in decline in $H_2O_2$ accumulation and lipid peroxidation in watermelon seedlings treated with melatonin under V stress. The results of this study suggested the use of melatonin to improve tolerance of watermelon seedlings against V stress (Nawaz et al. 2018). Melatonin application alleviated salinity stress in rice plants by decreasing lipid peroxidation, $H_2O_2$ accumulation, and increasing the activities of antioxidant enzymes (Li et al. 2017). The role of melatonin was explored in Bermuda grass under K deficiency. Melatonin pretreatment significantly enhanced the accumulation of K in plants. Besides, the efficiency of PSII was also enhanced in plants treated with melatonin (Chen et al. 2017). Melatonin effects on plant growth and photosynthesis were evaluated in nan-ZnO-stressed plants. Melatonin increased ATPase and RuBisCO activities, chlorophyll contents, and photosystems efficiency and thereby led to improved photosynthesis. Melatonin also protected plants from oxidative damage by enhancing the activities of antioxidant enzymes that safeguarded photosynthesis electron transport system. The results of the present investigation revealed the use of melatonin to protect wheat plants from the inhibitory effects of nano-ZnO stress (Zuo et al. 2017). Melatonin enhanced alkaline stress in *Malus hupehensis* Rehd. through the modulation of polyamine biosynthesis (Gong et al. 2017).

## 5 Chemical Priming with Polyamines

Polyamines application as seed treatment has been shown to improve plant tolerance to abiotic stress. In a study, polyamines (100 μM of spermine, Spm; putrescine, Put; and mixture of Spm and Put) application as seed priming and foliar application protected wheat plants under drought conditions. Drought-stressed plants displayed suppressed growth, changes in the accumulation of osmoprotectants, polyamines levels, expression of polyamines biosynthesis genes, and grain yield. Exogenous polyamines application enhanced osmoprotectant accumulation and cell water status and upregulated the expression of polyamines biosynthesis genes (Ebeed et al. 2017). Seed priming with 5 mM spermidine (Spd) modulated the seed polyamine metabolism that in turn improved chilling tolerance in rice plants. Chilling stress resulted in a significant decrease in seedling vigor index, seedling growth, and germination percentage. The key physiological attributes were significantly improved as a result of Spd priming. Priming with Spd increased glycinebetaine, flavonoids, and total phenolics accumulation. By contrast, priming decreased soluble proteins, sugars, and α-amylase activity in rice plants exposed to chilling stress. Activities of antioxidant enzymes (SOD, POD, APX, GPX) were higher in plants as a result of chilling stress. Priming with Spd resulted in further improvement in antioxidant enzyme activities. Chilling stress decreased the endogenous levels of Put and Spd, while the otherwise was true for Spm content. Polyamine biosynthesis genes expression was improved as a result of Spd priming. The ultrastructure of leaf cell alongside the grain structure was higher in primed plants compared with those unprimed plants (Sheteiwy et al. 2017). In another study, seed priming with Spd and Spm was investigated in rice (sensitive cv. IR-64 and tolerant cv. Nonabokra) subjected to salinity. The priming with polyamines mediated multiple metabolic pathways governing the salinity tolerance in rice seedlings. The transcript levels of osmolytes, ABA biosynthetic enzymes, nonenzymatic and enzymatic antioxidants, key transcription factors and ion transporters (NHX1), and polyamine metabolic enzymes were examined in roots and shoots of rice seedlings. Spm and Spd priming significantly enhanced the expression of antioxidant genes in different plant parts when compared with the nonprimed controls. The increase in gene expression was higher in case of Spm priming in salinity sensitive cv. IR-64 under salinity. The expression of genes related to osmolyte biosynthesis enzymes was higher as a result of seed priming with polyamines. Polyamines priming also enhanced the gene expression of ABA biosynthesis enzyme alongside the induced expression of NHX1 under salinity in seedlings. The endogenous levels of polyamines were also greater in primed plants (Paul and Roychoudhury 2017). Likewise, Spd seed pretreatment improved osmolyte concentration, anthocyanin contents, and activity of polyphenol oxidase in rice plants. Pretreatment enhanced the antioxidant properties of rice plant under salinity stress. Plants raised from primed seeds exhibited lower oxidative damage and stimulated antioxidant system and better osmotic adjustment (Paul et al. 2017).

Polyamines presowing seed treatment was studied to see changes in germination and seedling growth in rice. Seeds were soaked in Spm, Put, and Spd solutions (10 and 20 mg $L^{-1}$). It was seen that seed priming with polyamines enhanced germination. Significant increase in seedling fresh and dry weight was recorded as a result of polyamines pretreatment. The results revealed 10 mg $L^{-1}$ as the most effective dose that significantly improved germination and seedling growth in rice (Farooq et al. 2008). Seed priming in tomato with polyamines displayed significant effect on germination, antioxidant properties, and seedling vigor. Seeds were soaked in 50 mg $L^{-1}$ of Spm, Put, and Spd. Priming with Spm and Spd markedly enhanced seed vigor and stimulated antioxidant properties. However, Put priming did not influence seedling vigor and antioxidant properties (Afzal et al. 2009). In another study, seed priming with Put was investigated in tobacco under chilling stress. Chilling sensitive (MSk326) and tolerant (Honghuadajinyuan) varieties were used in the study. Seeds priming with 0.01 and 0.1 mM Put significantly enhanced seedling dry weight, seedling length, germination index, and germination percentage. When seedlings were subjected to chilling stress, priming with 0.1 mM significantly enhanced the activities of antioxidant enzymes (SOD, POD CAT, and APX) alongside the higher endogenous levels of polyamines (Spd, Spm, and Put). The results of the study recommended the use of 0.1 mM priming with Put could improve tobacco tolerance against chilling stress (Xu et al. 2011). In a study, the seed pretreatment with 2.5 and 5.0 mM Spd and Spm influence on ion accumulation and photosynthesis and growth was examined in wheat under salinity. Primed and unprimed seeds were grown under filed conditions at 15 dS $m^{-1}$ NaCl. Increase in growth and grain yield was seed as a result of all priming treatment; however, Spm priming was the most effective treatment in enhancing grain yield. The cultivars did not differ with respect to transpiration rate and photosynthesis. Spm priming decreased stomatal conductance in tolerant wheat cultivar. Polyamines priming mediated ion accumulation in sensitive (MH-97) and tolerant (Inqlab-91) wheat cultivars under salinity. Priming treatment significantly decreased accumulation of toxic Na and Cl besides the higher endogenous levels of K. Priming treatments had differential influence on growth, photosynthesis, and ion accumulation in two wheat cultivars (Iqbal and Ashraf 2005). Polyamine priming also increased drought tolerance in rice. Seeds were primed with 10 μM Put, Spd, and Spm besides the foliar application of polyamines. Presowing seed treatment significantly enhanced growth, water use efficiency, and photosynthesis in rice exposed to drought conditions. Accumulation of phenolics, anthocyanins, and proline was significantly higher in treated plants. Seed priming also improved membrane integrity under drought conditions. The results revealed higher antioxidant properties that protected plants from oxidative damage reflected in the form of lower cellular levels of $H_2O_2$ (Farooq et al. 2009b).

Plant growth and productivity is significantly deteriorated as a result of abiotic stresses. Besides conventional breeding and transgenic approaches, chemical priming offers an efficient way to combat these environmental hazards. Despite ample research on the physiological and biochemical responses of pretreated plants, there is lack of information on the expression of stress-related genes. The new chemicals are being discovered which exhibit the potential to improve plant growth and development under multiple abiotic stresses. The effective concentration of these chemicals may vary with plant species and type of stress. Therefore, further research is required to understand the role of these chemicals as presowing seed treatments.

## References

Abdel Latef AAH, Abu Alhmad MF, Abdelfattah KE (2017) The possible roles of priming with zno nanoparticles in mitigation of salinity stress in lupine (*Lupinus termis*) plants. J Plant Growth Regul 36:60–70

Abid M, Hakeem A, Shao Y, Liu Y, Zahoor R, Fan Y, Suyu J, Ata-Ul-Karim ST, Tian Z, Jiang D, Snider JL, Dai T (2018) Seed osmopriming invokes stress memory against post-germinative drought stress in wheat (*Triticum aestivum* L.). Environ Exp Bot 145:12–20

Afzal I, Munir F, Ayub C, Basra S, Hameed A, Nawaz A (2009) Changes in antioxidant enzymes, germination capacity and vigour of tomato seeds in response of priming with polyamines. Seed Sci Technol 37:765–770

Ahmad I, Ahmed Basra SM, Akram M, Wasaya A, Ansar M, Hussain S, Iqbal A, Azhar Hussain S (2017) Improvement of antioxidant activities and yield of spring maize through seed priming and foliar application of plant growth regulators under heat stress conditions. Semina: Ciências Agrárias 38:47

Akhtar N, Khan S, Malook I, Rehman SU, Jamil M (2017) Pb-induced changes in roots of two cultivated rice cultivars grown in lead-contaminated soil mediated by smoke. Environ Sci Pollut Res 24:21298–21310

Alasvandyari F, Mahdavi B, Hosseini SM (2017) Glycine betaine affects the antioxidant system and ion accumulation and reduces salinity-induced damage in safflower seedlings. Arch Biol Sci 69:139–147

Ali Q, Daud MK, Haider MZ, Ali S, Rizwan M, Aslam N, Noman A, Iqbal N, Shahzad F, Deeba F, Ali I, Zhu SJ (2017) Seed priming by sodium nitroprusside improves salt tolerance in wheat (*Triticum aestivum* L.) by enhancing physiological and biochemical parameters. Plant Physiol Biochem 119:50–58

Amooaghaie R, Nikzad K (2013) The role of nitric oxide in priming-induced low-temperature tolerance in two genotypes of tomato. Seed Sci Res 23:123–131

Antoniou C, Savvides A, Christou A, Fotopoulos V (2016) Unravelling chemical priming machinery in plants: the role of reactive oxygen–nitrogen–sulfur species in abiotic stress tolerance enhancement. Curr Opin Plant Biol 33:101–107

Ashraf MA, Rasheed R, Hussain I, Iqbal M, Haider MZ, Parveen S, Sajid MA (2015) Hydrogen peroxide modulates antioxidant system and nutrient relation in maize (*Zea mays* L.) under water-deficit conditions. Arch Agron Soil Sci 61:507–523

Bajwa AA, Farooq M, Nawaz A (2018) Seed priming with sorghum extracts and benzyl aminopurine improves the tolerance against salt stress in wheat (*Triticum aestivum* L.). Physiol Mol Biol Plants 24:239–249

Bibi A, Majid S, Ulfat A, Khatoon S, Munir A, Javed G (2017) Effect of nitric oxide seed priming on chilling induced water related physiological attributes in germinating wheat. J Anim Plant Sci 27:186

Chen L, Fan J, Hu Z, Huang X, Amombo E, Liu A, Bi A, Chen K, Xie Y, Fu J (2017) Melatonin is involved in regulation of bermudagrass growth and development and response to low $K^+$ stress. Front Plant Sci 8:2038

Chen Y-E, Mao J-J, Sun L-Q, Huang B, Ding C-B, Gu Y, Liao J-Q, Hu C, Zhang Z-W, Yuan S, Yuan M (2018) Exogenous melatonin enhances salt stress tolerance in maize seedlings by improving antioxidant and photosynthetic capacity. Physiol Plant 164:349–363

Christou A, Filippou P, Manganaris GA, Fotopoulos V (2014a) Sodium hydrosulfide induces systemic thermotolerance to strawberry plants through transcriptional regulation of heat shock proteins and aquaporin. BMC Plant Biol 14:42

Christou A, Manganaris GA, Fotopoulos V (2014b) Systemic mitigation of salt stress by hydrogen peroxide and sodium nitroprusside in strawberry plants via transcriptional regulation of enzymatic and non-enzymatic antioxidants. Environ Exp Bot 107:46–54

Cui G, Sun F, Gao X, Xie K, Zhang C, Liu S, Xi Y (2018) Proteomic analysis of melatonin-mediated osmotic tolerance by improving energy metabolism and autophagy in wheat (*Triticum aestivum* L.). Planta 248:69–87

Dai L-Y, Zhu H-D, Yin K-D, Du J-D, Zhang Y-X (2017) Seed priming mitigates the effects of saline-alkali stress in soybean seedlings. Chil J Agric Res 77:118–125

David BL, Christopher BF (2007) Global scale climate–crop yield relationships and the impacts of recent warming. Environ Res Lett 2:014002

Debnath B, Hussain M, Li M, Lu X, Sun Y, Qiu D (2018) Exogenous melatonin improves fruit quality features, health promoting antioxidant compounds and yield traits in tomato fruits under acid rain stress. Molecules 23:1868

Dianati Tilaki GA, Behtari B (2017) Alleviation of ionic and osmotic stress of salinity in seedling emergence of *Lolium perenne* L. with halopriming treatments growing in an hydroponic system. J Plant Nutr 40:219–226

Ding F, Wang G, Wang M, Zhang S (2018) Exogenous melatonin improves tolerance to water deficit by promoting cuticle formation in tomato plants. Molecules 23:1605

Duan P, Ding F, Wang F, Wang BS (2007) Priming of seeds with nitric oxide donor sodium nitroprusside (SNP) alleviates the inhibition on wheat seed germination by salt stress. J Plant Physiol Mol Biol 33:244–250

Ebeed HT, Hassan NM, Aljarani AM (2017) Exogenous applications of polyamines modulate drought responses in wheat through osmolytes accumulation, increasing free polyamine levels and regulation of polyamine biosynthetic genes. Plant Physiol Biochem 118:438–448

Ellouzi H, Sghayar S, Abdelly C (2017) $H_2O_2$ seed priming improves tolerance to salinity; drought and their combined effect more than mannitol in *Cakile maritima* when compared to *Eutrema salsugineum*. J Plant Physiol 210:38–50

Fan Q-J, Liu J-H (2012) Nitric oxide is involved in dehydration/drought tolerance in *Poncirus trifoliata* seedlings through regulation of antioxidant systems and stomatal response. Plant Cell Rep 31:145–154

Farooq M, Basra SMA, Rehman H, Hussain M (2008) Seed priming with polyamines improves the germination and early seedling growth in fine rice. J New Seeds 9:145–155

Farooq M, Basra SMA, Wahid A, Rehman H (2009a) Exogenously applied nitric oxide enhances the drought tolerance in fine grain aromatic rice (*Oryza sativa* L.). J Agron Crop Sci 195:254–261

Farooq M, Hussain M, Nawaz A, Lee D-J, Alghamdi SS, Siddique KHM (2017) Seed priming improves chilling tolerance in chickpea by modulating germination metabolism, trehalose accumulation and carbon assimilation. Plant Physiol Biochem 111:274–283

Farooq M, Wahid A, Lee D-J (2009b) Exogenously applied polyamines increase drought tolerance of rice by improving leaf water status, photosynthesis and membrane properties. Acta Physiol Plant 31:937–945

Filippou P, Tanou G, Molassiotis A, Fotopoulos V (2013) Plant acclimation to environmental stress using priming agents. In: Tuteja N, Singh Gill S (eds) Plant acclimation to environmental stress. Springer, New York, pp 1–27

Fotopoulos V, Christou A, Antoniou C, Manganaris GA (2015) Hydrogen sulphide: a versatile tool for the regulation of growth and defence responses in horticultural crops. J Hortic Sci Biotechnol 90:227–234

Gadelha CG, Miranda RS, Alencar NLM, Costa JH, Prisco JT, Gomes-Filho E (2017) Exogenous nitric oxide improves salt tolerance during establishment of *Jatropha curcas* seedlings by ameliorating oxidative damage and toxic ion accumulation. J Plant Physiol 212:69–79

Gao W, Zhang Y, Feng Z, Bai Q, He J, Wang Y (2018) Effects of melatonin on antioxidant capacity in naked oat seedlings under drought stress. Molecules 23:1580

Gao Y, Guo Y-K, Lin S-H, Fang Y-Y, Bai J-G (2010) Hydrogen peroxide pretreatment alters the activity of antioxidant enzymes and protects chloroplast ultrastructure in heat-stressed cucumber leaves. Sci Hortic 126:20–26

Ghassemi-Golezani K, Nikpour-Rashidabad N (2017) Seed pretreatment and salt tolerance of dill: Osmolyte accumulation, antioxidant enzymes activities and essence production. Biocatal Agric Biotechnol 12:30–35

Gong X, Shi S, Dou F, Song Y, Ma F (2017) Exogenous melatonin alleviates alkaline stress in *Malus hupehensis* Rehd by regulating the biosynthesis of polyamines. Molecules 22:1542

Habib N, Akram M, Javed M, Azeem M, Ali Q, Shaheen H, Ashraf M (2016) Nitric oxide regulated improvement in growth and yield of rice plants grown under salinity stress: antioxidant defense system. Appl Ecol Environ Res 14:91–105

Hameed A, Farooq T, Basra S, Sheikh M, Ibrahim M (2015) Wheat seed germination, antioxidant enzymes and biochemical enhancements by sodium nitroprusside priming. Agrochimica 59:93–107

Hasan MK, Liu C-X, Pan Y-T, Ahammed GJ, Qi Z-Y, Zhou J (2018) Melatonin alleviates low-sulfur stress by promoting sulfur homeostasis in tomato plants. Sci Rep 8:10182

Hassan M, Mansoor S (2017) Priming seeds with phytohormones alleviates cadmium toxicity in mung bean (*Vigna radiata* L. Wilczek) seedlings. Pak J Bot 49:2071–2078

He L, Gao Z (2009) Pretreatment of seed with $H_2O_2$ enhances drought tolerance of wheat (*Triticum aestivum* L.) seedlings. Afr J Biotechnol 8:6151

Hussain M, Farooq M, Lee D-J (2017) Evaluating the role of seed priming in improving drought tolerance of pigmented and non-pigmented rice. J Agron Crop Sci 203:269–276

Hussain S, Khaliq A, Tanveer M, Matloob A, Hussain HA (2018) Aspirin priming circumvents the salinity-induced effects on wheat emergence and seedling growth by regulating starch metabolism and antioxidant enzyme activities. Acta Physiol Plant 40:68

Ilyas N, Gull R, Mazhar R, Saeed M, Kanwal S, Shabir S, Bibi F (2017) Influence of salicylic acid and jasmonic acid on wheat under drought stress. Commun Soil Sci Plant 48:2715–2723

Imran M, Boelt B, Mühling K-H (2018) Zinc seed priming improves salt resistance in maize. J Agron Crop Sci 204:390–399

Iqbal M, Ashraf M (2005) Changes in growth, photosynthetic capacity and ionic relations in spring wheat (*Triticum aestivum* L.) due to pre-sowing seed treatment with polyamines. Plant Growth Regul 46:19–30

Jisha KC, Puthur JT (2016) Seed priming with beta-amino butyric acid improves abiotic stress tolerance in rice seedlings. Rice Sci 23:242–254

Kaczmarek M, Fedorowicz-Strońska O, Głowacka K, Waśkiewicz A, Sadowski J (2016) $CaCl_2$ treatment improves drought stress tolerance in barley (*Hordeum vulgare* L.). Acta Physiol Plant 39:41

Karadağ B, Yücel NC (2017) Salicylic acid and fish flour pre-treatments affect wheat phenolic and flavonoid compounds, lipid peroxidation levels under salt stress. Cereal Res Commun 45:192–201

Kaur H, Sirhindi G, Bhardwaj R, Alyemeni MN, Siddique KHM, Ahmad P (2018a) 28-homobrassinolide regulates antioxidant enzyme activities and gene expression in response to salt- and temperature-induced oxidative stress in *Brassica juncea*. Sci Rep 8:8735

Kaur K, Kaur K (2018) Nitric oxide improves thermotolerance in spring maize by inducing varied genotypic defense mechanisms. Acta Physiol Plant 40:55

Kaur R, Yadav P, Thukral AK, Sharma A, Bhardwaj R, Alyemeni MN, Wijaya L, Ahmad P (2018b) Castasterone and citric acid supplementation alleviates cadmium toxicity by modifying antioxidants and organic acids in *Brassica juncea*. J Plant Growth Regul 37:286–299

Ke Q, Ye J, Wang B, Ren J, Yin L, Deng X, Wang S (2018) Melatonin mitigates salt stress in wheat seedlings by modulating polyamine metabolism. Front Plant Sci 9:914

Khan F, Hussain S, Tanveer M, Khan S, Hussain HA, Iqbal B, Geng M (2018) Coordinated effects of lead toxicity and nutrient deprivation on growth, oxidative status, and elemental composition of primed and non-primed rice seedlings. Environ Sci Pollut Res 25:21185–21194

Li J, Zeng L, Cheng Y, Lu G, Fu G, Ma H, Liu Q, Zhang X, Zou X, Li C (2018a) Exogenous melatonin alleviates damage from drought stress in *Brassica napus* L. (rapeseed) seedlings. Acta Physiol Plant 40:43

Li L, Wang Y, Shen W (2012) Roles of hydrogen sulfide and nitric oxide in the alleviation of cadmium-induced oxidative damage in alfalfa seedling roots. Biometals 25:617–631

Li X, Jiang H, Liu F, Cai J, Dai T, Cao W, Jiang D (2013a) Induction of chilling tolerance in wheat during germination by pre-soaking seed with nitric oxide and gibberellin. Plant Growth Regul 71:31–40

Li X, Wei JP, Scott ER, Liu JW, Guo S, Li Y, Zhang L, Han WY (2018b) Exogenous melatonin alleviates cold stress by promoting antioxidant defense and redox homeostasis in *Camellia sinensis* L. Molecules 23:165

Li X, Yu B, Cui Y, Yin Y (2017) Melatonin application confers enhanced salt tolerance by regulating $Na^+$ and $cl^-$ accumulation in rice. Plant Growth Regul 83:441–454

Li Z-G, Yang S-Z, Long W-B, Yang G-X, Shen Z-Z (2013b) Hydrogen sulphide may be a novel downstream signal molecule in nitric oxide-induced heat tolerance of maize (*Zea mays* L.) seedlings. Plant Cell Environ 36:1564–1572

Liang D, Shen Y, Ni Z, Wang Q, Lei Z, Xu N, Deng Q, Lin L, Wang J, Lv X, Xia H (2018) Exogenous melatonin application delays senescence of kiwifruit leaves by regulating the antioxidant capacity and biosynthesis of flavonoids. Front Plant Sci 9:426

Lin L, Li J, Chen F, Liao MA, Tang Y, Liang D, Xia H, Lai Y, Wang X, Chen C, Ren W (2018) Effects of melatonin on the growth and cadmium characteristics of *Cyphomandra betacea* seedlings. Environ Monit Assess 190:119

Mahboob W, Khan M, Shirazi M, Faisal S, Asma MS (2018) Seed priming induced high temperature tolerance in wheat by regulating germination metabolism and physio-biochemical properties. Int J Agric Biol 20:2140–2148
Malook I, Shah G, Jan M, Shinwari KI, Aslam MM, Rehman SU, Jamil M (2017) Smoke priming regulates growth and the expression of myeloblastosis and zinc-finger genes in rice under salt stress. Arab J Sci Eng 42:2207–2215
Manjunatha G, Raj SN, Shetty NP, Shetty HS (2008) Nitric oxide donor seed priming enhances defense responses and induces resistance against pearl millet downy mildew disease. Pest Biochem Physiol 91:1–11
Mekawy AMM, Abdelaziz MN, Ueda A (2018) Apigenin pretreatment enhances growth and salinity tolerance of rice seedlings. Plant Physiol Biochem 130:94–104
Méndez AAE, Pena LB, Benavides MP, Gallego SM (2016) Priming with NO controls redox state and prevents cadmium-induced general up-regulation of methionine sulfoxide reductase gene family in Arabidopsis. Biochimie 131:128–136
Mohamed AKSH, Qayyum MF, Abdel-Hadi AM, Rehman RA, Ali S, Rizwan M (2017) Interactive effect of salinity and silver nanoparticles on photosynthetic and biochemical parameters of wheat. Arch Agron Soil Sci 63:1736–1747
Molassiotis A, Fotopoulos V (2011) Oxidative and nitrosative signaling in plants. Plant Signal Behav 6:210–214
Moreno C, Seal CE, Papenbrock J (2018) Seed priming improves germination in saline conditions for *Chenopodium quinoa* and *Amaranthus caudatus*. J Agron Crop Sci 204:40–48
Mouradi M, Bouizgaren A, Farissi M, Latrach L, Qaddoury A, Ghoulam C (2016) Seed osmopriming improves plant growth, nodulation, chlorophyll fluorescence and nutrient uptake in alfalfa (*Medicago sativa* L.) – rhizobia symbiosis under drought stress. Sci Hortic 213:232–242
Munir M, Shabbir G (2018) Salicylic acid mediated heat stress tolerance in selected bread wheat genotypes of Pakistan. Pak J Bot 50:2141–3146
Nagajyoti PC, Lee KD, Sreekanth TVM (2010) Heavy metals, occurrence and toxicity for plants: a review. Environ Chem Lett 8:199–216
Nawaz F, Naeem M, Akram A, Ashraf MY, Ahmad KS, Zulfiqar B, Sardar H, Shabbir RN, Majeed S, Shehzad MA, Anwar I (2017) Seed priming with KNO3 mediates biochemical processes to inhibit lead toxicity in maize (*Zea mays* L.). J Sci Food Agric 97:4780–4789
Nawaz MA, Jiao Y, Chen C, Shireen F, Zheng Z, Imtiaz M, Bie Z, Huang Y (2018) Melatonin pretreatment improves vanadium stress tolerance of watermelon seedlings by reducing vanadium concentration in the leaves and regulating melatonin biosynthesis and antioxidant-related gene expression. J Plant Physiol 220:115–127
Nejadalimoradi H, Nasibi F, Kalantari KM, Zanganeh R (2014) Effect of seed priming with L-arginine and sodium nitroprusside on some physiological parameters and antioxidant enzymes of sunflower plants exposed to salt stress. Agric Commun 2:23–30
Ni J, Wang Q, Shah FA, Liu W, Wang D, Huang S, Fu S, Wu L (2018) Exogenous melatonin confers cadmium tolerance by counterbalancing the hydrogen peroxide homeostasis in wheat seedlings. Molecules 23:799
Noman A, Ali Q, Maqsood J, Iqbal N, Javed MT, Rasool N, Naseem J (2018a) Deciphering physio-biochemical, yield, and nutritional quality attributes of water-stressed radish (*Raphanus sativus* L.) plants grown from Zn-Lys primed seeds. Chemosphere 195:175–189
Noman A, Ali Q, Naseem J, Javed MT, Kanwal H, Islam W, Aqeel M, Khalid N, Zafar S, Tayyeb M, Iqbal N, Buriro M, Maqsood J, Shahid S (2018b) Sugar beet extract acts as a natural biostimulant for physio-biochemical attributes in water stressed wheat (*Triticum aestivum* L.). Acta Physiol Plant 40:110
Paul S, Roychoudhury A (2017) Seed priming with spermine and spermidine regulates the expression of diverse groups of abiotic stress-responsive genes during salinity stress in the seedlings of indica rice varieties. Plant Gene 11:124–132

Paul S, Roychoudhury A, Banerjee A, Chaudhuri N, Ghosh P (2017) Seed pre-treatment with spermidine alleviates oxidative damages to different extent in the salt (NaCl)-stressed seedlings of three indica rice cultivars with contrasting level of salt tolerance. Plant Gene 11:112–123

Pehlivan N, Yesilyurt AM, Durmus N, Karaoglu SA (2017) *Trichoderma lixii* ID11D seed biopriming mitigates dose dependent salt toxicity in maize. Acta Physiol Plant 39:79

Perdomo JA, Conesa MÀ, Medrano H, Ribas-Carbó M, Galmés J (2015) Effects of long-term individual and combined water and temperature stress on the growth of rice, wheat and maize: relationship with morphological and physiological acclimation. Physiol Plant 155:149–165

Reddy PP (2015) Climate resilient agriculture for ensuring food security. Springer, New Delhi

Rinez I, Ghezal N, Rinez A, Farooq M, Dbara S, Saad I, Haouala R (2018) Improving salt tolerance in pepper by bio-priming with *Padina pavonica* and *Jania rubens* Aqueous Extracts. Int J Agric Biol 20:513–523

Samota MK, Sasi M, Awana M, Yadav OP, Amitha Mithra SV, Tyagi A, Kumar S, Singh A (2017) Elicitor-induced biochemical and molecular manifestations to improve drought tolerance in rice (*Oryza sativa* L.) through seed-priming. Front Plant Sci 8:934

Sani E, Herzyk P, Perrella G, Colot V, Amtmann A (2013) Hyperosmotic priming of Arabidopsis seedlings establishes a long-term somatic memory accompanied by specific changes of the epigenome. Genome Biol 14:R59

Sarwar M, Amjad M, Ayyub C (2017) Alleviation of salt stress in cucumber (*Cucumis sativus*) through seed priming with triacontanol. Int J Agric Biol 19:771

Sathiyaraj G, Srinivasan S, Kim Y-J, Lee OR, Parvin S, Balusamy SRD, Khorolragchaa A, Yang DC (2014) Acclimation of hydrogen peroxide enhances salt tolerance by activating defense-related proteins in *Panax ginseng* C.A. Meyer. Mol Biol Rep 41:3761–3771

Savvides A, Ali S, Tester M, Fotopoulos V (2016) Chemical priming of plants against multiple abiotic stresses: mission possible? Trends Plant Sci 21:329–340

Sepehri A, Rouhi HR (2016) Enhancement of seed vigor performance in aged groundnut (*Arachis hypogaea* L.) seeds by sodium nitroprusside under drought stress. Philipp Agric Sci 99:339–347

Shafiq F, Raza S, Bibi A, Khan I, Iqbal M (2018) Influence of proline priming on antioxidative potential and ionic distribution and its relationship with salt tolerance of wheat. Cereal Res Commun 46:287–300

Sheteiwy M, Shen H, Xu J, Guan Y, Song W, Hu J (2017) Seed polyamines metabolism induced by seed priming with spermidine and 5-aminolevulinic acid for chilling tolerance improvement in rice (*Oryza sativa* L.) seedlings. Environ Exp Bot 137:58–72

Sheteiwy MS, Gong D, Gao Y, Pan R, Hu J, Guan Y (2018) Priming with methyl jasmonate alleviates polyethylene glycol-induced osmotic stress in rice seeds by regulating the seed metabolic profile. Environ Exp Bot 153:236–248

Shukla P, Singh S, Dubey P, Singh A, Singh AK (2015) Nitric oxide mediated amelioration of arsenic toxicity which alters the alternative oxidase (Aox1) gene expression in Hordeum vulgare L. Ecotoxicol Environ Saf 120:59–65

Silva EN, Vieira SA, Ribeiro RV, Ponte LFA, Ferreira-Silva SL, Silveira JAG (2013) Contrasting physiological responses of jatropha curcas plants to single and combined stresses of salinity and heat. J Plant Growth Regul 32:159–169

Simlat M, Ptak A, Skrzypek E, Warchoł M, Morańska E, Piórkowska E (2018) Melatonin significantly influences seed germination and seedling growth of *Stevia rebaudiana Bertoni*. Peer J 6:e5009

Suzuki N, Rivero RM, Shulaev V, BlumwaldE MR (2014) Abiotic and biotic stress combinations. New Phytol 203:32–43

Tabassum T, Ahmad R, Farooq M, Basra SMA (2018) Improving the drought tolerance in barley by osmopriming and biopriming. Int J Agric Biol 20:1597–1606

Tanou G, Filippou P, Belghazi M, Job D, Diamantidis G, Fotopoulos V, Molassiotis A (2012) Oxidative and nitrosative-based signaling and associated post-translational modifications orchestrate the acclimation of citrus plants to salinity stress. Plant J 72:585–599

Uchida A, Jagendorf AT, Hibino T, Takabe T, Takabe T (2002) Effects of hydrogen peroxide and nitric oxide on both salt and heat stress tolerance in rice. Plant Sci 163:515–523
Wahid A, Perveen M, Gelani S, Basra SMA (2007) Pretreatment of seed with $H_2O_2$ improves salt tolerance of wheat seedlings by alleviation of oxidative damage and expression of stress proteins. J Plant Physiol 164:283–294
Wei Z, Gao T, Liang B, Zhao Q, Ma F, Li C (2018) Effects of exogenous melatonin on methyl viologen-mediated oxidative stress in apple leaf. Int J Mol Sci 19:316
Xiong J-L, Li J, Wang H-C, Zhang C-L, Naeem MS (2018) Fullerol improves seed germination, biomass accumulation, photosynthesis and antioxidant system in *Brassica napus* L. under water stress. Plant Physiol Biochem 129:130–140
Xu S, Hu J, Li Y, Ma W, Zheng Y, Zhu S (2011) Chilling tolerance in *Nicotiana tabacum* induced by seed priming with putrescine. Plant Growth Regul 63:279–290
Xu Y, Sun X, Jin J, Zhou H (2010) Protective effect of nitric oxide on light-induced oxidative damage in leaves of tall fescue. J Plant Physiol 167:512–518
Yang A, Akhtar SS, Iqbal S, Qi Z, Alandia G, Saddiq MS, Jacobsen S-E (2018a) Saponin seed priming improves salt tolerance in quinoa. J Agron Crop Sci 204:31–39
Yang X, Xu H, Li D, Gao X, Li T, Wang R (2018b) Effect of melatonin priming on photosynthetic capacity of tomato leaves under low-temperature stress. Photosynthetica:56(3)1–9
Zahra N, Mahmood S, Raza ZA (2018) Salinity stress on various physiological and biochemical attributes of two distinct maize (*Zea mays* L.) genotypes. J Plant Nutr 41:1368–1380
Zeng L, Cai J-S, Li J-J, Lu G-Y, Li C-S, Fu G-P, Zhang X-K, Ma H-Q, Liu Q-Y, Zou X-L, Cheng Y (2018) Exogenous application of a low concentration of melatonin enhances salt tolerance in rapeseed (*Brassica napus* L.) seedlings. J Integr Agric 17:328–335
Zhao F-Y, Liu W, Zhang S-Y (2009) Different responses of plant growth and antioxidant system to the combination of cadmium and heat stress in transgenic and non-transgenic rice. J Integr Plant Biol 51:942–950
Zheng Y, Yin X, Ma H (2018) Effects of hydrogen peroxide on seed germination, seedling growth and physiological characteristics of *Bombax ceiba* after heat shock. Pak J Bot 50:1327–1333
Zuo Z, Sun L, Wang T, Miao P, Zhu X, Liu S, Song F, Mao H, Li X (2017) Melatonin improves the photosynthetic carbon assimilation and antioxidant capacity in wheat exposed to nano-ZnO stress. Molecules 22:E1727

# Abiotic Stress Tolerance in Plants by Priming and Pretreatment with Hydrogen Peroxide

**Aditya Banerjee and Aryadeep Roychoudhury**

## Contents



**Abstract** Being a member of the reactive oxygen species, hydrogen peroxide ($H_2O_2$) is involved in signaling pathways associated with diverse antioxidant responses during abiotic stress. Thus, priming with $H_2O_2$ enables the seeds or seedlings to activate their antioxidant machineries and acclimatize prior to abiotic stress exposures. Hence, $H_2O_2$-priming actually hardens the plants to better cope with suboptimal conditions. This chapter discusses on the displayed mechanisms of $H_2O_2$-mediated tolerance in plants subjected to diverse abiotic stresses like salinity, drought, heat, cold and heavy metals.

**Keywords** Hydrogen peroxide · Reactive oxygen species · Priming · Signaling · Antioxidants · Abiotic stress · Tolerance

---

A. Banerjee · A. Roychoudhury (✉)
Post Graduate Department of Biotechnology, St. Xavier's College (Autonomous), Kolkata, West Bengal, India



M. Hasanuzzaman, V. Fotopoulos (eds.), *Priming and Pretreatment of Seeds and Seedlings*, https://doi.org/10.1007/978-981-13-8625-1_20

# 1 Introduction

Reactive oxygen species (ROS) like hydrogen peroxide ($H_2O_2$), hydroxyl, and superoxide radicals are toxic derivatives of oxygen, rapidly produced in plant cells subjected to almost any kind of abiotic stress (Banerjee and Roychoudhury 2018a). Abiotic stresses include salinity, drought, temperature, heavy metal toxicity, etc. (Banerjee and Roychoudhury 2018b, c, d, e). The uncontrolled production of ROS extensively degrades protein and nucleic acid structures, leading to cellular dysfunction and necrosis in plants (Banerjee et al. 2018a, b). Plants are intelligent organisms, which have evolved strategies to maintain low ROS titters even under abiotic stress. Many varieties show a tolerant phenotype under such suboptimal conditions (Banerjee et al. 2017; Banerjee and Roychoudhury 2016a, b).

Among the ROS, $H_2O_2$ has the longest half-life of 1 ms, owing to which it can traverse across various cellular compartments in order to initiate retrograde signaling (Noctor et al. 2014). The electron transport chains (ETCs) and redox reactions in the chloroplast and mitochondria are the sources of $H_2O_2$ within the cell (Miller et al. 2010). The NADPH-dependent oxidases, similar to respiratory burst oxidase homologs (RBOH), also produce $H_2O_2$. The peroxidases associated with the cell wall generate significant amounts of these molecules (Kapoor et al. 2015). The production of $H_2O_2$ under abiotic stress is balanced by the species-specific efficiency of the antioxidant machinery. Nonenzymatic antioxidants proline (Pro), glycinebetaine (GB), ascorbate (AsA), glutathione (GSH), polyamines (PAs), etc. and enzymatic antioxidants superoxide dismutase (SOD), ascorbate peroxidase (APX), glutathione peroxidase (GPX), catalase (CAT), etc. participate in the scavenging of ROS (Hossain et al. 2015).

Generation of abiotic stress-tolerant phenotype in plants has been achieved by several methods. Priming of seeds or seedlings or even exogenous treatment of the seedlings with a protective agent can be an economically cheap method for ameliorating the stress-induced injuries in plants (Nahar et al. 2014; Banerjee and Roychoudhury 2018f). In spite of being a toxic molecule, priming of plants with low doses of $H_2O_2$ interestingly confers tolerance to different abiotic stresses (Sathiyaraj et al. 2014). This process was found to increase basic physiological features like root length, shoot length, and chlorophyll (Chl) content and reduce the cytotoxic methylglyoxal (MG) level in stressed plants (Hossain and Fujita 2013). This chapter presents a brief discussion on the roles and instances of $H_2O_2$ acting as a signaling molecule to activate the antioxidant machinery during abiotic stresses.

# 2 $H_2O_2$-Induced Signaling Induces Stress Tolerance

Being an oxidant, $H_2O_2$ acts as a signaling molecule in coordination with GSH and Cys residues of protein. Being electron rich in nature, the Cys residues are the major sites of oxidant-mediated modifications (Akter et al. 2015). The heat shock factors

(HSFs) act as transcription factors (TFs) capable of regulating stress-inducible gene expression in $H_2O_2$-dependent fashion (Baniwal et al. 2004). Miller and Mittler (2006) reported that the HSFs directly sense ROS for their action during oxidative stress. Expression of cytosolic *APX1* and *APX2* is mediated by HSFs in ROS-dependent manner (Mazars et al. 2009).

A large signaling network comprising of $Ca^{2+}$-dependent signaling, protein kinases, and metabolic cross talks is thought to be associated with $H_2O_2$-mediated transduction processes. The protein kinases include $Ca^{2+}$-dependent protein kinases (CDPKs) and mitogen-activated protein kinases (MAPKs) (Zhou et al. 2014). A large number of MAPK and CDPK cascades are regulated by $H_2O_2$ during abiotic stresses (Roychoudhury and Banerjee 2017). Oxidative stress induces MAPKKK1, which together with $H_2O_2$ is required for ultimate MAPK4 activation (Mittler et al. 2011). $H_2O_2$ application also induced the expression of MAPK12 (Mittler et al. 2011). Cold stress increased $H_2O_2$ accumulation, which induced *RBOH1* expression and NADPH oxidase activity (Zhou et al. 2012). Pan et al. (2012) found that exogenous $H_2O_2$ application activated MAPK cascades in *Arabidopsis thaliana*, which ensured better ROS detoxification. *MAPK3/6* was found to be induced under $Cd^{2+}$-induced stress after accumulation of $H_2O_2$ (Liu et al. 2010).

$H_2O_2$ also regulated $Ca^{2+}$ homeostasis in order to dictate guard cell functioning during stomatal opening–closing under the conditions of abiotic stresses (Hossain et al. 2015). Bhattacharjee (2008) reported that the application of $Ca^{2+}$ inhibitor like trifluoperazine and $Ca^{2+}$ chelator like ethylene glycol tetraacetic acid (EGTA) in germinating *Amaranthus* seeds significantly decreased the endogenous $H_2O_2$ content. The expression of CDPKs was enhanced in tomato and wheat leaves treated with $H_2O_2$ (Wei et al. 2014). It was also reported that CAT scavenged $H_2O_2$ within the peroxisomes of *Arabidopsis* guard cells in a $Ca^{2+}$-dependent fashion (Costa et al. 2010).

The TFs like zinc fingers and WRKY are broadly involved in the regulation of ROS-detoxifying genes (Banerjee and Roychoudhury 2015, 2017). Application of $H_2O_2$ in *apx* knockout mutants of *Arabidopsis* resulted in accumulation of zinc finger proteins, ZAT7 and ZAT12 (Rizhsky 2004). $H_2O_2$ and nitric oxide (NO) together regulated stomatal closure in *Arabidopsis* leaves exposed to UV-B rays (He et al. 2005). Wang et al. (2010a, b) proposed that the $H_2O_2$-mediated NO synthesis in *Arabidopsis* might be regulated by MAPK6.

## 3 Mechanism of $H_2O_2$-Mediated Protection Against Abiotic Stress in Plants

It has been recently reviewed that $H_2O_2$ participates in retrograde signaling after its production from the chloroplast and mitochondria in order to induce the expression of stress-responsive genes encoded by the nucleus (Hossain et al. 2015). $H_2O_2$ priming enhances accumulation of heat shock proteins (Banerjee and Roychoudhury 2018b), activates the AsA–GSH cycle (Banerjee et al. 2016), and promotes the accumulation of Pro (Roychoudhury and Banerjee 2015; Hossain et al. 2015). This results in reduced ROS and MG accumulation along with enhanced rates of photosynthesis (Hossain et al. 2015). Huang et al. (2015) reported that pretreatment of seeds with $H_2O_2$ resulted in lowered accumulation of ROS and also conferred protection to the cellular organelles.

Correlation among $H_2O_2$ and other phytohormones has also been reported. Exogenous application of abscisic acid (ABA) resulted in inducing the expression of NADPH oxidase (Zhang et al. 2001). The ABA–ROS signaling cross talk was proposed to be integrated by MAPK9 and MAPK12 (Jammes et al. 2009). Kim et al. (2010) highlighted $H_2O_2$ as a crucial signaling mediator of ABA-induced stomatal closure during salinity and drought. The activities of $H_2O_2$-degrading enzymes like CAT and APX were suppressed by salicylic acid (SA) treatment (Yuan and Lin 2008). According to Rao and Davis (1999), $H_2O_2$ requires SA to trigger antioxidant- and defense-related responses in plants exposed to ozone stress. Vlot et al. (2009) inferred that SA and $H_2O_2$ form a self-amplifying feedback loop in plants subjected to both abiotic and biotic stresses. The interactions between $H_2O_2$ with jasmonic acid (JA) are also reported. Wang et al. (2012) showed that methyl jasmonate (MeJA) could be a crucial signal transducer and could regulate the levels of $H_2O_2$ and NO. JA interacted with ethylene, SA, and ABA during oxidative stresses triggered by desiccation (Brossa et al. 2011).

## 4 $H_2O_2$-Priming and Abiotic Stress Tolerance in Plants

### *4.1 Exogenous $H_2O_2$ Promotes Salt Stress Tolerance*

Salinity is also known as physiological drought stress due to toxic infiltration of $Na^+$ ions within tissues and altered osmotic pressure. Ellouzi et al. (2017) reported that seed priming by $H_2O_2$ promoted salt and drought tolerance in *Cakil maritima* compared to that by mannitol. $H_2O_2$ priming ameliorated growth and antioxidant defense during salinity and drought. The injuries caused by oxidative stress were drastically reduced due to high accumulation of AsA, GSH, and Pro. The redox balance was also well maintained, and $H_2O_2$ priming enabled the plants to memorize the stress-induced responses, so that they could rapidly activate the required signaling on later exposure to stress cues (Ellouzi et al. 2017). Treatment of *Panax ginseng* seedlings with 100 μM $H_2O_2$ enhances salt tolerance. This was accomplished due to increased activities of antioxidant enzymes like APX, CAT, and guaiacol peroxidase (POD). The endogenous accumulation of ROS and MDA was also suppressed (Sathiyaraj et al. 2014). $H_2O_2$ treatment increased seedling dry weight and content of photosynthetic pigments (Chl and carotenoids) and modulated the entire physiological metabolism. This led to better growth and development in the treated plants subjected to salt stress (Sathiyaraj et al. 2014). Treatment of wheat seedlings with 50–100 μM $H_2O_2$ mitigated the toxic effects of salt stress (Ashfaque et al. 2014). The water relations, Pro content, and nitrogen assimilation were improved in the treated plants compared to untreated seedlings under salt stress (Ashfaque et al. 2014). *Zea mays* plants treated with $H_2O_2$ exhibited higher relative water content (RWC), Chl, and decreased ROS accumulation. Interestingly AsA and GSH contents did not vary much after treatment, indicating that these might not be involved in the scavenging of ROS (Gondim et al. 2013). Li et al. (2011) also revealed that exogenous $H_2O_2$ increased APX, CAT, SOD, and POD activities in wheat seedlings grown in saline medium.

## 4.2 Exogenous $H_2O_2$ Promotes Drought Stress Tolerance

Physical drought stress is the prime abiotic factor behind desiccation in plant species across the globe. Priming with $H_2O_2$ alleviated the drought-induced injuries in soybean seedlings (Ishibashi et al. 2011). The RWC, photosynthetic rate and stomatal conductance along with the expression of genes like galactinol synthase and myo-inositol 3-phosphate synthase 2 increased in seedlings treated with $H_2O_2$ (Ishibashi et al. 2011). Exogenous application of $H_2O_2$ stimulated the production of the stress phytohormone, ABA in *Phaseolus vulgaris* seedlings. The photosynthetic pigment content and other growth parameters were all well maintained in the treated seedlings subjected to drought (Abass and Mohamed 2011). The cell membrane stability was increased, and the MDA production was suppressed in *Brassica juncea* seedlings treated with $H_2O_2$. The activities of antioxidant enzymes like APX, CAT, glutathione reductase (GR) and glutathione-*S*-transferase (GST) were all enhanced in these treated plants, which allowed lower accumulation of endogenous $H_2O_2$ (Hossain and Fujita 2013). Maize seedling pretreatment with 10 mM $H_2O_2$ ameliorated osmotic stress-induced injuries by decreasing the MDA level and promoting stomatal conductance (Terzi et al. 2014). The ABA concentration along with Pro, PAs, and soluble sugar contents also increased in these plants under drought (Terzi et al. 2014).

## 4.3 Exogenous $H_2O_2$ Promotes Temperature Stress Tolerance

The variations of temperature can result in both cold and heat stress in plant species. Priming with $H_2O_2$ has generated stress-tolerant phenotype in some plant species. Exogenous application of $H_2O_2$ induced acclimation to short-term cold stress in tomato seedlings (Iseri et al. 2013). The anthocyanin level in these plants was far higher than the non-treated seedlings subjected to 3 °C for 16 h. The acclimatized plants also exhibited high RWC and low MDA and lipid peroxidation. The activities of antioxidant enzymes like APX and CAT were enhanced in the roots of the treated plants exposed to cold stress (Iseri et al. 2013). Pretreatment of pea seeds with 70 mM $H_2O_2$ reduced the ROS-induced damages and increased Chl and carotenoid contents under cold stress (Moussa and Mohamed 2011). Foliar pretreatment of Mascarene grass and manila grass with 10 mM $H_2O_2$ increased the enzymatic activities of APX, GPX, CAT, GR, and POD during chilling stress (Wang et al. 2010a, b).

The exogenous application of 10 mM $H_2O_2$ promoted heat stress tolerance in *Lolium perenne* and *Festuca arundinacea* (Wang et al. 2014). The activities of POD, CAT, APX, GR, and GPX all increased in the treated plants under high temperature stress. Accumulation of osmolytes like AsA and GSH was also observed (Wang et al. 2014). The activities of SOD, CAT, APX, and GR, along with total thiol content, increased in the $H_2O_2$-treated rice cultivars SR 26B and Ratna, exposed to heat stress (Bhattacharjee 2012).

### 4.4 Exogenous $H_2O_2$ Alleviates Heavy Metal Toxicity

Application of $H_2O_2$ has also resulted in tolerance to heavy metal stress in plant species. Pretreatment with 100 μM $H_2O_2$ reduced the Cd-induced damages in rice seedlings (Hu et al. 2009). This was accomplished by an enhanced antioxidant system (Hu et al. 2009). Chao et al. (2009) reported that treatment with $H_2O_2$ increased the GSH content in rice plants which in turn regulated Cd-stress tolerance. Xu et al. (2010) highlighted that exogenous spraying of $H_2O_2$ induced the activities of CAT, POD, SOD, GPX, and GR and promoted tolerance to Al-induced phytotoxicity. Cr-induced stress in *Brassica napus* seedlings was mitigated upon $H_2O_2$ application (Yildiz et al. 2013). The thiol content and the activities of antioxidant enzymes were observed to be greater in these treated plants compared to the non-treated sets. The treated seedlings subjected to Cr stress also exhibited high accumulation of the metallothionein 1 protein and reduced content of MDA (Yildiz et al. 2013). Guzel and Terzi (2013) showed that the $H_2O_2$-pretreated seedlings of maize accumulated high levels of Pro and total soluble sugars. This enabled the treated plants to better tolerate Cu-mediated stress. The authors also suggested additional osmotic regulation in the mitigation of Cu-induced injuries in $H_2O_2$-primed maize seedlings (Guzel and Terzi 2013).

## 5 Conclusion

$H_2O_2$ is a member of the ROS family. Overaccumulation of this molecule occurs during almost all kinds of environmental stresses. $H_2O_2$ is involved in multiple signaling pathways associated with diverse antioxidant responses. This mechanism has been utilized by scientists to use $H_2O_2$ as a priming agent so that the endogenous antioxidant machinery remains active before encountering abiotic stress. As a result, the stress-induced injuries are less, and post-stress recovery is enhanced. Various instances show that seedling as well as seed priming with $H_2O_2$ have promoted tolerance to salinity, drought, cold, heat, and heavy metals in diverse plant species.

Such growth-promoting effects of $H_2O_2$ can be utilized for generating multiple abiotic stress tolerant traits, via an economically cheap and labor-free strategy.

## 6 Future Perspectives

The knowledge regarding the stress ameliorative effects of $H_2O_2$ has raised the necessity of understanding the associated genome-wide alterations and molecular signaling cascades. These aspects are less known in this field of research. Next-generation sequencing technologies including RNA-Seq and bisulfite sequencing can be used to identify the transcriptomic and epigenomic variations induced by $H_2O_2$. These technologies will better aid in the identification of molecular targets which can be utilized for transgenic studies. The development of a signaling blueprint operative during $H_2O_2$ priming in plant species would also be beneficial for serving the above purposes.

**Acknowledgments** Financial assistance from Council of Scientific and Industrial Research (CSIR), Government of India, through the research grant [38(1387)/14/EMR-II], Science and Engineering Research Board, Government of India, through the grant [EMR/2016/004799] and Department of Higher Education, Science and Technology and Biotechnology, Government of West Bengal, through the grant [264(Sanc.)/ST/P/S&T/1G-80/2017] to Dr. Aryadeep Roychoudhury is gratefully acknowledged. The authors acknowledge University Grants Commission, Government of India, for providing Junior Research Fellowship to Mr. Aditya Banerjee.

## References

Abass SM, Mohamed HI (2011) Alleviation of adverse effects of drought stress on common bean (*Phaseolus vulgaris* L.) by exogenous application of hydrogen peroxide. Bangladesh J Bot 41:75–83

Akter S, Huang J, Waszczak C, Jacques S, Gevaert K, Van Breusegem F et al (2015) Cysteines under ROS attack in plants: a proteomics view. J Exp Bot 66:2935–2944

Ashfaque F, Khan MIR, Khan NA (2014) Exogenously applied $H_2O_2$ promotes proline accumulation, water relations, photosynthetic efficiency and growth of wheat (*Triticum aestivum* L.) under salt stress. Annu Res Rev Biol 4:105–120

Banerjee A, Roychoudhury A (2015) WRKY proteins: signaling and regulation of expression during abiotic stress responses. Sci World J 2015:807560

Banerjee A, Roychoudhury A (2016a) Group II late embryogenesis abundant (LEA) proteins: structural and functional aspects in plant abiotic stress. Plant Growth Regul 79:1–17

Banerjee A, Roychoudhury A (2016b) Plant responses to light stress: oxidative damages, photoprotection and role of phytohormones. In: Ahammed GJ, Yu J-Q (eds) Plant hormones under challenging environmental factors. Springer, Dordrecht, pp 181–213

Banerjee A, Roychoudhury A (2017) Abscisic-acid-dependent basic leucine zipper (bZIP) transcription factors in plant abiotic stress. Protoplasma 254:3–16

Banerjee A, Roychoudhury A (2018a) Abiotic stress, generation of reactive oxygen species, and their consequences: an overview. In: Singh VP, Singh S, Tripathi D, Mohan Prasad S, Chauhan

DK (eds) Revisiting the role of reactive oxygen species (ROS) in plants: ROS boon or bane for plants? Wiley, Hoboken, pp 23–50

Banerjee A, Roychoudhury A (2018b) Small heat shock proteins: structural assembly and functional responses against heat stress in plants. In: Ahmad P, Ahanger MA, Singh VP, Tripathi DK, Alam P, Alyemeni MN (eds) Plant metabolites and regulation under abiotic stress. Academic/ Elsevier, London, pp 367–374

Banerjee A, Roychoudhury A (2018c) Effect of salinity stress on growth and physiology of medicinal plants. In: Ghorbanpour M et al (eds) Medicinal plants and environmental challenges. Springer, Cham, pp 177–188

Banerjee A, Roychoudhury A (2018d) Interactions of brassinosteroids with major phytohormones: antagonistic effects. J Plant Growth Regul. https://doi.org/10.1007/s00344-018-9828-5

Banerjee A, Roychoudhury A (2018e) Strigolactones: multi-level regulation of biosynthesis and diverse responses in plant abiotic stresses. Acta Physiol Plant 40:86

Banerjee A, Roychoudhury A (2018f) Seed priming technology in the amelioration of salinity stress in plants. In: Rakshit A, Singh HB (eds) Advances in seed priming. Springer Nature, Singapore, pp 81–93

Banerjee A, Roychoudhury A, Krishnamoorthi S (2016) Emerging techniques to decipher microRNAs (miRNAs) and their regulatory role in conferring abiotic stress tolerance in plants. Plant Biotechnol Rep 10:185–205

Banerjee A, Wani SH, Roychoudhury A (2017) Epigenetic control of plant cold responses. Front Plant Sci 8:1643

Banerjee A, Tripathi DK, Roychoudhury A (2018a) Hydrogen sulphide trapeze: environmental stress amelioration and phytohormone crosstalk. Plant Physiol Biochem 132:46–53

Banerjee A, Tripathi DK, Roychoudhury A (2018b) The karrikin 'callisthenics': can compounds derived from smoke help in stress tolerance? Physiol Plant. https://doi.org/10.1111/ppl.12836

Baniwal SK, Bharti K, Chan KY, Fauth M, Ganguli A, Kotak S et al (2004) Heat stress response in plants: a complex game with chaperones and more than twenty heat stress transcription factors. J Biosci 29:471–487

Bhattacharjee S (2008) Calcium-dependent signaling pathway in heat-induced oxidative injury in *Amaranthus lividus*. Biol Plant 52:1137–1140

Bhattacharjee S (2012) An inductive pulse of hydrogen peroxide pretreatment restores redox-homeostasis and mitigates oxidative membrane damage under extremes of temperature in two rice cultivars (*Oryza sativa* L., cultivars Ratna and SR26B). Plant Growth Regul 68:395–410

Brossa R, Lopez-Carbonell M, Jubany-Mari T, Alegre L (2011) Interplay between abscisic acid and jasmonic acid and its role in water-oxidative stress in wild-type, ABA-deficient, JA-deficient, and ascorbate deficient *Arabidopsis thaliana* plants. J Plant Growth Regul 30:322–333

Chao YY, Hsu YT, Kao CH (2009) Involvement of glutathione in heat shock-and hydrogen peroxide-induced cadmium tolerance of rice (*Oryza sativa* L.) seedlings. Plant Soil 318:37–45

Costa A, Drago I, Behera S, Zottini M, Pizzo P, Schroeder JI et al (2010) $H_2O_2$ in plant peroxisomes: an in vivo analysis uncovers a $Ca^{2+}$-dependent scavenging system. Plant J 62:760–772

Ellouzi H, Sghayar S, Abdelly C (2017) $H_2O_2$ seed priming improves tolerance to salinity; drought and their combined effect more than mannitol in *Cakile maritima* when compared to *Eutrema salsugineum*. J Plant Physiol 210:38–50

Gondim FA, Miranda RS, Gomes-Filho E, Prisco JT (2013) Enhanced salt tolerance in maize plants induced by $H_2O_2$ leaf spraying is associated with improved gas exchange rather than with non-enzymatic antioxidant system. Theor Exp Plant Physiol 25:251–260

Guzel S, Terzi R (2013) Exogenous hydrogen peroxide increases dry matter production, mineral content and level of osmotic solutes in young maize leaves and alleviates deleterious effects of copper stress. Bot Stud 54:26

He JM, Xu H, She XP, Song XG, Zhao WM (2005) The role and the interrelationship of hydrogen peroxide and nitric oxide in the UV-B- induced stomatal closure in broadbean. Funct Plant Biol 32:237–247

Hossain MA, Fujita M (2013) Hydrogen peroxide priming stimulates drought tolerance in mustard (*Brassica juncea* L.). Plant Gene Trait 4:109–123

Hossain MA, Bhattacharjee S, Armin S-M, Qian P, Xin W, Li H-Y, Burritt DJ, Fujita M, Tran L-SP (2015) Hydrogen peroxide priming modulates abiotic oxidative stress tolerance: insights from ROS detoxification and scavenging. Front Plant Sci 6:420

Hu Y, Ge Y, Zhang C, Ju T, Cheng W (2009) Cadmium toxicity and translocation in rice seedlings are reduced by hydrogen peroxide pretreatment. Plant Growth Regul 59:51–61

Huang YW, Zhou ZQ, Yang HX, Wei CX, Wan YY, Wang XJ et al (2015) Glucose application protects chloroplast ultrastructure in heat-stressed cucumber leaves through modifying antioxidant enzyme activity. Biol Plant 59:131–138

Iseri OD, Körpe DA, Sahin FI, Haberal M (2013) Hydrogen peroxide pretreatment of roots enhanced oxidative stress response of tomato under cold stress. Acta Physiol Plant 35:1905–1913

Ishibashi Y, Yamaguchi H, Yuasa T, Inwaya-Inoue M, Arima S, Zheng S (2011) Hydrogen peroxide spraying alleviates drought stress in soybean plants. J Plant Physiol 168:1562–1567

Jammes F, Song C, Shin D, Muncmasa S, Takeda K (2009) MAP kinases MPK9 and MPK12 are preferentially expressed in guard cells and positively regulate ROS-mediated ABA signaling. Proc Natl Acad Sci U S A 106:20520–20525

Kapoor D, Sharma R, Handa N, Kaur H, Rattan A, Yadav P et al (2015) Redox homeostasis in plants under abiotic stress: role of electron carriers, energy metabolism mediators and proteinaceous thiols. Front Environ Sci 3:13

Kim TH, Maik B, Hu HH, Noriyuki N, Julian IS (2010) Guard cell signal transduction network advances in understanding abscisic acid, $CO_2$, and $Ca^{2+}$ signaling. Annu Rev Plant Biol 61:56–91

Li JT, Qiu ZB, Zhang XW, Wang LS (2011) Exogenous hydrogen peroxide can enhance tolerance of wheat seedlings to salt stress. Acta Physiol Plant 33:835–842

Liu XM, Kim KE, Kim KC, Nguyen XC, Han HJ, Jung MS et al (2010) Cadmium activates *Arabidopsis* MPK3 and MPK6 via accumulation of reactive oxygen species. Photochemistry 71:614–618

Mazars C, Bourque S, Mithöfer A, Pugin A, Ranjeva R (2009) Calcium homeostasis in plant cell nuclei. New Phytol 181:261–274

Miller G, Mittler R (2006) Could heat shock transcription factors function as hydrogen peroxide sensors in plants? Ann Bot 98:279–288

Miller G, Suzuki N, Ciftci-Yilmaz S, Mittler R (2010) Reactive oxygen species homeostasis and signalling during drought and salinity stresses. Plant Cell Environ 33:453–467

Mittler R, Vanderauwera S, Suzuki N, Miller G, Tognetti VB, Vandepoele K et al (2011) ROS signaling: the new wave? Trends Plant Sci 16:300–309

Moussa HR, Mohamed MAEFH (2011) Role of nitric acid or $H_2O_2$ in antioxidant defense system of *Pisum sativum* L. under drought stress. Nat Sci 9:211–216

Nahar K, Hasanuzzaman M, Alam MM, Fujita M (2014) Exogenous glutathione confers high temperature stress tolerance in mungbean (*Vigna radiata* L.) by modulating antioxidant defense and methylglyoxal detoxification system. Environ Exp Bot 112:44–54

Noctor G, Mhamdi A, Foyer CH (2014) The roles of reactive oxygen metabolism in drought: not so cut and dried. Plant Physiol 164:1636–1648

Pan J, Zhang M, Kong X, Xing X, Liu Y, Zhou Y et al (2012) ZmMPK17,a novel maize group D MAP kinase gene, is involved in multiple stress responses. Planta 235:661–676

Rao MV, Davis KR (1999) Ozone-induced cell death occurs via two distinct mechanisms in *Arabidopsis*: the role of salicylic acid. Plant J 17:603–614

Rizhsky L (2004) The zinc finger protein Zat12 is required for cytosolic ascorbate peroxidase 1 expression during oxidative stress in *Arabidopsis*. J Biol Chem 279:11736–11743

Roychoudhury A, Banerjee A (2015) Transcriptome analysis of abiotic stress response in plants. Transcriptomics 3:2

Roychoudhury A, Banerjee A (2017) Abscisic acid signaling and involvement of mitogen activated protein kinases and calcium-dependent protein kinases during plant abiotic stress. In: Pandey G (ed) Mechanism of plant hormone signaling under stress, vol 1. Wiley, Hoboken, pp 197–241

Sathiyaraj G, Srinivasan S, Kim YJ, Lee OR, Balusamy SDR, Khorolaragchaa A et al (2014) Acclimation of hydrogen peroxide enhances salt tolerance by activating defense-related proteins in Panax ginseng CA. Meyer. Mol Biol Rep 41:3761–3771

Terzi R, Kadioglu A, Kalaycioglu E, Saglam A (2014) Hydrogen peroxide pretreatment induces osmotic stress tolerance by influencing osmolyte and abscisic acid levels in maize leaves. J Plant Interact 9:559–565

Vlot AC, Dempsey DMA, Klessig DF (2009) Salicylic acid, a multi- faceted hormone to combat disease. Annu Rev Phytopathol 47:177–206

Wang P, Du Y, Li Y, Ren D, Song CP (2010a) Hydrogen peroxide- mediated activation of map kinase 6 modulates nitric oxide biosynthesis and signal transduction in *Arabidopsis*. Plant Cell 22:2981–2998

Wang Y, Li J, Wang J, Li Z (2010b) Exogenous H2O2 improves the chilling tolerance of manila grass and mascarene grass by activating the antioxidative system. Plant Growth Regul 61:195–204

Wang KT, Zheng YH, Tang WC, Li TJ, Zhang Q, Shang HT (2012) Effects of methyl jasmonate treatment on levels of nitric oxide and hydrogen peroxide and phytoalexin synthesis in postharvest grapeberries. Acta Hort Sin 39:1559–1566

Wang Y, Zhang J, Li JL, Ma XR (2014) Exogenous hydrogen peroxide enhanced the thermotolerance of *Festuca arundinacea* and *Lolium perenne* by increasing the antioxidative capacity. Acta Physiol Plant 36:2915–2924

Wei S, Hu W, Deng X, Zhang Y, Liu X, Zhao X et al (2014) A rice calcium-dependent protein kinase OsCPK9 positively regulates drought stress tolerance and spikelet fertility. BMC Plant Biol 14:133

Xu FJ, Jin CW, Liu WJ, Zhang YS, Lin XY (2010) Pretreatment with $H_2O_2$ alleviates aluminum-induced oxidative stress in wheat seedlings. J Integr Plant Biol 54:44–53

Yildiz M, Terzi H, Bingül N (2013) Protective role of hydrogen peroxide pretreatment on defense systems and *BnMP1* gene expression in Cr(VI)-stressed canola seedlings. Ecotoxicology 22:1303–1312

Yuan S, Lin HH (2008) Role of salicylic acid in plant abiotic stress. Z Naturforsch C 63:313–320

Zhang X, Zhang L, Dong F, Gao J, Galbraith DW, Song C-P (2001) Hydrogen peroxide is involved in abscisic acid-induced stomatal closure in *Vicia faba*. Plant Physiol 126:1438–1448

Zhou J, Wang J, Shi K, Xia XJ, Zhou YH, Yu JQ (2012) Hydrogen peroxide is involved in the cold acclimation-induced chilling tolerance of tomato plants. Plant Physiol Biochem 60:141–149

Zhou J, Xia XJ, Zhou YH, Shi K, Chen Z, Yu JQ (2014) RBOH1- dependent $H_2O_2$ production and subsequent activation of MPK1/2 play an important role in acclimation-induced cross-tolerance in tomato. J Exp Bot 65:595–607

# Abiotic Stress Tolerance in Plants Through Pre-sowing Seed Treatments with Mineral Elements and Growth Regulators

**Aman Ullah, Babar Shahzad, Mohsin Tanveer, Faisal Nadeem, Anket Sharma, Dong Jin Lee, and Abdul Rehman**

## Contents



A. Ullah · F. Nadeem
Department of Agronomy, Faculty of Agriculture, University of Agriculture, Faisalabad, Pakistan

B. Shahzad (✉) · M. Tanveer
Tasmanian Institute of Agriculture, University of Tasmania, Hobart, TAS, Australia
e-mail: Babar.Shahzad@utas.edu.au

A. Sharma
Plant Stress Physiology Lab, Department of Botanical and Environmental Sciences, Guru Nanak Dev University, Amritsar, Punjab, India

D. J. Lee
Department of Crop Science and Biotechnology, Dankook University, Chungnam, Republic of Korea

A. Rehman
Department of Crop Science and Biotechnology, Dankook University, Chungnam, Republic of Korea

College of Agriculture, Bahauddin Zakariya University, Layyah, Pakistan



M. Hasanuzzaman, V. Fotopoulos (eds.), *Priming and Pretreatment of Seeds and Seedlings*, https://doi.org/10.1007/978-981-13-8625-1_21

**Abstract** Unfavorable environmental conditions and extreme abiotic stresses (high/low temperature, drought, salinity, and heavy metal stress) severely affect crop growth and productivity. Although different measures have been taken to counter the detrimental effects of climatic conditions through the development of drought-, salt-, and heat-tolerant varieties. Nevertheless, seed priming is considered one promising approach to get successful crop production due to improved seed germination and stand establishment ensuring food production and sustainability. Seed priming is a technique in which crop seeds are dipped in the water, nutrients, or plant growth regulators to allow the metabolic activities to occur without radicle protrusion. Priming of crop seeds with micronutrients has been shown to improve abiotic stress tolerance and quality of produce.

**Keywords** Abiotic stress tolerance · Nutrient priming · Oxidative stress · Crop productivity · Crop yield

## Abbreviations

| | |
|---|---|
| ABA | Abscisic acid |
| ATP | Adenosine triphosphate |
| BR | Brassinosteroid |
| CAT | Catalase |
| CK | Cytokinin |
| ET | Ethylene |
| GA | Gibberellin |
| HET | High-effective thiazole |
| IAA | Auxin |
| JA | Jasmonic acid |
| MDA | Malondialdehyde |
| NO | Nitric oxide |
| PEG | Polyethylene glycol |
| PGPR | Plant growth-promoting rhizobacteria |
| POX | Peroxidase |
| Put | Putrescine |
| SA | Salicylic acid |
| SNP | Sodium nitroprusside |
| SOD | Superoxide dismutase |
| Spd | Spermidine |
| Spm | Spermine |

# 1 Introduction

Abiotic stresses such as drought, heat stress, chilling, frost, heavy metal, waterlogging, and salinity cause severe yield reductions and deteriorate quality of crops. These stresses negatively affect plant growth and productivity (Jisha et al. 2013; Shahzad et al. 2018a, b) and cause unpredictable yield losses in crops (Jakab et al. 2005). More than 50% of yield reductions have been recorded under abiotic stresses in major field crops globally (Boyer 1982; Bray et al. 2000).

Seed priming is an effective approach to augment the uniform and rapid emergence and high seedling vigor and yield in many field crops under unfavorable environments (Paparella et al. 2015; Rehman et al. 2018a, b). The primed plants had quick and sturdy cellular defense against environmental stresses (Jisha et al. 2013). The salinity or drought stress causes diverse metabolic, biochemical, and physiological changes in crop plants (Xiong and Zhu 2002) leading to oxidative stress and results in poor plant performance (Shafi et al. 2009). However, primed seeds show improved germination rate, uniformity of germination, and total germination percentage under these stressed conditions (Basra et al. 2005).

# 2 Seed Priming and Abiotic Stress Tolerance

During priming, the cellular defense mechanism is activated in plants, which imparts tolerance against subsequent abiotic stresses under field conditions (Beckers and Conrath 2007). Seed priming with water, selenium, calcium chloride, hydrogen peroxide, and salicylic acid improves the chilling stress in rice seedlings owing to increase in peroxidase (POD), superoxide dismutase (SOD), catalase (CAT), glutathione (GT), and free proline contents (Hussain et al. 2016). Various priming methodologies and approaches have been developed so for including hydropriming, chemical priming, osmopriming, hormonal priming, etc., for imparting abiotic stress tolerance in plants (Table 1). It is evident that rate and uniformity of seedling emergence are positively improved with priming; however, its effectiveness varies greatly with different priming techniques and agents under different stresses (Parera and Cantliffe 1991; Iqbal and Ashraf 2005).

## *2.1 Hydropriming*

Hydropriming (HP) is a simple presoaking of seeds in sterilized water for some period before radical protrusion. It is a simple, economical, and safe approach towards stand establishment, osmotic adjustment, and crop production under suboptimal conditions (Kaur et al. 2002). The HP of chickpea seeds for 12 h has been shown to improve the germination of chickpea under low-temperature environment

**Table 1** Improvement in germination, grain yield, and stress tolerance by using various seed priming methods

| Crop | Methods | Salt/agent | Duration (h) | Germination (%) | Grain yield (%) | Enhanced tolerance for | References |
|---|---|---|---|---|---|---|---|
| Maize | Osmopriming | 5 g NaCl | 12 | 10 | – | Salt stress | Gebreegziabher and Qufa (2017) |
| Maize | Osmopriming | 5 g $CaCl_2$ | 12 | 17 | 139 | Salt stress | (Gebreegziabher and Qufa (2017) |
| Maize | Hydropriming | – | 24 | | 2.3 | Low root zone temperature | Imran et al. (2013) |
| Maize | Osmopriming | 8.5 mM Fe | 24 | | 14 | Low root zone temperature | Imran et al. (2013) |
| Maize | Osmopriming | 4 mM Zn + 2.5 mM Mn | 24 | | 15 | Low root zone temperature | Imran et al. (2013) |
| Maize | Hydropriming | – | – | 4 | – | Low temperature | Imran et al. (2013) |
| Maize | Hormonal priming | Abscisic acid (50 mg $L^{-1}$) | – | 15 | – | Low temperature | Imran et al. (2013) |
| Maize | Hormonal priming | Salicylic acid (50 mg $L^{-1}$) | – | 13 | – | Low temperature | Imran et al. (2013) |
| Maize | Biopriming | Sorghum water extract (10 mL $L^{-1}$) | – | 14 | – | Low temperature | Imran et al. (2013) |
| Maize | Biopriming | Moringa water extract (3%) | – | 15 | – | Low temperature | Imran et al. (2013) |
| Rice | Hydropriming | – | 48 | – | 83 | Terminal drought | Tilahun-Tadesse et al. (2013) |
| Rice | Chemical priming | 60 μM selenium | – | 2 | – | Submergence stress | Hussain et al. (2016) |
| Rice | Hormonal priming | 100 mg $L^{-1}$ salicylic acid | – | 3 | – | Submergence stress | Hussain et al. (2016) |
| Rice | Chemical priming | 50 μM selenium | – | 27 | – | Chilling stress | Wang et al. (2016) |
| Rice | Hormonal priming | 100 mg $L^{-1}$ salicylic acid | | 34 | – | Chilling stress | Wang et al. (2016) |
| Wheat | Hydropriming (pre-germinated) | – | – | 45 | 1 | Water stress | Meena et al. (2013) |
| Wheat | Hydropriming | – | 12 | – | 22 | Salinity stress | Jafar et al. (2012) |

| | | | | | | | |
|---|---|---|---|---|---|---|---|
| Wheat | Hormonal priming | Ascorbate (50 mg $L^{-1}$) | 12 | – | 38 | Salinity stress | Jafar et al. (2012) |
| Wheat | Hormonal priming | Salicylic acid (50 mg $L^{-1}$) | 12 | – | 25 | Salinity stress | Jafar et al. (2012) |
| Wheat | Hormonal priming | Kinetin (50 mg $L^{-1}$) | 12 | – | 9 | Salinity stress | Jafar et al. (2012) |
| Wheat | Osmopriming | $CaCl_2$ (50 mg $L^{-1}$) | 12 | – | 54 | Salinity stress | Jafar et al. (2012) |
| Mustard | Hydropriming | – | – | – | – | Salinity stress | Srivastava et al. (2010a, b) |
| Mustard | Osmopriming | $CaCl_2$ (100 µM) | – | – | – | Salinity stress | Srivastava et al. (2010a, b) |
| Mustard | Hormonal priming | Abscisic acid (100 µM) | – | – | – | Salinity stress | Srivastava et al. (2010a, b) |
| Sugarcane | Halopriming | NaCl (100 mM) | – | 27 | – | Drought stress | Patade et al. (2011) |
| Barley | On-farm priming | – | 12–16 | – | 53 | Salt stress | Rashid et al. (2006) |
| Wheat | Progeny seeds from Zn biopriming | 0.5 M ($ZnSO_4$) + PGPR | 12 | | 21 | Salt stress | |
| Wheat | Progeny seeds from Zn biopriming | 0.5 M ($ZnSO_4$) + PGPR | 12 | | 21.5 | Drought stress | Faran et al. 2019 |
| Wheat | Progeny seeds from Zn biopriming | 0.5 M ($ZnSO_4$) + PGPR | 12 | | 48 | Waterlogging | |

(Elkoca et al. 2007). HP also causes enhancement of biochemical and physiological aspects in the seeds during germination (Basra et al. 2003). For instance, HP increased root and shoot length by three- to four-fold under drought stress compared to unprimed seeds (Kaur et al. 2002). In a recent study, Farooq et al. (2018a, b) showed that terminal drought priming improves the drought tolerance in chickpea owing to accumulation of trehalose, proline, and total phenolics. Under saline conditions, HP of basil (*Ocimum basilicum* L.) improved seedling vigor, germination, and seedling dry weight (Farahani and Maroufi 2011). Similarly, HP enhanced the germination and seedling growth of mung bean at low temperature due to fast recovery of stress injuries (Posmyk and Janas 2007).

## 2.2 *Osmopriming*

Soaking of seeds in low water potential aerated solutions to restrict the amount and extent of imbibition is known as osmopriming. Osmopriming of seeds converts the seeds to physiologically active and hydrated state from dry and inactive state (Chen and Arora 2011).

### 2.2.1 Osmopriming with PEG

Several chemicals have been used to create low water potential during osmopriming such as $KNO_3$, $CaCl_2$, NaCl, KCl, $K_3PO_4$, $MgSO_4$, mannitol, $KH_2PO_4$, PEG, etc. (Farooq et al. 2005). However, polyethylene glycol (PEG) is the most commonly used priming agent due to its larger molecular size, nontoxic nature, and lower potential without penetrating into seeds during soaking (Thomas et al. 2000). Osmopriming (PEG-0.5 MPa) of chickpea seeds for 24 h enhanced the germination and showed vigorous growth under low temperature (Elkoca et al. 2007). Similarly, osmopriming of spinach seeds with PEG increased seed germination, and enhanced antioxidant activities alleviated stress acclimation (Chen and Arora 2011). In another study, PEG osmopriming improved seed germination of *Bromus inermis* under drought and salinity stress (Tavili et al. 2011). Yuan-Yuan et al. (2010) demonstrated that seed priming with PEG in a proper concentration was found effective in improving the germination and seedling growth under stress conditions.

Osmopriming of seeds with PEG has great potential to improve stand establishment and performance of crop plants under stressful environment (Zhang et al. 2015). Improvement in antioxidant activities like POX, SOD, and CAT has been observed in PEG osmoprimed sorghum (*Sorghum bicolor* L.) under suboptimal water stress (Zhang et al. 2015). Seed priming with PEG permits seeds to initiate the membrane repairing systems and metabolic preparations for germination through controlling the absorption of water in seed under abiotic stresses (Jisha et al. 2013). In addition, PEG has not any harmful effects on the germinated seed (Mehra et al. 2003). Under water deficit conditions, alfalfa (*Medicago sativa* L.) seeds treated with $PEG_{6000}$ (−0.6 MPa) showed improved germination and healthy seedling growth, improved root length and antioxidant defense mechanism, but reduced electrolyte leakage from cell and malondialdehyde (MDA) contents (Mouradi et al. 2016). Therefore, it is evident from several studies that seed priming with PEG helps in repairing damage of aged seeds or those experienced stress by increasing osmotic adjustment under drought and salinity stress as observed in rice and alfalfa (Butler et al. 2009; Sun et al. 2010; Amooaghaie 2013; Yacoubi et al. 2013). In another study, seed priming with PEG 6000 improved the germination and seedling growth of tomato under suboptimal temperature (Amooaghaie and Nikzad 2013). Moreover, osmoprimed seeds exhibited higher amylase (α and β) activities and sugar content under low-temperature stress (Amooaghaie and Nikzad 2013).

### 2.2.2 Osmopriming with NaCl

Salt stress tolerance in plants can be induced by priming seeds with NaCl as the seeds primed with salts may build osmotic adjustment and have high sugar contents owing to higher Na and Cl in roots (Cayuela et al. 1996). In a study, priming of capsicum seeds with normal water, warm water (40 °C), and NaCl (50 mM) increased germination percentage (7%–17%) exposed to subsequent stress (cold and salt). Only the seedlings primed with thiourea, hydrogen peroxide, or ABA showed higher tolerance under cold and salt stress with 100% survival for 10 days (Yadav et al. 2011).

Salt priming of sugarcane reduced the negative effects of salinity and drought (Patade et al. 2009). Some other studies showed similar results under saline environment as osmopriming induced improvement in germination/emergence and seedling growth of canola (Farhoudi and Sharifzadeh 2006) and chickpea (Sarwar et al. 2006). Seed priming with salt induced salt tolerance owing to improved seedling vigor, metabolism, enhanced accumulation of $Ca^{+2}$ and $K^{+}$, and reduced $Na^{+}$ contents in wheat (Afzal et al. 2005). Similarly, sunflower plants showed increased salt stress tolerance in primed seeds due to enhanced accumulation of $Ca^{+2}$, $K^{+}$, and proline (Bajehbaj 2010). In another study, Saha et al. (2010) observed that pretreatment of mung bean seeds with NaCl decreased the negative effects of NaCl by increasing the growth, photosynthetic pigments, and antioxidant activities.

### 2.2.3 Mannitol

Mannitol is an important osmolyte, and a significant proportion is synthesized in many plant species (Su et al. 1999; Mitoi et al. 2009). It plays crucial role in osmotic adjustment and scavenging of hydroxyl radicals ($OH^{•}$) (Shen et al. 1997; Srivastava et al. 2010b). Literature shows that seed priming with mannitol (4%) improves the activities of amylase and invertases (alkaline and acid), sucrose synthase, and sucrose phosphate synthase that enhanced the tolerance against drought and saline-stressed chickpea (Kaur et al. 2002; Sarwar et al. 2006). About two-fold increase in acid and alkaline invertases in the primed seedling was observed in the cotyledons of chickpea, while improved amylase activity enhances the quick hydrolysis of transitory starch that help to increase the glucose availability resulting in healthier seedling (Kaur et al. 2002). Contrarily, abiotic stresses decrease the starch mobilization in cotyledon and movement of sucrose toward embryonic axis from cotyledon (Gupta et al. 1993; Kaur et al. 2000) and reduce the sucrose accumulation because of low activities of enzymes related to breakdown of starch (Gonzalez et al. 1995). Amoghein et al. (2013) demonstrated that seed priming with mannitol (5%) improves the hypocotyl and coleoptile length, root length that helps plant to survive under saline stress.

### 2.2.4 Other Chemicals

The osmopriming with $CaCl_2$ and redox priming with $H_2O_2$ of rice seeds improves the chilling stress in rice seedlings (Hussain et al. 2016). Seed priming of tobacco with putrescine improved the cold stress tolerance in tobacco during germination and early seedling growth owing to activation of antioxidant system (Xu et al. 2011). Priming of maize seeds with chitosan enhanced the germination rate and seedling growth under chilling stress (Guan et al. 2009). Similarly, the priming of chickpea seeds with $CaCl_2$ (−1.25 MPa) improves the chilling stress owing to better germination metabolism and accumulation of trehalose contents (Farooq et al. 2017). Priming with $KNO_3$ enhanced the germination in watermelon (Demir and Mavi 2004); germination, root, and shoot length in tomato (Nawaz et al. 2011); and uniformity of emergence in leek (Brocklehurst et al. 1984), tomato (Ozbingol et al. 1998), and sorghum (Moradi and Younesi 2009). Salt priming with $KNO_3$ in cucumber and sunflower showed increased seedling vigor (Singh and Rao 1993; Ghassemi-Golezani and Esmaeilpour 2008). Osmopriming of rice seeds with KCl and $CaCl_2$ improved the emergence, emergence energy, and seedling emergence index (Farooq et al. 2006).

## *2.3 Nutrient Priming*

Nutrient priming is a technique in which seeds are soaked in solutions containing limiting nutrient instead of just water (Arif 2005). The mineral status of plants helps in resistance against environmental stress (Marschner 1995). Under stress conditions, potassium (K) plays a significant role in plant survival (Cakmak 2005). Seed priming with $Zn^{2+}$ improves the grain yield in several field crops like chickpea (Arif et al. 2007), rice (Farooq et al. 2018a, b), and wheat (Rehman et al. 2015, 2018a, b). Seed priming with Zn improves the Zn concentration in progeny seeds (Rehman et al. 2018a, b), which helped plants in improving resistance against salinity, drought, and waterlogging stress (Faran et al. 2019). Seed priming with $Mn^{2+}$ improves the grain yield and grain Mn concentration of wheat (Ullah et al. 2017b) and rice (Ullah et al. 2017a).

## 2.4 Hormonal Priming

Phytohormones such as cytokinin (CK), auxin (IAA), gibberellin (GA), abscisic acid (ABA), ethylene (ET), salicylic acid (SA), brassinosteroid (BR), jasmonic acid (JA), etc. are considered vital for the various processes related to growth and development of plants and also involved in signaling and cross talk under various internal and external stimulus. Hormonal priming involves the use of phytohormones or plant growth regulators in the priming medium. Plant growth regulators have been found effective in improving the crop performance under optimal and suboptimal growth conditions. For instance, ABA is a phytohormone involved in a number of abiotic stresses such as osmotic, low temperature, and drought stress (Fujita et al. 2006). The exogenous application of abscisic acid improves the salt tolerance in plants (Xiong and Zhu 2002). The Indian mustard seed priming with ABA increased the germination rate compared to non-primed seed (Srivastava et al. 2010a, b). The hormonal priming of wheat seed with IAA improved the hypocotyl length and dry weight of seedling under salt stress conditions (Akbari et al. 2007). The priming of *Agropyron elongatum* with ABA and gibberellin induced activities of CAT and SOD under drought stress (Eisvand et al. 2010). A decade ago, Tavili et al. (2009) demonstrated that seed priming with ABA (300 mg $L^{-1}$) improved the germination and seedling growth/vigor index of *Agropyron elongatum* (Tavili et al. 2009).

According to Gul and Khan (2003), priming of *Salicornia utahensis* with PGRs such as ethephon and kinetin improved the germination and alleviated the negative effects of salinity. Recently, Bajwa et al. (2018) demonstrated that seed priming with benzyl aminopurine (BAP) (5 mg $L^{-1}$) and sorghum water extract (SWE) alone or in combination improves the salt tolerance in bread wheat. The improved salt tolerance in BAP- and SWE-primed wheat seeds was attributed to the enhanced total phenolics, soluble sugars, photosynthetic pigment, and K uptake, while plants raised from hormonal primed seed exhibited lower Na accumulation. In a study, Hussain et al. (2016) reported that hormonal priming of rice seedlings with salicylic acid (100 mg $L^{-1}$ salicylic acid) improves the chilling stress in rice. The seed priming of sweet pepper with salicylic acid 0.3 mM improves the germination and early seedling growth under salinity stress (El-Afifi et al. 2016). Li et al. (2013) found that seed priming with $GA_3$ improved the germination and seedling growth in winter wheat with lower lipid peroxidation and ROS generation under chilling stress.

Hormonal priming has been found effective in inducing thermal stress (heat and cold stress) tolerance in crop plants. Few years back, Rehman et al. (2012) demonstrated that seed priming of maize with SA or JA helps to ameliorate the adversity of heat stress by hastening the seedling emergence and improving the seedling growth, water relations, and membrane stability. Recently, Górnik and Lahuta (2017) found that seed priming of sunflower with SA or JA followed by short-term exposure to high temperature (45 °C, 2 h) reduces the adverse effects of chilling stress by improving seedling growth, carbohydrate accumulation, sugar metabolism, and $F_v/F_m$ under low-temperature condition. In tomato, SA seed priming (0.25, 0.5, and 0.75 mM) improved the plant growth, flowering, fruit yield, and quality under supra-optimal temperature (Singh and Singh 2016). In a study, Guan et al. (2009) reported that seed priming of maize with chitosan improved the germination and seedling growth under low temperature. Pre-sowing seed soaking of rice seeds with selenium (Se) (50 μM) improves the chilling stress in rice seedlings (Hussain et al. 2016). At low concentrations, Se has a positive effect on crop growth and stress tolerance (Hasanuzzaman et al. 2010). The priming of bitter gourd with Se protects the plants under suboptimal temperature by increasing the antioxidant activities (Chen and Sung 2001).

## 2.5 *Seed Priming with Polyamines and Stress Signaling Compounds*

Polyamines (PAs) are low molecular weight nitrogenous compound, which play a vital role in plant developmental processes and stress tolerance. Most abundant polyamines are putrescine (Put), spermine (Spm), and Spermidine (Spd) (Alcázar et al. 2010). These PAs play a protective role against various biotic and abiotic stresses such as these bind to several negatively charged proteins and antioxidant enzymes and by modulating the protein activities (Bouchereau et al. 1999). Seed priming with Put (0.1 mM) substantially improved the germination and seedling growth in chilling tolerant and sensitive cultivars of tobacco. Tobacco plants raised from primed seeds exhibited higher accumulation of endogenous Put, Spd, and Spm and increased activities of APX, CAT, SOD, and POD activities on exposure to chilling stress (5 °C) with lower MDA level (Xu et al. 2011). In white clover, seed priming with Spd (30 μM) improved the germination-related traits, root development, and seedling dry weight under drought stress compared to non-primed seeds. Furthermore, Spd-treated seeds showed higher α- and β-amylase activities and higher transcript level of β-amylase gene, improved starch metabolism, and reducing sugar, glucose, and fructose content with lower MDA level during germination. Furthermore, Spd priming augments the antioxidant activities (APX, SOD, POD, and CAT) with higher expression of genes controlling these enzymes (Li et al. 2014). In another study, Chunthaburee et al. (2014) demonstrated that Spd (1 mM) primed seeds of color rice showed less reduction in growth, anthocyanin, and chlorophyll content under salt stress (150 mM) over control with lower ROS level and higher phenolics and antioxidant capacity under salt stress conditions.

Seed priming of winter wheat with sodium nitroprusside (SNP), a nitric oxide (NO) donor, substantially improved the seedling emergence, vigor index, and seedling growth through enhanced amylase activities, starch degradation, respiration, and better antioxidant defense capacity with lower MDA and $H_2O_2$ level (Li et al. 2013). A decade ago, Zheng et al. (2010) demonstrated that pre-sowing seed soaking in NO (0.1 mM) solution improved wheat performance under salt stress conditions (300 mM). The authors observed higher germination rate, radicle and coleoptile weight, respiration, ATP synthesis, amylase, CAT and SOD activities, and reduced

MDA, $H_2O_2$, and $O_2^{\bullet-}$ level. Amooaghaie and Nikzad (2013) showed that primed tomato seeds with SNP enhanced the germination index, seedling growth, sugar content, and α- and β-amylase activities under low-temperature stress which was associated with high NO level in primed seeds. Likewise, pre-sowing seed treatment with $H_2O_2$ improved the tolerance against drought stress in wheat by improving the germination, photosynthesis, leaf development, biomass production, water use efficiency, and proline content. Furthermore, $H_2O_2$ showed higher expression of CAT and APX with lower ROS damage (He and Gao 2009). Similarly, pre-sowing seed soaking in $H_2O_2$ solution (100 μM) mitigated the Cd-induced stress in rice by augmenting the activities of APX, CAT, SOD, guaiacol peroxidase (GPX), and glutathione *S*-transferase (GST) with substantial increase in ascorbic acid and reduced glutathione (GSH) level. Moreover, $H_2O_2$ lowered the ROS damage and Cd accumulation and translocation in rice plants (Hu et al. 2009).

In conclusion, redox priming and pre-sowing seed treatment with PAs and other signaling compounds help in mitigating the adverse effects of environmental stresses through enhanced cellular signaling, activation of antioxidant defense system, modulation of structure and function of proteins, and regulation of genes involved in stress responses.

## 2.6 *Biological Priming/Biopriming*

Different plant extracts can be used for seed priming to enhance plant's ability against abiotic stresses. Biopriming has been found effective in improving the growth, yield, nutrient solubilization, and uptake from soil and their translocation in seed (Rehman et al. 2018a, b). Biopriming refers to the induction of tolerance/resistance in tissue propagules by employing beneficial microorganism (PGPR, plant growth promoting rhizobacteria) under abiotic stresses during germination (Kavino et al. 2010). Soaking of pea (*Pisum sativum* L.) seeds in aqueous extract of *Typha angustifolia* improves root length, ameliorates cell metabolic activity (increase formazan concentration), increases the proline contents, decreases MDA contents, and modulates the membrane stability, ionic homeostasis, and photosynthetic pigments (Ghezal et al. 2016) that help in alleviating the inhibitory effects of salt stress. Moreover, seed priming increases the total soluble sugars and functioning of α-amylase activity; exhibits quick initiation of DNA, RNA synthesis, and protein repair (Tilahun-Tadesse et al. 2013; Soughir et al. 2013); and increases the cell division in apical meristem, leading to better germination and overall performance (Sakhabutdinova et al. 2003; Ghezal et al. 2016) of wheat, rice, and fenugreek and pea.

Addition of PGPR with seed priming benefits the growth of plant by increasing the root growth and germination, improves leaf area and chlorophyll contents, enhances uptake of mineral nutrients (including N, P and K), ameliorates hydraulic conductivity, modulates protein synthesis, and delays senescence under drought and salinity resulting in higher grain yield (Lucy et al. 2004; van Loon 2007). Application of *Bacillus cereus* in rice, chickpea, mung bean, *Pseudomonas putida*, and *Enterobacter cloacae* in wheat significantly promotes the growth and enhances the activities of CAT, POD, and SOD. Additionally, this process improves the functioning of β-1, 3 glucanase, chitinase, and phenyl alanine ammonia that helps the plant perform better under salinity stress (Chakraborty et al. 2013; Nadeem et al. 2013). Kasim et al. (2013) documented that biopriming with *A. brasilense* and *B. amyloliquefaciens* induces drought tolerance in wheat by upregulating the expression of stress-related genes.

## 2.7 Solid Matrix Priming

Solid matrix priming means to mix the seeds with a soil career, moistened with water to achieve the potential necessary for priming (Madsen et al. 2018); various reports indicated that sometimes solid matrix priming performs better than osmotic priming (Taylor et al. 1988; Harman and Taylor 1988). One thing that must be considered in practicing solid matrix priming is that it has properties like high water holding capacity, low osmotic potential (Khan et al. 1990), and low bilk density. In mung bean, soil matrix priming such as using celite matrix (10%) with various elicitors including ABA, kinetin, $GA_3$, chitosan, and naphthalene acetic acid improves tolerance against salt stress by increasing root and shoot tolerance index and maintaining relative water contents (Sen and Mandal 2018).

Another study reported that solid matrix priming with $O_2$ gave highest germination and lowest electrical conductivity and improved activities of enzymes (e.g., CAT) as observed in leek (*Allium ampeloprasum* L.; Ozden et al. 2018). Different substances such as volcanic cinder, ground charcoal, and sawdust used for priming in soybean had favorable impact under different incubation periods (Lorenzo 1991). However, longer incubation may have negative consequences because excessive moisture may lead to fungal development (Lorenzo 1991; Mercado and Fernandez 2002). The positive effect of solid matric priming has also been observed under thermal stress on germination of celery (Parera et al. 1993). In wheat, using press mud as solid matrix priming helps in significant increase in root and shoot length and enhances germination index resulting in better performance (Abbas et al. 2018).

## 2.8 Seed Coating and Pelleting

Seed coating is used to facilitate sowing, delivery of nutrients, PGRs, microbes, soil adjuvants, germination enhancers, and seed color (Pedrini et al. 2017), while seed pelleting is usually practiced for small grain crops to facilitate the seed handing during sowing. Seed coating with nutrients has been found effective (e.g., Zn) alone or in combination with PGPR (*Pseudomonas sp.*) and has been found effective in improving the water relations, chlorophyll, photosynthesis, grain yield, and Zn biofortification in wheat (Rehman and Farooq 2016; Rehman et al. 2018a, b) and chickpea (Ullah et al. 2018).

Hydro-absorber-coated seeds of wheat, rye, and barley exhibited >75% increase in seedling growth than uncoated seeds. They further reported that seeds of these cereals have high metabolic efficiencies resulting in more biomass production as the coated seeds might have hypoxic embryo which helps in efficient glucose utilization and reduced losses during respiration (Gorim and Asch 2015). In a recent study, Gorim and Asch (2017) tested two hydro-absorbers (Stockosorb and Geohumus) for seed coating of sorghum for their potential to improve sorghum tolerance to early season drought. They found that hydro-absorber coating improved the performance on sorghum under drought stress condition by improving early seedling growth, leaf development, and biomass production than uncoated seeds. Stockosorb was most effective coating in this regard.

Coated seeds exhibited improved crop performance under diverse climatic conditions. Chitosan seed coating improved the germination and induced resistance against salt stress in hybrid rice (Ruan and Xue 2002). Seed pelleting using poly (2-acrylamide-2-methyl propane sulfonic acid) (PAMPS) hydrogel along with SA improved the germination of tobacco under drought stress (Guan et al. 2014). In soybean, temperature-activated polymer coating improves the crop stand in no tilled soil under low soil temperature (Gesch et al. 2012). Zhang et al. (2007) documented that rice seed coating with high-effective thiazole (HET) (20 mg $L^{-1}$), YKJ (2000 mg $L^{-1}$), YKJ+ sodium nitro complex (SNC) (1000 mg $L^{-1}$), and ABA (10 mg $L^{-1}$) improved the resistance against chilling stress in rice by improving the root vigor, activities of antioxidant (CAT, SOD, and POD), photosynthetic pigment, soluble sugar, and proline accumulation with lower plant injury, electrolyte leakage, and MDA level than plants raised from uncoated seeds. Moreover, YKJ+ SNC seed coating was most effective in ameliorating the adverse effect of chilling stress in rice seedlings.

## 3 Summary

In the climate change scenario, the incidence of plant abiotic stresses is increasing. The possible solution to combat seed stress is development of stress-tolerant crop cultivars. However, it is a time-consuming strategy. Pre-sowing seed treatment with different salts, organic compounds, growth regulators, and PGPR help in improving the crop performance under suboptimal conditions. Seed priming induces stress tolerance by modulating gene expression, osmolyte production, and antioxidant activities due to the build-up of stress-related memory during seed priming which reciprocates on exposure to stress.

## References

Abbas MW, Khan M, Ahmad F, Nawaz H, Ahmad J, Ayub A, Amin H (2018) Germination and seedling growth of wheat as affected by seed priming and its duration. Agric Res Technol J 18(3):556062

Afzal I, Basra SMA, Ahmad N, Farooq M (2005) Optimization of hormonal priming techniques for alleviation of salinity stress in wheat (*Triticum aestivum* L.). Caderno de Pesquisa Ser. 17(1): 95–109

Akbari G, Sanavy SA, Yousefzadeh S (2007) Effect of auxin and salt stress (NaCl) on seed germination of wheat cultivars (*Triticum aestivum* L.). Pak J Biol Sci 10:2557–2561

Alcázar R, Altabella T, Marco F, Bortolotti C, Reymond M, Koncz C, Carrasco P, Tiburcio AF (2010) Polyamines: molecules with regulatory functions in plant abiotic stress tolerance. Planta 231(6):1237–1249

Amooaghaie R (2013) The effect of hydro and osmopriming on alfalfa seed germination and antioxidant defenses under salt stress. Afr J Biotechnol 10:6269–6275

Amooaghaie R, Nikzad K (2013) The role of nitric oxide in priming-induced low-temperature tolerance in two genotypes of tomato. Seed Sci Res 23:123–131

Amoghein MB, Amoghein RS, Tobeh A, Jamaati-e-Somarin S (2013) The effect of osmopriming and hydropriming on the different index of germination & early growth of wheat under salty stress. Int Res J Appl Basic Sci 4:1924–1931

Arif M (2005) Seed priming maize for improving emergence and seedling growth. Sarhad J Agric 21:239–243

Arif M, Waqas M, Nawab K, Shahid M (2007) Effect of seed priming in Zn solutions on chickpea and wheat. Afr Crop Sci Conf Proc 8:237–240

Bajehbaj AA (2010) The effects of NaCl priming on salt tolerance in sunflower germination and seedling grown under salinity conditions. Afr J Biotechnol 9(12):1764–1770

Bajwa AA, Farooq M, Nawaz A (2018) Seed priming with sorghum extracts and benzyl aminopurine improves the tolerance against salt stress in wheat (*Triticum aestivum* L.). Physiol Mol Biol Plants 24(2):239–249

Basra SMA, Pannu IA, Afzal I (2003) Evaluation of seedling vigor of hydro and matriprimed wheat (*Triticum aestivum* L.) seeds. Int Agric Biol 5:121–123

Basra S, Farooq M, Tabassam T, Ahmad N (2005) Physiological and biochemical aspects of pre-sowing seed treatments in fine rice (Oryza sativa L.). Seed Sci Technol 33:623–628

Beckers GJ, Conrath U (2007) Priming for stress resistance: from the lab to the field. Curr Opin Plant Biol 10:425–431

Bouchereau A, Aziz A, Larher F, Martin-Tanguy J (1999) Polyamines and environmental challenges: recent development. Plant Sci 140(2):103–125

Boyer JS (1982) Plant productivity and environment. Science 218:443–448

Bray EA, Bailey-Serres J, Weretilnyk E (2000) Responses to abiotic stresses. In: Gruissem W, Buchannan B, Jones R (eds) Biochemistry and molecular biology of plants. American Society of Plant Physiologists, Rockville, pp 1158–1249

Brocklehurst P, Dearman J, Drew R (1984) Effects of osmotic priming on seed germination and seedling growth in leek. Sci Hortic 24:201–210

Butler LH, Hay FR, Ellis RH, Smith RD, Murray TB (2009) Priming and re-drying improve the survival of mature seeds of *Digitalis purpurea* during storage. Ann Bot 103:1261–1270. Osmo

Cakmak I (2005) The role of potassium in alleviating detrimental effects of abiotic stresses in plants. J Plant Nutr Soil Sci 168:521–530

Cayuela E, Pérez-Alfocea F, Caro M, Bolarin M (1996) Priming of seeds with NaCl induces physiological changes in tomato plants grown under salt stress. Physiol Plant 96:231–236

Chakraborty U, Chakraborty BN, Chakraborty AP, Dey PL (2013) Water stress amelioration and plant growth promotion in wheat plants by osmotic stress tolerant bacteria. World J Microbiol Biotechnol 29:789–803

Chen K, Arora R (2011) Dynamics of the antioxidant system during seed osmopriming, post-priming germination, and seedling establishment in spinach (Spinacia oleracea). Plant Sci 180:212–220

Chen CC, Sung JM (2001) Priming bitter gourd seeds with selenium solution enhances germinability and antioxidative responses under sub-optimal temperature. Physiol Plant 111:9–16

Chunthaburee S, Sanitchon S, Pattanagul W, Theerakulpisut P (2014) Alleviation of salt stress in seedlings of black glutinous rice by seed priming with spermidine and gibberellic acid. Not Bot Hortic Agrobotan 42:405–413

Demir I, Mavi K (2004) The effect of priming on seedling emergence of differentially matured watermelon (Citrullus lanatus (Thunb.) Matsum and Nakai) seeds. Sci Hortic 102:467–473

Eisvand HR, Tavakkol-Afshari R, Sharifzadeh F, Maddah Arefi H, Hesamzadeh Hejazi SM (2010) Effects of hormonal priming and drought stress on activity and isozyme profiles of antioxidant enzymes in deteriorated seed of tall wheatgrass (*Agropyron elongatum* Host). Seed Sci Tech 38(2):280–297

El-Afifi ST, Zaghloul MM, Ahmed HM, Mohamed MA (2016) Alleviation of salinity stress during seed germination and early growth stage in sweet pepper by seed priming with acetyl salicylic acid. J Plant Prod 7:1255–1261

Elkoca E, Haliloglu K, Esitken A, Ercisli S (2007) Hydro-and osmopriming improve chickpea germination. Acta Agric Scand Sect B Soil Plant Sci 57:193–200

Farahani HA, Maroufi K (2011) Effect of hydropriming on seedling vigour in basil (*Ocimum basilicum* L.) under salinity conditions. Adv Environ Biol 5:828–833

Faran M, Farooq M, Rehman A, Nawaz A, Saleem MK, Ali N, Siddique KHM (2019) High intrinsic seed Zn concentration improves abiotic stress tolerance in wheat. Plant Soil. https://doi.org/10.1007/s11104-019-03977-3

Farhoudi R, Sharifzadeh F (2006) The effects of NaCl priming on salt tolerance in canola (Brassica napus L.) seedlings grown under saline conditions. Indian J Crop Sci 1:74–78

Farooq M, Basra SMA, Hafeez K (2005) Seed invigoration by osmohardening in Indica and Japonica rice. Seed Sci Technol 34:181–187

Farooq M, Basra SM, Rehman H, Mehmood T (2006) Germination and early seedling growth as affected by pre-sowing ethanol seed treatments in fine rice. Int J Agric Biol 8:19–22

Farooq M, Hussain M, Nawaz A, Lee D-J, Alghamdi SS, Siddique KH (2017) Seed priming improves chilling tolerance in chickpea by modulating germination metabolism, trehalose accumulation and carbon assimilation. Plant Physiol Biochem 111:274–283

Farooq M, Ullah A, Lee DJ, Alghamdi SS (2018a) Terminal drought-priming improves the drought tolerance in *desi* and *kabuli* chickpea. Int J Agric Biol 30:1129–1136

Farooq M, Ullah A, Rehman A, Nawaz A, Nadeem A, Wakeel A, Nadeem F, Siddique KHM (2018b) Application of zinc improves the productivity and biofortification of fine grain aromatic rice grown in dry seeded and puddled transplanted production systems. Field Crops Res 216:53–62

Fujita M, Fujita Y, Noutoshi Y, Takahashi F, Narusaka Y, Yamaguchi-Shinozaki K, Shinozak K (2006) Crosstalk between abiotic and biotic stress responses: a current view from the points of convergence in the stress signaling networks. Curr Opin Plant Biol 9:436–442

Gebreegziabher BG, Qufa CA (2017) Plant physiological stimulation by seeds salt priming in maize (Zea mays): prospect for salt tolerance. Afr J Biotechnol 16:209–223

Gesch RW, Archer DW, Spokas K (2012) Can using polymer-coated seed reduce the risk of poor soybean emergence in no-tillage soil? Field Crops Res 125:109–116

Ghassemi-Golezani K, Esmaeilpour B (2008) The effect of salt priming on the performance of differentially matured cucumber (Cucumis sativus) seeds. Not Bot Hortic Agrobotan Cluj-Napoca 36:67–70

Ghezal N, Rinez I, Sbai H, Saad I, Farooq M, Rinez A, Haouala R (2016) Improvement of Pisum sativum salt stress tolerance by bio-priming their seeds using Typha angustifolia leaves aqueous extract. S Afr J Bot 105:240–250

Gonzalez A, Gorden AJ, James CL, Arrese-Igor C (1995) The role of sucrose synthase in response of soybean nodules to drought. J Exp Bot 46:1515–1523
Gorim L, Asch F (2015) Seed coating reduces respiration losses and affects sugar metabolism during germination and early seedling growth in cereals. Funct Plant Biol 42:209–218
Gorim L, Asch F (2017) Seed coating with hydro-absorbers as potential mitigation of early season drought in sorghum (*Sorghum bicolor* L. Moench). Biology 6:3–33
Górnik K, Lahuta LB (2017) Application of phytohormones during seed hydropriming and heat shock treatment on sunflower (*Helianthus annuus* L.) chilling resistance and changes in soluble carbohydrates. Acta Physiol Plant 39(5):118
Guan YJ, Hu J, Wang XJ, Shao CX (2009) Seed priming with chitosan improves maize germination and seedling growth in relation to physiological changes under low temperature stress. J Zhejiang Univ Sci B 10:427–433
Guan Y, Cui H, Ma W, Zheng Y, Tian Y, Hu J (2014) An enhanced drought-tolerant method using SA-loaded PAMPS polymer materials applied on tobacco pelleted seeds. Sci World J 2014:9
Gul B, Khan MA (2003) Effect of growth regulators and osmotic in alleviating salinity effects on the germination of Salicornia utahensis. Pak J Bot 35:877–886
Gupta AK, Singh J, Kaur N, Singh R (1993) Effect of polyethylene glycol induced water stress on uptake and transport of sugars in chickpea seedlings. Plant Physiol Biochem 31:743–747
Harman GE, Taylor AG (1988) Improved seedling performance by integration of biological control methods at favorable pH levels with solid matrix priming. Phytopathology 78:520–525
Hasanuzzaman M, Hossain AM, Fujita M (2010) Selenium in higher plants: physiological role, antioxidant metabolism and abiotic stress tolerance. J Plant Sci 5:354–375
He L, Gao Z (2009) Pretreatment of seed with H2O2 enhances drought tolerance of wheat (*Triticum aestivum* L.) seedlings. Afr J Biotechnol 8:6151–6157
Hu Y, Ge Y, Zhang C, Ju T, Cheng W (2009) Cadmium toxicity and translocation in rice seedlings are reduced by hydrogen peroxide pretreatment. Plant Growth Regul 59:1–51
Hussain S, Khan F, Hussain HA, Nie L (2016) Physiological and biochemical mechanisms of seed priming-induced chilling tolerance in rice cultivars. Front Plant Sci 7:116
Imran M, Mahmood A, Römheld V, Neumann G (2013) Nutrient seed priming improves seedling development of maize exposed to low root zone temperatures during early growth. Eur J Agron 49:141–148
Iqbal M, Ashraf M (2005) Changes in growth, photosynthetic capacity and ionic relations in spring wheat (*Triticum aestivum* L.) due to presowing seed treatment with polyamines. Plant Growth Regul 46:19–30
Jafar MZ, Farooq M, Cheema MA, Afzal I, Basra SM, Wahid MA, Aziz T, Shahid M (2012) Improving the performance of wheat by seed priming under saline conditions. J Agron Crop Sci 198(1):38–45
Jakab G, Ton J, Flors V, Zimmerli L, Métraux JP, Mauch-Mani B (2005) Enhancing Arabidopsis salt and drought stress tolerance by chemical priming for its abscisic acid responses. Plant Physiol 139:267–274
Jisha KC, Vijayakumari K, Puthur JT (2013) Seed priming for abiotic stress tolerance: an overview. Acta Physiol Plant 35:1381–1396
Kasim WA, Osman ME, Omar MN, El-Daim IA, Bejai S, Meijer J (2013) Control of drought stress in wheat using plant-growth-promoting bacteria. J Plant Growth Regul 32(1):122–130
Kaur S, Gupta AK, Kaur N (2000) Effect of GA3, kinetin and indole acetic acid on carbohydrate metabolism in chickpea seedlings germinating under water stress. Plant Growth Regul 30:61–70
Kaur S, Gupta AK, Kaur N (2002) Effect of osmo-and hydropriming of chickpea seeds on seedling growth and carbohydrate metabolism under water deficit stress. Plant Growth Regul 37:17–22
Kavino M, Harish S, Saravanakumar D, Jeyakumar P, Kumar N, Samiyappan R (2010) Biological hardening-a new approach to enhance resistance against biotic and abiotic stresses in micropropagated plants. Tree For Sci Biotechnol 4:11–21

Khan AA, Miura H, Prusinski J, Ilyas S (1990) Matriconditioning of seeds to improve emergence. In: Proceedings of the Symposium on Stand Establishment of Horticultural Crops, pp 13–28
Li X, Jiang H, Liu F, Cai J, Dai T, Cao W, Jiang D (2013) Induction of chilling tolerance in wheat during germination by pre-soaking seed with nitric oxide and gibberellin. Plant Growth Regul 71:31–40
Li Z, Peng Y, Zhang XQ, Ma X, Huang LK, Yan YH (2014) Exogenous spermidine improves seed germination of white clover under water stress via involvement in starch metabolism, antioxidant defenses and relevant gene expression. Molecules 19:18003–18024
Lorenzo MCB (1991) Seed invigoration of soybean and corn through solid matrix priming. BS, Philippines
Lucy M, Reed E, Glick BR (2004) Applications of free living plant growth promoting rhizobacteria. Antonie Van Leeuwenhoek 86:1–25
Madsen MD, Svejcar L, Radke J, Hulet A (2018) Inducing rapid seed germination of native cool season grasses with solid matrix priming and seed extrusion technology. PLoS One 13:10
Marschner H (1995) Mineral nutrition of higher plants, 2nd edn. Academic, San Diego
Meena RP, Sendhil R, Tripathi SC, Chander S, Chhokar RS, Sharma RK (2013) Hydro-priming of seed improves the water use efficiency, grain yield and net economic return of wheat under different moisture regimes. SAARC J Agric 11:149–159
Mehra V, Tripathi J, Powell AA (2003) Aerated hydration treatment improves the response of Brassica juncea and Brassica campestris seeds to stress during germination. Seed Sci Technol 31:57–70
Mercado MFO, Fernandez PG (2002) Solid matrix priming of soybean seeds. Philipp J Crop Sci 27:27–35
Mitoi EN, Holobiuc I, Blindu R (2009) The effect of mannitol on antioxidative enzymes in vitro long term cultures of Dianthus tenuifolius and Dianthus spiculifolius. Rom J Biol Plant Biol 54:25–30
Moradi A, Younesi O (2009) Effects of osmo-and hydro-priming on seed parameters of grain sorghum (Sorghum bicolor L.). Aust J Basic Appl Sci 3:1696–1700
Mouradi M, Bouizgaren A, Farissi M, Makoudi B, Kabbadj A, Very AA, Sentenac H, Qaddoury A, Ghoulam C (2016) Osmopriming improves seeds germination, growth, antioxidant responses and membrane stability during early stage of Moroccan alfalfa populations under water deficit. Chil J Agric Res 76:265–272
Nadeem SM, Zaheer ZA, Naveed M (2013) Mitigation of salinity induced negative impact on the growth and yield of wheat by plant growth-promoting rhizobacteria in naturally saline conditions. Ann Microbiol 63:225–232
Nawaz A, Amjad M, Pervez MA, Afzal I (2011) Effect of halopriming on germination and seedling vigor of tomato. Afr J Agric Res 6:3551–3559
Ozbingol N, Corbineau F, Come D (1998) Responses of tomato seeds to osmoconditioning as related to temperature and oxygen. Seed Sci Res 8:377–384
Ozden E, Ermis S, Sahin O, Taskin MB, Demir I (2018) Solid matrix priming treatment with O2 enhanced quality of leek seed lots. Not Bot Hortic Agrobotan Cluj-Napoca 46:371–375
Paparella S, Araújo S, Rossi G, Wijayasinghe M, Carbonera D, Balestrazzi A (2015) Seed priming: state of the art and new perspectives. Plant Cell Rep 34:1281–1293
Parera CA, Cantliffe DJ (1991) Improved germination and modified imbibitions of shrunken-2 sweet corn by seed disinfection and solid matrix priming. J Am Soc Hortic Sci 116:942–945
Parera CA, Qiao P, Cantliffe DJ (1993) Enhanced celery germination at stress temperature via solid matrix priming. Hortic Sci 28:20–22
Patade VY, Bhargava S, Suprasanna P (2009) Halopriming imparts tolerance to salt and PEG induced drought stress in sugarcane. Agric Ecosyst Environ 134:24–28
Patade VY, Bhargava S, Suprasanna P (2011) Salt and drought tolerance of sugarcane under iso-osmotic salt and water stress: growth, osmolytes accumulation, and antioxidant defense. J Plant Interact 6(4):275–282

Pedrini S, Merritt DJ, Stevens J, Dixon K (2017) Seed coating: science or marketing spin?. Trend Plant Sci 22:106–116
Posmyk M, Janas K (2007) Effects of seed hydropriming in presence of exogenous proline on chilling injury limitation in Vigna radiata L. seedlings. Acta Physiol Plant 29:509–517
Rashid A, Hollington PA, Harris D, Khan P (2006) On-farm seed priming for barley on normal, saline and saline–sodic soils in north west Frontier Province, Pakistan. Eur J Agron 24:276–281
Rehman A, Farooq M (2016) Zinc seed coating improves the growth, grain yield and grain biofortification of bread wheat. Acta Physiol Plant 38:238
Rehman H, Afzal I, Farooq M, Aziz T, Ahmad SM (2012) Improving temperature stress resistance in spring maize by seed priming. In: Third International Conference-frontiers in agriculture, pp 28–32
Rehman A, Farooq M, Ahmad R, Basra SMA (2015) Seed priming with zinc improves the germination and early seedling growth of wheat. Seed Sci Technol 43:262–268
Rehman A, Farooq M, Naveed M, Ozturk L, Nawaz A (2018a) Pseudomonas-aided zinc application improves the productivity and biofortification of bread wheat. Crop Pasture Sci 69:659–672
Rehman A, Farooq M, Naveed M, Nawaz A, Shahzad B (2018b) Seed priming of Zn with endophytic bacteria improves the productivity and grain biofortification of bread wheat. Eur J Agron 94:98–107
Ruan SL, Xue QZ (2002) Effects of chitosan coating on seed germination and salt-tolerance of seedlings in hybrid rice (*Oryza sativa* L.). Acta Agron Sin 28:803–808
Saha P, Chatterjee P, Biswas AK (2010) NaCl pretreatment alleviates salt stress by enhancement of antioxidant defense system and osmolyte accumulation in mungbean (Vigna radiata L. Wilczek). Indian J Exp Biol 48:593–600
Sakhabutdinova AR, Fatkhutdinova DR, Bezrukova MV, Shakirova FM (2003) Salicylic acid prevents the damaging action of stress factors on wheat plants. Bulg J Plant Physiol. 21:314–319
Sarwar N, Yousaf S, Jamil FF (2006) Induction of salt tolerance in chickpea by using simple and safe chemicals. Pak J Bot 38:325
Sen SK, Mandal P (2018) Application of solid matrix priming to ameliorate salinity stress in mung bean ('Vigna radiata'). Aust J Crop Sci 12:458
Shafi M, Bakht J, Hassan MJ, Raziuddin M, Zhang G (2009) Effect of cadmium and salinity stresses on growth and antioxidant enzyme activities of wheat (*Triticum aestivum* L.). Bull Environ Contam Toxicol 82:772–776
Shahzad B, Tanveer M, Che Z, Rehman A, Cheema SA, Sharma A, Song H, Rehman S, Zhaorong D (2018a) Role of 24-epibrassinolide (EBL) in mediating heavy metal and pesticide induced oxidative stress in plants: a review. Ecotoxicol Environ Saf 147:935–944
Shahzad B, Tanveer M, Rehman A, Cheema SA, Fahad S, Rehman S, Sharma A (2018b) Nickel; whether toxic or essential for plants and environment – a review. Plant Physiol Biochem 132:641–651
Shen B, Jensen RG, Bohnert HJ (1997) Increased resistance to oxidative stress in transgenic plants by targeting mannitol biosynthesis to chloroplasts. Plant Physiol 113:1177–1183
Singh B, Rao G (1993) Effect of chemical soaking of sunflower (Helianthus-annuus) seed on vigor index. Indian J Agric Sci 63:232–233
Singh SK, Singh PK (2016) Effect of seed priming of tomato with salicylic acid on growth, flowering, yield and fruit quality under high temperature stress conditions. Int J Adv Res 4:723–727
Soughir M, Elouaer MA, Hannachi C (2013) The effect of NaCl priming on emergence, growth and yield of fenugreek under saline conditions. Cercetări Agron Moldova 2:73–83
Srivastava AK, Lokhande VH, Patade VY, Suprasanna P, Sjahril R, D'Souza SF (2010a) Comparative evaluation of hydro-, chemo-, and hormonal-priming methods for imparting salt and PEG stress tolerance in Indian mustard (*Brassica juncea* L.). Acta Physiol Plant 32:1135–1144
Srivastava AK, Ramaswamy NK, Suprasanna P, D'Souza SF (2010b) Genome-wide analysis of thiourea-modulated salinity stress responsive transcripts in seeds of Brassica juncea: identification of signalling and effector components of stress tolerance. Ann Bot 106:663–674

Su J, Chen PL, Wu R (1999) Transgene expression of mannitol-1-phosphate dehydrogenase enhanced the salt stress tolerance of the transgenic rice seedlings. Sci Agric Sin 32:101–103

Sun YY, Sun YJ, Wang MT, Li XY, Li X, Guo R (2010) Effects of seed priming on germination and seedling growth under water stress in rice. Acta Agron Sin 36:1931–1940

Tavili A, Zare S, Enayati A (2009) Hydropriming, ascorbic and salicylic acid influence on germination of *Agropyron elongatum* host. Seeds under salt stress. Res J Seed Sci 2:16–22

Tavili A, Zare S, Moosavi SA, Enayati A (2011) Effects of seed priming on germination characteristics of Bromus species under salt and drought conditions. Am Eur J Agric Environ Sci 10:163–168

Taylor AG, Klein DE, Whitlow TH (1988) Solid matrix priming of seeds. Sci Hortic 37:1–11

Thomas U, Varughese K, Thomas A, Sadanandan S (2000) Seed priming—for increased vigour, viability and productivity of upland rice. Leisa India 4:14

Tilahun-Tadesse F, Nigussie-Dechassa R, Bayu W, Gebeyehu S (2013) Effect of hydropriming and pre-germinating rice seed on the yield and terminal moisture stress mitigation of rain-fed lowland rice. Agric For Fish 2:89–97

Ullah A, Farooq M, Nadeem A, Rehman R, Asad SA, Nawaz A (2017a) Manganese nutrition improves the productivity and grain biofortification of fine grain aromatic rice in conventional and conservation production systems. Paddy Water Environ 15:563–572

Ullah A, Farooq M, Rehman A, Arshad MS, Shoukat H, Nadeem A, Nawaz A, Wakeel A, Nadeem F (2017b) Manganese nutrition improves the productivity and grain biofortification of bread wheat in alkaline calcareous soil. Exp Agric. https://doi.org/10.1017/S0014479717000369

Ullah A, Farooq M, Hussain M, Ahmad R, Wakeel A (2018) Zinc seed coating improves emergence and seedling growth in desi and kabuli chickpea types but shows toxicity at higher concentration. Int J Agric Biol 21:10–11

van Loon LC (2007) Plant responses to plant growth-promoting rhizobacteria. Eur J Plant Pathol 119:243–254

Wang W, Peng S, Chen Q, Mei J, Dong H, Nie L (2016) Effects of pre-sowing seed treatments on establishment of dry direct-seeded early rice under chilling stress. AoB Plants 8:plw074

Xiong L, Zhu JK (2002) Molecular and genetic aspects of plant responses to osmotic stress. Plant Cell Environ 25:131–139

Xu S, Hu J, Li Y, Ma W, Zheng Y, Zhu S (2011) Chilling tolerance in *Nicotiana tabacum* induced by seed priming with putrescine. Plant Growth Regul 63:279–290

Yacoubi R, Job C, Belghazi M, Chaibi W, Job D (2013) Proteomic analysis of the enhancement of seed vigour in osmoprimed alfalfa seeds germinated under salinity stress. Seed Sci Res 23:99–110

Yadav PV, Kumari M, Ahmed Z (2011) Seed priming mediated germination improvement and tolerance to subsequent exposure to cold and salt stress in capsicum. Res J Seed Sci 4:125–136

Yuan-Yuan S, Yong-Jian S, Ming-Tian W, Xu-Yi L, Xiang G, Rong H (2010) Effects of seed priming on germination and seedling growth under water stress in rice. Acta Agron Sin 36:1931–1940

Zhang H, Yingbin Z, Guochao X (2007) Effect and mechanism of cold tolerant seed coating agents on the cold tolerance of early Indica rice seedlings. Agric Sci China 6:792–801

Zhang F, Yu J, Johnston CR, Wang Y, Zhu K, Lu F (2015) Seed priming with polyethylene glycol induces physiological changes in sorghum (*Sorghum bicolor* L. Moench) seedlings under suboptimal soil moisture environments. PLoS ONE 10:e0140620

Zheng C, Jiang D, Dai T, Jing Q, Cao W (2010) Effects nitroprusside, a nitric oxide donor, on carbon and nitrogen metabolism and the activity of the antioxidation system in wheat seedlings under salt stress. Acta Ecol Sinica 30:1174–1183

# Abiotic Stress Tolerance in Plants by Priming and Pretreatments with Phytohormones

**Muhammad Sarfraz, Sajid Hussain, Muhammad Ijaz, Ahmad Nawaz, Tauqeer Ahmad Yasir, Ahmad Sher, Allah Wasaya, and Shakeel Ahmad**

## Contents



**Abstract** Seed priming and seedling pretreatments improve the stand establishment seedling growth by improving nutrient accumulation in roots/shoots, and the activity of antioxidant enzymes thus enhance tolerance against multiple abiotic stresses including drought, extreme temperature, salinity, toxic metal/metalloid(s), waterlogging, etc., which are threatening the productivity of agronomic and horticultural crops across the globe. The phytohormones involved in complex mechanisms including plant metabolic activities and signaling pathways, thus reducing generation of free radical and activation of defense mechanism against abiotic stresses. Many studies have reported that seed priming with plant hormones (especially gibberellic acid, cytokinins, abscisic acid, brassinosteroids, jasmonic acid, polyamines, salicylic acid) at low concentration might be useful to improve the crop establishment, leaf/shoot/root growth, and productivity of diverse crop biotypes under abiotic stresses. However, the role of seed priming with plant hormones in improving the crop performance under heat stress and waterlogging stress has rarely

M. Sarfraz · S. Hussain · M. Ijaz · A. Nawaz (✉) · T. A. Yasir · A. Sher · A. Wasaya
College of Agriculture, Bahauddin Zakariya University, Layyah, Pakistan

S. Ahmad (✉)
Department of Agronomy, Bahauddin Zakariya University, Multan, Punjab, Pakistan
e-mail: shakeelahmad@bzu.edu.pk



M. Hasanuzzaman, V. Fotopoulos (eds.), *Priming and Pretreatment of Seeds and Seedlings*, https://doi.org/10.1007/978-981-13-8625-1_22

been investigated, which needs the further attention of crop scientists. In this chapter, we have reviewed and accumulated the recent advances in plant science regarding seed priming and pretreatment with phytohormones to enhance crop tolerance against multiple abiotic stresses.

**Keywords** Abiotic stress · Salinity · Drought · Salicylic acid · Jasmonic acid · Abscisic acid

# 1 Introduction

The crop plants face a diverse array of biotic and abiotic stresses during their entire growth cycle. Among the major abiotic stresses, the salt stress, water stress, heat stress, heavy metals stress, waterlogging stress, and chilling stress are major stresses which cause reduction in the productivity of field and horticultural crops across the globe (Schutzendubel and Polle 2002; Zhu 2016; Shrivastava and Kumar 2015; Lamaoui et al. 2018). To cope with these abiotic stresses, most of the crop species have evolved complex mechanisms including subsequent variation in plant metabolic activities and signaling of plant hormone activation pathway (such as salicylic acid, jasmonate, abscisic acid, ethylene) (Farooq et al. 2009a; Hasanuzzaman et al. 2012, 2013; Lamaoui et al. 2018). The plants also produce various volatile organic compounds in response to abiotic stresses, which play key role in central defense system in the plant body (Suzuki et al. 2012). The released quantity of volatile compound depends upon the stress condition, duration of stress, and stress severity (Niinemets 2009).

In this scenario, seed priming might be an effective technique to increase the germination rate, seedling vigor, and yield under abiotic stresses through regulation of various growth cascades (Ashraf and Foolad 2005; Afzal et al. 2006; Farooq et al. 2008; Paparella et al. 2015). In this chapter, we have summarized the role of seed priming and seed pretreatments with plant hormones (e.g., auxin, strigolactone, ethylene, gibberellic acid, cytokinins, abscisic acid, brassinosteroids, jasmonic acid, polyamines, salicylic acid) in improving tolerance against drought stress, heat stress, low temperature stress, salinity stress, heavy metal stress, and waterlogging stress (Table 1) in agronomic and horticultural crops.

**Table 1** Improvement in the performance of crops due to seed priming with plant hormones under abiotic stresses

| Environmental constraints | Plant species | Priming agent | References |
|---|---|---|---|
| Salinity | *Triticum aestivum* | ABA + SA | Afzal et al. (2006) |
| Drought stress | *Oryza sativa* | Polyamines | Farooq et al. (2009b) |
| Drought stress | *Oryza sativa* | Salicylic acid | Li and Zhang (2012) |
| Low temperature stress | *Oryza sativa* | Salicylic acid | Pouramir-Dashtmian et al. (2014) |
| Low temperature stress | *Nicotiana tabacum* | Putrescine | Xu et al. (2011) |
| Heavy metal stress | *Poa pratensis* | Gibberellins | de Lespinay et al. (2010) |

## 2 Drought Stress

Drought stress negatively affects the plant growth, phenology, photosynthesis, nutrient and water relations, and respiration and assimilates partitioning in field crops (Yordanov et al. 2000). However, various studies showed that the seed priming with different plant hormones is effective for improving the drought tolerance in crop plants (Kubiś 2008; Farooq et al. 2009b; Davies 2010). In a study, seed priming with spermidine (0.5 mmol $L^{-1}$) in direct seeded rice improved the rice performance under drought stress in controlled and field conditions. The field emergence, shoot/root length, and root/shoot fresh weight were improved by 21%, 67%, 78%, 173%, and 232%, respectively, by seed priming with spermidine under drought stress (Zheng et al. 2016). Farooq et al. (2009b) also found that seed priming with three polyamines (spermidine, putrescine, and spermine) enhances tolerance against drought in fine grain rice. They found that seed priming with either polyamine (at 10 μM) was very beneficial for improvement in water use efficiency, net photosynthesis, water relations, accumulation of phenolics/proline, and membrane stability. Pretreatment of cucumber seedlings with spermidine cucumber (*Cucumis sativus* L.) improved the activity of guaiacol peroxidase and reduced the hydrogen peroxide and superoxide radical production under drought stress (Kubiś 2008). In a study, seed priming with salicylic acid improved the root length, photosynthetic pigments, stay green, water use efficiency, activities of peroxidase and superoxide dismutase, and accumulation of proline in drought-stressed rice (Li and Zhang 2012). In another study, application of salicylic acid (50–150 mg $L^{-1}$) under drought stress improved the tissue water potential, accumulation of compatible solutes, cell membrane stability, photosynthesis, and antioxidant system in rice (Farooq et al. 2009c). In a study, rice seed priming with two brassinolides, viz., 24-epibrassinolide and 28-homobrassinolide (each 0.01 μM), improved the water use efficiency, net assimilation of carbon dioxide, water relations, and accumulation of phenolics, proline, and anthocyanins under drought stress (Farooq et al. 2009d). In another study, Ansari et al. (2013) also reported an improvement in the seed germination and enzymatic activities due to pretreatment of rye seeds under drought stress conditions. In wheat, ascorbate seed priming (2 mm) enhanced the emergence of seedlings and improved the early growth, leaf development, stay green, and root growth under drought stress (Farooq et al. 2013).

## 3 Heat Stress

Heat stress is among the major abiotic stresses that affects the productivity of agricultural crops across the globe (Bray et al. 2000; Farooq et al. 2011; Nawaz et al. 2013; Zhao et al. 2017). In plants, heat stress damages can be mitigated through different types of scavenging mechanisms such as enzymatic (peroxidases and catalases) and non-enzymatic antioxidants and exogenous application of plant hormones

(Farooq et al. 2011; Ahmad et al. 2015). In a study, seed pretreatment with potassium nitrate and hydropriming improved the performance of *Gossypium hirsutum* under heat stress (Ashraf and Foolad 2005). Seed priming with plant hormones for improving heat stress tolerance in the agricultural crops has not been widely reported.

## 4 Low Temperature Stress

Low temperature stress (chilling or freezing stress) occurs at various phases of the crop growth, thus affecting the plant growth by disturbing the plant metabolic activities and losses in economic yield (Kang and Saltveit 2002; Cruz and Milach 2004; Cheng et al. 2007; Oliver et al. 2007; Ruelland et al. 2009; Kosova et al. 2013; Han et al. 2013).

In recent past, several studies have showed that seed priming with plant hormones has a positive effect on the performance of different crops under abiotic stresses. In a study, seed priming with methyl jasmonate (1–10 μM) or acetyl salicylic acid (0.05–1 mM) improved the germination percentage and the uniformity and progressive growth of sweet pepper under low temperature stress (Korkmaz 2005). Soaking of musk melon weeds in acetyl salicylic acid solution (0.1–1.0 mM) improved the dry weight of roots and leaf number under drought stress (Korkmaz et al. 2007). In another study, seed priming with potassium nitrate solution in combination with methyl jasmonic acid (1 or 3 μM) or spermine (1–5 mM) increased the germination percentage and uniformity of seedling emergence in water melon under low temperature stress (Korkmaz et al. 2004). In a study, maize seeds were primed with salicylic acid (20–40 mg $L^{-1}$) at low temperature stress. Seed priming improved the seedling growth by improving the stay green, nutrient accumulation in roots/shoots, and the activity of superoxide dismutase (Ahmad et al. 2015). In *Brassica napus*, gibberellic acid (100 μmol $L^{-1}$) seed priming improved the germination under low temperature stress (Yong-Ping et al. 2002).

In recent past, Farooq et al. (2008) conducted an experiment to evaluate the role of seed priming with salicylic acid in improving chilling tolerance in maize. Seeds of maize hybrid, viz., Hycorn 8288, were primed with aerated solution of salicylic acid (at 50, 100, and 150 mg $L^{-1}$). The results indicated that seed priming with salicylic acid at either rate increased the seedling emergence, seedling growth, root and leaf number, water relations, membrane stability, and activities of enzymatic antioxidants. In tobacco, seed priming using the putrescine was beneficial for improving the chilling tolerance through an improvement in seed germination, activities of antioxidants, and the seedling growth (Xu et al. 2011). In maize, Ahmad et al. (2012) found that seed priming with ascorbic acid or salicylic acid (eat at 20 and 40 mg $L^{-1}$) improved the speed and uniformity of emergence, root/shoot fresh and dry weights, and activities of antioxidants under low temperature stress. In another study on maize, seed priming with salicylic acid (at 20, 40, and 50 ppm) decreased mean emergence time, time to 50% emergence, and enhanced shoot length, root length, and relative water contents under low temperature stress (Singh and Pandurangam 2015). In rice, priming of seeds with salicylic acid improved the carbohydrate metabolism and activities of antioxidants and lowered the membrane damage under chilling stress (Hussain et al. 2016a).

## 5 Salinity Stress

Worldwide, soil salinity is one of the major problem of dryland agriculture, and it has became a significant hindrance for harvesting the better crop yield. The improper management of farming and irrigation practices is the main cause of salinity across the globe (Munns and Tester 2008; Abraha and Yohannes 2013). Salt stress affects the seed germination and subsequent growth through decreasing the osmotic potential and avoiding the uptake of water as well as by the toxic impact of $Na^+$ and $Cl^-$ ions on plant metabolic cascades (Shannon 1997; Shafi et al. 2009; Dkhil et al. 2014).

In this scenario, seed priming with plant hormones may improve the crop performance under salinity (Al-Hakimi and Hamada 2001; Carvalho et al. 2011; Ibrahim 2016). For example, seed priming with gibberellic acid (100 μmol $L^{-1}$) improved the germination under low salinity stress (Yong-Ping et al. 2002). In another study, seed priming with an ethylene releaser (i.e., chloroethylphosphonic acid) was beneficial for improvement in seedling biomass of maize under salinity stress (Carvalho et al. 2011). Parashar and Varma (1988) also reported an improvement in seed germination, plant height, and dry weight of wheat plants owing to seed priming with gibberellic acid at the rate of 50 mg $L^{-1}$. In similar studies on wheat, damaging effect of soil salinity on seed germination was mitigated by priming seeds with indole acetic acid or gibberellic acid (Balki and Padole 1982; Gulnaz et al. 1999). In wheat, soaking seed in sodium salicylate improved the crop growth by suppressing the salt-stress-induced proline accumulation (Al-Hakimi and Hamada 2001). In faba bean, seed priming with salicylic acid improved the activities of catalase, ascorbate peroxidase, glutathione reductase, and peroxidase under saline stress (Azooz 2009). In hot pepper, seed priming with salicylic acid and acetyl cyclic acid improved the germination percentage, time to complete 50% emergence, root/shoot length, and seedling dry weight under salt stress (Khan et al. 2009). In wheat, seed priming with salicylic acid (50 ppm) or ascorbic acid (50 ppm) improved the final germination percentage, shoot fresh/dry weight, and shoot length and decreased the germination time (Afzal et al. 2006).

In tomato, seed priming with three polyamines (spermine, putrescine, and spermidine each at 50 mg $L^{-1}$) improved the stand establishment, seedling vigor, and the activities of antioxidants under salt stress (Afzal et al. 2009). In another study on tomato, seed priming with salicylic acid and ascorbic acid (each at 50, 100, and 150 mg $L^{-1}$) improved the germination and seedling growth under saline stress (Ghoohestani et al. 2012). In two *Hedysarum* species, seed priming with salicylic acid improved the salinity tolerance by decreasing the lipid peroxidation (Dallali et al. 2012). In lucerne, seed priming with brassinolide (5 μM $L^{-1}$) improved seed germination, seedling vigor, root and shoot dry weight, the activities of enzymatic antioxidants and reduced lipid peroxidation under salt stress (Zhang et al. 2007).

Seed priming with gibberellic acid in okra (Vijayaraghavan 1999), tomato (Kang et al. 1996), sweet fennel, lettuce (Sedghi et al. 2010; Hela et al. 2012), and *Medicago sativa* (Younesi and Moradi 2014), with kinetin in pigeon pea (Jyotsna and Srivastava 1998), with 28-homobrassinolide in mung bean (Fariduddin et al. 2003), and with ascorbic acid in broad bean (Azooz et al. 2013) improved the germination of seeds and growth of seedlings of the respective crop species under salinity stress.

## 6 Heavy Metal Stress

The heavy metals such as lead, cadmium, mercury, copper, and chromium are regarded as major environmental pollutants in this industrial era. The accumulation of the heavy metals in the cropped land is of great concern as it may have an adverse effect on crop growth, soil, and environmental health and the safety and the marketability of foods (Nagajyoti et al. 2010). The influence of plants and their metabolic activities affects the geological and biological redistribution of heavy metals through pollution of the air, water, and soil. When plants are grown in the heavy metal-contaminated soils, the plant faces metabolic variations, growth reduction, disturbed water relations, and accumulation of heavy metals in edible plant parts (Barceló and Poschenrieder 1990; Nagajyoti et al. 2010).

However, studies have reported that seed priming with plant hormones may improve the crop performance under heavy metal stress with substantial reduction in uptake of these metals in plant parts. For example, the seeds of pigeon pea were primed with auxin, gibberellic acid, cytokinins, ethylene, and abscisic acid (each at 10 or 100 μM) under cadmium stress. Seed priming with either plant hormone at both concentrations improved the germination speed index, average germination time, final germination percentage, and time to 50% germination (Sneideris et al. 2015). In another study, seed priming with gibberellic acid (0.1 mM) improved the antioxidant system and light phase of photosynthesis in *Trifolium repens* L. in a heavy metal-contaminated soil (Galhaut et al. 2014). In safflower, seed priming with salicylic acid improved the germination rate, seedling dry weight, and specific leaf weight under calcium chloride stress (Jam et al. 2012).

## 7 Waterlogging Stress

Waterlogging stress causes hypoxia, subsequently promotes the leaf senescence and root death, and disturbs carbohydrate metabolism, stay green, and nutrient uptake, eventually reducing the productivity and profitability of agronomic crops (Tan et al. 2008; Watkins et al. 2014; Herzog et al. 2016). In a study, seed priming with salicylic acid under submerged conditions improved the germination and seedling growth and induced many genes which are involved in carbon metabolism, nitrogen metabolism, transcription, and oxidative stress responses (Hussain et al. 2016b). To the best of our knowledge, no other study is available on the role of seed priming with plant hormones in improving submergence tolerance in agronomic and horticultural crops. In an alternate study on direct seeded flooded rice, hydropriming and seed priming with 2% leaf extract of jamun in two rice lines (Swarna and Swarna-Sub1) improved the seedling establishment, accumulation of plant biomass, activities of alcohol dehydrogenase and total amylase, the yield parameters and seed yield. Grain yield was enhanced by 5.1%–16.8% in Swarna and 44.2%–55.3% in Swarna-Sub1 when seeds were sown after priming than non-primed seeds under waterlogging stress (Sarkar 2012). In another study on rice, the seeds were primed with potassium chloride solution or water. Seed priming improved the seedling emergence, activities of catalase and superoxide dismutase, amylase, while decreased the lipid peroxidation under direct seeded flooded conditions (Ella et al. 2011).

## 8 Conclusion

Abiotic stresses including drought, heat, low temperature, salinity, heavy metal, and waterlogging are threatening the productivity of agronomic and horticultural crops across the globe. Many studies have reported that seed priming with plant hormones (especially gibberellic acid, cytokinins, abscisic acid, brassinosteroids, jasmonic acid, polyamines, salicylic acid) at low concentration might be useful to improve the stand establishment, leaf/shoot/root growth, and productivity of diverse crop biotypes under abiotic stresses. However, the role of seed priming with plant hormones in improving the crop performance under heat stress and waterlogging stress has rarely been investigated, which needs the attention of crop scientists.

## References

Abraha B, Yohannes G (2013) The role of seed priming in improving seedling growth of maize (*Zea mays* L.) under salt stress at field conditions. Agric Sci 4:666–672

Afzal I, Basra SMA, Farooq M, Nawaz A (2006) Alleviation of salinity stress in spring wheat by hormonal priming with ABA, salicylic acid and ascorbic acid. Int J Agric Biol 8(1):23–28

Afzal I, Munir F, Ayub CM, Basra SMA, Hameed A, Nawaz A (2009) Changes in antioxidant enzymes, germination capacity and vigour of tomato seeds in response of priming with polyamines. Seed Sci Technol 37(3):765–770

Ahmad I, Ahmad TKA, Basra SMA, Hasnain Z, Ali A (2012) Effect of seed priming with ascorbic acid, salicylic acid and hydrogen peroxide on emergence, vigor and antioxidant activities of maize. Afr J Biotechnol 11(5):1127–1137

Ahmad I, Basra SMA, Hussain S, Hussain SA, Hafeez-ur-rehman RA, Ali A (2015) Priming with ascorbic acid, salicylic acid and hydrogen peroxide improves seedling growth of spring maize at suboptimal temperature. J Environ Agric Sci 3:14–22

Al-Hakimi AMA, Hamada AM (2001) Counteraction of salinity stress on wheat plants by grain soaking in ascorbic acid, thiamin or sodium salicylate. Biol Plant 44:253–261

Ansari O, Azadi MS, Sharif-zadeh F, Younesi E (2013) Effect of hormone priming on germination characteristics and enzyme activity of mountain rye (*Secalemontanum*) seeds under drought stress conditions. J Stress Physiol Biochem 9:61–71

Ashraf M, Foolad RM (2005) Pre-sowing seed treatment - a shotgun approach to improve germination, plant growth and crop yield under saline and non-saline conditions. Adv Agron 88:223–271

Azooz MM (2009) Salt stress mitigation by seed priming with salicylic acid in two faba bean genotypes differing in salt tolerance. Int J Agric Biol 11:343–350

Azooz MM, Alzahrani AM, Youssef MM (2013) The potential role of seed priming with ascorbic acid and nicotinamide and their interactions to enhance salt tolerance in broad bean (*Viciafaba* L.). Aust J Crop Sci 7(13):2091

Balki AS, Padole VR (1982) Effect of pre-soaking seed treatments with plant hormones on wheat under conditions of soil salinity. J Indian Soc Soil Sci 30:361–365

Barceló JUAN, Poschenrieder C (1990) Plant water relations as affected by heavy metal stress: a review. J Plant Nutr 13(1):1–37

Bray EA, Bailey-Serres J, Weretilnyk E (2000) Responses to abiotic stresses. In: Gruissem W, Buchannan B, Jones R (eds) Biochemistry and molecular biology of plants. American Society of Plant Physiologists, Rockville, MD, pp 1158–1249

Carvalho RF, Piotto FA, Schmidt D, Peters LP, Monteiro CC, Azevedo RA (2011) Seed priming with hormones does not alleviate induced oxidative stress in maize seedlings subjected to salt stress. Sci Agric 68(5):598–602

Cheng C, Yun KY, Ressom HW, Mohanty B, Bajic VB, Jia Y et al (2007) An early response regulatory cluster induced by low temperature and hydrogen peroxide in seedlings of chilling-tolerant japonica rice. BMC Genomics 8:175

Cruz R, Milach S (2004) Cold tolerance at the germination stage of rice: methods of evaluation and characterization of genotypes. Sci Agric 61:1–8

Dallali H, Maalej EM, Boughanmi NG, Haouala R (2012) Salicylic acid priming in hedysarum carnosum and hedysarum coronarium reinforces NACL tolerance at germination and the seedling growth stage. Aust J Crop Sci 6(3):407

Davies PJ (2010) The plant hormones: their nature, occurrence, and functions. In: Plant hormones. Springer, Dordrecht, pp 1–15

de Lespinay A, Lequeux H, Lambillotte B, Lutts S (2010) Protein synthesis is differentially required for germination in Poa pratensis and Trifolium repens in the absence or in the presence of cadmium. Plant Growth Regul 61:205–214

Dkhil BB, Issa A, Denden M (2014) Germination and seedling emergence of primed okra (Abelmoschus esculentus L.) seeds under salt stress and low temperature. Am J Plant Physiol 9:38–45

Ella ES, Dionisio-Sese ML, Ismail AM (2011) Seed pre-treatment in rice reduces damage, enhances carbohydrate mobilization and improves emergence and seedling establishment under flooded conditions. AoB Plants, 2011

Fariduddin Q, Ahmad A, Hayat S (2003) Photosynthetic Response of Vigna radiata to pre-sowing seed treatment with 28-homobrassinolide. Photosynthetica 41:307–310

Farooq M, Aziz T, Basra SMA, Cheema MA, Rehman H (2008) Chilling tolerance in hybrid maize induced by seed priming with salicylic acid. J Agron Crop Sci 194(2):161–168

Farooq M, Wahid A, Kobayashi N, Fujita D, Basra SMA (2009a) Plant drought stress: effects, mechanisms and management. Agron Sustain Dev 29:185–212

Farooq M, Wahid A, Lee DJ (2009b) Exogenously applied polyamines increase drought tolerance of rice by improving leaf water status, photosynthesis and membrane properties. Acta Physiologia Plantarum 31:937–945

Farooq M, Basra SMA, Wahid A, Ahmad N, Saleem BA (2009c) Improving the drought tolerance in rice (*Oryza sativa* L.) by exogenous application of salicylic acid. J Agron Crop Sci 195(4):237–246

Farooq M, Wahid A, Basra SMA (2009d) Improving water relations and gas exchange with brassinosteroids in rice under drought stress. J Agron Crop Sci 195(4):262–269

Farooq M, Bramley H, Palta JA, Siddique KH (2011) Heat stress in wheat during reproductive and grain-filling phases. Crit Rev Plant Sci 30(6):491–507

Farooq M, Irfan M, Aziz T, Ahmad I, Cheema SA (2013) Seed priming with ascorbic acid improves drought resistance of wheat. J Agron Crop Sci 199(1):12–22

Galhaut L, de Lespinay A, Walker DJ, Bernal MP, Correal E, Lutts S (2014) Seed priming of Trifolium repens L. improved germination and early seedling growth on heavy metal-contaminated soil. Water Air Soil Pollution 225(4):1905

Ghoohestani A, Gheisary H, Zahedi SM, Dolatkhahi A (2012) Effect of seed priming of tomato with salicylic acid, ascorbic acid and hydrogen peroxide on germination and plantlet growth in saline conditions. Int J Agron Plant Prod 3:700–704

Gulnaz A, Iqbal J, Azam F (1999) Seed treatment with growth regulators and crop productivity. II. Response of critical growth stages of wheat (*Triticum aestivum* L.) under salinity stress. Cereal Res Commun 27:419–426

Han Q, Kang G, Guo T (2013) Proteomic analysis of spring freeze-stress responsive proteins in leaves of bread wheat (*Triticum aestivum* L.). Plant Physiol Biochem 63:236–244

Hasanuzzaman M, Hossain MA, da Silva JAT, Fujita M (2012) Plant response and tolerance to abiotic oxidative stress: antioxidant defense is a key factor. In: Crop stress and its management: perspectives and strategies. Springer, Dordrecht, pp 261–315

Hasanuzzaman M, Nahar K, Alam M, Roychowdhury R, Fujita M (2013) Physiological, biochemical, and molecular mechanisms of heat stress tolerance in plants. Int J Mol Sci 14(5):9643–9684

Hela M, Zargouni H, Tarchoune I, Baatour O, Nasri N, Ben Massoud R et al (2012) Combined effect of hormonal priming and salt treatments on germination percentage and antioxidant activities in lettuce seedlings. Afr J Biotechnol 11:10373–10380

Herzog M, Striker GG, Colmer TD, Pedersen O (2016) Mechanisms of waterlogging tolerance in wheat - A review of root and shoot physiology. Plant Cell Environ 39:1068–1086

Hussain S, Khan F, Hussain HA, Nie L (2016a) Physiological and biochemical mechanisms of seed priming-induced chilling tolerance in rice cultivars. Front Plant Sci 7:116

Hussain S, Yin H, Peng S, Faheem A, Khan FA, Khan F, Sameeullah M, Hussain HA, Huang J, Kehui Cui K, Nie L (2016b) Comparative transcriptional prowling of primed and non-primed rice seedlings under submergence stress. Front Plant Sci 7:1125

Ibrahim EA (2016) Seed priming to alleviate salinity stress in germinating seeds. J Plant Physiol 192:38–46

Jam BJ, Shekari FARID, Azimi MR, Zangani ESMAEIL (2012) Effect of priming by salicylic acid on germination and seedling growth of safflower seeds under $CaCl_2$ stress. Int J Agri Res Rev 2:1097–1105

Jyotsna V, Srivastava AK (1998) Physiological basis of salt stress resistance in pigeon pea (*Cajanus cajan* L.) - II. Pre-sowing seed soaking treatment in regulating early seedling metabolism during seed germination. Plant Physiol Biochem (New Delhi) 25:89–94

Kang HM, Saltveit ME (2002) Chilling tolerance of maize, cucumber and rice seedling leaves and roots are differentially affected by salicylic acid. Physiol Plant 115:571–576

Kang J-S, J-L C, Y-O J (1996) Effect of seed priming on the germinability of tomato (*Lycopersicon esculentum* Mill.) seeds under water and saline stress. J Korean Soc Hort Sci 37:516–521

Khan HA, Pervez MA, Ayub CM, Ziaf K, Balal RM, Shahid MA, Akhtar N (2009) Hormonal priming alleviates salt stress in hot pepper (*Capsicum annuum* L.). Soil Environ 28(2):130–135

Korkmaz A (2005) Inclusion of acetyl salicylic acid and methyl jasmonate into the priming solution improves low-temperature germination and emergence of sweet pepper. Hort Sci 40(1):197–200

Korkmaz A, Tiryaki I, Nas MN, Ozbay N (2004) Inclusion of plant growth regulators into priming solution improves low-temperature germination and emergence of watermelon seeds. Can J Plant Sci 84(4):1161–1165

Korkmaz A, Uzunlu M, Demirkiran AR (2007) Treatment with acetyl salicylic acid protects muskmelon seedlings against drought stress. Acta Physiol Plant 29(6):503–508

Kosova K, Vitamvas P, Planchon S, Renaut J, Vankova R, Prasil IT (2013) Proteome analysis of cold response in spring and winter wheat (*Triticum aestivum*) crowns reveals similarities in

stress adaptation and differences in regulatory processes between the growth habits. J Proteome Res 12:4830–4845
Kubiś J (2008) Exogenous spermidine differentially alters activities of some scavenging system enzymes, H2O2 and superoxide radical levels in water-stressed cucumber leaves. J Plant Physiol 165(4):397–406
Lamaoui M, Jemo M, Datla R, Bekkaoui F (2018) Heat and drought stresses in crops and approaches for their mitigation. Front Chem 6:26
Li X, Zhang L (2012) SA and PEG-induced priming for water stress tolerance in rice seedling. In: Zhu E, Sambath S (eds) Information technology and agricultural engineering. Springer, Berlin, Heidelberg. 134: 881–887
Munns R, Tester M (2008) Mechanisms of salinity tolerance. Annu Rev Plant Biol 59:651–681
Nagajyoti PC, Lee KD, Sreekanth TVM (2010) Heavy metals, occurrence and toxicity for plants: a review. Environ Chem Lett 8(3):199–216
Nawaz A, Farooq M, Cheema SA, Wahid A (2013) Differential response of wheat cultivars to terminal heat stress. Int J Agric Biol 15:1354–1358
Niinemets U (2009) Mild versus severe stress and BVOCs: thresholds, priming and consequences. Trends Plant Sci 15:145–153
Oliver SN, Dennis ES, Dolferus R (2007) ABA regulates apoplastic sugar transport and is a potential signal for cold-induced pollen sterility in rice. Plant Cell Physiol 48:1319–1330
Paparella S, Araújo SS, Rossi G, Wijayasinghe M, Carbonera D, Balestrazzi A (2015) Seed priming: state of the art and new perspectives. Plant Cell Rep 34:1281–1293
Parashar A, Varma SK (1988) Effect of presowing seed soaking in gibberellic acid, duration of soaking, di Verent temperatures and their interaction on seed germination and early seedling growth of wheat under saline conditions. Plant Physiol Biochem (New Delhi) 15:189–197
Pouramir-Dashtmian F, Khajel-Hosseini M, Esfahani M (2014) Improving chilling tolerance of rice seedling by seed priming with salicylic acid. Arch Agron Soil Sci 60:1291–1302
Ruelland E, Vaultier MN, Zachowski A, Hurry V (2009) Cold signalling and cold acclimation in plants. Adv Bot Res 49:35–150
Sarkar RK (2012) Seed priming improves agronomic trait performance under flooding and non-flooding conditions in rice with QTL SUB1. Rice Sci 19(4):286–294
Schutzendubel A, Polle A (2002) Plant responses to abiotic stresses: heavy metal-induced oxidative stress and protection by mycorrhization. J Exp Bot 53(372):1351–1365
Sedghi M, Nemati A, Esmaielpour B (2010) Effect of seed priming on germination and seedling growth of two medicinal plants under salinity. Emir J Food Agr 22:130–139
Shafi M, Bakht J, Hassan MJ, Raziuddin M, Zhang G (2009) Effect of cadmium and salinity stresses on growth and antioxidant enzyme activities of wheat (*Triticum aestivum* L.). Bull Environ Contam Toxicol 82:772–776
Shannon MC (1997) Adaptation of plants to salinity. Adv Agron 60:75–120
Shrivastava P, Kumar R (2015) Soil salinity: a serious environmental issue and plant growth promoting bacteria as one of the tools for its alleviation. Saudi J Biol Sci 22(2):123–131
Singh P, Pandurangam V (2015) Evaluation of maize (*Zea mays* L.) inbred lines primed with salicylic acid under low temperature stress. Int J Agric Environ Biotechnol 8(4):867
Sneideris LC, Gavassi MA, Campos ML, D'Amico-Damiao V, Carvalho RF (2015) Effects of hormonal priming on seed germination of pigeon pea under cadmium stress. An Acad Bras Cienc 87(3):1847–1852
Suzuki N, Koussevitzky SHAI, Mittler RON, Miller GAD (2012) ROS and redox signalling in the response of plants to abiotic stress. Plant Cell Environ 35(2):259–270
Tan W, Liu J, Dai T, Jing Q, Cao W, Jiang D (2008) Alterations in photosynthesis and antioxidant enzyme activity in winter wheat subjected to post-anthesis water-logging. Photosynthetica 46:21–27
Vijayaraghavan H (1999) Effect of seed treatment with plant growth regulators on bhendi (*Abelmoschus esculentus* L.) grown under sodic soil conditions. Madras Agric J 86:247–249

Watkins JM, Hechler PJ, Muday GK (2014) Ethylene-induced flavonol accumulation in guard cells suppresses reactive oxygen species and moderates stomatal aperture. Plant Physiol 164:1707–1717

Xu S, Hu J, Li Y, Ma W, Zheng Y, Zhu S (2011) Chilling tolerance in Nicotiana tabacum induced by seed priming with putrescine. Plant Growth Regul 63:279–290

Yong-Ping G, Bonham-Smith PC, Gusta LV (2002) The role of peroxiredoxin antioxidant and calmodulin in ABA-primed seeds of Brassica napus exposed to abiotic stresses during germination. J Plant Physiol 159(9):951

Yordanov I, Velikova V, Tsonev T (2000) Plant responses to drought, acclimation, and stress tolerance. Photosynthetica 38:171–186

Younesi O, Moradi A (2014) Effect of priming of seeds of Medicago sativa 'bami' with gibberellic acid on germination, seedlings growth and antioxidant enzymes activity under salinity stress. J Hort Res 22(2):167–174

Zhang S, Hu J, Zhang Y, Xie XJ, Knapp A (2007) Seed priming with brassinolide improves lucerne (*Medicago sativa* L.) seed germination and seedling growth in relation to physiological changes under salinity stress. Aust J Agric Res 58(8):811–815

Zhao C, Liu B, Piao S, Wang X, Lobell DB, Huang Y, Huang M, Yao Y, Bassu S, Ciais P, Durand JL (2017) Temperature increase reduces global yields of major crops in four independent 507 estimates. Proc Nat Acad Sci USA 114:9326–9331

Zheng M, Tao Y, Hussain S, Jiang Q, Peng S, Huang J, Cui K, Nie L (2016) Seed priming in dry direct-seeded rice: consequences for emergence, seedling growth and associated metabolic events under drought stress. Plant Growth Regul 78:167–178

Zhu JK (2016) Abiotic stress signaling and responses in plants. Cell 167(2):313–324

# Abiotic Stress Tolerance in Plants by Priming and Pretreatments with Ascorbic Acid

**Sajid Ali, Aamir Nawaz, Sajjad Hussain, Samiya Mahmood Khan, Shaghef Ejaz, and Shakeel Ahmad**

## Contents



S. Ali (✉) · A. Nawaz · S. Hussain · S. Ejaz
Department of Horticulture, Bahauddin Zakariya University, Multan, Punjab, Pakistan
e-mail: sajidali@bzu.edu.pk

S. M. Khan
Department of Plant Pathology, Bahauddin Zakariya University, Multan, Pakistan

S. Ahmad (✉)
Department of Agronomy, Bahauddin Zakariya University, Multan, Punjab, Pakistan
e-mail: shakeelahmad@bzu.edu.pk



M. Hasanuzzaman, V. Fotopoulos (eds.), *Priming and Pretreatment of Seeds and Seedlings*, https://doi.org/10.1007/978-981-13-8625-1_23

**Abstract** Ascorbic acid (AA) has been found to play several imperative functions in the regulation of various responses of plants or tolerance to different abiotic stresses. The functions of the exogenous application of AA as a pretreatment or priming are well recognized ranging from the germination to accumulation of biomass or yield production under different types of abiotic stress conditions. AA acts as a signaling molecule, positively influencing vegetative growth and development. It also plays a critical role in oxidative stress amelioration by activating different enzymatic and non-enzymatic antioxidants that eventually helps in scavenging of various reactive oxygen species (ROS) produced during abiotic stresses. AA application ameliorates photosynthesis stress inhibition, impedes chlorophyll loss and leaf senescence, and conserves the integrity of cell membranes under abiotic stresses. In addition, it helps to maintain the ionic homeostasis and interacts with certain other phytohormones and metabolites during stress adaptation and regulation of plant growth under abiotic stresses. The present chapter provides a comprehensive overview about the role of AA in various biochemical, physiological, and molecular responses of plants in response to its pretreatments and short gun priming tool under abiotic stressful conditions. The upcoming research in the near future exploring proteomic, transcriptomic, and metabolic profiling would ensure advance insights regulating the possible mechanistic role of AA in the improvement of growth and amelioration of abiotic stresses of plants.

**Keywords** Ascorbic acid · Antioxidative system · Abiotic stresses · Photosynthesis · Reactive oxygen species · Signaling molecule

## Abbreviations

| | |
|---|---|
| AA | Ascorbic acid (ascorbate) |
| AAO | Ascorbic acid oxidase |
| ABA | Abscisic acid |
| APX | Ascorbate peroxidase |
| CAT | Catalase |
| Cu | Copper |
| Cd | Cadmium |
| DHA | Dehydroascorbic acid |
| DHAR | Dehydroascorbate reductase |
| EPR | Electron paramagnetic resonance spectroscopy |
| Fe | Iron |
| GSH | Glutathione |
| GR | Glutathione reductase |
| GSSG | Glutathione disulfide |
| GPX | Glutathione peroxidase |
| ET | Ethylene |

| | |
|---|---|
| $H_2O_2$ | Hydrogen peroxide |
| JA | Jasmonic acid |
| GA | Gibberellins |
| MAPK | Mitogen-activated protein kinase |
| MDHAR | Monodehydroascorbate reductase |
| MDA | Malondialdehyde |
| Mn | Manganese |
| NADPH | Nicotinamide adenine dinucleotide phosphate |
| Ni | Nickel |
| Pb | Lead |
| POD | Peroxidase |
| ROS | Reactive oxygen species |
| SA | Salicylic acid |
| SOD | Superoxide dismutase |
| sAPX | Stromal ascorbate peroxidase |
| tAPX | Thylakoid-bounded ascorbate peroxidase |
| Zn | Zinc |

## 1 Introduction

Different unfavorable environmental factors such as drought, heat, high temperature, cold, hypoxia, and saline stresses lead to a number of detrimental effects and cause significant economic losses to various crops around the world (Huang et al. 2013). All of the said abiotic stresses result in numerous complex responses at cellular level of the plants, leading to the overproduction of a number of unwanted ROS including hydroperoxyl, hydrogen peroxide, singlet oxygen, and superoxide anion radicals (Wang et al. 2003; Huang et al. 2013).

Excessive ROS production in the cells of plants has a general tendency to react with oxidation of protein contents, membrane lipids, and nucleic acids ultimately causing cellular damages subsequently leading to limited vegetative growth and biomass production of crop plants (Wang et al. 2003; Sharma et al. 2012). So, it is very critical to combat against these types of stresses to protect and sustain life in the plants. For this type of protection against the adverse and detrimental conditions, different plants have developed various defense mechanisms at cellular level. These include glutathione, tocopherols, and ascorbic acid. These are considered as effective ROS-scavenging non-enzymatic and enzymatic antioxidants (Veljović-Jovanović et al. 2017). The enzymatic antioxidants are ascorbate peroxidase (APX), catalase (CAT), and superoxide dismutase (SOD) (Inzé and Van Montagu 1995).

Among the antioxidants of plants, AA is known as one of the leading antioxidant compounds playing a critical function in excessive ROS mitigation through the action of non-enzymatic and enzymatic detoxification cascades (Mittler 2002; Veljović-Jovanović et al. 2017). Ascorbic acid also works as a signaling molecule in several cellular activities such as cell expansion, cell wall growth, and cell division

(Conklin and Barth 2004; Zhang et al. 2007). It also works as a key cofactor for numerous enzymes including 2-oxoacid-dependent dioxygenases (GA and ABA biosynthesis), 1-aminocyclopropane-1-carboxylic acid, as well as oxidase for ethylene biosynthesis (Davey et al. 2000; Smirnoff 2000). So, AA is an important antioxidant with signaling role and has enormous potential to alleviate various abiotic stresses in numerous crops of economic importance under global perspectives.

Plants with lower AA biosynthesis are considered potentially sensitive to certain types of environmental stresses influencing their overall growth and developmental phases (Gao and Zhang 2008). It has also been noted that ascorbic acid as a pretreatment or priming plays an imperative function in guarding the plants against numerous types of abiotic stresses including salinity (Venkatesh et al. 2012), low/high temperatures (Kwon et al. 2003; Larkindale et al. 2005), ozone (Feng et al. 2010), drought (Hemavathi et al. 2011), and increased light intensity (Talla et al. 2011). Besides pretreatment and priming, endogenous increased concentrations of AA in transgenic and/or mutants play a vital function in growth, developmental processes, and abiotic stress alleviation in various plants (Venkatesh and Park 2014). So, in the current chapter, the role of exogenous AA pretreatments and priming for alleviation of various abiotic stresses has been discussed.

## 2 Biosynthesis and Regulation of Ascorbic Acid

Plants being sessile have the ability to accumulate relatively higher levels of AA compared to animal cells (Foyer and Shigeoka 2011). Accumulation of AA in rapidly growing leaves might exceed 10% of the soluble carbohydrates. Initial work on AA biosynthetic pathway was carried out in the 1950s which advanced our understandings about carbon metabolisms in different plant species (Smirnoff and Wheeler 2000; Suekawa et al. 2017). Nevertheless, a detailed biosynthetic pathway was elucidated until 1999 by using various biochemical and genetics-related approaches (Conklin et al. 1999). Almost two decades ago, scientist characterized three AA biosynthetic pathways in plants with few exceptions in animals and some species of algae (Ishikawa and Shigeoka 2008).

The vast majority of animals except humans have the ability to synthesize AA in their bodies through a series of enzymatic reactions derived from UDP-glucose 6-phosphate. Similarly, a larger part of biosynthesis of AA is accomplished through two intricate and branchy pathways in plants, i.e., D-mannose pathway and D-galacturonic acid pathway (Nishikimi and Yagi 1996; Suekawa et al. 2017). The former pathway has a dominant role in leaf tissues, and the latter is operated in developing fruits. Following the D-glucuronate pathway, stereochemistry of D-glucuronate is inverted to AA through different carbon skeleton substrates including L-gulonate, D-glucuronate, and L-gulono-1,4-lactone (Nishikimi and Yagi 1996; Suekawa et al. 2017).

Although the counter-equilibrium between biosynthesis and degradations affects intracellular concentrations of AA, some plant tissues have much higher turnover rate of AA due to its multiple functional abilities (Pallanca and Smirnoff 2000; Suekawa et al. 2017). AA is not only limited to catalytic reactions (oxidation or reduction), but it is also itself an unstable metabolite that disturbs the balance of AA to dehydroascorbic acid (DHA). It has been demonstrated in several findings that AA also acts as a precursor of other metabolites including oxalates, L-tartrate, and L-threonate in various crop plants (Bánhegyi and Loewus 2004; Melino et al. 2009).

Structural cleavages of AA through a number of reduction, delactonization, and oxidation reactions at 2-, 3- or 3-, 6-carbon position atoms result in production of other useful metabolites. Cleavage at 1-, 2-carbon position yield oxalic acid, and 3-, 6-carbon atoms produce L-threonine, which is further oxidized to produce L-tartrate (Debolt et al. 2007). Active degradation of AA in different plant species has been elucidated to take place via the production of DHA, L-threonate, and oxalate (Green and Fry 2005). Apparently AA's cleavage pathways are plant species specific; hence, the comprehensive detail of degradation pathways remains to be elucidated until now (Bánhegyi and Loewus 2004; Suekawa et al. 2017).

Diverse plant species have different turnover rate of AA which is also supposed to be tissue specific. Turnover of AA in mature leaves of *Arabidopsis* is about 2.5% of the pool hour$^{-1}$, while the degradation rate has been demonstrated to be as high as 13% hour$^{-1}$ in developing axes of pea seedlings (Conklin et al. 1997; Melino et al. 2009). In addition to this, the degradation rate of AA was low at 1.41% hour$^{-1}$ of total concentration in the early ripening stages of black currant (Hancock et al. 2007). However, transformation rates of AA toward tartrate or oxalate accumulation in grape berries (and other related plants) are yet to be elucidated (Melino et al. 2009). In plants, the significance of L-tartrate accumulation is still unclear since its biosynthesis has not been demonstrated in a large variety of plants and needs further elucidation.

## 3 Ascorbic Acid-Mediated Functions in Growth and Development During Abiotic Stress

AA is known as a multifunctional molecule considered imperative in one or other way for the growth of plants under abiotic stress conditions. AA generally performs a number of critical functions as it is known as the leading cofactor as well as redox buffer and has been found to be involved in a wide variety of important processes such as respiration, photosynthesis, biosynthesis, and regeneration of certain antioxidants (Fig. 1) (Ortiz-Espín et al. 2017). In addition, AA influences cell cycle progression that importantly takes part in certain signal transduction cascades. Its role is also critical because it is important for the growth and development processes of tissues, embryogenesis, seed germination, and cell proliferation. Moreover, it is also imperative for the development of shoots, roots, and meristems and subsequent flowering of the plants (Ortiz-Espín et al. 2017). AA promoted root cell

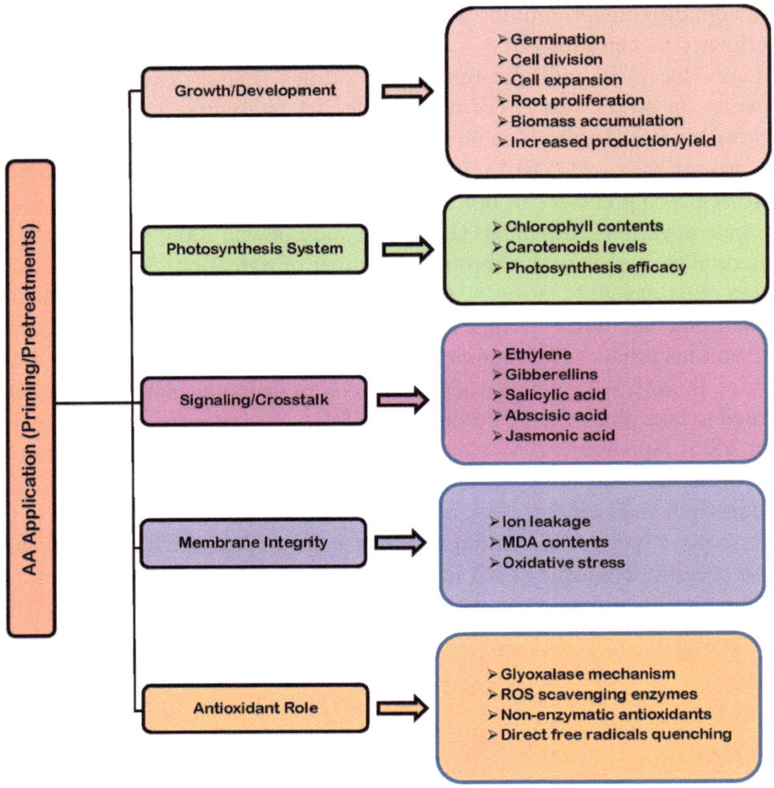


**Fig. 1** Physiological roles of ascorbic acid in different plants under abiotic stress

differentiation in onion by modulating division rate and larger phase of G1 (Liso et al. 1988). On the contrary, AA synthesis inhibition by the application of lycoporine resulted in blocking of cell cycles in G2 and G1 phase, specifying an important function of AA in the progression of cell cycles via G2/M and G1/S transitions (Innocenti et al. 1990). In the same way, similar changes were noted in tobacco BY-2 cell cycles (De Pinto et al. 2000), and pea root cells (Citterio et al. 1994). AA also plays a critical function in zygotic embryogenesis, root/shoot pole generation, and embryonic/cotyledon axis (West and Harada 1993). It directly or indirectly regulated auxiliary bud formation and apical dominance in tomato (Zhang 2012). It also takes part in certain critical aspects of seed germination during abiotic stresses. AA acts as a seed germination inductor under saline stress. It participates in a network of complex metabolic activities by integrating hormones, ROS, redox proteins, and antioxidant concentrations and modulations that eventually regulate the seed germination or developmental processes (Ortiz-Espín et al. 2017). The potential role of AA in shoot/root development has also been observed in onion roots. It regulates different phases of cell cycle that ultimately affect shoot/root development (Liso et al. 1988; Orman-Ligeza et al. 2016). More importantly, the potential part of

AA in the process of flowering has also been observed. It participates in various major cofactors in the biosynthesis of different important hormones that subsequently induce flowering in plants (Barth et al. 2006; Clifton et al. 2006).

## 4 Ascorbic Acid Signaling Under Abiotic Stress

Higher accumulation of ROS causes oxidative damage; however, lower ROS level activates distinctive signaling pathway by oxidation of specific metabolites such as signaling proteins. Posttranslational modifications of some proteins affect its structural solidity, catalytic activities, and interactions with certain important molecules (Dietz 2014, 2016; Mock and Dietz 2016). $H_2O_2$ modulates the transcriptional level of nuclear genes by its retrograde signaling mechanism (Bienert and Chaumont 2014). The tuning of the targeted signaling proteins is achieved through regulation of the spatial distribution and levels of primary as well as secondary oxidants that are putatively associated with redox regulations of thiol group of active proteins and nonproteins during stressful conditions (Dietz 2016). It has been suggested that AA coupled with ascorbate peroxidase (APX) reduced the contents of $H_2O_2$ inside the chloroplast. However, oxidized DHA drained electron via dehydroascorbate reductase (DHAR) of thiol–disulfide redox regulation networks. Recent studies have suggested that DHAR in the cytosol has a significant role for diverting the hydrogen peroxide ($H_2O_2$) metabolism toward oxidation of glutathione (GSH). The oxidized form of 2-CysPRXs and GSH/GSSG transmits further signaling toward transcriptional activity of the nucleus (Mock and Dietz 2016; Rahantaniaina et al. 2017). The 2-CysPRXs are known as the oxidized form which acts as redox sensors that are further reduced by thioredoxins and glutaredoxins. It has been demonstrated that 2-CysPRXs are involved in various metabolic processes inside the chloroplasts such as carbon, nitrogen, or sulfur metabolism, photosynthetic light reactions, certain antioxidant defense enzymes, and biosynthesis of secondary compounds (Liebthal et al. 2017). Levels of 2-CysPRXs significantly enhanced in high light acclimatized double tAPX–sAPX mutants than wild plants (Kangasjärvi et al. 2008). Awad et al. (2015) also observed that during high light stress in respective mutants, 2-CysPRX accumulation was more prevalent compared to tAPX. Exogenous application of AA as a pretreatment or priming reduces the transcriptional levels of PRX genes in chloroplast; however, the redox state of AA is an essential signal for upregulations of PRX gene expressions (Horling et al. 2003). So, AA acts as a repressor, and malondialdehyde (MDA) alone or coupled with DHA acts as a stimulator of PRX gene expression in the chloroplast under abiotic stress. Chloroplastic APX has a tendency to become inactivate in the deficiency of reduced AA; that can be considered as its signaling function (Exposito-Rodriguez et al. 2017).

The loosening of cell walls during drought, osmotic, and salinity stress is a vital phenomenon for cellular expansion (Tenhaken 2014). Accumulation of DHA is mainly pertinent in the loosening and reorganization of cell wall expansions

(Kukavica et al. 2008). DHA interacts with the side chain of cell wall protein arginine and lysine residues and prevents their bonding with polysaccharides and pectins. Apoplastic DHA is degraded to oxalates, L-threonate, and L-tartrate via multiple intermediates. These intermediates have a vital role in growth and developmental cascades of plants particularly under stress (Parsons and Fry 2012). Cell wall loosening is achieved by reducing pectin cross-linking and extraction of calcium from calcium–pectin complexes with the help of oxalate (Hocking et al. 2016). Conversely, it has been observed that an alternate mechanism for productions of hydroxyl radical in the AA reliant pathway in the presence of $Cu^{2+}$ or generation from $H_2O_2$ during degradation of AA and DHA exists in the cell wall (Fry 1998; Kärkönen and Fry 2006). The hydroxyl radical results in structural cleaving of xyloglucans and pectins in cell walls (Spasojević and Bogdanović-Pristov 2010). Ascorbic acid oxidase (AAO) has been found to be associated with active degradations of auxins that are imperative growth regulators mainly pertinent to respond to osmotic stresses levied by low temperature, drought, and salinity (Pignocchi et al. 2003). Among different plant growth regulators associated with protection against abiotic stresses, jasmonic acid (JA) is a positive regulator, while abscisic acid (ABA) and salicylic acid (SA) negatively regulate AAO expression levels (Sanmartin et al. 2007). In addition to this, AAO overexpressing tobacco contains improved levels of DHA than wild-type plants. Enhanced levels of DHA were found to be responsible for improved ABA and $H_2O_2$ levels, stomatal movement, and guard cell signaling through reversible redox regulations of particular proteins (Fotopoulos et al. 2008).

## 5 Mechanism of Ascorbic Acid-Induced Stress Tolerance

Plants being sessile are extensively exposed to hostile environmental stresses such as drought, salinity, heavy metal excess, high solar irradiation, heat, and nutrient deficiency. The significant effects of these factors are interrelated for plant and result in disturbance in redox homeostasis and reduced capacity of carbon fixation in photosynthesis and consequently stunt the growth magnitude (Cramer et al. 2011; Zhu 2017). During adverse situation, increased excitation pressure of free electrons in chloroplast surpasses the limit of cellular antioxidant capacity which causes oxidative damage of other organelles. The AA level fluctuates in plant cells due to various factors including diurnal rhythms, developmental stages, and light, and the redox status of AA is associated with the redox homeostasis of cells (Akram et al. 2017).

AA present in cytosol, peroxisome, and chloroplast has a key role in scavenging $H_2O_2$ through APX, and it is proficiently recycled by either ascorbate–glutathione cycle or by reductive activity of MDHAR directly (Akram et al. 2017). Co-localization of AA in apoplastic and vacuolar spaces advocates that these are major reducing agents of free radicals generated under oxidative stress. In addition to their antioxidative role, AA has a well-sophisticated role orchestrated with plant response to hostile environmental conditions such as regulation of enzymatic activities, biosynthesis of phytohormones, modulation of stress-related gene expressions, performing multiple tasks during redox signaling, and maintaining growth attributes (Fig. 1) (Akram et al. 2017). The magnitude of AA activity is strictly congruent to its distribution in cellular compartments. Hence, it is important to know about cellular distribution of ascorbate, which would determine the impact of signaling regulation for quenching ROS (Akram et al. 2017).

Climate change has been considered a major threat to the plants owing to abrupt and extreme heat shocks, floods, drought, and availability of energy resources. Plants employ a diverse array of adaptive feedback mechanisms that enable them to maintain phenotypic plasticity equilibrium against a continuously fluctuating hostile environment (Sultan 2000; Nicotra et al. 2010). The regular fluctuations in the environment bring different organisms close to hostile fluctuation turn into multiple stressors and thus act as determinants of plant performance and productivity of certain crops. Hence, a detailed understanding of stress magnitude and tolerance mechanism is the key for climate change predictions on different crop species (Koyro et al. 2012). Accumulation of ROS is a fundamental aspect of different abiotic stress types which disturb redox balance. If the magnitude of stress exceeds the antioxidation system and repairs capacity of any cell, the resulting accrued ROS leads to oxidative damages (Sharma et al. 2012). ROS accumulation is a hallmark of different molecular responses of plants to a wide range of hostile (heat, excess light, drought, heavy metal, chilling, and UV radiations) environmental conditions. Exogenous applied AA in the form of pretreatments or seed priming plays a critical role in ameliorating different abiotic stresses (Hamama and Murniati 2010; Penella et al. 2017). Overall, AA alleviates different abiotic stresses by playing an important role as a ROS scavenger (Fig. 2).

## 6 Ascorbic Acid–Glutathione Cycle and Abiotic Stress

Plants face continuous changes including biotic and abiotic stresses in the environment. The normal changes in temperature, irradiation, and other factors that influence various plant processes are respiration or photosynthesis (Miller et al. 2010). Increased levels of temperature and light may lead to over production of ROS that are considered very detrimental for the crop plants. So, some antioxidant defense system is required to combat against the said ROS. The non-enzymatic antioxidant includes AA and GSH (Noctor et al. 2014). Drought is considered highly destructive for plants. During drought conditions, closure of stomata is considered imperative to reduce loss of water through transpiration. AA–GSH takes part in alleviating ROS-induced damage during drought conditions (Fig. 2). The leaves of the wheat cultivars that had higher AA levels exhibited reduced oxidative damage compared to lower levels under drought conditions. Similar response was observed in salinity stress. The levels of AA–GSH and NADPH oxidase may be increased by exogenous applications of $H_2O_2$ that in turn scavenge harmful free radicals during salinity

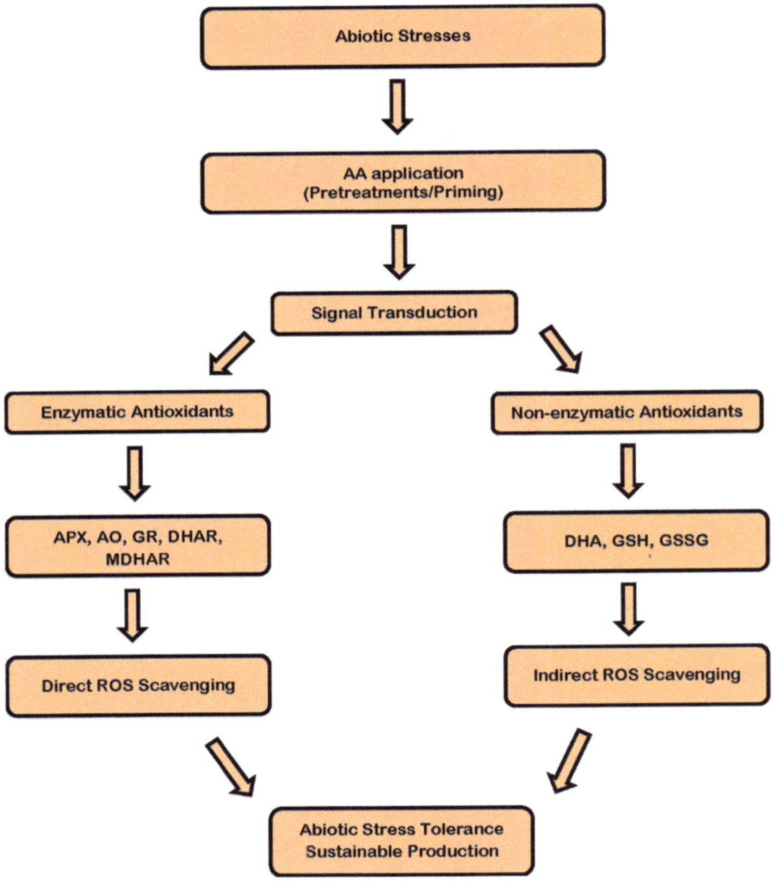


**Fig. 2** Generalized mechanism of ascorbic acid-induced abiotic stress tolerance in plants

stress (Schmidt et al. 2013). The plants lacking AA or APX enzyme are highly susceptible to photooxidation-induced damages. Moreover, high light intensity also leads to photosynthesis inhibition. It was observed that plants of tomato with lower AA concentration showed significant oxidative damage during higher light conditions. So, adequate AA–GSH level and higher potential to express APX enzyme are important to alleviate higher light stress in tomato plants (Baldet et al. 2013). In the same way, higher AA–GSH as well as APX enzyme is important to ameliorate oxidative stress during heat/high temperature and chilling stress as noted in spinach (Gómez et al. 2008; Duan et al. 2012).

## 7 Hydrogen Peroxide Scavenging and Ascorbic Acid Recycling During Abiotic Stress

ROS ($H_2O_2$) is produced as a result of numerous catabolic reactions; however, two key phenomena in plant cells, i.e., photosynthesis and photorespiration, during high temperature or light, drought, and chilling stress have the highest capacity to produce excessive ROS. During various abiotic stresses, ROS level is elevated in plant tissues (Cheeseman 2006; Slesak et al. 2007). It has also been observed that under abiotic stress, $H_2O_2$ accumulation in chloroplast chamber would also result in the disturbance of peroxidase, RuBisCO, Cu/Zn superoxide dismutase (SOD), and various thiol-sensitive enzymes of Calvin cycle (Weisiger and Fridovich 1973; Cheeseman 2007). Generally, it has been reported that the reduced state of AA, which is extracted by standard procedures from plant tissues, represents that the plant is healthy. The deviation of 0.90–0.95 in ratio of reduced AA to total AA indicates that ROS accumulation is outside of the antioxidative activity of the plant cells (Foyer and Noctor 2011). The total content of leaf AA is contingent on the plant growth and nutritional status, quality and quantity of light, diurnal rhythm, and rate of oxidative damage under environmental condition (Conklin and Barth 2004; Majer et al. 2016). APX, due to its high affinity for ROS, plays a crucial role for quenching $H_2O_2$ located in chloroplasts and utilizes AA as donor of electrons (Sharma and Dubey 2004). In addition to stromal ascorbate peroxidase (sAPX) and thylakoid-bounded ascorbate peroxidase (tAPX), $H_2O_2$ content in chloroplast is mainly controlled by peroxidase and glutathione peroxidases (GPXs). Similarly, $H_2O_2$ is scavenged by catalase in peroxisome, and $H_2O_2$ accumulation in apoplast, vacuoles, and cell wall is detoxified by class III peroxidases.

## 8 Free Radical Scavenging and Ascorbic Acid Under Abiotic Stress

Several environmental stressors such as salinity, water stress, heavy metals, ozone ($O_3$), and UV radiations contribute to disturbing the homeostatic balance in extracellular spaces. However, AA accumulation in apoplastic spaces act as the first antioxidative step to alleviate abiotic stress due to the said causes (Kangasjärvi and Kangasjärvi 2014). The major reducing agent in the extracellular compartment is AA whose redox state is mediated by AAO (De Tullio et al. 2013). In addition to this, AA plays a vibrant role as a reductant of $H_2O_2$ in a class III peroxidase/phenolic/AA system in apoplastic and vacuolar spaces (Takahama and Oniki 1997). During hydroxylic POD cycle in apoplast, the oxidation of ascorbate occurs during quenching of hydroxyl radicals (•OH) produced at the cell wall (Chen and Schopfer 1999). Exogenous application of AA enhanced the scavenging capacity of •OH (Veljović-Jovanović et al. 2005). Addition of AA significantly decreased the electron paramagnetic resonance signals due to the presence of •OH and quinhydrone in cell walls of pea roots (Kukavica et al. 2008). These results suggest that the redox state of AA pool in apoplast can also be a regulator of ${}^{\bullet}OH$, and consequently loosening or stiffening of plant cell wall may occur (Morina et al. 2010). It has been demonstrated by EPR analysis that exogenous treatment of excessive $Zn^{2+}$ and $Cu^{2+}$ leads to the accumulation of carbon-centered (${}^{\bullet}CH_3$) and free radicals (${}^{\bullet}OH$) in the cell wall of *Verbascum thapsus* plant's root which increase DHA content and downregulate the AA content (Morina et al. 2010). MDA and $H_2O_2$ are further metabolized by other enzymatic reactions which consequently regenerate reduced AA in the AA–GSH cycle. It has also been revealed that NADPH is seldom available in apoplastic tissues; hence, recycling of AA–GSH must take place in cytosol. AA degeneration into MDA in apoplast took place by the action of class III peroxidase/phenolic/AA systems (Takahama and Oniki 1992). Furthermore, generation of apoplastic MDA in the closest proximity of plasma membranes accepts electrons from the membranes enclosed cytochrome b and recovered as AA and DHA by metabolic activities of MDAR (Drazkiewicz et al. 2003). The obtained DHA is subsequently shifted toward cytosol where it is reduced to AA in AA–GSH cycles and scavenges free radicals (Horemans et al. 2000).

# 9 Interaction of Ascorbic Acid with Other Phytohormones Under Abiotic Stress

The levels of AA regulation can be witnessed in response to different abiotic stresses. In addition, different types of environmental stresses can also trigger various hormones including ethylene (ET), salicylic acid (SA), abscisic acid (ABA), gibberellins (GA), and jasmonic acid (JA) other than AA (Akram et al. 2017). AA cross talks with the abovementioned hormones in one or other way under abiotic stresses. The up- and/or downregulations of these hormones depend upon several factors (Fig. 1). For this purpose, one of the leading regulatory steps is protein interaction, transcription, and damage of the targeted proteins. In different plant species, mitogen-activated protein kinase (MAPK)-associated cascades play an imperative function in different abiotic stressful responses as well as responses of the phytohormones that consist of signaling of ROS molecules (Fujita et al. 2006). It is now a well-established fact that cross talks between different other hormones and AA do exist. The major signaling molecules integrated in abiotic stress regulation and environment of the plants include ABA and GA. GA biosynthesis regulated GA-20-oxidase needs AA for its characteristic activities (Coles et al. 1999). Furthermore, higher concentrations of ABA in vtc-1 apparently occurred due to AA up-relation needing biosynthetic enzyme of ABA such as NCED dioxygenase that might lead to late flowering phenotypes during short daytime periods. However, at the same time, this evidence appears to be slightly contradictory as well as supportive because the biosynthetic pathways of ABA need AA. In the same way, endogenous concentrations of AA positively influence various hormones and hormone-dependent signal transduction cascades under abiotic stresses (Pastori et al. 2003). For instance, the level of AA in tomato plant was noted to be positively associated with activation of the specific genes correlated with signaling of hormone dependent on oxidative stresses (Lima-Silva et al. 2012). It was also noted that exogenous AA treatment in combination with citric acid led to increased level of indole acetic acid, GA, brassinosteroids, and zeatin along with decreased contents of ABA in wheat under saline stress (Sadak et al. 2013). Moreover, biosynthesis regulation of ABA has also been found to be associated with redox signals together with DHA and AA level in plants (Chen et al. 2014). Likewise, AA is found to be associated with floral induction in indirect manner owing to its interface with GA and ABA biosynthesis (Barth et al. 2006). It has also been observed that GA enhances flowering (Xu et al. 2015); reduces senescence (Currey et al. 2013); plausibly antagonizes influence of JA, SA, and ABA; and exhibits a complicated interaction among various other hormones to intervene with the critical signaling cascades that might be a proper network for alleviation of certain abiotic stresses (He et al. 2001; Barth et al. 2006; Akram et al. 2017).

## 10 Ascorbic Acid and Drought Stress

### *10.1 Increased Endogenous Ascorbic Acid and Drought Stress*

Physiological functions such as stomatal closure in leaves, apoplastic fluctuations, and maintenance of xylem pH are regulated by AA-dependent ABA signaling cascades (Tombesi et al. 2015; Gloser et al. 2016). Upregulation of NCED3 gene enhances the biosynthesis of ABA during drought stress. Water deficit immediately triggered the increased level of ABA (McAdam and Brodribb 2016). Moreover, drought improved the circulation of ABA by de-conjugating ABA-glucose ester which is deposited in vacuole (Seiler et al. 2011). In addition to stomatal closures during drought stress, other physiological changes included the retarded growth, lignification, changes in antioxidant pool, accumulation of osmatic intermediates, and senescence induction cascade (Chaves et al. 2003). Pyngrope et al. (2013) reported two rice varieties including drought-tolerant (Dt) and drought-sensitive (Ds) cultivars and found that Ds cultivar downregulated AA and enhanced DHA. The authors concluded that Dt cultivar was more efficient in terms of maintenance of redox state during oxidative stress by water deficiency. The results were also consistent with lower $H_2O_2$ contents in the young seedling of Dt cultivar followed by exogenous treatment of glycol along with increased AA concentration. Several studies stressed on the significance of cytosolic isoforms of Cu/ZnSOD and APX enzymes in plants during water deficiency (Mittler and Zilinskas 1994; Sofo et al. 2005). Transgenic tobaccos (having enhanced endogenous AA) with duplicated cytosolic copies of APX as well as single copy of Cu/Zn SOD and APX in cytosol showed relatively higher tolerance to mild water deficiency stress than wild type (Faize et al. 2015). Lines of transgenic origin were less vulnerable to oxidative damage due to water deficiency due to less $H_2O_2$ content, MDA and electrolyte leakage, and higher water use efficiency along with enhanced growth. As the drought proceeded during the course of time, APX activity decreased initially and began to enhance with the passage of time in soybean young seedling roots (Kausar et al. 2012). Likewise, in four *Prunus* hybrids, after 3 weeks of commencement of drought stress, the activity of MDAR, APX, and DHAR enhanced as that of DHA and AA accumulation and continued to enhance unlike that of well-watered plants. The putative increment in GR activity was delayed which was consistent with accumulation of GSH content unlike to other antioxidant enzymes. The $H_2O_2$ level continually enhanced remarkably in drought treatments for up to 70 days, and after re-watering of plants, the AA–GSH cycle downregulated owing to reduced accretion of $H_2O_2$ and putative requirements for its decontamination (Sofo et al. 2005).

## 10.2 Exogenous Ascorbic Acid Application and Drought Stress

Drought is a very critical stress that negatively affects growth and productivity of crops. Its negative effects can be alleviated with exogenous AA pretreatments or seed priming. It has been observed that priming of sorghum seeds with AA showed increased fresh and dry weight and nodes per stem with improved vegetative growth under drought stress conditions (Arefi et al. 2016). The foliar pretreatment of young peach trees with 250 mg $L^{-1}$ AA resulted in increased $CO_2$ assimilation, stomatal conductance, ostiole length, proline contents, and reduced lipid peroxidation under water stress conditions (Penella et al. 2017). The foliar pretreatment with 300 mg $L^{-1}$ increased soluble proteins, cell membrane stability, and endogenous AA and APX activity with decreased lipid peroxidation in grapes under drought stress (Table 1) (Zonouri et al. 2014). Seed priming of maize with 55 mM AA showed increased germination speed, germination percentage, vigor index, root length, shoot length, seedling height, leaf number, and leaf area during drought stress (Hamama and Murniati 2010). Foliar pretreatment of AA in combination with 24-epibrassinolide showed increased oil content, reduced relative water content, enhanced antioxidative enzymes activities, and declined membrane leakage and $H_2O_2$ and MDA concentration of marigold under water stress (Hemmati et al. 2018). Seed pretreatment of cauliflower led to increased endogenous AA concentration, leaf proline, phenolics, root fresh and dry weight, chlorophyll, and root/shoot P and $K^+$ along with higher antioxidant enzyme (Latif et al. 2016). Priming of wheat seeds with AA resulted in increased shoot length, glycinebetaine concentration, and activities of GR, SOD, APX, CAT, and MDHAR with reduced MDA contents during drought conditions (Singh and Bhardwaj 2016). In the same way, priming of caper (*Capparis spinosa*) with 100 mg $L^{-1}$ AA showed increased germination percentage, germination rate, seedling length, root length, shoot length, and vigor index under drought stress (Heydariyan et al. 2014). Seed priming of wheat with AA resulted in higher chlorophyll, photosynthesis, and overall growth under drought stress (Malik and Ashraf 2012). In the same way, AA seed priming also showed higher chlorophyll, leaf emergence, specific leaf area, leaf area, seedling dry weight, and root length

**Table 1** Effect of ascorbic acid pretreatments on abiotic stress tolerance in plants

| Crop | Concentration | Stress type | Inferences | References |
|---|---|---|---|---|
| Pumpkin | 30 mg $L^{-1}$ | Salinity | Reduced salt stress severity, increased seedling fresh weight, protein concentration, and overall growth | Rafique et al. (2011) |
| Sugarcane | 0.5 mM | Salinity | Increased root length, soluble protein, and antioxidant enzyme activities under in vitro saline condition with reduced yellowing | Munir and Aftab (2011) |
| Rice | 5 mM | High temperature | Ameliorated high temperature stress by increasing potassium ions and antioxidant enzymes (CAT, POD, SOD) activities and lowering lipid peroxidation | Shah et al. (2011) |
| Pistachio | 0.5–1 mM | Salinity | Increased chlorophyll, photosynthesis capacity, and leaf proline and reduced electrolyte leakage | Bastam et al. (2013) |
| Common bean | 1 mM | Salinity | Alleviated oxidative damage, increased growth attributes, photosynthetic pigments, and lipid peroxidation | Saeidi-Sar et al. (2013) |
| Tomato | 50 mg $L^{-1}$ | Salinity | Increased germination rate, fresh and dry mass, leaves per plant, leaf area, chloroplast efficacy, photosynthesis pigments, inorganic minerals, and antioxidant enzyme activities | El-Sayed and El-Sayed (2013) |
| Maize | 40 mg $L^{-1}$ | Chilling/ cold | Increased growth, chlorophylls, leaf relative water, and membrane stability and enhanced enzymatic activities | Ahmad et al. (2014) |
| Strawberry | 3 mM | Heat | Increased AA contents and chlorophyll and reduced loss of turgidity | Ergin et al. (2014) |
| Cucumber | 750 µM | Chilling/ cold | Mitigated chilling stress, reduced oxidative damage, membrane leakage, and MDA production | Lukatkin and Anjum (2014) |
| Peach | 250 ppm | Drought | Increased $CO_2$ assimilation, stomatal conductance, ostiole length, and proline contents and reduced lipid peroxidation | Penella et al. (2017) |
| Pepino | 100 mM | Chilling | Improved AA, chlorophylls, carotenoids, total phenols, and proline. Decreased MDA contents | Sivaci et al. (2014) |
| Grapes | 300 mg $L^{-1}$ | Drought | Increased soluble proteins, cell membrane stability, endogenous AA, APX enzyme activity with decreased lipid peroxidation | Zonouri et al. (2014) |

(continued)

**Table 1** (continued)

| Crop | Concentration | Stress type | Inferences | References |
|---|---|---|---|---|
| Rice | 1 mM | Salinity | Increased $K^+$, chlorophyll, vegetative growth, fresh weight, dry weight, root length, and shoot length | Alhasnawi et al. (2015) |
| Tomato | 1 mM | Salinity | Higher chlorophyll contents, germination percentage, carotenoids, and shoot and root length with lower lipid peroxidation | Krupa-Małkiewicz et al. (2015) |
| Date palm | 100 mg $L^{-1}$ | Salinity | Increased shoot protein, chlorophyll, shoot length, root length, growth, and activity of APX and SOD enzymes | AL-Mayahi (2016) |
| Cauliflower | 75 or 150 mg $L^{-1}$ | Drought | Increased endogenous AA, leaf proline, phenolics, root fresh and dry weight, chlorophyll, root and shoot P and $K^+$ with improved antioxidant enzymes activities | Latif et al. (2016) |
| Rice | 10 mM | $O_3$ | Increased endogenous AA concentration but not alleviated $O_3$-induced photosynthesis inhibition | Kobayakawa and Imai (2017) |
| Marigold | 10 mM | Drought | Increased oil content, reduced relative water content, enhanced antioxidative enzymes activities, reduced membrane leakage, $H_2O_2$ and MDA concentration | Hemmati et al. (2018) |
| Tomato | 4 mM | Salinity | Increased germination, seedling biomass, seedling length, vigor index, and germination speed | Kaur and Gupta (2018) |

*AA* ascorbic acid, *APX* ascorbate peroxidase, *CAT* catalase, $CO_2$ carbon dioxide, $H_2O_2$ hydrogen peroxide, *MDA* malondialdehyde, $O_3$ ozone, *P* phosphorus, *SOD* superoxide dismutase

with reduced oxidative damage in wheat (Table 2) (Farooq et al. 2013). It has also been observed that AA priming also increased germination index, germination percentage, normal seedlings, seedling length, dry weight, and proline contents of sesame (Tabatabaei and Naghibalghora 2013). Similarly, priming of wheat seeds with AA led to increased vegetative growth, cell membrane stability, relative water content, and proline and antioxidant enzyme activity along with significantly suppressed oxidative damage (Table 2) (Malik et al. 2015).

**Table 2** Effect of ascorbic acid seed priming on abiotic stress tolerance in plants

| Crop | Concentration | Stress type | Inferences | References |
|---|---|---|---|---|
| Wheat | 50 mg $L^{-1}$ | Salinity | Reduced leakage rate and germination time, increased growth and biomass accumulation | Afzal et al. (2006) |
| Hot pepper | 50 mg $L^{-1}$ | Salinity | Improved germination, root/shoot length, seedling fresh weight, vigor index, and secondary roots | Amjad et al. (2007) |
| Maize | 55 mM | Drought | Increased germination speed, germination percentage, vigor index, root length, shoot length, seedling height, leaf number, and leaf area | Hamama and Murniati (2010) |
| Wheat | 1 mM | Drought | Higher chlorophyll, photosynthesis, and overall growth | Malik and Ashraf (2012) |
| Broad bean | 100 mg $L^{-1}$ | Salinity | Increased growth, soluble carbohydrates, proline, free amino acids, $K^+/Na^+$ ratio, and photosynthesis pigments | Azooz et al. (2013) |
| Wheat | 2 mM | Drought | Higher chlorophyll, leaf emergence, specific leaf area, leaf area, seedling dry weight, root length with reduced oxidative damage | Farooq et al. (2013) |
| Pumpkin | 500 μM | Salinity | Increased germination, root length, and activity of POD and CAT enzymes | Fazlali et al. (2013) |
| Sesame | 50 ppm | Drought | Increased germination index, germination percentage, normal seedlings, seedling length, dry weight, and proline contents | Tabatabaei and Naghibalghora (2013) |
| Sunflower | 2 mM | Salinity | Increased seed stamina index, germination speed, germination rate, hypocotyl dry weight, relative water content, and radical dry weight and reduced ion leakage | Fatemi (2014) |
| Caper | 100 mg $L^{-1}$ | Drought | Increased germination percentage, germination rate, seedling length, root length, shoot length, and vigor index | Heydariyan et al. (2014) |
| Wheat | 1 mM | Drought | Increased growth, oxidative damage, cell membrane stability, relative water content, and proline and increased antioxidant enzymes activities | Malik et al. (2015) |
| Sorghum | 100 mM | Drought | Increased fresh and dry weight and nodes per stem with improved growth | Arefi et al. (2016) |

(continued)

**Table 2** (continued)

| Crop | Concentration | Stress type | Inferences | References |
|---|---|---|---|---|
| Pea | 50 mg $L^{-1}$ | Salinity | Enhanced chlorophyll, stem length, stomatal conductance, and dry matter of roots | Balliu et al. (2014) |
| Eggplant | 100 or 200 mg $L^{-1}$ | Salinity | Increased germination, dry and fresh weights, seedling growth, carbohydrates, total chlorophyll, and proteins and reduced electrolyte leakage | Jan et al. (2016) |
| Wheat | 400 μM | Drought | Increased shoot length, glycinebetaine concentration, and activities of GR, SOD, APX, CAT, and MDHAR enzymes with reduced MDA contents | Singh and Bhardwaj (2016) |

*CAT* catalase, *POD* peroxidase, *APX* ascorbate peroxidase, *MDA* malondialdehyde, *MDHAR* monodehydroascorbate reductase, *SOD* superoxide dismutase

## 11 Ascorbic Acid and Low Temperature Stress

Chilling or cold stress is considered very detrimental for crop plants (Lukatkin and Anjum 2014). It damages the plants of economic importance by affecting the internal cell structure due to formation of crystals in the cytoplasm. So, it is a great concern with respect to commercial crop production and needs to be managed accordingly. It has been reported that cold or chilling stress can effectively be managed by priming or pretreatment with AA. Pretreatment of cucumber with 750 μM AA in combination with 100 μM glutathione alleviated chilling stress and reduced oxidative damage along with higher membrane integrity (Lukatkin and Anjum 2014). In the same way, priming or foliar pretreatment of maize with 40 mg $L^{-1}$ resulted in increased growth, chlorophylls, leaf relative water, and membrane stability and enhanced enzymatic activities subsequently developing in alleviated clod or low temperature stress under pot and filed conditions, respectively (Tables 1 and 2) (Ahmad et al. 2014). Similarly, exogenous pretreatment of pepino (*Solanum muricatum*) with 100 mM AA resulted in improved endogenous AA, chlorophyll contents, carotenoids, total phenols, and proline concentration along with decreased MDA contents that in turn efficiently ameliorated chilling stress (Sivaci et al. 2014). Besides exogenous pretreatment or priming, transgenic endogenous increased production of AA is also an effective approach to overcome chilling or cold stress in crop plants (Venkatesh and Park 2014). The increased endogenous production of AA showed alleviation of chilling stress in tomato (Wang et al. 2006). Likewise, OsAPX1 overexpression enhanced tolerance of chilling stress in rice (Sato et al., 2011). Similarly, the higher APX activity led to increased chilling tolerance in transgenic tobacco (Sun et al. 2010).

## 12 Ascorbic Acid and High Temperature Stress

High temperature or heat stress has been significantly increased during the recent years in the world. It not only negatively affects growth of crop plants but also reduces anticipated yield, eventually leading to the loss of food security. However, its impact can be alleviated with some exogenous AA pretreatments or priming strategies with varying effects (Tables 1 and 2). It has been reported that pretreatment of strawberry with AA efficiently alleviated the heat stress damage and resulted in increased AA contents and chlorophyll and reduced loss of turgidity (Table 1) (Ergin et al. 2014). Similarly, AA pretreatment also ameliorated high temperature stress by increasing potassium ions and antioxidant enzymes (CAT, POD, and SOD) activities and lowering lipid peroxidation in rice plants (Shah et al. 2011). Besides exogenous priming or pretreatments, overexpression of the endogenous concentrations of AA also helps in the amelioration of high temperature or heat stress in plants. It has been reported that cytosolic APX overexpression enhanced the heat stress tolerance in transgenic tomatoes (Wang et al. 2006). In the same way, overexpression of APX gene resulted in induced heat stress tolerance in transgenic potatoes due to *SWPA2* promoters (Tang et al. 2006). Incorporation of APX thylakoid-bound gene from tomatoes led to enhanced high temperature stress tolerance in the transgenic tobacco (Sun et al. 2010). It also resulted in the increased PS-II photochemical efficacy in the lines of transgenic plants (Wang et al. 2011).

## 13 Ascorbic Acid and Salinity Stress

Among different abiotic stresses, salinity is one of the major constraints that is considered as the leading challenge to agriculture and food security. It has been observed that different plants have developed certain defense mechanisms to combat saline stress conditions to balance cellular hyper-osmolarity and disequilibrium (Chandna et al. 2013). The negative effects of salinity may also be alleviated by using certain priming and pretreatments as well as transgenic increase of endogenous AA concentration (Tables 1 and 2). The pretreatment with AA led to increased $K^+$, chlorophyll, overall growth, fresh weight, dry weight, root length, and shoot length of rice under saline conditions (Alhasnawi et al. 2015). Pretreatment of date palm with AA resulted in higher shoot protein, chlorophyll concentration, shoot length, root length, growth, and activity of APX and SOD enzymes in response to saline environment under in vitro conditions (AL-Mayahi 2016). Foliar pretreatment of AA showed increased chlorophyll, photosynthesis capacity, and leaf proline and reduced electrolyte leakage in pistachio under salinity conditions (Bastam et al. 2013). Likewise, pretreatment of common bean with AA resulted in alleviation of ROS-induced oxidative damage and increased growth attributes, photosynthetic pigments, and lipid peroxidation under salinity stress (Table 1) (Saeidi-Sar et al., 2013). AA priming was highly effective to increase vegetative growth, soluble carbohydrates, proline, free amino acids, $K^+/Na^+$ ratio, and photosynthesis-related pigments of broad bean (Azooz et al. 2013). Similarly, pretreatment of tomato seeds with AA showed increased germination, seedling biomass, seedling length, vigor index, and germination speed under salinity conditions (Kaur and Gupta 2018). It was observed that pretreatment of sugarcane with AA resulted in increased root length, soluble protein, and antioxidant enzymes activities with reduced yellowing under in vitro saline conditions (Table 1) (Munir and Aftab 2011). Pretreatment of tomato seeds led to increased germination rate, fresh and dry mass, leaves per plant, leaf area, chloroplast efficacy, photosynthesis pigments, and inorganic minerals along with higher antioxidant enzymes activities (El-Sayed and El-Sayed 2013). Pre-sowing treatment of pumpkin with AA led to reduced salt stress severity and increased seedling fresh weight, protein concentration, and overall growth (Rafique et al. 2011). Similarly, tomato seed pretreatment with AA showed higher chlorophyll contents, germination percentage, and carotenoids and shoot and root length with lower lipid peroxidation (Krupa-Małkiewicz et al. 2015). Priming of eggplant with AA led to increased germination, dry and fresh weights, seedling growth, carbohydrates, total chlorophyll, and proteins and reduced electrolyte leakage under saline conditions (Table 2) (Jan et al. 2016). In the same way, hot pepper priming with AA also resulted in improved germination, root/shoot length, seedling fresh weight, vigor index, and secondary roots in response to saline conditions (Amjad et al. 2007). AA alleviated salinity-induced damage by showing enhanced chlorophyll concentration, stem length, stomatal conductance, and dry matter of pea plant roots (Balliu et al. 2014). Priming of naked pumpkin seeds with AA resulted in increased germination, root length, and activity of POD and CAT enzymes along with alleviation of salinity-induced damage (Fazlali et al. 2013). Likewise, AA priming treatment of wheat showed reduced leakage rate and germination time and increased growth and biomass accumulation under salinity stress (Afzal et al. 2006). Similarly, priming of sunflower with AA led to increased seed stamina index, germination speed, germination rate, hypocotyl dry weight, relative water content, and radical dry weight and reduced ion leakage under saline conditions (Fatemi 2014).

## 14 Ascorbic Acid and Heavy Metal Stress

The exploitation of environment by heavy metals and metalloids has been increased owing to their consistent presence and ability to bioaccumulate in plant tissues and deterioration of whole food chain. Excessive contents of heavy metals in different soil types are primarily due to expanding industrial wastes, mining activities, and urbanization. The bioavailability of metal ions depends on various factors including organic content, soil pH value, cation exchange capacity, and others. Certain metals are necessary component for growth and developments of plants including Cu, Zn, Mn, and Fe (called essential ions), while the rest of the ions (Cd, Pd, and Al) are non-essential, and these compete with other ions for their uptake and consequently cause metabolic disturbance. The mechanism of excess metal toxicity includes the enhanced generation of ROS that perturbs the redox state and wide range of modifications in antioxidative pool, destructive interaction with proteins (histidyl, thiol and carboxyl group), and displacement of essential cations (Sharma and Dietz 2009; Küpper and Andresen 2016). Metal toxicity causes severe damage to plant health by inhibiting photosynthesis as well as perturbing energy generation mechanisms, lipid peroxidation of organelles, and DNA damages eventually affecting growth and production. It has been noted that the effects of metal toxicity depend on the plant developmental stages and their concentration and exposure duration (Koffler et al. 2014; Anjum et al. 2014). Heavy metals, particularly redox active metals, play a key role for the generation of $O_2^{\bullet-}$, $H_2O_2$, and $^{\bullet}OH$ through Haber–Weiss and Fenton reaction inside the cells (Demidchik 2015). These metal ions bound to ligands with organic acids, essential amino acids, metallothioneins, and phytochelatins (Hossain et al. 2012). However, AA may act as proxidant when they are in direct contact with metal ions in the root apoplast. Conversely, redox-inactive heavy metal elements (Zn, Ni, Cd, and others) regenerate ROS by certain indirect mechanism, i.e., disabling antioxidant system and upregulating NADPH oxidase activity (Horemans et al. 2007; Anjum et al. 2014). Another alternate mechanism of redox-inactive heavy metals is Zn-induced oxidative stress which leads to accumulation and stabilization of phenoxyl radicals through charge transfer complex exhibiting prooxidant behavior (Morina et al. 2010).

It has been observed that exogenous application of Zn induced a more oxidizing state in root apoplast by decreasing AA/DHA ratio unlike leaves of *V. thasus* (Morina et al. 2010). The DHA accretion in roots is consistent with addition of QH that is localized in cell walls and inhibitive activities of most of the enzymes involved in AA–GSH cycle. Similarly, Zn reduced the AAO activity in the roots of *Zea mays*; however, Cu has no significant effect for the same crop (Vuletić et al. 2014). Cd significantly reduced the activity of AAO in roots of barley (Tamás et al. 2006). Moreover, exogenous pretreatment of AA alleviated salinity stress (Shalata and Neumann 2001). Fecht et al. (2003) suggested that peroxidases and AA in apoplast have improved the Mn tolerance in two cultivars of *Vigna unguiculata* metal-sensitive and metal-tolerant. It was noted that the effects of Mn toxicity significantly reduced the redox state and total AA content in apoplast in metal-sensitive cultivar but did not have any effect on the metal-tolerant cultivar. Moreover, DHA accumulated in apoplast is transported inside the cell to alter the cell cycle (Horemans et al. 2000) as consistently reported for BY-2 cell lines of tobacco (Potters et al. 2004). Horemans et al. (2000) also presumed that metal-induced oxidative stress or other hostile environmental conditions affect the redox signaling cascade by increasing cellular DHA content to cease the cell activities in order to allow the cell to adjust in different homeostatic range.

The transcriptional abundance of APX-1 was significantly induced in roots of *Arabidopsis* due to Cu toxicity-induced oxidative stress; however, Cd treatment had no effect on transcription levels of APX-1 gene (Opdenakker et al. 2012). Moreover, putative signaling pathway that is mediated by ZAT-12 transcription factor (sensitive to ROS accumulation) triggers APX-1 expression level, thus suggesting active role of APX-1 in cytosol during Cu-induced stress. Similarly, Zn accumulation induces the activity of MDAR and APX in young shoot tissues of common mullein (Morina et al. 2010). The biosynthesis of AA by enhanced activity of GalLGH and its efficient recycling through ascorbate–glutathione cycle was also stimulated. The sum of reduced and total GSH level was decreased due to Zn-induced oxidative stress in cell, indicating the affected AA recovery regardless of MDAR activity. Lucini and Bernardo (2015) elucidated that Zn toxicity stress exposure for a long time period enhanced the GalLDH activity in lettuce. Consistently depletion of AA content and elevated levels of DHA due to exogenous Cd stress have also been reported in the leaves of rice (Chao et al. 2010). During this duration, it has been demonstrated that oxidation of AA is closely associated with enhanced MDA content, suggesting its active role as an antioxidant. In addition to this, followed by Cd stress, the plants were treated with exogenous pretreatment of AA or L-galactono-1,4-lactone, oxidative stress of Cd was rescued, and there was reduced content of AA due to its putative consumption for alleviating stress.

The comprehensive role of AA in metal-induced stress has been investigated with mutant and transgenic studies. The increased expression of DHAR co-localized in cytosol of tobacco root cells enhanced the Al-induced stresses by reducing the ROS concentration, lipid accumulation and enhanced AA and APX enzymatic activity, than wild type. Unlike DHAR, overexpression of MDAR was not effective in contributing alleviation of oxidative stress induced by Al, though the AA level was higher than wild type (Yin et al. 2010). It has been observed that proficient AA recovery via cytosolic reduction of DHA provided the defense against oxidative stress in symplast and apoplast. However, this may imply to only availability of cellular GSH. Further work is needed to elucidate the redox reaction in cell wall and apoplast for better understanding of the signaling network of stress tolerance in plant cell under heavy metal stress.

## 15 Ascorbic Acid and UV Radiation Stress

Plants being sessile are generally exposed to UV irradiation which is about 7% of total solar radiations comprising UV-A (315–400 nm) and UV-B (290–315 nm) (Kerr and Fioletov 2008; Frohnmeyer and Staiger 2003). Preliminary investigations about UV-B irradiation have revealed its damaging effects for growth and developmental aspects of plants including detrimental effects on cellular membrane and proteins, regeneration of ROS, perturbing signaling network of D1 and D2 subunits of photosystem II, oxidative stress, and producing cyclobutane pyrimidine dimers in a DNA strands, affecting activities of RuBisCO (Hu et al. 2013; Lidon et al. 2012). Nevertheless, it has been observed that such damaging influences of UV-B irradiation are ascribed to increased fluxes as well as intensities of the artificial UV-A and UV-B exposure in growth chambers. Similarly, certain other unfavorable environmental conditions in the field are also detrimental for the growth and developmental processes of plants and yield including high temperature stress and drought conditions (Hideg et al. 2013).

In the same way, several reports have demonstrated the ecological importance of UV-B irradiation suggesting it to be vital for environmental cue for growth and development of plants (Heijde and Ulm 2012; Jenkins 2009). The UV responses on UVR-8-based signaling cascade is accountable for transcriptional abundance of multiple genes which are involved in flavonoid biosynthesis and regulation of antioxidant defense mechanism, along with activation of GSH metabolism cascades (Brown et al. 2005; Wu et al. 2012). Elevated concentrations of UV-B irradiation reduce the ascorbate redox state in pea and barley leaves due to enhanced levels of ROS accumulations (Hideg et al. 2006). However, UV-B irradiation also helps in activation of antioxidative defense via UVR-8 signaling pathway. This signaling pathway contains a number of components such as HY5-homologue (HYH), elongated hypocotyl-5 (HY5), and photomorphogenic-1 (COP1) (Vidović et al. 2017). However, during optimal circumstances, plants respond to the cumulative effect of UV-A, UA-B, and photosynthetically active radiation (PAR), and these overlapping responses may imply cross-tolerance (Behn et al. 2010; Majer and Hideg 2012; Klem et al. 2012).

Class III peroxidase/phenolic/AA and AA–APX systems are mainly responsible for efficient quenching of $H_2O_2$ and high AA redox state in leaf tissues of *P. coleoides* after high light (HL) exposure for 10 days (Vidović et al. 2015b). Plant exposure to UV-B irradiation and increased PAR diminished oxidative stress and thus didn't provoke enzymatic activities of POD and CAT, and hence no substantial changes in AA redox state occurred. This is in consistence with facts that UV-B-induced plant stresses in the fields are not much stronger (Ballaré et al. 2011; Hideg et al. 2013). Coupled exposure of higher intensity light along with UV-B irradiation provokes enhanced enrichment of AA, APX, and CAT activity compared to only high light, suggesting that two energy sources synergistically stimulate the antioxidant system in green zones of variegated leaves of *P. zonale* (Vidović et al. 2015a). In addition to this, UV-B acclimatization responses overlap with high light exposure; however, elevated AA redox state due to UV-B was only noted earlier under increased light acclimatized plants of tobacco (Majer and Hideg 2012).

## 16 Ascorbic Acid and High Light Stress

Plants are inevitably exposed to continuously varying PAR fluxes during natural environmental conditions (Lichtenthaler 2007). The PAR fluxes exceed the dissipation capacity and metabolic requirement of plant factory, and the rate of ROS generation enhances which sequentially inhibits the photosynthetic rate and assimilation of $CO_2$ (Li et al. 2010). These conditions are known as "photooxidative stress" (Foyer and Shigeoka 2011). It is very hard to classify the intensity of PAR which acts as a stressor for any specific plant species, since it depends on several factors such as salinity, drought, temperature stress, nutrient deficiency, developmental stages, and previous light acclimatization history (Li et al. 2010). It has been found that even optimal PAR flux might act as a stressor during limited $CO_2$ availability (Mittler et al. 2004).

During such situation, photoprotection mechanism is activated which downregulates the PSII in order to dissipate excessive photon energy by lowering the pH values and zeaxanthin accumulations (Karpinski et al. 1999). Molecular oxygen plays a vital function in dissipation of energy by accepting electron in both photosynthetic processes such as during photorespiration and Mehler reaction (Ort and Baker 2002).

AA content (Heyneke et al. 2013), cytosolic APX (Fryer et al. 2003), other essential components of ascorbate–glutathione cycle (Szechyńska and Karpiński 2013), anthocyanin and flavonoid contents (Pollastri and Tattini 2011), xanthophylls (Lichtenthaler 2007), and tocopherol (Munné et al. 2013) are all high light-induced components of anti-oxidative defense pool in plants under abiotic stresses. High light fluxes incite rapid biosynthesis of leaf AA content almost as directly proportional, supporting the role of AA in photo-protection mechanism (Noshi et al. 2016). Generally, high light induces 2–3 times of total AA content in different plant species, both after long-term exposure (10–15 days) (Page et al. 2012) and short-term exposure (< 24 h) (Laing et al. 2017). Subcellular fluctuations in AA content were studied in *Arabidopsis* leaves in response to high light exposure in mesophyll explicit that as the light irradiance enhances from 700 to 1500 μmol $m^{-2}$ $s^{-1}$, biosynthesis of AA level elevated progressively in vacuoles and chloroplasts (Heyneke et al. 2013). AA becomes prevalent in the lumen of thylakoids at very high PAR fluxes. Low PAR exposure at 700 μmol $m^{-2}$ $s^{-1}$ for 2 weeks produces more AA in all organelles. Significantly higher contents of AA were observed in vacuole (almost four times) and doubled in chloroplast. Conversely, relative ratio of AA was decreased in peroxisomes during high light exposure. The enhanced AA content in chloroplast is associated with photo-protection role, while such increment in vacuole is related to $H_2O_2$ quenching that also comprised of phenol compounds and PODs (Zipor and Oren 2013).

The detailed mechanism of $H_2O_2$ regulation in fully green (chloroplast) and white tissues regarding photosynthetic activity-dependent regeneration of ROS has not been elucidated. Recently, Vidović et al. (2016) exposed the *P. zonale* with variegated leaves to high intensity PAR (>1000 μmol $m^{-2}$ $s^{-1}$) and found that antioxidant pool in photosynthetic cells contains CAT, tAPX, and AA, while non-photosynthetic cells contain MnSOD, Cu/ZnSOD, GSH, and ascorbate–glutathione cycle enzymes. Sudden high light exposures to plants caused distinguish responses in those two cell types such as about 2.5-fold AA level increased in green sectors after 1 h exposure, while soluble cytosolic APX activity increased in white sectors of the same leaves. This was closely associated with 2.6-fold higher level of cytoplasmic AA of white sectors compared to green sectors. During certain treatments to induce more pro-oxidative conditions, AA level and activities of DHAR and MDAR were increased in non-photosynthetic cells. These results can be interpreted as high redox state level is maintained by effective recycling of AA (enhanced DHAR, APX, and GR enzymes activities and unchanged AA content) and not AA biosynthesis which is potentially adopted in photosynthetic cells. Such tissue-specific physiological adaptation for maintenance of AA redox status is ascribed to non-photosynthetic tissues dependency on AA precursors (Smirnoff 2011). The molecular mechanism of AA regulation and other enzymes of network coupled to the AA–GSH cycle in morphologically varied tissues of variegated leaves needs to be investigated further in the future.

## 17 Ascorbic Acid and Increased Ozone Stress

The increased concentration of ozone ($O_3$) is considered detrimental for the photosynthesis of crop plants. The $O_3$ enters through stomatal openings and reacts with symplastic and apoplastic cell components. The increased $O_3$ levels lead to overproduction of ROS. The higher levels of atmospheric $O_3$ decrease photosynthesis, ultimately leading to reduced overall yield of the crops (Long and Naidu 2002; Kobayakawa and Imai 2017). It has been observed that increased $O_3$ exposure badly influences guard cells and inhibits $K^+$ channel activity of plasma membranes (Torsethaugen et al. 1999). So, it is imperative to develop eco-friendly alternatives to reduce $O_3$-induced damages in the crop plants by using natural protective chemicals such as AA. Apoplastic AA is considered the first defense line against detrimental pollutants (Barnes et al. 2002). The increased apoplastic AA protects the plants by scavenging ROS. It was observed that the reduced apoplastic AA concentration significantly increased injury to foliage of tobacco in response to chronic $O_3$ exposure (Sanmartin et al. 2003). Besides endogenous increased apoplastic AA concentrations, exogenous priming or pretreatments are also considered potentially beneficial. The effect of AA pretreatment on $O_3$-induced photosynthesis was investigated on paddy rice under solution culture conditions. The increased concentration of $O_3$ resulted in photosynthesis inhibition and reduced AA levels in rice (Table 1). On the contrary, AA pretreatment led to increased endogenous AA concentration but did not efficiently ameliorated $O_3$-induced photosynthesis inhibition in paddy rice (Kobayakawa and Imai 2017).

## 18 Conclusion

In conclusion, biosynthesis of AA takes place via D-mannose/L-galactose pathways which has been known as an imperative ascorbate source in the higher plants. AA is considered imperative for various important physiological aspects of growth and development including cell division, root growth, cell proliferation, seed germination, shoot development, and stomatal movement of plants in combination with other hormones such as GA, ethylene, and ABA. It plays the leading function of the activities of ROS scavenging at cellular levels. Moreover, AA also affects numerous abiotic stress-related and/or stress-responsive activities of enzymes through the synergistic ways with certain important antioxidant compounds including $\alpha$-tocopherol and glutathione that eventually work in the very co-coordinative manner to alleviate oxidative damages to plant cells under abiotic stress conditions. The role of AA has also been recognized in the process of photosynthesis as it works as an alternative donor of electrons to PS-II under abiotic stresses and works for the protection of the apparatus of photosynthetic pigments located in the chloroplast by reducing the levels of ROS production. Application of AA through priming or pretreatments has been found effective to alleviate various abiotic stresses by scavenging different ROS, protecting photosynthetic apparatus, and increasing biomass accumulation of the treated crop plants under a wide variety of environmental conditions. In addition, various AA biosynthesizing pathways of the important transgenes have been incorporated in the plants of interest by using tools of genetic engineering to enhance the endogenous production levels of AA. Therefore, plants of transgenic origin provided with higher levels of endogenous AA resulted in the enhanced tolerance of abiotic stresses including high/low temperature, high light conditions, drought, hypoxia, ozone, salinity, chilling, heat, heavy metals, and UV radiations. However, it is needed to thoroughly study the role of AA in signal transduction cascades, stomatal movements, and upregulation of the defensive enzyme-related genes in the future. Moreover, study of proteomic changes in the AA primed and/or pretreated plants under various abiotic conditions is also suggested.

## References

Afzal I, Basra SM, Farooq M, Nawaz A (2006) Alleviation of salinity stress in spring wheat by hormonal priming with ABA, salicylic acid and ascorbic acid. Int J Agric Biol 8(1):23–28

Ahmad I, Basra SMA, Wahid A (2014) Exogenous application of ascorbic acid, salicylic acid and hydrogen peroxide improves the productivity of hybrid maize under at low temperature stress. Int J Agric Biol 16:825–830

Akram NA, Shafiq F, Ashraf M (2017) Ascorbic acid-a potential oxidant scavenger and its role in plant development and abiotic stress tolerance. Front Plant Sci 8:613. https://doi.org/10.3389/fpls.2017.00613

Alhasnawi AN, Kadhimi AA, Isahak A, Mohamad A, Yusoff WMW, Zain CRCM (2015) Exogenous application of ascorbic acid ameliorates detrimental effects of salt stress in rice (MRQ74 and MR269) seedlings. Asian J Crop Sci 7(3):186–196

AL-Mayahi AMW (2016) Influence of salicylic acid (SA) and ascorbic acid (ASA) on in vitro propagation and salt tolerance of date palm (*Phoenix dactylifera* L.) cv. 'Nersy'. Aust J Crop Sci 10:969–976

Amjad M, Ziaf K, Iqbal Q, Ahmad I, Riaz MA, Saqib ZA (2007) Effect of seed priming on seed vigour and salt tolerance in hot pepper. Pak J Agric Sci 44(3):408–416

Anjum NA, Gill SS, Gill R, Hasanuzzaman M, Duarte AC, Pereira E, Ahmad I, Tuteja R, Tuteja N (2014) Metal/metalloid stress tolerance in plants: role of ascorbate, its redox couple, and associated enzymes. Protoplasma 251:1265–1283

Arefi F, Ganjali HR, Rad MRN (2016) Influence of drought stress and ascorbic acid on some characteristics of sorghum. Int J Agric Biosci 5(3):113–115

Awad J, Stotz HU, Fekete A, Krischke M, Engert C, Havaux M, Berger S, Mueller MJ (2015) 2-cysteine peroxiredoxins and thylakoid ascorbate peroxidase create a water-water cycle that is essential to protect the photosynthetic apparatus under high light stress conditions. Plant Physiol 167:1592–1603

Azooz MM, Alzahrani AM, Youssef MM (2013) The potential role of seed priming with ascorbic acid and nicotinamide and their interactions to enhance salt tolerance in broad bean (*Vicia faba* L.). Aust J Crop Sci 7(13):2091–2100

Baldet P, Bres C, Okabe Y, Mauxion JP, Just D, Bournonville C, Ferrand C, Mori K, Ezure H, Rothan C (2013) Investigating the role of vitamin C in tomato through TILLING identification of ascorbate-deficient tomato mutants. Plant Biotechnol 30:308–314

Ballaré CL, Caldwell MM, Flint SD, Robinson SA, Bornman JF (2011) Effects of solar ultraviolet radiation on terrestrial ecosystems. Patterns, mechanisms, and interactions with climate change. Photochem Photobiol Sci 10:226–241

Balliu A, Sallaku G, Nasto T (2014) The effects of some osmoprotectant compounds on growth parameters of pea plants (*Pisum sativum*) under saline conditions. Acta Hortic 1142:377–382

Bánhegyi G, Loewus FA (2004) Ascorbic acid catabolism: breakdown pathways in animals and plants. In: Asard H, May JM, Smirnoff N (eds) Vitamin C: functions and biochemistry in animals and plants. BIOS Scientific Publishers Ltd, Oxford, pp 31–48

Barnes JD, Zheng Y, Lyons TM (2002) Plant resistance to ozone: the role of ascorbate. In: OmAA K, Saji H, Youssefian S, Kondo N (eds) Air pollution and plant biotechnology. Springer, Tokyo, pp 235–254

Barth C, De Tullio M, Conklin PL (2006) The role of ascorbic acid in the control of flowering time and the onset of senescence. J Exp Bot 57:1657–1665

Bastam N, Banin AAb B, Ghobadi C (2013) Interactive effects of ascorbic acid and salinity stress on the growth and photosynthetic capacity of pistachio (*Pistacia vera* L.) seedlings. J Hortic Sci Biotechnol 88(5):610–616

Behn H, Albert A, Marx F, Noga G, Ulbrich A (2010) Ultraviolet-B and photosynthetically active radiation interactively affect yield and pattern of monoterpenes in leaves of peppermint (Mentha × piperita L.). J Agric Food Chem 58:7361–7367

Bienert GP, Chaumont F (2014) Aquaporin-facilitated transmembrane diffusion of hydrogen peroxide. Biochim Biophys Acta 1840:1596–1604

Brown BA, Cloix C, Jiang GH, Kaiserli E, Herzyk P, Kliebenstein DJ, Jenkins GI (2005) A UV-B-specific signaling component orchestrates plant UV protection. Proc Natl Acad Sci USA 102:18225–18230

Chandna R, Azooz MM, Ahmad P (2013) Recent advances of metabolomics to reveal plant response during salt stress. In: Ahmad P, Azooz MM, Prasad MNV (eds) Signaling, omics and adaptations. Springer, New York, pp 1–14

Chao YY, Hong CY, Kao CH (2010) The decline in ascorbic acid content is associated with cadmium toxicity of rice seedlings. Plant Physiol Biochem 48:374–381

Chaves MM, Maroco JP, Pereira JS (2003) Understanding plant responses to drought-from genes to the whole plant. Funct Plant Biol 30:239–264

Cheeseman JM (2006) Hydrogen peroxide concentrations in leaves under natural conditions. J Exp Bot 57:2435–2444

Cheeseman JM (2007) Hydrogen peroxide and plant stress: a challenging relationship. Plant Stress 1:4–15

Chen SX, Schopfer P (1999) Hydroxyl-radical production in physiological reactions. FEBS J 260:726–735

Chen C, Letnik I, Hacham Y, Dobrev P, Ben-Daniel BH, Vanková R (2014) Ascorbate peroxidase protects Arabidopsis desiccating and germinating seeds from stress and mediates cross talk between reactive oxygen species, abscisic acid, and auxin. Plant Physiol 166:370–383

Citterio S, Sgorbati S, Scippa S, Sparvoli E (1994) Ascorbic acid effect on the onset of cell proliferation in pea root. Physiol Plant 92:601–607

Clifton IJ, McDonough MA, Ehrismann D (2006) Structural studies on 2-oxoglutarate oxygenases and related double-stranded B-helix fold proteins. J Inorg Biochem 100:644–669

Coles JP, Phillips AL, Croker SJ, Garcia-Lepe R, Lewis MJ, Hedden P (1999) Modification of gibberellin production and plant development in Arabidopsis by sense and antisense expression of gibberellin 20-oxidase genes. Plant J 17:547–556

Conklin PL, Barth C (2004) Ascorbic acid, a familiar small molecule intertwined in the response of plants to ozone, pathogens, and the onset of senescence. Plant Cell Environ 27:959–970

Conklin PL, Pallanca JE, Last RL, Smirnoff N (1997) L-ascorbic acid metabolism in the ascorbate-deficient arabidopsis mutant vtc1. Plant Physiol 115:1277–1285

Conklin PL, Norris SR, Wheeler GL, Williams EH, Smirnoff N, Last RL (1999) Genetic evidence for the role of GDP-mannose in plant ascorbic acid (vitamin C) biosynthesis. Proc Natl Acad Sci U S A 96:4198–4203

Cramer GR, Urano K, Delrot S, Pezzotti M, Shinozaki K (2011) Effects of abiotic stress on plants: a systems biology perspective. BMC Plant Biol 11:163

Currey CJ, Lopez RG, Rapaka VK, Faust JE, Runkle ES (2013) Exogenous applications of benzyladenine and gibberellic acid inhibit lower-leaf senescence of geraniums during propagation. Hortic Sci 48:1352–1357

Davey MW, Van Montagu M, Inze D, Sanmartin M, Kanellis A, Smirnoff N, Benzie IJJ, Strain JJ, Favell D, Fletcher J (2000) Plant l – ascorbic acid: chemistry, function, metabolism, bioavailability and effects of processing. J Sci Food Agric 80:825–860

De Pinto MC, Tommasi F, De Gara L (2000) Enzymes of the ascorbate biosynthesis and ascorbate-glutathione cycle in cultured cells of tobacco Bright Yellow 2. Plant Physiol Biochem 38:541–550

De Tullio M, Guether M, Balestrini R (2013) Ascorbate oxidase is the potential conductor of a symphony of signaling pathways. Plant Signal Behav 8:e23213

Debolt S, Melino V, Ford CM (2007) Ascorbate as a biosynthetic precursor in plants. Ann Bot 99:3–8

Demidchik V (2015) Mechanisms of oxidative stress in plants: From classical chemistry to cell biology. Environ Exp Bot 109:212–228

Dietz K-J (2014) Redox regulation of transcription factors in plant stress acclimation and development. Antioxid Redox Signal 21:1356–1372

Dietz K-J (2016) Thiol-based peroxidases and ascorbate peroxidases: why plants rely on multiple peroxidase systems in the photosynthesizing chloroplast? Mol Cells 39:20–25

Drazkiewicz M, Skórzyńska-Polit E, Krupa Z (2003) Response of the ascorbate-glutathione cycle to excess copper in Arabidopsis thaliana (L). Plant Sci 164:195–202

Duan M, Ma N-N, Li D, Deng Y-S, Kong F-Y, Lv W, Meng Q-W (2012) Antisense-mediated suppression of tomato thylakoidal ascorbate peroxidase influences anti-oxidant network during chilling stress. Plant Physiol Biochem 58:37–45

El-Sayed HEA, El-Sayed A (2013) Exogenous application of ascorbic acid for improve germination, growth, water relations, organic and inorganic components in tomato (*Lycopersicon esculentum* Mill.) plant under salt-stress. NY Sci J 6:123–139

Ergin S, Aydogan C, Ozturk N, Turhan E (2014) Effects of ascorbic acid application in strawberry plants during heat stress. Turk J Agric Nat Sci 7(7):1486–1491

Exposito-Rodriguez M, Laissue PP, Yvon-Durocher G, Smirnoff N, Mullineaux PM (2017) Photosynthesis-dependent $H_2O_2$ transfer from chloroplasts to nuclei provides a high-light signalling mechanism. Nat Commun 8:49

Faize M, Nicolás E, Faize L, Díaz-Vivancos P, Burgos L, Hernández JA (2015) Cytosolic ascorbate peroxidase and Cu/Zn-superoxide dismutase improve seed germination, plant growth, nutrient uptake and drought tolerance in tobacco. Theor Exp Plant Physiol 27:215–226

Farooq M, Irfan M, Aziz T, Ahmad I, Cheema SA (2013) Seed priming with ascorbic acid improves drought resistance of wheat. J Agron Crop Sci 199(1):12–22

Fatemi SN (2014) Ascorbic acid and its effects on alleviation of salt stress in sunflower. Ann Res Rev Biol 4(24):3656–3665

Fazlali R, Asli DE, Moradi P (2013) The effect of seed priming by ascorbic acid on bioactive compounds of naked seed pumpkin (*Cucurbita pepo* var. styriaca) under salinity stress. Int J Farm Allied Sci 2(17):587–590

Fecht CMM, Maier P, Horst WJ (2003) Apoplastic peroxidases and ascorbate are involved in manganese toxicity and tolerance of *Vigna unguiculata*. Physiol Plant 117:237–244

Feng Z, Pang J, Nouchi I, Kobayashi K, Yamakawa T, Zhu J (2010) Apoplastic ascorbate contributes to the differential ozone sensitivity in two varieties of winter wheat under fully open – air field conditions. Environ Pollut 158:3539–3545

Fotopoulos V, De Tullio MC, Barnes J, Kanellis AK (2008) Altered stomatal dynamics in ascorbate oxidase over-expressing tobacco plants suggest a role for dehydroascorbate signalling. J Exp Bot 59:729–737

Foyer CH, Noctor G (2011) Ascorbate and glutathione: the heart of the redox hub. Plant Physiol 155:2–18

Foyer CH, Shigeoka S (2011) Understanding oxidative stress and antioxidant functions to enhance photosynthesis. Plant Physiol 155:93–100

Frohnmeyer H, Staiger D (2003) Ultraviolet-B radiation-mediated responses in plants. Balancing damage and protection. Plant Physiol 133:1420–1428

Fry SC (1998) Oxidative scission of plant cell wall polysaccharides by ascorbate-induced hydroxyl radicals. Biochem J 332:507–515

Fryer MJ, Ball L, Oxborough K, Karpinski S, Mullineaux PM, Baker NR (2003) Control of Ascorbate Peroxidase 2 expression by hydrogen peroxide and leaf water status during excess light stress reveals a functional organization of Arabidopsis leaves. Plant J 33:691–705

Fujita M, Fujita Y, Noutoshi Y, Takahashi F, Narusaka Y, Yamaguchi-Shinozaki K, Shinozaki K (2006) Crosstalk between abiotic and biotic stress responses: A current view from the points of convergence in the stress signalling networks. Curr Opin Plant Biol 9:436–442

Gao Q, Zhang L (2008) Ultraviolet-B-induced oxidative stress and antioxidant defense system responses in ascorbate-deficient vtc1 mutants of Arabidopsis thaliana. J Plant Physiol 165:138–148

Gloser V, Korovetska H, Martín-Vertedor AI, Hájíčková M, Prokop Z, Wilkinson S, Davies W (2016) The dynamics of xylem sap pH under drought: A universal response in herbs? Plant Soil 409:259–272

Gómez F, Fernández L, Gergoff G, Guiamet JJ, Chaves A, Bartoli CG (2008) Heat shock increases mitochondrial $H_2O_2$ production and extends postharvest life of spinach leaves. Postharvest Biol Technol 49:229–234

Green MA, Fry SC (2005) Vitamin C degradation in plant cells via enzymatic hydrolysis of 4-O-oxalyl-L-threonate. Nature 433:83–87

Hamama H, Murniati E (2010) The effect of ascorbic acid treatment on viability and vigor maize (Zea mays L.) seedling under drought stress. Hayati J Biosci 17(3):105–109

Hancock RD, Walker PG, Pont SDA, Marquis N, Vivera S, Gordon SL, Brennan RM, Viola R (2007) L-Ascorbic acid accumulation in fruit of *Ribes nigrum* occurs by in situ biosynthesis via the L-galactose pathway. Funct Plant Biol 34:1080–1091

He Y, Tang W, Swain JD, Green AL, Jack TP, Gan S (2001) Networking senescence-regulating pathways by using Arabidopsis enhancer trap lines. Plant Physiol 126:707–716

Heijde M, Ulm R (2012) UV-B photoreceptor-mediated signalling in plants. Trends Plant Sci 17:230–237

Hemavathi CP, Upadhyaya N, Akula HS, Kim JH, Jeon MH, Oh SC, Chun DHK, Park SW (2011) Biochemical analysis of enhanced tolerance in transgenic potato plants overexpressing d – galacturonicacidreductase gene in response to various abiotic stresses. Mol Breed 28:105–115

Hemmati K, Ebadi A, Khomari S, Sedghi M (2018) Influence of ascorbic acid and 24-epibrassinolide on physiological characteristics of pot marigold under water-stress condition. J Plant Interact 13(1):364–372

Heydariyan M, Basirani N, Sharifi-Rad M, Khmmari I, Rafat Poor S (2014) Effect of seed priming on germination and seedling growth of the caper (*Capparis spinosa*) under drought stress. Int J Adv Biol Biomed Res 2(8):2381–2389

Heyneke E, Luschin-Ebengreuth N, Krajcer I, Wolkinger V, Müller M, Zechmann B (2013) Dynamic compartment specific changes in glutathione and ascorbate levels in Arabidopsis plants exposed to different light intensities. BMC Plant Biol 13:104–123

Hideg É, Rosenqvist E, Váradi G, Bornman J, Vincze É (2006) A comparison of UV-B induced stress responses in three barley cultivars. Func Plant Biol 33:77–90

Hideg É, Jansen MAK, Strid Å (2013) UV-B exposure, ROS, and stress: inseparable companions or loosely linked associates? Trends Plant Sci 18:107–115

Hocking B, Tyerman SD, Burton RA, Gilliham M (2016) Fruit calcium: transport and physiology. Front Plant Sci 7:569

Horemans N, Foyer CH, Asard H (2000) Transport and action of ascorbate at the plant plasma membrane. Trends Plant Sci 5:263–267

Horemans N, Raeymaekers T, Van Beek K, Nowocin A, Blust R, Broos K, Cuypers A, Vangronsveld J, Guisez Y (2007) Dehydroascorbate uptake is impaired in the early response of Arabidopsis plant cell cultures to cadmium. J Exp Bot 58:4307–4317

Horling F, Lamkemeyer P, König J, Finkemeier I, Kandlbinder A, Baier M, Dietz KJ (2003) Divergent light-, ascorbate-, and oxidative stress-dependent regulation of expression of the peroxiredoxin gene family in Arabidopsis. Plant Physiol 131:317–325

Hossain MA, Piyatida P, da Silva JAT, Fujita M (2012) Molecular mechanism of heavy metal toxicity and tolerance in plants: central role of glutathione in detoxification of reactive oxygen species and methylglyoxal and in heavy metal chelation. J Bot 2012:872875

Hu Z, Li H, Chen S, Yang Y (2013) Chlorophyll content and photosystem II efficiency in soybean exposed to supplemental ultraviolet-B radiation. Photosynthetica 51:151–157

Huang J, Levine A, Wang Z (2013) Plant abiotic stress. Scientific World J. https://doi.org/10.1155/2013/432836

Innocenti AM, Bitonti MB, Arrigoni O, Liso R (1990) The size of quiescent centre in roots of Allium cepa L. grown with ascorbic acid. New Phytol 114:507–509

Inzé D, Van Montagu M (1995) Oxidative stress in plants. Curr Opin Biotechnol 6:153–158

Ishikawa T, Shigeoka S (2008) Recent advances in ascorbate biosynthesis and the physiological significance of ascorbate peroxidase in photosynthesizing organisms. Biosci Biotechnol Biochem 72:1143–1154

Jan S, Hamayun M, Wali S, Bibi A, Gul H, Rahim F (2016) Foliar application of ascorbic acid mitigates sodium chloride induced stress in eggplant (*Solanum melongena* L.). Pak J Bot 48(3):869–876

Jenkins GI (2009) Signal transduction in responses to UV-B radiation. Annu Rev Plant Biol 60:407–431

Kangasjärvi S, Kangasjärvi J (2014) Towards understanding extracellular ROS sensory and signaling systems in plants. Adv Bot 2014:538946

Kangasjärvi S, Lepistö A, Hännikäinen K, Piippo M, Luomala EM, Aro EM, Rintamäki E (2008) Diverse roles for chloroplast stromal and thylakoid bound ascorbate peroxidases in plant stress responses. Biochem J 412:275–285

Kärkönen A, Fry SC (2006) Effect of ascorbate and its oxidation products on $H_2O_2$ production in cell-suspension cultures of Piceaabies and in the absence of cells. J Exp Bot 57:1633–1644

Karpinski S, Reynolds H, Karpinska B, Wingsle G, Creissen G, Mullineaux P (1999) Systemic signaling and acclimation in response to excess excitation energy in Arabidopsis. Science 284:654–657

Kaur H, Gupta N (2018) Ameliorative Effect of proline and ascorbic acid on seed germination and vigour parameters of tomato (*Solanum lycopersicum* L.) under salt stress. Int J Curr Microbiol App Sci 7(1):3523–3532

Kausar R, Hossain Z, Makino T, Komatsu S (2012) Characterization of ascorbate peroxidase in soybean under flooding and drought stresses. Mol Biol Rep 39:10573–10579

Kerr JB, Fioletov VE (2008) Surface ultraviolet radiation. Atmosphere-Ocean 46:159–184

Klem K, Ač A, Holub P, Kovác D, Špunda V, Robson TM, Urban O (2012) Interactive effects of PAR and UV radiation on the physiology, morphology and leaf optical properties of two barley varieties. Environ Exp Bot 75:52–64

Kobayakawa H, Imai K (2017) Exogenous ascorbic acid scarcely ameliorates inhibition of photosynthesis in rice leaves by $O_3$. Plant Prod Sci 20(1):83–89

Koffler BE, Polanschütz L, Zechmann B (2014) Higher sensitivity of pad2-1 and vtc2-1 mutants to cadmium is related to lower subcellular glutathione rather than ascorbate contents. Protoplasma 251:755–769

Koyro HW, Ahmad P, Geissler N (2012) Abiotic stress responses in plants: an overview. In: Ahmad P, Prasad MNV (eds) Environmental adaptations and stress tolerance of plants in the era of climate change. Springer, New York, pp 1–28

Krupa-Małkiewicz M, Smolik B, Ostojski D, Sędzik M, Pelc J (2015) Effect of ascorbic acid on morphological and biochemical parameters in tomato seedling exposure to salt stress. Och Srod Zasob Nat 26(2):21–25

Kukavica B, Mojović M, Vučinić Ž, Maksimović V, Takahama U, Veljović-Jovanović S (2008) Generation of hydroxyl radical in isolated pea root cell wall, and the role of cell wall-bound peroxidase, Mn-SOD and phenolics in their production. Plant Cell Physiol 50:304–317

Küpper H, Andresen E (2016) Mechanisms of metal toxicity in plants. Metallomics 8:269–285

Kwon SY, Choi SM, Ahn YO, Lee HS, Lee HB, Park YM, Kwak SS (2003) Enhanced stress – tolerance of transgenic plants expressing a human dehydroascorbate reductase gene. J Plant Physiol 160:347–353

Laing W, Norling C, Brewster D, Wright M, Bulley S (2017) Ascorbate concentration in Arabidopsis thaliana and expression of ascorbate related genes using RNAseq in response to light and the diurnal cycle. BioRxiv:138008

Larkindale J, Hall JD, Knightand MR, Vierling E (2005) Heat stress phenotypes of Arabidopsis mutants implicate multiple signaling pathways in the acquisition of thermotolerance. Plant Physiol 138:882–897

Latif M, Akram NA, Ashraf M (2016) Regulation of some biochemical attributes in drought-stressed cauliflower (*Brassica oleracea* L.) by seed pre-treatment with ascorbic acid. J Hortic Sci Biotechnol 91(2):129–137

Li F, Wu QY, Sun YL, Wang LY, Yang XH, Meng QW (2010) Overexpression of chloroplastic monodehydroascorbate reductase enhanced tolerance to temperature and methyl viologen-mediated oxidative stresses. Physiol Plant 139:421–434

Lichtenthaler HK (2007) Biosynthesis, accumulation and emission of carotenoids, alpha-tocopherol, plastoquinone, and isoprene in leaves under high photosynthetic irradiance. Photosynth Res 92:163–179

Lidon FJ, Reboredo FH, Leitã AE, Silva MMA, Duarte MP, Ramalho JC (2012) Impact of UV-B radiation on photosynthesis-an overview. Emirates J Food Agric 24:546–556

Liebthal M, Maynard D, Dietz K-J (2017) Peroxiredoxins and redox signaling in plants. Antioxid Redox Signal. https://doi.org/10.1089/ars.2017.7164

Lima-Silva V, Rosado A, Amorim-Silva V (2012) Genetic and genome-wide transcriptomic analyses identify co-regulation of oxidative response and hormone transcript abundance with vitamin C content in tomato fruit. BMC Genomics 13:187. https://doi.org/10.1186/1471-2164-13-187

Liso R, Innocenti AM, Bitonti MB, Arrigoni O (1988) Ascorbic acid-induced progression of quiescent centre cells from G1 to S phase. New Phytol 110:469–471

Long SP, Naidu SL (2002) Effects of oxidants at the biochemical, cell and physiological levels, with particular reference to ozone. In: Bell JNB, Treshow M (eds) Air pollution and plant life. Wiley, Chichester, pp 69–88

Lucini L, Bernardo L (2015) Comparison of proteome response to saline and zinc stress in lettuce. Front Plant Sci 6:240

Lukatkin AS, Anjum NA (2014) Control of cucumber (*Cucumis sativus* L.) tolerance to chilling stress—evaluating the role of ascorbic acid and glutathione. Front Environ Sci 2:62

Majer P, Hideg É (2012) Existing antioxidant levels are more important in acclimation to supplemental UV-B irradiation than inducible ones: studies with high light pretreated tobacco leaves. Emirates J Food Agric 24:598–606

Majer P, Vidović M, Czégény G, Veljović-Jovanović S, Strid Å, Hideg E (2016) Evaluation of procedures for assessing anti- and pro-oxidants in plant samples. Anal Methods 8:5569–5580

Malik S, Ashraf M (2012) Exogenous application of ascorbic acid stimulates growth and photosynthesis of wheat (*Triticum aestivum* L.) under drought. Soil Environ 31(1):72–77

Malik S, Ashraf M, Arshad M, Malik TA (2015) Effect of ascorbic acid application on physiology of wheat under drought stress. Pak J Agric Sci 52(1):209–217

McAdam SA, Brodribb TJ (2016) Linking turgor with ABA biosynthesis: implications for stomatal responses to vapour pressure deficit across land plants. Plant Physiol 171:2008–2016

Melino VJ, Soole KL, Ford CM (2009) Ascorbate metabolism and the developmental demand for tartaric and oxalic acids in ripening grape berries. BMC Plant Biol 9:145

Miller G, Suzuki N, Ciftci-Yilmaz S, Mittler R (2010) Reactive oxygen species homeostasis and signalling during drought and salinity stresses. Plant Cell Environ 33:453–467

Mittler R (2002) Oxidative stress, antioxidants and stress tolerance. Trends Plant Sci 7:405–410

Mittler R, Zilinskas BA (1994) Regulation of pea cytosolic ascorbate peroxidase and other antioxidant enzymes during the progression of drought stress and following recovery from drought. Plant J 5:397–405

Mittler R, Vanderauwera S, Gollery M, Van Breusegem F (2004) Reactive oxygen gene network of plants. Trends Plant Sci 9:490–498

Mock HP, Dietz K-J (2016) Redox proteomics for the assessment of redox-related posttranslational regulation in plants. Biochim Biophys Acta 1864:967–973

Morina F, Jovanović LJ, Mojović M, Vidović M, Panković D, Veljović-Jovanović S (2010) Zinc-induced oxidative stress in *Verbascum thapsus* is caused by an accumulation of reactive oxygen species and quinhydrone in the cell wall. Physiol Plant 140:209–224

Munir N, Aftab F (2011) Enhancement of salt tolerance in sugarcane by ascorbic acid pretreatment. Afr J Biotechnol 10(80):18362–18370

Munné BS, Queval G, Foyer CH (2013) The impact of global change factors on redox signaling underpinning stress tolerance. Plant Physiol 161:5–19

Nicotra AB, Atkin OK, Bonser SP, Davidson AM, Finnegan EJ, Mathesius U, Poot P, Purugganan MD, Richards CL, Valladares F, van Kleunen M (2010) Plant phenotypic plasticity in a changing climate. Trends Plant Sci 15:684–692

Nishikimi M, Yagi K (1996) Biochemistry and molecular biology of ascorbic acid biosynthesis. Subcell Biochem 25:17–39

Noctor G, Mhamdi A, Foyer CH (2014) The roles of reactive oxygen metabolism in drought: Not so cut and dried. Plant Physiol 164:1636–1648

Noshi M, Hatanaka R, Tanabe N, Terai Y, Maruta T, Shigeoka S (2016) Redox regulation of ascorbate and glutathione by a chloroplastic dehydroascorbate reductase is required for high-light stress tolerance in Arabidopsis. Biosci Biotechnol Biochem 80:870–877

Opdenakker K, Remans T, Keunen E, Vangronsveld J, Cuypers A (2012) Exposure of Arabidopsis thaliana to Cd or Cu excess leads to oxidative stress mediated alterations in MAP Kinase transcript levels. Environ Exp Bot 83:53–61

Orman-Ligeza B, Parizot B, de Rycke R (2016) RBOH-mediated ROS production facilitates lateral root emergence in Arabidopsis. Development 143:3328–3339

Ort DR, Baker NR (2002) A photoprotective role for $O_2$ as an alternative electron sink in photosynthesis? Curr Opin Plant Biol 5:193–198

Ortiz-Espín A, Sánchez-Guerrero A, Sevilla F, Jiménez A (2017) The role of ascorbate in plant growth and development. In: Hossain MA, MunneBosch S, Burritt DJ, Dian-Vivancos P, Fujita M, Lorence A (eds) Ascorbic acid in plant growth, development and stress tolerance. Springer, Cham, pp 24–45

Page M, Sultana N, Paszkiewicz K, Florance H, Smirnoff N (2012) The influence of ascorbate on anthocyanin accumulation during high light acclimation in *Arabidopsis thaliana*: further evidence for redox control of anthocyanin synthesis. Plant Cell Environ 35:388–404

Pallanca JE, Smirnoff N (2000) The control of ascorbic acid synthesis and turnover in pea seedlings. J Exp Bot 51:669–674

Parsons HT, Fry SC (2012) Oxidation of dehydroascorbic acid and 2,3-diketogulonate under plant apoplastic conditions. Phytochemistry 75:41–49

Pastori GM, Kiddle G, Antoniw J, Bernard S, Veljovic-Jovanovic S, Verrier PJ, Noctor G, Foyer CH (2003) Leaf vitamin C contents modulate plant defense transcripts and regulate genes that control development through hormone signalling. Plant Cell 15:939–951

Penella C, Calatayud Á, Melgar JC (2017) Ascorbic acid alleviates water stress in young peach trees and improves their performance after rewatering. Front Plant Sci 8:1627

Pignocchi C, Fletcher JM, Wilkinson JE, Barnes JD, Foyer CH (2003) The function of ascorbate oxidase in tobacco. Plant Physiol 132:1631–1641

Pollastri S, Tattini M (2011) Flavonols: old compounds for old roles. Ann Bot 108:1225–1233

Potters G, Horemans N, Bellone S, Caubergs RJ, Trost P, Guisez Y, Asard H (2004) Dehydroascorbate influences the plant cell cycle through a glutathione-independent reduction mechanism. Plant Physiol 134:1479–1487

Pyngrope S, Bhoomika K, Dubey RS (2013) Reactive oxygen species, ascorbate-glutathione pool, and enzymes of their metabolism in drought-sensitive and tolerant indica rice (*Oryza sativa* L.) seedlings subjected to progressing levels of water deficit. Protoplasma 250:585–600

Rafique N, Raza SH, Qasim M, Iqbal N (2011) Pre-sowing application of ascorbic acid and salicylic acid to seed of pumpkin and seedling response to salt. Pak J Bot 43(6):2677–2682

Rahantaniaina MS, Li S, Chatel-Innocenti G, Tuzet A, Issakidis-Bourguet E, Mhamdi A, Noctor G (2017) Cytosolic and chloroplastic DHARs cooperate in oxidative stress-driven activation of the salicylic acid pathway. Plant Physiol 174:956–971

Sadak MS, Elhamid EMA, Mostafa HM (2013) Alleviation of adverse effects of salt stress in wheat cultivars by foliar treatment with antioxidants I. changes in growth, some biochemical aspects and yield quantity and quality. Am Eur J Agric Environ Sci 13:1476–1487

Saeidi-Sar S, Abbaspour H, Afshari H, Yaghoobi SR (2013) Effects of ascorbic acid and gibberellin A 3 on alleviation of salt stress in common bean (*Phaseolus vulgaris* L.) seedlings. Acta Physiol Plant 35(3):667–677

Sanmartin M, Drogoudi PA, Lyons T, Pateraki I, Barnes J, Kanellis AK (2003) Over-expression of ascorbate oxidase in the apoplast of transgenic tobacco results in altered ascorbate and glutathione redox states and increased sensitivity to ozone. Planta 216:918–928

Sanmartin M, Pateraki I, Chatzopoulou F, Kanellis AK (2007) Differential expression of the ascorbate oxidase multigene family during fruit development and in response to stress. Planta 225:873–885

Sato Y, Masuta Y, Saito K, Murayama S, Ozawa K (2011) Enhanced chilling tolerance at the booting stage in rice by transgenic overexpression of the ascorbate peroxidase gene. OsAPXa Plant Cell Rep 30:399–406

Schmidt R, Mieulet D, Hubberten H-M, Obata T, Hoefgen R, Fernie AR, Fisahn J, San Segundo B, Guiderdoni E, Schippers JHM, Mueller-Roeber B (2013) Salt-responsive ERF1 regulates reactive oxygen species-dependent signaling during the initial response to salt stress in rice. Plant Cell 25:2115–2131

Seiler C, Harshavardhan VT, Rajesh K, Reddy PS, Strickert M, Rolletschek H, Scholz U, Wobus U, Sreenivasulu N (2011) ABA biosynthesis and degradation contributing to ABA homeostasis during barley seed development under control and terminal drought-stress conditions. J Exp Bot 62:2615–2632

Shah F, Huang J, Cui K, Nie L, Shah T, Wu W, Chen C (2011) Physiological and biochemical changes in rice associated with high night temperature stress and their amelioration by exogenous application of ascorbic acid (vitamin C). Aust J Crop Sci 5(13):1810–1816

Shalata A, Neumann PM (2001) Exogenous ascorbic acid (vitamin C) increases resistance to salt stress and reduces lipid peroxidation. J Exp Bot 524:2207–2211

Sharma SS, Dietz K-J (2009) The relationship between metal toxicity and cellular redox imbalance. Trends Plant Sci 14:43–50

Sharma P, Dubey RS (2004) Ascorbate peroxidase from rice seedlings: Properties of enzyme isoforms, effects of stresses and protective roles of osmolytes. Plant Sci 167:541–550

Sharma P, Jha AB, Dubey RS, Pessarakli M (2012) Reactive oxygen species, oxidative damage, and antioxidative defense mechanism in plants under stressful conditions. J Bot 217037

Singh N, Bhardwaj RD (2016) Ascorbic acid alleviates water deficit induced growth inhibition in wheat seedlings by modulating levels of endogenous antioxidants. Biologia 71(4):402–413

Sivaci A, Kaya A, Duman S (2014) Effects of ascorbic acid on some physiological changes of pepino (*Solanum muricatum* Ait.) under chilling stress. Acta Biol Hung 65(3):305–318

Slesak I, Libik M, Karpinska B, Karpinski S, Miszalski Z (2007) The role of hydrogen peroxide in regulation of plant metabolism and cellular signalling in response to environmental stresses. Acta Biochim Pol 54:39–50
Smirnoff N (2000) Ascorbate biosynthesis and function in photoprotection. Philos Trans R Soc Lond Ser B Biol Sci 355:1455–1464
Smirnoff N (2011) Vitamin C: The metabolism and functions of ascorbic acid in plants. Adv Bot Res 59:107–177
Smirnoff N, Wheeler GL (2000) Ascorbic acid in plants: biosynthesis and function. Crit Rev Biochem Mol Biol 35:291–314
Sofo A, Tuzio AC, Dichio B, Xiloyannis C (2005) Influence of water deficit and rewatering on the components of the ascorbate–glutathione cycle in four interspecific Prunus hybrids. Plant Sci 169(2):403–412
Spasojević I, Bogdanović-Pristov J (2010) The potential physiological implications of polygalacturonic acid-mediated production of superoxide. Plant Signal Behav 5:1525–1529
Suekawa M, Kondo T, Fujikawa Y, Esaka M (2017) Regulation of ascorbic acid biosynthesis in plants. In: Hossain MA, MunneBosch S, Burritt DJ, Dian-Vivancos P, Fujita M, Lorence A (eds) Ascorbic acid in plant growth, development and stress tolerance. Springer. Cham. Springer, Cham, pp 157–176
Sultan SE (2000) Phenotypic plasticity for plant development, function and life history. Trends Plant Sci 5:537–542
Sun WH, Duan M, Li F, Shu DF, Yang S, Meng QW (2010) Overexpression of tomato tAPX gene in tobacco improves tolerance to high or low temperature stress. Biol Plant 54:614–620
Szechyńska HM, Karpiński S (2013) Light intensity-dependent retrograde signalling in higher plants. J Plant Physiol 170:1501–1516
Tabatabaei SA, Naghibalghora SM (2013) The effect of ascorbic acid on germination characteristics and proline of sesame seeds under drought stress. Int J Agric Crop Sci 6(4):208–212
Takahama U, Oniki T (1992) Regulation of peroxidase-dependent oxidation of phenolics in the apoplast of spinach leaves by ascorbate. Plant Cell Physiol 33:379–387
Takahama U, Oniki T (1997) A peroxidase/phenolics/ascorbate system can scavenge hydrogen peroxide in plant cells. Physiol Plant 101:845–852
Talla S, Riazunnisa K, Padmavathi L, Sunil B, Rajsheel P, Raghavendra AS (2011) Ascorbic acid is a key participant during the interactions between chloroplasts and mitochondria to optimize photosynthesis and protect against photoinhibition. J Biosci 36:163–173
Tamás L, Bočová B, Huttová J, Mistrík I, Ollé M (2006) Cadmium-induced inhibition of apoplastic ascorbate oxidase in barley roots. Plant Growth Regul 48:41–49
Tang L, Kwon SY, Kim SH, Kim JS, Choi JS, Cho KY, Sung CK, Kwak SS, Lee HS (2006) Enhanced tolerance of transgenic potato plants expressing both superoxide dismutase and ascorbate peroxidase in chloroplasts against oxidative stress and high temperature. Plant Cell Rep 25:1380–1386
Tenhaken R (2014) Cell wall remodeling under abiotic stress. Front Plant Sci 5:771
Tombesi S, Nardini A, Frioni T, Soccolini M, Zadra C, Farinelli D, Poni S, Palliotti A (2015) Stomatal closure is induced by hydraulic signals and maintained by ABA in drought-stressed grapevine. Sci Rep 5:12449
Torsethaugen G, Pell EJ, Assmann SM (1999) Ozone inhibits guard cell K1 channels implicated in stomatal opening. Proc Natl Acad Sci U S A 96:13577–13582
Veljović-Jovanović S, Kukavica B, Cvetić T, Mojović M, Vučinić Ž (2005) Ascorbic acid and the oxidative processes in pea root cell wall isolates: characterization by fluorescence and EPR spectroscopy. Ann N Y Acad Sci 1048:500–504
Veljović-Jovanović S, Vidović M, Morina F (2017) Ascorbate as a key player in plant abiotic stress response and tolerance. In: Hossain MA, MunneBosch S, Burritt DJ, Dian-Vivancos P, Fujita M, Lorence A (eds) Ascorbic acid in plant growth, development and stress tolerance. Springer, Cham, pp 47–109
Venkatesh J, Park SW (2014) Role of L-ascorbate in alleviating abiotic stresses in crop plants. Bot Stud 55:38. http://www.as-botanicalstudies.com/content/55/1/38

Venkatesh J, Upadhyaya CP, Yu JW, Hemavathi A, Kim DH, Strasser RJ, Park SW (2012) Chlorophyll a fluorescence transient analysis of transgenic potato overexpressing D-galacturonic acid reductase gene for salinity stress tolerance. Hortic Environ Biotechnol 53:320–328

Vidović M, Morina F, Milić S, Albert A, Zechmann B, Tosti T, Winkler JB, Jovanović SV (2015a) Carbon allocation from source to sink leaf tissue in relation to flavonoid biosynthesis in variegated *Pelargonium zonale* under UV-B radiation and high PAR intensity. Plant Physiol Biochem 93:44–55

Vidović M, Morina F, Milić S, Zechmann B, Albert A, Winkler JB, Veljović-Jovanović S (2015b) Ultraviolet-B component of sunlight stimulates photosynthesis and flavonoid accumulation in variegated leaves depending on background light. Plant Cell Environ 38(5):968–979

Vidović M, Morina F, Lj P, Milić-Komić S, Živanović B, Veljović-Jovanović S (2016) Antioxidative response in variegated Pelargonium zonale leaves and generation of extracellular $H_2O_2$ in (peri) vascular tissue induced by sunlight and paraquat. J Plant Physiol 206:25–39

Vidović M, Morina F, Veljović-Jovanović S (2017) Stimulation of various phenolics in plants under ambient UV-B radiation. In: Singh VP, Singh S, Prasad SM, Parihar P (eds) UV-B Radiation: from environmental stressor to regulator of plant growth. Wiley-Blackwell, Chichester, pp 9–56

Vuletić M, Hadži-Tašković Šukalović V, Marković K, Kravić N, Vučinić Ž, Maksimović V (2014) Differential response of antioxidative systems of maize (*Zea mays* L.) roots cell walls to osmotic and heavy metal stress. Plant Biol 16:88–96

Wang W, Vinocur B, Altman A (2003) Plant responses to drought, salinity and extreme temperatures: towards genetic engineering for stress tolerance. Planta 218:1–14

Wang Y, Wisniewski M, Meilan R, Cui M, Fuchigami L (2006) Transgenic tomato (*Lycopersicon esculentum*) overexpressing cAPX exhibits enhanced tolerance to UV–B and heat stress. J Appl Hortic 8:87–90

Wang HS, Yu C, Zhu ZJ, Yu XC (2011) Overexpression in tobacco of a tomato GMPase gene improves tolerance to both low and high temperature stress by enhancing antioxidation capacity. Plant Cell Rep 30:1029–1040

Weisiger RA, Fridovich I (1973) Mitochondrial superoxide dismutase site of synthesis and intramitochondrial localization. J Biol Chem 248:4793–4796

West M, Harada JJ (1993) Embryogenesis in higher plants: An overview. Plant Cell 5:1361–1369

Wu D, Hu Q, Yan Z, Chen W, Yan C, Huang X, Zhang J, Yang P, Deng H, Wang J, Deng X (2012) Structural basis of ultraviolet-B perception by UVR8. Nature 484:214–219

Xu M, Lu Y, Yang H, He J, Hu Z, Hu X et al (2015) ZmGRF, a GA regulatory factor from maize, promotes flowering and plant growth in Arabidopsis. Plant Mol Biol 87:157–167

Yin L, Wang S, Eltayeb AE, Uddin MI, Yamamoto Y, Tsuji W, Takeuchi Y, Tanaka K (2010) Overexpression of dehydroascorbate reductase, but not monodehydroascorbate reductase, confers tolerance to aluminum stress in transgenic tobacco. Planta 231:609–621

Zhang Y (2012) Ascorbic acid in plants. Biosynthesis, regulation and enhancement. Springer, Boston. ISSN 2192-1210

Zhang L, Wang Z, Xia Y, Kai G, Chen W, Tang K (2007) Metabolic engineering of plant L – ascorbic acid biosynthesis: recent trends and applications. Crit Rev Biotechnol 27:173–182

Zhu JK (2017) Abiotic stress signaling and responses in plants. Cell 167(2):313–324

Zipor G, Oren SM (2013) Do vacuolar peroxidases act as plant caretakers? Plant Sci 199:41–47

Zonouri M, Javadi T, Ghaderi N (2014) Effect of foliar spraying of ascorbic acid on cell membrane stability, lipid peroxidation, total soluble protein, ascorbate peroxidase and leaf ascorbic acid under drought stress in grapes. Int J Adv Biol Biomed Res 2:349–354

# An Assessment of Efficiency of Zinc Priming in Rice (cv. MTU-7029) During Germination and Early Seedling Growth

**Debojyoti Moulick, Subhas Chandra Santra, Dibakar Ghosh, and Sanjib Kumar Panda**

## Contents



**Abstract** We laid down this experiment with two objectives. First to evaluate the germination and early seedling growth behaviour of rice seeds primed with water (hydroprimed or HP) and primed with zinc (Zn) using zinc sulphate under arsenic

D. Moulick (✉)
Department of Environmental Science, University of Kalyani, Nadia, West Bengal, India

Plant Molecular Biotechnology Laboratory, Department of Life Science, Assam University, Silchar, Assam, India

S. C. Santra
Department of Environmental Science, University of Kalyani, Nadia, West Bengal, India

D. Ghosh
ICAR Directorate of Weed Research, Jabalpur, Madhya Pradesh, India

S. K. Panda
Plant Molecular Biotechnology Laboratory, Department of Life Science, Assam University, Silchar, Assam, India

 
M. Hasanuzzaman, V. Fotopoulos (eds.), *Priming and Pretreatment of Seeds and Seedlings*, https://doi.org/10.1007/978-981-13-8625-1_24

free (As-) and As spiked (As+) condition. Whereas, our second intention was to asses whether seed priming with Zn in rice can ameliorate As induced phytotoxicity or not. Here in this experiment, total six priming treatments (0.0, HP, 0.5,1.0,1.5 and 2.0 % Zn priming) and three stress regimes {0.0, 8.0 ($As^{3+}$) and 8.0 ($As^{5+}$) mg $L^{-1}$} were considered and arranged these treatment combinations in complete randomized design (CRD) in 90.0 mm glass petriplates. Results from the current investigation suggest that priming rice seeds with Zn significantly (at $p \leq 0.0001$) improves germination and seedling growth and minimize the stress-induced biochemical markers under normal condition than those primed with water and unprimed seeds at 10 days after sowing (DAS). Whereas, under As stress germination and seedling growth were inhibited in a noteworthy fashion, irrespective of seed priming treatments. Among the stressor $As^{3+}$ seems to be more phytotoxic than $As^{5+}$ on germination and seedling. Seeds which were primed had greater germination and longer root and shoot growth when compared with the seedlings of hydroprimed and unprimed seeds. Our findings also indicate that, may be due to a significant interactive aspect (at $p \leq 0.001$) among Zn and As (Zn × As), Zn primed seedlings accumulated lesser amount of As at 10 DAS than the unprimed and hydroprimed seedlings, irrespective of nature of stress.

**Keywords** Seed priming · Rice · Zinc · Arsenic · Interaction

## 1 Introduction

More than 30% of soils are deficient in phyto-available Zn around the globe (Alloway 2008). Among the field crops, cereals are considered more prone to Zn deficiency resulting in considerable decrease in grain yield as well as nutritional quality (Cakmak et al. 1999). Moreover, in rice agroecosystem Zn deficiency may reach up to 50% (Fageria et al. 2002; Quijano-Guerta et al. 2002). In Asia, Zn deficiency has been regarded as one of the most important nutritional stress factors for irrigated rice agroecosystem (Quijano-Guerta et al. 2002). In the soil, Zn generally exists in varied number of forms, differs in solubility and chemistry such as soluble (water-soluble form as solutions), co-exists along with organic matter or associated with other minerals, is an integral part of primary minerals, etc. Whereas among the various forms, the water-soluble form is accessed by plants for uptake (Marschner 1995; Shuman 1991; Almendros et al. 2008), which in turn depends on other factors like acidity or alkalinity of soil, organic carbon and redox potential (Alloway 2009). Rice is generally transplanted and cultivated in submerged paddy field environments. In the paddy fields, various levels of biological as well as physicochemical alternations take place on continuous basis under waterlogged condition that ultimately emerges to be an important factor in maintaining sustainability of rice production (De Datta 1981). Findings of Forno et al. (1975) and Mikkelsen and Kuo (1976) suggest that a reduction in Zn content in soil solution is associated soon after flooding of paddy fields but at the same time an effort to restore by 0.3–0.5 μM (micro molar) has also been observed. Few other factors associated with paddy field soil such as zinc carbonate (found mostly in calcareous soil) tendency of Zn to precipitate as zinc oxide are also found to govern phyto-availability of Zn. Besides these, formation of franklinite (an insoluble form of zinc, iron, and oxygen) along with zinc sulphide (in acidic soil mostly) may attribute behind depletion of Zn content in paddy field soil (Sajwan and Lindsay 1986; Kittrick 1976; Bostick et al. 2001). Authors like Johnson-Beebout et al. (2009) and Beebout et al. (2011) found that redox potential and sulphur content are also responsible for low phytoavailability of Zn and further added that with lowering in redox potential, precipitation of Zn increases and subsequently cuts down in phytoavailable Zn concentration.

Besides deficiency of essential minerals, presence of toxicant(s) like arsenic (As) is also a crucial factor or threat to rice agroecosystem. Earth crust is considered as the major source of As in the environment, with the presence of ≥200 As-bearing minerals (arsenopyrite being the major source). According to the observation made by Nriagu and Pacyna (1988) based on a decade-long study, it was found that anthropogenic activity has contributed as much as 52,000–112,000 tons of As on a yearly basis in soil environment. Authors like Cozzolino et al. (2010), Mondal et al. (2006) and Verbruggen et al. (2009) reported that besides geogenic activities, application of various agents such as phosphate fertilizers, herbicides and insecticides in agricultural sector as well as wastes from semiconductor industry are the major sources of As in environment. But most prevalent and adverse consequences of As contamination can be seen in the arsenic "hotspot area" where As-rich (leaching As naturally into the ground water) ground water has been used to irrigate paddy fields and drinking, from countless past (Santra et al. 2013; Kim et al. 2009).

Previous findings suggest that supplementation of mineral nutrients though foliar and soil application could have been an effective way to reduce heavy metal and / metalloid phytoavailability and their uptake as well as toxicity (Naeem et al. 2015; Saifullah et al. 2014, 2016). Earlier observations made by authors like Bakhat et al. (2017) and Islam et al. (2016) have also explained various strategies to decrease As-induced toxicity and subsequent accumulation in rice. Keeping in mind the above possibilities we laid down the current experiment to investigate the potential of supplementing rice seeds with Zn (by using seed priming technology) on germination and early seedling growth under arsenic stress condition.

## 2 Materials and Methods

### *2.1 Rice Variety and Seed Priming Treatments*

In the current experiment, *MTU-7029* rice variety was chosen. This variety is popularly known as *Swarna*, having dark brown seed coat colour with short and bold grain morphology. The seeds of Swarna had initial moisture content (on dry weight basis) of 10.1%, and length to width ratio was 2.62 ± 0.03. We collected this variety from Regional Rice Research Station located in Hooghly district of West Bengal, India. Seeds are first surface sterilized with 0.1% $HgCl_2$ solution and divided into six parts followed by priming with solution containing 0.5%, 1.0%, 1.5% and 2.0% Zn as $ZnSO_4.7H_2O$ salt in weight by volume ratio. During the course of priming treatment, a portion of seed was also primed with double distilled water and designated as hydroprimed (HP), and remaining one portion was kept as control (Moulick et al. 2016a, 2017).

## 2.2 Experimental Layout

We conducted the entire experiment under laboratory condition in 90.0 mm glass petriplates (Borosil). First the seeds from each treatment were placed in glass petriplates lined with cotton and filter paper. These seeds are either treated with double distilled water or equal amount of, that is, 8.0 mg $L^{-1}$ of $As^{5+}$ (sodium arsenate; having molecular weight (MW)—312.01, purchased from MERCK) and $As^{3+}$ (sodium arsenite salt; with MW—130, also purchased from MERCK) and covered with another filter paper. Then these petriplates were kept inside an incubator at $25 \pm 0.8$ °C in the absence of light for 10 days. In this experiment 6 priming treatment and 3 stress regimes were arranged in accordance with complete randomized design (CRD) replicated five times in 90 different petriplates (Moulick et al. 2016a).

## 2.3 Determination of Morphometric and Biochemical Attributes

At 10 DAS (days after sowing) seeds were regarded as germinated, only when they have radical length $\geq$ 2.0 mm, and their respective attributes like root and shoot length and dry weights were noted down by previously established protocol (Moulick et al. 2016a). The estimation of lipid peroxidation as MDA content using thiobarbituric acid reactive substances (TBARS) and proline content was carried out according to the method described by Heath and Packer (1968) and Bates et al. (1973), respectively.

## 2.4 Quantification of Total Arsenic Content in Intact Rice Seedlings and Quality Assurance of Process

Oven-dried intact rice seedlings (0.5 g) were grinded into fine powder form then were acid digested using a tri-acid mixture having perchloric acid (70%), concentrate nitric acid and sulphuric acid (all ACS grade MERCK), according to the method described by Das et al. (2004). Before the acid digestion process all the glassware and stainless steel instruments were dipped in 2 M chromic acid (freshly prepared) solution and washed adequately and oven dried. During the course of acid digestion, pure reagent blank (only the tri-acid mixture without sample) and SRM or Standard Reference Material [rice flour (Item No. 1568A), obtained from National Institute of Science and Technology (NIST), USA] were also digested. Total As content of intact rice seedlings was then analyzed by following the methodology given by Welsch et al. (1990) using FI-HG-AAS or flow injection hydride generation atomic absorption spectrometer (Perkin Elmer A Analyst 400). The certified value of SRM was 0.29 ± 0.03, and our observed value was 0.282 ± 0.011, respectively.

### 2.5 Statistical Analysis

Here we expressed all the data as a mean value of five replications (n = 5). By using the statistical analysis software (SAS Windows Version 9.4) and using two-way analyses of variance (*TWO-WAY ANOVA*) and adopting GLM or general linear model procedure, the effects of various priming treatments and As stress (alone or in combinations) were separated. Further the means of treatments (alone or in combinations) were separated with the help of *Tukey's Honest Significant Difference* (HSD) test at a 5%.

## 3 Results

### 3.1 Consequences of Priming Rice Seeds with Zn and As Stress on Seed Germination and Seedling Growth

During germination, As stress (irrespective of nature, i.e. $As^{3+}/As^{5+}$) was found to inhibit the seed germination of the tested variety in a significant manner when compared with the control (Table 1). Among the two different forms of stressor (s) considered here, 8.0 mg As $L^{-1}$ ($As^{3+}$) seems to be more inhibitory than 8.0 mg As $L^{-1}$ ($As^{5+}$) at 10 DAS. When the primed seeds of the tested rice variety germinated under similar As stress regimes were compared with the unprimed seeds, a noteworthy (significant at $\leq 0.05$ level) enhancement in FGP, i.e. restoration in germination%, can be seen (Table 1). When the priming treatments were further compared among themselves, the hydroprimed seeds though had a marginal edge than that of unprimed seeds under both the stress regimes, but those that were primed with Zn had a significant effect. Among the different doses of Zn that were considered here, those seeds primed with 1.0% Zn had significantly greater FGP under As stress than the remaining priming treatments. Furthermore, another important aspect was also emerging out from the current experiment, which the primed seeds had significantly (at $\leq 0.0001$ level) greater score (FGP) over the unprimed seeds when germinated

**Table 1** Effects of various seed priming and As stress on seed germination and seedling growth of rice genotype at 10 DAS

| Seed priming treatments | As (stress) mg $L^{-1}$ | Final germination (%) | Root length (cm) | Shoot length (cm) | Dry weight (g) |
|---|---|---|---|---|---|
| 0 | 0 | [(1)]94.44 ab | 6.07 d | 2.63 b | 0.13 a |
| 0 | 8.0 $As^{3+}$ | 26.66 e | 1.93 i | 0.93 d | 0.075 a |
| 0 | 8.0 $As^{5+}$ | 52.22 c-e | 2.57 g-i | 1.2 cd | 0.084 a |
| HP | 0 | 68.89 a-d | 6.57 cd | 3.27 ab | 0.14 a |
| HP | 8.0 $As^{3+}$ | 38.89 de | 2.1 hi | 1.23 cd | 0.090 a |
| HP | 8.0 $As^{5+}$ | 55.55 b-e | 2.73 gh | 1.43 cd | 0.101 a |
| 0.5% Zn | 0 | 100 a | 6.9 bc | 3.53 a | 0.16 a |
| 0.5% Zn | 8.0 $As^{3+}$ | 51.11 c-e | 3.17 fg | 1.43 cd | 0.112 a |
| 0.5% Zn | 8.0 $As^{5+}$ | 65.55 a-e | 3.6 ef | 1.53 cd | 0.112 a |
| 1.0% Zn | 0 | 100 a | 7.43 b | 3.27 ab | 0.172 a |
| 1.0% Zn | 8.0 $As^{3+}$ | 64.44 a-e | 3.5 ef | 1.57 cd | 0.097 a |
| 1.0% Zn | 8.0 $As^{5+}$ | 64.44 a-e | 4.1 e | 1.73 c | 0.112 a |
| 1.5% Zn | 0 | 93.33 ab | 8.27 a | 3.87 a | 0.164 a |
| 1.5% Zn | 8.0 $As^{3+}$ | 51.11 c-e | 4.13 e | 1.83 c | 0.112 a |
| 1.5% Zn | 8.0 $As^{5+}$ | 65.55 a-e | 4.13 e | 1.73 c | 0.127 a |
| 2.0% Zn | 0 | 82.22 a-c | 6.23 cd | 2.73 b | 0.118 a |
| 2.0% Zn | 8.0 $As^{3+}$ | 34.44 de | 2.4 hi | 0.93 d | 0.112 a |
| 2.0% Zn | 8.0 $As^{5+}$ | 38.89 de | 2.6 g-i | 1.03 d | 0.082 a |
| ***Sources of Variation*** | | ***P-Values*** | | | |
| Zn | | *≤0.0001* | *≤0.0001* | *≤0.0001* | *Ns* |
| As | | *≤0.0001* | *≤0.0001* | *≤0.0001* | *Ns* |
| Zn × AS | | *≤0.05* | *≤0.05* | *Ns* | *Ns* |

[(1)]Values referring to the mean value (n = 5) followed by the identical letter case(s) in a column were not significantly different at $P < 0.05$
*HP* hydro-primed, *DAS* days after sowing

alike the control, i.e. in As-free environment. Besides these, when the germination pattern was carefully observed, it suggests that irrespective of As stress there is a decline in germination% that can also be seen beyond 1.5% priming with Zn.

Results from the present investigation also indicate that seedlings primed with water (HP) and different doses of Zn and grown similar to the control had relatively larger root and shoot length (significant at $p \leq 0.0001$ level) than the control (unprimed seed). Our findings point out towards a positive impact of seed priming with Zn as growth-promoting agent (Table 1). Whereas a noteworthy reduction in root and shoot growth can also be seen among the seedlings under As stress. At 10 DAS unprimed seedlings having 1.93 cm long root and 0.93 cm long shoot under $As^{3+}$ stress along with 2.57 cm root and 1.2 cm shoot can be seen in case of $As^{5+}$ stress, when compared to the seedlings having 6.07 cm long root and 2.67 cm long shoot of the control. The inhibitory effects of As (irrespective of species) were more pronounced in root elongation process (62.93%) than in shoot growth (59.51%). The results also revealed that among the two As species considered here, $As^{3+}$ stress had 24.90% and 22.5% greater phytotoxic effects on root and shoot elongation compared to the $As^{5+}$ stress at 10 DAS in the tested variety. Besides these, substantial interactive effects among Zn and As on the root and shoot length of rice seedlings were also visible (Table 1). Seedlings of rice seeds when primed with 1.5% mg Zn $L^{-1}$ had longer root (by 1.85 times) and shoot length (by 1.71 fold) than the seedlings of unprimed seeds grown under similar As-stressed condition at 10 DAS.

When the seedlings were further compared for dry weight or biomass accumulation, three different scenarios can be seen. First a noteworthy reduction in dry weight content in intact seedlings of unprimed seeds grown in As stress condition can be seen. When compared with the dry weight of intact seedlings of unprimed seeds without As stress, i.e. control a 42.30 and 35.39% reduction in dry weight can be seen in case of the seedlings grown in $As^{3+}$ and $As^{5+}$ stresses, respectively. The results also show that $As^{3+}$ stress was 16.37% more phytotoxic than $As^{5+}$ in terms of biomass accumulation. In the second case, there is a sharp enhancement in biomass of seedlings of primed seeds grown alike the control in As-free environment at 10 DAS. A 0.91 to 1.32% increase in biomass accumulation of primed seedlings can be detected, compared to the control. In the third case, an enhancement in biomass of intact seedlings of primed seedlings can be seen in the seedlings of primed seedlings grown in As stress (irrespective of nature) than the unprimed seedlings (Table 1).

## 3.2 *Effect of Zn Supplementation Through Seed Priming and As Stress on MDA and Proline Contents*

The presence of As (as $As^{3+}/As^{5+}$) to unprimed seeds enhanced the total MDA content by 1.64 times compared to the control at 10 DAS (Table 2). On the other hand, the supplementation of Zn has decreased the MDA content of intact rice seedlings by 15.56% as compared to the control, grown alike in As-free medium in a significant manner (at $p \leq 0.0001$ level). Besides these, a noteworthy difference among the Zn doses on MDA content was observed at 10 DAS. However, a significant interactive aspect among Zn and As (Zn × As) was also observed among the seedlings of Zn primed seeds grown in As stressed condition.

Similar to the above situation an enhancement (by 1.71 time) in proline content in the seedlings of unprimed seeds exposed to As stress was also visible, when compared with the control. When proline content of unprimed seedlings experienced As stress was compared with the control (no as stress), a twofold greater proline content was noted. Furthermore, the results also suggest that priming rice seeds prior to sow with Zn and double distilled water (HP) significantly reduce proline content by 1.21 and 1.13 times in case exposed to $As^{3+}$ and $As^{5+}$ stress, respectively, than the proline content of seedlings of unprimed seeds exposed to similar As stress at 10 DAS (Table 2).

## 3.3 *Effect of Priming with Zn and As Stress on Total Arsenic Accumulation*

Seedlings grown under $As^{3+}$ stress had 13.81% greater As content than the seedlings grown under $As^{5+}$ stress at 10 DAS. Priming rice seeds with Zn effectively decreased the As concentration by 1.13-fold (in case of $As^{3+}$ stress) and 1.19 times (in case of $As^{5+}$ stress) in seedlings, than the unprimed seedlings exposed to identical As stress regimes. It is evident from the findings that seed priming with 0.5 to 1.5 mg Zn $L^{-1}$ (though are not statistically different) were effective in minimizing the As content in rice seedlings. A noteworthy interactive effect among Zn and As on As build-up ($p < 0.001$) in rice seedlings can be seen (Table 2).

**Table 2** Effects of various seed priming and As stress on selected biochemical attributes and As content of rice seedling at 10 DAS

| Seed priming treatments | As (stress) mg $L^{-1}$ | MDA (nmol/g FW) | Proline (m mole/g FW) | tAS (mg$Kg^{-1}$DW) |
|---|---|---|---|---|
| 0 | 0 | 36.75 i | 0.54 fg | 0 f |
| 0 | 8.0 $As^{3+}$ | 62.58 a | 0.91 a | 5.14 a |
| 0 | 8.0 $As^{5+}$ | 57.6 b | 0.80 bc | 4.43 b-d |
| HP | 0 | 33.66 jk | 0.45 hi | 0 f |
| HP | 8.0 $As^{3+}$ | 53.16 c | 0.82 a-c | 4.89 ab |
| HP | 8.0 $As^{5+}$ | 48.13 ef | 0.72 c-e | 4.23 cd |
| 0.5% Zn | 0 | 29.47 lm | 0.42 hi | 0 f |
| 0.5% Zn | 8.0 $As^{3+}$ | 50.8 d | 0.69 de | 4.61 a-c |
| 0.5% Zn | 8.0 $As^{5+}$ | 42.8 7 g | 0.63 ef | 3.9 de |
| 1.0% Zn | 0 | 27.43 m | 0.38 i | 0 f |
| 1.0% Zn | 8.0 $As^{3+}$ | 46.5 f | 0.67 e | 4.74 a-c |
| 1.0% Zn | 8.0 $As^{5+}$ | 39.4 h | 0.66 e | 3.54 e |
| 1.5% Zn | 0 | 31.37 kl | 0.52 fg | 0 f |
| 1.5% Zn | 8.0 $As^{3+}$ | 54.5 c | 0.78 b-d | 4.27 b-d |
| 1.5% Zn | 8.0 $As^{5+}$ | 46.2 f | 0.78 b-d | 3.83 de |
| 2.0% Zn | 0 | 33.23 j | 0.53 fg | 0 f |
| 2.0% Zn | 8.0 $As^{3+}$ | 54.57 c | 0.84 ab | 4.72 a-c |
| 2.0% Zn | 8.0 $As^{5+}$ | 50.13 de | 0.78 b-d | 3.95 de |
| ***Sources of variation*** | | ***P-values*** | | |
| Zn | | *≤0.0001* | *≤0.0001* | *≤0.0001* |
| As | | *≤0.0001* | *≤0.0001* | *≤0.0001* |
| Zn × AS | | *≤0.0001* | *≤0.05* | *≤0.001* |

[(1)]Values referring to the mean value (n = 5) followed by the identical letter case(s) in a column were not significantly different at $P < 0.05$
*HP* hydro-primed, *DAS* days after sowing

## 4 Discussion

Rice plant (*Oryza sativa*. L) has been regarded as one of the most efficient As accumulators. It is an established fact that rice plants have capability to accumulate As mainly in root from paddy field soil and then translocate it into its shoot (aerial part). Arsenic content in rice grain after maturation is determined by the capability of rice plants to assimilate As into root from contaminated field (irrigated with As contaminated ground water), followed by its (rice plant) ability to translocate. According to the opinion of Chen et al. (2016), Dai et al. (2016) and Sahoo and Kim (2013), the whole process of As accumulation and translocation has been influenced by a large number of soil physicochemical para meters such as pH, conductivity (EC), texture, etc. From countless past, rice seedlings are uprooted (from seed bed) by the farmers and then manually transplanted into puddled submerged soil (with 3–4 cm of stagnant water) under anaerobic condition. Such anaerobic environment favours the conversion of $As^{5+}$ to $As^{3+}$ (Ghosh et al. 2016; Takahashi et al. 2004) or by simply broadcasting the seeds to the field itself (known as direct seeding). Once the seeds or seedlings come in contact with As, ROS or reactive oxygen species generation amplifies within the cell and in general stimulates the lipid peroxidation process, which may finally result to cell death (Farooq et al. 2016a, b). Various other phytotoxic effects of As in rice plant have been documented so far. Among the noteworthy adverse consequences of As stress on rice plant such as inhibition of germination process, seedling development (Moulick et al. 2018c; Moulick et al. 2017; Moulick et al. 2016a; Shri et al. 2009) to marked reduction in chlorophyll content, tiller number, decrease in flag leaf area have been reported. In addition to considerable reduction in test weight, biomass as well as grain quality traits etc. have been documented by authors from time to time (Moulick et al. 2018a, b, d; Moulick et al. 2016b).

Findings from our current investigation support the previous findings regarding marked reduction in germination and early seedling growth reported by Moulick et al. (2016a, 2017, 2018c). There is no doubt that the presence of As can significantly inhibit germination and seedling growth process in both soil-less and soil-based media. The inhibition of germination may have attributed to the susceptibility of seeds (irrespective of priming treatment) to As stress. Germination is the crucial stage of plant's life cycle where for the first time it comes in contact with the environment. Whereas, Zn is considered as an essential element for plant growth. Though its (Zn) synergistic and/or antagonistic aspect with As has not been addressed in detail so far. The present study was laid down to explore the effects of Zn supplementation (using seed priming technique) in rice seed of *Swarna* variety on germination and seedlings growth under As stress {as $As^{3+}$(dominant in pre monsoon season) and $As^{5+}$ stress (dominant in monsoon season)} of equal strength (Table 1).

The current study revealed that when Zn is added (fortified) in seeds of the tested variety, by seed priming technology, and permitted to germinate under normal condition (absence of As stress), the seeds had longer root and shoot length and greater seedling biomass accumulation, demonstrating the growth promoting aspect of Zn. The findings from the current experiment also show that supplementation of Zn through seed priming can detoxify As-induced toxicity and promote seed germination and elongation of root and shoot length and facilitate greater seedling biomass of rice, when compared with the seedlings of unprimed seeds. However, this improvement process might be dependent on the dosages of Zn and forms of inorganic As species ($As^{3+}$ / $As^{5+}$). Furthermore, seed priming with Zn (with 1.0 and 1.5%) seems to be more beneficial for seed germination and seedling growth. Results also reveal that the $As^{3+}$ stress is more phytotoxic on seed germination and seedlings growth of rice, than that of $As^{5+}$ species. The augmentation of seed germination and seedlings growth by Zn seed priming was higher in $As^{5+}$ stress, compared to the $As^{3+}$ stress. The results above matched well with our previous observation (Moulick et al. 2016a, b), which describes the effects of supplementing of selenium in rice on germination and seedling growth under As stress (Table 2).

The MDA level is a vital stress induced biomarker represents significant cellular injury due to inadequate ability to quench the excess ROS production. The addition of As promotes the MDA production in 10-day-old rice seedlings. From the results, it is evident that As triggered more membrane damages. Furthermore, among the two species considered here, the potency of inducing damage was more in $As^{3+}$ than $As^{5+}$. A decrease in MDA contents in rice seedlings (primed with Zn) when exposed to As stress was also found from the results; it indicates the potential of Zn as an effective antioxidant (Chen et al. 2009). The decrease in MDA contents may be due to the lessening of As-induced oxidative stress and ROS-induced damages and a significant interaction among Zn and As. Proline, an amino acid and a well-established osmoticum, is capable to protect the protein against denaturation (Kavi Kishor et al. 2005). The elevated content of proline is generally used as a stress-related biomarker for ROS-induced damage. The enhancement in proline content along with reduced seedling growth under As stress showed the adverse role of As on rice seedlings. The supplementation of Zn (before sowing, through seed priming technology) showed a reduction in the proline content of rice seedlings grown under As stress. Among the doses considered here, lower dose of Zn (0.5 and 1.0%) was more effective in dropping the proline content even under As stress in rice (Table 2).

The total As content was found to be much lower in Zn primed rice seedlings, when compared to unprimed seedlings grown under similar As stress at 10 DAS (Table 2). The trend of having reduction in As content by Zn primed seedlings than unprimed one though reported for the first time, yet our findings support partially a study published by Das et al. (2005) who reported the effects of supplementing Zn in soil on As accumulation in mature rice plant, found similar reduction in As content. But a few other literatures suggest that Zn can ameliorate cadmium-induced toxicity and accumulation in wheat (Naeem et al. 2015).

## 5 Conclusion

In our opinion, if Zn supplementation using seed priming technology (before sowing) at concentration ranging from 0.5% to 1.5% can be effective in minimizing the adverse effects of As on seed germination and seedling growth in rice by decreasing the As uptake, modulating the oxidative damage. Thus, seed priming with Zn may have probability to improve seedling emergence and luxurious growth of rice plant by diminishing the As toxicity. Whereas the present study also reveals that in the absence of As stress, priming rice seeds with Zn can also be beneficial, too. This is a laboratory-based study, and further field-based investigation should be carried out to elaborate the potential of seed priming with Zn, which can be assessed in a much better way.

## References

Alloway BJ (2008) Zinc in soils and crop nutrition, 2nd edn. IZA Brussels, Belgium

Alloway BJ (2009) Soils factors associated with zinc deficiency in crops and humans. Environ Geochem Health 31(5):537–548

Almendros P, Gonzalez D, Obrador A, Alvarez JM (2008) Residual zinc forms in weakly acidic and calcareous soils after an oilseed flax crop. Geophysical Research Abstracts EGU General Assembly 10, EGU2008-A-12479

Bakhat HF, Zia Z, Fahad S, Abbas S, Hammad HM, Shahzad AN, Abbas F, Alharby H, Shahid M (2017) Arsenic uptake, accumulation and toxicity in rice plants: possible remedies for its detoxification: a review. Environ Sci Pollut Res 24(10):9142–9158

Bates LS, Waldren RP, Teare ID (1973) Rapid determination of free proline for water-stress studies. Plant Soil 39(1):205–207

Beebout SJ, Francis HCR, Dennis SJT, Ranee CM (2011) Reasons for variation in rice (Oryza sativa) grain zinc response to zinc fertilization. In: 3rd International Zinc Symposium 10–14 October 2011, Hyderabad, India

Bostick BC, Hansel CM, La Force MJ, Fendorf S (2001) Seasonal fluctuations in Zinc speciation within a contaminated wetland. Environ Sci Technol 35:3823–3829

Cakmak I, Kalayci M, Ekiz H, Braun HJ, Yilmaz A (1999) Zinc deficiency as an actual problem in plant and human nutrition in Turkey: a NATO Science for Stability Project. Field Crop Res 60:175–188

Chen W, He ZL, Yang X, Feng Y (2009) Zinc efficiency is correlated with root morphology, ultrastructure, and antioxidative enzymes in rice. J Plant Nutr 32:287–305

Chen H, Yuan X, Li T, Hu S, Ji J, Wang C (2016) Characteristics of heavy metal transfer and their influencing factors in different soil–crop systems of the industrialization region, China. Ecotoxicol Environ Saf 126:193–201

Cozzolino V, Pigna M, Di Meo V, Caporale AG, Violante A (2010) Effects of arbuscular mycorrhizal inoculation and phosphorus supply on the growth of Lactuca sativa L. and arsenic and phosphorus availability in an arsenic polluted soil under nonsterile conditions. Appl Soil Ecol 45:262–268

Dai Y, Lv J, Liu K, Zhao X, Cao Y (2016) Major controlling factors and prediction models for arsenic uptake from soil to wheat plants. Ecotoxicol Environ Saf 130:256–262

Das HK, Mitra AK, Sengupta PK, Hossain A, Islam F, Rabbani GH (2004) Arsenic concentrations in rice, vegetables, and fish in Bangladesh: a preliminary study. Environ Int 30(3):383–387

Das DK, Garai TK, Sarkar S, Sur P (2005) Interaction of Arsenic with Zinc and Organics in a Rice (Oryza sativa L.)–Cultivated Field in India. Sci World J 5:646–651

De Datta SK (1981) Principles and practices of rice production. Wiley, New York

Fageria NK, Baligar VC, Clark RB (2002) Micronutrients in crop production. Adv Agron 77:185–268

Farooq MA, Gill RA, Ali B, Wang J, Islam F, Ali S, Zhou WJ (2016a) Subcellular distribution, modulation of antioxidant and stress-related genes response to As in Brassica napus L. Ecotoxicology 25:350–366

Farooq MA, Gill RA, Islam F, Ali B, Liu H, Xu J, He S, Zhou WJ (2016b) Methyl jasmonate regulates antioxidant defense and suppresses As uptake in Brassica napus L. Front Plant Sci 7:468. https://doi.org/10.3389/fpls.2016.00468

Forno DA, Yoshida S, Asher CJ (1975) Zinc deficiency in rice I. Soil factors associated with the deficiency. Plant Soil 42:537–550

Ghosh D, Singh UP, Brahmachari K et al (2016) An integrated approach to weed management practices in direct-seeded rice under zero-tilled rice wheat cropping system. Int J Pest Manag. https://doi.org/10.1080/09670874.2016.1213460

Heath RL, Packer L (1968) Photoperoxidation in isolated chloroplasts: I. Kinetics and stoichiometry of fatty acid peroxidation. Arch Biochem Biophys 125(1):189–198

Islam S, Rahman MM, Islam MR, Naidu R (2016) Arsenic accumulation in rice: consequences of rice genotypes and management practices to reduce human health risk. Environ Int 96:139–155

Johnson-Beebout SE, Angeles OR, Alberto MCR, Buresh RJ (2009) Simultaneous minimization of nitrous oxide and methane emission from rice paddy soils is improbable due to redox potential changes with depth in a greenhouse experiment without plants. Geoderma 149:45–53

Kim K-W, Bang S, Zhu Y, Meharg AA, Bhattacharya P (2009) Arsenic geochemistry, transport mechanism in the soil-plant system, human and animal health issues. Environ Int 35:453–454

Kishor PK, Sangam S, Amrutha RN, Laxmi PS, Naidu KR, Rao KR, Rao S, Reddy KJ, Theriappan P, Sreenivasulu N (2005) Regulation of proline biosynthesis, degradation, uptake and transport in higher plants: its implications in plant growth and abiotic stress tolerance. Curr Sci 88:424–438

Kittrick JA (1976) Control of Zn2+ in soil solution by sphalerite. Soil Sci Soc Am J 40:314–317

Marschner H (1995) Mineral nutrition of higher plants. Academic, San Diego, p 889

Mikkelsen DS, Kuo S (1976) Zinc fertilization and behavior in flooded soils. In: The fertility of paddy soils and fertilizer application of rice. Food and Fertilizer Technology Centre, Taipei, Taiwan, pp 170–196

Mondal P, Majumdar CB, Mohanty B (2006) Laboratory based approaches for arsenic remediation from contaminated water; recent developments. J Hazard Mater 137:464–479

Moulick D, Ghosh D, Santra SC (2016a) Evaluation of effectiveness of seed priming with selenium in rice during germination under arsenic stress. Plant Physiol Biochem 109:571–578

Moulick D, Ghosh D, Santra SC (2016b) An assessment of some physicochemical properties and cooking characteristics of milled rice and associated health risk in two rice varieties of arsenic contaminated areas of West Bengal, India. Int J Res Agric Food Sci 6:44–55

Moulick D, Santra SC, Ghosh D (2017) Seed priming with Se alleviate As induced phytotoxicity during germination and seedling growth by restricting As translocation in rice (*Oryza sativa* L. cv IET-4094). Ecotoxicol Environ Saf 145:449–456

Moulick D, Santra SC, Ghosh D (2018a) Effect of selenium induced seed priming on arsenic accumulation in rice plant and subsequent transmission in human food chain. Ecotoxicol Environ Saf 152:67–77

Moulick D, Santra SC, Ghosh D (2018b) Rice seed priming with se: a novel approach to mitigate as induced adverse consequences on growth, yield and as load in brown rice. J Hazard Mater 355:187–196

Moulick D, Santra SC, Ghosh D (2018c) Seed priming with se mitigates as-induced phytotoxicity in rice seedlings by enhancing essential micronutrient uptake and translocation and reducing as translocation. Environ Sci Pollut Res. https://doi.org/10.1007/s11356-018-2711-x

Moulick D, Santra SC, Ghosh D (2018d) Consequences of paddy cultivation in arsenic-contaminated paddy fields of lower Indo-Gangetic plain on arsenic accumulation pattern and selected grain quality traits: A preliminary assessment. In: Mechanisms of arsenic toxicity and tolerance in plants. Springer, Singapore, pp 49–78

Naeem A, Saifullah, Ghafoor A, Farooq M (2015) Suppression of cadmium concentration in wheat grains by silicon is related to its application rate and cadmium accumulating abilities of cultivars. J Sci Food Agric 95:2467–2472

Nriagu JO, Pacyna JM (1988) Quantitative assessment of worldwide contamination of air, water and soils by trace metals. Nature 333:134–139

Quijano-Guerta C, Kirk GJD, Portugal AM, Bartolome VI, McLaren GC (2002) Tolerance of rice germplasm to zinc deficiency. Field Crop Res 76:123–130

Sahoo PK, Kim K (2013) A review of the arsenic concentration in paddy rice from the perspective of geoscience. Geosci J 17:107–122

Saifullah, Sarwar N, Bibi S, Ahmad M, Ok YS (2014) Effectiveness of zinc application to minimize cadmium toxicity and accumulation in wheat (Triticum aestivum L.). Environ Earth Sci 71:1663–1672

Saifullah, Javed H, Naeem A, Rengel Z, Dahlawi S (2016) Timing of foliar Zn application plays a vital role in minimizing Cd accumulation in wheat. Environ Sci Pollut Res 23:16,432–16,439

Sajwan KS, Lindsay WL (1986) Effects of redox (rē'dŏks): see oxidation and reduction and zinc deficiency in paddy rice. Soil Sci Soc Am J 50:1264–1269

Santra SC, Samal AC, Bhattacharya P, Banerjee S, Biswas A, Majumdar J (2013) Arsenic in foodchain and community health risk: a study in Gangetic West Bengal. Procedia Environ Sci 1(18):2–13

Shri M, Kumar S, Chakrabarty D, Trivedi PK, Mallick S, Misra P, Shukla D, Mishra S, Srivastava S, Tripathi RD, Tuli R (2009) Effect of As on growth, oxidative stress, and antioxidant system in rice seedlings. Ecotoxicol Environ Saf 72:1102–1110

Shuman LM (1991) Chemical forms of micronutrients in soils. In: Mortvedt JJ et al (eds) Micronutrients in agriculture. ASA, CSSA, and SSSA, Madison, WI, pp 113–144

Takahashi Y, Minamikawa R, Hattori KH, Kurishima K, Kihou N, Yuita K (2004) As behavior in paddy fields during the cycle of flooded and non-flooded periods. Environ Sci Technol 38:1038–1044

Verbruggen N, Hermans C, Schat H (2009) Mechanisms to cope with arsenic or cadmium in plants. Curr Opin Plant Biol 12:1–9

Welsch FP, Crock JG, Sanzolone R (1990) Trace elements determination of arsenic and selenium using continuous-flow hydride generation atomic absorption spectrophotometry (HG-AAS). In: Arbogast BF (ed) Quality Assurance Manual for the Branch of Geochemistry. U.S. Geological Survey, Denver, pp 38–45

# Application of Endophytes Through Seed Priming

**Ahmad Mahmood and Ryota Kataoka**

## Contents



**Abstract** The nutrition requirement for the intensive agriculture has pressurized non-renewable resources and demands the shift to renewable resources, "biofertilizers," being the most important in case of soil fertility. The application of such biofertilizers has been carried out through conventional methods, mainly seed inoculation. However, with the increasing focus on endophytic microbes due to their edge over rhizosphere microbes, the application methods need to be modified accordingly. The conventional methods usually do not offer proper conditions for the bacteria to enter the seed before sowing. The microbial seed priming, biopriming, potentially has been explored for possible application of desired microbes to the seeds and resulted positively. The biopriming with/without subsequent incubation at microbe-favorable temperatures allows sufficient time for the endophytes to enter and colonize the seeds which is seldom the case in other treatments. Diversity of biopriming methods has been employed; here, we evaluate the methods for endophytes besides giving insights into possible methods which can be employed for successful application of endophytes to the seeds.

**Keywords** Plant growth-promoting endophytes · Application · Biopriming · Inoculation

A. Mahmood · R. Kataoka (✉)
Faculty of Life & Environmental Sciences, Department of Environmental Sciences, University of Yamanashi, Yamanashi, Japan
e-mail: rkataoka@yamanashi.ac.jp



M. Hasanuzzaman, V. Fotopoulos (eds.), *Priming and Pretreatment of Seeds and Seedlings*, https://doi.org/10.1007/978-981-13-8625-1_25

# 1 Introduction

The food shortage, decreasing farmland, genotypic and phenotypic issues of the plants, and inputs-related limitations need to be resolved to feed the ever-increasing population. Intensive agriculture and fertilizer responsive crop varieties ask for huge amount of inputs especially fertilizers. Resultantly, the fertilizers like phosphorus and potassium being prepared from available forms have put pressure on nonrenewable resources, requiring sustainable solutions. Another issue related to chemical fertilizers is the linked problem of pollution, which encompasses runoff and leaching of the nutrients to the water bodies. Similarly, the use of natural gas in preparation of fertilizers also increases the price, thus making it difficult for the farming community to afford.

These issues thus bring forward the need for alternate nutrient sources like organic and biofertilizers. The organic fertilizers although being an economical choice ask for more labor and pose application difficulties, besides being bulky. Additionally, the problem of weed seed dissemination and other pathogen transfer remain major concerns, proposing the composting of such organic materials before application. The composting, however, causes the nutrient loss. Under such circumstances, biofertilizers become feasible option for their multiple roles and sustainability. The biofertilizers encompassing plant growth-promoting microorganisms (PGPM), bacteria (PGPB), fungi (PGPF), and cyanobacteria (PGPC) have been applied for enhancing the plant growth through variety of methods (Mahmood and Kataoka 2018). Bacteria dominate the farming application, followed by mycorrhiza, and other fungi. Cyanobacterial application studied mainly in rice remains least explored (Karthikeyan et al. 2007). Bacteria mainly have been applied through seed inoculation, mycorrhiza and other fungi through spores, and cyanobacteria as the cell extracts (Shariatmadari et al. 2013). Mycorrhizal inoculum has also been applied involving mycorrhizal cut roots, hyphae, and spores mixed together (Jäderlund et al. 2008). Several other methods at different levels of use have been employed (Mahmood et al. 2016), bringing forward the problems in application. The problems as discussed in later parts include limited survival of the microbes, macro and micro-environmental limitations, and insufficient colonization.

Seed priming can be an attractive alternative for enhancing the seed germination, vigor, and seedling establishment. It offers sufficient time for the endophytes to enter the seeds and start the initial colonization. The possibility of occupying the space by desired microbes would also pose competition to the pathogens, which utilize the same path. The bottlenecks propose the integration of seed priming with the application of microbes, also at various instances termed as biopriming, for efficient application of endophytes. The rhizosphere microbes have been applied through different methods; however, the endophytes need to be incorporated through improved methods. This chapter will hence discuss the application methods for endophytes, besides brief introduction of endophytes, and the past methods.

## 2 What Are Endophytes?

The microbes, which spend at least part of their life within the plant body without showing symptoms of disease and help the plants grow better with multitude of mechanisms, are termed as endophytes. The endophytes can be categorized as plant growth-promoting endophytic bacteria (PGPEB) and fungi (PGPEF). The endophytes are ubiquitous in nature and colonize on all type of plants (Santoyo et al. 2016). The earliest reports from the nearly 200 years ago to the last decade of the twentieth century (reviewed by Hardoim et al. 2015) bring forward the concept of interactions between plants and microbes beyond just rhizosphere. The endophytes help the plant grow better, like that of rhizosphere microbes, through several mechanisms encompassing production and competition. The widely accepted direct mechanisms include release of certain hormones, fixing nitrogen, solubilizing phosphorus, and sequestering iron. Among the indirect mechanisms, physical and chemical competition with the pathogenic microbes, induced systemic resistance, 1-aminocy clopropane-1-carboxylate (ACC) deaminase, and siderophore production (Santoyo et al. 2016) can be reported.

Several genera of endophytes from diverse group of plants have been reported: *Acetobacter* from pineapple (Tapia-Hernández et al. 2000) and sugarcane (Lee et al. 2000), *Acinetobacter* from sugar beet (Shi et al. 2011) and soybean (Zhao et al. 2018), *Agrobacterium* from common bean (Chihaoui et al. 2015), *Bacillus* from maize (Gond et al. 2015) and chickpea (Abd_Allah et al. 2018), *Burkholderia* from grapes (Compant et al. 2005), *Enterobacter* from maize (Naveed et al. 2014), *Pantoea* from *Alhagi sparsifolia* (Chen et al. 2017), *Pseudomonas* from ginger (Jasim et al. 2014), and *Serratia* from *Solanum nigrum* (Khan et al. 2017) for bacteria, and *Alternaria* from Phalsa (Deshidi et al. 2017), *Aspergillus* from maize (Potshangbam et al. 2017), *Epicoccum* from *Entadaabyssinica* (Dzoyem et al. 2017), *Fusarium* from *Nothapodytes foetida* (Combès et al. 2012), *Paraconiothyrium* from *Cephalotaxus harringtonia* (Combès et al. 2012), *Penicillium* from *Fagonia cretica* (Meng et al. 2011), *Talaromyces* from cotton (Yuan et al. 2017), and *Trichoderma* from *Panax notoginseng* (Potshangbam et al. 2017) for fungi, quoting a few. The diversity of plants inhabiting endophytes and their mutual relationships offer insights to the vast adaptations by both.

The endophytic environment provides stability as compared to phyllosphere and rhizosphere, thus ensuring stress-free functionality of the microbes. The microbes applied to the soil by any of the methods must cope with the competition posed by indigenous microbial community (Mahmood et al. 2016); thus, the endophytes have a plus. For instance, the diversity of soil environments, being polluted with organics, or inorganics, or the hurdles of carbon to nitrogen ratios, limit the functionality of microbes. Simultaneous problems of desiccation and exposure to extreme climatic parameters also limit the functioning and survival of microbes so the endophytes become advantageous.

## 3 Colonization and Host Specificity

The attraction of the endophytes toward the plant roots, same as the rhizosphere microbes, is facilitated by the affinity toward the root exudates (Kandel et al. 2017). The competition for the attachment to the root surface is thus also helped by the microbial exopolysaccharides (Meneses et al. 2011). Similar to plant pathogens, the endophytes enter the plant through germinating radicles, stomata, root cracks or openings, and injured plant parts (reviewed by Mahmood et al. 2016), where the main entry point is root zone (Ibáñez et al. 2017). The entry through aerial tissues is also possible; thus, the reports of endophytes have been from flower, fruits, and leaves, as observed in rose plant parts (El-Deeb et al. 2012). Likewise, for the endophytes, the secretion of certain enzymes for cell wall modifications also helps in priority entry to the plants.

Once inside the plants, the endophytes move toward intercellular spaces, where they reside, and have been reported from the intercellular spaces of seeds, flowers, leaves, stem, and roots (Mattos et al. 2008). The colonization further varies within the plant, also depending on the role of microbes. Similarly, the colonization of the forming seeds by endophytes for the possibility of vertical transfer guarantees presence in the subsequent plant population (Shahzad et al. 2018).

Different factors are responsible for successful colonization by the endophytic community. These include the host and microbe genotype, plant part, and environmental conditions (Hardoim et al. 2015). The diversity of endophytic microbes has brought forward different approaches, functions, and mechanisms being utilized by them. For example, *Aspergillus oryzae* besides promoting the growth of radish also conferred herbivore resistance (Sun et al. 2018). The dynamics suggest involvement of endophytes at different levels, changing the physiology of the endosphere for various purposes (Hardoim et al. 2015). *Piriformospora indica* (*Serendipita indica*), another interesting example, induces programmed cell death besides promoting the plant growth as observed in lettuce (Fig. 1). The cell death being usual response of plants to pathogens (Hoeberichts and Woltering 2003) is also triggered by the fungal colonization. Qiang et al. (2012) postulate this cell death by *P. indica* to facilitate

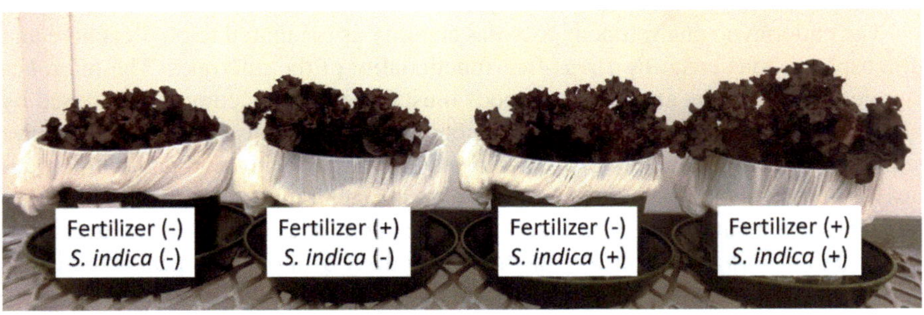


**Fig. 1** Effect of *Serendipita indica* inoculation on growth of lettuce with and without fertilizer application. The fungal application was carried out following Fig. 3

the spore formation. The variety of endophytic functions and their adaptability to certain environments result into facultative colonization of the novel hosts (Gladieux 2018; Inoue et al. 2017).

The available reports on endophyte from diverse habitats and plants propose the widespread relationship; however, the host specificity prevails, as observed in fungi in trees (Collado et al. 2000). The studies regarding specificity occurrence on level of the taxa remain in initial stages. The interaction is mainly influenced by the functions of the microbes, plants being the employers, and playing an active role (Rosenblueth and Martínez-Romero 2006). With the attachment of the microbes to the plant surface, the host plant starts to shortlist the possible candidates, and with the desired purpose, certain chemicals are released for the recruitment (Ibáñez et al. 2017). The plants release certain exudates for the recruitment of desired microbes (Kandel et al. 2017), yet there is ambiguity in understanding at what level plant influences the microbes, which are allowed to enter the plant.

# 4 The Methods Used, Bottlenecks, and Prospects

The widely used inoculation method by the farming community encompasses the use of bacterial culture, facilitated by a carrier, mixed with the artificially made adhesive seeds, and sown. Certain commercial formulations also include liquid cultures applied directly or with the seed or granular fertilizer. Several other methods, e.g., seed coating, pelleting, root dipping, direct soil application, and foliar spray, have also been employed, yet not utilized for commercial purposes (O'Callaghan 2016). Given the endophytes, seed inoculation can serve the purpose, but the engaging time, subsequent physical abrasion, and further competition with the surrounding microbes in soil limit the efficiency. The film- or slurry-based seed coating method, employed mainly for experimental purposes, remains uncommon among the farmers due to insufficient colonization time and application difficulties. Root dipping and/or seedling treated with microbial suspension also faces the later handling and contamination problems; however, it can be beneficial for fruit trees when the number of plants is not very high per unit area. The application of seedlings in microbial suspension seems good option due to presence of root hair, which would offer easy entry for the crops grown in nursery. The pelleting and/or the direct soil application of phytobeneficial microbes offers the application in bulk population, yet the lack of uniformity in the field and exposure to environmental adversities remain a problem. Among all the conventional methods, the microbes need to survive for several weeks to gain entry inside the plant upon the appearance of root hair. Although, the facultative nature of some endophytes indicates the possibility of subsequent colonization if they can survive in the rhizosphere, yet the ecological limitations exist.

On the other hand, the seed priming, with predetermined duration, can help the microbes gain entry during the imbibition. This length of priming must consider the soaking of the seeds until the emergence of the radicle, which besides preserving the delicate radicle, will also ensure a head start to the treated seeds. The seed priming has been utilized for various purposes, including the osmotic adjustments, quick germination, and stand establishment in many crops (Yan 2015; Hussain et al. 2016; Moreno et al. 2018; Zhang et al. 2015), but the introduction of microbes through seed priming, alternatively termed as biopriming, remains primitive.

The review of application of endophytes through seed priming brings forward few approaches being used. The soaking of seeds, and rarely seedling, has been employed, where time of soaking ranges from 10 minutes to 4 hours (Table 1). Zhu and She (2018) also practiced the incision of the seeds prior to soaking. Such practices have been observed in terms of bacteria where the endophytic fungi have been applied to the plants as liquid cultures (You et al. 2012; Waqas et al. 2015).

The application of microbes to the plants through seed priming thus comes as attractive alternative. For bacteria, the seed priming can successfully help them gain entry in the seeds during the imbibition process (Fig. 2). The predetermined soaking durations, different for different crops, would provide sufficient time for the absorption of the suspension involving bacteria. The bulk population will also provide the desired microbes to overcome the pathogens already present in seeds, and with the growing plant parts, such microbes can get additional advantage and quick colonization. Besides, the earlier interaction with the primordial plant parts also offer speedy recruitment. The methods already used show colonization (Table 1); however, increasing the priming duration would offer the radicle emergence besides the entry of endophytes. Similarly, the broth culture at the lab scale can be converted to microbial suspension in water, or saline buffer to make it available commercially.

The fungal application through seed priming using liquid media/suspension is hindered by certain growth properties of fungus. However, seed priming using conidial solution has been practiced for rhizosphere fungi (Jogaiah et al. 2013). Hence, the lab-scale illustration (Fig. 3) involves the agar media followed by incubation or co-cultivation. For the commercial application, the fungus can be grown on organic material, rice bran as an example, followed by mixing the seed and incubation. However, the mixture poses difficulties in planting; thus, application through spores and/or liquid suspension stands out. The application of endophytes is evolving, and further research for successful colonization would strengthen science as well as the farming community.

## 5 Conclusion

Application of endophytes through seed priming offers sufficient time, entry during the imbibition, and speedy colonization besides the established benefits of seed priming. The bulk entry and gaining the ground quickly as compared to subsequent pathogens would also limit the biotic stress to the plants besides the functional support from desired endophytes. Besides the variety of functions, the stability of endophytic environment, reduced or no competition with other microbes, and closer interaction with the host prioritize the use of such microbes.

**Table 1** Reports of endophytes applied through seed priming

| Species | Isolated from/applied to | Mechanisms shown | Crop responses | Method used | Priming duration | Confirmation of infection | Reference |
|---|---|---|---|---|---|---|---|
| *Bacillus*, *Staphylococcus* and *Arthrobacter* | *Ammodendron bifolium* | Indole acetic acid, ACC-deaminase, phosphorus solubilization, extracellular enzyme production | Increased seed germination and radicle elongation | Soaking the seeds | 4 h (incised seeds) | No info given | Zhu and She (2018) |
| *Bacillus subtilis* | Chickpea | Indole acetic acid, phosphorus solubilization, hydrogen cyanide, siderophore, exo-enzyme activities, antagonistic activity | Nodule number, root dry weight and shoot dry weight | Soaking the seeds | 10 min | Yes (10 out of 40 were re-isolated as endophytes) | Egamberdieva et al. (2017) |
| *Bacillus cereus,* and *Bacillus subtilis*[a] | *Teucrium polium*/maize | Ammonia, indole acetic acid, and extracellular enzymes production, phosphorus solubilization, | Increased root length (consortium of the microbes) | Soaking the seedlings | 2 h | No info given | Hassan (2017) |
| *Pseudomonas* sp. and *Burkholderia* sp. | *Echinacea purpurea* and *Lonicera japonica* | Phosphorus solubilization, siderophore, indole acetic acid, hydrogen cyanide production, fixation of atmospheric nitrogen, and antifungal activities | Root, shoot length, fresh and dry weight, number of lateral roots | Soaking the seeds | 10 min | No info given | Gupta et al. (2016) |

(continued)

**Table 1** (continued)

| Species | Isolated from/applied to | Mechanisms shown | Crop responses | Method used | Priming duration | Confirmation of infection | Reference |
|---|---|---|---|---|---|---|---|
| *Bacillus endophyticus*, *Bacillus tequilensis*, *Planococcus rifietoensis*, *Variovorax paradoxus*, and *Arthrobacter agilis* | *Salicornia europaea* | ACC-deaminase, indole acetic acid, phosphorus solubilization | Increased germination, and seedling growth | Soaking the seeds | 2 h | Yes | Zhao et al. (2016) |
| *Pseudomonas* sp. | *Alyssum serpyllifolium*/*A. serpyllifolium* and *Brassica juncea* | ACC-deaminase, phosphorus solubilization, indole acetic acid, siderophore | Increased biomass | Soaking the seeds | 2 h | Yes | Ma et al. (2011) |

[a]Also included fungus

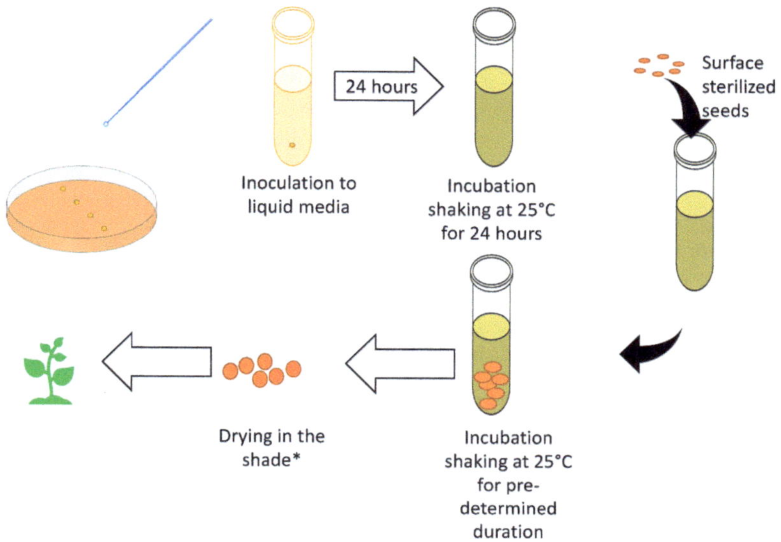


**Fig. 2** Schematic representation of lab-scale endophytic bacterial application through seed priming. The liquid suspension for large-scale application can include saline buffer or water. *The subsequent seed drying can be included for delayed sowings

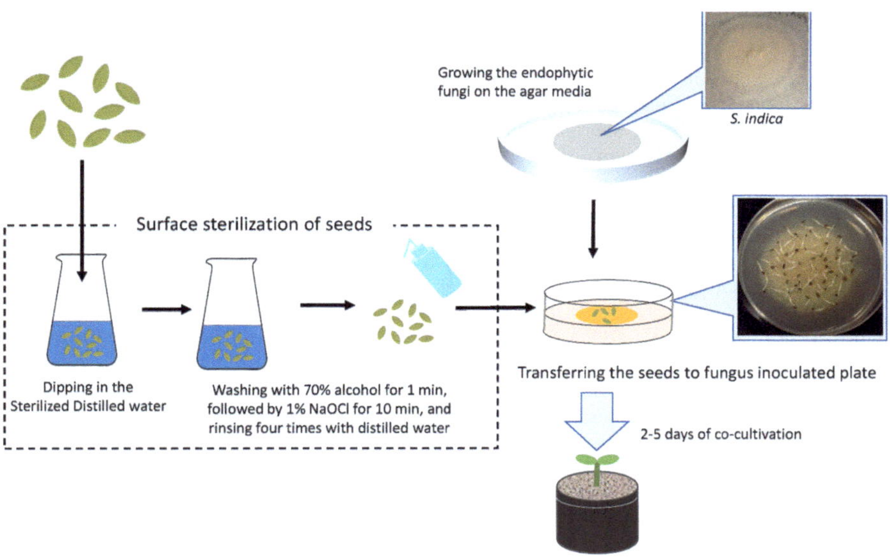


**Fig. 3** Schematic representation of lab-scale endophytic fungal application

**Acknowledgments** The first author would like to thank Rotary Yoneyama Memorial Foundation for the scholarship.

## References

Abd_Allah EF, Alqarawi AA, Hashem A, Radhakrishnan R, Al-Huqail AA, Al-Otibi FON, Malik JA, Alharbi RI, Egamberdieva D (2018) Endophytic bacterium Bacillus subtilis (BERA 71) improves salt tolerance in chickpea plants by regulating the plant defense mechanisms. J Plant Interact 13(1):37–44. https://doi.org/10.1080/17429145.2017.1414321

Chen C, Xin K, Liu H, Cheng J, Shen X, Wang Y, Zhang L (2017) Pantoea alhagi, a novel endophytic bacterium with ability to improve growth and drought tolerance in wheat. Sci Rep 7:41564. https://doi.org/10.1038/srep41564

Chihaoui S-A, Trabelsi D, Jdey A, Mhadhbi H, Mhamdi R (2015) Inoculation of Phaseolus vulgaris with the nodule-endophyte agrobacterium sp. 10C2 affects richness and structure of rhizosphere bacterial communities and enhances nodulation and growth. Arch Microbiol 197(6):805–813. https://doi.org/10.1007/s00203-015-1118-z

Collado J, Platas G, Peláez F (2000) Host specificity in fungal endophytic populations of Quercus ilex and Quercus faginea from Central Spain. Nova Hedwigia 71(3/4):421–430

Combès A, Ndoye I, Bance C, Bruzaud J, Djediat C, Dupont J, Nay B, Prado S (2012) Chemical communication between the endophytic fungus Paraconiothyrium Variabile and the Phytopathogen Fusarium oxysporum. PLoS One 7(10):e47313. https://doi.org/10.1371/journal.pone.0047313

Compant S, Reiter B, Sessitsch A, Nowak J, Clément C, Ait Barka E (2005) Endophytic colonization of Vitis vinifera L. by plant growth-promoting bacterium Burkholderia sp. strain PsJN. Appl Environ Microbiol 71(4):1685–1693. https://doi.org/10.1128/AEM.71.4.1685-1693.2005

Deshidi R, Devari S, Kushwaha M, Gupta AP, Sharma R, Chib R, Khan IA, Jaglan S, Shah BA (2017) Isolation and quantification of Alternariols from endophytic fungus, Alternaria alternata: LC-ESI-MS/MS. Analysis ChemistrySelect 2(1):364–368. https://doi.org/10.1002/slct.201601649

Dzoyem JP, Melong R, Tsamo AT, Maffo T, Kapche DGWF, Ngadjui BT, McGaw LJ, Eloff JN (2017) Cytotoxicity, antioxidant and antibacterial activity of four compounds produced by an endophytic fungus Epicoccum nigrum associated with Entada abyssinica. Rev Bras 27:251–253

Egamberdieva D, Wirth SJ, Shurigin VV, Hashem A, Abd_Allah EF (2017) Endophytic Bacteria improve plant growth, symbiotic performance of chickpea (Cicer arietinum L.) and induce suppression of root rot caused by Fusarium solani under salt stress. Front Microbiol 8:1887. https://doi.org/10.3389/fmicb.2017.01887

El-Deeb B, Bazaid S, Gherbawy Y, Elhariry H (2012) Characterization of endophytic bacteria associated with rose plant (Rosa damascena trigintipeta) during flowering stage and their plant growth promoting traits. J Plant Interact 7(3):248–253. https://doi.org/10.1080/17429145.2011.637161

Gladieux P (2018) What makes a specialized endophyte special? Mol Ecol 27(15):3037–3039. https://doi.org/10.1111/mec.14775

Gond SK, Bergen MS, Torres MS, White JF Jr (2015) Endophytic Bacillus spp. produce antifungal lipopeptides and induce host defence gene expression in maize. Microbiol Res 172:79–87. https://doi.org/10.1016/j.micres.2014.11.004

Gupta H, Saini RV, Pagadala V, Kumar N, Sharma DK, Saini AK (2016) Analysis of plant growth promoting potential of endophytes isolated from Echinacea purpurea and Lonicera japonica. J Soil Sci Plant Nutr 16:558–577

Hardoim PR, van Overbeek LS, Berg G, Pirttilä AM, Compant S, Campisano A, Döring M, Sessitsch A (2015) The hidden world within plants: ecological and evolutionary considerations for defining functioning of microbial endophytes. Microbiol Mol Biol Rev 79(3):293–320. https://doi.org/10.1128/MMBR.00050-14

Hassan SE-D (2017) Plant growth-promoting activities for bacterial and fungal endophytes isolated from medicinal plant of Teucrium polium L. J Adv Res 8(6):687–695. https://doi.org/10.1016/j.jare.2017.09.001

Hoeberichts FA, Woltering EJ (2003) Multiple mediators of plant programmed cell death: interplay of conserved cell death mechanisms and plant-specific regulators. BioEssays 25(1):47–57. https://doi.org/10.1002/bies.10175

Hussain S, Khan F, Hussain HA, Nie L (2016) Physiological and biochemical mechanisms of seed priming-induced chilling tolerance in Rice cultivars. Front Plant Sci 7:116. https://doi.org/10.3389/fpls.2016.00116

Ibáñez F, Tonelli ML, Muñoz V, Figueredo MS, Fabra A (2017) Bacterial endophytes of plants: diversity, invasion mechanisms and effects on the host. In: Endophytes: biology and biotechnology. Springer, Cham, pp 25–40

Inoue Y, Vy TTP, Yoshida K, Asano H, Mitsuoka C, Asuke S, Anh VL, Cumagun CJR, Chuma I, Terauchi R, Kato K, Mitchell T, Valent B, Farman M, Tosa Y (2017) Evolution of the wheat blast fungus through functional losses in a host specificity determinant. Science 357(6346):80–83. https://doi.org/10.1126/science.aam9654

Jäderlund L, Arthurson V, Granhall U, Jansson JK (2008) Specific interactions between arbuscular mycorrhizal fungi and plant growth-promoting bacteria: as revealed by different combinations. FEMS Microbiol Lett 287(2):174–180. https://doi.org/10.1111/j.1574-6968.2008.01318.x

Jasim B, Joseph AA, John CJ, Mathew J, Radhakrishnan EK (2014) Isolation and characterization of plant growth promoting endophytic bacteria from the rhizome of Zingiber officinale. 3 Biotech 4(2):197–204. https://doi.org/10.1007/s13205-013-0143-3

Jogaiah S, Abdelrahman M, Tran L-SP, Shin-ichi I (2013) Characterization of rhizosphere fungi that mediate resistance in tomato against bacterial wilt disease. J Exp Bot 64(12):3829–3842. https://doi.org/10.1093/jxb/ert212

Kandel SL, Joubert PM, Doty SL (2017) Bacterial endophyte colonization and distribution within plants. Microorganisms 5(4):77. https://doi.org/10.3390/microorganisms5040077

Karthikeyan N, Prasanna R, Nain L, Kaushik BD (2007) Evaluating the potential of plant growth promoting cyanobacteria as inoculants for wheat. Eur J Soil Biol 43(1):23–30. https://doi.org/10.1016/j.ejsobi.2006.11.001

Khan AR, Park G-S, Asaf S, Hong S-J, Jung BK, Shin J-H (2017) Complete genome analysis of Serratia marcescens RSC-14: a plant growth-promoting bacterium that alleviates cadmium stress in host plants. PLoS One 12(2):e0171534. https://doi.org/10.1371/journal.pone.0171534

Lee S, Reth A, Meletzus D, Sevilla M, Kennedy C (2000) Characterization of a major cluster of nif, fix, and associated genes in a sugarcane endophyte, Acetobacter diazotrophicus. J Bacteriol 182(24):7088–7091

Ma Y, Rajkumar M, Luo Y, Freitas H (2011) Inoculation of endophytic bacteria on host and non-host plants—effects on plant growth and Ni uptake. J Hazard Mater 195:230–237. https://doi.org/10.1016/j.jhazmat.2011.08.034

Mahmood A, Kataoka R (2018) Potential of biopriming in enhancing crop productivity and stress tolerance. In: Rakshit A, Singh HB (eds) Advances in seed priming. Springer, Singapore, pp 127–145. https://doi.org/10.1007/978-981-13-0032-5_9

Mahmood A, Turgay OC, Farooq M, Hayat R (2016) Seed biopriming with plant growth promoting rhizobacteria: a review. FEMS Microbiol Ecol 92(8):fiw112–fiw112. https://doi.org/10.1093/femsec/fiw112

Mattos KA, Pádua VLM, Romeiro A, Hallack LF, Neves BC, Ulisses TMU, Barros CF, Todeschini AR, Previato JO, Mendonça-Previato L (2008) Endophytic colonization of rice (Oryza sativa L.) by the diazotrophic bacterium Burkholderia kururiensis and its ability to enhance plant growth. An Acad Bras Cienc 80:477–493

Meneses CHSG, Rouws LFM, Simões-Araújo JL, Vidal MS, Baldani JI (2011) Exopolysaccharide production is required for biofilm formation and plant colonization by the nitrogen-fixing endophyte Gluconacetobacter diazotrophicus. Mol Plant-Microbe Interact 24(12):1448–1458. https://doi.org/10.1094/MPMI-05-11-0127

Meng L, Sun P, Tang H, Li L, Draeger S, Schulz B, Krohn K, Hussain H, Zhang W, Yi Y (2011) Endophytic fungus Penicillium chrysogenum, a new source of hypocrellins. Biochem Syst Ecol 39(2):163–165

Moreno C, Seal CE, Papenbrock J (2018) Seed priming improves germination in saline conditions for Chenopodium quinoa and Amaranthus caudatus. J Agron Crop Sci 204(1):40–48. https://doi.org/10.1111/jac.12242

Naveed M, Mitter B, Yousaf S, Pastar M, Afzal M, Sessitsch A (2014) The endophyte Enterobacter sp. FD17: a maize growth enhancer selected based on rigorous testing of plant beneficial traits and colonization characteristics. Biol Fertil Soils 50(2):249–262. https://doi.org/10.1007/s00374-013-0854-y

O'Callaghan M (2016) Microbial inoculation of seed for improved crop performance: issues and opportunities. Appl Microbiol Biotechnol 100:5729–5746. https://doi.org/10.1007/s00253-016-7590-9

Potshangbam M, Devi SI, Sahoo D, Strobel GA (2017) Functional characterization of endophytic fungal community associated with Oryza sativa L. and Zea mays L. Front Microbiol 8:325. https://doi.org/10.3389/fmicb.2017.00325

Qiang X, Weiss M, Kogel K-H, Schäfer P (2012) Piriformospora indica—a mutualistic basidiomycete with an exceptionally large plant host range. Mol Plant Pathol 13(5):508–518. https://doi.org/10.1111/j.1364-3703.2011.00764.x

Rosenblueth M, Martínez-Romero E (2006) Bacterial endophytes and their interactions with hosts. Mol Plant-Microbe Interact 19(8):827–837. https://doi.org/10.1094/MPMI-19-0827

Santoyo G, Moreno-Hagelsieb G, del Carmen Orozco-Mosqueda M, Glick BR (2016) Plant growth-promoting bacterial endophytes. Microbiol Res 183:92–99. https://doi.org/10.1016/j.micres.2015.11.008

Shahzad R, Khan AL, Bilal S, Asaf S, Lee I-J (2018) What is there in seeds? Vertically transmitted endophytic resources for sustainable improvement in plant growth. Front Plant Sci 9:24. https://doi.org/10.3389/fpls.2018.00024

Shariatmadari Z, Riahi H, Seyed Hashtroudi M, Ghassempour A, Aghashariatmadary Z (2013) Plant growth promoting cyanobacteria and their distribution in terrestrial habitats of Iran. Soil Sci Plant Nutr 59(4):535–547. https://doi.org/10.1080/00380768.2013.782253

Shi Y, Lou K, Li C (2011) Growth promotion effects of the endophyte Acinetobacter johnsonii strain 3–1 on sugar beet. Symbiosis 54(3):159–166. https://doi.org/10.1007/s13199-011-0139-x

Sun B-T, Akutse KS, Xia X-F, Chen J-H, Ai X, Tang Y, Wang Q, Feng B-W, Goettel MS, You M-S (2018) Endophytic effects of Aspergillus oryzae on radish (Raphanus sativus) and its herbivore, Plutella xylostella. Planta 248(3):705–714. https://doi.org/10.1007/s00425-018-2928-4

Tapia-Hernández A, Bustillos-Cristales M, Jimenez-Salgado T, Caballero-Mellado J, Fuentes-Ramirez L (2000) Natural endophytic occurrence of Acetobacter diazotrophicus in pineapple plants. Microb Ecol 39(1):49–55

Waqas M, Khan AL, Hamayun M, Shahzad R, Kang S-M, Kim J-G, Lee I-J (2015) Endophytic fungi promote plant growth and mitigate the adverse effects of stem rot: an example of Penicillium citrinum and Aspergillus terreus. J Plant Interact 10(1):280–287. https://doi.org/10.1080/17429145.2015.1079743

Yan M (2015) Seed priming stimulate germination and early seedling growth of Chinese cabbage under drought stress. S Afr J Bot 99:88–92. https://doi.org/10.1016/j.sajb.2015.03.195

You Y-H, Yoon H, Kang S-M, Shin J-H, Choo Y-S, Lee I-J, Lee J-M, Kim J-G (2012) Fungal diversity and plant growth promotion of endophytic fungi from six halophytes in Suncheon Bay. J Microbiol Biotechnol 22(11):1549–1556

Yuan Y, Feng H, Wang L, Li Z, Shi Y, Zhao L, Feng Z, Zhu H (2017) Potential of endophytic Fungi isolated from cotton roots for biological control against Verticillium wilt disease. PLoS One 12(1):e0170557. https://doi.org/10.1371/journal.pone.0170557

Zhang F, Yu J, Johnston CR, Wang Y, Zhu K, Lu F, Zhang Z, Zou J (2015) Seed priming with polyethylene glycol induces physiological changes in Sorghum (Sorghum bicolor L. Moench) seedlings under suboptimal soil moisture environments. PLoS One 10(10):e0140620. https://doi.org/10.1371/journal.pone.0140620

Zhao S, Zhou N, Zhao Z-Y, Zhang K, Wu G-H, Tian C-Y (2016) Isolation of endophytic plant growth-promoting Bacteria associated with the halophyte Salicornia europaea and evaluation of their promoting activity under salt stress. Curr Microbiol 73(4):574–581. https://doi.org/10.1007/s00284-016-1096-7

Zhao L, Xu Y, Lai X (2018) Antagonistic endophytic bacteria associated with nodules of soybean (Glycine max L.) and plant growth-promoting properties. Braz J Microbiol 49(2):269–278. https://doi.org/10.1016/j.bjm.2017.06.007

Zhu Y, She X (2018) Evaluation of the plant-growth-promoting abilities of endophytic bacteria from the psammophyte Ammodendron bifolium. Can J Microbiol 64(4):253–264. https://doi.org/10.1139/cjm-2017-0529

# Microbial Inoculation of Seeds for Better Plant Growth and Productivity

**Muhammad Arif Ali, Fariha Ilyas, Muhammad Arshad, Sajjad Hussain, Mazhar Iqbal, Shakeel Ahmad, Abdul Saboor, Ghulam Mustafa, and Niaz Ahmed**

## Contents



M. A. Ali · F. Ilyas · A. Saboor · G. Mustafa · N. Ahmed (✉)
Department of Soil Science, Bahauddin Zakariya University, Multan, Punjab, Pakistan
e-mail: niaz.ahmad@bzu.edu.pk

M. Arshad
Institute of Environmental Sciences and Engineering, School of Civil and Environmental Engineering, National University of Sciences and Technology, Islamabad, Pakistan

S. Hussain
Department of Horticulture, Bahauddin Zakariya University, Multan, Punjab, Pakistan

M. Iqbal
Department of Environmental Sciences, Quaid-i-Azam University, Islamabad, Pakistan

S. Ahmad (✉)
Department of Agronomy, Bahauddin Zakariya University, Multan, Punjab, Pakistan
e-mail: shakeelahmad@bzu.edu.pk



M. Hasanuzzaman, V. Fotopoulos (eds.), *Priming and Pretreatment of Seeds and Seedlings*, https://doi.org/10.1007/978-981-13-8625-1_26

**Abstract** Microorganisms are beneficial for crop growth and productivity. These are the best alternatives to replace the chemical fertilizers and pesticides. Microorganisms are known to promote plant growth and produce resistance against the pathogen attached on plants. Various methods of microbial applications have been tested and evaluated for crops. Among them seed treatment of beneficial microorganism is an efficient method to incorporate microbial inocula into the soil. Once inoculated, the microorganisms start colonizing the roots of young seedlings. The roots, which are colonized by microorganisms are protected from the attack of many soilborne diseases and pests. These inoculants can also (1) promote plant growth by secretion of growth-promoting hormones, (2) make nutrients soluble and available for plants, (3) help plant to acquire nutrients from soil, and (4) improve soil physical and biochemical properties. Despite of many advantages of seed application of the microorganisms, there are challenges to maintain the viability of microorganisms on seeds throughout commercial seed treatment processes and storage. Future research in this field will have great potential for ecosystem restoration and biodegradation. Advances in development of new inoculants that can maintain viability of microorganisms and seeds while in storage for extended period of time will be highly rewarding for plant growth and production.

**Keywords** Microorganism · Bacteria · Fungi · Seed inoculation · Mycorrhizae · Plant growth

## 1 Introduction

Soil is a complex and heterogeneous material, which consists of living and nonliving entities. Living things include plants, micro and macro-organisms, whereas non-living components include soil organic matter in various forms, minerals, and soil fractions (sand, silt, and clay). Soil microorganisms are comprised of soil flora (algae, bacteria, fungi) and fauna (protozoa and nematodes). Globally, it is thought that microorganisms contain about 3% nitrogen (N) and 1.5% carbon (C) of total earth ecosystem (Wardle 1992; Chaparro et al. 2012). Although these proportions are small, the decomposition and respiratory activities of microorganisms are responsible for nutrient mineralization and release of gases in the soil atmosphere.

By weight, > 0.5% of soil mass is comprised of microorganisms, and about 60–80% of metabolism is due to microflora. In 1 gram of soil (Table 1), there are more than 1 billion bacteria, up to 100 million actinomycetes, 1 million fungi, and nearly about 100 nematodes (Balasubramanian 2017). Even microbial biomass of nitrogen and carbon ranges from 2 g $m^{-2}$, in desert grassland, to 134 g/$m^2$ in natural grassland prairie (Zak et al. 1994). Bacterial numbers are greater in magnitude of orders than fungi and other microorganisms in the soil as illustrated in Table 1. Soil organisms are very essential for soil health, structure, and maintenance. They play an important role in controlling environmental and economic issues, as they play a central role in nutrient recycling.

Microorganisms not only influence soil health but also improve plant growth. This is because they relocate and decompose organic material and protect plant from many diseases and pathogens. More organisms reside in topsoil layer as compared to subsoil layers; however the magnitude of these microorganisms varies from crop to crop. Microorganisms are especially abundant near the root (Table 2) surface area (termed as rhizosphere), as there are sloughed-off root cells and exudates, released from roots. These materials provide a readily available source of food for microorganisms. A portion of these microorganisms possess the ability to suppress the plant diseases and improve plant growth by supplying them nutrients and optimum moisture (mycorrhizal fungi) (Coombs et al. 2003: Bloemberg and Lugtenberg 2001: Robert et al. 2001).

Some endophytic microorganisms (residing in roots) offer more benefits to plants in terms of their colonization ability, improving microbe effectiveness (Iqbal et al. 2018), and this shows that there is a strong bond or relation in between endophytic microorganisms and plant roots. This ability of microorganisms is sometimes beneficial for plants, or sometimes, it causes antagonistic effects on

**Table 1** Microbial groups with representative size and numbers

| Microbial group | Example | Size (μm) | Numbers ($g^{-1}$ of soil) |
|---|---|---|---|
| Virus | *Tobacco mosaic* | 0.02 × 0.3 | $10^{10}$–$10^{11}$ |
| Bacteria | *Pseudomonas* | 0.5 × 1.5 | $10^8$–$10^9$ |
| Actinomycetes | *Streptomyces* | 0.5 × 2.0 | $10^7$–$10^8$ |
| Fungi | *Mucor* | 8.0 | $10^5$–$10^6$ |
| Algae | *Chlorella* | 5 × 13 | $10^3$–$10^6$ |
| Protozoa | *Euglena* | 15 × 50 | $10^3$–$10^5$ |
| Nematodes | *Pratylenchus* | 1000 | $10^1$–$10^2$ |

Source: Balasubramanian (2017)

**Table 2** Quantification of microorganisms present in rhizosphere and in non-rhizosphere soil

| Days after sowing | Wheat rhizosphere soil | Oat rhizosphere soil | Non-rhizosphere soil |
|---|---|---|---|
| 3 | 250 | 250 | 10 |
| 6 | 1100 | 240 | 12 |
| 10 | 400 | 300 | 30 |
| 16 | 580 | 398 | 12 |

plants. However, this ability proves that microorganisms' releases would not always impart desired effects (Okon and Hadar 1987).

Some microorganisms are already present in soil termed as indigenous microorganism (Zahid et al. 2016), while some microorganisms are inoculant. Many different species of microorganisms are used as inoculant, and trend will shift toward the inoculation of beneficial microorganisms for better plant productivity. It is estimated that these inoculants would replace the pesticides and fertilizers. Globally, increased food production in a few decades increased the use of pesticides and fertilizers, which are no longer sustainable (Oerke 2006). This leads to the development of many synthetic chemicals on customer demands, but some of them have environmental concerns. However, cost of development and registration of synthetic pesticides is increasing, which has led to the significant reduction of new chemistries (Glare et al. 2012). Alternative solutions must be developed in such crisis, which will be easy to use, cost-effective, long-term sustainable, and environment-friendly in nature. Use of soil microorganisms is a good approach not only for plant health but also for use as biopesticide. They are used as liquid or dry formulations, applied at the time of planting of seeds or used as seed priming technique. Application of microorganisms at the time of planting is not feasible for large scale, as large amount of inoculant will be needed, which is sometimes very expensive.

Application of microbial inoculants to seeds is an efficient method for the introduction of microorganisms into soil system, where it will explore and colonize the plant roots. This not only proves beneficial for plants but also projects on soil structure. The inoculation of beneficial microorganism is not a new idea, there is a long history of rhizobium bacteria that colonize legume roots, which underpins the importance of legumes (Graham and Vance 2003). Despite of the long history of legumes being inoculated, there are a few commercially available seed microbial inoculants. The use of these microbial inoculants is economically feasible in modern agriculture as it replaced the use of expensive fertilizers and pesticides in some cases. In this chapter, the types and species of microorganisms used as inoculants, their carrier material, and effect on plant growth and productivity are discussed.

## 2 Microbial Interaction with Plants

The living portion of soil comprises of several microorganisms and roots of many plants, both use the organo-mineral part of soil for their growth and survival (Spence and Bais 2013). That part of soil where their interaction occurs is termed as rhizosphere. Rhizosphere is the place from where plants get their nutrients, and there they excrete exudates from their roots (Zahra et al. 2015; Arshad et al. 2016). Due to these functions of plants, they modify soil physical and chemical properties and microbial populations. Microbial activities in the vicinity of root zone also affect the plants greatly (Vessey 2003).

Plants are constantly in interaction with these microorganisms, either beneficial or pathogenic (Barea 2015). Effect on plants may vary with plant species, plant-microbe interaction, microbe species, soil properties, and environmental conditions. Plant interaction with microbes can be proved beneficial for both or negative when microbes compete with plants or invade in plant form completing their life cycles. The first step for developing any interaction is the colonization of roots by microorganisms present in soil or inoculated with seeds, as described by Hiltner (1904). The microorganisms are attracted to root exudates for nutrient source. Root exudates also develope signals, which attract microorganisms; they recognize those signals and start colonizing the roots. Chemotaxis is a well-known mechanism for attraction of microorganisms by plants. It is also believed that plant roots produce electrogenic ion transport mechanism, for attraction of spores and small microbes to surface of roots. These signals mask chemical signals during short range of attraction. Biotic and abiotic factors also influence microbe-plant interaction; biotic conditions include plant species, cultivar, developmental/growth stage, and plant health conditions. Abiotic conditions include temperature, nutritional status of soil and plant, moisture contents, season, soil structure, and geological factors. Now, we will describe about positive and negative interaction in between plant and microorganisms.

## 2.1 Negative Interaction

Soilborne microorganisms also proved to be lethal for plants, and sometimes plant root exudates become toxic for some soilborne pathogens/microbes (Fig. 1). Plants secrete phytoalexins or phenolic compounds, which inhibit the growth of pathogens. Rosmarinic acid (RA) and caffeic acid product proved toxic for an array of soilborne pathogens/microorganism, e.g., *Pseudomonas aeruginosa* (Bais et al. 2002). *Lithospermum erythrorhizon* roots secrete nephthoquinones, which hinders

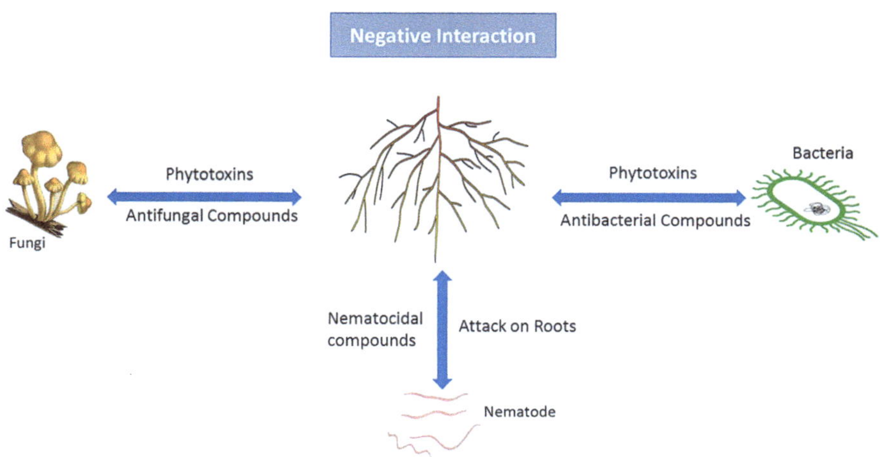


**Fig. 1** Common negative interaction occurs in between plants and soil microorganisms

growth of most pathogenic bacteria and fungi (Brigham et al. 1999). Other examples of phytoalexins include phenylpropanoid, indolic acid, glyceollins, ferulic acid, vanillic acids, trans-cinnamic acid, lubimin, aethione, sesquiterpenoid solavetivone and 6,7-dimethoxybenzoxazolinone (Yoneyama and Natsume 2010).

Plants also get adversely affected by soilborne pathogens. When elicitors and receptors of pathogen and host, respectively, recognized each other, pathogens then proliferate in the host body and cause disease. Pathogens secrete effective proteins and enzymes, which destroy cell structure, and pathogens enter in the plant system (Borah et al. 2018). Adam et al. (2018) found that phytopathogenic and phytoassociated fungi, belonging to ascomycetes, encode special genes, which make them identify their host. Pathogen may be highly selective in their host or may not be. Plant pathogenic fungi belong to *Aspergillus*, *Fusarium*, *Verticillium*, *Uncinula*, and *Thielaviopsis*. Bacterial species belonging to genus *Agrobacterium*, *Burkholderia*, *Clavibacter*, *Erwinia*, *Phytoplasma*, *Pseudomonas*, *Spiroplasma*, and *Xanthomonas* could cause diseases in plants (Wells and Varel 2011).

## 2.2 Positive Interaction

Positive interaction is found between plant and microorganisms, in which both interact with each other and form mutual relationship called as symbiosis. Symbiosis is a Greek word, which means "living together"; it is a kind of biological interaction, which occurs in between specific organisms, from which both organisms get benefit from this interaction (Table 3). This interaction occurs between two specific organisms, as interaction is compulsory for both, and occasionally they cannot survive in the absence of each other, or this is obligatory interaction. Both species take advantages from each other. Interaction may be highly obligate, sometimes that they cannot survive alone, e.g., mycorrhizal fungi and rhizobium. Plants attract microorganisms by providing signals via root exudates. These exudates are composed of organic acids, oxygen, mucilage, and mucigels. Many examples of beneficial interaction exist. As plants are the only producers on earth and only source of carbon for microorganisms, in return, plants obtain their nutrient (required for their proper

**Table 3** Habitually occurring bacterial species colonize plant roots symbiotically

| Species | Host plant | Species | Host plant |
|---|---|---|---|
| *Sinorhizobium meliloti* | Medicago, alfalfa, yellow sweet clover | *Rhizobium tropici* | Common bean |
| *Mesorhizobium cicero* | Chickpea | *Rhizobium leguminosarum* | Faba bean, lentil, pea |
| *Rhizobium etli* | Beans | *Bradyrhizobium elkanii* | Soybean |
| *Bradyrhizobium japonicum* | Soybean | *Pseudomonas fluorescens* | *Chlorophytum*, maize |
| *Azotobacter chroococum* | Grasses, forage crops | *Pseudomonas syringae* | Maize |

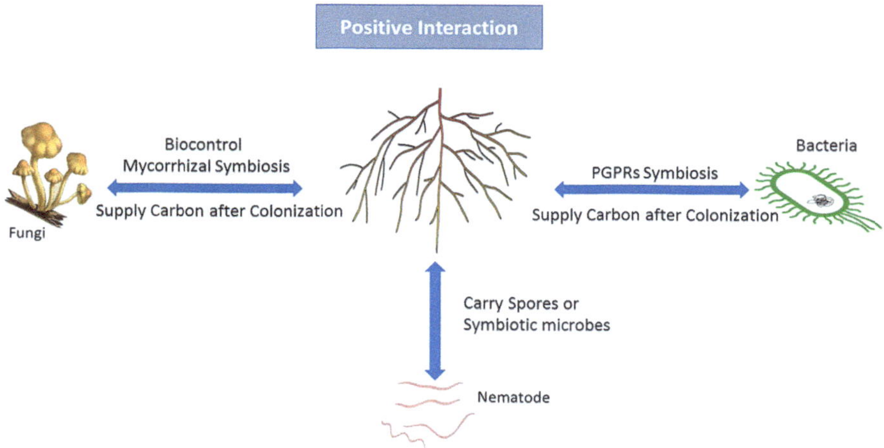


**Fig. 2** Common positive interaction occurs in between plants and soil microorganisms

functioning) from soil. These nutrients are sometimes depleted from soil or become deficient (N and P especially), and plants cannot uptake them; in such case, plant growth is reduced. Plants found their solution in such cases, in that they form association with microorganisms. According to Barea (2015), beneficial saprophytic microorganisms improve plant health by acting in following ways:

- Decomposers of organic detritus
- Plant growth promotion
- Antagonist to the pathogens

There are some main plant-microbe positive interactions explained briefly in Fig. 2.

## 3 Rhizobium

Rhizobium (bacteria) is rod-shaped gram-negative bacteria that form interaction with leguminous crop roots. Free living rhizobium can enter the root hairs by a series of infection process and root colonization. This leads to the formation of a tube-like infection thread that spread along the root hair, which contain a large number of rhizobia cells. Rhizobia release into the cytoplasm of root cells from the infection thread tip, and multiplication occurs (Somasegaran and Hoben 1994). It is also considered that plants other than Fabaceae can also form rhizobium association. These bacteria form root nodules and fix atmospheric nitrogen ($N_2$) there and account for more than 50% of biologically fixed nitrogen (Long 2001). However, rhizobium is host specific that only specific strains of rhizobia can nodulate specific host species (Fig. 3). For example, *Sinorhizobium* spp. can colonize only *Medicago*,

**Fig. 3** Root nodules formed with roots of *Lotus japonicus* plant (©Ali and Gamper)

*Melilotus*, and *Trigonella* genera. Some rhizobium strain can also colonize more than 200 species of plants from 112 genera. For example, *Parasponia andersonii* can nodulate more than 130 plant roots (Pueppke and Broughton 1999). Plants strictly regulate the number of nodules because they take photosynthates from host plants as source of carbon. For maintaining this nodulation system, plants have an autoregulation mechanism of nodules (AON), which involves signal transfer from root to shoot and vice versa, genetically controlled, e.g., HAR1 gene in *Lotus japonicas* (Oka-Kira and Kawaguchi 2006, Ferguson et al. 2010). Roots secrete flavonoid molecules, which are named as Nod factors, coded by special host nodulation gene primarily composed from lipochito-oligosaccharides.

These signaling molecules are recognized by rhizobium and cause internalization followed by nodulation. After the initiation of the infection, nitrogenase enzyme production starts. Nitrogen fixation is an energy-consuming process and requires high amount of oxygen, and nodules have low oxygen. Nodulated plants produce leghemoglobin, an oxygen carrier protein that carries oxygen toward nodules (Spaink 1995; Wells and Varel 2011). Rhizobium species falls in genus *Bradyrhizobium*, *Azorhizobium*, *Mesorhizobium*, and *Rhizobium*.

## 4 Plant Growth-Promoting Bacteria

Plant growth-promoting bacteria are naturally occurring bacteria, which immediately colonize the root and enhance plant growth. Bacteria are found in abundant quantity near the root zones. Root exudates (carbohydrates and amino acids) attract plant growth-promoting bacteria (PGPB). Chemotaxis proves important for successful and competitive interaction. The outer membrane of bacteria secretes a protein, involved in host signal detection and attachment of bacteria with root surface (Burdman et al. 2001). PGPB stimulate plant growth and are involved in nitrogen fixation. PGPB secrete plant growth hormones like cytokinins and gibberellin (Steenhoudt and Vanderleyden 2000). So, the inoculation of the PGPB strains at early stage of plant development stimulates growth of the plants.

PGPB have the ability to produce siderophores, which are low-molecular-weight organic compound. Production and regulation of siderophores is controlled by genes and some environmental factors. Genetic factors involved are Fur protein, e.g., GacA and GacS, and sigma factors RpoS, PvdS, and FpvI (Ravel and Cornelis 2003). Soil environment factors include pH of soil, iron level, soil form, and also the availability of the nutrients especially carbon, nitrogen, and phosphorous (Duffy and Defago 1999). Plants inoculated with PGPB strains show 20–40%, 60–75%, and 40% increase in the uptake of N, P and K, respectively. Enhanced uptake of nutrients is due to their solubilizing power of nutrients present in soil (in various forms). They are of different types like P-solubilizing PGPB, K-solubilizing PGPB, etc. (Rafique et al. 2018). PGPB are comprised of *Azotobacter*, *Azospirillum*, *Bacillus*, *Clostridium*, *Pseudomonas*, and *Streptomyces* species.

## 5 Mycorrhizae

Mycorrhizae are literally translated to fungus root association. It is one of the most important symbioses on earth. It is believed that mycorrhizae evolved 450 million years ago. More than 80% plants on earth can colonize with this unique fungus. The mycorrhizal fungi provide nutrients, like nitrogen, phosphorous, and other micronutrients. Due to the filamentous nature of fungi, they can exploit large volume of their substrate for their nutrition. It can also increase the plant tolerance level of biotic (pathogen and weeds) and abiotic stress (drought, salinity, nutrients, and heavy metals). Mycorrhiza, in return, obtains 4–20% of plant photosynthates as carbon source (food), whereas some plants of families bryophytes, pteridophytes, and angiosperms parasitize mycorrhizae and obtain their food from them. In that case, mycorrhizae are also associated with neighboring photosynthetic host plant (Wright et al. 1998). Based on its structural diversity and nature of symbiotic association, mycorrhiza is further divided into different categories, which are briefly summarized in Table 4.

In ectomycorrhiza fungus can penetrate intracellularly in the root cortex of plants. Ectomycorrhizae can colonize plants of high economic values, woody, perennial, temperate, cold, and forest trees (Taylor and Alexander 2005). This fungus has saprotrophic mood of action and belongs to *Basidiomycetes*. Some common example of ectomycorrhiza is provided in Table 4. ECM fungi can grow in axenic culture in the absence of their host. They are highly competitive in acquisition of their food as they have dual lifestyle. They secrete a number of hydrolytic enzymes, which can degrade little polymers, so they can use organic substrates also. They can

**Table 4** Mycorrhizal species that frequently colonize terrestrial plant roots

| | | | |
|---|---|---|---|
| Mycorrhiza | Ectomycorrhiza | *Amanita muscaria, Hebeloma cylindrosporum, Laccaria bicolor, Paxillus involutus, Pisolithus tinctorius, Suillus bovinus, Xerocomus badius* | Basidiomycota |
| | | *involutus, Pisolithus tinctorius, Suillus bovinus, Xerocomus badius* | Ascomycota |
| | Arbuscular mycorrhiza | *Glomus intraradices, G. geosporum, G. deserticola, G. mosseae, G. claroideum, G. etunicatum* | |
| | | *Gigaspora albida, G. margarita, G. rosea* | |
| | | *Paraglomus laccatum, P. occultum* | |
| | | *Archaeospora trappei, Acaulospora laevis, Scutellospora calospora, Entrophospora infrequens* | |
| | Ericaceous mycorrhiza | Ericoid mycorrhiza | *Cairneyella variabilis* |
| | | | *Oidiodendron maius* |
| | | Arbutoid mycorrhiza | *Arctostaphylos virgata* |
| | | Monotropoid mycorrhiza | *Russula brevipes* |
| | | | *Boletus edulis* |
| | Orchidaceous mycorrhiza | – | |

colonize the plant roots, at the same time, as symbiotic association. ECM fungi no longer have the ability to degrade plants' cell wall components, such as cellulose, polysaccharides, and pectin; due to this reason, ECM fungi cannot penetrate in root cells (Finlay 2008; Martin et al. 2008).

After establishment of symbiosis, ECM form interwoven mantle of hyphae around roots; this mantle is termed as hartig net. Penetration of hartig net varies in between angiosperm and gymnosperms (confined to epidermis and extends up to endodermis, respectively) (Smith and Read 2008). This sheath is involved in nutrient exchange and water transport. The mantle also protects plant roots from the surrounding environment and creates a closed interfacial apoplast (Ashford et al. 1988; Bucking et al. 2002). External hyphae of fungus act as extension in root system and enhance absorbing area of roots. It is estimated that *Pisolithus tinctorius* hyphae can absorb 99% of pine tree nutrients. These absorbed nutrients immediately transfer across the membrane. Plant photosynthates transfer from source to sink; ECM act as a strong sink of C (sucrose), which is hydrolyzed to make glucose and fructose.

In endomycorrhizae (also known as vesicular arbuscular mycorrhiza), hyphae can penetrate within the root cortical cells and have intercellular growth of vesicles and arbuscules. It is the most occurring form of mycorrhiza and colonize 65% of all known plants including agricultural crops like maize, wheat, rice, and soybean and horticultural crops. It is further divided into arbuscular, orchidaceous, and ericaceous mycorrhiza.

All orchid plants need orchidaceous mycorrhiza at some point of their growth, mostly at their seedling stage. As orchid seeds are very small, they have not enough nutrients to support seedling growth. Mycorrhiza infection occurs after embryo stage, when seed coat ruptures. Mycorrhizal plant uptake nutrients via both pathways, i.e., plant pathway and mycorrhizal pathway. Nutrients or water gets absorbed by ERM and then transferred to hartig net or IRM, and plants get these materials from interfacial apoplast. Nowadays, we use seed inoculation technique for better growth of our seedling plants.

## 6 Nematode-Bacteria Interaction

Nematodes are diversified microorganisms and present in abundance in soil environment. Beneficial nematodes are used to control pest and insects, and more than 200 families of insects are susceptible to these beneficial nematodes. Seven nematode families, i.e., *Mermithidae*, *Allantonematidae*, *Sphaerularidae*, *Tetradonematidae*, *Phaenopsitylenchidae*, *Steinernematidae*, and *Heterorhabditidae*, are important in biological point of view (Kaya and Stock 1997). Nematode-bacteria symbiosis is widely reported in protecting plants from many pest and insect attacks. Entomopathogenic nematodes carry gram-negative and non-spore-forming bacteria (Table 5) *Xenorhabdus* and *Photorhabdus*, which don't have any resistant stage. Their life cycle is completed in two stages, i.e., (i) free stage, bacteria live within the nematodes, and (ii) parasitic stage, bacteria are released in the host body and reproduce there. Nematodes get entry in host (pest or insect) from their piercing mouth parts, and after getting entry into host, nematodes release juvenile bacteria. Bacteria degrade the body parts of the host through their enzymatic activity and provide a nutritive medium for growth of nematode. On the other hand, nematodes provide safety to these bacteria from external atmosphere and sometimes suppress the defense mechanism of host against bacteria (Askary 2010). Jan et al. (2008) found that bacterial strain *Xanthomonas nematophila* isolated from *S. carpocapsae* (nematode) has the ability to suppress the growth of wax moth; the host becomes flaccid and dies. This mechanism is useful in controlling pest/insects of plants.

**Table 5** Entomopathogenic nematodes and symbiotically interacted bacterial species

| Nematodes | Bacteria |
|---|---|
| *Steinernema carpocapsae* | *Xenorhabdus nematophila* |
| *S. feltiae, S. affine, S. intermedium* | *Xenorhabdus bovienii* |
| *S. rarum* | *Xenorhabdus szentirmaii* |
| *S. thermophilum* | *Xenorhabdus indica* |
| *Heterorhabditis megidis Palaearctic group* | *Photorhabdus temperata temperate* |
| *H. indica* | *Photorhabdus luminescens akhurstii* |

Source: Ganguly (2006)

# 7 Types of Microbial Inoculation

Land with long-term fallow period or continuously tilled usually contained less microorganisms especially beneficial fungus. There is a need to re-establish the effective microorganism population in less cost and by effective method. Methods for the rhizobium inoculum preparation have been widely overviewed. The most reliable technique to build up these microbial species is application of commercially available microbial inoculum directly to seed or root or indirectly in soil. Many inoculums have shelf life of up to 2–3 months to years. An inoculum should carry a range of viable cells from 10,000 to 1 million per seed; this number of viable cells is necessary for successful infection to the developing root of the plants. Commercial inoculums are available in powder, liquid, and granular forms (Armanthus et al. 2012). There may be two types of inoculants used:

1. Single strain (Uni strain)
   Inoculum would have only a single type of strain.
2. Several strains (Multi strain)
   This type of inoculants has more than one type of strain. Consortium of species is usually used because of positive microbe-microbe interaction. For example, bacteria enhance growth of germ tube and branches of AM fungi and promote mutual or symbiotic interaction (Lojan et al. 2016).

- **Inoculation**: Application of microorganisms to seeds (Elegba and Rennie 1984).
- **Seed Priming:** It is the soaking of seeds in a solution containing required agent for priming followed by re-drying of seed (McDonald 1999).
- **Biopriming**: It is the soaking of seed in microbial solution, which allows the microbial attachment and imbibition into the seed (Mahmood et al. 2016; Abuamsha et al. 2011).
- **Consortium**: Two or more microbial species, living or working together symbiotically. First introduced by Johannes Reinke in 1872.
- **Strain**: Sub-type of microorganisms. This term is usually used to denote a pure culture (Dijkshoorn et al. 2000).

## *7.1 Sources of Inoculum*

Rhizosphere soil of plant, hosting required microorganisms, can be used as microbial inoculum for the next plant. Soil inoculum is comprised of soil, plant roots, spores, and indigenous microorganism. The significance of this method is that soil may also contain pathogens and weed seeds, which hampers plant growth. Another issue relating to the use of this inoculum is the quantity of soil, which will be used as inoculum, as quantification of soil microbes is often difficult and uncertain. So, the use of this source should be less preferred when other sources are available.

Viable microorganisms (spores and bacteria) can also be extracted from soil and known host plant; such inoculum is known as crude inoculum. It consists of infected roots and viable microorganisms, which can be isolated by various methods. Infected roots of plants can also be used as inoculum, which is called root inoculum. Crude inoculum potential depends upon several factors: state of starter culture, type of host plant, medium, and growth environment.

Inoculum quantity is the main aspect in microbial inoculation. In case of fungal inoculum, generally four to eight propagules, 4000 cm of infected roots required per gram of inoculum. There are two methods of inoculum application, i.e., direct and indirect application. The following types of microbial inoculations are used.

## *7.2 Microbial Soil Inoculation*

Soil application is preferred when indigenous microbes of antagonist species are present in excess to harm the plant tissue or when seed is already treated with fungicides (Mahmood et al. 2016). Inoculum can also be used on the soil directly using drip irrigation system or spray on the soil. This type of inoculum may be in different types, i.e., liquid or frozen concentrates, or it may be in small granules like structure; these granules may have a small amount of marble, silica, or calcite particles, which are moistened with an adhesive material and mixed with peat powder (Podile and Kishore 2006). After drying, these granules are ready for field applications; mostly, broadcast method is suggested for granule application. Inoculum is also placed in soil below the seeds (up to 30 mm). Inoculum calculation will be based upon the presence of the number of viable microorganisms and spores in case of fungi, in 10 g inoculum. Usually 500 spores per plant are recommended in arbuscular mycorrhizal fungi (Orates and Bykova 2018). In case of nematode application, mix them with water and spray on ground surface, and continue mixing inoculum during the process to avoid sinking nematodes in bucket or drum. Its application is recommended only after transplanting the plant.

## *7.3 Microbial Seed Inoculation*

Direct soil application of microbes is preferred when the conditions are favorable and is the most recommended method; when conditions are dry, hot, and highly acidic, seed coating of inoculants is restricted. This is another kind of seed priming, in which seeds are treated with microorganisms before planting (Fig. 4). The advantage of this method is that seed will quickly colonize after germination. Direct seed application is recommended in the following situations:

1. Dry, hot, and acidic conditions of soil.
2. Weather conditions are adverse.
3. It is essential to coat seed with chemical toxic to rhizobia.
4. Rhizobia flora is present in soil to compete with introduced strain of bacteria.

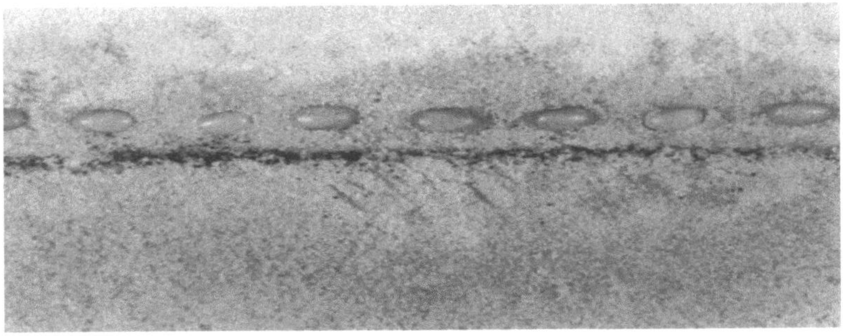

**Fig. 4** Inoculum placed beneath the seeds of bean in soil

## *7.4 Root Inoculation*

Plant roots may also be treated with inoculum, after establishment of plants. In this technique, inoculum is placed in deep soil, adjacent with growing roots. Roots become colonized when they encounter inoculum. This method is not recommended for vegetables or sensitive crops. Perennial crops such as alfalfa and grasslands treated with this method get benefit over several years. Much greater amount of inoculum is required in this method, so this method is not used commercially as it is a cost-expensive technique (Bashan 1998).

Survival of microorganisms on seed is influenced by temperature, pH, and seed coat toxicity, which is ameliorated by selection of tolerant strains (Deaker et al. 2004). Selection of above microbial inoculation method depends upon available inoculum and equipment, plant species, labor availability, cost, and soil and atmospheric conditions.

## *7.5 Carriers for Inoculants*

For successful inoculation, a cheap and effective carrier material is required. Shelf life of carrier containing microorganisms is an important factor in the selection of carrier material. Carrier should have the following properties (Somasegaran and Hoben 1994; Kaljeet et al. 2011):

- Nontoxic for microorganisms.
- Absorption quality should be good.
- Can be easily sterilized or autoclaved.
- Adhesion power with seed.
- Easily available.
- Cost-effective.

Commonly used carrier materials include peat, rice husk, farmyard manure, clay, kaolinite, organic matter, charcoal, rock phosphate, soil-charcoal mixture, lignite, and bentonite (Tarley and Arruda 2004; Murray 2002). Physical and chemical characteristics of carrier material should be determined before use, e.g., water holding capacity, pH, considered nutrient concentration, and organic matter contents.

Peat is one of most common and generally used carriers for the inoculants, and it is also preferred in commercial use (Kaljeet et al. 2011). Peat is a nutritive media for the microorganisms, e.g., bacteria, and is essential for them during distribution time. Peat, which must be used in inoculation for coating of seeds, should be highly adsorptive, having high organic matter contents, nontoxic for microorganisms, easy to be sterilized, and easily available at low cost. Peat would be sterilized using autoclaving and ultraviolet radiations before use.

## 8 Methods of Inoculation

There could be several methods of microbial inoculation including single-seed to single-microbe infectivity to a large-scale commercial microbial application. Inoculation can be performed by the following methods and as summarized in Table 6 along with the brief significance of the methods.

**Table 6** Significance of different microbial inoculation methods

| Inoculation methods significance | Slurry method |
|---|---|
| | Less costly and inoculant can cope up with fertilizers |
| | Cost required on labor and requires coatings (lime) |
| | Non-flexible seeding |
| | Can't tolerate fertilizers and pesticides |
| | **Granular** |
| | Less labor and time required |
| | Can tolerate fertilizers and chemicals |
| | Flexible seeding |
| | Higher cost and bulk volume |
| | **Seed priming** |
| | Tolerates pesticide, fertilizers, and fungicides |
| | Flexible seeding |
| | High cost and most of the microbial species have low life span |

## 8.1 Slurry Method

The most common method of seed coating is slurry method. In this method, inoculum is diluted in water to cover the seeds and to increase the adhesion gums, and sugar solution is added to it (Fig. 5). Surface sterilized the seeds by soaking them in 5% sodium hypochlorite solution for 2 min and then in ethanol and distilled water. Seeds will then be dipped in suspension containing microbial cells at room temperature for 12 h. Before sowing, seeds would be properly coated with this slurry solution (Afzal et al. 2012; Berdeni et al. 2018). Seeds should be lime coated after inoculation. After inoculation seeds should be sown into a moist soil within 12 h, and in case of lime coating, use it within 24 h. Another technique is pouring of inoculum suspension after placement of seed in soil, above or below the seed or onto the soil surface (Afzal et al. 2012).

Seed-coated rhizobium strains are affected by the pH very badly. At low pH the rhizobial population decreases very rapidly. Similar effect also occurs when seeds are applied to the fertilizer-treated soil. If we need to inoculate this type of soil, seeds should be coated with peat layer having lime in it, and this lime layer will make pallet around the seed.

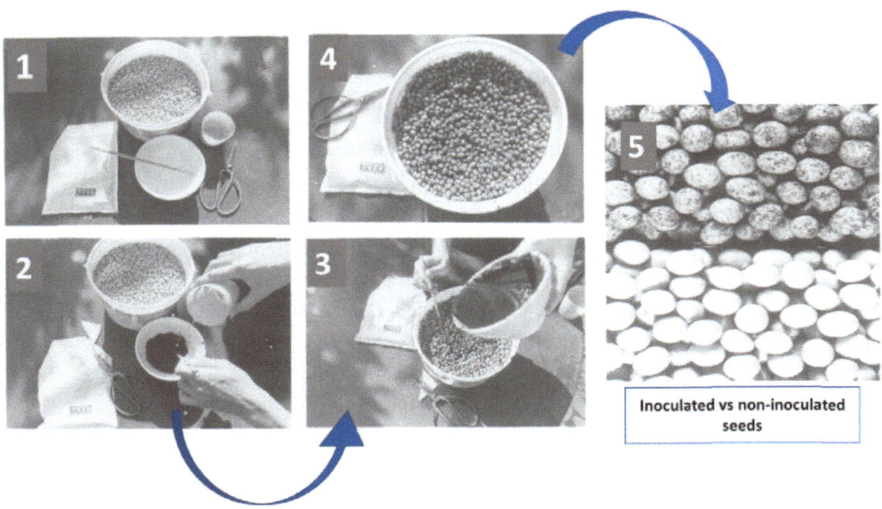


**Fig. 5** Slurry inoculation method 1: Required equipment, 2: add water into peat-based inoculant, 3: mix slurry of peat and inoculum with seeds, 4: let them dry 5: inoculated vs non-inoculated seeds appearance

### 8.2 Granular Method

Granular material can be purchased and used as stand-alone product and inoculate on the sides of seed. This material can also be mixed with soil or inoculated with seeds, through drill box. Inoculum is mixed in calcite or silica grains along with adhesive material. After drying, inoculum is broadcasted over soil surface or can be drilled in soil easily. This enables the equal or uniform distribution of granular inoculum. Due to physical separation of inoculum and seed in this method, seeds can also be pretreated with pesticides or insecticides as it will not damage the inoculum material. Granular method is preferable where indigenous microbes are in abundance. This method requires a large amount of inoculum, so it is the less frequently used method.

### 8.3 Pre-coated Seeds

Coated seeds are available in markets on which inoculum is applied before sale. Along with desired microbes, seeds will be treated with insecticides, pesticides, and fertilizers. With exception of clover, serradella and biserrula, seeds treated with this method have less shelf life as compared to other methods. Exception of these crops may be because of microbial species (used for such crops have short life span).

### 8.4 Freeze-Dried

A suspension is prepared by mixing inoculum powder with an effective liquid. This liquid will activate the microbes and a protective agent is also mixed with this suspension for making possible the survival of microorganisms till sowing. For microbial inoculation on seeds, this suspension can be sprayed on seeds, and then these seeds will be air dried. Lime coating will not be required in this method. Seeds should be sown within 5 h of inoculation.

## 9 Seed Inoculation with Microorganisms

This is a technique in which seeds will be soaked in a solution containing required microorganisms, which aids in germination and radical emergence (Ashraf and Foolad 2005). The primary objective of this treatment is to provide enough and effective amounts of microorganisms to rhizosphere. Seed priming shows better and effective results in inoculation methods. Liquid or powdered inoculum should be used for seed priming as powder form inoculum is used in hairy texture seed such as wheat, barley, and other grasses. Adhesion of inoculum must be insured for effective colonization. Liquid inoculum is preferred in smooth surface seeds because of better adhesion ability, e.g., corn, alfalfa, and beans. A sticky material should be used with inoculum to keep inoculum attached with seed surface. On commercial scale, seeds should be treated with some cement mixer or by spraying inoculum on seeds.

## 9.1 Thin Coating

A single coat of dormant microbial inoculum is applied on seed along with any sticky material, e.g., methyl cellulose or sugar, and then the seeds will be dried. Carrier of the inoculant would have a neutral pH, e.g., clay, kaolinite, or montmorillonite (Kosanke et al. 1992). Microbes survive for a very short duration, and technique is usually used for experimental purposes.

## 9.2 Slurry Coated Seeds

In this method inoculum is added to the peat solution with the addition of moisture, to allow the adhesion of inoculum with seeds. In this technique, inoculum will grow in a peat or other solid medium, and this medium will then be used as secondary inoculum for seed priming/inoculation along with some sticky material (Afzal et al. 2012). For sowing of inoculated seeds in acidic soils, seeds must be coated with lime. After coating with peat-based inoculum, seeds are quickly mixed with pulverized limestone; this step is termed as lime-pelleting.

Inoculant slurry must be prepared with some adhesive material to bind the lime coating with seeds, e.g., Arabic gum or synthetic gum. Gum solution for making slurry solution has moderate viscosity. A diagrammatic procedure of slurry inoculation method is described in the above diagram.

## 9.3 Pelleted Seeds

In this technique, inoculum is applied on seeds along with additives. Seed is coated with peat-based inoculum when large quantity of microorganisms will be required. One-third inoculum will be applied along with gum or peat, and two-thirds inoculum will be applied after application of peat-based inoculum. Pelleting technique is applied on seeds having an irregular surface. Pelleting makes the seed surface regular, smooth, large, and heavy. Pellet increases the size of seed up to 2–50-folds or more. Pelleting allows layers of inoculum at different and desired points as layers are coated in this technique. For instance, pesticides would be coated on the outer surface for seed protection, and keep these chemicals away from seed.

For acidic soil lime ($CaCO_3$), pelleting technique is introduced for seed microbial inoculation/priming for reducing the effect of soil pH on microbe survival, e.g., *R. leguminosarum* bv. *trifolii* (Jensen 1943) and *Sinorhizobium meliloti* (Amarger 1980) are sensitive to low pH. Loneragan et al. (1955) found that treating of pelleted seeds with limestone powder is effective.

### *9.4 Bioprimed Seeds*

This method is the latest and diligent for inducing endophytic microbial effects in seeds. It is quite effective and provides no harm to plants. This technique provides seed a favorable environment for their growth with desired composition and properties of endophytic microorganisms. In this technique, the flower of a growing plant contacted with a source of endophytic microorganisms (e.g., be spraying). These microorganisms get natural entry in plant from the route of flowers and conveyed to new producing seeds. These seeds are then collected as bioprimed seeds of microbes. This technique facilitates the growth and productivity of plant, seed vigor, and biomass production (Mitter et al. 2018). Microbial inoculation in flowers can also be fulfilled by using honeybees or other pollen-feeding insects (butterfly, wasp, and flies). By planting these naturally inoculated seeds, endophytic microorganisms get proliferated and colonize the daughter plant.

## 10 Microbial Inoculation and Plant Growth

Microbial inoculation to seeds delivered the microorganisms directly to soil that surrounds the roots of plants (rhizosphere), where plants interact directly with microorganisms and invertebrates affect biogeochemical cycling, plant growth, and abiotic and biotic stress factors (Philippot et al. 2013). Most of the beneficial microorganisms are found in rhizosphere, where they have different mechanisms of accomplishing growth of plants (Babalola 2010).

### *10.1 Influence on Nutrient Availability*

Best known example of microbial seed inoculation is legume inoculation with *rhizobia*, which aims to maximize the yield potential of plant by enhancing the number *rhizobia* in rhizosphere zone. Biological nitrogen fixation by rhizobia-legume symbiosis is vital for N-input soil agroecosystem. Bacterial species of genera *Allorhizobium*, *Azorhizobium*, *Bradyrhizobium*, *Mesorhizobium*, *Rhizobium* and *Sinorhizobium* form nodulation with plant roots and fix $N_2$ in legumes. Plant growth enhancement related to their microbial association is due their ability to act as

**Fig. 6** Pot screenhouse experiment in the presence (+) and absence (–) of *Striga hermonthica* (PGPR strain) in maize plant. (Source: Babalola et al. 2007)

"biofertilizer"; they play a role in nutrient availability (Vessey 2003), as it eliminates the use of N-fertilizer. In case indigenous microorganisms are present in soil, the yield will increase up to one-third or one-half of the previous yield (Fig. 6).

Sometimes higher yield can also be expected from other species. Many EM fungi have the ability to mineralize and utilize amino acids and amide (glutamine, glutamate, and alanine), present abundantly in soil organic matter pool. Mycorrhizal plant can obtain more or less 75% of their N from mycorrhizal pathway. VAM fungi typically increase plant nutrient uptake especially P and that of micronutrients. Li et al. (2006) found that sometimes, AM wheat plants obtain their 50% of P from mycorrhiza, whereas in tomato 100% of their P come from the mycorrhizal source. This shows that nutrient acquisition depends upon fungal and plant species. Microbial inoculation proved to be an effective replacement of chemical fertilizers. In a very recent study, the role of microbial inoculation on P, K, and C/N uptake by plant in soil has been shown (Fig. 7). Plant growth-promoting bacteria have th ability to solubilize and utilize the ferric ion through the production of low-molecular-weight compounds, siderophores (Whipps 2001).

## *10.2 Abiotic Stress Tolerance*

Various adverse abiotic conditions affect plant growth negatively and contributed about 50% losses in agriculture. Abiotic stress conditions include drought, temperature, and heavy metal stress. These conditions affect plant physiology, morphology, seed physiology, and yield of plants (Chodak et al. 2015). In such worse condition, microbial seed inoculation benefits plant by direct and indirect mechanisms. Zolla et al. (2013) manifested that microorganisms cause changes in morphology and physiology of plants, altering the transporter activity and composition of root

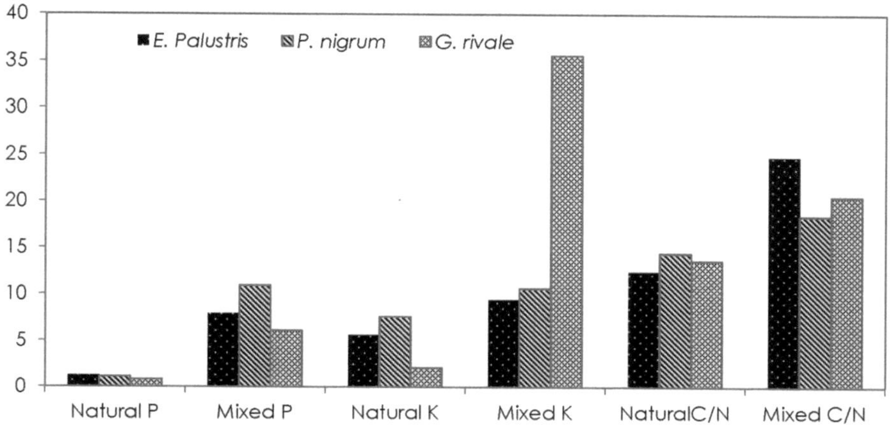


**Fig. 7** Mean soil nutrient concentration (mg/100 g soil) and C/N ratio comparison of *Euphorbia palustris*, *Phyteuma nigrum*, and *Geum rivale* bacterial species response in natural and inoculated conditions. (Source: Michaelis and Diekmann 2018)

exudates and develop some regular mechanisms to maintain the crop productivity under stress. Microorganisms evolved mechanisms for their survival in such conditions by modifying their structures. Irrespective of stress conditions in rhizosphere, they colonize roots and supply nutrients to plants. Mechanisms involved in tolerance include production of phytohormones (IAA, cytokinins, abscisic acid), secretion of exopolysaccharides and ACC deaminase, and improvement of plant systematic mechanisms (Porcel et al. 2014). Microbial inoculation decreased antioxidant activity and enhanced the production of amino acids, sugar, and proline in plant under drought stress conditions (Staudinger et al. 2016; Tiwari et al. 2016; Bano et al. 2013). Under salinity stress, microorganisms produce siderophores, fix nitrogen, produce phytohormones and mobilize nutrients to improve the growth of plants and root activity (Hayat et al. 2010). Vardharajula et al. (2011) demonstrated that microbes bind cations and make them unavailable for plant use. For example, seeds inoculated with PGPB strains *Rhizobium* and *Pseudomonas* show better growth as compared to non-inoculated seeds in saline soils.

## *10.3 Resist Plant from Attack of Pathogens*

Seed inoculation with microorganism antagonizes the soilborne pathogens and insects, and it becomes a more popular and safe method than the treatment of seeds with fungicides and insecticides. Use of these chemical causes environmental hazards, and pathogens also acquire resistance to these chemicals. More than 400 fungal species can retard the attack of insects and mites on plant, e.g., *Verticillium lecanii* control aphids and *Hirsutella thompsonii* control mites. In Table 7, the role

**Table 7** Beneficial microorganisms used in research and commercial trails against different pathogens

| Microorganisms | Disease/pathogen | Plant | References |
|---|---|---|---|
| *Funneliformis mosseae* and *Rhizophagus irregularis* | Citrus canker | Citrus | Berdeni et al. (2018) |
| *Bacillus tequilensis, Bacillus flexus, Trichoderma harzianum, and Glomus aggregatum* | Root knot | Sweet basil | Tiwari et al. (2017) |
| *Trichoderma harzianum* | *Fusarium verticillioides* | Maize | Nayaka et al. (2010) |
| *Pseudomonas fluorescens* | Downy mildew | Millet | Raj et al. (2004) |
| *Pseudomonas chlororaphis* | *Tilletia caries* | Wheat, barley, peas | BioAgri AB (2016) |
| *Bacillus firmus* | *Plant parasitic nematodes* | Corn, cotton, sorghum, soybean | Bayer Crop Science (2016) |
| Bacterial consortia—*Bacillus* spp., *Pseudomonas aeruginosa, Streptomyces* sp. | Sunflower necrosis virus disease | Sunflower | Srinivasan and Mathivanan (2011) |
| *Fusarium oxysporum* f. sp. *strigae* | Root parasitic weed | Witch weed | Elzein et al. (2010) |
| *B. cepacia strain OSU-7* | Fusarium dry rot | Potatoes | Recep et al. (2009) |
| *Streptomyces spororaveus* | *Fusarium udum* | Wheat | Al Sahli and Abdulkhair (2012) |

of microorganisms to control plant diseases has been summarized. Application of beneficial microbe to rhizosphere protects the seeds from seed rot and seedling dumping due to soilborne pathogens such as *Pythium* and *Rhizoctonia*. It is observed that disease effect was reduced up to 50–70% due to biopriming of seeds (Muller and Berg 2008).

After successful experiments on protecting seeds from pathogens, seed coating technique with microorganisms is gaining attention nowadays. *Bacillus subtilis* retards crown rot of shoot disease in strawberry plants and reduces the number of oospores found in root (Vestberg et al. 2004). *Bacillus*, *Trichoderma*, and *Glomus* species are currently used as liquid slurry in seed treatment along with some pesticides and insecticides as biocontrol agents (Tiwari et al. 2017). Another example describing the advantage of seed priming method is the inoculation of endophytic bacterium *Clavibacter xyli*, which is transformed to produce protein toxic for European corn borer. After germination of inoculated seeds, toxin will be released in the xylem fluid, which help in resisting plant from insect attack. Results of several studies indicate that consortia (mixture of fungi, AM, and bacteria) application provides effective results and environment-friendly results for controlling pests, as compared to the use of single strain.

### *10.4 Insect-Plant Interaction*

Microorganisms seem to play an important role in insect-plant interaction such as host range (Chu et al. 2013), efficiency of plant in feeding insects (Brune and Dietrich 2015), and symbiotic interaction of insect and microbe by altering insect and plant physiology (Giron et al. 2016). Microorganisms present in rhizosphere positively improve plant resistance to most of the herbivore insect (prevent their attack) and prove much better in their role as compared to genotypic resistance. There is no influence of genotypic traits in interaction of microbe-insect- plants. Hubbard et al. (2018) found that seed priming with *Nitrobacter* spp. improves plant stress to herbivores, as $NO_3^-$ availability to plants increases, and thus concentration of primary and secondary metabolites also relatively increases, which is involved in host resistance to pathogen and insect attack. All plant microbial seed inoculations with respect to controlling insect pest need to be understood completely.

Beneficial nematodes control over 100 families of insects. Nematodes can attack on soil-dwelling insects (cucumber beetle, grubs, gall midge, root weevil, beetles, masked chafer, flea, strawberry root and black vine weevils, chafer, squash bugs, termite, cutworms, white grubs, algae gnats, black fly, potato tubeworm, meal worm, bark beetle, corn root weevil, fire ant, sting bugs, gall gnats, gypsy moth, billbug, thrips, ants, and termites). Nematodes enter the larvae and spread specific bacteria which can degrade insect host tissues. Degraded material is further utilized easily by nematodes for their growth, and host insect will die soon.

## 11 Conclusion and Future Prospective

To address multiple challenges faced by seed, a technique is established widely known as seed priming with microorganisms or biopriming to improve germination and development. Various studies are available in the literature on the beneficial use of microorganisms for agricultural crops, but little attention has been paid to their use as seed inoculation that could help in sustainable agriculture. Another advantage of seed inoculation is that it provides enough quantity of microorganisms to seed and subsequent development. Despite their importance being well recognized, their role in rhizosphere community and their use in nature conservation demand more focus and detailed studies. Emphasis should be paid on determining better methods and formulations of microbial inoculation and improving application methods and cost of this microorganism inoculum as it is a serious hurdle. There is a need to fill in the gap between the knowledge of our understandings and biotechnical strategies to ameliorate the plant adaptation and elevate the capability of microorganisms to thrive in stress situations, as mechanisms of interaction in between plants and microbes under stress conditions are poorly understood.

# References


Abuamsha R, Salman M, Ehlers R (2011) Effect of seed priming with Serratia plymuthica and pseudomonas chlororaphis to control *Leptosphaeria maculans* in different oilseed rape cultivars. Eur J Plant Pathol 130:287–295

Adam A, Deimel S, Pardo-Medina J, Garcia-Martinez J, Konte T, Limon M, Avalos J, Terpitz U (2018) Protein activity of the *Fusarium fujikuroi* rhodopsins CarO and OpsA and their relation to fungus–plant interaction. Int J Mol Sci 19:215

Afzal M, Yousaf S, Reichenauer TG, Sessitsch S (2012) The inoculation method affects colonization and performance of bacterial inoculant strains in the phytoremediation of soil contaminated with diesel oil. Int J Phytoremediation 14:35–47

Al Sahli AA, Abdulkhair WM (2012) Biocontrol of *Fusarium udum* diseases for some wheat cultivars by *Streptomyces spororaveus*. Afr J Microbiol Res 6:190–196

Amarger N (1980) Aspect microbiologique de la culture des legumineuses. Le Selectionneur Francais 28:61–66

Armanthus M, Simpson L, Malajczuk N (2012) Inoculate with mycorrhizae. It's as easy as A-B-Seeds ACRES 42(1)

Arshad M, Merlina G, Uzu G, Sobanska S, Sarret G, Dumat C, Silvestre J, Pinelli E, Kallerhoff J (2016) Phytoavailability of lead altered by two *Pelargonium* cultivars grown on contrasting lead-spiked soils. J Soils Sediments 16:581–591

Ashford AE, Peterson CA, Carpenter JL, Cairney JWG, Allaway WG (1988) Structure and permeability of the fungal sheath in the *Pisonia* mycorrhiza. Protoplasma 147:149–161

Ashraf M, Foolad MR (2005) Pre-sowing seed treatment-a shotgun approach to improve germination growth and crop yield under saline and none-saline conditions. Adv Agron 88:223–271

Askary TH (2010) Nematodes as biocontrol agents. In: Lichtfouse E (ed) Sociology, organic farming, climate change and soil Science. Springer, Dordrecht/Heidelberg/London/New York, pp 347–378

Babalola OO (2010) Beneficial bacteria of agricultural importance. Biotechnol Lett 32:1559–1570

Babalola OO, Berner DK, Amusa NA (2007) Evaluation of some bacterial isolates as germination stimulants of *Striga hermonthica*. Afr J Agric Res 2(1):27–30

Bais HP, Walker TS, Schweizer HP, Vivanco JM (2002) Root specific elicitation and antimicrobial activity of rosmarinic acid in hairy root cultures of sweet basil (*Ocimum basilicum L.*). plant Physiol. Biochemist 40:983–995

Balasubramanian A (2017) Soil microorganisms technical report

Bano Q, Ilyas N, Bano A, Zafar N, Akrami A, Hassan FU (2013) Effect of *Azospirillum* inoculation on maize *(Zea mays L.)* under drought stress. Pak J Bot 45:13–20

Barea JM (2015) Future challenges and perspectives for applying microbial biotechnology in sustainable agriculture based on a better understanding of plan-microbiome interaction. J Soil Sci Plant Nutr 15(2):261–282

Bashan Y (1998) Inoculants for plant growth-promoting bacteria in agriculture. Biotechnol Adv 16:729–770

Bayer Crop Science (2016). www.cropscience.bayer.us/products/seedgrowth/poncho-votivo/the-votivo-solution. Accessed 4 Nov 2018

Berdeni D, Cotton TEA, Daniell TJ, Bidartondo M, Cameron DD, Evans KL (2018) The effects of arbuscular mycorrhizal fungal colonisation on nutrient status, growth, productivity and canker resistance of apple (*Malus pumila*). Frontiers microbiol 9(1461):1–30

BioAgri AB (2016). www.bioagri.se/?p=30933&m=4353&topic=products. Accessed 4 Nov 2018

Bloemberg GV, Lugtenberg BJ (2001) Molecular basis of plant growth promotion and biocontrol by rhizobacteria. Curr Opin Plant Biol 4:343–350

Borah N, Albarouki E, Schirawski J (2018) Comparative methods for molecular determination of host-specificity factors in plant-pathogenic fungi. Int J Mol Sci 19(3):1–19

Brigham LA, Michaels PJ, Flores HE (1999) Cell-specific production and antimicrobial activity of naphthoquinones in roots of *Lithospermum erythrorhizon*. Plant Physiol 119:417–428

Brune A, Dietrich C (2015) The gut microbiota of termites: digesting the diversity in the light of ecology and evolution. Ann Rev Microbiol 69:145–166

Bucking H, Kuhn AJ, Sehroder WH, Heyser W (2002) The fungal sheath of ectomycorrhizal pine roots: an apoplastic barrier for the entry of calcium, magnesium, and potassium into the root cortex. J Exp Bot 53(374):1659–1669

Burdman S, Dulguerova G, Okon Y, Jurkevitch E (2001) Purification of the major outer membrane protein of *Azospirillum brasilense,* its affinity to plant roots, and its involvement in cell aggregation. Mol Plant-Microbe Interact 14:555–558

Chaparro JM, Sheflin AM, Manter DK, Vivanco JM (2012) Manipulating the soil microbiome to increase soil health and plant fertility. Biology & Fertility of Soils 48(5):489–499

Chodak M, Gołębiewski M, Ploskonka JM, Kuduk K, Niklinska M (2015) Soil chemical properties affect the reaction of forest soil bacteria to drought and rewetting stress. Annal Microbiol 65:1627–1637

Chu CC, Spencer JL, Curzi MJ, Zavala JA, Seufferheld MJ (2013) Gut bacteria facilitate adaptation to crop rotation in the western corn rootworm. Proc Natl Acad Sci U S A 110:11917–11922

Coombs JT, Michelsen PP, Franco CMM (2003) Evaluation of endophytic actinobacteria as antagonists of *Gaeumannomyces graminis var. tritici* in wheat. Biol Control 29:359–366

Deaker R, Roughley RJ, Kennedy IR (2004) Legume seed inoculation technology—a review. Soil Biol Biotech 36:1275–1288

Dijkshoorn J, Ursing BM, Ursing JB (2000) Train, clone and species: comments on three basic concepts of bacteriology. J Med Microbiol 49:397–401

Duffy BK, Defago G (1999) Environmental factors modulating antibiotic and siderophore biosynthesis by Pseudomonas fluorescens biocontrol strains. Appl Environ Microbiol 65:2429–2438

Elegba MS, Rennie RJ (1984) Effect of different inoculant adhesive agents on survival, nodulation, and nitrogenase (acetylene-reducing) activity of soybeans [*Glycine max* (L.) Merrill]. Cana J Soil Sci 64(4):631–636

Elzein A, Heller A, Ndambi B, de Mol M, Kroschel J, Cadisch G (2010) Cytological investigations on colonization of sorghum roots by the mycoherbicide *Fusarium oxysporum* f. sp. *strigae* and its implications for Striga control using a seed treatment delivery system. Biol Control 53:249–257

Ferguson BJ, Indrasumunar A, Hayashi S, Lin MH, Lin YH, Reid DE, Gresshoff PM (2010) Molecular analysis of legume nodule development and autoregulation. J Integr Plant Biol 52:61–76

Finlay RD (2008) Ecological aspects of mycorrhizal Symbiosis: with special emphasis on the functional diversity of interactions involving the Extraradial mycelium. J Exp Bot 59(5):1115–1126

Ganguly S (2006) Recent taxonomic status of entomopathogenic nematodes: a review. Int J Nematol 36:158–176

Giron D, Huguet E, Stone GN, Body M (2016) Insect-induced effects on plants and possible effectors used by galling and leaf-mining insects to manipulate their hostplant. J Insect Physiol 84:70–89

Glare T, Caradus J, Gelernter W, Jackson T, Keyhani N, Kohl J, Marrone P, Morin L, Stewart A (2012) Have biopesticides come of age. Trends Biotechnol 30:250–258

Graham PH, Vance CP (2003) Legumes: importance and constraints to greater use. Plant Physiol 131:872–877

Hayat R, Ali S, Amara U, Khalid R, Ahmad I (2010) Soil beneficial bacteria and their role in plant growth promotion: a review. Ann of Microbiol 60:579–598

Hiltner L (1904) Uberneure Erfahrungen und probleme auf demgebeit der boden back teriologieund unter besondererberuck sichtigung der grundungung und brache. Arb Deut LandwirschGes 98:59–78

Hubbard CJ, Li B, McMin R, Brock MT, Maignien L, Ewers BE, Kliebenstein D, Weinig C (2018) Rhizosphere microbes and host plant genotype influence the plant metabolome and reduce insect herbivory. Biorxiv 1-30

Iqbal A, Arshad M, Hashmi H, Karthikeyan R, Gentry TJ, Schwab AP (2018) Biodegradation of phenol and benzene by endophytic bacterial strains isolated from refinery wastewater-fed *Cannabis sativa*. Environ Technol 39:1705–1714

Jan ND, Mahar GM, Mahar AN, Hullio MH, Lanjar AG, Gower SR (2008) Susceptibility of different insect pupae to the bacterial symbiont, *Xenorhabdus nematophila*, isolated from the entomopathogenic nematode, *Steinernema carpocapsae*. Pak J Nematol 26:59–67
Jensen HL (1943) The influence of reaction on the formation of root nodules in *Medicago* and *Trifolium*. In: Nitrogen fixation in leguminous plants, vol 68. Proceedings of the Linnean Society, NSW, pp 207–213
Kaljeet S, Keyeo F, Amir HG (2011) Influence of carrier materials and storage temperature on survivability of Rhizobial inoculant. Asian J Plant Sci 10(6):331–337
Kaya HK, Stock SP (1997) Techniques in insects nematology. In: Lacey L (ed) Manual of techniques in insect pathology, biological techniques series. Academic, San Diego, CA, pp 281–324
Kosanke JW, Osburn RM, Shuppe GI, Smith RS (1992) Slow rehydration improves the recovery of dried bacterial populations. Can J Microbiol 38:520–525
Li HY, Smith SE, Holloway RE, Zhu YG, Smith FA (2006) Arbuscular mycorrhizal fungi contribute to phosphorus uptake by wheat grown in a phosphorus-fixing soil even in the absence of positive growth responses. New Phytol 172:536–543
Lojan P, Demortier M, Velivelli SL, Pfeiffer S, Suarez JP, deVos P, Prestwich BD, Sessitsch A, Declerck S (2016) Impact of plant growth-promoting rhizobacteria on root colonization potential and life cycle of *Rhizophagus irregularis* following co-entrapment into alginate beads. J Appl Microbiol 122:429–444
Loneragan JF, Meyer D, Fawcett RC, Anderson AJ (1955) Lime pelleted clover seeds for nodulation on acid soils. Aust J Agri Res 21:264–265
Long SR (2001) Genes and signals in the rhizobium-legume symbiosis. Plant Physiol 125:69–72
Mahmood A, Turgay OC, Farooq M, Hayat R (2016) Seed biopriming with plant growth promoting rhizobacteria: a review. FEMS Microbiol. Ecol 92(8):1–14
Martin F, Aerts A, Ahren D, Brun A, Danchin EGJ, Duchaussoy F, Gibon J, Kohler A, Lindquist E, Pereda V, Salamov A, Shapiro HJ, Wuyts J, Blaudez D, Buee M, Brokstein P, Canback B, Cohen D, Courty PE, Coutinho PM, Delaruelle C, Detter JC, Deveau A, DiFazio S, Duplessis S, Fraissinet-Tachet L, Lucic E, Frey-Klett P, Fourrey C, Feussner I, Gay G, Grimwood J, Hoegger PJ, Jain P, Kilaru S, Labbe J, Lin YC, Legue V, Tacon FL, Marmeisse R, Melayah D, Montanini B, Muratet M, Nehls U, Niculita-Hirzel H, Oudot-Le Secq MP, Peter M, Quesneville H, Rajashekar B, Reich M, Rouhier N, Schmutz J, Yin T, Chalot M, Henrissat B, Kues U, Lucas S, Van de Peer Y, Podila GK, Polle A, Pukkila PJ, Richardson PM, Rouze P, Sanders IR, Stajich JE, Tunlid A, Tuskan G, Grigoriev IV (2008) The genome of Laccaria bicolor provides insights into mycorrhizal Symbiosis. Nature 452(7183):88–U7
McDonald MB (1999) Seed deterioration: physiology, repair and assessment. Seed Sci Technol 27:177–237
Michaelis J, Diekmann M (2018) Effects of soil types and bacteria inoculum on the cultivation and reintroduction success of rare plant species. Plant Ecol 219(4):441–453
Mitter B, Sessitsch A, Naveed M (2018) Method for producing plant seed containing endophytic microorganisms. U.S. patent app. Pub. US 2018 / 0132486 A1
Muller H, Berg G (2008) Impact of formulation procedures on the effect of the biocontrol agent *Serratia plymuthica* HRO-C48 on Verticillium wilt in oilseed rape. Biol Control 53:905–916
Murray H (2002) Industrial clays case study. Indiana University. http://pubs.iied.org/pdfs/G00556.pdf
Nayaka SC, Niranjana SR, Shankar ACU, Raj SN, Reddy MS, Prakash HS, Mortensen CN (2010) Seed biopriming with novel strain of *Trichoderma harzianum* for the control of toxigenic *Fusarium verticillioides* and *fumonisins* in maize. Arch Phytopathol Plant Prot 43:264–282
Oerke EC (2006) Crop losses to pests. J Agric Sci 144:31–43
Oka-Kira E, Kawaguchi (2006) Long-distance signaling to control root nodule number. Curr Opin Plant Biol 9:496–502
Okon Y, Hadar Y (1987) Microbial inoculants as crop-yield enhancers. Crit Rev Biotechnol 6(1):61–85
Orates I, Bykova A (2018) The effect of mycorrhiza inoculation and phosphorus application on phosphorus efficiency of wheat plants. Commun Soil Sci Plant Anal 49(10):1199–1207

Philippot L, Raaijmakers JM, Lemanceau P, Putten WHVD (2013) Going back to the roots: the microbial ecology of the rhizosphere. Nat. Rev Microbiol 11:789–799

Podile AR, Kishore GK (2006) Plant growth promoting rhizobacteria. In: Gnanamanickam SS (ed) Plant associated bacteria. Springer, Dordrecht, pp 195–230

Porcel R, Zamarreno AM, Mina JMG, Aroca R (2014) Involvement of plant endogenous ABA in bacillus megaterium PGPR activity in tomato plants. BMC Plant Biol 14(36):1–12

Pueppke SG, Broughton WJ (1999) Rhizobium sp. strain NGR234 and R. fredii USDA257 share exceptionally broad, nested host ranges. Mol. Plant-Microbe Interact 12:293–318

Rafique M, Riaz A, Anjum A, Qureshi MA, Mujeeb F (2018) Role of bioinoculants for improving growth and yield of okra (*Abelmoshuses culentum*). Uni J Agri Res 6(3):105–112

Raj SN, Shetty NP, Shetty HS (2004) Proline—an inducer of resistance against pearl millet downy mildew disease caused by *Sclerospora graminicola*. Phytoparasitica 32:523–527

Ravel J, Cornelis P (2003) Genomics of pyoverdine-mediated iron uptake in pseudomonads. Trends Microbiol 11:195–200

Recep K, Fikrettin S, Erkol D, Cafer E (2009) Biological control of the potato dry rot caused by *Fusarium* species using PGPR strains. Biol Control 50(2):194–198

Robert MA, Stodola AJW, Tims JE, Saxton AM (2001) Moisture retention properties of a mycorrhizal soil. Plant Soil 230:87–97

Smith SE, Read DJ (2008) Mycorrhizal Symbiosis, 3rd edn. Academic Press, Amsterdam

Somasegaran, P. and H. J. Hoben. 1994. Handbook for rhizobia: methods in legume rhizobium technology. p. 217–2178. Springer, Berlin/Heidelberg/New York

Spaink HP (1995) The molecular basis of infection and nodulation by rhizobia: the ins and outs of sympathogenesis. Annu Rev Phytopathol 33:345–368

Spence C, Bais H (2013) Probiotics for plants: rhizospheric microbiome and plant fitness. In: de Bruijn FJ (ed) Molecular microbial ecology of the rhizosphere, vol 2. Wiley, Hoboken, NJ, pp 713–721

Srinivasan K, Mathivanan N (2011) Plant growth promoting microbial consortia mediated classical biocontrol of sunflower necrosis virus disease. J Biopest 4:65–72

Staudinger C, Tershani VM, Quintana EG, Gonzalez EM, Hofhansl F, Bachmann G, Wienkoop S et al (2016) Evidence for a rhizobia-induced drought stress response strategy in *Medicago truncatula*. J Proteome 136:202–213

Steenhoudt O, Vanderleyden J (2000) *Azospirillum*, a free-living nitrogen-fixing bacterium closely associated with grasses: genetic, biochemical and ecological aspects. FEMS Microbiol Rev 24:487–506

Tarley CRT, Arruda MAZ (2004) Biosorption of heavy metals using rice milling by-products: characterization and application for removal of metals from aqueous effluents. Chemosphere 54:987–995

Taylor AFS, Alexander I (2005) The ectomycorrhizal Symbiosis: life in the real world. Mycol 19:102–112

Tiwari S, Lata C, Chauhan PS, Nautiyal CS (2016) Pseudomonas putida attunes morphophysiological: biochemical and molecular responses in *Cicer arietinum L.* during drought stress and recovery. Plant Physiol. Biochemist 99:108–117

Tiwari S, Pandey SW, Chauhanb PS, Pandey R (2017) Biocontrol agents in co-inoculation manages root knot nematode [*Meloidogyne incognita* (Kofoid & White) Chitwood] and enhances essential oil contents in *Ocimum basilicum* L. Ind Crops Prod 97:292–301

Vardharajula S, Ali SZ, Grover M, Reddy G, Bandi V (2011) Drought-tolerant plant growth promoting bacillus spp.: effect on growth osmolytes, and antioxidant status of maize under drought stress. J Plant Interact 6:1–14

Vessey JK (2003) Plant growth promoting rhizobacteria as biofertilizers. Plant Soil 255:571–586

Vestberg M, Kukkonen S, Saari K, Parikka P, Huttunen J, Tainio L, Devos N, Weekers F, Kevers C, Thonart P, Lemoine MC (2004) Microbial inoculation for improving the growth and health of micro propagated strawberry. Appl Soil Ecol 27(3):243–258

Wardle DA (1992) A comparative assessment of the factors which influence microbial biomass carbon and nitrogen levels in soil. Biol Rev 67:321–358

Wells J, Varel V (2011) Food safety issues in animal source foods related to animal health and welfare. In: Pond WG, Bazer FW, Rollen BE (eds) Animal welfare in animal agriculture: husbandry, stewardship and sustainability in animal production. CRC Press, New York, pp 185–203

Whipps JM (2001) Microbial interactions and biocontrol in the rhizosphere. J Exp Bot 52:487–511

Wright DP, Read DJ, Scholes JD (1998) Mycorrhizal sink strength influences whole plant carbon balance of *Trifolium repens* L. Plant Cell Environ 21:881–891

Yoneyama K, Natsume M (2010) Comprehensive natural product II. Chem Biol 4:539–561

Zahid MS, Iqbal A, Arshad M (2016) Benzene degradation with bacterial strains isolated from rhizosphere of *Cannabis sativa* being irrigated with petroleum refinery wastewater. Desalin Water Treat 57:17579–17584

Zahra Z, Arshad M, Rafique R, Mahmood A, Habib A, Qazi IA, Khan SA (2015) Metallic nanoparticle ($TiO_2$ and $Fe_3O_4$) application modifies rhizosphere phosphorus availability and uptake by *Lactuca sativa*. J Agric Food Chem 63:6876–6882

Zak RD, Tilman D, Permenter RR, Rice CW, Fisher FM, Vose J, Milchunas D, Martin CW (1994) Plant production and soil microorganisms in late-successional ecosystem: a continental-scale study. Ecol Soci America 75(8):2333–2347

Zolla G, Bakker MG, Badri DV, Chaparro JM, Sheflin AM, Manter DK, Vivanco J (2013) Understanding root-microbiome interactions. In: de Bruijn FJ (ed) Molecular microbial ecology of the Rhizosphere, vol 2. Wiley, New York, pp 745–754

# Role of Seed Priming and Plant Growth-Promoting Rhizobacteria in Modulating Crops' Responses to Salinity Stress

**Abdul Majeed, Zahir Muhammad, and Rehmanullah**

## Contents



**Abstract** Salinity, a leading problem in agriculture, has been regarded as a major limiting factor in growth performance of crops with substantial yield losses. Although different crops respond differently to salt stress, nearly all physiological, biochemical, developmental, and growth processes of all crops are affected by salinity to some extent. Major effects of salinity on crops include reduced uptake of water and nutrient, the altered rate of photosynthesis, respiration and transpiration, stomatal conductance, and ion toxicity, which collectively reduce their growth and yield potentials. The problem of salinity shows its adversities in those areas where evaporation rate exceed the precipitation rate. Some other factors corresponding to salinity emergence in agricultural lands include natural weathering, nonsystemic irrigation practices, and intensification of agriculture. Development of salt tolerant crops through applied biotechnology seems an ideal approach; however, research in that domain is a novice quest. One of the best sustainable approaches in salt stress management is the employment of pre-sowing seed priming techniques involving different priming agents because they do not pose any ecological or environmental problems. Similarly, several strains of plant growth-promoting rhizobacteria (PGPR) have the abilities to recover soil quality degraded by salinity and directly or

---

A. Majeed (✉)
Department of Botany, Government Degree College Pabbi,
Nowshera, Khyber Pakhtunkhwa, Pakistan

Z. Muhammad · Rehmanullah
Department of Botany, University of Peshawar, Peshawar, Khyber Pakhtunkhwa, Pakistan

 
M. Hasanuzzaman, V. Fotopoulos (eds.), *Priming and Pretreatment of Seeds and Seedlings*, https://doi.org/10.1007/978-981-13-8625-1_27

indirectly improve crops' growth in a saline environment. In this chapter, we review the effects of salinity on major cultivated crops and the potential role of seed priming and PGPR in helping them to adjust to salinity stress.

**Keywords** Salinity tolerance · Electrical conductivity · Ion exchange · Physiological stress · Rhizosphere · Yield

## 1 Introduction

Cultivated crops are major drivers in agricultural outputs and meeting food requirements of millions of people throughout the world. Increase in crop production and yield is emergent because the world's population is growing at a faster rate, which will require more food and hence greater production of crops. The productivity of agricultural crops is threatened by several biological and environmental constraints among which salinity remains a significant one. It is estimated that global irrigated land affected by salinity is about 20%, which is anticipated to elevate to 50% by the next three decades (Pitman and Läuchli 2002; Soda et al. 2016). Drastic effects of salinity on crops include poor seed germination and seedling growth or even death which depends on the concentration of salinity and concerned tolerance level of crop species. If crops survived under salinity stress, still they will show poor growth; abnormal physiological functions such as altered rate of photosynthesis, respiration, and transpiration; and reduced yields (Muhammad and Hussain 2010a, b; Khalilzadeh et al. 2018; Mumtaz et al. 2018).

In order to make cultivated crops adaptable to the threat of salinity, molecular breeding, utilization of functional genomics and metabolomics, and development of transgenic crops with salinity tolerance traits are widely worked on tools (Kumari and Parida 2018; Patel et al. 2018); however, these methods have yet not resulted in durable solution to ever-growing salinity problems which cultivated crops are challenged with. Moreover, high costs and technical expertise in molecular methods for the development of salt tolerant crop varieties are additional hurdles in achieving salt stress management targets. Thus easy, economical and sustainable methods are crucially desired to address the issue of salinity stress and make crops more adaptable to saline conditions.

Seed priming, which involves soaking seeds with some chemical agents before sowing, can work better in achieving desirable responses of cultivated crops toward salinity stress. Potential priming agents include water, different concentration of salts, polyethylene glycol (PEG), superoxides of sulfur (S) and nitrogen (N), salicylic acid, proline, and several other natural metabolites (Salah et al. 2015; Savvides et al. 2016; Sano et al. 2017). The basic mechanism involved in priming is stimulation of metabolic activities at pre-germination state. Many studies suggest a promising role of seed priming in improving crops' response to stress conditions and better growth and production. In our previous work (Muhammad et al. 2015), we achieved improved salinity tolerance in different genotypes of wheat treated with halopriming agents. Salah et al. (2015) have also revealed that priming with PEG improved growth and yield attributes of rice under zinc oxide stress. Better seedling growth in maize was observed in alkalinity stress when seeds were treated with jasmonic acid prior to germination (Mir et al. 2018). Yang et al. (2018) demonstrated that seed pretreatment with different concentrations of saponin significantly elevated germination percentage while decreasing mean germination time of quinoa at 400 mM salinity stress, owing to the significance of priming techniques in plant stress physiology.

Some microorganisms, those present in the rhizosphere either free-living or associated with plants, have extraordinary potentials of stimulating plant's growth, soil fertility, and protection of hosts from a wide range of biotic and abiotic stresses. These microorganisms are collectively termed as plant growth-promoting rhizobacteria (PGPR). PGPR offer excellent choice as a sustainable measure to reduce the drastic consequences of salinity on crops. Several studies have outlined the active role of rhizobacteria in mitigating the salinity stress, and consequent growth improvement of crops under salt stress has been well established (Mayak et al. 2004; Han and Lee 2005; Tank and Saraf 2010; Upadhyay et al. 2011). Nadeem et al. (2006) documented that inoculation of maize seedling with rhizobacteria significantly improved the growth performance under salinity stress. Peanut exposed to 100 mM NaCl salinity level exhibited salt tolerance and improved growth in the presence of PGPR *Pseudomonas* and *Bacillus* sp. (Shukla et al. 2012). Gururani et al. (2013) showed that different strains of *Bacillus* induced tolerance to potato against different levels of salinity and other stresses. Some strains of *Bacillus* were reported to improve growth of rice under imposed salinity stress of 200 mM (Nautiyal et al. 2013). In a recent study, Chatterjee et al. (2018) revealed that *Brevibacterium* sp. confer salinity tolerance to the rice crop. The aim of this chapter is to overview the role of seed priming and PGPR in salinity tolerance to crops, which will further strengthen our knowledge about salinity mitigation potential of these approaches in agriculture.

## 2 Salinity and Its Effect on Field Crops

Salinity is referred to a state of soil where electrical conductivity of the saturated paste equals or increases 4 $dSm^{-1}$ (~40 mM NaCl or other salts) resulting osmotic disturbances and difficulty for crops to grow (Munns and Tester 2008). It is estimated that salt-affected land throughout the world exceeds 800 mha which represents almost 6% of the global net arable land (Munns and Tester 2008). Causes of salinity are diverse, which range from natural flooding and salt accumulation to man-triggered irrigation of soil with substandard water and agricultural intensification.

Saline conditions impose several physiological, biochemical, and growth-inhibitory challenges upon crops. Salinity stress causes primarily excessive generation of reactive oxygen species (ROS) in addition to imparting negative influence on nutrient uptake and their transport, ion toxicity, water availability in soil, and damage of root hairs (Rady and Mohamed 2015; Kumari and Parida 2018). Reduction in transpiration rate occurs when crops get exposed to salinity, which is generally correlated with "ionic component" and "osmotic component" – phenomena which accumulate $Na^+$ and $Cl^-$ in tissues, whereby toxicity of ions results in premature senescence and anticipated yield losses (Al-Tamimi et al. 2016). Increased osmotic stress and ion toxicity developed to the high level of salinity often cause stomata to open and close in an unpredictable manner leading to reduced photosynthesis and consequent growth and yield reduction of crops (Jiang et al. 2017). Effect of different concentrations of NaCl on growth, yield, and physiological activities of major food crops has been presented in Table 1. Amirjani (2011) observed retarded shoot length, chlorophyll *b*, the dry biomass of plants, but significant increase in root length of rice (*Oryza sativa*) at 200 mM NaCl stress. In other similar studies, decline in shoot growth, yield, grain weight and quality, rate of transpiration, nutrient uptake, photosynthetic activities, and chlorophyll pigments in rice crop were caused by different concentrations of salinity (Hakim et al. 2014; Thitisaksakul et al. 2015; Mumtaz et al. 2018). A dose concentration of salinity ranging between 2 and 108 $dSm^{-1}$ significantly lowered germination, radicle and plumule growth, leaf area, and dry biomass, while 30–90 mM NaCl stress resulted in deficient grain yield, stomatal conductance, and water uptake in wheat cultivars (Muhammad and Hussain 2012; Muhammad et al. 2015; Siddiqui et al. 2017; Khalilzadeh et al. 2018). In other crops such as maize, soybean, common bean, tomato, and cucumber, decreased growth, biomass, leaf area, and chlorophyll content and yield were documented under different salinity levels (Hamayun et al. 2010; Dolatabadian et al. 2011; Babu et al. 2012; Sali et al. 2015; Rady and Mohamed 2015; Taïbi et al. 2016; Jiang et al. 2017; Luo et al. 2018; Youssef et al. 2018). Here only results of few studies are presented. In fact, salinity adversely affects almost all cultivated crops with more growth and yield depression correlated with higher salinity concentrations.

## 3 Seed Priming: A Promising Approach to Salinity Tolerance Induction in Crops

Seed priming is an applied technique, which implicates soaking of seeds of different crops in different concentrations of priming agents which range from water to natural metabolites and chemicals of diverse classes. The most promising and widely employed priming agents are $H_2O$ (hydropriming); different concentrations of salts such as NaCl, $MgCl_2$, $MgSO_4$, etc. (halopriming); polyethylene glycol (PEG); reactive oxygen species of S and N; hormones like salicylic acid, jasmonic acid, auxin, gibberellins, and proline; leaf extracts; and several other natural metabolites (Salah et al. 2015; Savvides et al. 2016; Sano et al. 2017). Seeds are soaked in priming agents of different concentrations for different durations. Primed seeds are dried

**Table 1** Effect of different salinity levels on growth, yield, and physiological and biochemical attributes of important crops

| Crops | Salinity concentration | Effects | References |
|---|---|---|---|
| Rice | 200 mM | Decreased shoot length and dry biomass but increased root length, decreased chlorophyll *b* | Amirjani (2011) |
| Rice | 4–12 $dSm^{-1}$ | Increased $N^+$ accumulation, decreased shoot growth, dry biomass, and yield | Hakim et al. (2014) |
| Rice | 2–4 $dSm^{-1}$ | Decreased seed fertility and grain weight | Thitisaksakul et al. (2015) |
| Rice | 150 mM | Decreased growth and transpiration rate | Al-Tamimi et al. (2016) |
| Rice | 0.24–10.8 $dSm^{-1}$ | Drastic effects on growth, yield, grain quality, and ionic uptake | Mumtaz et al. (2018) |
| Wheat | 2–8 $dSm^{-1}$ | Decreased seedling growth at 8 $dSm^{-1}$ | Muhammad and Hussain (2012) |
| Wheat | 2–8 $dSm^{-1}$ | Lowered germination, radicle and plumule length, dry biomass, and moisture content | Muhammad et al. (2015) |
| Wheat | 8–108 $dSm^{-1}$ | Decreased growth, total dry biomass, yield, leaf area | Siddiqui et al. (2017) |
| Wheat | 30–90 mM | Decreased grain yield, stomatal conductance, water uptake | Khalilzadeh et al. (2018) |
| Maize | 11 $dSm^{-1}$ | Reduced shoot length, leaf area, and transpiration | Hütsch et al. (2015) |
| Maize | 50–400 mM | Declined seed germination, germination index, and chlorophyll pigments | Sali et al. (2015) |
| Maize | 0–100 mM | Reduced shoot length and dry weight | Jiang et al. (2017) |
| Maize | 0–100 mM | Declined growth rate, root-shoot length, fresh and dry weight | Luo et al. (2018) |
| Soybean | 70–140 mM | Reduced growth, dry weight, and chlorophyll contents | Hamayun et al. (2010) |
| Soybean | 25–100 mM | Retardation in plant height, dry weight, leaf number and area | Dolatabadian et al. (2011) |
| Tomato | 25–200 mM | Decrease bio-matter, leaf area, $N^+/K^+$ ratio | Babu et al. (2012) |
| Common bean | 6.23–6.28 $dSm^{-1}$ | Reduction in shoot length, leaf no., dry weight, and moisture contents | Rady and Mohamed (2015) |
| Common bean | 50–200 mM | Reduction in growth, biomass, and pigments at higher salt concentration | Taïbi et al. (2016) |
| Cucumber | 50–100 mM | Drastic effects on growth, yield, leaf water content, and chlorophyll pigments | Youssef et al. (2018) |

and then sown under stressed environment to evaluate the influence of priming techniques on growth and yield performance of crops. Nature of the priming agents and soaking duration are key factors involved in salt stress induction although variation may be evident in different crop species in their response toward priming. Various researchers prefer different priming techniques such as hydropriming (use of $H_2O$), halopriming (different salts at variable concentrations), osmopriming (chemicals with osmotic gradient), hormones, and other natural compounds, which correspond to different effects on germination, growth, and establishment of test seeds exposed to priming (Sano et al. 2017; Song et al. 2017).

Basic mechanism involved in priming-induced stress tolerance in crops is the activation of pre-germination "memory," which involves several metabolic and physiological pathways ranging from hormonal communication, enzymes activities, signaling, assembly of transcription factors, synthesis of specialized proteins, gene expression to alterations in dormancy, and embryonic state of seeds exposed to priming (Bruce et al. 2007; Varier et al. 2010; Savvides et al. 2016). Kubala et al. (2015) stated that osmopriming induces stress tolerance in seeds by activating antioxidative system and buildup of dehydrins, which results in the awakening of "stress-related memory," which seeds might experience during and after germination. Growth regulatory substances and hormones, e.g., auxins, gibberellins, abscisic acid, ethylene, salicylic acid, nitric oxide, and jasmonic acid, are believed to be correlated with signaling traits and stress tolerance responses in plants (Bruce et al. 2007), and seed priming can alter these substances and overall signaling processes by adjusting crops to stressful environment. Negative influences of salinity and other abiotic stresses on germination, growth, and physiological functions can be attributed to the production of reactive oxygen species (ROS) and membranes abnormalities of seeds. Seed priming may impart healthy effects on growth and germination by regulating endogenous substances, destruction of ROS, homeostatic regulation, and production of stress-related transcription factors and proteins, which will consequently induce salinity stress tolerance in exposed plants (Savvides et al. 2016) (Fig. 1).

## 4 Role of Seed Priming in Improving Salinity Tolerance of Major Crops

The efficiency of seed priming techniques in developing resistance to salinity and other stressful conditions is well established (Table 2). Pre-sowing seed treatment with water, osmolytes, salts, hormones, and plant extracts in different studies have been found effective in inducing salt tolerance to different crops (Ali et al. 2017; Bajwa et al. 2018; Farooq et al. 2018). In previous work, Basra et al. (2005, 2006) achieved promising results in inducing salinity tolerance in wheat to a high level of salinity (15 $dSm^{-1}$) when seeds were pretreated with different priming agents. Studies carried out by different researchers revealed the efficacy of different priming agents at different concentrations on germination, growth potential, yield, nutrient uptake, and physiological functions of wheat genotypes (Afzal et al. 2005, 2006; Wahid et al. 2007; Jamal et al. 2011; Jafar et al. 2012; Bajwa et al. 2018). Similarly

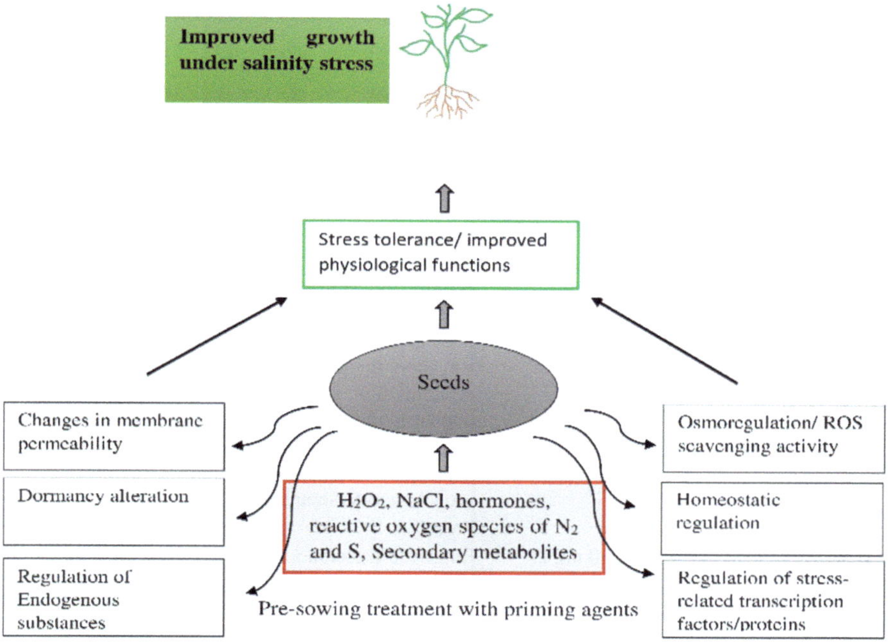


**Fig. 1** An illustration of the possible mechanism of seed priming in mitigating salinity stress in plants

in rice, an important agricultural crop, salt tolerance has been successfully induced by the application of $H_2O$, $ZnSO_4$, $CaCl_2$, spermidine, etc. with improved germination, growth, production, yield, and biomass under saline conditions (Afzal et al. 2012; Chunthaburee et al. 2015; Wei et al. 2017; Bajwa et al. 2018). Priming techniques have also contributed to induction of salinity resistance in maize in pot experiments and field trials. Seeds treated with 200 mM NaCl, KCl, and $H_2O$ for 8 h soaking duration improved germination, root-shoot elongation, and fresh/dry weight of seedling at higher salinity (Ashraf and Rauf 2001). In another study, Janmohammadi et al. (2008) reported that hydropriming of maize seeds for 36 h, while osmopriming with 1.2 MPa urea for 96 h, resulted in better germination, vigor, and seedling growth at the saline environment. Similar results were obtained in different researches when maize seeds were exposed to hydropriming, NaCl, thiamin, and melatonin at different concentrations and soaking duration, which yielded enhanced salt tolerance in maize crops and improved germination, growth, yield, and physicochemical activities under salty conditions (Bakht et al. 2011; Kaya et al. 2015; Jiang et al. 2016).

**Table 2** Effect of different priming agents on growth, yield, and physicochemical performance of major crops under different salinity concentrations

| Crops | Priming agent and soaking duration | Response of crops under saline environment | References |
|---|---|---|---|
| Maize | 0–200 mM NaCl, $CaCl_2$, KCl, $H_2O$ (8 h) | Better germination, fresh and dry weight of radical and plumule at elevated salinity stress | Ashraf and Rauf (2001) |
| Wheat | $H_2O$; 25 mM $Ca(NO_3)_2$; 50 mM NaCl; 100 mM $CaCl_2$ (24 h for each priming agent) | Significantly improved emergence time, seedling length and dry weight at salt stress (15 $dSm^{-1}$) | Basra et al. (2005) |
| Wheat | $H_2O$ (12 h), 50 mm $NaCl_2$, 50 mM $CaCl_2.H_2O$, 0.56 MPa PEG (24 h) | Seed germination, vigor, root and shoot length and dry weight, accumulation of $N^+$ and $K^+$ improved in salinity (15 $dSm^{-1}$) | Basra et al. (2006) |
| Wheat | Kinetin, $GA_3$, IAA (500 ppm) | Kinetin improved germination, mean germination time, seedling growth, and fresh and dry weight at 15 $dSm^{-1}$ salinity level | Afzal et al. (2005) |
| Wheat | 50 mM $CaCl_2$, 50 mg $L^{-1}$ ascrobate (12 h) | Improved germination and seedling growth at 15 $dSm^{-1}$ | Afzal et al. (2006) |
| Wheat | 1–120 μM $H_2O_2$ (5 h) | Early germination, better photosynthesis, stomatal conductance, growth 150 mM | Wahid et al. (2007) |
| Maize | Hydropriming (36 h); osmopriming with 1.2 MPa urea (96 h) | Better germination, vigor, seedling growth achieved with hydropriming at 21.8 $dSm^{-1}$ salt stress | Janmohammadi et al. (2008) |
| Wheat | 1–8 $dSm^{-1}$ | Decrease in total water uptake and grain yield | Gowing et al. (2009) |
| Canola | 5–20 $dSm^{-1}$ | Decreased germination, seedling growth and biomass | Bybordi and Tabatabaei (2009) |
| Soybean | 50–200 mM | Plant height, biomass, and nutrients decreased significantly | Amirjani (2010) |
| Wheat | 30 mM NaCl (12 h) | Higher K/N ration, shoot dry weight at 120 mM | Jamal et al. (2011) |
| Maize | 6–8 $dSm^{-1}$ | Delayed germination, growth and developmental aspects | Bakht et al. (2011) |
| Wheat | 50 mg $L^{-1}$ $CaCl_2$, kinetin, salicylic acid, ascorbate (12 h) | Improvement in germination, productive and yield traits, and enzymatic activity at 10.65 $dSm^{-1}$ | Jafar et al. (2012) |
| Rice | 50 mM $H_2O_2$ and $CaCl_2$ (36 h) | Enhanced germination and growth parameters achieved at priming with $CaCl_2$ under 80 mM NaCl stress | Afzal et al. (2012) |
| Maize | $H_2O$, 100 mM NaCl+150 mg $L^{-1}$ of thiamin (24 h) | Enhanced germination, growth and biochemical parameters at 100 mM NaCl stress | Kaya et al. (2015) |

(continued)

**Table 2** (continued)

| Crops | Priming agent and soaking duration | Response of crops under saline environment | References |
|---|---|---|---|
| Wheat | 2–8 $dSm^{-1}$ | Germination and seedling growth linearly decreased with increase in NaCl concentration | Muhammad et al. (2015) |
| Rice | 4 $dSm^{-1}$ | Reduction in grain weight and seed fertility | Thitisaksakul et al. (2015) |
| Mung bean | 75 $dSm^{-1}$ | Abnormalities in growth and yield attributes | Sehrawat et al. (2015) |
| Maize | 50–400 mM NaCl and $CaCl_2$ | Germination and chlorophyll contents reduced | Sali et al.(2015) |
| Rice | $H_2O$, 1 mM spermidine (12 h) | Improved growth and anti-oxidative activity at 150 mM NaCl stress | Chunthaburee et al. (2015) |
| Leek | 7 $dSm^{-1}$ | Decrease in shoot length, stem diameter, dry biomass | Kiremit and Arslan (2016) |
| Wheat | $H_2O$, 1.5% $CaCl_2$ (12 h) | Enhanced growth, yield, and physiological responses at 100 mM NaCl | Tabassum et al. (2017) |
| Maize | 0.8 mM melatonin (24 h) | Improvement in germination, growth, and physiological attributes at 150 mM NaCl stress | Jiang et al. (2016) |
| Rice | 200 mM | Shoot and root lengths, biomass, chlorophyll pigments, and proline content decreased | Jogawat et al. (2016) |
| Chickpea | 40–150 mM | Reduced plant height, biomass, pod filling, and pod abortion | Atieno et al. (2017) |
| Rice | 10 μM ABA (24 h) | Elevated growth performance under saline and alkaline soils | Wei et al. (2017) |
| Wheat | 10–100 mM | Reduction in plant height and yield and water contents | Tabassum et al. (2017) |
| Alfalfa | 18.3–24.5 $dSm^{-1}$ | Reduced leaf area, stomatal conductance, decreased mineral ions in leaves | Sandhu et al. (2017) |
| Faba bean | 50–150 mM | Reduced growth, yield, and chlorophyll contents | Hussein et al. (2017) |
| Wheat | 0.5 mM $ZnSO_4$ (12 h) | Improvement in productivity and grain yield | Rehman et al. (2018) |
| Wheat | $H_2O$, 5% sorghum water extract, 5 $mgl^{-1}$ benzyl aminopurine (12 h) | Enhanced growth and biochemical activities at 10 $dSm^{-1}$ salt stress | Bajwa et al. (2018) |

(continued)

**Table 2** (continued)

| Crops | Priming agent and soaking duration | Response of crops under saline environment | References |
|---|---|---|---|
| Rice | 0.5 M $ZnSO_4 \cdot 7H_2O$ (24 h) | Enhanced productive and yield traits | Farooq et al. (2018) |
| *Brassica* species | 8–16 $dSm^{-1}$ | Complete failure of growth at salinity concentration exceeding 15 $dSm^{-1}$ | Shirazi et al. (2018) |
| Forage legumes (*Lotus* sp.) | 50–250 mM | Germination reduced by 47% | Hajri et al. (2018) |

## 5 PGPR: Induction of Salinity Tolerance in Crops and Governing Mechanisms

The rhizosphere provides an excellent interface to plants and millions of microorganisms to interact with each other in both beneficial and harmful directions. The microorganism inhabiting the rhizosphere may be pathogens, neutral or beneficial in their actions toward interacting plants. Among beneficial ones, those bacteria, which have the potential to stimulate plant growth either directly or indirectly through several mechanisms, are known as plant growth-promoting rhizobacteria (PGPR) (Lugtenberg and Kamilova 2009). Plant growth-stimulating characters of PGPR are widely attributed to their potentials of producing plant hormones and volatile organic compounds, nitrogen fixation, phosphate solubilization and enhancing nutrient acquisition, and suppression of diseases and stressful environment, to which plants are exposed (Lugtenberg and Kamilova 2009; Vacheron et al. 2013). The PGPR may be free-living or symbionts and comprise diverse strains. Different strains of *Pseudomonas*, *Bacillus*, *Azospirillum*, *Azotobacter*, and several species have been well studied PGPR, which have multiple activities regarding plant growth enhancement, soil fertility, and a wide array of biotic and abiotic stress management (Ahmad et al. 2008; Beneduzi et al. 2012; Shameer and Prasad 2018). The PGPR may be endophytes (living in tissues of plants), or they may form an association with root hairs and root surfaces, which in either case is facilitated by root exudates (Ilangumaran and Smith 2017).

In the ever-growing concern of agricultural salinity, mitigating effects of PGPR and induction of salinity tolerance to non-tolerant crops are promising, and several strains of these beneficial microbes have been identified, which can be used as "salt stress alleviators" (Shrivastava and Kumar 2015; Islam et al. 2016; Numan et al. 2018). A list of rhizobacterial flora, which has salt mitigating effects, is presented in Table 3. In earlier studies, specific strains of PGPR such as *Achromobacter piechaudii*, *Pseudomonas* sp., and *Enterobacter* sp. have been documented for conferring salinity tolerance to tomato (Mayak et al. 2004; Ali et al. 2014; Kim et al. 2014; Sapre et al. 2018). Wheat exposed to different salinity levels exhibited improved

**Table 3** A list of different rhizobacteria which have salt mitigating and crops growth-promoting effects

| Crops | PGPB/ strains | Salinity stress | Response | References |
|---|---|---|---|---|
| Tomato | Strains of *Achromobacter piechaudii* | 172 mM | Fresh and dry weight of challenged plant improved under salinity stress | Mayak et al. (2004) |
| Canola | *Pseudomonas putida* | 150 mM | Induced salt resistance and improved growth characters of the crop | Cheng et al. (2007) |
| Groundnut | *Pseudomonas fluorescens* (TDK1) | 120 mM | Salinity resistances, improvement in general growth parameters | Saravanakumar and Samiyappan (2007) |
| Lettuce | *Pseudomonas mendocina* (Palleroni) | 0.34–0.92 $dSm^{-1}$ | Enhanced root and shoot biomass and water content | Kohler et al. (2009) |
| Mung bean | *Pseudomonas* and *Rhizobium* sp. | 4–6 $dSm^{-1}$ | Salinity resistance, improvement in growth and dry biomass | Ahmad et al. (2011) |
| Wheat | *Nitrinicola* sp. (SL 11), *Halomonas* sp. (SL 9), *Arthrobacter* sp. (AS 18), *Pseudomonas* sp. (AS 40), *Bacillus* sp. (AS 121) | ≥4.6 $dSm^{-1}$ | Improved growth, dry weight, total proteins and sugars, chlorophyll pigments | Tiwari et al. (2011) |
| Red pepper | *Brevibacterium iodinum*, *Bacillus licheniformis*, and *Zhihengliuela alba* | 100–200 mM | Improved growth, biomass, root-shoot ratio | Siddikee et al. (2011) |
| Peanut | *Brachybacterium saurashtrense* (JG-06), *Brevibacterium casei* (JG-08), and *Haererohalobacter* (JG-11) | 100 mM | Induction of salt tolerance, better growth potentials, improved Na/K ratios, nutrient uptake | Shukla et al. (2012) |
| Rice | *Azotobacter strains* (C5, C7, C8,C9) | 2.93–5.85 $gl^{-1}$ | Ion regulation, improvement in biomass, chlorophyll content | Rojas-Tapias et al. (2012) |
| Wheat | *Bacillus subtilis* SU47 and *Arthrobacter* sp. SU18 | 2–6 $dSm^{-1}$ | Dry biomass, proline content, and soluble sugars elevated at salinity stress | Upadhyay et al. (2012) |
| Wheat | *Planococcus rifietoensis* (strain SAL-15) | 300 $dSm^{-1}$ | Better growth performance in salinity than under control | Rajput et al. (2013) |

(continued)

**Table 3** (continued)

| Crops | PGPB/ strains | Salinity stress | Response | References |
|---|---|---|---|---|
| Tomato | *Pseudomonas fluorescens* (YsS6), *Pseudomonas migulae* (8R6) | 165–185 mM | Growth, reproductive characters and chlorophyll contents improved | Ali et al. (2014) |
| Cucumber | *Burkholderia cepacia* (SE4), *Promicromonospora* sp. (SE188), and *Acinetobacter calcoaceticus* (SE370) | 120 mM | Induction of salt tolerance, reduced oxidative damage, and improved growth in response to salinity | Kang et al. (2014) |
| Tomato | *Enterobacter* sp. (EJ01) | 200 mM | Enhance growth and biomass production under salinity | Kim et al. (2014) |
| Rice | *Pseudomonas* (PDMCd 0501, PDMCd2007) | 16 $dSm^{-1}$ | Stimulatory effects on growth and germination depending on strains | Nakbanpote et al. (2014) |
| Rice | *Pseudomonas* sp. (TDK1 and PF1) | 100 mM | Increase in seedling length and dry weight | Sen and Chandrasekhar (2014) |
| Maize | *Enterobacter* sp. (FD17), *Burkholderia phytofirmans* (PsJN) | 25 mM | Regulation of ions and nutrient uptake | Akhtar et al. (2015) |
| Soybean | *Pseudomonas* sp. (AK-1), *Bacillus* sp. (SJ-5) | 40–200 mM | Overall growth and biomass improvement in saline conditions | Kumari et al. (2015) |
| Barley | *Curtobacterium flaccumfaciens* (E108) | 250 mM | Salt tolerance and growth improvement | Cardinale et al. (2015) |
| Wheat | *Dietzia* sp. (STR1) | 150 mM | Overall growth and biomass improvement | Bharti et al. (2016) |
| Wheat | *Bacillus* sp., *Staphylococcus* sp., and many other strains | 200 mM | Increased growth by 62–78% under salinity stress | Orhan (2016) |
| Maize | *Pseudomonas fluorescens* (002) | 150 mM | Root number, growth and biomass enhanced | Zerrouk et al. (2016) |
| Sunflower | *Chryseobacterium humi* (ECP37), *Ochrobactrum haematophilum* (ZR3–5) | – | Improved N/K ratio, dry biomass and enzymatic activity | Pereira et al. (2016) |
| Mung bean | *Enterobacter cloacae* and *Bacillus drentensis* | 3.12–7.81 $dSm^{-1}$ | Improved plant height, leaf area, biomass and chlorophyll contents | Mahmood et al. (2016) |

(continued)

**Table 3** (continued)

| Crops | PGPB/ strains | Salinity stress | Response | References |
|---|---|---|---|---|
| Maize | *Bacillus aquimaris* (DY-3) | – | Salinity resistance inducted | Li and Jiang (2017) |
| Wheat | *Arthrobacter* sp. (SA3), *Dietzia* sp. (STR1) | 100 mM | Enhanced seedling growth, dry weight, and transpiration | Barnawal et al. (2017) |
| Wheat | *Enterobacter cloacae* (SBP-8) | 200 mM | Enhanced metabolic activities under salinity | Singh et al. (2017) |
| Wheat | *P. fluorescens* (PGU2–79) | 6 $dSm^{-1}$ | Improvement in germination and growth parameters of seedling | Safari et al. (2018) |
| Rice | *Enterobacter* sp. (P23) | 200 mM | Growth promotion in saline conditions | Sarkar et al. (2018) |
| Rice | *Streptomyces* sp. (GMKU 336) | 150 mM | Enhanced growth and chlorophyll contents | Jaemsaeng et al. (2018) |
| Black gram | *Pseudomonas fluorescens* | 7 $dSm^{-1}$ | Improvement in leaf water content, chlorophyll content, root/shoot growth and biomass | Yasin et al. (2018a) |
| Chili | *Bacillus fortis* (SSB 21) | 1–2 NaCl $kg^{-1}$ | Enhanced growth, biomass and biochemical parameters in salt stress | Yasin et al. (2018b) |
| Tomato | *Pseudomonas* sp. (OFT2, OFT5) | 75 mM | Enhanced ionic balance, physiological activity, growth and total biomass | Win et al. (2018) |
| Oat | *Klebsiella* sp. (IG3) | 100 mM | Improved shoot and root growths | Sapre et al. (2018) |

growth and physiological activities and better nutrient acquisition when different strains of *Nitrinicola* sp., *Halomonas* sp., *Arthrobacter* sp., *Pseudomonas* sp., *Bacillus* sp., *Planococcus rifietoensis*, *Dietzia* sp., *Staphylococcus* sp., *Enterobacter cloacae*, and *Pseudomonas* sp. were used as inoculants or bioformulants (Tiwari et al. 2011; Upadhyay et al. 2012; Rajput et al. 2013; Bharti et al. 2016; Orhan 2016; Barnawal et al. 2017; Singh et al. 2017; Safari et al. 2018). There are many reports about the positive influences of diverse bacterial strains on maize (Akhtar et al. 2015; Zerrouk et al. 2016; Li and Jiang 2017), rice (Rojas-Tapias et al. 2012; Nakbanpote et al. 2014; Sen and Chandrasekhar 2014; Sarkar et al. 2018; Jaemsaeng et al. 2018), canola (Cheng et al. 2007), groundnut (Saravanakumar and Samiyappan

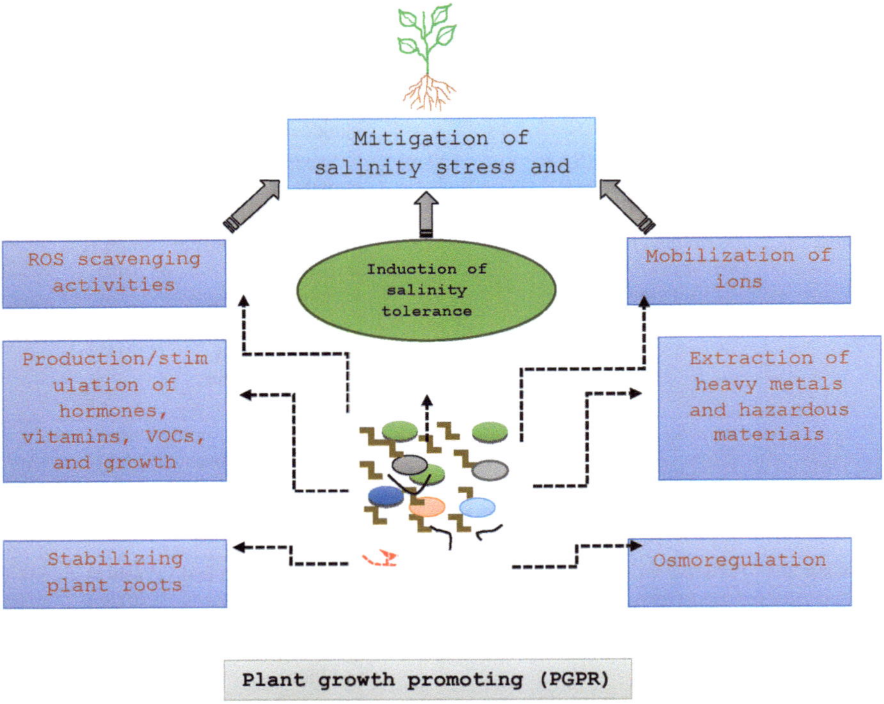


**Fig. 2** Role of plant growth-promoting bacteria in conferring salt tolerance in plants

2007; Shukla et al. 2012), mung bean (Ahmad et al. 2011; Mahmood et al. 2016), and soybean (Kumari et al. 2015), which exhibited improved growth, biomass, nutrient uptake, ionic ratio, and physiological parameters under saline conditions.

The induction of salt tolerance to crops and alleviation of salinity toxicity by PGPR may be attributed to diverse mechanisms, which include the production of plant growth-stimulating hormones, detoxification of reactive oxygen species produced as a result of salinity stress, balancing of $Na^+$ and $K^+$ ratio, homeostatic regulations, and regulation of several stress pathways (Ilangumaran and Smith 2017; Numan et al. 2018) (Fig. 2). Growth abnormalities of plants, which are exposed to above threshold levels of salinity, occur due to the production of reactive oxygen species (ROS), accumulation of certain ions in different tissues, and malfunctioning of stress-regulating system. ROS, ionic imbalance, and deficit running of stress regulatory pathways result in tissue damage and abnormalities in biochemical and physiological processes, which lead to hindered growth and development of the challenged plants. PGPR help in accommodating these abnormalities either directly or indirectly. In a comprehensive review, Ilangumaran and Smith (2017) outlined the underlined mechanisms involved in stress tolerance to plants by PGPR. They provided evidence about the influential role of PGPR in salinity stress induction by modulation of the rate of transpiration and hydraulic conductivity in maize. The production and stimulation of hormones and growth regulatory substances (auxins, ethylene, abscisic acid) and other extracellular molecules triggered by PGPR were also correlated with induced salinity tolerance to crops. Ethylene, a growth regulator, in higher concentration can potentially damage the root development and overall growth of crops; thus regulation of ethylene's production is crucially required, which in many studies has been confirmed to be positively linked with PGPR (Bhattacharyya and Jha 2012; Siddikee et al. 2012; Ahemad and Kibret 2014). Yang et al. (2009) concluded in their review that ethylene is produced in response to stressed conditions, while PGPR reverses its harmful effects on crops by producing 1-aminocyclopropane-1-carboxylate (ACC) deaminase. In many other studies, 1-aminocyclopropane-1-carboxylate (ACC) deaminase producing PGPR have been reported to confer resistance to crops, which are exposed to salinity and other stressful conditions (Gontia-Mishra et al. 2014; Akhgar et al. 2014; Heydarian et al. 2016; Wang et al. 2016; Habib et al. 2016; Jaemsaeng et al. 2018; Sarkar et al. 2018). Several other mechanisms have also been reported for PGPR-induced salinity tolerance in crops. Farooq et al. (2015) asserted that the imbalanced presence of $Na^+$ and $Cl^-$ in soil drastically affects the availability of required nutrients to crops which certain PGPR have the ability to recover and make them available to crops under stressed environment. They further elaborated that PGPR has an active role in regulating ionic homeostasis, osmoprotection, and nitrogen fixation and enhancing the anti-oxidant defense system. In a recent review, Etesami and Beattie (2018) recommended the use of halotolerant PGPR because of their ability to induce systemic tolerance to crops against salinity and other biotic/abiotic challenges via nitrogen fixation, hormones production and regulation, siderophore production, phosphate solubilization, and augmenting plants growth by controlling disease damages under stresses.

# 6 Conclusions

In summary, seed priming proves to be an effective, suitable, easy-to-manage, and cost-effective strategy to enhance salinity resistance in different crops. Major improvements have been achieved in germination, growth, and yield potentials and physiological and biochemical response of a variety of crops to salinity stress when pre-sowing priming was performed. Since wheat, maize, and rice are important food and staple crops and they are challenged with salinity threat along with other several abiotic and biotic stresses, employment of seed priming in these crops would substantially improve their yield and production under increasing salinity problems. Besides traditional seed priming approaches which employ mostly $H_2O$, NaCl, PEG, hormones, and growth regulatory substances, extensive research is needed to test different plant extracts and nano-molecules as priming agent as it will elaborate our understanding of the basic mechanisms of seed priming and potential utilization of such techniques on a wide scale to induce salinity resistance in agronomic crops. Moreover, PGPR, bacteria in the rhizosphere with potential characteristics of stimulating plant growth, are ideal candidates for using in agriculture to help crops cope with salinity stress. Bacterial strains in genera *Rhizobium*, *Pseudomonas*, *Achromobacter*, *Arthrobacter*, *Halomonas*, *Bacillus*, *Azotobacter*, *Burkholderia*, *Enterobacter*, *Dietzia*, and *Klebsiella* are some of the well-known PGPR, which can be successfully employed in agriculture for management of salinity problem. The PGPR induce systemic salinity stress to crops in several ways. Production of hormones, 1-aminocyclopropane-1-carboxylate (ACC) deaminase, balancing of toxic ions in the rhizosphere by producing compatible molecules, and homeostatic regulations triggered by PGPR result in better growth and physiological activities of crops in a stressed environment. Wide employment of seed priming and PGPR in agriculture can benefit the environment, ecosystem, and agriculture in a more sustainable manner. There is also a potential space for priming and PGPR in tolerance induction in crops to other stress conditions such as drought, heat, and cold stress.

## References

Afzal I, Basra SA, Iqbal A (2005) The effects of seed soaking with plant growth regulators on seedling vigor of wheat under salinity stress. J Stress Physiol Biochem 1:6–14

Afzal I, Basra SM, Farooq M, Nawaz A (2006) Alleviation of salinity stress in spring wheat by hormonal priming with ABA, salicylic acid and ascorbic acid. Int J Agric Biol 8(1):23–28

Afzal I, Butt A, Rehman H, Basra AB, Afzal A (2012) Alleviation of salt stress in fine aromatic rice by seed priming. Aust J Crop Sci 6:1401–1407

Ahemad M, Kibret M (2014) Mechanisms and applications of plant growth promoting rhizobacteria: current perspective. J King Saud Univ Sci 26:1–20

Ahmad F, Ahmad I, Khan MS (2008) Screening of free-living rhizospheric bacteria for their multiple plant growth promoting activities. Microbiol Res 163:173–181

Ahmad M, Zahir ZA, Asghar HN, Asghar M (2011) Inducing salt tolerance in mung bean through coinoculation with rhizobia and plant-growth-promoting rhizobacteria containing 1-aminocyclopropane-1-carboxylate deaminase. Can J Microbiol 57:578–589

Akhgar AR, Arzanlou M, Bakker P, Hamidpour M (2014) Characterization of 1-aminocyclopropane-1-carboxylate (ACC) deaminase-containing *Pseudomonas* spp. in the rhizosphere of salt-stressed canola. Pedosphere 24:461–468

Akhtar SS, Andersen MN, Naveed M, Zahir ZA, Liu F (2015) Interactive effect of biochar and plant growth-promoting bacterial endophytes on ameliorating salinity stress in maize. Funct Plant Biol 42:770–781

Ali S, Charles TC, Glick BR (2014) Amelioration of high salinity stress damage by plant growth-promoting bacterial endophytes that contain ACC deaminase. Plant Physiol Biochem 80:160–167

Ali Q, Daud MK, Haider MZ, Ali S, Rizwan M, Aslam N, Ali T (2017) Seed priming by sodium nitroprusside improves salt tolerance in wheat (*Triticum aestivum* L.) by enhancing physiological and biochemical parameters. Plant Physiol. Biochemist 119:50–58

Al-Tamimi N, Brien C, Oakey H, Berger B, Saade S, Ho YS, Schmöckel SM, Tester M, Negrão S (2016) Salinity tolerance loci revealed in rice using high-throughput non-invasive phenotyping. Nat Commun 7(1)

Amirjani MR (2010) Effect of salinity stress on growth, mineral composition, proline content, antioxidant enzymes of soybean. Am J Plant Physiol 5(6):350–360

Amirjani MR (2011) Effect of salinity stress on growth, sugar content, pigments and enzyme activity of rice. Int J Bot 7(1):73–81

Ashraf M, Rauf H (2001) Inducing salt tolerance in maize (Zea mays L.) through seed priming with chloride salts: growth and ion transport at early growth stages. Acta Physiol Plant 23(4):407–414

Atieno J, Li Y, Langridge P, Dowling K, Brien C, Berger B et al (2017) Exploring genetic variation for salinity tolerance in chickpea using image-based phenotyping. Sci Rep 7(1):1300

Babu MA, Singh D, Gothandam KM (2012) The effect of salinity on growth, hormones and mineral elements in leaf and fruit of tomato cultivar PKM1. J Anim Plant Sci 22(1):159–164

Bajwa AA, Farooq M, Nawaz A (2018) Seed priming with sorghum extracts and benzyl aminopurine improves the tolerance against salt stress in wheat (Triticum aestivum L.). Physiol Mol Biol Plants 24(2):239–249

Bakht J, Shafi M, Jamal Y, Sher H (2011) Response of maize (Zea mays L.) to seed priming with NaCl and salinity stress. Span J Agric Res 9(1):252–261

Barnawal D, Bharti N, Pandey SS, Pandey A, Chanotiya CS, Kalra A (2017) Plant growth promoting rhizobacteria enhances wheat salt and drought stress tolerance by altering endogenous phytohormone levels and TaCTR1/TaDREB2 expression. Physiol Plant 161(4):502–514

Basra SMA, Farooq M, Tabassam R, Ahmad N (2005) Physiological and biochemical aspects of pre-sowing seed treatments in fine rice (Oryza sativa L.). Seed Sci Technol 33(3):623–628

Basra SMA, Farooq M, Wahid A, Khan MB (2006) Rice seed invigoration by hormonal and vitamin priming. Seed Sci Technol 34(3):753–758

Beneduzi A, Ambrosini A, Passaglia LM (2012) Plant growth-promoting rhizobacteria (PGPR): their potential as antagonists and biocontrol agents. Genet Mol Biol 35(4):1044–1051

Bharti N, Pandey SS, Barnawal D, Patel VK, Kalra A (2016) Plant growth promoting rhizobacteria Dietzia natronolimnaea modulates the expression of stress responsive genes providing protection of wheat from salinity stress. Sci Rep 6:34768

Bhattacharyya PN, Jha DK (2012) Plant growth-promoting rhizobacteria (PGPR): emergence in agriculture. World J Microbiol Biotechnol 28(4):1327–1350

Bruce TJ, Matthes MC, Napier JA, Pickett JA (2007) Stressful "memories" of plants: evidence and possible mechanisms. Plant Sci 173(6):603–608

Bybordi A, Tabatabaei J (2009) Effect of salinity stress on germination and seedling properties in canola cultivars (Brassica napus L.). Notulae Botanicae Horti Agrobotanici Cluj-Napoca 37(2):71–76

Cardinale M, Ratering S, Suarez C, Montoya AMZ, Geissler-Plaum R, Schnell S (2015) Paradox of plant growth promotion potential of rhizobacteria and their actual promotion effect on growth of barley (Hordeum vulgare L.) under salt stress. Microbiol Res 181:22–32

Chatterjee P, Samaddar S, Niinemets Ü, Sa TM (2018) Brevibacterium linens RS16 confers salt tolerance to Oryza sativa genotypes by regulating antioxidant defense and H+ ATPase activity. Microbiol Res 215:89–101

Cheng Z, Park E, Glick BR (2007) 1-Aminocyclopropane-1-carboxylate deaminase from Pseudomonas putida UW4 facilitates the growth of canola in the presence of salt. Can J Microbiol 53(7):912–918

Chunthaburee S, Sanitchon J, Pattanagul W, Theerakulpisut P (2015) Application of exogenous spermidine (Spd) improved salt tolerance of rice at the seedling and reproductive stages. Procedia Environ Sci 29:134

Dolatabadian A, Sanavy S, Ghanati F (2011) Effect of salinity on growth, xylem structure and anatomical characteristics of soybean. Not Sci Biol 3(1):41

Etesami H, Beattie GA (2018) Mining halophytes for plant growth-promoting halotolerant bacteria to enhance the salinity tolerance of non-halophytic crops. Front Microbiol 9:148

Farooq M, Hussain M, Wakeel A, Siddique KH (2015) Salt stress in maize: effects, resistance mechanisms, and management. A Rev Agron Sustain Dev 35(2):461–481

Farooq M, Ullah A, Rehman A, Nawaz A, Nadeem A, Wakeel A, Siddique KH (2018) Application of zinc improves the productivity and biofortification of fine grain aromatic rice grown in dry seeded and puddled transplanted production systems. Field Crops Res 216:53–62

Gontia-Mishra I, Sasidharan S, Tiwari S (2014) Recent developments in use of 1-aminocyclop ropane-1-carboxylate (ACC) deaminase for conferring tolerance to biotic and abiotic stress. Biotechnol Lett 36(5):889–898

Gowing JW, Rose DA, Ghamarnia H (2009) The effect of salinity on water productivity of wheat under deficit irrigation above shallow groundwater. Agric Water Manag 96(3):517–524

Gururani MA, Upadhyaya CP, Baskar V, Venkatesh J, Nookaraju A, Park SW (2013) Plant growth-promoting rhizobacteria enhance abiotic stress tolerance in Solanum tuberosum through inducing changes in the expression of ROS-scavenging enzymes and improved photosynthetic performance. J Plant Growth Regul 32(2):245–258

Habib SH, Kausar H, Saud HM (2016) Plant growth-promoting rhizobacteria enhance salinity stress tolerance in okra through ROS-scavenging enzymes. BioMed Res Int 2016:1–10

Hajri R, Ouhibi C, Mechri M, Kourda H, Younes MB (2018) Salinity and water deficit effects on seed germination and recovery of lotus populations from northern Tunisia. Pak J Bot 50(6):2085–2090

Hakim MA, Juraimi AS, Hanafi MM, Ismail MR, Rafii MY, Islam MM, Selamat A (2014) The effect of salinity on growth, ion accumulation and yield of rice varieties. J Anim Plant Sci 24(3):874–885

Hamayun M, Khan SA, Khan AL, Shinwari ZK, Hussain J, Sohn EY, Lee IE (2010) Effect of salt stress on growth attributes and endogenous growth hormones of soybean cultivar. Pak J Bot 42(5):3103–3112

Han HS, Lee KD (2005) Plant growth promoting rhizobacteria effect on antioxidant status, photosynthesis, mineral uptake and growth of lettuce under soil salinity. Res J Agric Biol Sci 1(3):210–215

Heydarian Z, Yu M, Gruber M, Glick BR, Zhou R, Hegedus DD (2016) Inoculation of soil with plant growth promoting bacteria producing 1-aminocyclopropane-1-carboxylate deaminase or expression of the corresponding acdS gene in transgenic plants increases salinity tolerance in Camelina sativa. Front Microbiol 7:1966

Hussein M, Embiale A, Husen A, Aref IM, Iqbal M (2017) Salinity-induced modulation of plant growth and photosynthetic parameters in faba bean (Vicia faba) cultivars. Pak J Bot 49(3):867–877

Hütsch BW, Jung SA, Schubert S (2015) Comparison of salt and drought-stress effects on maize growth and yield formation with regard to acid invertase activity in the kernels. J Agron Crop Sci 201(5):353–367

Ilangumaran G, Smith DL (2017) Plant growth promoting rhizobacteria in amelioration of salinity stress: a systems biology perspective. Front Plant Sci 8:1768

Islam F, Yasmeen T, Arif MS, Ali S, Ali B, Hameed S, Zhou W (2016) Plant growth promoting bacteria confer salt tolerance in Vigna radiata by up-regulating antioxidant defense and biological soil fertility. Plant Growth Regul 80(1):23–36

Jaemsaeng R, Jantasuriyarat C, Thamchaipenet A (2018) Molecular interaction of 1-aminocyclopr opane-1-carboxylate deaminase (ACCD)-producing endophytic Streptomyces sp. GMKU 336 towards salt-stress resistance of Oryza sativa L. cv. KDML105. Sci Rep 8(1):1950

Jafar MZ, Farooq M, Cheema MA, Afzal I, Basra SMA, Wahid MA, Shahid M (2012) Improving the performance of wheat by seed priming under saline conditions. J Agron Crop Sci 198(1):38–45

Jamal Y, Shafi M, Arif M (2011) Seed priming improves salinity tolerance of wheat varieties. Pak J Bot 43(6):2683–2686

Janmohammadi M, Dezfuli PM, Sharifzadeh F (2008) Seed invigoration techniques to improve germination and early growth of inbred line of maize under salinity and drought stress. Gen Appl Plant Physiol 34(3–4):215–226

Jiang X, Li H, Song X (2016) Seed priming with melatonin effects on seed germination and seedling growth in maize under salinity stress. Pak J Bot 48:1345–1352

Jiang C, Zu C, Lu D, Zheng Q, Shen J, Wang H, Li D (2017) Effect of exogenous selenium supply on photosynthesis, Na+ accumulation and antioxidative capacity of maize (Zea mays L.) under salinity stress. Sci Rep 7:42039

Jogawat A, Vadassery J, Verma N, Oelmüller R, Dua M, Nevo E, Johri AK (2016) PiHOG1, a stress regulator MAP kinase from the root endophyte fungus Piriformospora indica, confers salinity stress tolerance in rice plants. Sci Rep 6:36765

Kang SM, Khan AL, Waqas M, You YH, Kim JH, Kim JG et al (2014) Plant growth-promoting rhizobacteria reduce adverse effects of salinity and osmotic stress by regulating phytohormones and antioxidants in Cucumis sativus. J Plant Interact 9(1):673–682

Kaya C, Ashraf M, Sonmez O, Tuna AL, Polat T, Aydemir S (2015) Exogenous application of thiamin promotes growth and antioxidative defense system at initial phases of development in salt-stressed plants of two maize cultivars differing in salinity tolerance. Acta Physiol Plant 37(1):1741

Khalilzadeh R, Seyed Sharifi R, Jalilian J (2018) Growth, physiological status, and yield of salt-stressed wheat (Triticum aestivum L.) plants affected by biofertilizer and cycocel applications. Arid Land Res Manag 32(1):71–90

Kim K, Jang YJ, Lee SM, Oh BT, Chae JC, Lee KJ (2014) Alleviation of salt stress by Enterobacter sp. EJ01 in tomato and Arabidopsis is accompanied by up-regulation of conserved salinity responsive factors in plants. Mol Cells 37(2):109

Kiremit MS, Arslan H (2016) Effects of irrigation water salinity on drainage water salinity, evapotranspiration and other leek (Allium porrum L.) plant parameters. Sci Hortic 201:211–217

Kohler J, Hernández JA, Caravaca F, Roldán A (2009) Induction of antioxidant enzymes is involved in the greater effectiveness of a PGPR versus AM fungi with respect to increasing the tolerance of lettuce to severe salt stress. Environ Exp Bot 65(2–3):245–252

Kubala S, Wojtyla L, Quinet M, Lechowska K, Lutts S, Garnczarska M (2015) Enhanced expression of the proline synthesis gene P5CSA in relation to seed osmopriming improvement of Brassica napus germination under salinity stress. J Plant Physiol 183:1–12

Kumari A, Parida AK (2018) Metabolomics and network analysis reveal the potential metabolites and biological pathways involved in salinity tolerance of the halophyte Salvadora persica. Environ Exp Bot 148:85–99

Kumari S, Vaishnav A, Jain S, Varma A, Choudhary DK (2015) Bacterial-mediated induction of systemic tolerance to salinity with expression of stress alleviating enzymes in soybean (Glycine max L. Merrill). J Plant Growth Regul 34(3):558–573

Li HQ, Jiang XW (2017) Inoculation with plant growth-promoting bacteria (PGPB) improves salt tolerance of maize seedling. Russ J Plant Physiol 64(2):235–241

Lugtenberg B, Kamilova F (2009) Plant-growth-promoting rhizobacteria. Annu Rev Microbiol 63:541–556

Luo M, Zhao Y, Weisong W, Zhang R, Su A, Li C, Zhao J (2018) Effect of saline stress on the physiology and growth of maize hybrids and their related inbred lines. Maydica 62(2):1–8

Mahmood S, Daur I, Al-Solaimani SG, Ahmad S, Madkour MH, Yasir M et al (2016) Plant growth promoting rhizobacteria and silicon synergistically enhance salinity tolerance of mung bean. Front Plant Sci 7:876

Mayak S, Tirosh T, Glick BR (2004) Plant growth-promoting bacteria confer resistance in tomato plants to salt stress. Plant Physiol Biochem 42(6):565–572

Mir MA, John R, Alyemeni MN, Alam P, Ahmad P (2018) Jasmonic acid ameliorates alkaline stress by improving growth performance, ascorbate glutathione cycle and glyoxylase system in maize seedlings. Sci Rep 8(1):2831

Muhammad Z, Hussain F (2010a) Vegetative growth performance of five medicinal plants under NaCl salt stress. Pak J Bot 42(1):303–316

Muhammad Z, Hussain F (2010b) Effect of NaCl salinity on the germination and seedling growth of some medicinal plants. Pak J Bot 42(2):889–897

Muhammad Z, Hussain F (2012) Effect of NaCl salinity on the germination and seedling growth of seven wheat genotypes. Pak J Bot 44(6):1845–1850

Muhammad Z, Hussain F, Rehmanullah M, Majeed A (2015) Effect of halopriming on the induction of NaCl salt tolerance in different wheat genotypes. Pak J Bot 47(5):1613–1620

Mumtaz MZ, Saqib M, Abbas G, Akhtar J, Qamar ZU (2018) Genotypic variation in rice for grain yield and quality as affected by salt-affected field conditions. J Plant Nutr 41(2):233–242
Munns R, Tester M (2008) Mechanisms of salinity tolerance. Annu Rev Plant Biol 59:651–681
Nadeem SM, Hussain I, Naveed M, Asghar HN, Zahir ZA, Arshad M (2006) Performance of plant growth promoting rhizobacteria containing ACC-deaminase activity for improving growth of maize under salt-stressed conditions. Pak J Agric Sci 43(3–4):114–121
Nakbanpote W, Panitlurtumpai N, Sangdee A, Sakulpone N, Sirisom P, Pimthong A (2014) Salt-tolerant and plant growth-promoting bacteria isolated from Zn/Cd contaminated soil: identification and effect on rice under saline conditions. J Plant Interact 9(1):379–387
Nautiyal CS, Srivastava S, Chauhan PS, Seem K, Mishra A, Sopory SK (2013) Plant growth-promoting bacteria Bacillus amyloliquefaciens NBRISN13 modulates gene expression profile of leaf and rhizosphere community in rice during salt stress. Plant Physiol Biochem 66:1–9
Navarro JM, Flores P, Garrido C, Martinez V (2006) Changes in the contents of antioxidant compounds in pepper fruits at different ripening stages, as affected by salinity. Food Chem 96(1):66–73
Numan M, Bashir S, Khan Y, Mumtaz R, Shinwari ZK, Khan AL et al (2018) Plant growth promoting bacteria as an alternative strategy for salt tolerance in plants: a review. Microbiol Res 209:21–32
Orhan F (2016) Alleviation of salt stress by halotolerant and halophilic plant growth-promoting bacteria in wheat (Triticum aestivum). Braz J Microbiol 47(3):621–627
Patel VK, Srivastava R, Sharma A, Srivastava AK, Singh S, Srivastava AK, Saxena AK (2018) Halotolerant Exiguobacterium profundum PHM11 tolerate salinity by accumulating L-Proline and fine-tuning gene expression profiles of related metabolic pathways. Front Microbiol 9:423
Pereira SI, Moreira H, Argyras K, Castro PM, Marques AP (2016) Promotion of sunflower growth under saline water irrigation by the inoculation of beneficial microorganisms. Appl Soil Ecol 105:36–47
Pitman MG, Läuchli A (2002) Global impact of salinity and agricultural ecosystems. In: Salinity: environment-plants-molecules. Springer, Dordrecht, pp 3–20
Rady MM, Mohamed GF (2015) Modulation of salt stress effects on the growth, physio-chemical attributes and yields of Phaseolus vulgaris L. plants by the combined application of salicylic acid and Moringa oleifera leaf extract. Sci Hortic 193:105–113
Rajput L, Imran A, Mubeen F, Hafeez FY (2013) Salt-tolerant PGPR strain Planococcus rifietoensis promotes the growth and yield of wheat (Triticum aestivum L.) cultivated in saline soil. Pak J Bot 45(6):1955–1962
Rehman A, Farooq M, Naveed M, Nawaz A, Shahzad B (2018) Seed priming of Zn with endophytic bacteria improves the productivity and grain biofortification of bread wheat. Eur J Agron 94:98–107
Rojas-Tapias D, Moreno-Galván A, Pardo-Díaz S, Obando M, Rivera D, Bonilla R (2012) Effect of inoculation with plant growth-promoting bacteria (PGPB) on amelioration of saline stress in maize (Zea mays). Appl Soil Ecol 61:264–272
Safari D, Jamali F, Nooryazdan HR, Bayat F (2018) Evaluation of ACC deaminase producing Pseudomonas fluorescens strains for their effects on seed germination and early growth of wheat under salt stress. Aust J Crop Sci 12:413–421
Salah SM, Yajing G, Dongdong C, Jie L, Aamir N, Qijuan H et al (2015) Seed priming with polyethylene glycol regulating the physiological and molecular mechanism in rice (Oryza sativa L.) under nano-ZnO stress. Sci Rep 5:14278
Sali A, Rusinovci I, Fetahu S, Gashi B, Simeonovska E, Rozman L (2015) The effect of salt stress on the germination of maize (Zea mays L.) seeds and photosynthetic pigments. Acta Agric Slov 105(1):85–94
Sandhu D, Cornacchione MV, Ferreira JF, Suarez DL (2017) Variable salinity responses of 12 alfalfa genotypes and comparative expression analyses of salt-response genes. Sci Rep 7:42958
Sano N, Kim JS, Onda Y, Nomura T, Mochida K, Okamoto M, Seo M (2017) RNA-Seq using bulked recombinant inbred line populations uncovers the importance of brassinosteroid for seed longevity after priming treatments. Sci Rep 7(1):8095

Sapre S, Gontia-Mishra I, Tiwari S (2018) Klebsiella sp. confers enhanced tolerance to salinity and plant growth promotion in oat seedlings (Avena sativa). Microbiol Res 206:25–32
Saravanakumar D, Samiyappan R (2007) ACC deaminase from Pseudomonas fluorescens mediated saline resistance in groundnut (Arachis hypogea) plants. J Appl Microbiol 102(5):1283–1292
Sarkar A, Ghosh PK, Pramanik K, Mitra S, Soren T, Pandey S et al (2018) A halotolerant Enterobacter sp. displaying ACC deaminase activity promotes rice seedling growth under salt stress. Res Microbiol 169(1):20–32
Savvides A, Ali S, Tester M, Fotopoulos V (2016) Chemical priming of plants against multiple abiotic stresses: Mission possible? Trends Plant Sci 21(4):329–340
Sehrawat N, Yadav M, Bhat KV, Sairam RK, Jaiwal PK (2015) Effect of salinity stress on mungbean [Vigna radiata (L.) Wilczek] during consecutive summer and spring seasons. J Agric Sci 60(1):23–32
Sen S, Chandrasekhar CN (2014) Effect of PGPR on growth promotion of rice (Oryza sativa L.) under salt stress. Asian J Plant Sci Res 4:62–67
Shameer S, Prasad T (2018) Plant growth promoting rhizobacteria for sustainable agricultural practices with special reference to biotic and abiotic stresses. Plant Growth Regul 84(3):603–615
Shirazi M, Ansari R, Khan M, Tahir SS, Rajput MT, Shereen A, Ali M (2018) Salinity response of some amphidiploids species of brassica (Brassicaceae) at Seedling and maturity stages. Pak J Bot 50(6):2113–2122
Shrivastava P, Kumar R (2015) Soil salinity: a serious environmental issue and plant growth promoting bacteria as one of the tools for its alleviation. Saudi J Biol Sci 22(2):123–131
Shukla PS, Agarwal PK, Jha B (2012) Improved salinity tolerance of Arachis hypogaea (L.) by the interaction of halotolerant plant-growth-promoting rhizobacteria. J Plant Growth Regul 31(2):195–206
Siddikee MA, Glick BR, Chauhan PS, Jong Yim W, Sa T (2011) Enhancement of growth and salt tolerance of red pepper seedlings (Capsicum annuum L.) by regulating stress ethylene synthesis with halotolerant bacteria containing 1-aminocyclopropane-1-carboxylic acid deaminase activity. Plant Physiol Biochem 49(4):427–434
Siddikee MA, Chauhan PS, Sa T (2012) Regulation of ethylene biosynthesis under salt stress in red pepper (Capsicum annuum L.) by 1-aminocyclopropane-1-carboxylic acid (ACC) deaminase-producing halotolerant bacteria. J Plant Growth Regul 31(2):265–272
Siddiqui MN, Mostofa MG, Akter MM, Srivastava AK, Sayed MA, Hasan MS, Tran LSP (2017) Impact of salt-induced toxicity on growth and yield-potential of local wheat cultivars: oxidative stress and ion toxicity are among the major determinants of salt-tolerant capacity. Chemosphere 187:385–394
Singh RP, Runthala A, Khan S, Jha PN (2017) Quantitative proteomics analysis reveals the tolerance of wheat to salt stress in response to Enterobacter cloacae SBP-8. PLoS One 12(9):e0183513
Soda N, Sharan A, Gupta BK, Singla-Pareek SL, Pareek A (2016) Evidence for nuclear interaction of a cytoskeleton protein (OsIFL) with metallothionein and its role in salinity stress tolerance. Sci Rep 6:34762
Song GC, Choi HK, Kim YS, Choi JS, Ryu CM (2017) Seed defense biopriming with bacterial cyclodipeptides triggers immunity in cucumber and pepper. Sci Rep 7(1):14209
Tabassum T, Farooq M, Ahmad R, Zohaib A, Wahid A (2017) Seed priming and transgenerational drought memory improves tolerance against salt stress in bread wheat. Plant Physiol Biochem 118:362–369
Taïbi K, Taïbi F, Abderrahim LA, Ennajah A, Belkhodja M, Mulet JM (2016) Effect of salt stress on growth, chlorophyll content, lipid peroxidation and antioxidant defence systems in Phaseolus vulgaris L. S Afr J Bot 105:306–312
Tank N, Saraf M (2010) Salinity-resistant plant growth promoting rhizobacteria ameliorates sodium chloride stress on tomato plants. J Plant Interact 5(1):51–58
Thitisaksakul M, Tananuwong K, Shoemaker CF, Chun A, Tanadul OUM, Labavitch JM, Beckles DM (2015) Effects of timing and severity of salinity stress on rice (Oryza sativa L.) yield, grain composition, and starch functionality. J Agric Food Chem 63(8):2296–2304

Tiwari S, Singh P, Tiwari R, Meena KK, Yandigeri M, Singh DP, Arora DK (2011) Salt-tolerant rhizobacteria-mediated induced tolerance in wheat (Triticum aestivum) and chemical diversity in rhizosphere enhance plant growth. Biol Fertil Soils 47(8):907

Upadhyay SK, Singh JS, Singh DP (2011) Exopolysaccharide-producing plant growth-promoting rhizobacteria under salinity condition. Pedosphere 21(2):214–222

Upadhyay SK, Singh JS, Saxena AK, Singh DP (2012) Impact of PGPR inoculation on growth and antioxidant status of wheat under saline conditions. Plant Biol 14(4):605–611

Vacheron J, Desbrosses G, Bouffaud ML, Touraine B, Moënne-Loccoz Y, Muller D et al (2013) Plant growth-promoting rhizobacteria and root system functioning. Front Plant Sci 4:356

Varier A, Vari AK, Dadlani M (2010) The subcellular basis of seed priming. Curr Sci 99:450–456

Wahid A, Perveen M, Gelani S, Basra SM (2007) Pretreatment of seed with H2O2 improves salt tolerance of wheat seedlings by alleviation of oxidative damage and expression of stress proteins. J Plant Physiol 164(3):283–294

Wang Q, Dodd IC, Belimov AA, Jiang F (2016) Rhizosphere bacteria containing 1-aminocyclop ropane-1-carboxylate deaminase increase growth and photosynthesis of pea plants under salt stress by limiting Na+ accumulation. Funct Plant Biol 43(2):161–172

Wei LX, Lv BS, Li X, Wang MM, Ma HY, Jiang CJ (2017) Priming of rice (Oryza sativa L.) seedlings with abscisic acid enhances seedling survival, plant growth, and grain yield in saline-alkaline paddy fields. Field Crops Res 203:86–93

Win KT, Tanaka F, Okazaki K, Ohwaki Y (2018) The ACC deaminase expressing endophyte Pseudomonas spp. Enhances NaCl stress tolerance by reducing stress-related ethylene production, resulting in improved growth, photosynthetic performance, and ionic balance in tomato plants. Plant Physiol Biochem 127:599–607

Yang J, Kloepper JW, Ryu CM (2009) Rhizosphere bacteria help plants tolerate abiotic stress. Trends Plant Sci 14(1):1–4

Yang A, Akhtar SS, Iqbal S, Qi Z, Alandia G, Saddiq MS, Jacobsen SE (2018) Saponin seed priming improves salt tolerance in quinoa. J Agron Crop Sci 204(1):31–39

Yasin NA, Khan WU, Ahmad SR, Ali A, Ahmad A, Akram W (2018a) Imperative roles of halotolerant plant growth-promoting rhizobacteria and kinetin in improving salt tolerance and growth of black gram (Phaseolus mungo). Environ Sci Pollut Res 25(5):4491–4505

Yasin NA, Akram W, Khan WU, Ahmad SR, Ahmad A, Ali A (2018b) Halotolerant plant-growth promoting rhizobacteria modulate gene expression and osmolyte production to improve salinity tolerance and growth in Capsicum annum L. Environ Sci Pollut Res:1–15

Youssef SM, Elhady SA, Aref RM, Riad GS (2018) Salicylic acid attenuates the adverse effects of salinity on growth and yield and enhances peroxidase isozymes expression more competently than proline and glycine betaine in cucumber plants. Gesunde Pflanzen:1–16

Zerrouk IZ, Benchabane M, Khelifi L, Yokawa K, Ludwig-Müller J, Baluska F (2016) A Pseudomonas strain isolated from date-palm rhizospheres improves root growth and promotes root formation in maize exposed to salt and aluminum stress. J Plant Physiol 191:111–119

# Pretreatment of Seedlings with Exogenous Protectants for Abiotic Stress Tolerance

**Shaghef Ejaz, Muhammad Akbar Anjum, Sajjad Hussain, Muhammad Azam, Sajid Ali, and Shakeel Ahmad**

## Contents



S. Ejaz (✉) · M. A. Anjum · S. Hussain · S. Ali
Department of Horticulture, Bahauddin Zakariya University, Multan, Punjab, Pakistan
e-mail: shaghef.ejaz@bzu.edu.pk

M. Azam
Institute of Horticultural Sciences, University of Agriculture Faisalabad, Faisalabad, Pakistan

S. Ahmad
Department of Agronomy, Bahauddin Zakariya University, Multan, Punjab, Pakistan



M. Hasanuzzaman, V. Fotopoulos (eds.), *Priming and Pretreatment of Seeds and Seedlings*, https://doi.org/10.1007/978-981-13-8625-1_28

**Abstract** Abiotic stresses are one of the most plant growth- and yield-limiting factors in agricultural production system. Drought, salinity, chilling temperature, heat, freezing, and heavy metal toxicity are ubiquitously present around the globe. Plants have evolved different adaptive mechanisms to sustain their growth and reproduction in adverse environmental conditions. Accumulation of phytoprotectants is one such strategy, in which bioactive molecules that are effective against stress are accumulated by plants. The most effective phytoprotectants include polyamines, hormones, osmolytes, signaling molecules, antioxidants, and trace elements. In recent few decades, researchers have established procedures for exogenously applying these phytoprotectants to facilitate plants to sustain growth and development under abiotic stresses. Seedling pretreatment is one such method, where seedlings are exogenously supplied (through roots or leaves) with phytoprotectants prior to exposing the seedlings to abiotic stress. Seedling pretreatment with phytoprotectants has shown the potential to improve plant growth and stress tolerance during stressful conditions. In this chapter, we have summarized, which phytoprotectants in recent times, have been studied for seedling pretreatment to mitigate various abiotic stresses. Moreover, this chapter is focused on polyamines, signaling molecules, antioxidants, amino acids, quaternary ammonium compounds, sugars, and sugar alcohols.

**Keywords** Stress mitigation · Exogenous application · Osmolytes · Polyamines · Antioxidant defense mechanism · Climate change · Osmoprotectant

## 1 Introduction

Extremes in environment conditions initiate responses within plants to adapt the changes. This environmental condition is called abiotic stress for plants. With the change in climate patterns, abiotic stresses are becoming common and occurring around the globe. The most common and threatening abiotic stresses for plant production are water scarcity, salinity, low and high temperature (chilling and heat), freezing, and heavy metal toxicity. Abiotic stresses modulate diversity and distribution of plants around the globe and plant productivity in agricultural systems. Generally, these stresses adversely impact growth and reproduction in plants. At cellular levels, the production and accumulation of reactive oxygen species during stress is one of the most harmful adversaries for plants. Reactive oxygen species (ROS) reduce photosynthetic efficiency by damaging photosystems and destabilizing membrane structures, DNA, and lipids, consequently reducing the repairing ability of cell organelles leading to cell or plant death. Almost half of the cultivable land in the world is facing similar kind of yield-limiting stressful conditions affecting a huge chunk of human population. In this scenario, the capacity of a plant's defense mechanism defines its chances of survival and to remain productive. For this purpose, nature has equipped plants with several tools and strategies. Plants have evolved these strategies to preserve their cellular systems from harmful effects of abiotic stresses. One such tool is the production and accumulation of phytoprotectants.

Phytoprotectants are endogenously produced bioactive compounds that are effective against stresses. Exogenous application of these compounds has been found to facilitate plant to carry on its normal metabolic functions during stress and to increase tolerance to various stresses. The class of phytoprotectants or osmoprotectants is broad with notable categories of polyamines, antioxidants, signaling molecules, amino acids, quaternary ammonium compounds, sugars, and sugar alcohols.

For decades, researchers have been exogenously applying phytoprotectants to fortify plants with these bioactive compounds to better cope with adverse environmental conditions. The most common methods adapted for this purpose are seed priming, seedling pretreatments, and foliar spray. The exogenous application might be applied before or after the induction of stress, both methods having their own merits and demerits. Seedling pretreatment is one such method, where seedlings are supplied with phytoprotectant before exposing the seedlings to abiotic stress. This strategy has been used in numerous studies and has shown promising findings, whereby seedlings pretreated with various phytoprotectants withstood well against stress and continued their growth and development contrary to the seedlings that were not pretreated.

## 2 Phytoprotectants and Their Role in Normal Functioning of Plant Cell

In recent decades, a lot of bioactive compounds have been categorized as phytoprotectant owing to their effectiveness in facilitating a plant to carry on its normal metabolic activities during stress. Not only these compounds are produced endogenously by plants during stress, but also the exogenous application of the phytoprotectants has been found to strengthen plant response to stress. Phytoprotectants have evidently shown that if applied exogenously, these protectants increase plant's tolerance to various stresses, growth, and yield. Exogenously applied phytoprotectants or osmoprotectants include polyamines, antioxidants, signaling molecules, amino acids, quaternary ammonium compounds, sugars, and sugar alcohols.

Polyamines are ubiquitous in all living organisms including plants. Polyamines regulate various basic metabolic activities at cellular or tissue levels including DNA replication, cell division, activities of enzymes, stability of membranes, embryogenesis, plant morphology, dormancy breaking and seed germination, flower bud initiation and setting, and ripening of fruits (Gill and Tuteja 2010). The most commonly occurring polyamines include putrescine, spermidine, and spermine. During various basic metabolic functions like respiration, production of reactive oxygen species (ROS) is a natural phenomenon. Plants use certain bioactive compounds called antioxidants to protect its vital organelles like chloroplast, mitochondria, and other membranous structures from ROS, which otherwise can destabilize these organelles. The antioxidants compounds include ascorbic acid, glutathione, α-tocopherol, flavonoids, etc. Ascorbic acid is the most abundant antioxidant molecule in plants that also activates plant defense mechanism. It is one of the primary metabolites involved in hydrogen peroxide detoxification (Munir and Aftab 2011).

The earliest detection of abiotic stress by plants is a key for stimulating defense mechanism and adapting the changing conditions. Each plant cell performs immune function, and there are certain molecules that are part of plant signaling system. Receptors on cell membrane receive the stress stimulus and then trigger the signaling system leading to activation of plant defense responses to that specific stress. Some of the well-recognized signaling molecules in plants include hydrogen sulfide, hydrogen peroxide (an ROS), hormones, and nitric oxide.

Osmolytes are a class of phytoprotectant that are water-soluble, neutral and non-toxic compounds with no interaction with normal metabolic activities (Slama et al. 2015). Concentration of osmolyte increases in cells with the decrease in the cellular water potential. Osmolytes can alter the solubility of water, adjust osmotic potential, stabilize the folded proteins, and protect membrane structures when climatic conditions are unfavorable (Yancey, 2005). Major categories of osmolytes include amino acids (proline, glutamine, and ornithine), quaternary ammonium compounds (glycinebetaine), sugars (sucrose and trehalose), and sugar alcohol—also called polyols—including mannitol and myoinositol.

## 3 Seedling Pretreatments with Phytoprotectants to Mitigate Abiotic Stress

### *3.1 Polyamines*

Polyamines are a group of aliphatic amine compounds that contain more than two amino groups. Polyamines are present in almost all living organisms including plants. The well-recognized polyamines, since the discovery of first polyamine 300 years ago, are putrescine (a diamine), spermidine (a triamine), and spermine (a tetraamine). Polyamines regulate many elementary and vital cellular activities such as transcription, translation and DNA replication, cell division and enlargement, activities of enzymes, cation-anion balance within cells, and stability of membrane structures. Further functional roles of polyamines include seed germination, breaking of dormancy in tubers, embryogenesis, development of flower buds, fruit setting, and development and ripening of fruits (Gill and Tuteja 2010).

The response of different polyamines to same abiotic stress is variable. Generally, polyamines have been shown to stabilize cellular membrane structures, induce expression of stress retarding genes, scavenge ROS, and protect plants from growth reduction and altered development (Alcázar et al. 2010; Hussain et al. 2011). In acquisition of tolerance against stresses, polyamines are suggested to be potent stress alleviators during cold stress, salt stress, hyperosmosis, heavy metal stress, pollutant stress, and hypoxia (Liu et al. 2007; Garcia-Jimenez et al. 2007).

One of the ways polyamine mitigate drought stress is attaining the stability of membranes in plant roots during drought by inducing the accumulation of proline. Kubis et al. (2014) studied this phenomenon in roots of cucumber seedling whereby roots were pretreated with polyamines (i.e., putrescine, spermidine, and spermine). Dehydration caused injury to membrane and increase in the activity of lipoxygenase, lipid peroxidation, and proline content. However, polyamine pretreatment reduced membrane injury, thus regulating membrane permeability along with the reduction in the activity of lipoxygenase and lipid peroxidation.

It is considered that spermine has a role to offer in mitigating drought and high temperature stress. Similarly, the response of spermine pretreatment to nullify the combined effect of these two stresses has shown that spermine upregulates the stress-mitigating genes, activates antioxidant enzymes, and increases ROS scavenging capacity (Fu et al. 2014). Pretreatment of seedling of trifoliate orange with spermine resulted in low accumulation of hydrogen peroxide and superoxide radicals, whereas superoxide dismutase (SOD), peroxidase (POD), and catalase (CAT) were noticeably more active in pretreated seedlings than in non-pretreated seedlings (Fu et al. 2014). Seedling pretreated with spermine also showed that spermine acted as a signaling molecule that resulted in higher expression of proteins related to heat shock, abscisic acid-responsive binding factor, and 9-cis-epoxycarotenoid dioxygenase. This suggests that spermine plays a key role in conferring the tolerance to trifoliate oranges by counteracting simultaneous stresses of drought and heat stress. Like spermine, spermidine also induced enzymes that scavenge ROS. Barley seedlings were pretreated with spermidine prior to subjecting the seedling to insufficient water stress (Kubis 2003). Where water deficiency caused a significant increase in the activity of CAT and glutathione peroxidase (GPX), pretreatment with spermidine markedly reduced the activities of both enzymes. Moreover, the accumulation of overall polyamines increased in pretreated seedlings.

## *3.2 Amino Acids*

### 3.2.1 Proline

In salinity-stressed plants, a well-observed adaptive mechanism is the accumulation of phytoprotectant resulting in stabilization of cellular membrane system, adjustment of cell osmotic potential, upregulation of the expression of responsive proteins, and activation of enzymatic antioxidant system (Ashraf and Foolad 2007; Khedr et al. 2003). Accumulation of proline was first reported in ryegrass growing under drought condition (Kemble and MacPherson 1954). Since then, proline has been widely recognized as a compatible phytoprotectant and a stress resistant biochemical marker in plants. Besides conferring osmoprotection to plant tissues, proline also mitigates oxidative stress by scavenging radical oxygen species such as hydroxyl radicals (Sharma and Dietz 2006). During adverse environmental conditions, especially stresses like salinity and drought, synthesis and accumulation of proline increase to a noticeable level in higher plants that play a major role in stabilizing proteins and enzymes (RuBisCO) and protecting electron transport complex II (Hamilton and Heckathorn 2001).

Proline and glycinebetaine mitigate salinity-induced oxidative stress by modifying antioxidant mechanism and methylglyoxal detoxification system in salinity-stressed plants. In mung bean seedlings, salt stress increased the concentration of oxidized glutathione and decreased the ascorbic acid content and reduced/oxidized glutathione ratio (Hossain et al. 2011). Similarly, salt stress has shown noticeable effects on enzymes, whereby there was an increase in the activities of glutathione reductase (GR), GPX, glutathione *S*-transferase (GST), and glyoxalase II (Gly II) activities. Moreover, a sharp decrease in the enzymatic activities of monodehydroascorbate reductase (MDHAR), dehydroascorbate reductase (DHAR), CAT, and glyoxalase I (Gly I) was observed. Increase in the concentration of hydrogen peroxide and malondialdehyde was also noticed. However, the seedlings pretreated with proline and glycinebetaine showed significantly enhanced levels of ascorbic acid, reduced glutathione and oxidized glutathione and reduced/oxidized glutathione ratio, and decreased levels of hydrogen peroxide and malondialdehyde content. These seedlings also maintained higher activities of enzymes involved in ROS scavenging and methylglyoxal detoxification system that includes ascorbate peroxidase (APX), DHAR, GR, GST, GPX, CAT, Gly I, and Gly II.

Alleviation of heavy metal stress by pretreatment of seedlings with phytoprotectants has been a useful strategy. Chromium induced lipid peroxidation and its mitigation though proline pretreatment was studied in rice seedlings by Yu et al. (2017). Rice seedlings stressed with chromium toxicity showed classic symptoms of abiotic stress such as accumulation of malondialdehyde and high cell death rate in roots along with retarded growth rate. Pretreatment with proline significantly reduced lipid peroxidation, malondialdehyde accumulation, and root cell death showing that seedlings could be effectively pretreated with proline to reduce chromium-induced oxidative damages.

### 3.2.2 γ-Aminobutyric Acid

γ-Aminobutyric acid, generally known as GABA, is a nonprotein amino acid and is produced by both types of organisms, i.e., prokaryotes and eukaryotes. Endogenous level of GABA in plants is comparatively low where glutamate decarboxylase catalyzes the synthesis of GABA using glutamate as a primary precursor. However, adverse environmental conditions cause the accumulation of GABA in higher concentration, even higher than proline in some plants. Upregulation of gene (GAD1) regulating GABA biosynthesis pathway along with inhibitory effect of GABA on ROS ($O_2^{\bullet -}$, $H_2O_2$ and $^1O_2$) has been observed in tobacco under drought stress (Liu et al. 2011).

Seed germination and early seedling growth are generally halted by cold temperature during these phases of growth, and, consequently, it affects overall productivity of plants. However, pretreatment of seedling has been found to reverse chilling temperature-induced growth inhibition. Tomato is one such plant that is sensitive to cold stress at seedling stage. Malekzadeh et al. (2014) pretreated tomato seedling with GABA 3 days before exposing the seedlings to 2.0 ± 0.05 °C (chilling stress). Pretreatment of tomato seedlings with GABA significantly reduced electrolyte leakage, malondialdehyde accumulation, and activities of antioxidant enzymes and enhanced sugar and proline concentration. These findings encourage the use of pretreatment of seedling with GABA as chilling stress mitigator.

Similar observations were made by Malekzadeh et al. (2012) in wheat seedlings that were subjected to chilling stress. Inhibition of elongation of roots in wheat seedling was observed in non-pretreated seedling contrary to pretreated seedling, whereby GABA mitigated this growth inhibitory effect of low temperature stress. Likewise, GABA reduced malondialdehyde content and alleviated oxidative damages, otherwise, caused by cold stress.

### 3.2.3 β-Aminobutyric Acid

Like GABA, β-aminobutyric acid (BABA) also plays a protective role against abiotic stresses. Hossain et al. (2012) pretreated soybean seedling with 200 μM β-aminobutyric acid to alleviate cadmium-induced stress damage to soybean seedlings. After 24 h of pretreatment, the seedlings were exposed to cadmium stress. The results showed that BABA restored the relative intensity levels of heat-shock cognate protein and activity of enzymes including carbonic anhydrase, methionine synthase, and glycine dehydrogenase. In addition, BABA pretreatment in cadmium-stressed plants also increased the content of protein peroxidation and glycolytic enzymes involved in plant defense. Moreover, low biophoton emission (an indicator of less oxidative stress) was observed in BABA pretreated soybean seedlings.

### 3.2.4 Arginine

Arginine has many diverse functions in living cells such as contributing to cell structure and acting as a precursor to important biological molecules. Arginine is a component of many types of proteins and acts as a primary substrate for nitric oxide synthase-catalyzed pathway, giving nitric oxide and citrulline, and arginase-catalyzed pathway, giving ornithine, a precursor for proline and putrescine (Chen et al. 2004). In plants, nitric oxide acts as a functional metabolite, and arginine-dependent activity of nitric oxide synthase has been observed in plants (Arasimowicz and Floryszak-Wieczorek 2007). Nitric oxide synthase catalyzes the production of nitric oxide from L-arginine. Inhibiting this enzyme also retards the formation of nitric oxide by blocking the conversion of arginine to nitric oxide. One of such nitric oxide inhibitors used in studies is Nω-nitro-l-arginine methyl ester (LNAM). Nasibi et al. (2011) pretreated tomato seedlings with arginine and then subsequently subjected to drought stress. The pretreatment caused the alleviation of oxidative stress due to drought by reducing lipid peroxidation within cells and the activities of CAT and guaiacol peroxidase, whereas increasing the activities of SOD, APX, and GR. Arginine also led to higher ascorbate and glutathione content and lower dehydroascorbate level. Nasibi et al. (2011) also applied Nω-nitro-l-arginine methyl ester along with arginine to inhibit the conversion of arginine to nitric oxide. The findings suggested that combined pretreatment with both, arginine and Nω-nitro-l-arginine methyl ester, reduced the accumulation of malondialdehyde showing that other arginine involving pathway is more effective than nitric oxide synthase pathway.

#### 3.2.5 5-Aminolevulinic Acid

5-Aminolevulinic acid is a nonprotein endogenous amino acid and is also known as D-aminolevulinic acid or D-amino ketone acid (Liu et al. 2016). 5-Aminolevulinic acid is a hydrocarbon containing oxygen and nitrogen. It modulates the growth, development, yield, and numerous physiological traits in plants. In crops, 5-aminolevulinic acid enhances growth and yield of plants by increasing the levels of total chlorophyll, ratio of chlorophyll *a* to *b*, and photosynthesis-associated activities of photosystems, PS I and II (Xu et al. 2010).

Liu et al. (2016) exogenously applied 5-aminolevulinic acid through pretreating the seedlings of *Leymus chinensis* grass and then subjected the seedlings to drought stress using 50% soil relative water content. The findings showed that pretreatment with 5-aminolevulinic acid could facilitate the growth by increasing the height of a plant, moisture level of plant tissues, leaf area, vegetative biomass, carotenoids, and chlorophyll *a* and *b*. Moreover, 5-aminolevulinic acid triggers the enzymatic antioxidant system, improves the activities of POD, SOD, CAT, GR, and APX and decreases the levels of malondialdehyde.

Numerous attempts have been made to promote the application of 5-aminolevulinic acid in production system of agriculture. Li (2011) pretreated cucumber seedling with 5-aminolevulinic acid and later exposed the seedlings to polyethylene glycol-induced water deficiency with an aim to examining the role of 5-aminolevulinic acid in modulating the enzymatic antioxidant defense system. 5-aminolevulinic acid not only increased the activities of APX, GPX, GR, and MDHAR but also accumulated higher content of reduced glutathione and ascorbate in the leaves of drought-stressed cucumber seedlings. 5-Aminolevulinic acid also caused the alleviation of stress-induced growth inhibition and concentration of ROS like superoxide radical and hydrogen peroxide.

### *3.3 Quaternary Ammonium Compounds*

#### 3.3.1 Glycinebetaine

All organisms including plants, animals, and microorganisms produce glycinebetaine. However, plants synthesize and accumulate higher levels of glycinebetaine when exposed to abiotic stresses. Accumulation of glycinebetaine increases resistance to abiotic stress in plants by protecting proteins and membranes, regulating transcription and translation processes, decreasing malondialdehyde levels, scavenging ROS, and enhancing antioxidative enzyme activity, for example, activities of catalase, superoxide dismutase, peroxidase, and ascorbate peroxidase (Bharwana et al. 2014; Hu et al. 2012; Park et al. 2006). During salt stress, glycinebetaine confers protection to plant cells by adjusting cellular osmotic potential, protecting photosynthesis apparatus, stabilizing proteins and enzymes, and scavenging ROS like hydrogen peroxide (Chen and Murata 2008; Cha-Um and Kirdmanee 2010).

Proline and glycinebetaine mitigate salinity-induced oxidative stress by modifying antioxidant mechanism and methylglyoxal detoxification system in salinity-stressed plants. In mung bean seedlings, salt stress increased the concentration of oxidized glutathione and decreased ascorbic acid content and reduced/oxidized glutathione ratio (Hossain et al. 2011). Similarly, salt stress has shown noticeable effects on enzymes, whereby there was an increase in the activities of glutathione reductase, glutathione peroxidase, glutathione *S*-transferase and glyoxalase II. Moreover, a sharp decrease in the enzymatic activities of monodehydroascorbate reductase, dehydroascorbate reductase, catalase, and glyoxalase I was observed. Increase in the concentration of hydrogen peroxide and malondialdehyde was also noticed. However, the seedlings pretreated with proline and glycinebetaine showed significantly enhanced levels of ascorbic acid, reduced and oxidized glutathione and reduced/oxidized glutathione ratio, and decreased levels of hydrogen peroxide and malondialdehyde content. These seedlings also maintained higher activities of enzymes involved in ROS scavenging and methylglyoxal detoxification system that includes APX, DHAR, GR, GST, GPX CAT, Gly I, and Gly II.

Besides inducing antioxidant defense mechanism, glycinebetaine also plays a significant role in mitigating abiotic stress-induced ultrastructural damages to chloroplast and mitochondria. It is suggested that, to cope with salt stress, glycinebetaine enhances the production of vacuoles in roots cells that may accumulate higher concentrations of sodium, thus diminishing its levels in shoot cells (Rahman et al. 2002). Such changes in response to glycinebetaine pretreatments also result in greening of, otherwise growth inhibited, plants. Rahman et al. (2002) studied this protective role of glycinebetaine against salt-induced ultrastructural damages in cells of rice seedling. Salt stress, generally, led to inflammation of thylakoid membranes along with collapse of grana stacking and lamellae. Similar symptoms including low numbers of cristae, vacuolation, and swelling were observed in mitochondria. The number of mitochondria also increased with their aggregation localized mostly in the cytoplasm of the cells of the root tip and root cap. Exogenous application of glycinebetaine on the salinity-stressed rice seedlings, on the one hand, reinforced the inhibitory effects of salt stress on root growth, while, on the other hand, it improved the stress-induced retarded shoot growth. Further, an increase in the concentrations of $K^+$ and a decrease in the concentration of $Na^+$ were noticed.

When the plant is under stress, the most important physiological function to maintain for its survival is photosynthesis. Mostly, any abiotic stress reduces the photosynthetic efficiency or photosynthetic rate of plant cells either by targeting photosynthesis-related membrane structures or decreasing the content of photosynthetic pigments. In this scenario, glycinebetaine protects photosynthetic machinery and maintains or increases the concentration of photosynthetic pigments under stress.

Demiral and Türkan (2006) exogenously applied glycinebetaine on NaCl-stressed rice seedling to investigate the mitigating effects of glycinebetaine on photosynthetic efficiency and pigments. They found that the growth of rice seedling improved during salt stress if the plants were pretreated with glycinebetaine compared to the growth of non-pretreated plants. However, effects of exogenous glycinebetaine application on photosystem II efficiency and photosynthetic pigments were discrepant, and two cultivars showed different behaviors for these parameters. Under salinity, the content of chlorophyll *b* and carotenoids decreased in cultivar IR-28, whereas those of cultivar Pokkali remained unchanged. Further, even glycinebetaine pretreated IR-28 plants lost photosynthetic pigments during stress.

Another role of glycinebetaine in the fight against stresses is that it facilitates the accumulation of growth regulators like abscisic acid and jasmonic acid in addition to a well-known osmoregulatory compatible solute, proline (Mahouachi et al. 2012). Glycinebetaine does so by regulating the movement of stomata and the availability of phytoprotectants, thus enhancing relative leaf water content, vegetative growth, and photosynthetic efficiency. Papaya seedlings were pretreated with glycinebetaine and then exposed to drought, rehydration and again drought, respectively (Mahouachi et al. 2012). Although water scarcity resulted in inefficient photosynthesis, less leaf biomass, low stomatal conductance, low relative water content, and higher leaf abscission, glycinebetaine pretreatment decreased the impact of drought by increasing the accumulation of abscisic acid, jasmonic acid, and proline. Moreover, higher seedling growth, photosynthetic rate, and relative water content of leaves were also observed in pretreated papaya seedlings. This suggests that glycinebetaine regulates stomatal movement, photosynthesis, and phytoprotectant accumulation in drought-stressed plants.

## 3.4 Sugars

Carbohydrates are vital for cell structure, signaling, and photosynthate transport and storage. Soluble sugars are recognized phytoprotectants that increase stress tolerance in plants by helping them to continue their regular cellular metabolic activities (Peshev et al. 2013). Sugars also contribute to antioxidant scavenging system during stress by producing NADPH needed in glutathione synthesis (Bolouri-Moghaddam et al. 2010).

### 3.4.1 Trehalose

Although present in scarcely detectable amount in most plants, trehalose, a disaccharide sugar, is one of the major sugar signaling molecules within plant cells. During stressful conditions, trehalose stabilizes macromolecules within cells (Karim et al. 2007). Trehalose also activates the enzymatic antioxidant defense system consisting of superoxide dismutase, peroxidase, and catalase and nonenzymatic antioxidants such as total phenolics and ascorbic acid (Shafiq et al. 2015).

The role of trehalose in the process of regulating response of a plant against salinity stress has been scarcely studied. Trehalose reduces accumulation of reactive oxygen species and activates enzymatic and nonenzymatic antioxidant system in salinity-stressed plants. Mostofa et al. (2015) applied 10 mM trehalose for 48 h to salt-stressed hydroponically grown rice seedlings with an aim to alleviate oxidative stress induced by salinity. NaCl application resulted in the appearance of symptoms related to salt toxicity including growth inhibition, yellowing and rolling of leaves at higher NaCl concentrations, increase in lipoxygenase activity and contents of malondialdehyde, hydrogen peroxide, and proline with a decrease in the content of ascorbic acid, total chlorophyll, and relative water content. Pretreatment with trehalose caused a decrease in the activities of SOD and MDHAR and an increase in the activities of CAT, GST, APX, and DHAR compared to no pretreatment, whereas showing similar GPX and GR activities. In salt-stressed seedlings, trehalose was found to be further enhancing the activities of Gly I and Gly II enzymes involved in methylglyoxal detoxifying mechanism and, thus, reducing oxidative stress.

## *3.5 Sugar Alcohol (Polyols)*

Sugar alcohols or polyols including inositol, mannitol, and sorbitol have shown ROS scavenging activities (Smirnoff and Cumbes 1989).

### 3.5.1 Mannitol

Mannitol is a multifunctional molecule showing the capacity to be an osmoprotectant, a radical scavenger, a protein and membrane stabilizer, and a protector of photosynthesis apparatus under abiotic stress (Seckin et al. 2009). Mannitol-accumulating transgenic tomato plants, when exposed to low temperature of 4 °C, showed a significant increase in the enzymatic activities of superoxide dismutase and catalase (Khare et al. 2010). Similarly, during drought and salinity, these transgenic tomato plants exhibited an antioxidative response. Shen et al. (1997) suggested that transgenic lines accumulate mannitol in high concentrations in chloroplast that confer ability to scavenge endogenous $OH^{\bullet}$ and mitigate cellular oxidative stress due to ROS. Considering the role of mannitol in osmoregulation, ROS scavenging, membranes and protein stabilizing, protection of photosynthetic apparatus, its exogenous application to the plants –which do not produce mannitol– might be beneficial. Wheat is one of such plants that is not able to synthesize mannitol. Therefore, wheat is a very good model plant to study the effect of exogenously applied mannitol. Seckin et al. (2009) pretreated seedlings of salt-sensitive wheat cultivar with mannitol to investigate its role in moderating salinity stress in wheat seedlings. The seedlings were pretreated with mannitol for 24 h and then transferred to NaCl solution and grew for 5 days. The researchers observed an increase in the activities of antioxidant enzymes such as SOD, POD, CAT, APX, and glutathione reductase in pretreated seedlings. There were 11 SOD activity bands observed in mannitol pretreated seedlings compared to only 9 in non-pretreated seedlings. Moreover, mannitol reduced the otherwise enhanced lipid peroxidation of membranes.

### 3.5.2 Myoinositol

Myoinositol indirectly scavenges ROS by regulating the synthesis of myoinositol oxygenase, which is responsible for the conversion of myoinositol to ascorbic acid (Lorence et al. 2004). Similar increase in ascorbic acid content in plants over expressing myoinositol oxygenase gene has been observed by Lisko et al. (2013) in *Arabidopsis*. Myoinositol induces antioxidant defense system in plants and regulate $Na^+$ to $K^+$ ratio during stress, thus minimizing the inhibition of cellular metabolic activities.

Pretreatment of *Malus hupehensis* seedlings with myoinositol has been shown to markedly alleviate net photosynthesis rate and upregulate the size of stomatal aperture during salt stress (Hu et al. 2018). This was also observed that exogenous myoinositol pretreatment reduced the accumulation of ROS and uptake of $Na^+$ ions, whereas increased the accumulation of sugars or polyols contributing to the cells osmoprotection.

## 3.6 *Signaling Molecules*

### 3.6.1 Hydrogen Peroxide

Various adverse environmental conditions cause the formation of $H_2O_2$ in plant cells. Hydrogen peroxide, on the one hand, is a harmful molecule triggering oxidative stress if accumulated at high concentration, while on the other hand, it takes part in plant cell signaling at low concentrations and influences growth and development of plant (Quan et al. 2008). Recently, researchers have been pretreating plant seedlings with $H_2O_2$ at low concentrations to induce tolerance against abiotic stress. $H_2O_2$ has been found to be involved in several protective pathways to maintain high plant survival percentage and effectively improve the growth of plants during stresses like salinity, drought, flooding, and high and low temperature (Andrade et al. 2018; Sarwar et al. 2017; Kong et al. 2017; Sun et al. 2016; Yu et al. 2003).

$H_2O_2$ induces the expression of glutathione *S*-transferase1 gene and accumulation of $Ca^{2+}$ ions (Rentel and Knight 2004). Exposure of *Arabidopsis* seedling to $H_2O_2$ causes a biphasic elevation of $Ca^{2+}$, the early $Ca^{2+}$rise being localized in the cotyledons, and late peak within the roots. $H_2O_2$ also triggers the activities of antioxidant enzymes and accumulation of reduced glutathione and ascorbate if pretreated at low concentrations (Liu et al. 2010). Similarly, $H_2O_2$ treatment also facilitates the transcription of heat-shock proteins (Volkov et al. 2006).

$H_2O_2$ coordinates with CAT and APX and enhances nonenzymatic antioxidant in salt-stressed plants to mitigate the oxidative damages of abiotic stress. Bagheri et al. (2019) pretreated pistachio rootstock seedlings with $H_2O_2$ and then subjected the seedlings to salt stress for 7 days. Salinity stress reduced vegetative growth and ascorbate and carotenoid levels, whereas pretreated seedlings showed improvement in growth and maintained higher ascorbate and carotenoids content than non-pretreated seedlings. Similarly, $H_2O_2$ pretreatment induced the activities of APX and CAT and elevated glutathione content. Overall, pretreated pistachio seedlings showed better vigor against salt stress that may be attributed to better osmoregulation of $H_2O_2$ in harmony with APX and CAT and enhanced ascorbate, glutathione, and carotenoid levels.

Heat stress is becoming a global reality in the wake of changing climate. Heat stress induces alterations in DNA methylation. $H_2O_2$ changes such alteration in cytosine methylation levels in genes. Cao et al. (2013) exposed cucumber seedlings to heat stress after pretreatment with $H_2O_2$ and detected the cytosine methylation. Heat suppressed seedling growth and caused methylation of multiple loci. However, the suppression in growth was reversed by $H_2O_2$ pretreatment with demethylation of certain loci.

Other than pretreatment of seedling by feeding through the root, another strategy of pretreatment is foliar application. $H_2O_2$ pretreatment via foliar spray has shown the similar mitigation on salt-stressed plant such as improved growth, reduction in lipid peroxidation, and activation of plant antioxidant mechanism. Maize seedlings were leaf sprayed with $H_2O_2$ to stimulate enzymatic antioxidant system for better plant growth under salt stress (Gondim et al. 2012). As expected, salinity reduced maize seedling growth, but $H_2O_2$ pretreatment minimized this effect and stimulated growth in salt-stressed plants. The pretreated maize seedlings showed higher CAT activity within 48 h, whereas APX and glutathione peroxidase showed late response after 240 h. Higher CAT activity was accompanied with low malondialdehyde content. This suggests that $H_2O_2$ pretreatment through foliar application stimulates growth and induces antioxidant enzymes to counter the adverse effect of stress.

### 3.6.2 Nitric Oxide

Nitric oxide is a small and simple gaseous molecule ubiquitously present throughout the kingdom of plants mostly in radical form. It is an instrumental signaling molecule that modulates growth and development of plants during germination of seeds, seedling development, flowering, stomatal responses, and senescence. Its bioactivity has been mainly observed during adverse environmental condition faced by plants. Plants can absorb nitric oxide from the atmosphere and endogenously synthesize it in an adequate amount as well. Nitric oxide is synthesized in plants both enzymatically and nonenzymatically. Enzymatically, nitric oxide synthase catalyzes the dominant pathway to form nitric oxide using L-arginine as primary substrate. However, it is synthesized through nitrate nonenzymatically (Palavan-Unsal and Arisan 2009). Nitric oxide has been widely applied in seed priming and seedling pretreatment techniques to facilitate plants to cope with various abiotic stress. Exogenous application of nitric oxide has shown improvement in growth of plants seedlings in response to various abiotic stresses (Uchida et al. 2002; Zhang et al. 2006; Zhou et al. 2012; Liu et al. 2015).

### 3.6.3 Hydrogen Sulfide

Although initially thought as a toxic gaseous molecule, multifunctional roles of hydrogen sulfide ($H_2S$) within cells have been discovered such as its role as an anti-oxidative signaling molecule (Ali et al. 2019). In addition to carbon monoxide and nitric oxide, $H_2S$ is now considered as an important gaseous signaling molecule. $H_2S$ facilitates the biogenesis of chloroplast, expression of enzymes involved in photosynthesis, and modification of thiol redox system, all this translating into high photosynthetic efficiency (Chen et al. 2011). Christou et al. (2013) also observed that $H_2S$ modifies ROS biosynthesis and regulates transcriptional factors involved in multiple defense mechanisms to confer tolerance to strawberry plants during salinity and nonionic osmotic stresses. $H_2S$ induces growth, germination, and lateral root formation (Dooley et al. 2013; Fang et al. 2014).

Heavy metal toxicity is one of the emerging stresses that greatly influence the quality of agricultural food products. Phytoprotectants play an important role in minimizing the adverse effects of heavy metal toxicity (Islam et al. 2009). Mercury toxicity is one of the major yield- and quality-limiting factors in contaminated soils. Pretreatment of rice seedling with hydrogen sulfide, a multifunctional molecule, has been tested to investigate its role in reducing mercury toxicity damages to rice seedlings (Chen et al. 2017). Rice seedlings were pretreated with sodium hydrosulfide, a hydrogen sulfide donor, prior to exposing the seedlings to mercuric chloride. The pretreatment inhibited the mercury transport to shoots and held the mercury in roots. This was achieved due to improvement in transcription of membrane-associated transcription factors and expression of thiols and metallothioneins. Hydrogen sulfide pretreatment, in the presence of mercury and SOD and CAT inhibitors (diethyldithiocarbamate or 3-amino-1,2,4-triazole), also noticeably improved the seedlings growth. Hydrogen sulfide prevents oxidative stress-induced damages as it might inhibit or scavenge ROS and suppress malondialdehyde and $H_2O_2$ accumulation. The findings of this study suggested that hydrogen sulfide retarded mercury toxicity by holding it in roots or by scavenging ROS in pretreated seedlings.

Tian et al. (2016) also investigated the involvement of hydrogen sulfide in defense mechanism against heavy metal toxicity in millet. Pretreatment with sodium hydrosulfide stabilized the membrane integrity and decreased the leakage of electrolytes in seedling stressed with cadmium toxicity. Further, levels of malondialdehyde and $H_2O_2$ were also found to be reduced in pretreated seedling along with higher photosynthetic efficiency in cadmium-stressed millet seedlings. Activities and transcript levels of pyrroline-5-carboxylate reductase and proline dehydrogenase, enzymes involved in proline biosynthesis pathway, were markedly influenced by sodium hydrosulfide pretreatment. With using inhibitors for endogenous hydrogen sulfide, cadmium-related increase in proline content shrunk. Proline and hydrogen sulfide showed synergistic effect, whereby parallel pretreatment with proline and hydrogen sulfide reversed cadmium-stressed suppression of growth, stomatal movement, and oxidative injury of millet seedling.

## *3.7 Antioxidants*

### 3.7.1 Ascorbic Acid

Ascorbic acid is the most abundant antioxidant molecule in plants. Ascorbic acid also takes part in activating plant defense mechanism. Ascorbic acid is the primary metabolite; other metabolites include glutathione and NADPH in the hydrogen peroxide detoxification cycle and, therefore, if needed, it might activate plant defense system (Munir and Aftab 2011). Indirect mitigation of oxidative stress by ascorbic acid through regenerating and protecting tocopherol and carotene on the surface of thylakoid is also important for plant response to stress (Tappel 1977).

Ascorbic acid has been scarcely studied for its role in alleviation of stress induces oxidative damages through its seedling pretreatment. Munir and Aftab (2011) pretreated nearly 8-week-old in vitro-grown sugarcane seedlings with 0.5 mM ascorbic acid containing MS basal media. Afterward, the pretreated seedlings were transferred to NaCl containing MS basal media for salt stress induction. Pretreatment with ascorbic acid led to decrease in yellowing of leaves in NaCl-stressed seedlings. Ascorbic acid moderated the enzymatic plant defense mechanism, whereby, generally, the activities of POD and CAT increased, and activity of SOD decreased. The pretreatment also modified plant growth characters like root length and concentrations of osmoprotectants such as soluble proteins.

### 3.7.2 Melatonin

N-Acetyl-5-methoxytryptamine (melatonin) is a low molecular weight tryptophan-derived molecule, which is recognized as an important and multifunctional antioxidant and signaling molecule in plants (Erland et al. 2017). It is present almost in all organisms. During the early life forms, melatonin primarily was a ROS scavenger acting as a potent antioxidant such as in unicellular organisms; however, its secondary role evolved as a signaling molecule (Tan et al. 2015). Melatonin is induced even during moderate oxidative stress, and its secondary and tertiary derivatives are also potent ROS scavenger; one melatonin molecule is considered to reduce up to 10 ROS compared to $\leq 1$ by other antioxidants (Tan et al. 2015). Therefore, melatonin is more efficient in protecting cells under oxidative stress than other antioxidants. Melatonin is also considered to enhance resistance against heavy metal toxicity in plants. Vanadium is one such heavy metal which, although present universally on earth, has not been explored enough for its effects on plants. Nawaz et al. (2018)

investigated the pretreatment of watermelon seedlings with melatonin to sustain the growth and development of seedlings when these were subsequently subjected to vanadium toxicity. The pretreatment with melatonin led to higher levels of relative chlorophyll, photosynthetic assimilation, and better plant growth compared with non-pretreated seedlings. The pretreatment reduced the vanadium levels in seedling leaf and stem and reduced its transport from the root to shoot system (Nawaz et al. 2018). As for enzymatic antioxidant system, melatonin stimulated the activities of SOD and CAT and reduced malondialdehyde and $H_2O_2$ content. These effects of pretreatment were attributed to ability of melatonin to regulate gene expression for SOD, POD, APX, glutathione peroxidase, and glutathione *S*-transferase.

Melatonin is found naturally in plants and plays important role in maintaining the growth and development of plant during adverse environmental conditions. Especially, it aids in the development of roots, scavenges ROS, accumulates osmoregulatory substances, and increases the biomass of salinity-stressed seedling (Zeng et al. 2018). If rapeseed seedlings are pretreated with melatonin before exposing to salt stress, it diminishes the inhibitory effects of salinity on growth by facilitating the increase in fresh and dry weights of shoots and roots (Zeng et al. 2018). Moreover, antioxidant enzyme activities of POD, CAT, and APX enhanced, and hydrogen peroxide levels decreased in the rapeseed seedlings pretreated exogenously with melatonin. In pretreated seedling, accumulation of stress retarding phytoprotectants including soluble proteins, proline, and water-soluble glucan was also encouraged.

Melatonin has shown potential to reverse the growth inhibition caused by other abiotic stresses such as salt stress. Zeng et al. (2018) pretreated rapeseed seedling with 30 $\mu mol\ L^{-1}$ melatonin and then subjected the seedlings to salt-induced stress. The pretreatment stimulated the growth that was otherwise inhibited in non-pretreated seedlings. This was indicated by the increase in fresh and dry weights of the shoot and root of pretreated seedlings. Similarly, the activities of enzymes involved in antioxidant defense mechanism such as POD, CAT, and APX enhanced, whereas $H_2O_2$ content decreased in pretreated seedlings compared to in non-pretreated seedlings. One of the important advantages of melatonin pretreatment is its osmoregulation through accumulation of phytoprotectants as shown in this study, whereby concentration of proline, soluble sugars, and soluble proteins increased. In conclusion, melatonin facilitates scavenging of $H_2O_2$, enhancement of antioxidant enzymatic activities, production of osmoregulatory substances, and improvement of root and shoot growth under salt stress.

Chilling injury to the plants is one of the drawbacks of changing environmental conditions with extremes of temperatures that limits the yield of agricultural plants. Pretreatment with phytoprotectants is one of the effective strategies to counter chilling injury and sustain growth and yield of plants. In one of such studies, melatonin was used to pretreat tomato seedlings to ameliorate cold damage caused by low temperature (Ding et al. 2017). With less electrolyte leakage and malondialdehyde levels and higher antioxidant enzyme activities, nonenzymatic reducing agents were noted in melatonin-pretreated seedlings under low temperature stress. Similarly, in pretreated seedlings, cold-response-associated genes were upregulated, and the gene encoding sedoheptulose-1,7-bisphosphatase (a Calvin cycle enzyme) was noticeably expressed in pretreated seedling during low temperature stress which was also confirmed by increase in assimilation of photosynthetic carbon. Further, levels of other phytoprotectants such as proline, sucrose, and polyamines also increased, all this providing evidence of the mitigating ability of melatonin for low temperature-stressed plants.

## 4 Conclusion

Phytoprotectants are the class of bioactive compounds that have shown the capacity to enable plant to withstand against abiotic stress. Seedling pretreatment prepares the seedlings for abiotic stress and has been proven to be an effective strategy that can be utilized in the scenario of changing environmental patterns. Various categories of phytoprotectants including polyamines, signaling molecules, antioxidants, amino acids, quaternary ammonium compounds, sugars, and sugar alcohols have shown that these compounds are species specific and in some cases stress specific. Further, testing a phytoprotectant for its suitability for a certain species under a specific stress will extend the knowledge of this already well-explored research area.

## References

Alcázar R, Altabella T, Marco F, Bortolotti C, Reymond M, Koncz C, Carrasco P, Tiburcio AF (2010) Polyamines: molecules with regulatory functions in plant abiotic stress tolerance. Planta 231:1237–1249

Ali S, Nawaz A, Ejaz S, Haider STA, Alam MW, Javed HU (2019) Effects of hydrogen sulfide on postharvest physiology of fruits and vegetables: an overview. Sci Hortic 243:290–299

Andrade CA, de Souza KRD, de Oliveira Santos M, da Silva DM, Alves JD (2018) Hydrogen peroxide promotes the tolerance of soybeans to waterlogging. Sci Hortic 232:40–45

Arasimowicz M, Floryszak-Wieczorek J (2007) Nitric oxide as a bioactive signaling molecule in plant stress responses. Plant Sci 172:876–887

Ashraf M, Foolad MR (2007) Roles of glycine betaine and proline in improving plant abiotic stress resistance. Environ Exp Bot 59:206–216

Bagheri M, Gholami M, Baninasab B (2019) Hydrogen peroxide-induced salt tolerance in relation to antioxidant systems in pistachio seedlings. Sci Hortic 243:207–213

Bharwana SA, Ali S, Farooq MA, Iqbal N, Hameed A, Abbas F, Ahmad MSA (2014) Glycine betaine-induced lead toxicity tolerance related to elevated photosynthesis, antioxidant enzymes suppressed lead uptake and oxidative stress in cotton. Turk J Bot 38(2):281–292

Bolouri-Moghaddam MR, Le Roy K, Xiang L, Rolland F, Van den Ende W (2010) Sugar signalling and antioxidant network connections in plant cells. FEBS J 277(9):2022–2037

Cao YY, Gao Y, Sun WJ, Huang YW, Zhang J, Bai JG (2013) Role of hydrogen peroxide pretreatment in heat-induced alteration of DNA methylation in cucumber leaves. Sci Hortic 151:173–183

Cha-Um S, Kirdmanee C (2010) Effect of glycinebetaine on proline, water use, and photosynthetic efficiencies, and growth of rice seedlings under salt stress. Turk J Agric For 34:517–527

Chen TH, Murata N (2008) Glycinebetaine: an effective protectant against abiotic stress in plants. Trends Plant Sci 13(9):499–505

Chen H, McCarig B, Melotto M, Yang HS, Howe GA (2004) Regulation of plant arginase by wounding, jasmonate and the phytotoxin coronatine. J Biol Chem 279:45998–46007
Chen J, Wu FH, Wang WH, Zheng CJ, Lin GH, Dong XJ, He JX, Pei ZM, Zheng HL (2011) Hydrogen sulphide enhances photosynthesis through promoting chloroplast biogenesis, photosynthetic enzyme expression, and thiol redox modification in *Spinacia oleracea* seedlings. J Exp Bot 62:4481–4493
Chen Z, Chen M, Jiang M (2017) Hydrogen sulfide alleviates mercury toxicity by sequestering it in roots or regulating reactive oxygen species productions in rice seedlings. Plant Physiol Biochem 111:179–192
Christou A, Manganaris GA, Papadopoulos I, Fotopoulos V (2013) Hydrogen sulfide induces systemic tolerance to salinity and non-ionic osmotic stress in strawberry plants through modification of reactive species biosynthesis and transcriptional regulation of multiple defence pathways. J Exp Bot 64:1953–1966
Demiral T, Türkan I (2006) Exogenous glycinebetaine affects growth and proline accumulation and retards senescence in two rice cultivars under NaCl stress. Environ Exp Bot 56:72–79
Ding F, Liu B, Zhang S (2017) Exogenous melatonin ameliorates cold-induced damage in tomato plants. Sci Hortic 219:264–271
Dooley FD, Nair SP, Ward PD (2013) Increased growth and germination success in plants following hydrogen sulfide administration. PLoS One 8:e62048
Erland LAE, Saxena PK, Murch SJ (2017) Melatonin in plant signalling and behaviour. Funct Plant Biol 45(2):58–69
Fang T, Cao Z, Li J, Shen W, Huang L (2014) Auxin-induced hydrogen sulfide generation is involved in lateral root formation in tomato. Plant Physiol Biochem 76:44–51
Fu XZ, Xing F, Wang NQ, Peng LZ, Chun CP, Cao L, Ling LL, Jiang CL (2014) Exogenous spermine pretreatment confers tolerance to combined high-temperature and drought stress in vitro in trifoliate orange seedlings via modulation of antioxidative capacity and expression of stress-related genes. Biotechnol Biotechnol Equip 28:192–198
Garcia-Jimenez P, Just PM, Delgado AM, Robaina RR (2007) Transglutaminase activity decrease during acclimation to hyposaline conditions in marine seaweed *Grateloupia doryphora* (*Rhodophyta*, Halymeniaceae). J Plant Physiol 164:367–370
Gill SS, Tuteja N (2010) Reactive oxygen species and antioxidant machinery in abiotic stress tolerance in crop plants. Plant Physiol Biochem 48:909–930
Gondim FA, Gomes-Filho E, Costa JH, Mendes Alencar NL, Prisco JT (2012) Catalase plays a key role in salt stress acclimation induced by hydrogen peroxide pretreatment in maize. Plant Physiol Biochem 56:62–71
Hamilton EW, Heckathorn SA (2001) Mitochondrial adaptation to NaCl. Complex I is protected by antioxidants and small heat shock proteins, whereas complex II is protected by proline and betaine. Plant Physiol 126:1266–1274
Hossain MA, Hasanuzzaman M, Fujita M (2011) Coordinate induction of antioxidant defense and glyoxalase system by exogenous proline and glycinebetaine is correlated with salt tolerance in mung bean. Front Agric China 5:1–14
Hossain Z, Makino T, Komatsu S (2012) Proteomic study of β-aminobutyric acid-mediated cadmium stress alleviation in soybean. J Proteome 75:4151–4164
Hu L, Hu T, Zhang X, Pang H, Fu J (2012) Exogenous glycine betaine ameliorates the adverse effect of salt stress on perennial ryegrass. J Am Soc Hortic Sci 137(1):38–46
Hu L, Zhou K, Li Y, Chen X, Liu B, Li C, Gong X, Ma F (2018) Exogenous myo-inositol alleviates salinity-induced stress in *Malus hupehensis* Rehd. Plant Physiol Biochem 133:116–126
Hussain SS, Ali M, Ahmad M, Siddique KHM (2011) Polyamines: natural and engineered abiotic and biotic stress tolerance in plants. Biotechnol Adv 29:300–311
Islam MM, Hoque MA, Okuma E, Banu MN, Shimoishi Y, Nakamura Y, Murata Y (2009) Exogenous proline and glycinebetaine increase antioxidant enzyme activities and confer tolerance to cadmium stress in cultured tobacco cells. J Plant Physiol 166(15):1587–1597

Karim S, Aronsson H, Ericson H, Pirhonen M, Leyman B, Welin B, Mantyla E, Palva ET, Van Dijck P, Holmstrom KO (2007) Improved drought tolerance without undesired side effects in transgenic plants producing trehalose. Plant Mol Biol 64:371–386

Kemble AR, MacPherson HT (1954) Liberation of amino acids in perennial rye grass during wilting. Biochem J 58(1):46–49

Khare N, Goyary D, Singh NK, Shah P, Rathore M, Anandhan S, Sharma D, Arif M, Ahmed Z (2010) Transgenic tomato cv. Pusa Uphar expressing a bacterial mannitol-1-phosphate dehydrogenase gene confers abiotic stress tolerance. Plant Cell Tissue Organ Cult 103:267–277

Khedr AHA, Abbas MA, Wahid AAA, Quick WP, Abogadallah GM (2003) Proline induces the expression of salt-stress-responsive proteins and may improve the adaptation of *Pancratium maritimum* L. to salt-stress. J Exp Bot 54(392):2553–2562

Kong C-C, Ren C-G, Li R-Z, Xie Z-H, Wang J-P (2017) Hydrogen peroxide and strigolactones signaling are involved in alleviation of salt stress induced by arbuscular mycorrhizal fungus in *Sesbania cannabina* seedlings. J Plant Growth Regul 3:734–742

Kubis J (2003) Polyamines and "scavenging system": influence of exogenous spermidine on catalase and guaiacol peroxidase activities, and free polyamine level in barley leaves under water deficit. Acta Physiol Plant 4:337–343

Kubis J, Floryszak-Wieczorek J, Arasimowicz-Jelonek M (2014) Polamines induce adaptive responses in water deficit stressed cucumber roots. J Plant Res 127:151–158

Li D-M, Zhang J, Sun W-J, Li Q, Dai A-H, Bai J-G (2011) 5-Aminolevulinic acid pretreatment mitigates drought stress of cucumber leaves through altering antioxidant enzyme activity. Sci Hortic 130(4):820–828

Lisko KA, Torres R, Harris RS, Belisle M, Vaughan MM, Jullian B, Chevone BI, Mendes P, Nessler CL, Lorence A (2013) Elevating vitamin C content via overexpression of myo-inositol oxygenase and L-gulono-1,4-lactone oxidase in Arabidopsis leads to enhanced biomass and tolerance to abiotic stresses. In Vitro Cell Dev Biol Plant 49:643–655

Liu J-H, Kitashiba H, Wang J, Ban Y, Moriguchi T (2007) Polyamines and their ability to provide environmental stress tolerance to plants. Plant Biotechnol 24:117–126

Liu Z-J, Guo Y-K, Bai J-G (2010) Exogenous hydrogen peroxide changes antioxidant enzyme activity and protects ultrastructure in leaves of two cucumber ecotypes under osmotic stress. J Plant Growth Regul 29:171–183

Liu C, Zhao L, Yu G (2011) The dominant glutamic acid metabolic flux to produce γ-amino butyric acid over proline in Nicotiana tabacum leaves under water stress relates to its significant role in antioxidant activity. J Integr Plant Biol 53(8):608–618

Liu N, Gong B, Jin Z, Wang X, Wei M, Yang F, Li Y, Shi Q (2015) Sodic alkaline stress mitigation by exogenous melatonin in tomato needs nitric oxide as a downstream signal. J Plant Physiol 186–187:68–77

Liu M, Li J, Niu J, Wang R, Song J, Lv J, Zong X, Wang S (2016) Interaction of drought and 5-aminolevulinic acid on growth and drought resistance of *Leymus chinensis* seedlings. Acta Ecol Sin 36:180–188

Lorence A, Chevone BI, Mendes P, Nessler CL (2004) Myo-inositol oxygenase offers a possible entry point into plant ascorbate biosynthesis. Plant Physiol 134:1200–1205

Mahouachi J, Argamasilla R, Gómez-Cadenas A (2012) Influence of exogenous glycine betaine and abscisic acid on papaya in responses to water-deficit stress. J Plant Growth Regul 31:1–10

Malekzadeh P, Khara J, Heidari R (2012) Effect of exogenous gamma-aminobutyric acid on physiological tolerance of wheat seedlings exposed to chilling stress. Iran J Plant Physiol 3:611–618

Malekzadeh P, Khara J, Heydari R (2014) Alleviating effects of exogenous gamma-aminobutyric acid on tomato seedling under chilling stress. Physiol Mol Biol Plants 20:133–137

Mostofa MG, Hossain MA, Fujita M (2015) Trehalose pretreatment induces salt tolerance in rice (*Oryza sativa* L.) seedlings: oxidative damage and co-induction of antioxidant defense and glyoxalase systems. Protoplasma 252:461–475

Munir N, Aftab F (2011) Enhancement of salt tolerance in sugarcane by ascorbic acid pretreatment. Afr J Biotechnol 10:18362–18370

Nasibi F, Yaghoobi MM, Kalantari KM (2011) Effect of exogenous arginine on alleviation of oxidative damage in tomato plant underwater stress. J Plant Interact 6:291–296
Nawaz MA, Jiao Y, Chen C, Shireen F, Zheng Z, Imtiaz M, Bie Z, Huang Y (2018) Melatonin pretreatment improves vanadium stress tolerance of watermelon seedlings by reducing vanadium concentration in the leaves and regulating melatonin biosynthesis and antioxidant-related gene expression. J Plant Physiol 220:115–127
Palavan-Unsal N, Arisan D (2009) Nitric oxide signalling in plants. Bor Rev 75(2):203–229
Park E-J, Jeknic Z, Chen TH (2006) Exogenous application of glycinebetaine increases chilling tolerance in tomato plants. Plant Cell Physiol 47(6):706–714
Peshev D, Vergauwen R, Moglia A, Hideg É, Van den Ende W (2013) Towards understanding vacuolar antioxidant mechanisms: a role for fructans? J Exp Bot 64(4):1025–1038
Quan L, Zhang B, Shi W, Li H (2008) Hydrogen peroxide in plants: a versatile molecule of the reactive oxygen species network. J Integr Plant Biol 50:2–18
Rahman MS, Miyake H, Takeoka Y (2002) Effects of exogenous glycinebetaine on growth and ultrastructure of salt-stressed rice seedlings (*Oryza sativa* L.). Plant Prod Sci 1:33–44
Rentel MC, Knight MR (2004) Oxidative stress-induced calcium signaling in Arabidopsis. Plant Physiol 135:1471–1479
Sarwar M, Saleem MF, Najeeb U, Shakeel A, Ali S, Bilal MF (2017) Hydrogen peroxide reduces heat-induced yield losses in cotton (*Gossypium hirsutum* L.) by protecting cellular membrane damage. J Agron Crop Sci 203:429–441
Seckin B, Sekmen AH, Türkan I (2009) An enhancing effect of exogenous mannitol on the antioxidant enzyme activities in roots of wheat under salt stress. J Plant Growth Regul 28:12–20
Shafiq S, Akram NA, Ashraf M (2015) Does exogenously-applied trehalose alter oxidative defense system in the edible part of radish (*Raphanus sativus* L.) under water-deficit conditions? Sci Hortic 185:68–75
Sharma SS, Dietz KJ (2006) The significance of amino acids and amino acid-derived molecules in plant responses and adaptation to heavy metal stress. J Exp Bot 57:711–726
Shen BO, Jensen RG, Bohnert HJ (1997) Increased resistance to oxidative stress in transgenic plants by targeting mannitol biosynthesis to chloroplasts. Plant Physiol 113(4):1177–1183
Slama I, Abdelly C, Bouchereau A, Flowers T, Savouré A (2015) Diversity, distribution and roles of osmoprotective compounds accumulated in halophytes under abiotic stress. Ann Bot 115:433–447
Smirnoff N, Cumbes QJ (1989) Hydroxyl radical scavenging activity of compatible solutes. Phytochemistry 28:1057–1060
Sun Y, Wang H, Liu S, Peng X (2016) Exogenous application of hydrogen peroxide alleviates drought stress in cucumber seedlings. S Afr J Bot 106:23–28
Tan D-X, Manchester LC, Esteban-Zubero E, Zhou Z, Reiter RJ (2015) Melatonin as a potent and inducible endogenous antioxidant: synthesis and metabolism. Molecules 20:18886–18906
Tappel AL (1977) Protection against free radical lipid peroxidation reactions. In: Roberts J, Adelman RC, Cristofalo V (eds) Pharmacological intervention in the aging process. Plenum Press, New York, pp 111–113
Tian B, Qiao Z, Zhang L, LiH PY (2016) Hydrogen sulfide and proline cooperate to alleviate cadmium stress in foxtail millet seedlings. Plant Physiol Biochem 109:293–299
Uchida A, Jagendorf AT, Hibino T, Takabe T, Takabe T (2002) Effects of hydrogen peroxide and nitric oxide on both salt and heat stress tolerance in rice. Plant Sci 163:515–523
Volkov RA, Panchuk II, Mullineaux PM, Schöffl F (2006) Heat stress-induced $H_2O_2$ is required for effective expression of heat shock genes in Arabidopsis. Plant Mol Biol 61:733–746
Xu F, Zhu J, Cheng SY, Zhang WW, Wang Y (2010) Effect of 5-aminolevulinic acid on photosynthesis, yield, nutrition and medicinal values of kudzu (*Pueraria phaseoloides*). Trop Grasslands 44:260–265
Yancey PH (2005) Organic osmolytes as compatible, metabolic and counteracting cytoprotectants in high osmolarity and other stresses. J Exp Biol 208:2819–2830

Yu C-W, Murphy TM, Lin C-H (2003) Hydrogen peroxide-induced chilling tolerance in mung beans mediated through ABA-independent glutathione accumulation. Funct Plant Biol 30:955–963

Yu XZ, Lin YJ, Fan WJ, Lu MR (2017) The role of exogenous proline in amelioration of lipid peroxidation in rice seedlings exposed to Cr (VI). Int Biodeterior Biodegrad 123:106–112

Zeng L, Cai J-S, Li J-P, Lu G-Y, Li C-S, Fu G-P, Zhang X-K, Ma H-Q, Liu Q-Y, Zou X-L, Cheng Y (2018) Exogenous application of a low concentration of melatonin enhances salt tolerance in rapeseed (*Brassica napus* L.) seedlings. J Integr Agric 17:328–335

Zhang YY, Wang LL, Liu YL, Zhang Q, Wei QP, Zhang WH (2006) Nitric oxide enhances salt tolerance in maize seedlings through increasing activities of proton-pump and Na+/H+ antiport in the tonoplast. Planta 224:545–555

Zhou WH, Shi SL, Kou JT (2012) Effects of exogenous nitric oxide on the growth and nitrogen metabolism of alfalfa seedlings under salt stress. Ying Yong Sheng Tai Xue Bao 23:3003–3008

# Improving Antioxidant Defense in Plants Through Seed Priming and Seedling Pretreatment

**Idrees Haider, Muhammad Akmal, Muhammad Taimoor Shakeel, Shakeel Ahmad, Niaz Ahmad, Sajjad Hussain, Muhammad Bilal, and Muhammad Arif Ali**

## Contents



**Abstract** Under abiotic and biotic stress conditions, the first and foremost effect on crop is reduction/poor seed germination, which results in poor crop stand and ultimately results in poor crop production. Seed priming proved to be a promising approach to alter the seed hydration and nutrition level for better germination, improved and synchronized crop stand, and better crop yield in normal and stress

I. Haider · N. Ahmad · M. A. Ali (✉)
Department of Soil Science, Bahauddin Zakariya University, Multan, Punjab, Pakistan
e-mail: arif1056@bzu.edu.pk

M. Akmal
Institute of Soil Science, Pir Mehr Ali Shah Arid Agriculture University, Rawalpindi, Punjab, Pakistan

M. T. Shakeel
Department of Plant Pathology, Bahauddin Zakariya University, Multan, Punjab, Pakistan

S. Ahmad (✉)
Department of Agronomy, Bahauddin Zakariya University, Multan, Punjab, Pakistan
e-mail: shakeelahmad@bzu.edu.pk

S. Hussain
Department of Horticulture, Bahauddin Zakariya University, Multan, Punjab, Pakistan

M. Bilal
Department of Environmental Sciences, COMSATS University Islamabad, Islamabad, Pakistan



M. Hasanuzzaman, V. Fotopoulos (eds.), *Priming and Pretreatment of Seeds and Seedlings*, https://doi.org/10.1007/978-981-13-8625-1_29

conditions as well. Abiotic and biotic stress rejuvenates the oxidative stress in different plant cellular compartments due to production of reactive oxygen species (ROS) as by-product of plant metabolic processes. To overcome the damaging effects of ROS, plants naturally develop antioxidant defense system within the plant body through production of enzymatic and nonenzymatic antioxidants to regulate oxidant and antioxidant balance in plants. Different types of seed priming techniques like hydropriming, nutrient priming, thermopriming, osmo-priming, etc. help protect the plants from oxidative stress by production of antioxidants for ROS scavenging to control lipid, DNA, and protein oxidation within the plant body. Seed priming found to be an operative approach to improve antioxidant defense system in the plant body and had promotive advantages to advance seed germination, plant vigor, and crop stand and overcome nutrient deficiency and minimize the negative impacts of stress conditions. In this chapter, we have reviewed promotive effects of seed priming and seedling pretreatment to improve antioxidant defense system in the plant body under stress conditions and role of seed priming in reducing the ROS production and ROS scavenging to maintain oxidants and antioxidant balance in plant's cellular compartments.

**Keywords** Seed priming · Seedling pretreatment · Oxidative stress · Antioxidant defense system

## 1 Introduction to Seed Priming

Manipulation/alteration and management of seed hydration level in controlled conditions to enhance the seed sprouting is known as seed priming (Murungu et al. 2004). It is a technique which may help the seed to better germinate in soil under unfavorable moisture and temperature conditions. Seed priming initiates metabolic processes within the seed for germination that results in uniform and rapid plant stand. Priming synchronizes the germination process to improve crop stand in suboptimal temperature conditions and under low temperature may effectively improve seed germination and seedling establishment (Afzal et al. 2008). There are different types of seed priming like hydropriming, osmo-priming, hormonal priming, halopriming, thermopriming, chemical priming, redox priming, biopriming, and nutrient priming (Jisha et al. 2013; Paparella et al. 2015) that can be employed for speedy, enhanced, and synchronized seed germination (Ozden et al. 2017).

Under stress conditions of drought, salt and/or nutrients, production of cereal crops is limited due to poor germination and poor crop stand. The negative effects of such stress conditions on seed germination can be minimized through seed priming. The type of seed priming is also dependent on the type of stress delaying the seed germination. Generally, seed priming is the hydration of seeds to a certain level for easy germination; and in some conditions, treated or hydro-primed seeds are dried to their original moisture contents, resulting in rapid germination in field conditions, i.e., normal or stress conditions (Ashraf and Foolad 2005). If any seed lot is found to have deficiency of one nutrient or certain nutrients, then the seed is primed or pretreated with nutrient solutions containing the deficient nutrient(s). Nutrient priming has supplemented advantage of improved nutrient supply from seed for seedlings along with primed/hydrated seed (Asgedom and Becker 2001). Studies have shown that different nutrient primings have considerable results. Like, priming of seeds with phosphorous (P) has shown support at early growth stages and stimulated seedling growth in calcareous soils (Peltonen-Sainio et al. 2006; Abdulrahmani et al. 2007). Zinc solution priming and Zn dressing of seeds for Zn-deficient soils has enhanced root and shoot biomass and crop production (Yilmaz et al. 1998). Similarly, seed hydropriming in drought conditions has minimized crop failure risks.

## 2 Effects and Causes of Seed Priming

Prevalence of biotic and abiotic stresses poses adverse effects on seed germination, seedling establishment, plant growth and development, and crop yield. Such stress conditions are proved to be unfavorable growth conditions for plant growth and result in overproduction of ROS that causes oxidation of cellular components like protein, lipids, and nucleic acid. These unfavorable growth conditions determine the need for seed and seedling treatment that have promotive effect on seed germination, seedling establishment, and plant growth and development to enhance crop yield. Different types of seed treatments are in practice to enhance the tolerance to abiotic stresses and environmental conditions that include seed priming, presoaking hydration, seed coating techniques, seed pelleting and rounding, seed conditioning, etc. (Hoseini et al. 2013). Seed priming is a practical and facile approach to enhance the tolerance to biotic and abiotic stress conditions (Paparella et al. 2015). Studies have shown multiple effects of seed priming on crops including improvement in seedling establishment in rice sown in waterlogged conditions, improvement in ROS scavenging in seeds by the activation of SOD and CAT enzymes, and rapid carbohydrate metabolism (Ella et al. 2011). Hydropriming or presoaking of seeds and then drying after soaking resulted in better germination percentage, seedling emergence, and plant biomass (fresh and dry weight) (Ozden et al. 2017). Multiple effects of different seed priming techniques were considered in several studies, and major priming effects include enhanced, rapid, and synchronized germination, healthy seedling establishment, better plant vigor, and increased crop yield in stressed environments.

## 3 Seed Priming in Relation to Antioxidant Defense in Plants

Plants naturally develop a number of defense mechanisms against biotic and abiotic stresses by producing antioxidants inside the plant body. Antioxidant refers to any substance that significantly delays or avoids cell components' (DNA, proteins, lipids, etc.) oxidation from reactive oxygen species (ROS) or free radicals when present in low concentrations (Gupta and Sharma 2006; Fig. 1), and at higher concentration, it controls/minimizes the generation of ROS. Due to biotic and abiotic stress, as a result of metabolic processes, ROS are produced in different parts of the plant body and impose oxidative stress due to imbalance between oxidant and antioxidant in favor of the oxidant inside the plant body (Valko et al. 2006). When plants are exposed to different biotic or abiotic stress conditions like nutrient deficiency, salt stress, drought/water deficiency, air pollution, or high/low temperature, the level of ROS generation in different plant's cellular compartments (like chloroplast, peroxisomes, or mitochondria) increases, resulting in the activation of the antioxidant defense system (Gill and Tuteja 2010). To minimize/cope with unfavorable effects of ROS and oxidative stress, plants display enzymatic and nonenzymatic low molecular weight antioxidant defense system for scavenging of ROS and free radicals inside plant's cellular compartments (Ahmad et al. 2015).

A number of studies were conducted to assess the effects of seed priming on germination, crop stand, crop vigor, and changes in antioxidant defense mechanism after priming the seeds with different priming techniques as production and detoxification of ROS have its key importance in plant response to different stress conditions. Hydropriming of seeds proved to be an effective approach that results in enhancement of seedling emergence from seed, seedling health, and enhanced level of tolerance to environmental and other stress conditions (Paparella et al. 2015).

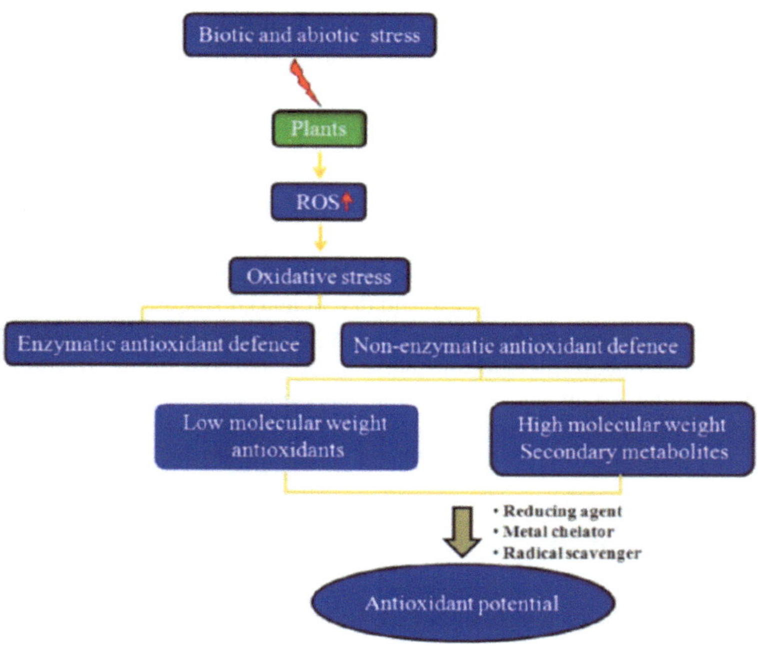


**Fig. 1** ROS-induced oxidative stress and antioxidant potential in plants. (Adopted from Kasote et al. 2015)

Primed seeds also resulted in the vigorous plant seedling and enhanced tolerance to prevailing stress conditions because of the regulated osmotic adjustments, improved energy balance of plants, large size of the embryo, faster and efficient enzyme activation, and speedy antioxidant defense in response to stress or enhanced production of ROS (Jisha et al. 2013). Seed priming is also proved to be an effective and promising approach to generate plant tolerance to chilling or freezing stress, as chilling stress favors the oxidative conditions inside cellular compartments and seed priming results in activation of antioxidant defense in plant cells by rapid activation of cellular enzymes (Xu et al. 2011; Hussain et al. 2016a, b).

In a recent study, Hussain et al. (2016a, b) examined the response of rice seedlings in different nutrient regimes and chilling stress with seed thermopriming and nutrient priming to analyze the production and detoxification of ROS along with seedling growth. They found fluctuations in enzymatic and nonenzymatic antioxidant production within rice seedling, and total antioxidant capability in rice leaves was activated in response to nutrient deficiency or deprivation condition. They concluded that seed priming regulated the total antioxidant capability in chilling and nutrient deficiency stress conditions and found significantly higher antioxidant capability because of nutrient priming. Seed priming proved to be an effective approach to improve seed germination and plant vigor through activation of enzymes that play a key role in plant growth and development, e.g., amylases, proteases, and lipases (Ashraf and Foolad 2005).

## 4 Effect of Seed Priming and Seedling Treatment on Plant Growth and Yield Parameters

Life cycle of crops plants is the combination of different growth and development stages of plant that includes germination stage, seedling stage, vegetative growth stage, flowering stage, reproduction stage, and regeneration stage. After regeneration, plants repeat its life cycle from newly produced seeds from first generation to second generation and so on. Germination stage is considered as one of the most important stages in plant life cycle. The nature and type of germination determine the quality of seedling establishment, growth, and development of plants. Germination is thought to be a complex process which starts from water uptake, and by utilizing the dry seed mass, a metabolically active seedling emerges. Once embryonic axis emerges from the embryo and that stage is considered as the final stage of germination (Bewley et al. 2013). Germination including seedling emergence and plant growth/establishment is a very important feature of agricultural, horticultural, and higher plants and also has great significance for seed and seedling vigor.

Different priming techniques are being used nowadays, and each priming type may have varying results depending on various factors including the plant type, plant development stage, concentration of priming agent, and period of application.

But researchers are more focused to uniform and rapid germination and seedling emergence and plant establishment (Tzortzakis 2009). Priming controls the metabolite leakage from seed, allows DNA replication, fosters protein and RNA synthesis, and ultimately promotes seedling growth and establishment (McDonald 2000). Seed priming with plant growth regulators is one of the effective methods of seed treatment and found to create resistance against plant stress response (Pedranzani et al. 2003) and promote cell elongation and cell division to enhance plant growth and plays its role in germination, vegetative growth, flowering, and seed development (Golenberg and West 2013). Seed treatment with salicylic acid enhances the plant defense mechanism against biotic and abiotic stress which is well understood (Ghassemi-Golezani and Hosseinzadeh-Mahootchi 2015). Under osmotic and non-osmotic stress conditions, hormonal priming and halopriming improve all germination traits, and if no stress prevails, then such priming results in maximum germination percentage, healthy seed vigor, vegetative growth, and conversion efficiency of seed reserve to seedling tissues (Balouchi et al. 2015). Halopriming of seeds (treatment with $KNO_3$) and treatment with gibberellic acid ($GA_3$) proved to be an economic and easy application that results in rapid and uniform germination and seedling emergence and proved to be a great initiative for plant development in greenhouses and nurseries. Such technique could be favorable for farmers producing mass plant stock (Tzortzakis 2009).

Studies have also found that different concentrations of priming agents and priming/treatment time also have varying effects on seed germination ratio, germination time, plant growth, development, and vigor. Pulok et al. (2014) conducted a study with different types of priming agents (gibberellin, lemon juice, and tamarind leaf extract) in varying concentrations with different priming/pretreatment durations and found that gibberellin at 30 ppm and 12 h of treatment time produced high germination index, while gibberellin at 30 ppm with 6 h treatment duration resulted in high germination speed for cowpea crop. Such studies indicate that type of priming agent, concentration and duration of priming are specific to crop and type of stress (biotic and abiotic stress). Thus, same type of priming concentration and duration may not produce the desired results and in some instances may have harmful effects on germination, seedling establishment, plant growth and development.

Seed priming proved to be a facile technique to ensure fast and uniform germination, improved seedling vigor, and enhanced crop yield of multiple field crops under stress conditions (Paparella et al. 2015). Seed priming results in higher crop yield through better crop stand by improving germination, promoting the repair of metabolic functions during seed imbibition (Farooq et al. 2006; Hussain et al. 2015) and osmotic adjustments during unfavorable environmental conditions. Several studies have indicated that plants grown through primed seeds exhibit better tolerance to biotic and abiotic stress conditions through the activation of the cellular defense system (Jisha et al. 2013).

## 5 Seed Priming and Antioxidant Production in Plants

Enzymatic antioxidants like catalase (CAT), ascorbate peroxidase (APX), guaiacol peroxidase (GPOX), and nonenzymatic antioxidants like ascorbic acid production inside the plant body proved to be enhanced as a result of seed priming or pre-sowing seed treatment to minimize the detrimental effects of oxygen-generated ROS (Scandalios 2005). Seed treatment with ascorbate (AsA) is found to have reduction in ROS-generated oxidative stress through the activation of antioxidant enzymes including catalase (CAT), ascorbate peroxidase (APX), and glutathione reductase (GR) in different cellular compartments like mitochondria and cytoplasm. Ascorbate (AsA) seed treatment also had reductive effect on activity of glutathione peroxidase (GPOX) (Alcântara et al. 2015). Samota et al. (2017) conducted a study on drought tolerance of drought-tolerant and drought-susceptible rice cultivars and found that antioxidant enzymes SOD, CAT, and POD activity in both types of varieties increased sufficiently with seed priming in the presence of drought stress. Enhanced activity of antioxidant enzymes and proline contents increased the tolerance of rice cultivars to drought.

## 6 Promotive Effects of Seed Priming on Antioxidant Defense System in Plants

Abiotic stresses like salinity, drought, chilling/freezing, heating, and flooding and biotic stresses including stress due to bacteria, viruses, fungi, parasites, insects, weeds, and other native or invading plants cause oxidative damage to plants. Several studies were conducted globally to assess the level of these biotic and abiotic stress conditions on oxidative damage to plants and activation of antioxidant defense system in plants as individual or cumulative effect of such stresses in plants' body. Abiotic stresses produce ROS in higher concentrations and ROS cause oxidation of lipids, proteins, and DNA and cause oxidative stress (Gill and Tuteja 2010). Antioxidant defense system in plants comprises of enzymatic antioxidant defense through activation of superoxide dismutase (SOD), catalase (CAT), guaiacol peroxidase (GPX), ascorbate peroxidase (APX), glutathione reductase (GR), etc., while nonenzymatic antioxidant defense includes production of ascorbate (AsA), glutathione (GSH), and tocopherols, proline, carotenoids, phenols, tannins, etc. to protect the plants from uncontrolled oxidation and avoid the oxidative damage due to ROS-induced oxidative stress (Anjum et al. 2015; Chen et al. 2015).

Hydropriming, chemical priming, and hormonal priming are found to have stimulatory effect or at least result to maintain antioxidant defense system in rice seedlings through significant reduction in production of ROS (Hussain et al. 2016a, b). Presence of abiotic stresses like salinity negatively alters the antioxidant defense in plants, and hormonal priming with salicylic acid improves the activities of SOD, CAT, and AsA contents in plants cells in wheat cultivars that positively contribute in plants' antioxidant defense system (Afzal et al. 2011). Under water deficit conditions, hydro-primed seed with hydrogen peroxide ($H_2O_2$) results in improved antioxidant defense system and reported higher levels of CAT, APX activity, and proline contents (He et al. 2009). Chilling stress usually results in inhibitory effects on antioxidant defense system due to overproduction of ROS. Osmo-priming with calcium chloride, redox priming with $H_2O_2$, chemical priming with selenium, and hormonal priming with salicylic acid have proved triggering effects on antioxidant enzymes like SOD, POD, CAT activity, and GSH contents in rice seedlings under chilling stress (Hussain et al. 2016a, b). The increased production of antioxidants under abiotic stress conditions due to seed priming enhances the level of ROS scavenging and helps in reduction of ROS concentration in plant cellular compartments, which help improves the naturally developed antioxidant defense system in the plant body.

## References

Abdulrahmani B, Ghassemi-Golezani K, Valizadeh M, FeiziAsl V (2007) Seed priming and seedling establishment of barley (Hordeum vulgare L.). J Food Agric Environ 5(3&4):179–184

Afzal I, Basra SMA, Shahid M, Saleem M (2008) Priming enhances germination of spring maize in cool conditions. Seed Sci Technol 36:497–503

Afzal I, Basra SMA, Ahmad N, Cheema MA, Haq MA, Kazmi MH, Irfan S (2011) Enhancement of antioxidant defense system induced by hormonal priming in wheat. Cereal Res Commun 39(3):334–342. https://doi.org/10.1556/CRC.39.2011.3.3

Ahmad I, Basra SMA, Hussain S, Hussain SA, Hafeez-ur-Rehman, Rehman A, Ali A (2015) Priming with ascorbic acid, salicylic acid and hydrogen peroxide improves seedling growth of spring maize at suboptimal temperature. J Environ Agric Sci Sarghoda 3:14–22

Alcântara BK, Machemer-Noonan K, Silva-Júnior FG, Azevedo RA (2015) Dry priming of maize seeds reduces aluminum stress. PLoS One 10(12):e0145742. https://doi.org/10.1371/journal.pone.0145742

Anjum SA, Tanveer M, Hussain S, Bao M, Wang L, Khan I (2015) Cadmium toxicity in maize (Zeamays L.): consequences on antioxidative systems, reactive oxygen species and cadmium accumulation. Environ Sci Pollut Res 22:17022–17030. https://doi.org/10.1007/s11356-015-4882-z

Asgedom H, Becker M (2001) Effects of seed priming with nutrient solutions on germination, seedling growth and weed competitiveness of cereals in Eritrea. In: Proceedings of Deutscher Tropentag 2001, University of Bonn and ATSAF, Margraf Publishers Press, Weickersheim, p 282

Ashraf M, Foolad MR (2005) Pre-sowing seed treatment – a shotgun approach to improve germination, plant growth and crop yield under saline and non-saline conditions. Adv Agron 88:223–271

Balouchi H, Dehkordi SA, Dehnavi MM, Behzadi B (2015) Effect of priming types on germination of Nigella sative under osmotic stress. South Western J Hortic Biol Environ 6(1):1–20

Bewley JD, Bradford KJ, Hilhorst HMW, Nonogaki H (2013) Seeds: physiology of development, germination and dormancy. Springer, New York

Chen W, Guo C, Hussain S, Zhu B, Deng F, Xue Y (2015) Role of xylo-oligosaccharides in protection against salinity-induced adversities in Chinese cabbage. Environ Sci Pollut Res 23:1254–1264. https://doi.org/10.1007/s11356-015-5361-2

Ella ES, Dionisio-Sese ML, Ismail AM (2011) Seed pre-treatment in rice reduces damage, enhances carbohydrate mobilization and improves emergence and seedling establishment under flooded conditions, AoB Plants (Open Access Research Article). Oxford University Press, Oxford. https://doi.org/10.1093/aobpla/plr007

Farooq M, Basra SMA, Hafeez K (2006) Seed invigoration by osmo-hardening in coarse and fine rice. Seed Sci Technol 34:181–187. https://doi.org/10.15258/sst.2006.34.1.19

Ghassemi-Golezani K, Hosseinzadeh-Mahootchi A (2015) Improving physiological performance of safflower under salt stress by application of salicylic acid and jasmonic acid. Walia J 31:104–109

Gill SS, Tuteja N (2010) Reactive oxygen species and antioxidant machinery in abiotic stress tolerance in crop plants (Review). Plant Physiol Biochem 48:909–930

Golenberg EM, West NW (2013) Hormonal interactions and gene regulation can link monoecy and environmental plasticity to the evolution of dioecy in plants. Am J Bot 100:1022–1037

Gupta VK, Sharma SK (2006) Plants as natural antioxidants. Nat Prod Radiance 5(4):326–334

He L, Gao Z, Li R (2009) Pretreatment of seed with H2O2 enhances drought tolerance of wheat (*Triticum aestivum* L.) seedlings. Afr J Biotechnol 8(22):6151–6157

Hoseini M, Feqenabi F, Tajbakhsh M, Babazadeh-Igdir H (2013) Introduction of seed treatment techniques (seed priming). Int J Biosci 3(5):1–12

Hussain S, Zheng M, Khan F, Khaliq A, Fahad S, Peng S (2015) Benefits of rice seed priming are offset permanently by prolonged storage and the storage conditions. Sci Rep 5:8101. https://doi.org/10.1038/srep08101

Hussain S, Khan F, Cao W, Wu L, Geng M (2016a) Seed priming alters the production and detoxification of reactive oxygen intermediates in rice seedlings grown under sub-optimal temperature and nutrient supply. Front Plant Sci 7:439. https://doi.org/10.3389/fpls.2016.00439

Hussain S, Khan F, Hussain HA, Nie L (2016b) Physiological and biochemical mechanisms of seed priming-induced chilling tolerance in rice cultivars. Front Plant Sci 7:116. https://doi.org/10.3389/fpls.2016.00116

Jisha KC, Vijayakumari K, Puthur JT (2013) Seed priming for abiotic stress tolerance: an overview. Acta Physiol Plant 35:1381–1396. https://doi.org/10.1007/s11738-012-1186-5

Kasote DM, Katyare SS, Hegde MV, Bae H (2015) Significance of antioxidant potential of plants and its relevance to therapeutic applications. Int J Biol Sci 11(8):982–991. https://doi.org/10.7150/ijbs.12096

McDonald MB (2000) Seed pre-treatments. In: Black M, Bewley JD (eds) Seed technology and its biological basis. Sheffield Academic Press, Sheffield, pp 287–325

Murungu FS, Chiduza C, Nyamugafata P, Clark LJ, Whalley WR, Finch-Savage WE (2004) Effects of on-farm seed priming on consecutive daily sowing occasions on the emergence and growth of maize in semi-arid Zimbabwe. Field Crop Res 89(1):49–57

Ozden E, Ermis S, Demir I (2017) Seed priming increases germination and seedling quality in Antirrhinum, Dahlia, Impatiens, Salvia and Zinnia seeds. J Ornamental Plants 7(3):171–176

Paparella S, Arau'jo SS, Rossi G, Wijayasinghe M, Carbonera D, Balestrazzi A (2015) Seed priming: state of the art and new perspectives. Plant Cell Rep 34:1281–1293. https://doi.org/10.1007/s00299-015-1784-y

Pedranzani H, Racagni G, Alemano S, Miersch O, Ramírez I, Peña-Cortés H, Taleisnik E, Machado-Domenech E, Abdala G (2003) Salt tolerant tomato plants show increased levels of jasmonic acid. Plant Growth Regul 41:149–158

Peltonen-Sainio P, Kontturi M, Peltonen J (2006) Phosphorus seed coating enhancement on early growth and yield components in oat. Agron J 98:206–211

Pulok MAI, Roy TS, Sarker S, Islam MS, Rahima U (2014) Different types of priming effect on the seed germination, vigour and seedling parameters of cowpea. Bangladesh Res Publication J 10(2):209–213

Samota MK, Sasi M, Singh A (2017) Impact of seed priming on proline content and antioxidant enzymes to mitigate drought stress in Rice genotype. Int J Curr Microbiol App Sci 6(5):2459–2466

Scandalios JG (2005) Oxidative stress: molecular perception and transduction of signals triggering antioxidant gene defenses. Braz J Med Biol Res 38:995–1014. PMID: 16007271

Tzortzakis NG (2009) Effect of pre-sowing treatment on seed germination and seedling vigour in endive and chicory. Hortic Sci (Prague) 36(3):117–125

Valko M, Rhodes CJ, Moncol J, Izakovic M, Mazur M (2006) Free radicals, metals and antioxidants in oxidative stress-induced cancer. Chem Biol Interact 160:1–40

Xu S, Hu J, Li Y, Ma W, Zheng Y, Zhu S (2011) Chilling tolerance in Nicotiana tabacum induced by seed priming with putrescine. Plant Growth Regul 63:279–290. https://doi.org/10.1007/s10725-010-9528-z

Yilmaz A, Ekiz H, Gultekin I, Torun B, Barut H, Karanlik S, Cakmak I (1998) Effect of seed zinc content on grain yield and zinc concentration of wheat grown in zinc-deficient calcareous soils. J Plant Nutr 21:2257–2264

MIX
Papier aus verantwortungsvollen Quellen
Paper from responsible sources
FSC® C105338

If you have any concerns about our products,
you can contact us on
**ProductSafety@springernature.com**

In case Publisher is established outside the EU,
the EU authorized representative is:
**Springer Nature Customer Service Center GmbH**
**Europaplatz 3, 69115 Heidelberg, Germany**

Printed by Libri Plureos GmbH
in Hamburg, Germany